Monte Carlo and Quasi-Monte Carlo Methods 2006

Alexander Keller • Stefan Heinrich
Harald Niederreiter
Editors

Monte Carlo and Quasi-Monte Carlo Methods 2006

Springer

Alexander Keller
Department of Computer Science
Ulm University
Albert-Einstein-Allee 11
89069 Ulm
Germany
alexander.keller@uni-ulm.de

Stefan Heinrich
Department of Computer Science
University of Kaiserslautern
67653 Kaiserslautern
Germany
heinrich@informatik.uni-kl.de

Harald Niederreiter
Department of Mathematics
National University of Singapore
2 Science Drive 2
Singapore 117543
Republic of Singapore
nied@math.nus.edu.sg

ISBN 978-3-540-74495-5 e-ISBN 978-3-540-74496-2

Library of Congress Control Number: 2007936240

Mathematics Subject Classification (2000): Primary: 11K45, 65-06, 65C05, 65C10
Secondary: 11K38, 65D18, 65D30, 65D32, 65R20, 91B28

Cover Design: WMXDesign GmbH, Heidelberg

Printed on acid-free paper

9 8 7 6 5 4 3 2 1

spinger.com

Preface

This volume represents the refereed proceedings of the Seventh International Conference on Monte Carlo and Quasi-Monte Carlo Methods in Scientific Computing, which was held at Ulm University, Germany, from 14–18 August 2006. The program of this conference was arranged by an international committee consisting of Ivan Dimov (Bulgarian Academy of Sciences), Henri Faure (CNRS Marseille), Paul Glasserman (Columbia University), Stefan Heinrich (co-chair, University of Kaiserslautern), Fred Hickernell (Illinois Institute of Technology), Alexander Keller (co-chair, University of Ulm), Pierre L'Ecuyer (Université de Montréal), Michael Mascagni (The Florida State University), Peter Mathé (Weierstrass Institute for Applied Analysis and Stochastics), Harald Niederreiter (co-chair, National University of Singapore), Erich Novak (Friedrich-Schiller-Universität Jena), Art Owen (Stanford University), Klaus Ritter (TU Darmstadt), Ian Sloan (University of New South Wales), Denis Talay (INRIA Sophia Antipolis), and Henryk Woźniakowski (Columbia University).

The local arrangements were in the hands of the PhD students Sabrina Dammertz, Holger Dammertz, Matthias Raab, Carsten Wächter, the students Bernhard Finkbeiner, Leonhard Grünschloss, Daniela Hauser, Johannes Hanika, Christian Kempter, Manuel Kugelmann, Sehera Nawaz, Daniel Seibert, and our secretary Claudia Wainczyk, all at the computer graphics group of Ulm University.

This conference continued the tradition of biennial MCQMC conferences which was begun at the University of Nevada in Las Vegas, Nevada, USA, in June 1994 and followed by conferences at the University of Salzburg, Austria, in July 1996, the Claremont Colleges in Claremont, California, USA, in June 1998, Hong Kong Baptist University in Hong Kong, China, in November 2000, the National University of Singapore, Republic of Singapore, in November 2002, and at the Palais des Congrès in Juan-les-Pins, France, in June 2004.

The proceedings of these previous conferences were all published by Springer-Verlag, under the titles *Monte Carlo and Quasi-Monte Carlo Methods in Scientific Computing* (H. Niederreiter and P.J.-S. Shiue, eds.), *Monte*

Carlo and Quasi-Monte Carlo Methods 1996 (H. Niederreiter, P. Hellekalek, G. Larcher and P. Zinterhof, eds.), *Monte Carlo and Quasi-Monte Carlo Methods 1998* (H. Niederreiter and J. Spanier, eds.), *Monte Carlo and Quasi-Monte Carlo Methods 2000* (K.-T. Fang, F.J. Hickernell and H. Niederreiter, eds.), *Monte Carlo and Quasi-Monte Carlo Methods 2002* (H. Niederreiter, ed.), and *Monte Carlo and Quasi-Monte Carlo Methods 2004* (H. Niederreiter and D. Talay, eds.). The next MCQMC conference will be held in Montréal, Canada, in July 2008.

The program of the conference was rich and varied with over 120 talks being presented. Highlights were the invited plenary talks given by Ronald Cools (Katholieke Universiteit Leuven), Sergei Mikhailovitch Ermakov (Saint-Petersburg State University), Alan Genz (Washington State University), Frances Kuo (The University of New South Wales), Thomas Müller-Gronbach (Otto-von-Guericke-Universität Magdeburg), Harald Niederreiter (National University of Singapore), Gilles Pagès (Universités Paris VI et VII - CNRS), Karl Sabelfeld (Weierstraßinstitut für angewandte Analysis und Stochastik), and Peter Shirley (University of Utah) as well as the special sessions that were organized by designated chairpersons. The papers in this volume were carefully screened and cover both the theory and the applications of Monte Carlo and quasi-Monte Carlo methods.

We gratefully acknowledge generous financial support of the conference by the German Science Foundation (DFG) and the financial backup by the mental images GmbH, Berlin.

We also thank the anonymous reviewers for their reports and many others who contributed enormously to the excellent quality of the conference presentations and to the high standards for publication in these proceedings by careful review of the abstracts and manuscripts that were submitted.

Leonhard Grünschloß provided invaluable help with editing the proceedings. Finally, we want to express our gratitude to Springer-Verlag for publishing this volume.

Ulm,
August 2007

Alexander Keller
Stefan Heinrich
Harald Niederreiter

Contents

Part I Invited Articles

A Belgian View on Lattice Rules
Ronald Cools and Dirk Nuyens 3

MCQMC Algorithms for Solving some Classes of Equations
Sergej Ermakov 23

MCQMC Methods for Multivariate Statistical Distributions
Alan Genz 35

Minimal Errors for Strong and Weak Approximation of Stochastic Differential Equations
Thomas Müller-Gronbach and Klaus Ritter 53

Nets, (t, s)-Sequences, and Codes
Harald Niederreiter 83

Quadratic Optimal Functional Quantization of Stochastic Processes and Numerical Applications
Gilles Pagès 101

Random Field Simulation and Applications
Karl Sabelfeld 143

Monte Carlo and Quasi-Monte Carlo Methods for Computer Graphics
Peter Shirley, Dave Edwards, and Solomon Boulos 167

Part II Contributed Articles

Random Walk Algorithm for Estimating the Derivatives of Solution to the Elliptic BVP
Alexander Burmistrov 181

Free-Knot Spline Approximation of Fractional Brownian Motion
Jakob Creutzig and Mikhail Lifshits 195

Simulation on Rank-1 Lattices
Holger Dammertz, Alexander Keller, and Sabrina Dammertz 205

Image Synthesis by Rank-1 Lattices
Sabrina Dammertz and Alexander Keller 217

Continuous Runge-Kutta Methods for Stratonovich Stochastic Differential Equations
Kristian Debrabant and Andreas Rößler 237

Issues on Computer Search for Large Order Multiple Recursive Generators
Lih-Yuan Deng 251

Design and Implementation of Efficient and Portable Multiple Recursive Generators with Few Zero Coefficients
Lih-Yuan Deng, Huajiang Li, Jyh-Jen Horng Shiau, and Gwei-Hung Tsai 263

Approximation of Functions Using Digital Nets
Josef Dick, Peter Kritzer, and Frances Y. Kuo 275

Construction of Low-Discrepancy Point Sets of Small Size by Bracketing Covers and Dependent Randomized Rounding
Benjamin Doerr and Michael Gnewuch 299

A Coding Theoretic Approach to Building Nets with Well-Equidistributed Projections
Yves Edel and Pierre L'Ecuyer 313

Improvements on Low Discrepancy One-Dimensional Sequences and Two-Dimensional Point Sets
Henri Faure 327

Improved Multilevel Monte Carlo Convergence using the Milstein Scheme
Mike Giles 343

Generalized Tractability for Linear Functionals
Michael Gnewuch and Henryk Woźniakowski 359

An Improved Implementation of Stochastic Particle Methods and Applications to Coagulation Equations
Flavius Guiaş 383

(t, m, s)-Nets and Maximized Minimum Distance
Leonhard Grünschloß, Johannes Hanika, Ronnie Schwede, and Alexander Keller 397

Quasi-Monte Carlo Simulation of Discrete-Time Markov Chains on Multidimensional State Spaces
Rami El Haddad, Christian Lécot, and Pierre L'Ecuyer 413

Computational Engine for a Virtual Tissue Simulator
Carole Hayakawa, Jerome Spanier, and Vasan Venugopalan 431

Randomized Approximation of Sobolev Embeddings
Stefan Heinrich 445

Tractability of Linear Multivariate Problems in the Average Case Setting
Fred Hickernell, Greg Wasilkowski, and Henryk Woźniakowski 461

Zinterhof Sequences in GRID-Based Numerical Integration
Heinz Hofbauer, Andreas Uhl, and Peter Zinterhof 495

A Pragmatic View on Numerical Integration of Unbounded Functions
Heinz Hofbauer, Andreas Uhl, and Peter Zinterhof 511

Assessment of Genetic Association using Haplotypes Inferred with Uncertainty via Markov Chain Monte Carlo
Raquel Iniesta and Victor Moreno 529

The Generalized Gibbs Sampler and the Neighborhood Sampler
Jonathan Keith, George Sofronov, and Dirk Kroese 537

The Weighted Dyadic Diaphony of Digital Sequences
Peter Kritzer and Friedrich Pillichshammer 549

A New Criterion for Finiteness of Weight Estimator Variance in Statistical Simulation
Ilya Medvedev and Gennadii Mikhailov 561

Optimal Pointwise Approximation of a Linear Stochastic Heat Equation with Additive Space-Time White Noise
Thomas Müller-Gronbach, Klaus Ritter, and Tim Wagner 577

Unbiased Global Illumination with Participating Media
Matthias Raab, Daniel Seibert, and Alexander Keller 591

SIMD-Oriented Fast Mersenne Twister: a 128-bit Pseudorandom Number Generator
Mutsuo Saito and Makoto Matsumoto 607

A New Lower Bound on the t-Parameter of (t, s)-Sequences
Rudolf Schürer 623

Walk-on-Spheres Algorithm for Solving Boundary-Value Problems with Continuity Flux Conditions
Nikolai Simonov 633

Good Lattice Rules with a Composite Number of Points Based on the Product Weighted Star Discrepancy
Vasile Sinescu and Stephen Joe 645

Ergodic Simulations for Diffusion in Random Velocity Fields
Nicolae Suciu, Călin Vamoş, and Karl Sabelfeld 659

Efficient Simultaneous Simulation of Markov Chains
Carsten Wächter and Alexander Keller 669

Part III Appendix

Conference Participants 687

Author Index 697

Part I

Invited Articles

A Belgian View on Lattice Rules

Ronald Cools[1] and Dirk Nuyens[2]

[1] Dept. of Computer Science, K.U.Leuven, Celestijnenlaan 200A, B-3001 Heverlee, Belgium
Ronald.Cools@cs.kuleuven.be

[2] Dept. of Computer Science, K.U.Leuven, Celestijnenlaan 200A, B-3001 Heverlee, Belgium
Dirk.Nuyens@cs.kuleuven.be

1 Introduction

The problem we consider is the approximation of multivariate integrals over the s-dimensional unit cube

$$I[f] := \int_0^1 \cdots \int_0^1 f(x_1, \ldots, x_s)\, \mathrm{d}x_1 \cdots \mathrm{d}x_s = \int_{[0,1)^s} f(\mathbf{x})\, \mathrm{d}\mathbf{x}.$$

We are interested in approximations of the form

$$Q[f] := \sum_{j=1}^{n} w_j f(\mathbf{y}^{(j)}) \tag{1}$$

with weights $w_j \in \mathbb{R}$ and points $\mathbf{y}^{(j)} \in [0,1)^s$.

Many people call this a *quadrature problem*, although strictly speaking the word "quadrature" refers to the 1-dimensional case, i.e., measuring an area. By only using this key word in a search, one misses a whole world of relevant literature. The more appropriate word is "cubature". In written English, it appears already in the 17th century to refer to measuring a volume.[3] Because one speaks about an s-dimensional cube, it is natural to use the same word in connection with measuring s-dimensional volumes, i.e., integrals. So, if $s = 1$ then Q is called a *quadrature formula* and if $s \geq 2$ then Q is called a *cubature formula*.

We are particularly interested in cubature formulas where the points $\mathbf{y}^{(j)}$ and weights w_j are chosen independent of the integrand f. It is usually difficult and time consuming to construct such cubature formulas, but the result is usually hard coded in programs or tables.

[3] An equivalent exists in other languages, e.g., in German "Kubatur" and in Dutch "kubatuur".

In the taxonomy of cubature formulas one can distinguish two major classes: polynomial based methods (e.g., methods exact for algebraic or trigonometric polynomials) and number theoretic methods (e.g., quasi-Monte Carlo methods and even Monte Carlo methods based on pseudo random number generators). As in zoology, some species are difficult to classify. Lattice rules are a family of cubature formulas that are studied as members of both classes, depending on the background of the researcher. They are in the focus of this text.

Definition 1. *An s-dimensional lattice rule is a cubature formula which can be expressed in the form*

$$Q[f] = \frac{1}{d_1 d_2 \dots d_t} \sum_{j_1=1}^{d_1} \sum_{j_2=1}^{d_2} \cdots \sum_{j_t=1}^{d_t} f\left(\left\{\frac{j_1 \mathbf{z}_1}{d_1} + \frac{j_2 \mathbf{z}_2}{d_2} + \dots + \frac{j_t \mathbf{z}_t}{d_t}\right\}\right),$$

where t and $d_i \in \mathbb{N} \setminus \{0\}$ and $\mathbf{z}_i \in \mathbb{Z}^s$ for all i.

The notation $\{\cdot\}$ denotes to take the fractional part componentwise.

An alternative definition is given below. This already shows that lattice rules can be approached in different ways.

Definition 2. *A* multiple integration lattice Λ *is a subset of $\mathbb{R}^s$ which is discrete and closed under addition and subtraction and which contains $\mathbb{Z}^s$ as a subset. A* lattice rule *is a cubature formula where the n points are the points of a multiple integration lattice Λ that lie in $[0, 1)^s$ and the weights are all equal to $1/n$.*

We must emphasize that a lattice rule has different representations of the form given in Definition 1. The minimal number of sums (i.e., the minimal number of generating vectors $\mathbf{z}_i$) required is called the *rank* of the lattice rule. Even if the number of generators is fixed, the rules can still be represented using different generating vectors. Many papers only consider lattice rules of rank 1. A *rank-1* lattice rule is generated by one vector $\mathbf{z}$ and has the form

$$Q[f] = \frac{1}{n} \sum_{j=1}^{n} f\left(\left\{\frac{j\mathbf{z}}{n}\right\}\right).$$

The view on lattice rules presented in this text is strongly biased. It reflects how the first author got into contact with lattice rules, and how he started looking at them from the view on multivariate integration he had at that time. (For a different view on lattice rules, which also includes other kinds of quasi-Monte Carlo point sets, we refer to [LL02].) In Section 2 an overview of quality and construction criteria for lattice rules is given, biased towards what is less known in the qMC-world, i.e., the target audience of this volume. In Section 3 we will briefly describe recent approaches for constructing lattice rules, making it clear that the choice of quality criterion determines the required construction effort. In Section 4 we will point to techniques to make lattice rules work in practice and in Section 5 we will illustrate that lattice rules are used from 2-dimensions to high dimensions. Final remarks are given in Section 6.

2 Quality Criteria

2.1 Rules Exact for Polynomials

There are many quality criteria to specify and classify cubature formulas in general, and lattice rules in particular. Trigonometric polynomials play an important role in the world of lattice rules. Algebraic polynomials play a role in connection with more "classical" cubature formulas. In this section we will point to some similarities.

Let $\mathbf{h} = (h_1, h_2, \ldots, h_s) \in \mathbb{Z}^s$ and $|\mathbf{h}| := \sum_{j=1}^{s} |h_j|$. An *algebraic polynomial* is a finite sum of the form

$$p(\mathbf{x}) = \sum_{\mathbf{h}\in\mathbb{Z}^s} a_{\mathbf{h}} \mathbf{x}^{\mathbf{h}} = \sum_{\mathbf{h}\in\mathbb{Z}^s} a_{\mathbf{h}} \prod_{j=1}^{s} x_j^{h_j}, \quad \text{with } h_j \geq 0.$$

A *trigonometric polynomial* is a finite sum of the form

$$t(\mathbf{x}) = \sum_{\mathbf{h}\in\mathbb{Z}^s} a_{\mathbf{h}} e^{2\pi i \mathbf{h}\cdot\mathbf{x}} = \sum_{\mathbf{h}\in\mathbb{Z}^s} a_{\mathbf{h}} \prod_{j=1}^{s} e^{2\pi i x_j h_j}.$$

The degree of a polynomial is defined as $\max_{a_{\mathbf{h}}\neq 0} |\mathbf{h}|$. The space of all algebraic polynomials in s variables of degree at most d is denoted by $\mathbb{P}_d^s$. The space of all trigonometric polynomials in s variables of degree at most d is denoted by $\mathbb{T}_d^s$. We will use the symbol $\mathbb{V}_d^s$ to refer to one of the vector spaces $\mathbb{P}_d^s$ or $\mathbb{T}_d^s$.

The dimensions of the vector spaces of polynomials are

$$\dim \mathbb{P}_d^s = \binom{s+d}{d} \quad \text{and} \quad \dim \mathbb{T}_d^s = \sum_{j=0}^{s} \binom{s}{j} \binom{d}{j} 2^j.$$

The right hand sides are polynomials in d of degree s.

A very old quality criterion for cubature formulas comes from demanding that the formula gives the exact value of the integral for polynomials.

Definition 3. *A cubature formula Q has algebraic (trigonometric) degree d if it is exact for all polynomials of algebraic (trigonometric) degree at most d.*

Once this criterion is put forward, it is natural to ask how many points are needed in a cubature formula to obtain a specified degree of precision. This is obviously related to the dimension of the space for which the formula reproduces the exact value of the integral.

Theorem 1. *If a cubature formula is exact for all polynomials of $\mathbb{V}_{2k}^s$, then the number of points $n \geq \dim \mathbb{V}_k^s$.*

A proof of this result for algebraic degree is given in [Rad48] for $s = 2$ and in [Str60] for general s. For trigonometric degree it is presented in [Mys87]. So, the required number of points increases exponentially with the dimension. Furthermore a "large" part of the weights in a cubature formula have to be positive.

Theorem 2. *If a cubature formula is exact for all polynomials of* $\mathbb{V}_d^s$ *and has only real points and weights, then it has at least* $\dim \mathbb{V}_k^s$ *positive weights,* $k = \lfloor \frac{d}{2} \rfloor$.

This result is proven in [Mys81] for algebraic degree and [Coo97] for trigonometric degree. The combination of the two previous theorems implies that formulas attaining the lower bound of Theorem 1 have only positive weights. For trigonometric degree, we even know more [BC93].

Corollary 1. *If a cubature formula of trigonometric degree* $2k$ *has* $n = \dim \mathbb{T}_k^s$ *points, then all weights are equal.*

One cannot expect that the lower bound of Theorem 1 can be attained for odd degrees $2k+1$, since in that case it is equal to the bound for degree $2k$. For algebraic degree, there exists an improved lower bound for odd degrees that takes into account information on the symmetry of the integration region. The first such result was derived for centrally symmetric regions such as a cube. For surveys of achievements in this particular area we refer to [Coo97, CMS01]. A similar result holds for the trigonometric degree case. Let G_k be the span of trigonometric monomials of degree $\leq k$ with the same parity as k.

Theorem 3. *The number of points* n *of a cubature formula for the integral over* $[0,1)^s$ *which is exact for all trigonometric polynomials of degree at most* $d = 2k+1$ *satisfies*

$$n \geq 2 \dim G_k.$$

This result is mentioned in [Nos85] and a complete proof appears in [Mys87].

Structures do not only play a role in the derivation of lower bounds; their role in constructing cubature formulas is even more important. Imposing structure on the points and weights is used since the beginning of history to reduce the complexity of the construction problem for cubature formulas. The basic structure for lattice rules is "shift symmetry". In the trigonometric case this structure plays the same role as "central symmetry" in the algebraic case.

Definition 4. *A cubature formula is called* shift symmetric *if it is invariant with respect to the group of transformations*

$$\left\{ \mathbf{x} \mapsto \mathbf{x}, \mathbf{x} \mapsto \left\{ \mathbf{x} + \left(\frac{1}{2}, \dots, \frac{1}{2} \right) \right\} \right\}.$$

Hence, the multiple integration lattice Λ *of a shift symmetric cubature formula satisfies*

$$\left\{ \mathbf{x} + \left(\frac{1}{2}, \dots, \frac{1}{2} \right) \mid \mathbf{x} \in \Lambda \right\} = \Lambda.$$

This structure was exploited to derive the following result [BC93].

Theorem 4. *If a shift symmetric cubature formula of degree* $2k+1$ *has* $n = 2 \dim G_k$ *points, then all weights are equal.*

In the algebraic case it is proven that formulas attaining the lower bound for odd degrees for centrally symmetric regions are also centrally symmetric. For the trigonometric case it was conjectured in [Coo97].

The results of Corollary 1 and Theorem 4 motivate us to restrict searches for cubature formulas of trigonometric degree to equal weight cubature formulas. Hence the general form (1) simplifies to

$$Q[f] = \frac{1}{n} \sum_{j=1}^{n} f(\mathbf{y}^{(j)}). \tag{2}$$

Formulas for which the lower bounds in Theorems 1 and 3 are sharp, are only known for degrees 1, 2 and 3 in all dimensions, for all degrees in 2 dimensions, and for degree 5 in 3 dimensions. We refer to [Coo97, Lyn03] for a detailed survey. Almost all known formulas of trigonometric degree that attain these lower bounds are (shifted) lattice rules. The only exceptions are derived in [CS96].

Theorem 5. *The following points*

$$\left(C_p + \frac{j}{2(k+1)}, \; C_p + \frac{j+2p}{2(k+1)} \right) \qquad \text{for } \; j = 0, \ldots, 2k+1, \quad p = 0, \ldots, k,$$

with $C_0 = 0$ and $C_1, \ldots, C_k$ arbitrary, are the points of a cubature formula for the integral over $[0,1)^2$ of trigonometric degree $2k+1$.

We are not aware of successful efforts to construct cubature formulas of trigonometric degree that are not (shifted) lattice rules. In Section 3 we will mention recent construction methods for lattice rules with the trigonometric degree criterion, not necessarily attaining the known lower bounds.

Most of the results summarized above were obtained using *reproducing kernels*, see [Aro50]. A reproducing kernel K is in general a function of two s-dimensional variables with the property that an evaluation of a function f can be written as the inner product of f with K. If we work in a finite dimensional space of polynomials, then a reproducing kernel can be written using orthogonal polynomials. The trigonometric case is easier to work with than the algebraic case because orthonormal polynomials are readily available. Indeed, the trigonometric monomials form an orthonormal sequence. A further simplifying aspect of the trigonometric case is that the reproducing kernel can be written as a function of one s-dimensional variable. For $s = 2$ and $\mathbb{T}_d^s$ it has a simple form which was exploited in [CS96] to obtain formulas with the lowest possible number of points, including lattice rules and others (see Theorem 5).

2.2 On Route to Other Quality Criteria

So far, we focused on integrating polynomials. What do we know if we apply a cubature formula to a function that is not a polynomial? To answer this question, let us assume the integrand function f can be expanded into an absolutely convergent multiple Fourier series:

$$f(\mathbf{x}) = \sum_{\mathbf{h}\in\mathbb{Z}^s} \hat{f}(\mathbf{h})\, e^{2\pi i \mathbf{h}\cdot\mathbf{x}} \quad \text{with} \quad \hat{f}(\mathbf{h}) := \int_{[0,1)^s} f(\mathbf{x})\, e^{-2\pi i \mathbf{h}\cdot\mathbf{x}}\, \mathrm{d}\mathbf{x}.$$

Then the approximation error of an equal weight cubature formula (2) is given by

$$\begin{aligned} Q[f] - I[f] &= \frac{1}{n} \sum_{j=1}^{n} \left(\sum_{\mathbf{h}\in\mathbb{Z}^s\setminus\{\mathbf{0}\}} \hat{f}(\mathbf{h})\, e^{2\pi i \mathbf{h}\cdot\mathbf{y}^{(j)}} \right) \\ &= \sum_{\mathbf{h}\in\mathbb{Z}^s\setminus\{\mathbf{0}\}} \left(\hat{f}(\mathbf{h})\, \frac{1}{n} \sum_{j=1}^{n} e^{2\pi i \mathbf{h}\cdot\mathbf{y}^{(j)}} \right). \end{aligned}$$

Observe that

$$\frac{1}{n}\sum_{j=1}^{n} e^{2\pi i \mathbf{h}\cdot\mathbf{y}^{(j)}} = \begin{cases} 1 & \text{if } \mathbf{h}\cdot\mathbf{y}^{(j)} \in \mathbb{Z}, \\ 0 & \text{if } \mathbf{h}\cdot\mathbf{y}^{(j)} \notin \mathbb{Z}. \end{cases}$$

So, if our equal weight cubature formula is a lattice rule, many terms in the expression for the error vanish. This brings us to a very important tool to investigate the error of a lattice rule and a well known theorem by Sloan and Kachoyan [SK87].

Definition 5. *The* dual *of the multiple integration lattice Λ is*

$$\Lambda^\perp := \{\mathbf{h} \in \mathbb{Z}^s : \mathbf{h}\cdot\mathbf{x} \in \mathbb{Z}\ \ \forall \mathbf{x} \in \Lambda\}.$$

Theorem 6. *Let Λ be a multiple integration lattice. Then the corresponding lattice rule Q has an error*

$$Q[f] - I[f] = \sum_{\mathbf{h}\in\Lambda^\perp\setminus\{\mathbf{0}\}} \hat{f}(\mathbf{h}).$$

Remember that our analysis in this section assumes that the integrand can be expanded in an absolutely convergent multiple Fourier series. So, lattice rules look interesting for periodic functions. Not surprisingly, the trigonometric degree can be defined in terms of the dual lattice.

Definition 6. *The* trigonometric degree *of a lattice rule Q is*

$$d(Q) := \min_{\mathbf{h}\in\Lambda^\perp\setminus\{\mathbf{0}\}} \left(\sum_{j=1}^{s} |h_j| \right) - 1.$$

For many years this criterion was only used in Russia for construction. Some references are [Mys85, Mys90, Rez90, Nos85, Nos88, Tem91, Sem96, Sem97, Osi04, OP04].

Another popular criterion for lattice rules that can also be defined in terms of the dual lattice is the *Zaremba index* or *figure of merit.*

Definition 7. *The* Zaremba index *or* figure of merit *is*

$$\rho(Q) := \min_{\mathbf{h}\in\Lambda^\perp\setminus\{\mathbf{0}\}} (\overline{h}_1 \overline{h}_2 \cdots \overline{h}_s) \qquad \textit{with} \quad \overline{h} := \max(1, |h|).$$

The Zaremba index was used in a computer search for good lattice rules in three and four dimensions by Maisonneuve [Mai72], and also in, e.g., [BP85]. The now classical survey [Lyn89] already presented both the Zaremba index and the trigonometric degree (there it is called "overall degree") in the form of the above definitions.

We will now sketch the origin of the Zaremba index. For $c > 0$ and fixed $\alpha > 1$, let $E_s^\alpha(c)$ be the class of functions f whose Fourier coefficients satisfy

$$|\hat{f}(\mathbf{h})| \leq \frac{c}{(\overline{h}_1 \overline{h}_2 \cdots \overline{h}_s)^\alpha}.$$

This is essentially a class of functions of a certain smoothness, given by α. The worst possible function in class $E_s^\alpha(1)$ is

$$f_\alpha(\mathbf{x}) := \sum_{\mathbf{h}\in\mathbb{Z}^s} \frac{1}{(\overline{h}_1 \overline{h}_2 \cdots \overline{h}_s)^\alpha} e^{2\pi i \mathbf{h}\cdot\mathbf{x}}.$$

Now define $P_\alpha(Q)$ as the error of the lattice rule Q for the function f_α:

$$P_\alpha(Q) := \sum_{\mathbf{h}\in\Lambda^\perp\setminus\{\mathbf{0}\}} \frac{1}{(\overline{h}_1 \overline{h}_2 \cdots \overline{h}_s)^\alpha}. \tag{3}$$

When α is an even integer $P_\alpha(Q)$ is easy to compute because in that case f_α can be written as a product of Bernoulli polynomials. It was introduced by Korobov [Kor59] who showed the existence of lattice rules for which $P_\alpha(Q)$ is $O(n^{-\alpha+\epsilon})$, $\epsilon > 0$, or $O(n^{-\alpha}(\log(n))^{\alpha s})$ in [Kor60].

It follows easily that the larger $\rho(Q)$ is, the smaller we expect $P_\alpha(Q)$ to be; the $\mathbf{h}$ which achieve the minimum in the definition of $\rho(Q)$ make up the largest value in the sum for $P_\alpha(Q)$. A lower bound on $P_\alpha(Q)$ can easily be derived from the definitions as

$$\frac{2}{\rho(Q)^\alpha} \leq P_\alpha(Q)$$

but the real use of $\rho(Q)$ is in deriving upper bounds, see, e.g., [Nie78, Nie92] for an overview.

Another related criterion is given by

$$R(Q) := \sum_{\substack{\mathbf{h} \in \Lambda^{\perp} \setminus \{\mathbf{0}\} \\ -\frac{n}{2} < h_j \leq \frac{n}{2}}} \frac{1}{(\overline{h}_1 \overline{h}_2 \cdots \overline{h}_s)}.$$

Here, Fourier coefficients which are already at a certain distance from the origin are not considered anymore. This has the benefit that no smoothness parameter α has to be chosen. In other words: $R(Q)$ is a modified version of P_α, chosen in such a way that α can be set to one. A similar lower bound as for $P_\alpha(Q)$ is given by

$$\frac{1}{\rho(Q)} \leq R(Q).$$

Recent searches based on $R(Q)$ were done by Joe [Joe04] and Sinescu and Joe [JS07]. These searches are in fact searches for the "star discrepancy" by using a nice relationship in terms of $R(Q)$. (Loosely speaking, a point set has low discrepancy if the points are fairly well uniformly distributed in relation to the number of points used, see, e.g., [Nie78, Nie92].) For large n it can be inferred that large values of $\rho(Q)$ will give small values of both $R(Q)$ and $P_\alpha(Q)$.

In a Korobov space with smoothness α the value of $P_\alpha(Q)$ is the square of the *worst-case error*. The worst-case error of a cubature rule Q in a space $\mathcal{F}$ is given by

$$e(Q, \mathcal{F}) = \sup_{\substack{f \in \mathcal{F} \\ \|f\|_{\mathcal{F}} \leq 1}} |I(f) - Q(f)|.$$

Such a Korobov space is a reproducing kernel Hilbert space (see [Aro50]). As was already mentioned, a reproducing kernel Hilbert space is a function space for which the evaluation of a function can be written as the inner product with the reproducing kernel K. This reproducing kernel is in general a function of two variables, but when the function space is periodic the kernel can in fact be written as a function in one variable. Such a kernel is then called *shift-invariant*. For a shift-invariant kernel K and a rank-1 lattice the squared worst-case error is given by

$$e^2(Q, K) = -\int_{[0,1)^s} K(\mathbf{x}, \mathbf{0})\, \mathrm{d}\mathbf{x} + \frac{1}{n} \sum_{j=1}^{n} K\left(\left\{\frac{j\mathbf{z}}{n}\right\}, \mathbf{0}\right),$$

see, e.g., [Hic98a]. So if one knows the reproducing kernel, one can obtain an explicit formula for the worst-case error in the space under consideration.

A very important and recent ingredient in these reproducing kernel Hilbert spaces are weights which are used to denote the importance of certain sets of variables. (Note that these weights are different from the ones in the cubature formula (1).) The most simple and useful form of the kernel is the kernel for a shift-invariant and tensor-product weighted reproducing kernel Hilbert space. In this case the kernel can be written as

$$K(\mathbf{x}, \mathbf{y}) = \prod_{k=1}^{s} (1 + \gamma_k \, \omega(\{x_k - y_k\})).$$

The weights $\gamma_k \geq 0$ are used to denote the importance of the different dimensions. For a rank-1 rule the typical form for the squared worst-case error in such a weighted space is then

$$e_s^2(\mathbf{z}) = -1 + \frac{1}{n} \sum_{j=1}^{n} \prod_{k=1}^{s} \left(1 + \gamma_k \, \omega \left(\left\{ \frac{j z_k}{n} \right\} \right) \right), \tag{4}$$

where we assumed that $\int_0^1 \omega(x) \, \mathrm{d}x = 0$.

The use of reproducing kernel Hilbert spaces has created a very elegant theory in which all kinds of discrepancies can be defined in terms of the worst-case error in a certain space (see, e.g., [Hic98a, Hic98b]). Moreover, it enables the study of the error in non-periodic spaces, see, e.g., [SKJ02a, SKJ02b].

3 Recent Constructions

In Section 2 we presented lower bounds for the number of points that is required in a cubature formula to attain a specified trigonometric degree. The theorems are not constructive and—as mentioned in Section 2—formulas that attain the known bounds are only known for small s or d. The construction of lattice rules is done by searches. The parameters in such a search are the number of points, the number of generating vectors and the components of these vectors. Obviously the complete search space is huge. Furthermore, the cost to verify that a lattice rule has trigonometric degree d is proportional to d^s. Consequently only "moderate" dimensions are feasible for this criterion for this reason only.

Practical constructions of lattice rules start with restricting the search space. A popular restriction is to consider only rank-1 lattice rules with one generating vector, hence only s components have to be determined. Actually, most authors only consider so-called rank-1 simple rules, where the first component of the generating vector is equal to 1. Then only $s - 1$ components have to be determined.

The search space can be even further reduced by considering only generator vectors of the form

$$\mathbf{z}(\ell) = (1, \ell, \ell^2 \bmod n, \ldots, \ell^{s-1} \bmod n), \quad 1 < \ell < n. \tag{5}$$

This is the form of so-called Korobov rules [Kor60].

In the remainder of this section we will sketch two recent successful types of searches. They use a different quality criterion and have their own way to restrict the search space.

3.1 Rules of Exact Trigonometric Degree

Many searches for lattice rules use the generator matrix of the dual lattice. We will first properly introduce this concept. Recent searches impose some structure on this matrix.

Any s-dimensional lattice Λ can be specified in terms of s linearly independent vectors $\{\mathbf{a}_1, \mathbf{a}_2, \ldots, \mathbf{a}_s\}$. These vectors are known as generators of Λ. (In addition to the t vectors in Definition 1, one can always take $s-t$ unit vectors.) Associated with the generators is an $s \times s$ *generator matrix* A whose rows are $\mathbf{a}_1, \mathbf{a}_2, \ldots, \mathbf{a}_s$. All lattice points $\mathbf{x}$ are of the form $\mathbf{x} = \sum_{i=1}^{s} \lambda_i \mathbf{a}_i = \boldsymbol{\lambda} A$ for some $\boldsymbol{\lambda} \in \mathbb{Z}^s$.

The dual lattice $\Lambda^\perp$ has generator matrix $B = (A^{-1})^T$. Some authors use this as the definition of a dual lattice, instead of Definition 5. They are equivalent. It can be shown that the number of points $n = |\det A|^{-1} = |\det B|$.

Recent searches in low dimensions are based on the following argument by [CL01]: "*It is reasonable to believe that the lattice Λ of an* optimal *lattice rule will have $\Lambda^\perp$ with many elements on the boundary of $S(O_s, d+1)$*". Here $S(O_s, d+1)$ denotes a magnification of the unit octahedron O_s by a factor $d+1$. Their searches only consider lattice rules whose dual has s generators lying on the boundary of $S(O_s, d+1)$. The corresponding lattice rules are called *K-optimal rules.*

The cost for searching this space mainly depends on the number of generator matrices that must be considered; this is $O(d^{s^2-s})$. Most of these can be eliminated quickly but for a minute proportion one has to verify their trigonometric degree, at a cost over d^{s-1}. This leads to a complexity bounded above by d^{s^2-1}. This is a pessimistic bound, but it indicates the fundamental problem of this approach. In [CL01] it was used for 3 and 4 dimensions. In [LS06] the results of a 'Seti@home'-type of search is described for 5 dimensions, limited to $d \leq 11$.

One can impose structure on the generator matrix of the dual lattice to reduce the number of free parameters in the search. In [LS04, CG03] the search was restricted to (skew-)circulant generator matrices. This reduces the cost to $O(d^{2s-2})$ and was very successful in 4–6 dimensions. This approach also lead to closed expressions for lattice rules of arbitrary degrees. A more detailed summary of this approach is presented in [CN06]

We will conclude this part with a digression: linking the search for lattice rules to the field known as "*geometry of numbers*". To compare the number of points of different lattice rules of the same degree we require a proper scaling. The *packing factor* provides this.

Definition 8. *The* packing factor *is*

$$\hat{\rho}(n) := \frac{(d+1)^s}{s!n}.$$

The packing factor is a measure of the efficiency of a rule and provides a convenient way for making pictures because $0 \leq \hat{\rho}(n) \leq 1$.

Actually, $\hat{\rho}(n)$ is bounded above by what people working in the area of "geometry of numbers" call the *density of the densest lattice packing* of the crosspolytope (octahedron) $\theta(O_s)$ [GL87]. This provides a better lower bound for lattice rules for trigonometric degree than those of Theorems 1 and 3:

$$n \geq \frac{(d+1)^s}{s!\theta(O_s)}.$$

The problem is however that $\theta(O_s)$ is only known for $s = 1, 2$ and 3: $\theta(O_1) = \theta(O_2) = 1$, $\theta(O_3) = \frac{18}{19}$. This last result is due to Minkowski [Min67] and was already used in [Fro77] to construct lattice rules.

Lattice rules provide constructive lower bounds for $\theta(O_s)$. From a lattice rule with n points having degree d follows

$$\theta(O_s) \geq \frac{(d+1)^s}{s!n}.$$

The currently best known bounds for $\theta(O_s), s = 4, 5$ and 6 all follow from known lattice rules [OP04, Coo06].

3.2 Rules Minimizing a Worst-Case Error

The introduction of weights in the function space, e.g., as in (4), makes it practically impossible to hard code the cubature rules in tables since there are an infinite number of weighted function spaces to choose from. The weights give the flexibility to tune the function space, but at a price. Luckily, for shift-invariant spaces we are able to construct lattice rules just in time by a fast algorithm.

If one wants to search lattice rules which minimize $P_\alpha(Q)$, $R(Q)$ or any other weighted worst-case error for a given function space, then again the search space has to be limited in one way or another. In this section the focus will be on rank-1 rules. A traditional approach was to consider Korobov rules (5), but more recently, the component-by-component construction [SR02] has opened many more possibilities. Since the publication of [SR02] a lot of results concerning component-by-component construction were obtained, both on the existence and on the construction side, see, e.g., [SKJ02a, SKJ02b, DK04a, Kuo03, CKN06].

Instead of trying to find an optimal generating vector of a predefined form, the components of the generating vector are now searched, and fixed, component by component. In this way the complexity of the search is reduced from $O(n^s\, \kappa(n,s))$ to $O(sn\, \kappa(n,s))$ where $\kappa(n,s)$ is the cost of calculating the worst-case error by formula (4). By inspection we find that $\kappa(n,s) = O(sn)$ and thus the total cost is $O(s^2n^2)$. However, simply considering the product as a cumulative product, since the previous components of $\mathbf{z}$ are fixed, reduces the construction cost to $O(sn^2)$ at the expense of $O(n)$ memory. This allows for moderately larger values of s and n than for the exactness criteria, but really large values are still infeasible unless more advanced arguments are used.

In [DK04a, DK04b] Dick and Kuo conceive a modified method to find lattice rules with "millions of points", for which n needs to be a product of few primes. But even without modifying the search it turns out to be possible to construct lattice rules with millions of points and in thousands of dimensions. This was first shown by the authors for n prime in [NC06a] and later extended for any composite n [NC06b]. This fast algorithm allows for the construction of lattice rules on a just in time basis.

The fast algorithm works by exploiting some structural properties in the worst-case error formula. Starting from (4) it can be observed that the ω function is evaluated on a multiplicative algebraic structure modulo n

$$\omega\left(\left\{\frac{jz}{n}\right\}\right) = \omega\left(\frac{j \cdot z \bmod n}{n}\right).$$

By rewriting (4) as a matrix-vector product it can be shown that a matrix-vector multiplication with a matrix with the above structure can be done in $O(n \log n)$ using fast Fourier transforms, see [NC06b]. Therefore, construction takes only $O(sn \log(n))$ using $O(n)$ memory.

4 Toward Using Lattice Rules

Many texts start with saying that lattice rules are for integrating periodic functions. The different quality criteria we mentioned before make that clear. The traditional line of thought is that one first has to transform the region to the unit cube and then periodize the function. Periodization is further discussed in Section 4.1.

However, it is nowadays known that lattice rules can successfully be applied to non-periodic functions as well, see, e.g., [SKJ02a, SKJ02b]. In Section 4.2 we describe a recent trend in which lattice rules are even used as a sequence. Both these new insights reduce the historical differences between low discrepancy sequences and lattice rules.

4.1 Periodizing Transformations

A non-periodic function on the unit cube can be transformed by a periodizing transform ϕ:

$$\int_{[0,1)^s} f(x_1, \ldots, x_s)\, \mathrm{d}\mathbf{x} = \int_{[0,1)^s} f(\phi(x_1), \ldots, \phi(x_s))\, \phi'(x_1) \cdots \phi'(x_s)\, \mathrm{d}\mathbf{x}.$$

Using a periodizing transformation is equivalent to using a transformed point set, $\mathbf{y}^{(j)} \mapsto (\phi(y_1^{(j)}), \ldots, \phi(y_s^{(j)}))$, with weights $w_j = \prod_{k=1}^{s} \phi'(y_k^{(j)})$ in (1).

There are several practical problems with periodization. Many periodizing transformations exist. They are mainly used in one dimension and selecting the right transform for a given function is not trivial. It seems that the factor $(\log(n))^{\alpha s}$ in the theoretical convergence, as mentioned in Section 2.2, is often very well visible and this gets worse for higher s. Consequently more initial points are needed to achieve the $O(n^{-\alpha})$.

The periodizing transformations lead to machine dependent cubature rules. When n gets larger, calculations have to be done in higher precision. Indeed, IEEE double precision is not enough since different points map to the same floating point representation even for relatively small n; the floating point cube $[0, 1]^s$ is not symmetric. Furthermore, when s gets higher the weights at the boundaries get very small. Insiders know these problems already a long time. In the recent paper [HR06] this is nicely analyzed.

It follows that periodization is only applicable in low dimensions and with few points. But even then the transformation can give a transformed integrand which is much harder to integrate than before, see [Hic02] for a theoretical discussion and an alternative, and [HR06] for an example. Summarized, periodization is a powerful tool in the hands of an expert but in the hands of the unwary it is a dangerous tool!

4.2 Lattice Sequences

In practice one wants to have an error estimate for the approximation. The traditional approach is to use multiple randomly shifted copies of one lattice rule and then using the standard error of the multiple results as a stochastic error estimate [CP76]. However, recent interest is in lattice sequences. These are, not surprisingly, sequences of lattice rules, of which the points are embedded. That is

$$\Lambda_0 \subset \Lambda_1 \subset \cdots \subset \Lambda_\ell \subset \cdots . \tag{6}$$

In this way it becomes possible to obtain an error estimate, e.g., by using the difference of two successive approximations.

Different schemes for this embedding exist. Joe and Sloan [JS92, SJ94] introduced so-called copy rules for this purpose. This idea has been extended in low dimensions to so-called augmented lattice sequences, see, e.g. [HR99, RH02]. A different approach is by using a number of points which is a power of a given integer base. This was done by [HHLL01] and the theoretical existence of good extensible lattice rules was given in [HN03]. Also in [CKN06] and [DPW] such good lattice sequences were successfully constructed.

For such lattice sequences one uses the property that a lattice rule with b^m points consists of b smaller lattice rules with b^{m-1} points, which in turn all consist of b smaller lattice rules with b^{m-2} points, and so on. By ordering the points of the biggest lattice rule in a specific way, while keeping the embedding (6) it is even possible to stop anywhere and still have a reasonable good uniform distribution. This fixes one of the historical problems with lattice rules: one can keep on adding points until the error estimate is sufficiently small. An example in two dimensions is given in Figure 1.

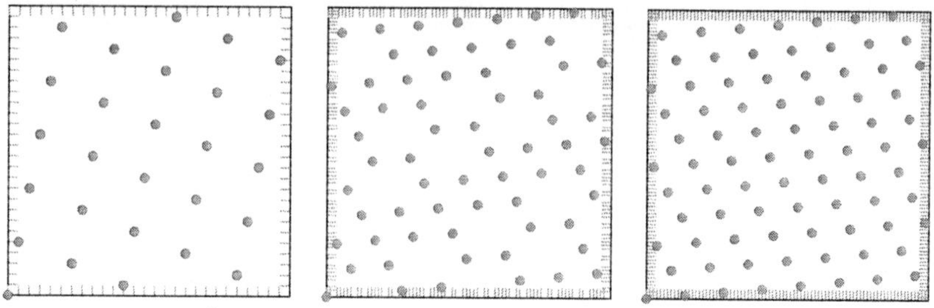

$n = 27 = 3^3$ $n = 64 = 2 \times 3^3 + 3^2 + 1$ $n = 81 = 3^4$

Fig. 1. A lattice sequence in base 3. The first image shows a full lattice using 27 points. The next image shows an extension of this lattice to 64 points, since this is not a power of 3, the resulting point set is not a full lattice. If we keep on adding points we arrive at the last figure with 81 points, again a full lattice. For a good lattice sequence the intermediate points are well distributed.

5 Lattice Rules in Action

In most recent papers on lattice rules, the emphasis is in high dimensions. Let us point out that they are useful and indeed used also in low dimensions, starting from two.

Several general purpose, black-box integration routines for 2-dimensional integration are based on lattice rules. DITAMO [RdD81] is based on the product rectangle rule in combination with the IMT periodizing transformation. d2lri and r2d2lri [HR99, RH02] use augmented lattice rules combined with a periodizing transformation. All these routines are based on sequences of embedded rules and error estimators derived from these. A nice application of 2-dimensional lattice rules is described in [Rev95]. Lattice rules also found applications in the area of computer graphics, see, e.g., [Kel04, DKD08, DK08].

An example of lattice rules in action on a 5-dimensional example is presented in [CN06]. There the result of a lattice rule of high trigonometric degree, constructed along the lines described in Section 3.1, is compared to the result of a lattice rule minimizing some worst-case error, constructed along the lines described in Section 3.2 (and constructed to be a good lattice sequence as described in Section 4). Good results were obtained without the use of periodization and both rules were used as a sequence. Lattice rules and sequences also find applications in much higher dimensions, see, e.g., [CKN06, KDSWW] for examples in 100 and more dimensions.

During the conference several speakers presented results on lattice rules in low and high dimensions. Some of these are included in this volume, e.g., [DKD08, DK08, SJ08].

6 Final Remarks

The situation of construction methods for lattice rules can be summarized as follows. Searches for lattice rules using the "classical" criteria are doomed to fail for increasing dimensions, not only because the search space is too big but also because the cost for evaluating these criteria is too high. The component-by-component algorithm, relying on the worst-case error for a reproducing kernel Hilbert space beats this curse of dimensionality. It allows the construction of lattice rules very quickly even if n and s are large.

But work remains to be done. For the component-by-component construction, tuning of the function space using the weights must be done so that a given problem belongs to (or is close to) the underlying reproducing kernel Hilbert space. More experience with reliable, cheap and deterministic error estimators for sequences, especially in high dimensions, would be interesting. Currently the usage of a low number of randomizations seems to be the preferred method. We are not aware of any extensive tests for error estimation in high dimensions.

Note that lattice rules are useful for low and high dimensions, and are not only for integrating periodic functions. Furthermore different quality criteria can be useful. Finally the difference between lattice rules and "classical" low discrepancy sequences evaporates. Lattice rules with large n can be constructed easily and can be used as low discrepancy sequences.

We would like to express our hope that some readers want to apply lattice rules in practical problems. We hope that their experiences are positive and that their reports find their way in the growing literature on lattice rules.

7 Tourist Information

It is beyond any doubt that the biggest monument in the world devoted to a lattice is the Atomium[4] in Brussels, Belgium. This monument was designed for the Brussels World's Fair that took place in 1958 (Expo '58). The Atomium consist of 9 balls symbolizing a unit cell of the body centered cubic lattice crystal structure of iron magnified 165×10^9 times.

References

[Aro50] N. Aronszajn. Theory of reproducing kernels. *Trans. Amer. Math. Soc.*, 68:337–404, 1950.

[BC93] M. Beckers and R. Cools. A relation between cubature formulae of trigonometric degree and lattice rules. In H. Brass and G. Hämmerlin, editors, *Numerical Integration IV*, pages 13–24, Basel, 1993. Birkhäuser Verlag.

[4] See www.atomium.be.

[BP85] M. Bourdeau and A. Pitre. Tables of good lattices in four and five dimensions. *Numer. Math.*, 47:39–43, 1985.

[Coo97] R. Cools. *Constructing cubature formulae: the science behind the art*, volume 6 of *Acta Numerica*, pages 1–54. Cambridge University Press, 1997.

[Coo06] R. Cools. More about cubature formulas and densest lattice packings. *East Journal on Approximations*, 12(1):37–42, 2006.

[CG03] R. Cools and H. Govaert. Five- and six-dimensional lattice rules generated by structured matrices. *J. Complexity*, 19(6):715–729, 2003.

[CKN06] R. Cools, F. Y. Kuo and D. Nuyens. Constructing embedded lattice rules for multivariate integration. *SIAM J. Sci. Comput.*, 28(6):2162–2188, 2006.

[CL01] R. Cools and J. Lyness. Three- and four-dimensional K-optimal lattice rules of moderate trigonometric degree. *Math. Comp.*, 70(236):1549–1567, 2001.

[CMS01] R. Cools, I. Mysovskikh and H. Schmid. Cubature Formulae and Orthogonal Polynomials. *J. Comput. Appl. Math.*, 127:121–152, 2001.

[CN06] R. Cools and D. Nuyens. The role of structured matrices for the construction of integration lattices. *JNAIAM J. Numer. Anal. Ind. Appl. Math.*, 1(3):257–272, 2006.

[CP76] R. Cranley and T. Patterson. Randomization of number theoretic methods for multiple integration. *SIAM J. Numer. Anal.*, 13:904–914, 1976.

[CS96] R. Cools and I. H. Sloan. Minimal cubature formulae of trigonometric degree. *Math. Comp.*, 65(216):1583–1600, 1996.

[DKD08] H. Dammertz, A. Keller and S. Dammertz. Simulation on rank-1 lattices. In this volume, pages 205–212.

[DK08] S. Dammertz and A. Keller. Image synthesis by rank-1 lattices. In this volume, pages 217–236.

[DK04a] J. Dick and F. Y. Kuo. Constructing good lattice rules with millions of points. In Niederreiter [Nie04], pages 181–197.

[DK04b] J. Dick and F. Y. Kuo. Reducing the construction cost of the component-by-component construction of good lattice rules. *Math. Comp.*, 73(248):1967–1988, 2004.

[DPW] J. Dick, F. Pillichshammer and B. Waterhouse. The construction of good extensible rank-1 lattices. *Math. Comp.* To appear.

[Fro77] K. Frolov. On the connection between quadrature formulas and sublattices of the lattice of integral vectors. *Dokl. Akad. Nauk SSSR*, 232:40–43, 1977. (Russian) Soviet Math. Dokl. 18: 37–41, 1977 (English).

[GL87] P. Gruber and C. Lekkerkerker. *Geometry of numbers*. North Holland, 1987.

[Hic98a] F. J. Hickernell. Lattice rules: How well do they measure up? In P. Hellekalek and G. Larcher, editors, *Random and Quasi-Random Point Sets*, volume 138 of *Lecture Notes in Statistics*, pages 109–166. Springer-Verlag, 1998.

[Hic98b] F. Hickernell. A generalized discrepancy and quadrature error bound. *Math. Comp.*, 67(221):299–322, 1998.

[Hic02] F. J. Hickernell. Obtaining $O(n^{-2+\epsilon})$ convergence for lattice quadrature rules. In K. T. Fang, F. J. Hickernell and H. Niederreiter, editors, *Monte Carlo and Quasi-Monte Carlo Methods 2000*, pages 274–289. Springer-Verlag, 2002.

[HHLL01] F. J. Hickernell, H. S. Hong, P. L'Écuyer and C. Lemieux. Extensible lattice sequences for quasi-Monte Carlo quadrature. *SIAM J. Sci. Comput.*, 22:1117–1138, 2001.

[HN03] F. J. Hickernell and H. Niederreiter. The existence of good extensible rank-1 lattices. *J. Complexity*, 19(3):286–300, 2003.

[HR99] M. Hill and I. Robinson. d2lri: A nonadaptive algorithm for two-dimensional cubature. *J. Comput. Appl. Math.*, 112(1–2):121–145, 1999.

[HR06] M. Hill and I. Robinson. Quadrature using 64-bit IEEE arithmetic for integrands over $[0, 1]$ with a singularity at 1. *Theoret. Comput. Sci.*, 351(1):82–100, 2006.

[Joe04] S. Joe. Component by component construction of rank-1 lattice rules having $O(n^{-1}(\ln(n))^d)$ star discrepancy. In Niederreiter [Nie04], pages 293–298.

[JS92] S. Joe and I. H. Sloan. Embedded lattice rules for multidimensional integration. *SIAM J. Numer. Anal.*, 29:1119–1154, 1992.

[JS07] S. Joe and V. Sinescu. Good lattice rules based on the general weighted star discrepancy. *Math. Comp.*, 76(258):989–1004, 2007.

[Kel04] A. Keller. Stratification by rank-1 lattices. In Niederreiter [Nie04], pages 299–313.

[Kor59] N. Korobov. On approximate calculation of multiple integrals. *Dokl. Akad. Nauk SSSR*, 124:1207–1210, 1959. (Russian).

[Kor60] N. Korobov. Properties and calculation of optimal coefficients. *Dokl. Akad. Nauk SSSR*, 132:1009–1012, 1960. (Russian) Soviet Math. Dokl. 1: 696–700, 1960 (English).

[Kuo03] F. Y. Kuo. Component-by-component constructions achieve the optimal rate of convergence for multivariate integration in weighted Korobov and Sobolev spaces. *J. Complexity*, 19:301–320, 2003.

[KDSWW] F. Y. Kuo, W. T. M. Dunsmuir, I. H. Sloan, M. P. Wand and R. S. Womersley. Quasi-Monte Carlo for highly structured generalised response models. *Methodology and Computing in Applied Probability*. To appear.

[LL02] P. L'Écuyer and C. Lemieux. Recent advances in randomized quasi-Monte Carlo methods. In M. Dror, P. L'Ecuyer and F. Szidarovszki, editors, *Modeling Uncertainty: An Examination of Stochastic Theory, Methods, and Applications*, pages 419–474. Kluwer Academic Publishers, 2002.

[Lyn89] J. Lyness. An introduction to lattice rules and their generator matrices. *IMA J. Numer. Anal.*, 9:405–419, 1989.

[Lyn03] J. Lyness. Notes on lattice rules. *J. Complexity*, 19(3):321–331, 2003.

[LS04] J. Lyness and T. Sørevik. Four-dimensional lattice rules generated by skew-circulant matrices. *Math. Comp.*, 73(245):279–295, 2004.

[LS06] J. Lyness and T. Sørevik. Five-dimensional k-optimal lattice rules. *Math. Comp.*, 75(255): 1467–1480, 2006.

[Mai72] D. Maisonneuve. Recherche et utilisation des "bons treillis". Programmation et résultats numerériques. In S. Zaremba, editor, *Applications of Number Theory to Numerical Analysis*, pages 121–201. Academic Press, 1972.

[Min67] H. Minkowski. *Gesammelte Abhandlungen.* Chelsea Publishing Company, New York, Reprinted (originally published, in 2 volumes, Leipzig, 1911) edition, 1967.

[Mys81] I. Mysovskikh. *Interpolatory Cubature Formulas.* Izdat. 'Nauka', Moscow-Leningrad, 1981. (Russian).

[Mys85] I. Mysovskikh. Quadrature formulae of the highest trigonometric degree of accuracy. *Zh. vychisl. Mat. mat. Fiz.*, 25:1246–1252, 1985. (Russian) *U.S.S.R. Comput. Maths. Math. Phys.* 25:180–184, 1985 (English).

[Mys87] I. Mysovskikh. On cubature formulas that are exact for trigonometric polynomials. *Dokl. Akad. Nauk SSSR*, 296:28–31, 1987. (Russian) *Soviet Math. Dokl.* 36:229–232, 1988 (English).

[Mys90] I. Mysovskikh. On the construction of cubature formulas that are exact for trigonometric polynomials. In A. Wakulicz, editor, *Numerical Analysis and Mathematical Modelling*, volume 24 of *Banach Center Publications*, pages 29–38. PWN - Polish Scientific Publishers, Warsaw, 1990. (Russian).

[NC06a] D. Nuyens and R. Cools. Fast algorithms for component-by-component construction of rank-1 lattice rules in shift-invariant reproducing kernel Hilbert spaces. *Math. Comp.*, 75(2):903–920, 2006.

[NC06b] D. Nuyens and R. Cools. Fast component-by-component construction of rank-1 lattice rules with a non-prime number of points. *J. Complexity*, 22(1):4–28, 2006.

[Nie78] H. Niederreiter. Quasi-Monte Carlo methods and pseudo-random numbers. *Bull. Amer. Math. Soc.*, 84(6):957–1041, 1978.

[Nie92] H. Niederreiter. *Random Number Generation and Quasi-Monte Carlo Methods*, volume 63 of *CBMS-NSF regional conference series in applied mathematics.* SIAM, Philadelphia, 1992.

[Nie04] H. Niederreiter, editor. *Monte-Carlo and Quasi-Monte Carlo Methods - 2002.* Springer-Verlag, 2004.

[Nos85] M. Noskov. Cubature formulae for the approximate integration of periodic functions. *Metody Vychisl.*, 14:15–23, 1985. (Russian).

[Nos88] M. Noskov. Formulas for the approximate integration of periodic functions. *Metody Vychisl.*, 15:19–22, 1988. (Russian).

[Osi04] N. N. Osipov. *Cubature formulas for periodic functions.* Ph.D. thesis, Krasnoyarsk State Technical University, 2004. (Russian).

[OP04] N. N. Osipov and A. V. Petrov. Construction of sequences of lattice rules which are exact for trigonometric polynomials in four variables. *Vychisl. Tekhnol.*, 9, Spec. Iss. 1:102–110, 2004. (Russian).

[Rad48] J. Radon. Zur mechanischen Kubatur. *Monatsh. Math.*, 52:286–300, 1948.

[RdD81] I. Robinson and E. deDoncker. Algorithm 45: Automatic computation of improper integrals over a bounded or unbounded planar region. *Computing*, 27:253–284, 1981.

[Rev95] M. Revers. Numerical integration of the Radon transform on classes E_s^α in multiple finite dimensions. *Computing*, 54(2):147–165, 1995.

[Rez90] A. Reztsov. On cubature formulas of Gaussian type with an asymptotic minimal number of nodes. *Mathematicheskie Zametki*, 48:151–152, 1990.

[RH02] I. Robinson and M. Hill. Algorithm 816: r2d2lri: an algorithm for automatic two-dimensional cubature. *ACM Trans. Math. Software*, 28(1):75–100, 2002.

[Sem96] A. Semenova. Computing experiments for construction of cubature formulae of high trigonometric accuracy. In M. Ramazanov, editor, *Cubature Formulas and their Applications (Russian)*, pages 105–115, Ufa, 1996.

[Sem97] A. Semenova. An algorithm for the construction of cubature formulas of high trigonometric accuracy. In C. Shoynjurov, editor, *Cubature Formulas and their Applications (Russian)*, pages 93–105, Ulan-Ude, 1997.

[SJ08] V. Sinescu and S. Joe. Good lattice rules with a composite number of points based on the product weighted star discrepancy. In this volume, pages 645–658.

[SJ94] I. H. Sloan and S. Joe. *Lattice Methods for Multiple Integration*. Oxford University Press, 1994.

[SK87] I. H. Sloan and P. Kachoyan. Lattice mathods for multiple integration: theory, error analysis and examples. *SIAM J. Numer. Anal.*, 24:116–128, 1987.

[SKJ02a] I. H. Sloan, F. Y. Kuo and S. Joe. Constructing randomly shifted lattice rules in weighted Sobolev spaces. *SIAM J. Numer. Anal.*, 40(5):1650–1665, 2002.

[SKJ02b] I. H. Sloan, F. Y. Kuo and S. Joe. On the step-by-step construction of quasi-Monte Carlo integration rules that achieve strong tractability error bounds in weighted Sobolev spaces. *Math. Comp.*, 71(240):1609–1640, 2002.

[SR02] I. H. Sloan and A. V. Reztsov. Component-by-component construction of good lattice rules. *Math. Comp.*, 71(237):263–273, 2002.

[Str60] A. Stroud. Quadrature methods for functions of more than one variable. *New York Acad. Sci.*, 86:776–791, 1960.

[Tem91] N. Temirgaliev. Application of divisor theory to the numerical integration of periodic functions of several variables. *Math. USSR Sbornik*, 69(2):527–542, 1991.

MCQMC Algorithms for Solving some Classes of Equations*

Sergej Ermakov

Saint-Petersburg State University faculty of Mathematics and Mechanics, Universitetsky prospekt, 28, 198504, Peterhof, St. Petersburg, Russia
Sergej.Ermakov@pobox.spbu.ru

1 Introduction

The Monte-Carlo method is known to be used for solving problems of very different nature. Equation solving constitutes one of the very important classes of problems. A stochastic process which can be effectively simulated by computer is usually associated with the equation under consideration. Then some functional of process trajectories is constructed in order to obtain an unbiased estimation of the required value which can be either solution of the equation or some functional of the solution. And finally, one of the laws of large numbers or limit theorems is used. Stochastic methods usually permit to apply a simple software implementation, they are easily adapted for parallel computer systems and can also effectively use a priori information about the exact problem's solution (i.e. methods of variance reduction). The well-known disadvantage of the stochastic methods is a comparatively low speed of the error decrease as the number of independent process realizations grows. There are a lot of works aimed at overcoming this disadvantage. The works concerning application of the deterministic methods in computational schemes (Quasi Monte-Carlo Method - QMC) are among them. It is important to notice that the QMC methods preserve the parallel structure of classical stochastic algorithms. It seems that the parallelism of algorithms is one of the most important problems in the modern theory of the Monte-Carlo methods. Another important problem is in comparison of computational complexities of stochastic algorithms and similar deterministic algorithms. Investigations in these fields of MC theory might be important to find out the structure of modern computational systems. This article includes a brief revue of the author's and his colleague's investigations in these and related fields. Generalizations of some results and their analysis from the point of view of parallelism are presented for the first time.

* The work is supported by the RFBJ grant No01-05-00865a

2 Classes of Discussed Equations

The following equations are considered

$$\varphi = f + \sum_{l=1}^{\infty} \mathcal{K}_l(\varphi, \ldots, \varphi) \quad (mod \mu), \tag{1}$$

where φ and f are functions of $x \in X$, they are integrable with respect to measure μ at $x \in \mathbb{X} \in R$, μ is probability measure defined on σ-algebra of subsets of $\mathbb{X}$. Operator $\mathcal{K}$ is defined by expression

$$\mathcal{K}_l(\varphi, \ldots, \varphi)(x) = \int \mu^l(dx_1, \ldots, dx_l) k_l(x, x_1, \ldots, x_l) \prod_{j=1}^{l} \varphi(x_j), \tag{2}$$

where $\mu^l(dx_1, \ldots, dx_l) = \otimes_{j=1}^{l} \mu(dx_j)$.

$(mod \mu)$ in equality (1) means that this equality takes place for every $x \in supp\ \mu$. Further $\overline{\mathcal{K}}_l(\varphi, \ldots, \varphi)$ denotes operator similar to one defined in (2) but with kernel $|k_l|$, where $|k_l|$ is absolute value of k_l.

Further we will discuss in details several particular cases of equation (1). In particular linear equation will be concerned.

3 Linear Case

In this section we'll consider

$$\varphi = f + \mathcal{K}\varphi \quad (mod \mu) \tag{3}$$

that in the case of discrete measure μ turns out to be a system of linear algebraic equations (S.L.A.E). The problem of the S.L.A.E solving is known to be one of the most important tasks of computational mathematics. We'll see further that the problem of solving (1) can be reduced to solving of (3). Due to this fact equation (3) deserves special attention.

The most widespread methods for solving (3) based on the Markov chain modeling were suggested by J. v. Neumann and Ulam [For50]. The scheme of the Markov chain connection with (3) we will further call the N.U.Scheme. The simplest variant of this scheme consists of the following operations. Fist we choose parameters of the Markov chain: the density of initial distribution $p^0(x)$ (correspondingly to measure μ) and transition density $p(x, y)$ that satisfies condition

$$\int p(x, y)\mu(dy) = 1 - g(x) \quad (mod \mu), \ 0 \leq g(x) \leq 1.$$

The chosen densities p^0 and p must also satisfy the concordance conditions:

- $p^0(x) > 0, \quad if\ h(x) \neq 0$
- $p(x,y) > 0, \quad if\ k(x,y) \neq 0$, where k is a kernel of operator $\mathcal{K}$
- $g(x) > 0, \quad if\ f(x) \neq 0$

It is obvious that $g(x)$ can be chosen in such a way that almost all trajectories will be finite. Then we simulate the Markov chain with chosen parameters and compute the functional estimate by collisions:

$$J_s(w_\tau) = \frac{h_0 k_{0,1} \dots k_{\tau-1,\tau} f_\tau}{p_0^0 p_{0,1} \dots p_{\tau-1,\tau} g_\tau}, \tag{4}$$

where $h(x_0) = h_0$, $k(x_i, x_j) = k_{i,j}$, $f(x_i) = f_i$ and similar denotations are used in the case of p_0^0 and $p_{i,j}$. We also presume $k_{-1,0} = p_{-1,0} = 1$; $\tau = 0, 1, \dots$, $w_\tau = x_0 \rightarrow x_1 \rightarrow \dots \rightarrow x_\tau$ is the chain's trajectory (see [Erm75]). Then the following theorem is valid under assumption of integral $(h, \varphi) = \int h(x)\varphi(x)\mu(dx)$ existence. Here $h = h(x)$ is a given integrable at X function.

Theorem 1. *In order that the $J_s(w_\tau)$ be an unbiased estimate of functional (h, φ) the following conditions are necessary and sufficient:*

1. *Concordance conditions hold true,*
2. *Convergence of majorant iterative process*

$$\varphi_{n+1} = \overline{\mathcal{K}\varphi_{n+1}} + \overline{|f|} \tag{5}$$

takes place.

Remark 1. In case when μ is a discrete measure p^0 is a vector and $p = P = \|p_{i,j}\|$ is a matrix of transition probabilities.

Remark 2. $J_s(w_\tau)$ is not the unique unbiased estimate. There are infinitely many of such estimates. We'll point out another estimate that is comparatively simple in computation:

$$J_s(w'_\tau) = \sum_{l=1}^{\tau} \frac{h_0}{p_0^0} \frac{k_{0,1} \dots k_{l-1,l}}{p_{0,1} \dots p_{l-1,l}} \cdot f_l, \tag{6}$$

Noticing that $(h, \varphi) = (\varphi^*, f)$, where φ^* is the solution of equation $\varphi^* = \mathcal{K}^*\varphi^* + h$, $\mathcal{K}^*$ and $\mathcal{K}$ are conjugate operators in the sense of Lagrange (in the given case $k^*(x,y) = k(y,x)$), we state that conjugate estimates for $J_s(w_\tau)$ and $J_s(w'_\tau)$ are also unbiased. For example the conjugate estimate to (6) is:

$$J_s(w''_\tau) = \sum_{l=1}^{\tau} \frac{f_0}{p_0^0} \frac{k_{0,1} \dots k_{l-1,l}}{p_{0,1} \dots p_{l-1,l}} \cdot h_l \tag{7}$$

Concordance conditions are different for every type of estimate and therefore the densities (p^0, p) that define corresponding Markov chains are different for estimates (4), (6), (7) in the general case.

4 Estimates of Comparative Complexity in the Simplest Case

It is wellknown that the stochastic methods can be more useful for estimation of separate functionals (h, φ) and for solving of equations in complicated space regions and so on. A number of papers ([Dan95] - [Erm01a]) have shown that the Monte-Carlo method can be more effective than the deterministic methods also in case of soling S.L.A.E, but only for high-dimensional systems. We'll prove several new statements confirming this point of view below. Let's consider system

$$X = AX + F,\ X = (x_1, \ldots, x_n)^T,\ F = (f_1, \ldots, f_n)^T,\ A = \|a_{i,j}\|_{i,j=1}^n.$$

We presume $\rho(|A|) < 1$, where ρ is spectral radius, and suppose $X_m = \sum\limits_{l=1}^{m} A^l F$. When m goes to infinity $X_m \to X_T$ – exact estimate of the system. Let's fix m and vector $H = (h_1, \ldots, h_n)$ and notice that

$$J_{s,m}^* = \sum_{l=1}^{r} \frac{f_{i_0}}{p_{i_0}^0} \frac{a_{i_0,i_1} \cdots a_{i_{l-1},i_l}}{p_{i_0}^0 p_{i_0,i_1} \cdots p_{i_{l-1},i_l}} \cdot h_{i_l}, \tag{8}$$

where $r = m$ if $\tau > m$ and $r = \tau$ if $\tau \leq m$, p^0 and P are an initial distribution and a matrix of transition probabilities both satisfying concordance conditions. Thus (8) is an unbiased estimate of (H, X_m). Let's choose such confidence level p and such number of independent trajectory chains N that the mean error $\frac{1}{N} \sum\limits_{j=1}^{N} J_{s,m}^*(j) - (X_m, H) = \delta(m, N)$ will belong the confidence interval $|\delta(m, N)| < \varepsilon$, where $\varepsilon = |(H, X_T - X_m)|$. Then we estimate the number of computational operations $d(n)$ needed for direct computation of value $(H, \sum\limits_{l=0}^{m} A^l F)$, and the number of computational operations $p(n)$ needed for its estimation with the Monte-Carlo method.

Here we suppose that modeling of the discrete distributions is performed with the use of

a) the bisectional method
b) the Walker method with some necessary preliminary calculations. The method demands $O(\nu(n))^2$ operations [Wal97].

Concerning matrix A we suppose that the average number of nonzero elements in row is $\nu(n)$. Then the next theorem is valid.

Theorem 2. *It exits such constants C_1, C_2 and C_3 that the next inequality takes place: In the case a)*

$$\frac{p(n)}{d(n)} \leq C_1 \frac{N log_2 \nu(n)}{n \nu(n)} \tag{9}$$

In the case b)

$$\frac{p(n)}{d(n)} = C_2 \frac{\nu(n)}{m} + C_3 \frac{N}{n\nu(n)} \tag{10}$$

The theorem's proof:

To proceed from $\sum_{k=0}^{l} A^k F$ to $\sum_{k=0}^{l+1} A^k F$ one obviously needs $2\nu(n) \cdot n + n$ operations. Thus one needs $m(2\nu(n) \cdot n + n)$ operations to compute X_m. Then we can conclude that to compute the average of estimates (8) one needs:

in case a) $(log_2\nu(n) + 3)\tau \cdot N \leq mNlog_2\nu(n)$ operations (we neglect logical and transmit operations). Thus

$$\frac{p(n)}{d(n)} = \frac{mNlog_2\nu(n)}{mn(2\nu(n)+1)} = \frac{N}{n} \frac{log_2\nu(n)}{2\nu(n)+1}$$

in case b) it is known [Wal97] that the working time for efficient modeling of each distribution takes $(\nu(n))^2 c_1$ operations, and therefore $c_1 n(\nu(n))^2$ operations as a whole. Then one need one operation of the use of a random number's generator and one operation of comparison – c_2 operations. Hence, $p(n) = c_1 n(\nu(n))^2 + c_2 mN$ and

$$\frac{p(n)}{d(n)} = \frac{c_1 n(\nu(n))^2 + c_2 mN}{mn(2\nu(n)+1)} = \frac{c_1(\nu(n))^2}{m(2\nu(n)+1)} + \frac{c_2 N}{(2\nu(n)+1)n}$$

Then (9) and (10) are evident.

Corollary 1. *Suppose N and m are fixed values. Then fraction $p(n)/d(n)$ tends to zero when n tends to infinity in case 'a'. In case 'b' this fraction tends to c/m where c is some absolute constant.*

In such a way, when sufficiently large systems are under consideration, the Monte-Carlo method can be less laborious than the iterations method. From expressions (9) and (10) one can deduct, that the previous statement is true if n is commensurable with N. The well-known proportion $N \sim 1/\varepsilon^2$ obviously remains valid.

One might find detailed analysis of these results under other assumptions about iterations processes and error's ε behavior in the papers [Dan95] - [Erm01a].

5 Parallelism

We have just considered a problem of comparison of complexity in some classes of deterministic and stochastic algorithms. It is natural to deduce, that one can recommend stochastic algorithms for estimation of solution of systems with rather high dimensions and when no high accuracy is required. But for all that we have not took into account such a important property of the Monte-Carlo algorithms as natural parallelism. This property becomes more and more important in view of the fast computer engineering development. The property of parallelism, which we mention as natural, is peculiar to algorithms of integral's evaluation with the use of quadrature formulas. In our case it is

$$(u,h) \sim \frac{1}{N}\sum_{j=1}^{N} J(\omega_\tau^j),$$

where $J(\omega_\tau^j)$ is one of estimates (4) - (8) calculated at j-th independent realization of the Markov chain. But this case requires absolute integrability of the estimate or existence of the iterative solution of majorant equation (5). This condition can be rather strict. If iterative process $\overline{\varphi}_{n+1} = \overline{\mathcal{K}}\overline{\varphi}_n + |f|$ is divergent while the process $\varphi_{n+1} = \mathcal{K}\varphi_n + f$ converge, the following approach could be used. At first, we calculate step by step $\varphi_1 = Kf + f$, $\varphi_2 = K(Kf + f) + f$, etc.; the calculation must be done with sufficient accuracy by the Monte-Carlo method. Convergence of the majorant equation isn't required in this case, but one needs two other restrictions:

1. the value of φ_m must be stored before calculation of φ_{m+1} for every m (synchronization)
2. the number of realizations of estimate φ_m can be very large.

Below we will discuss several problems connected with the mentioned points. Let's consider the case of S.L.A.E. The general case of linear operator $\mathcal{K}$ can be considered analogically provided we take into account works [Erm01b],[Erm02], [Ada04]. Suppose $X_{m+1} = AX_m + F$ (as in the previous case), but one uses the Monte-Carlo method to multiply matrix A by the vector for every m. Then one obtains:

$$\Xi_{m+1} = \widetilde{A}\Xi_m + F \tag{11}$$

where $\widetilde{A}$ is random matrix independent from Ξ_m, $E(\widetilde{A}\Xi_m \mid \Xi_m) = A\Xi_m$ and $\widetilde{A}\Xi_m$ is the average of N independent results of simulations. In the general case N can depend on m but we are considering more simple situation. For error's vector $\varepsilon_{m+1} = \Xi_{m+1} - X_{m+1}$ one obtains

$$\varepsilon_{m+1} = A\varepsilon_m + \Delta(X_m + \varepsilon_m)$$

where $\Delta = \widetilde{A} - A$. Provided equality $E\varepsilon_m = 0, m = 0, 1, \ldots$ it is easy to obtain a recurrent dependence for covariance of matrix ε_m. Namely

$$Cov\varepsilon_{m+1} = E(\varepsilon_{m+1}, \varepsilon_{m+1}^T) = ACov\varepsilon_m A^T + E(\Delta Cov\varepsilon_m \Delta^T) + \Delta X_m X_m^T \Delta^T \tag{12}$$

We call the algorithm of $\widetilde{A}\Xi_m$ computation stochastically stable if the norm of matrix $Cov\varepsilon_m$ remains bounded with the growth of m. The following theorem is a natural conclusion from the previous notices, especially from equality (12). Note that equality (12) one can interpret as equality $\overrightarrow{C_{m+1}} = \mathcal{A}\overrightarrow{C_m} + \overrightarrow{D}$, where $\overrightarrow{C_m}$ is n^2-vector composed from columns of $Cov\varepsilon_m$, $\mathcal{A}$ is $n^2 \times n^2$ matrix determined by (12) and $\overrightarrow{D}$ is a vector independent from $\overrightarrow{C_m}$.

Theorem 3. *If a spectral radius of matrix A is strictly less than one then such integer number N exists that the algorithm of Ξ_{m+1} computation is stochastically stable.*

The proof is an obvious consequence of two facts. The first one consists that elements of matrix $E(\Delta Cov\varepsilon_m \Delta^T)$ are of the same order as $\frac{1}{N}$ and $\widetilde{A}\Xi_n$ is calculated as an average of N simulated estimates of Ξ_n. The second fact consists that the spectral radius of matrix continuously depends on its elements.

Let's suppose now that one knows N_0 that provides stochastic stability of the algorithm. If one repeats independent calculations (with the use of different random numbers) of estimates X_m while m is sufficiently large then one will be able to estimate the final result as an average of these independent values.

Algorithm (11) also possesses the property of largegranular natural parallelism which is characterized by parameter N_0. It should be noticed that the *Theorem 5.1* is rather close to the results described in works [Erm01b] - [Ada04], where the property of stochastic stability was examined in application to difference schemes. These works suggested some methods of N_0 estimation. One succeeds in construction of an effective algorithm with parallel structure for difference schemes connected with hyperbolic equations in partial derivatives.

6 Biased and Unbiased Estimates

As we have just observed, the parallelism of the algorithms considered in the previous point is strongly connected with unbiasedness of the solution estimates. Here the unbiasedness implies the possibility of solution representation as an average of some distribution (an integral with discrete measure) which is equal to the possibility of approximate representation as a sum. Construction of the unbiased estimated is more complicated in nonlinear case. The difficulties are connected with the following circumstances. Suppose $\mathcal{F}(x)$ is a nonlinear function, $\xi_N = \frac{1}{N}\sum_{j=1}^{N}\xi_j$, where ξ_j are independent realizations of random value ξ and $a = E\xi$, $\sigma^2 = Var\xi$. Then $E\mathcal{F}(\xi_N) \neq \mathcal{F}(a)$. Presuming necessary smoothness of function $\mathcal{F}$ one obtain

$$\mathcal{F}(a) + \mathcal{F}'(a)(\xi_N - a) + \frac{\mathcal{F}''(a)}{2!}(\xi_N - a)^2 + \cdots$$

and

$$E\mathcal{F}(\xi_N) = \mathcal{F}(a) + \frac{\mathcal{F}''(a)}{2!} \cdot \frac{\sigma^2}{N} + O\left(N^{-3/2}\right) \tag{13}$$

One can neglect unbiasedness in equation (13) with the growth of N since

$$E\left(\mathcal{F}(\xi_N) - \mathcal{F}(a)\right)^2 \sim (\mathcal{F}'(a))^2 \frac{\sigma^2}{N} + O\left(N^{-3/2}\right).$$

Let's suppose however that we dispose of N_1 independent processors for calculation of $\xi_{N_1}^{(l)}, l = 1, \ldots, N_1$ and $\mathcal{F}^{(l)} = \mathcal{F}(\xi_{N_1}^{(l)})$, then average $\frac{1}{N_1}\mathcal{F}^{(l)}$ will have the following bias:

$$\frac{\mathcal{F}''(a)}{2} E\left(N_1^{-3/2} \sum_{l=1}^{N_1} (\xi_{N_1}^{(l)} - a)^2)\right) + O\left(N_1^{-3/2}\right) = \frac{\mathcal{F}''(a)}{2!} \cdot \frac{\sigma^2}{N_1} + O\left(N_1^{-3/2}\right)$$

This fact implies that the calculations with the small degree of accuracy but with the use of great number (N_1) of independent processors might lead to the great value of bias. One can also conclude importance of construction of unbiased estimates. Further we will discuss the methods of their construction for solving of equations with polynomial nonlinearity in the form (1).

7 Equations with the Polynomial Nonlinearity

For equations written in the form

$$\varphi(x) = f(x) + \sum_{l=1}^{M} \int k_l(x, y_1, \ldots, y_l) \prod_{j=1}^{l} (\varphi(y_j)\mu(dy_j)) \quad (mod \mu)$$

one can construct an analogue of estimate (4) provided existence of the iterative solution of the majorant equation

$$\overline{\varphi}(x) = |f(x)| + \sum_{l=1}^{M} \int |k_l(x, y_1, \ldots, y_l)| \prod_{j=1}^{l} (\overline{\varphi}(y_j)\mu(dy_j))$$

The corresponding estimate is constructed over a trajectory of the branching Markov chain, time is considered discrete and every particle might possess $0, 1, \ldots, M$ descendants at every consequent moment. Description of the corresponding computational algorithm can be found in book [Erm89]. The analogue of dual estimate (6) in nonlinear case is also discussed in this book. The corresponding algorithm also has properties of natural parallelism similar to the linear case.

Unfortunately, the analogue of the linear case with direct use of branching processes takes place only for the simplest estimate (4).

The problem can be solved by another way of construction of branching trajectory. Let's discuss the simplest case $M = 2$ in details. One can notice that equation

$$\varphi = f + \mathcal{K}_1\varphi + \mathcal{K}_2\varphi\varphi \tag{14}$$

can be rewritten in form

$$\begin{aligned}\psi_{l+1}(x, x_1, \ldots, x_l) = {} & f(x)\psi_l(x_1, \ldots, x_l) + \int k_1(x, y)\psi_{l+1}(y, x_1, \ldots, x_l)\mu(dy) + \\ & + \int\int k_2(x, y, z)\psi_{l+2}(y, z, x_1, \ldots, x_l)\mu(dy)\mu(dz),\end{aligned} \tag{15}$$

where $\psi_l(x_1, \ldots, x_l) = \prod_{j=1}^{l} \varphi(x_j)$, $l = 0, 1, 2, \ldots$. It follows from the definition of ψ that

$$\psi_l(x_1, \ldots, x_l) = \psi_l(x_{i_1}, \ldots, x_{i_l}), \tag{16}$$

where $i_1, .., i_l$ is an arbitrary permutation of indexes $1, \ldots, l$.

Linear system (15) has matrix

$$\mathcal{A} = \begin{pmatrix} \mathcal{K}_1 & \mathcal{K}_2 & 0 & 0 & . \\ f & \mathcal{K}_1 & \mathcal{K}_2 & 0 & . \\ 0 & f & \mathcal{K}_1 & \mathcal{K}_2 & . \\ . & . & . & . & . \end{pmatrix}$$

and right part $(f, 0, 0, \ldots)^T$.

The N.U.scheme applications determines the form of the Markov chain transition probability matrix:

$$\mathcal{P} = \begin{pmatrix} p_1(x,y) & p_2(x,y,z) & 0 & 0 & . \\ p_0(x) & p_1(x,y) & p_2(x,y,z) & 0 & . \\ 0 & p_0(x) & p_1(x,y) & p_2(x,y,z) & . \\ 0 & 0 & p_0 & . & . \end{pmatrix} \tag{17}$$

Elements of this matrix satisfy concordance conditions and equation

$$p_0(x) + \int p_1(x, y)\mu(dy) + \int\int p_2(x, y, z)\mu(dy)\mu(dz) = 1.$$

The process under consideration can be described as the process of particles "birth-death". A row of matrix (17) describes behavior of L particle group in point of time t. One of the particles is randomly chosen with equal probabilities. If it is located in point x in point of time t, then in the point of time $t+1$ it can

(1) either die with probability $p_0(x)$,
(2) or go to point y with probability $p_1(x) = \int p_1(x, y)\mu(dy)$,

(3) or give birth to two particles in points y and z with probability $p_2(x) = 1 - p_0(x) - p_1(x)$. In this case the particles y, z are distributed with density $p_2(x, y, z)/p_2(x)$.

The other particles don't change their coordinates while time changes from t to $t + 1$. One can obviously choose such p_0 that all particles will die with probability 1 after a finite number of iterations.

Since $h(x)$ is given, the estimates of functional (h, φ) are constructed analogously to the above-mentioned linear case (system (15) is linear).

The same reasoning can be applied to a system dual to (15), this system will be characterized by matrix $\mathcal{A}^T$ and right part $(h(x), 0, 0, \ldots)$. Similar considerations concerning $\mathcal{A}^T$ lead to a model of "collision" process with two particles which can stick together.

Since our constructed estimated are unbiased, they can be used in a scheme analogical to the one considered in section 4 of this article. This approach permits construction of "large granular" parallel algorithm.

8 Some Notes about the Quasi-Monte Carlo

The recent results concerning the modified method's application to the very simple case of solving of linea algebraic systems are presented below.

It should be noticed that quasi-random sequences could be used in the scheme of the Monte-Carlo methods and in this case they preserve parallelism which is a very important advantage the methods.

Let's notice several peculiar properties of the Quasi Monte-carlo methods application to solving of linear algebraic systems.

Suppose we have fixed some integer $s \geq 1$. Then estimate (6) can be presented in form

$$\sum_{k=0}^{min(s,\tau)} \frac{h_{i_0}}{p^0_{i_0}} \frac{a_{i_0,i_1} \ldots a_{i_{k-1},i_k}}{p_{i_0,i_1} \ldots p_{i_{k-1},i_k}} \cdot f_{i_k} + \sum_{k=s+1}^{\tau} \frac{h_{i_0}}{p^0_{i_0}} \frac{a_{i_0,i_1} \ldots a_{i_{k-1},i_k}}{p_{i_0,i_1} \ldots p_{i_{k-1},i_k}} \cdot f_{i_k}.$$

Let's denote the first sum by S_1 and the second by S_2. Due to the definition the second sum will be considered equal to zero if $s = \tau$. Then one obtains $ES_1 = \sum\limits_{k=1}^{s} HA^kF$ and $ES_2 = \sum\limits_{k=s+1}^{\infty} HA^kF$. The last component (ES_1) can be considered as an integral over trajectories with length s. Therefore the uniformly distributed in the s-dimensional hypercube quasi-random numbers can be used foe modeling the first s points of the Markov chain's trajectory. If the trajectory length is greater than s then pseudorandom numbers can be used for modeling other points of the trajectory. In this case the error decreases like $c_1 \frac{ln^s N}{N} + \frac{c_2}{\sqrt{N}}$. This approach was suggested before in a number o papers concerning the Quasi Monte-Carlo application to solving of the integral equations.

Researches in article [Erm06] showed that one can improve the results by $ln^s N$ times with the use of the "large granular" modification of the methods for linear system's solving (section (4) of this article). Indeed we use only one-dimensional quasirandom sequences for solving of S.L.A.E. In the general case of integral equation one needs r-dimensional points if $X \in \mathcal{R}^r$ Note in conclusion that properties of parallelism and possibility of use of the QMC methods made the MC method a very powerful tool in computational mathematics (even for solving some classes of classical computational problems).

References

[Ada04] A. V. Adamov and S. M. Ermakov. Stochastic Stability of the Monte Carlo Method (the Case of Operators). *Vestnik St. State Petersburg Univ.*, 37(2), 2004.

[Dan95] D. L. Danilov and S. M. Ermakov. On Complexity of Neumann-Ulam Scheme for the Multidemensional Dirichlet Problem for Nets Equations. *Vestnik of S-Petersburg State University*, 1, 1995.

[Erm75] S. M. Ermakov. *The Monte Carlo method and related topics*. Moscow, Nauka, 1975.

[Erm89] S. M. Ermakov, V. V. Nekrutkin, and A. S Sipin. *Random Processes for Classical Equations of Mathematical Physics*. Kluwer Publ., 1989.

[Erm01a] S. M. Ermakov. Addendum to the Article on the Monte-Carlo Method. *Vestnik of S-Petersburg State University*, 41(6), 2001.

[Erm01b] S. M. Ermakov and W. Wagner. Stochastic Stability and Parallelism of the Monte Carlo Method. *Doklady Rus. Akad. Nauk*, 379(4):431–439, 2001.

[Erm02] S. M. Ermakov and W. Wagner. Monte Carlo Difference Schemes for Wave Equations. *Monte Carlo Methods and Applications*, 8(1):1–29, 2002.

[Erm06] S. M. Ermakov and A. I. Rukavishnikova. Quasi Monte-Carlo Algorithms for Solving Linear Algebraic Equation. *Monte Carlo Methods and Applications*, 12(5):363–384, 2006.

[For50] G. E. Forsithe and R. Z. Leibler. Matrix Inversion by the Monte Carlo methods. *Math. tables and other aids to computations*, 4:127–129, 1950.

[Wal97] A. J. Walker. An Efficient Method for Generating Discrete Random Variables with General Distribution. *ASM Trans. Math. Software*, 3:253–256, 1997.

MCQMC Methods for Multivariate Statistical Distributions

Alan Genz

Washington State University, Mathematics Department, Pullman, WA 99164-3113
alangenz@wsu.edu

Summary. A review and comparison is presented for the use of Monte Carlo and Quasi-Monte Carlo methods for multivariate Normal and multivariate t distribution computation problems. Spherical-radial transformations, and separation-of-variables transformations for these problems are considered. The use of various Monte Carlo methods, Quasi-Monte Carlo methods and randomized Quasi-Monte Carlo methods are discussed for the different problem formulations and test results are summarized.

1 Introduction

Modern statistical computations often require multivariate probabilities. For continuous distributions, these probabilities are defined by multivariate integrals of multivariate density functions over application specific integration regions. The most important multivariate continuous distribution is the multivariate normal distribution. This review will consider the multivariate normal (MVN) distribution for hyper-rectangular integration regions. This type of MVN distribution is defined [To90] by

$$\boldsymbol{\Phi}(\mathbf{a},\mathbf{b},\boldsymbol{\Sigma}) = \frac{1}{\sqrt{|\boldsymbol{\Sigma}|(2\pi)^m}} \int_{a_1}^{b_1}\int_{a_2}^{b_2}\cdots\int_{a_m}^{b_m} e^{-\frac{1}{2}\mathbf{x}^t\boldsymbol{\Sigma}^{-1}\mathbf{x}} d\mathbf{x}, \tag{1}$$

where $\mathbf{x} = (x_1, x_2, \ldots, x_m)^t$, $-\infty \le \mathbf{a} < \mathbf{b} \le \infty$, $d\mathbf{x} = dx_m dx_{m-1}\cdots dx_1$, and $\boldsymbol{\Sigma}$ is a symmetric positive definite $m \times m$ (covariance) matrix.

A second multivariate distribution which will be considered is the multivariate t (MVT) distribution. For a hyper-rectangular integration region, the MVT distribution is defined [To90] by

$$\mathbf{T}(\mathbf{a},\mathbf{b},\boldsymbol{\Sigma},\nu) = \frac{\Gamma(\frac{\nu+m}{2})}{\Gamma(\frac{\nu}{2})(\nu\pi)^{\frac{m}{2}}\sqrt{|\boldsymbol{\Sigma}|}} \int_{a_1}^{b_1}\int_{a_2}^{b_2}\cdots\int_{a_m}^{b_m} (1+\frac{\mathbf{x}^t\boldsymbol{\Sigma}^{-1}\mathbf{x}}{\nu})^{-\frac{\nu+m}{2}} d\mathbf{x}. \tag{2}$$

An equivalent MVT form (see [Co54]), which is useful for numerical computation, is given as an integral of an MVN distribution, by

$$\mathbf{T}(\mathbf{a}, \mathbf{b}, \boldsymbol{\Sigma}, \nu) = \frac{2^{1-\frac{\nu}{2}}}{\Gamma(\frac{\nu}{2})} \int_0^\infty s^{\nu-1} e^{-\frac{s^2}{2}} \boldsymbol{\Phi}(\frac{s\mathbf{a}}{\sqrt{\nu}}, \frac{s\mathbf{b}}{\sqrt{\nu}}, \boldsymbol{\Sigma}) ds. \tag{3}$$

There are now highly accurate and fast methods available for MVN and MVT probability computations when $m < 4$, but simulation methods are usually required for higher dimensional problems. The purpose of this paper is to describe how Monte Carlo (MC), quasi-Monte Carlo (QMC), and randomized QMC (MCQMC) methods can be used for efficient approximation of MVN and MVT probabilities. Most currently available simulation methods are designed for approximate integration over the unit hyper-cube $C_m = [0, 1]^m$. Therefore, an important aspect of the construction of efficient MCQMC methods for MVN and MVT computations is the transformation of the MVN and MVT problems in the standard forms (1), (2), (3), into integrals over C_m. A significant part of this paper will be the description of transformation methods for these problems. The discussion will begin with a description of transformations based on an initial transformation to a spherical-radial coordinate system. This will be followed a discussion of transformations that result in a separation of the variables.

All of the transformations described in this paper begin with a transformation which uses the Cholesky decomposition of the covariance matrix. Let $\boldsymbol{\Sigma} = LL^t$, where is L is the $m \times m$ lower triangular Cholesky factor for $\boldsymbol{\Sigma}$. This Cholesky factor is then used to change variables from $\mathbf{x}$ to $\mathbf{y}$ with $\mathbf{x} = L\mathbf{y}$, so that equation (1) becomes

$$\boldsymbol{\Phi}(\mathbf{a}, \mathbf{b}, \boldsymbol{\Sigma}) = \frac{1}{\sqrt{(2\pi)^m}} \int_{\mathbf{a} \le L\mathbf{y} \le \mathbf{b}} e^{-\frac{\mathbf{y}^t\mathbf{y}}{2}} d\mathbf{y} \tag{4}$$

where $\mathbf{x}^t \boldsymbol{\Sigma}^{-1} \mathbf{x} = \mathbf{y}^t\mathbf{y}$, with $d\mathbf{x} = Ld\mathbf{y} = \sqrt{|\Sigma|}\mathbf{y}$. There is a similar expression for the MVT distribution (2) which becomes

$$\mathbf{T}(\mathbf{a}, \mathbf{b}, \boldsymbol{\Sigma}, \nu) = \frac{\Gamma(\frac{\nu+m}{2})}{\Gamma(\frac{\nu}{2})(\nu\pi)^{\frac{m}{2}}} \int_{\mathbf{a} \le L\mathbf{y} \le \mathbf{b}} (1 + \frac{\mathbf{y}^t\mathbf{y}}{\nu})^{-\frac{\nu+m}{2}} d\mathbf{x}. \tag{5}$$

2 Spherical-Radial Methods

This section considers the use of an additional transformation which will be called a "spherical-radial" transformation. The MVN case will be considered first. The "spherical-radial" transformation methods for MVN problems were first described and analyzed in detail by Deák (see [De80, De86, De90]).

2.1 The MVN Spherical-Radial Transformation

By letting $\mathbf{y} = r\mathbf{z}$, with $||\mathbf{z}||_2 = 1$, so that $\mathbf{y}^t\mathbf{y} = r^2$ and $d\mathbf{y} = r^{m-1}drd\mathbf{z}$, and reorganizing the normalization constant, equation (4) becomes

$$\begin{aligned}\boldsymbol{\Phi}(\mathbf{a},\mathbf{b},\boldsymbol{\Sigma}) &= \frac{\Gamma(\frac{m}{2})}{2\pi^{\frac{m}{2}}} \int\limits_{||\mathbf{z}||_2=1} \frac{2^{1-\frac{m}{2}}}{\Gamma(\frac{m}{2})} \int\limits_{\mathbf{a}\le rL\mathbf{z}\le\mathbf{b}} r^{m-1}e^{-\frac{r^2}{2}}drd\mathbf{z} \\ &= \frac{\Gamma(\frac{m}{2})}{2\pi^{\frac{m}{2}}} \int\limits_{||\mathbf{z}||_2=1} \frac{2^{1-\frac{m}{2}}}{\Gamma(\frac{m}{2})} \int\limits_{\rho_l(\mathbf{a},\mathbf{b},L,\mathbf{z})}^{\rho_u(\mathbf{a},\mathbf{b},L,\mathbf{z})} r^{m-1}e^{-\frac{r^2}{2}}drd\mathbf{z},\end{aligned}$$

where $\rho_l(\mathbf{a},\mathbf{b},L,\mathbf{z})$ and $\rho_u(\mathbf{a},\mathbf{b},L,\mathbf{z})$ are distances from the origin to points where a vector in the $\mathbf{z}$ direction intersects the integration region. If we let $\mathbf{v} = C\mathbf{z}$, the limits for the r-variable integration are given by

$$\rho_l(\mathbf{z}) = \max\{0, \max_{v_i>0}\{a_i/v_i\}, \max_{v_i<0}\{b_i/v_i\}\}$$

and

$$\rho_u(\mathbf{z}) = \max\{0, \min\{\min_{v_i>0}\{b_i/v_i\}, \min_{v_i<0}\{a_i/v_i\}\}\}.$$

If the origin is inside the integration region then $\rho_l(\mathbf{a},\mathbf{b},L,\mathbf{z}) = 0$. The equation for $\boldsymbol{\Phi}(\mathbf{a},\mathbf{b},\boldsymbol{\Sigma})$ can be rewritten as

$$\boldsymbol{\Phi}(\mathbf{a},\mathbf{b},\boldsymbol{\Sigma}) = \frac{\Gamma(\frac{m}{2})}{2\pi^{\frac{m}{2}}} \int\limits_{||\mathbf{z}||_2=1} F(\mathbf{a},\mathbf{b},L,\mathbf{z})d\mathbf{z}, \tag{6}$$

where $F(\mathbf{a},\mathbf{b},L,\mathbf{z})$ is given by

$$F(\mathbf{a},\mathbf{b},L,\mathbf{z}) = \frac{2^{1-\frac{m}{2}}}{\Gamma(\frac{m}{2})} \int\limits_{\rho_l(\mathbf{a},\mathbf{b},L,\mathbf{z})}^{\rho_u(\mathbf{a},\mathbf{b},L,\mathbf{z})} r^{m-1}e^{-\frac{r^2}{2}}dr.$$

Given $\mathbf{a}$, $\mathbf{b}$, L and $\mathbf{z}$, $F(\mathbf{a},\mathbf{b},L,\mathbf{z})$ can be computed using differences of univariate χ distribution function values, with $\chi_\nu(u)$ defined by

$$\chi_\nu(u) = \frac{2^{1-\frac{m}{2}}}{\Gamma(\frac{m}{2})} \int\limits_0^u s^{\nu-1}e^{-\frac{s^2}{2}}ds.$$

This is a standard statistical distribution which can be computed with standard statistical software. The problem of estimating $\boldsymbol{\Phi}(\mathbf{a},\mathbf{b},\boldsymbol{\Sigma})$ reduces to the problem of integration of F over U_m, the surface of the unit m-sphere. A simple N-point Monte-Carlo method would use

$$\boldsymbol{\Phi}(\mathbf{a},\mathbf{b},\boldsymbol{\Sigma}) \approx \frac{1}{N}\sum_{i=1}^{N} F(\mathbf{a},\mathbf{b},L,\mathbf{z}_i),$$

where the points $\mathbf{z}_i$ are chosen randomly (see [Dv86]) from the surface of U_m.

In order to easily use quasi-random point sets, a transformation from C_{m-1} to U_m is needed. A good transformation for this is given in the book by Fang and Wang [FW94]. This transformation from a point $\mathbf{w} \in C_{m-1}$ to a $\mathbf{z} \in U_m$, is defined by

$$z_{m-2i+2}(\mathbf{w}) = \sin(2\pi w_{m-2i+1})\sqrt{1 - w_{m-2i}^{\frac{2}{m-2i}}} \prod_{k=1}^{i-1} w_{m-2k}^{\frac{1}{m-2k}}$$

$$z_{m-2i+1}(\mathbf{w}) = \cos(2\pi w_{m-2i+1})\sqrt{1 - w_{m-2i}^{\frac{2}{m-2i}}} \prod_{k=1}^{i-1} w_{m-2k}^{\frac{1}{m-2k}}$$

for $i = 1, 2, \ldots l$, where $l = \lfloor \frac{m}{2} \rfloor - 1$, and ending with

$$z_2(\mathbf{w}) = \sin(2\pi w_1) \prod_{k=1}^{l} w_{m-2k}^{\frac{1}{m-2k}} \text{ and } z_1(\mathbf{w}) = \cos(2\pi w_1) \prod_{k=1}^{l} w_{m-2k}^{\frac{1}{m-2k}},$$

when m is even, or ending with

$$z_3(\mathbf{w}) = (2w_1 - 1) \prod_{k=1}^{l} w_{m-2k}^{\frac{1}{m-2k}},$$

$$z_2(\mathbf{w}) = 2\sin(2\pi w_2)\sqrt{w_1(1-w_1)} \prod_{k=1}^{l} w_{m-2k}^{\frac{1}{m-2k}}$$

and

$$z_1(\mathbf{w}) = 2\cos(2\pi w_2)\sqrt{w_1(1-w_1)} \prod_{k=1}^{l} w_{m-2k}^{\frac{1}{m-2k}},$$

when m is odd. This transformation has a constant Jacobian, so an MC algorithm for the MVN problem, based on uniform C_{m-1} points, uses

$$\boldsymbol{\Phi}(\mathbf{a}, \mathbf{b}, \boldsymbol{\Sigma}) \approx \frac{1}{N} \sum_{k=1}^{N} F(\mathbf{a}, \mathbf{b}, L, \mathbf{z}(\mathbf{w}_k)), \tag{7}$$

with all $w_{i,k} \sim$ Uniform(0, 1). A QMC algorithm for the MVN problem replaces the MC $\{w_k\}$ point set with an appropriately chosen QMC point set. Another transformation, with a simpler formula, but an extra integration variable and a higher computational cost, uses $\mathbf{w} \in C_m$ followed by $\mathbf{z} = \Phi^{-1}(\mathbf{w})/||\Phi^{-1}(\mathbf{w})||_2$, with $\Phi^{-1}(\mathbf{w})$ applied component-wise (see [Dv86]).

2.2 MVT Spherical-Radial Transformation

The use of spherical-radial transformations for MVT computations have been discussed and analyzed by Somerville ([So97, So98a, So98b, So99]) and Genz and Bretz ([GB00, GB01, GB02]). After the spherical-radial transformation $\mathbf{y} = r\mathbf{z}$, with $||\mathbf{z}||_2 = 1$, equation (2) becomes

$$\mathbf{T}(\mathbf{a},\mathbf{b},\boldsymbol{\Sigma},\nu) = \frac{\Gamma(\frac{m}{2})}{2\pi^{\frac{m}{2}}} \int\limits_{||\mathbf{z}||_2=1} \frac{2\Gamma(\frac{\nu+m}{2})}{\Gamma(\frac{m}{2})\Gamma(\frac{\nu}{2})\nu^{\frac{m}{2}}} \int\limits_{\mathbf{a}\le rL\mathbf{z}\le\mathbf{b}} \frac{r^{m-1}}{(1+\frac{r^2}{\nu})^{\frac{\nu+m}{2}}} dr d\mathbf{z}$$

$$= \frac{\Gamma(\frac{m}{2})}{2\pi^{\frac{m}{2}}} \int\limits_{||\mathbf{z}||_2=1} \frac{2\Gamma(\frac{\nu+m}{2})}{\Gamma(\frac{m}{2})\Gamma(\frac{\nu}{2})\nu^{\frac{m}{2}}} \int\limits_{\rho_l(\mathbf{a},\mathbf{b},L,\mathbf{z})}^{\rho_u(\mathbf{a},\mathbf{b},L,\mathbf{z})} \frac{r^{m-1}}{(1+\frac{r^2}{\nu})^{\frac{\nu+m}{2}}} dr d\mathbf{z},$$

which can be rewritten as

$$\mathbf{T}(\mathbf{a},\mathbf{b},\boldsymbol{\Sigma},\nu) = \frac{\Gamma(\frac{m}{2})}{2\pi^{\frac{m}{2}}} \int\limits_{||\mathbf{z}||_2=1} G(\mathbf{a},\mathbf{b},L,\nu,\mathbf{z}) d\mathbf{z}.$$

with $G(\mathbf{a},\mathbf{b},L,\nu,\mathbf{z})$ defined by

$$G(\mathbf{a},\mathbf{b},L,\nu,\mathbf{z}) = \frac{2\Gamma(\frac{\nu+m}{2})}{\Gamma(\frac{m}{2})\Gamma(\frac{\nu}{2})\nu^{\frac{m}{2}}} \int\limits_{\rho_l(\mathbf{a},\mathbf{b},L,\mathbf{z})}^{\rho_u(\mathbf{a},\mathbf{b},L,\mathbf{z})} \frac{r^{m-1}}{(1+\frac{r^2}{\nu})^{\frac{\nu+m}{2}}} dr. \tag{8}$$

This function can be computed using the univariate distribution

$$g(u) = \frac{2\Gamma(\frac{\nu+m}{2})}{\Gamma(\frac{m}{2})\Gamma(\frac{\nu}{2})\nu^{\frac{m}{2}}} \int_0^u \frac{r^{m-1}}{(1+\frac{r^2}{\nu})^{\frac{\nu+m}{2}}} dr.$$

This is a standard statistical distribution (related to the "F" distribution, and often transformed to a Beta distribution), which can be computed using standard statistical software.

If the transformation from C_{m-1} given in the previous section is used, then MC or QMC approximations to $\mathbf{T}(\mathbf{a},\mathbf{b},\boldsymbol{\Sigma},\nu)$ are given by

$$\mathbf{T}(\mathbf{a},\mathbf{b},\boldsymbol{\Sigma},\nu) \approx \frac{1}{N}\sum_{k=1}^{N} G(\mathbf{a},\mathbf{b},L,\nu,\mathbf{z}(\mathbf{w}_k)), \tag{9}$$

for a selected set of MC or QMC points $\{\mathbf{w}_k\}$, with each $w_k \in C_{m-1}$.

An alternate spherical radial approximation method can be based on the definition for $\mathbf{T}(\mathbf{a},\mathbf{b},\boldsymbol{\Sigma},\nu)$ given by equation (3). After the spherical-radial transformation is used with this equation,

$$\mathbf{T}(\mathbf{a},\mathbf{b},\boldsymbol{\Sigma},\nu) = \frac{2^{1-\frac{\nu}{2}}}{\Gamma(\frac{\nu}{2})} \int\limits_0^\infty s^{\nu-1} e^{-\frac{s^2}{2}} \frac{\Gamma(\frac{m}{2})}{2\pi^{\frac{m}{2}}} \int\limits_{||\mathbf{z}||_2=1} F(\frac{s\mathbf{a}}{\sqrt{\nu}}, \frac{s\mathbf{b}}{\sqrt{\nu}}, L, \mathbf{z}) d\mathbf{z} ds.$$

Then MC or QMC approximations to $\mathbf{T}(\mathbf{a}, \mathbf{b}, \boldsymbol{\Sigma}, \nu)$ are given by

$$\mathbf{T}(\mathbf{a}, \mathbf{b}, \boldsymbol{\Sigma}, \nu) \approx \frac{1}{N} \sum_{k=1}^{N} F\left(\frac{\chi_\nu^{-1}(w_{km})\mathbf{a}}{\sqrt{\nu}}, \frac{\chi_\nu^{-1}(w_{km})\mathbf{b}}{\sqrt{\nu}}, L, \nu, \mathbf{z}(\mathbf{w}_k)\right). \quad (10)$$

Because of the extra s integration variable, the w_k points in this formula have an extra component $w_{km} \in [0,1]$, which is transformed to the s interval $[0, \infty)$ using the inverse χ_ν transformation.

The use of various types of antithetic variates, developed by Deák [De90], and described in the previous section for MVN problems, can also be used for spherical radial computations for MVT problems if the G and F functions in equations (9) and (10) are replaced by the appropriate antithetic variable sums.

3 Separated-Variable Methods

Separated-variable methods for MVN problems were first studied by Genz [Ge92], Geweke [Ge91] and Hajivassiliou [Ha93, HMR96], and later by Vijverberg [Vi96, Vi97, Vi00]. These methods start by using the Cholesky decomposition coefficients $l_{i,j}$ to produce explicit integration limits for the y variables, given by

$$b_i'(y_1, \ldots, y_{i-1}) = (b_i - \sum_{j=1}^{i-1} l_{i,j} y_j)/l_{i,i},$$

and

$$a_i'(y_1, \ldots, y_{i-1}) = (a_i - \sum_{j=1}^{i-1} l_{i,j} y_j)/l_{i,i},$$

for $i = 1, 2, \ldots, m$.

3.1 MVN Separated-Variable Methods

In the MVN case, equation (4) becomes

$$\boldsymbol{\Phi}(\mathbf{a}, \mathbf{b}, \boldsymbol{\Sigma}) = \frac{1}{\sqrt{(2\pi)^m}} \int_{a_1'}^{b_1'} e^{-\frac{y_1^2}{2}} \int_{a_2'(y_1)}^{b_2'(y_1)} e^{-\frac{y_2^2}{2}} \cdots \int_{a_m'(y_1,\ldots,y_{m-1})}^{b_m'(y_1,\ldots,y_{m-1})} e^{-\frac{y_m^2}{2}} d\mathbf{y}.$$

Then, the univariate normal distribution $\Phi(y)$, defined by

$$\Phi(y) = \frac{1}{\sqrt{2\pi}} \int_{-\infty}^{y} e^{-\frac{s^2}{2}} ds,$$

is used to transform the $\mathbf{y}$ variables to $\mathbf{z}$ variables using $y_i = \Phi^{-1}(z_i)$ for $i = 1, \ldots, m$. Now,

$$\boldsymbol{\Phi}(\mathbf{a}, \mathbf{b}, \boldsymbol{\Sigma}) = \int\limits_{\Phi(a'_1)}^{\Phi(b'_1)} \int\limits_{\Phi(a'_2(\Phi^{-1}(z_1)))}^{\Phi(b'_2(\Phi^{-1}(z_1)))} \cdots \int\limits_{\Phi(a'_m(\Phi^{-1}(z_1),\ldots,\Phi^{-1}(z_{m-1})))}^{\Phi(b'_m(\Phi^{-1}(z_1),\ldots,\Phi^{-1}(z_{m-1})))} d\mathbf{z}.$$

A final set of transformations to $[0, 1]$ $\mathbf{w}$ variables is defined by

$$z_i = d_i(w_1, \ldots, w_{i-1}) + (e_i(w_1, \ldots, w_{i-1}) - d_i(w_1, \ldots, w_{i-1}))w_i,$$

so that $dz_i = (e_i - d_i)dw_i$, with

$$e_i(w_1, \ldots, w_{i-1}) = \Phi((b_i - \sum_{j=1}^{i-1} l_{i,j}\Phi^{-1}(d_j + (e_j - d_j)w_j))/l_{i,i})$$

and

$$d_i(w_1, \ldots, w_{i-1}) = \Phi((a_i - \sum_{j=1}^{i-1} l_{i,j}\Phi^{-1}(d_j + (e_j - d_j)w_j))/l_{i,i}),$$

for $i = 1, 2, \ldots, m$. The result is a formula for $\boldsymbol{\Phi}(\mathbf{a}, \mathbf{b}, \boldsymbol{\Sigma})$ as an integral over C_m given by

$$\boldsymbol{\Phi}(\mathbf{a}, \mathbf{b}, \boldsymbol{\Sigma}) = (e_1 - d_1) \int_0^1 (e_2(w_1) - d_2(w_1))$$
$$\cdots \int_0^1 (e_m(w_1, \ldots, w_{m-1}) - d_m(w_1, \ldots, w_{m-1})) \int_0^1 d\mathbf{w}.$$

Note: the innermost integral is 1 exactly, so this is an m-1-dimensional integration problem. MC or QMC approximations to $\boldsymbol{\Phi}(\mathbf{a}, \mathbf{b}, \boldsymbol{\Sigma})$ can computed using

$$\boldsymbol{\Phi}(\mathbf{a}, \mathbf{b}, \boldsymbol{\Sigma}) \approx \frac{1}{N} \sum_{i=k}^{N} D(\mathbf{w}_k), \tag{11}$$

where

$$D(\mathbf{w}) = (e_1 - d_1) \cdots (e_m(w_1, \ldots, w_{m-1}) - d_m(w_1, \ldots, w_{m-1})),$$

and the $\{\mathbf{w}_k\}$ point set is an MC or QMC point set from C_{m-1}.

3.2 MVT Separated-Variable Methods

Separated-variables for MVT problems were first carefully studied by Genz and Bretz [GB99, GB01, GB02]. In this case, if explicit a' and b' limits are introduced in equation (5),

$$\mathbf{T}(\mathbf{a},\mathbf{b},\boldsymbol{\Sigma},\nu) = K_\nu^{(m)} \int_{a_1'}^{b_1'} \int_{a_2'(y_1)}^{b_2'(y_1)} \cdots \int_{a_m'(y_1,\ldots,y_{m-1})}^{b_m'(y_1,\ldots,y_{m-1})} (1+\frac{\mathbf{y}^t\mathbf{y}}{\nu})^{-\frac{\nu+m}{2}} d\mathbf{y},$$

with $K_\nu^{(m)} = \Gamma(\frac{\nu+m}{2})/\Gamma(\frac{\nu}{2})(\nu\pi)^{\frac{m}{2}}$. The variables are still not separated, but the formula

$$(1+\frac{\sum_{j=1}^m y_j^2}{\nu}) = (1+\frac{y_1^2}{\nu})(1+\frac{y_2^2}{\nu+y_1^2})\cdots(1+\frac{y_m^2}{\nu+\sum_{j=1}^{m-1} y_j^2})$$

can be used to rewrite $\mathbf{T}(\mathbf{a},\mathbf{b},\boldsymbol{\Sigma},\nu)$ as

$$\mathbf{T}(\mathbf{a},\mathbf{b},\boldsymbol{\Sigma},\nu) = K_\nu^{(m)} \int_{a_1'}^{b_1'} \frac{1}{(1+\frac{y_1^2}{\nu})^{\frac{m+\nu}{2}}} \int_{a_2'(y_1)}^{b_2'(y_1)} \frac{1}{(1+\frac{y_2^2}{\nu+y_1^2})^{\frac{m+\nu}{2}}}$$
$$\cdots \int_{a_m'(y_1,\ldots,y_{m-1})}^{b_m'(y_1,\ldots,y_{m-1})} \frac{1}{(1+\frac{y_m^2}{\nu+\sum_{j=1}^{m-1} y_j^2})^{\frac{m+\nu}{2}}} d\mathbf{y}$$

This formula motivates (see [GB99]) an extra set of transformations

$$y_i = u_i\sqrt{\frac{\nu+\sum_{j=1}^{i-1} y_j^2}{\nu+i-1}}, \text{ with } dy_i = du_i\sqrt{\frac{\nu+\sum_{j=1}^{i-1} y_j^2}{\nu+i-1}}, \quad i=1,\ldots,m.$$

After defining the rescaled limits

$$\tilde{a}_i = a_i'\sqrt{\frac{\nu+i-1}{\nu+\sum_{j=1}^{i-1} y_j^2}}, \text{ and } \tilde{b}_i = b_i'\sqrt{\frac{\nu+i-1}{\nu+\sum_{j=1}^{i-1} y_j^2}}, \text{ for } i=1,\ldots,m,$$

the resulting separated-variable form equation for $\mathbf{T}(\mathbf{a},\mathbf{b},\boldsymbol{\Sigma},\nu)$ is

$$\mathbf{T}(\mathbf{a},\mathbf{b},\boldsymbol{\Sigma},\nu) = \int_{\tilde{a}_1}^{\tilde{b}_1} \frac{K_\nu^{(1)}}{(1+\frac{u_1^2}{\nu})^{\frac{1+\nu}{2}}} \cdots \int_{\tilde{a}_m(y_1,\ldots,y_{m-1})}^{\tilde{b}_m(y_1,\ldots,y_{m-1})} \frac{K_{\nu+m-1}^{(1)}}{(1+\frac{u_m^2}{m+\nu-1})^{\frac{m+\nu}{2}}} d\mathbf{u}.$$

The transformations $u_i = t_{\nu+i-1}^{-1}(z_i)$, for $i = 1, \ldots, m$, where $t_\nu(u)$ is the standard univariate Student-t distribution defined by

$$t_\nu(u) = K_\nu^{(1)} \int_{-\infty}^{u} (1 + \frac{s^2}{\nu})^{-\frac{1+\nu}{2}} ds.$$

can then be used to transform to $[0, 1]$ variables. Then, $\mathbf{T}(\mathbf{a}, \mathbf{b}, \boldsymbol{\Sigma}, \nu)$ becomes

$$\mathbf{T}(\mathbf{a}, \mathbf{b}, \boldsymbol{\Sigma}, \nu) = \int_{t_\nu(\tilde{a}_1)}^{t_\nu(\tilde{b}_1)} \cdots \int_{t_{\nu+m-1}(\tilde{a}_m(t_\nu^{-1}(z_1),\ldots,t_{\nu+m-2}^{-1}(z_{m-1})))}^{t_{\nu+m-1}(\tilde{b}_m(t_\nu^{-1}(z_1),\ldots,t_{\nu+m-2}^{-1}(z_{m-1})))} d\mathbf{z}.$$

Finally, the transformations to the w_i's are given by $z_i = \tilde{d}_i + (\tilde{e}_i - \tilde{d}_i)w_i$, with

$$\tilde{d}_i(w_1, \ldots, w_{i-1}) = t_{\nu+m-1}(\tilde{a}_m(t_\nu^{-1}(z_1(w_1)), \ldots, t_{\nu+i-2}^{-1}(z_{i-1}(w_{i-1}))))$$

and

$$\tilde{e}_i(w_1, \ldots, w_{i-1}) = t_{\nu+m-1}(\tilde{b}_m(t_\nu^{-1}(z_1(w_1)), \ldots, t_{\nu+i-2}^{-1}(z_{i-1}(w_{i-1})))),$$

for $i = 1, 2, \ldots, m$, so that

$$\mathbf{T}(\mathbf{a}, \mathbf{b}, \boldsymbol{\Sigma}, \nu) = (\tilde{e}_1 - \tilde{d}_1) \int_0^1 (\tilde{e}_2(w_1) - \tilde{d}_2(w_1))$$
$$\cdots (\tilde{e}_m(w_1 \ldots w_{m-1}) - \tilde{d}_m(w_1 \ldots w_{m-1})) \int_0^1 d\mathbf{w}.$$

Note: the innermost integral is again 1 exactly, so this is also an m-1-dimensional integration problem. MC or QMC approximations to $\mathbf{T}(\mathbf{a}, \mathbf{b}, \boldsymbol{\Sigma}, \nu)$ can computed using

$$\mathbf{T}(\mathbf{a}, \mathbf{b}, \boldsymbol{\Sigma}, \nu) \approx \frac{1}{N} \sum_{i=k}^{N} E(\mathbf{w}_k), \qquad (12)$$

where

$$E(\mathbf{w}) = (\tilde{e}_1 - \tilde{d}_1) \cdots (\tilde{e}_m(w_1, \ldots, w_{m-1}) - \tilde{d}_m(w_1, \ldots, w_{m-1})),$$

and $\{\mathbf{w}_k\}$ is an MC or QMC point set from C_{m-1}.

An alternate separated-variable approximation method, based on alternate definition for $\mathbf{T}(\mathbf{a}, \mathbf{b}, \boldsymbol{\Sigma}, \nu)$ given by equation (3), can also be described. After the separated-variable transformation is used with this equation,

$$\mathbf{T}(\mathbf{a}, \mathbf{b}, \boldsymbol{\Sigma}, \nu) = \frac{2^{1-\frac{\nu}{2}}}{\Gamma(\frac{\nu}{2})} \int_0^\infty s^{\nu-1} e^{-\frac{s^2}{2}} \int_{\hat{a}_1(s)}^{\hat{b}_1(s)} e^{-\frac{y_1^2}{2}} \cdots \int_{\hat{a}_m(s,y_1,\ldots,y_{m-1})}^{\hat{b}_m(s,y_1,\ldots,y_{m-1})} e^{-\frac{y_m^2}{2}} d\mathbf{y} ds,$$

with

$$\hat{a}_i(s, y_1. \ldots, y_{i-1}) = \frac{s}{\sqrt{\nu}}(a_i - \sum_{j=1}^{i-1} c_{i,j} y_j)/c_{i,i},$$

and

$$\hat{b}_i(s, y_1. \ldots, y_{i-1}) = \frac{s}{\sqrt{\nu}}(b_i - \sum_{j=1}^{i-1} c_{i,j} y_j)/c_{i,i}.$$

Then, using $\hat{z}_i = \hat{d}_i + (\hat{e}_i - \hat{d}_i) w_i$,

$$\mathbf{T}(\mathbf{a}, \mathbf{b}, \boldsymbol{\Sigma}, \nu) = \frac{2^{1-\frac{\nu}{2}}}{\Gamma(\frac{\nu}{2})} \int_0^\infty s^{\nu-1} e^{-\frac{s^2}{2}} (\hat{e}_1(s, \mathbf{w}) - \hat{d}_1(s, \mathbf{w})) \int_0^1 \cdots (\hat{e}_m(s, \mathbf{w}) - \hat{d}_m(s, \mathbf{w})) \int_0^1 d\mathbf{w} ds,$$

where

$$\hat{d}_i(s, \mathbf{w}) = \Phi(\hat{a}_m(s, \Phi^{-1}(\hat{z}_1(w_1)), \ldots, \Phi^{-1}(\hat{z}_{i-1}(w_{i-1})))),$$

and

$$\hat{e}_i(s, \mathbf{w}) = \Phi(\hat{b}_m(s, \Phi^{-1}(\hat{z}_1(w_1)), \ldots, \Phi^{-1}(\hat{z}_{i-1}(w_{i-1})))).$$

for $i = 1, 2, \ldots, m$. The innermost $\mathbf{w}$ component integral has value one, so the $\mathbf{T}$ integral is determined as an m-dimensional integral. An MC or QMC algorithm for this formulation of the MVT problem uses

$$\mathbf{T}(\mathbf{a}, \mathbf{b}, \boldsymbol{\Sigma}, \nu) \approx \frac{1}{N} \sum_{k=1}^{N} \prod_{i=1}^{m} (\hat{e}_i(\chi^{-1}(w_{m,k}), \mathbf{w}) - \hat{d}_i(\chi^{-1}(w_{m,k}), \mathbf{w})) \qquad (13)$$

4 MCQMC Algorithms for MVN and MVT

There have been many types of Monte Carlo and quasi-Monte Carlo methods that have been used for the transformed integrands described in the previous sections. The use of Monte Carlo methods will be discussed first.

4.1 MC Algorithms for MVN and MVT Computations

All of the transformed problems discussed so far, can be approximated using MC methods for integrals of the form

$$I(f) = \int_0^1 \int_0^1 \dots \int_0^1 f(\mathbf{v}) d\mathbf{v}, \tag{14}$$

with f is appropriately chosen, and $\mathbf{v}$ has length m or $m-1$. Simple antithetic variate methods, where $f(\mathbf{v})$ is replaced by $(f(\mathbf{v}) + f(\mathbf{1} - \mathbf{v}))/2$ will usually improve convergence.

The use of more sophisticated types of antithetic variates was described Deák [De90] for the spherical-radial transformed MVN algorithms. Let Z be an $m \times m$ uniformly random (with Haar measure, see Stewart [St90]) orthogonal matrix with columns $\{\mathbf{z}_j\}$, and define

$$S_n(Z) = \frac{1}{2^n \binom{m}{n}} \sum_{\mathbf{s}} \sum_{1 \le j_1 < \cdots < j_n \le m} F(\mathbf{a}, \mathbf{b}, L, \frac{\sum_{l=1}^{n} s_l \mathbf{z}_{j_l}}{\sqrt{n}}),$$

where $\mathbf{s} = (s_1, s_2, \dots, s_n) = (\pm 1, \dots, \pm 1)$ and the outer sum is taken over the 2^n possible sign combinations for the components of $\mathbf{s}$. The sample points used by $S_n(Z)$ are very evenly spread over the surface of the unit m-sphere. For MVN problems, Deák found that the larger values for the parameter n (which must satisfy $n \le m$) can provide values for S_n with significantly smaller variances. But the larger the n value, the higher the computational cost for S_n, so these two features of the S_n sums must be balanced, for practical computations. Deák recommended values of $n = 1, 2$ or 3 for typical computations. MVN estimates based on S_n are obtained using

$$\Phi(\mathbf{a}, \mathbf{b}, \Sigma) \approx \frac{1}{N} \sum_{k=1}^{N} S_n(Z_k). \tag{15}$$

These methods can also be written in the form given by equation (14), because each random orthogonal matrix can be generated from a sequence of $\frac{m(m-1)}{2}$ Uniform$(0, 1)$ numbers (see [FW94]), but $\mathbf{v}$ will have length $\frac{m(m-1)}{2}$. These Deák antithetic variate methods can also be used for MVT spherical-radial transformed algorithms.

The standard error can be used to provide an error estimate for all of the algorithms for MC methods. To standardize notation, let $\hat{\mathbf{S}}_N$ denote an approximation to Φ or $\mathbf{T}$ obtained using N integrand values $\{f(\mathbf{v}_k)\}_{k=1}^{N}$,with

$$\hat{\mathbf{S}}_N = \frac{1}{N} \sum_{k=1}^{N} f(\mathbf{v}_k).$$

The standard error σ_N^2, for $\hat{\mathbf{S}}_N$, is defined using

$$\sigma_N^2 = \frac{1}{N(N-1)}\sum_{k=1}^{N}(f(\mathbf{v}_k) - \hat{\mathbf{S}}_N)^2.$$

The error estimate $\hat{\epsilon} = 3\sigma_N$ provides an approximate confidence level of 99%.

One problem with the SR transformation algorithms is that $F(\mathbf{a}, \mathbf{b}, L, \nu, \mathbf{z})$ (or $G(\mathbf{a}, \mathbf{b}, L, \nu, \mathbf{z})$) as a function of $\mathbf{z}$, although continuous, is not very smooth, because of sharp corners of the integration region defined by $\mathbf{a} \le rL\mathbf{z} \le \mathbf{b}$. This problem can produce approximations with large variation and slower convergence for SR algorithms for MVN and MVT problems (see [Ge93]), [GB02]). For some combinations of $\mathbf{a}$ and $\mathbf{b}$, (e.g. if $\mathbf{0} < \mathbf{a} < \mathbf{b}$) many F or G values will be zero, and this can cause further reductions in the efficiency of the SR algorithms.

Prioritization of Variables

The MVN and MVT problems defined initially by equations (1), (2), and (3), actually have $m!$ equivalent definitions each, based on the $m!$ possible ways that the variables can be permuted. The choice of a particular permutation determines the Cholesky factor L for $\boldsymbol{\Sigma}$ and, after all of the transformations have be completed for a particular method, could affect the overall variation in the function that is used for simulation in the equations (7), (9), (10), (11), (12), and (13). Schervish [Sc84] originally suggested that the computation of MVN probabilities should be easier for numerical integration methods if the variables are reordered (and appropriate rows and columns of $\boldsymbol{\Sigma}$ are permuted) so that the innermost integrals have the larger integration intervals. This sorting heuristic often has the effect that the innermost integrals have expected value closer to one, thereby reducing the overall variation in the integrand. Gibson, Glasbey and Elston [GG94], suggested an improved prioritization of the variables, which will now be briefly described.

Using the MVN problem definition given by equation ((4)), the first (outermost) integration variable is chosen by selecting a variable i where $\min_{1\le i\le m}\{\Phi(b_i/\sqrt{\sigma_{i,i}}) - \Phi(a_i/\sqrt{\sigma_{i,i}})\}$ is achieved. The limits and rows and columns of $\boldsymbol{\Sigma}$ for variables 1 and i are interchanged. Then the first column of the Cholesky decomposition L of $\boldsymbol{\Sigma}$ is computed using $l_{1,1} = \sqrt{\sigma_{1,1}}$ and $l_{i,1} = \sigma_{i,1}/l_{1,1}$, for $i = 2, \ldots, m$, and the expected value for the y_1 is determined using

$$y_1 = \Big(\int_{a_1}^{b_1} \frac{se^{-\frac{s^2}{2}}}{\sqrt{2\pi}} ds\Big) / \Big(\Phi(b_1) - \Phi(a_1)\Big).$$

Note: this weighted 1-d Normal integral has an easy analytic evaluation using $\int_a^b se^{-\frac{s^2}{2}} ds = e^{-\frac{a^2}{2}} - e^{-\frac{b^2}{2}}$. Given this expected value for y_1, the second integration variable is chosen by selecting a variable i where

$$\min_{2\le i\le m}\left\{\Phi\Big(\frac{b_i-l_{i,1}y_1}{\sqrt{\sigma_{i,i}-l_{i,1}^2}}\Big)-\Phi\Big(\frac{a_i-l_{i,1}y_1}{\sqrt{\sigma_{i,i}-l_{i,1}^2}}\Big)\right\}$$

is achieved. The integration limits, rows and columns of $\boldsymbol{\Sigma}$, and rows of L for variables 2 and i are interchanged. Then the second column of L is computed using $l_{2,2}=\sqrt{\sigma_{2,2}-l_{2,1}^2}$ and $l_{i,2}=(\sigma_{i,2}-l_{2,1}l_{i,1})/l_{2,2}$, for $i=3,\ldots,m$, and the expected value for y_2 is computed using

$$y_2=\Big(\int_{\tilde{a}_2}^{\tilde{b}_2}\frac{se^{-\frac{s^2}{2}}}{\sqrt{2\pi}}ds\Big)/\Big(\Phi(b_2')-\Phi(a_2')\Big).$$

At a general stage j, given the expected values for $y_1,y_2,\ldots,y_{j-1}$, the jth integration variable is chosen by selecting a variable i, where

$$\min_{j\le i\le m}\left\{\Phi\Big(\frac{b_i-\sum_{k=1}^{j-1}l_{i,k}y_k}{\sqrt{\sigma_{i,i}-\sum_{k=1}^{j-1}l_{i,k}^2}}\Big)-\Phi\Big(\frac{a_i-\sum_{k=1}^{j-1}l_{i,k}y_k}{\sqrt{\sigma_{i,i}-\sum_{k=1}^{j-1}l_{i,k}^2}}\Big)\right\}$$

is achieved. The integration limits, rows and columns of $\boldsymbol{\Sigma}$, and rows of L for variables j and i are interchanged. Then the jth column of L is computed using $l_{j,j}=\sqrt{\sigma_{j,j}-\sum_{k=1}^{j-1}l_{j,k}^2}$ and $l_{i,j}=(\sigma_{i,j}-\sum_{k=1}^{j-1}l_{j,k}l_{i,k})/l_{j,j}$, for $i=j+1,\ldots,m$, and y_j is computed using

$$y_j=\Big(\int_{\tilde{a}_2}^{\tilde{b}_2}\frac{se^{-\frac{s^2}{2}}}{\sqrt{2\pi}}ds\Big)/\Big(\Phi(b_j')-\Phi(a_j')\Big).$$

The complete $m-1$ stage process has overall cost $O(m^3)$, which is not significant compared to the rest of the computation cost for the methods discussed here, and is therefore a relatively cheap preconditioning step that can be used with the algorithms.

With this variable sorting method, the variables are sorted so that the innermost integrals have the largest expected integration intervals. This method uses **a**, **b** and $\boldsymbol{\Sigma}$ in the sorting process, and it should therefore further increases the likelihood that the innermost integrals have values close to one and improve the convergence of the numerical integration methods. Numerical experiments have shown that variable permutations can often significantly reduce the overall variation of the final integrand for many MVN problems. The Gibson, Glasbey and Elston method can be generalized to MVT problems, and this generalization was described in detail in [GB02].

4.2 QMC Algorithms for MVN and MVT Computations

Tests by Beckers and Haegemans [BH92], and by Genz [Ge93], for MVN problems demonstrated that the performance of MC MVN methods could usually be improved if the sets of (pseudo-)random numbers used by the MC methods were replaced by appropriate sets of quasi-random numbers. In order to construct QMC MVN and MVT methods, the Uniform(0, 1) random numbers for the MC methods are replaced by appropriately chosen sets of $Quasi(0,1)$ random numbers. However, simple QMC methods do not provide the statistically robust (standard) error estimates that MC methods provide, so some type of randomization is needed for practical QMC algorithms. There have been tests by many researchers investigating the combination of different QMC algorithms with different randomizations (see [BH92, Ge93, GB02, GDS02, HH97, HMR96, HH97, SA04], also [LL00] for finance problems). The most comprehensive tests comparing different QMC methods for MVN problems were completed by Sandor and Andras [SA04], mostly for small sample size, but m as large as 50, where lattice rule QMC methods had the best overall performance.

The extensive Genz and Bretz [GB02] MVT tests focussed on comparing separated-variable and spherical-radial methods with various types of MC and QMC algorithms. The MCQMC(!) methods that generally had the best overall performance over the range of dimensions 5-20, used approximations to $I(f)$ in the form

$$\begin{aligned}\hat{\mathbf{S}}_{N,P} &= \frac{1}{N}\sum_{i=1}^{N}\frac{1}{2P}\sum_{j=1}^{P}(f(|2\{\mathbf{p}_j+\mathbf{w}_i\}-\mathbf{1}|)+f(\mathbf{1}-|2\{\mathbf{p}_j+\mathbf{w}_i\}-\mathbf{1}|))\\ &\equiv \frac{1}{N}\sum_{i=1}^{N}Q_P(\mathbf{w}_i). \qquad (16)\end{aligned}$$

In this definition, $\{\mathbf{x}\}$ denotes the vector obtained by taking the fractional part of each of the components of $\mathbf{x}$, the random shifts $\mathbf{w}_i$ have components $w_{k,i} \sim$ Uniform(0, 1) and the point set $\{\mathbf{p}_j\}_{j=1}^{P}$, for prime P, is a set of good lattice rule points (see [SJ94]). The formula (16) uses simple antithetic variates, and the "periodizing" transformation $|2\mathbf{x}-\mathbf{1}|$, where $\mathbf{1} = (1, 1, \ldots, 1)^t$, (because lattice rules have better convergence properties for periodic integrands. The standard error σ_N for these approximation (16) can be determined using

$$\sigma_N^2 = \frac{1}{N(N-1)}\sum_{i=1}^{N}(Q_P(\mathbf{w}_i) - \hat{\mathbf{S}}_{N,P})^2.$$

N is usually small (say 8-10), because the most significant performance improvement for the $\hat{\mathbf{S}}_{N,P}$ type of approximation comes from the use of the QMC method, where an approximate $O(1/P)$ error behavior is predicted and often observed. With this choice for N, $\alpha\sigma_N$, with α in the 3-4 range, usually provides a robust error estimate for $\hat{\mathbf{S}}_{N,P}$.

The MVN tests by Genz [Ge93] and MVT tests by Genz and Bretz [GB02] showed that the performance of MCQMC methods based on separated-variable transformations have much better performance than the methods that use the spherical-radial transformations, although Somerville [So99] has provided some spherical-radial implementations which are very effective for some specialized (multiple comparison problem) MVT computations. The [GB02] tests used sets of 100 random MVT problems with randomly generated correlation matrices, random integration limits (where $-\infty < \mathbf{a} < \mathbf{0} < \mathbf{b} < \infty$), random integer ν's $\in [1, 10\sqrt{m}]$ and $m = 4 - 20$. The tests with "automatic" algorithms based on (16) (with prioritized variables) requested an accuracy level ϵ and then determined the time needed for a particular algorithm to produce a result at that accuracy level (as estimated by the algorithm). The overall result was that MCQMC separated-variable algorithms were usually significantly more efficient than spherical-radial MCQMC algorithms, with differences that became more pronounced with increasing m and decreasing ϵ. For example, at accuracy level $\epsilon = 10^{-3}$, the separated-variable-normal transformation (13) algorithm was approximately twice as fast for $m = 6$ compared to the spherical-radial (8) algorithm, and for $m = 20$, the separated-variable-normal algorithm was approximately 10 times faster. These algorithms used $3.5\sigma_8$ for error estimates, and the reliability of the algorithms was also tested, with average correct digits approximately 30% more than requested correct digits. Spherical-radial algorithms with more complicated symmetrizations based on (15), with $n = 1$ and $n = 2$, were significantly less efficient than those which used equation (8) for the m's tested ($m = 4 - 20$). The Genz and Bretz [GB02] MVT tests also showed that the MCQMC algorithms which used the alternate MVT definition (3) (instead of (2)) were more efficient, in spite of the extra integration variable, because the m-1 different Φ^{-1} values for each integration point for (3) require significantly less time (1/2-1/5) than the t^{-1} values for the (2) form of the MVT problem. Similar results (but limited to $m \leq 14$) were reported for the MVN problem by Genz [Ge93]. Results for a simple acceptance-rejection ("hit or miss") algorithm were also provided with both sets of tests ([Ge93] and [GB02]) but this method was very inefficient compared to other methods tested. There have not been any significant tests completed for problems where m is very large (e.g. $m = 100 - 1000$), where it would be interesting to investigate whether the predicted approximate $O(1/P)$ time complexity is achieved. With currently available software it is now possible to compute typical MVN and MVT probabilities with $m \leq 20$ to 3-5 digit accuracy in less than a second of computer workstation time.

Software implementations in Matlab, Fortran 77 and Fortran 90 for MVN and MVT probabilities for MCQMC separated-variable algorithms which use the formula (16) are available from the Genz website at

www.math.wsu.edu/faculty/genz .

This includes software for problems where the covariance matrix $\boldsymbol{\Sigma}$ is singular, and related problems where the integration region is defined by a set of linear inequalities. These problems were not considered in this paper, but efficient separated-variable methods can also be applied to these problems [GK99].

5 Concluding Remarks

Two methods, spherical-radial and separated-variable, were considered for the transformation of MVN and MVT probabilities for hyper-rectangular regions to problems that could be posed as multi-dimensional integrals over a unit hyper-cube. Test results were reviewed, which showed that the separated-variable transformed MVN and MVT probabilities can be most efficiently approximated with QMC methods using lattice rules. For practical calculation, which require robust error estimates, these QMC methods must be combined with MC methods; the results are efficient MCQMC methods for MVN and MVT probabilities. An interesting question for current research is whether recent work on efficient methods for the construction of lattice rules for weighted function spaces (see e.g. [Sl02, NC06]) could lead to the construction of better lattice rules for MVN and MVT problems.

References

[BH92] M. Beckers and A. Haegemans. 'Comparison of Numerical Integration Techniques for Multivariate Normal Integrals', Computer Science Department Preprint, Catholic University of Leuven, Belgium, 1992.

[Co54] E. A. Cornish. 'The Multivariate t-Distribution Associated with a Set of Normal Sample Deviates', *Australian Journal of Physics* **7**, pp. 531–542, 1954.

[De80] I. Deák. 'Three Digit Accurate Multiple Normal Probabilities' *Numer. Math.* **35**, pp. 369–380, 1980.

[De86] I. Deák. 'Computing Probabilities of Rectangles in Case of Multinormal Distribution' *J. Statist. Comput. Simul.* **26**, pp. 101–114, 1986.

[De90] I. Deák. *Random Number Generation and Simulation*, Akadémiai Kiadó, Budapest, Chapter 7, 1990.

[Dv86] L. Devroye. 'Non-Uniform Random Variate Generation' Springer-Verlag, Berlin, 1986.

[FW94] K.-T. Fang and Y. Wang. *Number-Theoretic Methods in Statistics*, Chapman and Hall, London, pp. 167–170, 1994.

[GDS02] H. I. Gassmann, I. Deák, and T. Szántai. 'Computing Multivariate Normal Probabilities: a New Look', *J. Comp. Graph. Stat.* **11**, pp. 920–949, 2002.

[Ge92] A. Genz. 'Numerical Computation of the Multivariate Normal Probabilities', *J. Comput. Graph. Stat.* **1**, pp. 141–150, 1992.

[Ge93] A. Genz. 'A Comparison of Methods for Numerical Computation of Multivariate Normal Probabilities', *Computing Science and Statistics* **25**, pp. 400–405, 1993.

[GB99] A. Genz and F. Bretz. Numerical Computation of Multivariate t Probabilities with Application to Power Calculation of Multiple Contrasts, *J. Stat. Comp. Simul.* **63**, pp. 361–378, 1999.

[GB00] A. Genz and F. Bretz. 'Numerical Computation of Critical Values for Multiple Comparison Problems', in *Proceedings of the Statistical Computing Section*, American Statistical Association, Alexandria, VA, pp. 84–87, 2000.

[GB01] A. Genz and F. Bretz. 'Critical Point and Power Calculations for the Studentised Range Test', *J. Stat. Comp. Simul.* **71**, pp. 85–97, 2001.

[GB02] A. Genz and F. Bretz. 'Comparison of Methods for the Computation of Multivariate t Probabilities', *J. Comp. Graph. Stat.* **11**, pp. 950–971.

[GK99] A. Genz and K. S. Kwong. 'Numerical Evaluation of Singular Multivariate Normal Distributions', *J. Stat. Comp. Simul.* **68**, pp. 1–21, 1999.

[Ge91] J. Geweke. 'Efficient Simulation from the Multivariate Normal and Student-*t* Distributions Subject to Linear Constraints', *Computing Science and Statistics* **23**, pp. 571–578, 1991.

[GG94] G. J. Gibson, C. A. Glasbey, and D. A. Elston. 'Monte-Carlo Evaluation of Multivariate Normal Integrals and Sensitivity to Variate Ordering', in *Proceedings of the Third International Conference in Numerical Methods and Applications*, World Scientific, Singapore, pp. 120–126, 1994.

[Ha93] V. Hajivassiliou. 'Simulating Normal Rectangle Probabilities and Their Derivatives: The effects of Vectorization', *The International Journal of Supercomputer Applications* **7**, pp. 231–253, 1993.

[HMR96] V. Hajivassiliou, D. McFadden, and O. Rudd. 'Simulation of Multivariate Normal Rectangle Probabilities and Their Derivatives: Theoretical and Computational Results', *Journal of Econometrics*, **72**, pp. 85–134, 1996

[HH97] F. J. Hickernell and H. S. Hong. 'Computing Multivariate Normal Probabilities Using Rank-1 Lattice Sequences', in *Proceedings of the Workshop on Scientific Computing (Hong Kong)*, (G. H. Golub, S. H. Lui, F. T. Luk, and R. J. Plemmons, eds.), Springer-Verlag, Singapore, pp. 209–215, 1997.

[HHLL00] F. J. Hickernell, H. S. Hong, P. L'Ecuyer, and C. Lemieux. 'Extensible Lattice Sequences for QMC Quadrature' *SIAM Journal of Scientific and Statistical Computing* **22**, pp. 1117–1138, 2000.

[LL00] C. Lemieux and P. L'Ecuyer. 'A Comparison of Monte Carlo, Lattice Rules and Other Low-Discrepancy Point Sets', in *Monte Carlo and Quasi-Monte Carlo methods 1998*, (H. Niederreiter and J. Spanier Eds.), Springer, Berlin, pp. 326–340, 2000.

[NC06] D. Nuyens and R. Cools. 'Fast Component-By-Component Construction of Rank-1 Lattice Rules in Shift-Invariant Reproducing Kernel Hilbert Spaces' *Math. Comp.* **75**, pp. 903–920, 2006.

[SA04] Z. Sandor and P. Andras. 'Alternative Sampling Methods for Estimating Multivariate Normal Probabilities', Journal of Econometrics, **120**, pp. 207–234, 2004.

[Sc84] M. Schervish. 'Multivariate Normal Probabilities with Error Bound', *Applied Statistics* **33**, pp. 81–87, 1984.

[Sl02] I. H. Sloan. 'QMC Integration – Beating Intractability by Weighting the Co-ordinate Directions' in *Monte Carlo and Quasi-Monte Carlo Methods 2000* (K. T. Fang, F. J. Hickernell, and H. Niederreiter, eds.), Springer-Verlag, Berlin, pp. 103–123, 2002.

[SJ94] I. H. Sloan and S. Joe. *Lattice Methods for Multiple Integration*, Oxford University Press, Oxford, 1994.

[So97] P. N. Somerville. 'Multiple Testing and Simultaneous Confidence Intervals: Calculation of Constants' *Comp. Stat. & Data Analysis* **25**, pp. 217–223, 1997.

[So98a] P. N. Somerville. 'Numerical Computation of Multivariate Normal and Multivariate-t Probabilities Over Convex Regions', *J. Comput. Graph. Stat.* **7**, pp. 529–545, 1998.

[So98b] P. N. Somerville. 'A Fortran 90 Program for Evaluation of Multivariate Normal and Multivariate t Integral over Convex Regions', *Journal of Statistical Software* **3**, 1998, available at http://www.jstatsoft.org.

[So99] P. N. Somerville. 'Critical Values for Multiple Testing and Comparisons: One Step and Step Down Procedures' *J. Stat. Plan. & Inf.*, **82**, pp. 129–138, 1999.

[So01] P. N. Somerville. 'Numerical Computation of Multivariate Normal and Multivariate t Probabilities over Ellipsoidal regions. *Journal of Statistical Software* **6**, 2001, available at http://www.jstatsoft.org.

[St90] G. W. Stewart. 'The Efficient Generation of Random Orthogonal Matrices with An Application to Condition Estimation', *SIAM J. Numer. Anal.* **17**, pp. 403–409, 1980.

[To90] Y. L. Tong. *The Multivariate Normal Distribution*, Springer-Verlag, New York, 1990.

[Vi96] W. P. M. Vijverberg. 'Monte Carlo Evaluation of Multivariate Student's t Probabilities', *Economics Letters* **52**, pp. 1–6, 1996.

[Vi97] W. P. M. Vijverberg. 'Monte Carlo Evaluation of Multivariate Normal Probabilities',*Journal of Econometrics* **76**, pp. 281–307, 1997.

[Vi00] W. P. M. Vijverberg. 'Rectangular and Wedge-shaped Multivariate Normal Probabilities', *Economics Letters* **68**, pp. 13–20, 2000.

Minimal Errors for Strong and Weak Approximation of Stochastic Differential Equations

Thomas Müller-Gronbach[1] and Klaus Ritter[2]

[1] Fakultät für Mathematik und Informatik, FernUniversität Hagen, Lützowstraße 125, 58084 Hagen, Germany
Thomas.Mueller-Gronbach@FernUni-Hagen.de

[2] Fachbereich Mathematik, Technische Universität Darmstadt, Schloßgartenstraße 7, 64289 Darmstadt, Germany
ritter@mathematik.tu-darmstadt.de

Summary. We present a survey of results on minimal errors and optimality of algorithms for strong and weak approximation of systems of stochastic differential equations. For strong approximation, emphasis lies on the analysis of algorithms that are based on point evaluations of the driving Brownian motion and on the impact of non-commutativity, if present. Furthermore, we relate strong approximation to weighted integration and reconstruction of Brownian motion, and we demonstrate that the analysis of minimal errors leads to new algorithms that perform asymptotically optimal. In particular, these algorithms use a path-dependent step-size control. For weak approximation we consider the problem of computing the expected value of a functional of the solution, and we concentrate on recent results for a worst-case analysis either with respect to the functional or with respect to the coefficients of the system. Moreover, we relate weak approximation problems to average Kolmogorov widths and quantization numbers as well as to high-dimensional tensor product problems.

1 Introduction

Construction and analysis of algorithms for stochastic differential equations started with the work of Maruyama [Mar55] in 1955, and by now it is a very active field of research at the intersection of numerical analysis and stochastic processes with numerous applications in different areas. We refer to the monographs [KP99, Mil95] as well as to the survey papers [Pla99, Tal95].

A partial list of recent developments includes equations that are driven by fractional noise, see [BT05, Lin95, Neu06, NN06], stochastic delay differential equations, see, e.g., [BB00, BS05, HMG06, HMY04, KP00, TT87], jump-diffusions, see, e.g., [Gar04, GM04, HK06, KP02, LL00, Mag98], conditional sampling, see [HSV06, HSVW05, SVW04], and quantization, see, e.g., [Der03, Der04, DFMS03, DMGR06, DS06, LP02, LP04, LP06, Pag07, PP05]. Furthermore, stochastic partial differential equations are studied since about 10 years from an algorithmic point of view, see, e.g., [DG01, GK96, GN97, Hau03, MGR07a, MGR07b, MGRW07].

As a typical result, upper bounds for the error $e(A)$ of specific algorithms A in terms of their computational cost $c(A)$ are obtained. In this paper, however, we focus on minimal errors

$$e_N(\mathcal{A}) = \inf\{e(A) : A \in \mathcal{A},\ c(A) \leq N\}$$

that are achievable for broad classes $\mathcal{A}$ of algorithms. The sequence of minimal errors $e_N(\mathcal{A})$ quantifies the intrinsic difficulty of the computational problem under investigation, and it only depends on the class $\mathcal{A}$ and the definition of the error and the cost. We refer to [TWW88] for a general abstract theory together with many applications to specific numerical problems.

Asymptotic results for $(e_N(\mathcal{A}))_{N\in\mathbb{N}}$, say[1]

$$e_N(\mathcal{A}) \asymp N^{-\alpha}$$

for some exponent $\alpha > 0$, consist of an upper and a lower bound. The upper bound states that for a suitable sequence of algorithms $A_N \in \mathcal{A}$ there are constants $\gamma_i \geq 0$ such that

$$c(A_N) \leq \gamma_1 \cdot N \qquad \text{and} \qquad e(A_N) \leq \gamma_2 \cdot N^{-\alpha}$$

for every $N \in \mathbb{N}$. The lower bound states that there exists a constant $\gamma > 0$ such that

$$c(A) \leq N \qquad \Rightarrow \qquad e(A) \geq \gamma \cdot N^{-\alpha}$$

holds for every algorithm $A \in \mathcal{A}$ and every $N \in \mathbb{N}$.

Minimal errors constitute a benchmark for any specific algorithm $A \in \mathcal{A}$ by comparing $e(A)$ with $e_N(\mathcal{A})$ for $N = \lfloor c(A) \rfloor$. Furthermore, the definition of optimality of algorithms as well as weak or strong asymptotic optimality of sequences of algorithms is based on this concept. We add that the analysis of minimal errors sometimes leads to new algorithms, too. See, e.g., Section 3.3.

Actually, two different kinds of numerical problems arise for stochastic differential equations

$$dX(t) = a(X(t))\,dt + b(X(t))\,dW(t). \tag{1}$$

Strong approximation deals with approximation of the trajectories of the solution process X, i.e., approximation of a random function, while weak

[1] By definition, $x_N \asymp y_N$ for sequences of positive real numbers x_N and y_N, if $\gamma_1 \cdot x_N \leq y_N \leq \gamma_2 \cdot x_N$ holds for every $N \in \mathbb{N}$ with constants $\gamma_i > 0$.

approximation aims at approximation of deterministic quantities that only depend on the distribution of X. As an important instance of the latter problem we study the approximation of expectations $E(h(X))$ for functionals $h : C([0,1],\mathbb{R}^d) \to \mathbb{R}$.

The proof of lower bounds for strong approximation relies on the fact that every algorithm may only use partial information about the trajectories of the driving Brownian motion W, e.g., its values at a finite number of points. For weak approximation it is partial information about the equation, i.e., about the drift coefficient a and the diffusion coefficient b, or partial information about the functional h that enables the proof of lower bounds.

For illustration we discuss a strong approximation problem, where minimal errors have been studied for the first time in the context of stochastic differential equations. Consider a scalar equation

$$dX(t) = a(X(t))\,dt + dW(t) \tag{2}$$

with a deterministic initial value $X(0) = x_0 \in \mathbb{R}$ and a scalar Brownian motion W. Suppose that we wish to approximate the solution X at $t = 1$, and assume that the drift coefficient $a : \mathbb{R} \to \mathbb{R}$ satisfies (at least) a global Lipschitz condition.

The simplest approximation is provided by the strong Euler scheme $A_N^{\mathrm{E}}(W)$ with constant step-size $1/N$, where

$$A_N^{\mathrm{E}} : C([0,1],\mathbb{R}) \to \mathbb{R}$$

is defined by $A_N^{\mathrm{E}}(w) = x_N$ with

$$x_n = x_{n-1} + a(x_{n-1})/N + w(n/N) - w((n-1)/N)$$

for $n = 1,\ldots,N$ and any trajectory $w \in C([0,1],\mathbb{R})$ of W. Since (2) is a stochastic differential equation with additive noise, the Euler scheme with step-size $1/N$ satisfies

$$\left(E|X(1) - A_N^{\mathrm{E}}(W)|^2\right)^{1/2} \leq \gamma \cdot N^{-1} \tag{3}$$

for some unspecified constant $\gamma = \gamma(a, x_0) \geq 0$.

Clearly,

$$A_N^{\mathrm{E}}(W) = \phi_N(W(1/N),\ldots,W(1))$$

with a mapping $\phi_N : \mathbb{R}^N \to \mathbb{R}$, and the question arises whether a different choice of ϕ_N may reduce the error. In the sequel we thus consider every mapping $A_N : C([0,1],\mathbb{R}) \to \mathbb{R}$ of the form

$$A_N(w) = \phi_N(w(1/N),\ldots,w(1))$$

with an arbitrary measurable mapping $\phi_N : \mathbb{R}^N \to \mathbb{R}$ as an algorithm for approximation of $X(1)$, which s a whole build the class $\mathcal{A}^{\mathrm{equi}}$. The error and the cost of A_N are defined by

$$e(A_N) = \left(E|X(1) - A_N(W)|^2\right)^{1/2}, \tag{4}$$

as for the Euler scheme, and

$$c(A_N) = N. \tag{5}$$

Then the optimal choice of ϕ_N is given by the conditional expectation

$$\phi_N^{\mathrm{c}}(y) = E(X(1) \mid (W(1/N), \ldots, W(1)) = y),$$

i.e.,

$$e(A_N^{\mathrm{c}}) = e_N(\mathcal{A}^{\mathrm{equi}})$$

for $A_N^{\mathrm{c}}(w) = \phi_N^{\mathrm{c}}(w(1/N), \ldots, w(1))$.

In 1980 Clark and Cameron determined the strong asymptotic behaviour of the minimal errors for the class $\mathcal{A}^{\mathrm{equi}}$, see also Section 3.1.

Theorem 1 ([CC80]). *Suppose that the drift coefficient in equation* (2) *satisfies $a \in C^3(\mathbb{R})$ with bounded derivatives a', a'', a'''. Put*

$$C = \left(\frac{1}{12} \cdot \int_0^1 E\left(a'(X(t)) \cdot \exp\left(\int_t^1 a'(X(u))\,du\right)\right)^2 dt\right)^{1/2}.$$

If $C = C(a, x_0) > 0$ then[2]

$$e_N(\mathcal{A}^{\mathrm{equi}}) \approx C \cdot N^{-1}.$$

Remark 1. From Theorem 1 and (3) we immediately get the weak asymptotic optimality

$$e(A_N^{\mathrm{E}}) \asymp e_N(\mathcal{A}^{\mathrm{equi}})$$

of the Euler scheme with constant step-size $1/N$, if $C > 0$. We add that the constant C is in fact positive in all non-trivial cases; more precisely, $C = 0$ iff $X(1) = g(W(1))$ for some measurable mapping $g : \mathbb{R} \to \mathbb{R}$.

Remark 2. The class $\mathcal{A}^{\mathrm{equi}}$ is defined by the requirement that all trajectories of W may only be evaluated at equidistant points. Since no restriction except measurability is imposed on ϕ_N, we do not care whether algorithms from $\mathcal{A}^{\mathrm{equi}}$ are actually implementable on a computer. Moreover, if so, our definition of cost does not take into account any computational overhead besides evaluation of W. However, smaller classes $\mathcal{A}^{\mathrm{equi}}$ or a more detailed definition of cost could

[2] By definition, $x_N \approx y_N$ for sequences of positive real numbers x_N and y_N, if $\lim_{N\to\infty} x_N/y_N = 1$.

only yield larger minimal errors, and the asymptotic lower bound $C \cdot N^{-1}$ for the minimal error would still be valid.

On the other hand, if we also include the number of evaluations of the drift coefficient a and the total number of all further arithmetic operations in the definition of the cost, then the cost of the Euler scheme would only be changed by a small multiplicative constant. Consequently, the same holds true for the upper bound $\gamma \cdot N^{-1}$ for the minimal error, which follows from (3).

2 Deterministic and Randomized Algorithms

The analysis of minimal errors requires a formal definition of the class of algorithms under investigation. To this end we essentially consider the real number model of computation, which is used at least implicitly in most considerations of computational problems for stochastic differential equations; for convenience we proceed slightly more general. We refer to [Nov95] for a general study of the real number model in numerical analysis.

For both problems, strong and weak approximation, there is some underlying class F of functions, which constitutes the problem instances, and a mapping $S : F \to G$, which takes $f \in F$ to the corresponding solution $S(f) \in G$ of the computational problem. For strong approximation, $F = C([0,1], \mathbb{R}^m)$ is the class of trajectories of the driving Brownian motion W, and we use $G = \mathbb{R}^d$ or $G = L_p([0,1], \mathbb{R}^d)$ depending on whether we want to approximate the trajectories of X at a single point, say $t = 1$, or globally on $[0,1]$. For weak approximation, F is a class of drift and diffusion coefficients or a class of functionals on $C([0,1], \mathbb{R}^d)$, and we take $G = \mathbb{R}$. To cover both cases we consider any class F of functions on some set $\mathfrak{X}$ with values in a finite-dimensional real vector space $\mathfrak{Y}$.

We assume that every algorithm may evaluate the elements $f \in F$ at a finite number of sequentially chosen points from a subset $\mathfrak{X}_0 \subseteq \mathfrak{X}$, which is modeled by an oracle in the real number model. A deterministic sequential evaluation is formally defined by a point

$$\psi_1 \in \mathfrak{X}_0$$

and a sequence of mappings

$$\psi_n : \mathfrak{Y}^{n-1} \to \mathfrak{X}_0, \qquad n \geq 2.$$

For every $f \in F$ the evaluation starts at the point ψ_1, and after n evaluations the values

$$y_1 = f(\psi_1)$$

and

$$y_\ell = f(\psi_\ell(y_1, \ldots, y_{\ell-1})), \qquad \ell = 2, \ldots, n,$$

are known. A decision to stop or to further evaluate f is made after each step, and this is formally described by a sequence of mappings

$$\chi_n : \mathfrak{Y}^n \to \{\mathrm{STOP}, \mathrm{GO}\}, \qquad n \geq 1.$$

The total number of evaluations is given by

$$\nu(f) = \min\{n \in \mathbb{N} : \chi_n(y_1, \ldots, y_n) = \mathrm{STOP}\},$$

which is finite for every $f \in F$ by assumption. Finally, an output

$$A(f) = \phi_{\nu(f)}(y_1, \ldots, y_{\nu(f)})$$

is defined by a sequence of mappings

$$\phi_n : \mathfrak{Y}^n \to G, \qquad n \geq 1.$$

Up to some measurability assumptions, which will be stated in Sections 3 and 4, every such mapping $A : F \to G$ will be considered as a deterministic algorithm, with algorithm being understood in a broad sense, and the resulting class of mappings is denoted by $\mathcal{A}^{\mathrm{det}}(\mathfrak{X}_0)$. For convenience, we identify A with the point ψ_1 and the sequences of mappings ψ_n, χ_n, and ϕ_n.

The computational cost $c(A, f)$ of applying A to f is defined by

$$c(A, f) = \nu(f) \cdot s(\mathfrak{X}_0). \tag{6}$$

In most cases we simply take

$$s(\mathfrak{X}_0) = 1,$$

which means that we only count the number of evaluations of f. However, for the weak approximation problem that is studied in Section 4.1 we have $\mathfrak{X} = C([0,1], \mathbb{R}^d)$ and $\mathfrak{X}_0 \subsetneq \mathfrak{X}$ is any finite-dimensional subspace. In this case it is reasonable to take

$$s(\mathfrak{X}_0) = \dim(\mathfrak{X}_0),$$

which captures that a basis representation of the points $\psi_\ell(y_1, \ldots, y_{\ell-1}) \in \mathfrak{X}_0$ is submitted to the oracle.

For $G = \mathbb{R}$ a randomized (or Monte Carlo) broad sense algorithm based on sequential evaluation is formally defined by a probability space $(\Omega, \mathfrak{A}, P)$ and a mapping

$$A : \Omega \times F \to \mathbb{R}$$

such that

(i) $A(\omega, \cdot) \in \mathcal{A}^{\mathrm{det}}(\mathfrak{X}_0)$ for every $\omega \in \Omega$,
(ii) $A(\cdot, f)$ is measurable for every $f \in F$,
(iii) $\omega \mapsto c(A(\omega, \cdot), f)$ is measurable for every $f \in F$.

We refer to [NY83, Was89] for this and an equivalent definition of randomized algorithms, which captures the assumption that randomized algorithms have access to perfect random number generators. By $\mathcal{A}^{\mathrm{ran}}(\mathfrak{X}_0)$ we denote the class of all mappings A with properties (i)–(iii) on any probability space. Clearly,

$$\mathcal{A}^{\mathrm{det}}(\mathfrak{X}_0) \subsetneq \mathcal{A}^{\mathrm{ran}}(\mathfrak{X}_0).$$

It might seem that these classes of algorithms are too large and that $c(A, f)$ is too small, as it only takes into account a fraction of the actual cost of a computation. See, however, Remarks 2 and 12 and Section 3.3.

3 Strong Approximation

In this section we consider d-dimensional autonomous systems of stochastic differential equations (1) with drift and diffusion coefficients

$$a : \mathbb{R}^d \to \mathbb{R}^d \qquad \text{and} \qquad b : \mathbb{R}^d \to \mathbb{R}^{d\times m},$$

and initial value $x_0 \in \mathbb{R}^d$. Hence the solution X is a d-dimensional process and we have an m-dimensional driving Brownian motion W. For simplicity, we assume throughout that

(I) both a and b are Lipschitz continuous,
(II) all components a_i of a and $b_{i,j}$ of b are differentiable with Lipschitz continuous derivatives

$$\nabla a_i = \left(\frac{\partial}{\partial x_1} a_i, \dots, \frac{\partial}{\partial x_d} a_i\right) \quad \text{and} \quad \nabla b_{i,j} = \left(\frac{\partial}{\partial x_1} b_{i,j}, \dots, \frac{\partial}{\partial x_d} b_{i,j}\right),$$

respectively.

Condition (I) assures the existence and uniqueness of a strong solution of (1), and condition (II) is a standard assumption for the analysis of strong approximation problems. These assumptions suffice to derive all of the stated results except for Theorem 2, which deals with one-point approximation of scalar equations.

We study approximation of $X(1)$ in Section 3.1 and global approximation of X on the interval $[0, 1]$ in Section 3.2. For both problems we drop the assumption from the introductory example in Section 1 that all trajectories of the Brownian motion W must be evaluated with a fixed step-size. Instead, we consider all algorithms that are based on a finite number of sequential evaluations of these trajectories. Except for measurability conditions we do not impose any further restrictions, so that we cover any step-size control used in practice. The discussion of minimal errors in Sections 3.1 and 3.2 is complemented by remarks on asymptotically optimal algorithms in Section 3.3.

3.1 One-Point Approximation

Deterministic and randomized algorithms have been introduced in Section 2 in a general setting. Specifically for one-point approximation we take

$$F = C(\mathfrak{X}, \mathfrak{Y})$$

with

$$\mathfrak{X} = [0,1] \qquad \text{and} \qquad \mathfrak{Y} = \mathbb{R}^m,$$

because the trajectories w of W in equation (1) are elements of this space F. Since $w(0) = 0$ we put

$$\mathfrak{X}_0 =]0,1]\,.$$

Furthermore, we take

$$G = \mathbb{R}^d,$$

since we aim at approximation of $S(W) = X(1)$.

We consider the corresponding class $\mathcal{A}^{\text{det}} = \mathcal{A}^{\text{det}}(\mathfrak{X}_0)$ of algorithms. Hence the sequence $(\psi_n)_{n\in\mathbb{N}}$ determines the evaluation sites for every $w \in F$, and the total number of evaluations to be made is determined by the sequence $(\chi_n)_{n\in\mathbb{N}}$ of stopping rules. Finally, $(\phi_n)_{n\in\mathbb{N}}$ is used to obtain the $\mathbb{R}^d$-valued approximation $A(w)$ to the corresponding trajectory of the solution X at $t = 1$. For technical reasons we require measurability of all mappings ψ_n, χ_n, and ϕ_n.

Analogously to (4), the error of $A \in \mathcal{A}^{\text{det}}$ is defined by

$$e(A) = \left(E|X(1) - A(W)|^2\right)^{1/2},$$

where $|\cdot|$ denotes the euclidean norm, and for the definition of the cost we take the expected number of evaluations of W, i.e.,

$$c(A) = E(c(A,W))$$

with $s(\mathfrak{X}_0) = 1$, see (6) and compare (5). Thus we perform an average-case analysis of deterministic algorithms, see [Rit00, TWW88].

Motivated by the introductory example on one-point approximation in Section 1 we first treat the case of scalar autonomous equations (1), i.e., $d = m = 1$. Put

$$M(t) = \exp\left(\int_t^1 \left(a' - 1/2 \cdot (b')^2\right)(X(u))\,du + \int_t^1 b'(X(u))\,dW(u)\right)$$

and

$$Y_1(t) = \left(ba' - ab' - 1/2 \cdot b^2 b''\right)(X(t)) \cdot M(t),$$

provided that b is sufficiently smooth. Note that $Y_1(t)$ is the product of an Itô-Taylor coefficient function, evaluated at $X(t)$, and the mean-square derivative $M(t)$ of $X(1)$ w.r.t. the state at time t. Furthermore, $b(X(t)) \cdot M(t)$ coincides with the first order Malliavin derivative $\mathcal{D}_t X(1)$, see, e.g., [Nua06]. In particular, $Y_1(t) = \frac{d}{dt}\mathcal{D}_t X(1)$ holds in the case of equation (2).

Theorem 2 ([MG04]). *Assume $d = m = 1$ in (1) and suppose that the drift coefficient $a : \mathbb{R} \to \mathbb{R}$ and the diffusion coefficient $b : \mathbb{R} \to \mathbb{R}$ are bounded and have continuous bounded derivatives up to order three. Put*

$$C_1 = \frac{1}{\sqrt{12}} \cdot \left(\int_0^1 E|Y_1(t)|^{2/3}\, dt \right)^{3/2}.$$

If $C_1 = C_1(a, b, x_0) > 0$ then

$$e_N(\mathcal{A}^{\mathrm{det}}) \approx C_1 \cdot N^{-1}.$$

For details concerning the following two remarks we also refer to [MG04]. We add that Theorem 2 holds true under weaker assumptions concerning a and b, and the strong asymptotic behaviour of the minimal errors is known in the non-autonomous case, too.

Remark 3. Algorithms from the class $\mathcal{A}^{\mathrm{equi}} \subsetneq \mathcal{A}^{\mathrm{det}}$, which was already discussed in Section 1 for equation (2), are defined by constant mappings $\psi_n = n/N$, $\chi_1 = \ldots = \chi_{N-1} = \mathrm{GO}$, and $\chi_N = \mathrm{STOP}$. It turns out that

$$e_N(\mathcal{A}^{\mathrm{equi}}) \approx C_1^{\mathrm{equi}} \cdot N^{-1}$$

with

$$C_1^{\mathrm{equi}} = \frac{1}{\sqrt{12}} \cdot \left(\int_0^1 E|Y_1(t)|^2\, dt \right)^{1/2}$$

if the latter constant is positive. In the particular case of $b = 1$ we recover the constant from Theorem 1.

An intermediate case $\mathcal{A}^{\mathrm{equi}} \subsetneq \mathcal{A}^{\mathrm{fix}} \subsetneq \mathcal{A}^{\mathrm{det}}$ was considered by Cambanis and Hu in 1996, who essentially studied all algorithms of the form

$$A_N(W) = \phi_N(W(t_1), \ldots, W(t_N))$$

for any fixed choice of knots $0 < t_1 < \cdots < t_N \leq 1$, i.e., for χ as previously and any choice of constant mappings $\psi_1, \ldots, \psi_N$, see [CH96]. Here we have

$$e_N(\mathcal{A}^{\mathrm{fix}}) \approx C_1^{\mathrm{fix}} \cdot N^{-1}$$

with

$$C_1^{\mathrm{fix}} = \frac{1}{\sqrt{12}} \cdot \left(\int_0^1 \left(E|Y_1(t)|^2\right)^{1/3} dt \right)^{3/2}$$

if this constant is positive.

The superiority of $\mathcal{A}^{\mathrm{det}}$ over $\mathcal{A}^{\mathrm{equi}}$ and $\mathcal{A}^{\mathrm{fix}}$ is thus expressed by the ratio of the respective asymptotic constants, and this ratio quantifies in particular the potential of an optimal path-dependent step-size control, see also Section 3.3. In most cases we have

$$C_1^{\mathrm{equi}} > C_1^{\mathrm{fix}} > C_1 > 0,$$

and sometimes the ratio C_1^{fix}/C_1 is large. For instance, if $a = 1$, $b(x) = \sigma \cdot x$, and $x_0 = 1$, we obtain

$$
\begin{aligned}
C_1^{\mathrm{equi}} &= 1/\sqrt{12} \cdot (\exp(\sigma^2) - 1)^{1/2}, \\
C_1^{\mathrm{fix}} &= 1/\sqrt{12} \cdot \sqrt{27}/\sigma^2 \cdot (\exp(\sigma^2/3) - 1)^{3/2}, \\
C_1 &= 1/\sqrt{12} \cdot 27/\sigma^2 \cdot (1 - \exp(-\sigma^2/9))^{3/2},
\end{aligned}
$$

so that C_1^{fix}/C_1 grows exponentially as a function of σ^2.

Remark 4. It turns out that approximation of $X(1)$ is strongly connected to an integration problem for the Brownian motion W with the random weight $(Y_1(t))_{t\in[0,1]}$. Here we only give a precise formulation of this fact in the elementary case of a Langevin equation, namely for $a(x) = x$, $b = 1$, and $x_0 = 0$, where

$$Y_1(t) = \exp(1 - t)$$

is actually deterministic and

$$X(1) = W(1) - \int_0^1 Y_1(t) \cdot W(t)\, dt.$$

Integration problems for stochastic processes are well-studied in the case of deterministic weight functions, see [Rit00] for results and references. Basic ideas carry over to the case of random weights, and this allows the construction of easily implementable algorithms that enjoy strong asymptotic optimality, see Section 3.3.

Under the assumptions of Theorem 2, the order of convergence of the minimal errors $e_N(\mathcal{A}^{\mathrm{det}})$ is at least N^{-1} for scalar equations (1). This is no longer true, in general, for systems of equations. If the dimension m of the driving Brownian motion W in (1) is larger than one then the asymptotic behaviour of the minimal errors $e_N(\mathcal{A}^{\mathrm{det}})$ for one-point approximation crucially depends on whether the diffusion coefficient b satisfies the so-called commutativity condition.

To be more precise, put

$$b^{(j)} = \begin{pmatrix} b_{1,j} \\ \vdots \\ b_{d,j} \end{pmatrix} \quad \text{and} \quad \nabla b^{(j)} = \begin{pmatrix} \nabla b_{1,j} \\ \vdots \\ \nabla b_{d,j} \end{pmatrix}$$

for $j = 1, \dots, m$. Then b has the commutativity property if

$$\nabla b^{(j_1)} \cdot b^{(j_2)} = \nabla b^{(j_2)} \cdot b^{(j_1)} \tag{7}$$

holds for all $1 \le j_1 < j_2 \le m$. Roughly speaking, this condition assures that the trajectories of the solution X depend continuously on the trajectories of the driving Brownian motion W. Clearly, (7) holds if W is one-dimensional, i.e., if $m = 1$.

To study the impact of non-commutativity we consider the $\mathbb{R}^d$-valued processes given by

$$\mathrm{dev}_{j_1,j_2}(t) = \left(\nabla b^{(j_1)} \cdot b^{(j_2)} - \nabla b^{(j_2)} \cdot b^{(j_1)}\right)(X(t))$$

as well as the $d \times d$-dimensional random field

$$\Phi = (\Phi(t,s))_{0 \le t \le s \le 1},$$

that satisfies the matrix stochastic differential equations

$$d\Phi(t,s) = \nabla a(X(s)) \cdot \Phi(t,s)\, ds + \sum_{j=1}^{m} \nabla b^{(j)}(X(s)) \cdot \Phi(t,s)\, dW_j(s)$$

with initial values $\Phi(t,t)$ equal to the $d \times d$-dimensional identity matrix Id_d. The processes dev_{j_1,j_2} measure the deviation from commutativity along the trajectories of X, and the field Φ serves as a measure of the variability of the solution X on $[t,1]$ w.r.t. its state at time t. Note that the process $(\Phi(t,1))_{t\in[0,1]}$ coincides with the process M in Theorem 2 in the case $d = m = 1$.

We quantify the effect of deviation from commutativity by the random field

$$\vartheta(t,s) = \left(\sum_{j_1<j_2} |\Phi(t,s) \cdot \mathrm{dev}_{j_1,j_2}(t)|^2\right)^{1/2}, \quad 0 \le t \le s \le 1.$$

Theorem 3 ([MG02a]). *Put* $Y_2(t) = \vartheta(t,1)$ *and*

$$C_2 = \frac{1}{2} \cdot \int_0^1 E(Y_2(t))\, dt.$$

If $C_2 = C_2(a,b,x_0) > 0$ *then*[3]

$$1/\sqrt{3} \cdot C_2 \cdot N^{-1/2} \lesssim e_N(\mathcal{A}^{\mathrm{det}}) \lesssim C_2 \cdot N^{-1/2}.$$

For details of the following remarks we refer to [MG02a].

Remark 5. For one-point approximation in the non-commutative case, the asymptotic behaviour of the minimal errors for subclasses of $\mathcal{A}^{\mathrm{det}}$ is studied as well. We briefly discuss the class $\mathcal{A}^{\mathrm{equi}}$ of algorithms that use the same

---

3 By definition, $x_N \lesssim y_N$ for sequences of positive real numbers x_N and y_N, if $\limsup_{N\to\infty} x_N/y_N \le 1$.

equidistant evaluation sites for every trajectory of W, see Remark 3 for the formal definition. We have

$$e_N(\mathcal{A}^{\mathrm{equi}}) \approx C_2^{\mathrm{equi}} \cdot N^{-1/2}$$

with

$$C_2^{\mathrm{equi}} = \frac{1}{2} \cdot \left(\int_0^1 E(Y_2^2(t))\, dt \right)^{1/2}$$

if the latter constant is positive.

Minimal errors for one-point approximation in the non-commutative case have first been studied in [CC80], where the class $\mathcal{A}^{\mathrm{equi}}$ is considered for the specific 2-dimensional system

$$dX_1(t) = dW_1(t), \qquad dX_2(t) = X_1(t)\, dW_2(t),$$

with initial value $x_0 = 0$. Here the corresponding random field ϑ satisfies $\vartheta(t,s) = 1$ for all $0 \le t \le s \le 1$, and we obtain $C_2 = C_2^{\mathrm{equi}} = 1/2$.

We further mention [Rum82], where the order of convergence $N^{-1/2}$ of the minimal errors for a subclass $\mathcal{A} \subsetneq \mathcal{A}^{\mathrm{equi}}$ of Runge-Kutta algorithms is derived in the case of systems (1) with $C_2^{\mathrm{equi}} > 0$.

Remark 6. If the constant C_2 in Theorem 3 is positive, then the order of convergence of the minimal errors is only $N^{-1/2}$, and, consequently, one-point approximation is as hard as L_2-approximation in this case, see Theorem 4.

On the other hand, assume that the diffusion coefficient b satisfies the commutativity condition (7). Then the field ϑ vanishes, which implies $C_2 = 0$, and the order of convergence of the minimal errors turns out to be at least N^{-1}. In fact, (7) guarantees that the strong Milstein scheme with constant step-size $1/N$ yields an algorithm $A_N^{\mathrm{M}} \in \mathcal{A}^{\mathrm{equi}}$ with

$$c(A_N^{\mathrm{M}}) = N \qquad \text{and} \qquad e(A_N^{\mathrm{M}}) \le \gamma \cdot N^{-1}$$

for some unspecified constant $\gamma = \gamma(a, b, x_0) > 0$. Lower bounds for the minimal errors are unknown in this situation, except for the particular case $d = m = 1$, see Theorem 2.

3.2 Global Approximation

In addition to assumptions (I) and (II) from page 59 we assume that

(III) $P(b(X(t)) \neq 0) > 0$ for some $t \in [0, 1]$

to exclude deterministic equations.

As in the case of one-point approximation we take

$$F = C([0,1], \mathbb{R}^m) \qquad \text{and} \qquad \mathfrak{X}_0 =]0, 1]$$

in the definition of the class of algorithms $\mathcal{A}^{\det} = \mathcal{A}^{\det}(\mathfrak{X}_0)$. However, in contrast to one-point approximation we now aim at approximation of $S(W) = X$ globally on the interval $[0, 1]$. Here we restrict considerations to the problem of L_2-approximation, i.e., we take

$$G = L_2([0, 1], \mathbb{R}^d)$$

equipped with the norm

$$\|g\|_{L_2} = \left(\int_0^1 |g(t)|^2 \, dt \right)^{1/2}.$$

Hence the measurable sequences $(\psi_n)_{n\in\mathbb{N}}$ and $(\chi_n)_{n\in\mathbb{N}}$ associated with an algorithm $A \in \mathcal{A}^{\det}$ determine the location and the number of evaluation sites for every trajectory $w \in F$ of W, as previously, while the measurable sequence $(\phi_n)_{n\in\mathbb{N}}$ yields a function $A(w) \in G$, which serves as an approximation to the corresponding trajectory of X on the interval $[0, 1]$.

For the definition of the error of A we consider the pathwise L_2-distance $\|X - A(W)\|_{L_2}$ and average over all trajectories, i.e.,

$$e(A) = \left(E\|X - A(W)\|_{L_2}^2 \right)^{1/2}.$$

The cost of A is given by the expected number of evaluations of W, as in the case of one-point approximation.

There are two factors that determine the asymptotic behaviour of the minimal errors $e_N(\mathcal{A}^{\det})$ for L_2-approximation, namely the smoothness of the solution X in the mean square sense and the impact of non-commutativity, if present.

To be more precise with respect to the first issue, we note that the components X_i of X satisfy

$$E\left(|X_i(t + \delta) - X_i(t)|^2 \,\middle|\, X(t) = x\right) = |b_i|^2(x) \cdot \delta + o(\delta), \tag{8}$$

where $b_i = (b_{i,1}, \ldots, b_{i,m})$ denotes the i-th row of the diffusion coefficient b. Hence X is Hölder continuous of order $1/2$ in the mean square sense, and locally, in time and space, the smoothness of X is determined by

$$\|b\|_2(X(t)) = \left(\sum_{i=1}^d |b_i|^2(X(t)) \right)^{1/2}.$$

To take non-commutativity into account, we use the random field ϑ, see Section 3.1, to define the process Ψ by

$$\Psi(t) = \int_t^1 \vartheta^2(t, s) \, dt,$$

which accumulates the effect of a deviation from commutativity at time t over the interval $[t, 1]$.

Theorem 4 ([HMGR01, MG02a]). *Put*

$$C_3 = \frac{1}{\sqrt{6}} \cdot \int_0^1 E(Y_3(t))\,dt,$$

where

$$Y_3(t) = \left(\|b\|_2^2(X(t)) + 3/2 \cdot \Psi(t)\right)^{1/2},$$

Then

$$1/\sqrt{3} \cdot C_3 \cdot N^{-1/2} \lesssim e_N(\mathcal{A}^{\text{det}}) \lesssim C_3 \cdot N^{-1/2}.$$

Moreover, if the diffusion coefficient b satisfies the commutativity condition (7) *then $\Psi = 0$ and*

$$e_N(\mathcal{A}^{\text{det}}) \approx C_3 \cdot N^{-1/2}.$$

For details concerning the following three remarks we refer to [HMGR01, MG02a, MG02b]

Remark 7. As in the case of one-point approximation we compare the classes of algorithms $\mathcal{A}^{\text{det}}$ and $\mathcal{A}^{\text{equi}}$ for L_2-approximation with respect to the asymptotic behaviour of the corresponding minimal errors. For the latter class we have

$$e_N(\mathcal{A}^{\text{equi}}) \approx C_3^{\text{equi}} \cdot N^{-1/2}$$

with

$$C_3^{\text{equi}} = \frac{1}{\sqrt{6}} \cdot \left(\int_0^1 E(Y_3^2(t))\,dt\right)^{1/2}.$$

Consider, for instance, the equation

$$dX(t) = \sum_{j=1}^{m} \sigma_j \cdot X(t)\,dW_j(t)$$

with $\sigma = (\sigma_1, \ldots, \sigma_m) \in \mathbb{R}^m$ and initial value $x_0 = 1$. Hence $d = 1$ and $a = 0$ and the diffusion coefficient $b(x) = x \cdot \sigma$ satisfies the commutativity condition (7). Here we have

$$Y_3(t) = |\sigma| \cdot X(t) = |\sigma| \cdot \exp(-|\sigma|^2 \cdot t/2 + \sigma \cdot W(t)),$$

which yields

$$C_3^{\text{equi}} = \frac{1}{\sqrt{6}} \cdot (\exp(|\sigma|^2) - 1)^{1/2} \qquad \text{and} \qquad C_3 = \frac{1}{\sqrt{6}} \cdot |\sigma|.$$

Thus, C_3^{equi}/C_3 grows exponentially as a function of $|\sigma|^2$.

Remark 8. Due to Theorem 4 the order of convergence of the minimal errors for L_2-approximation is not affected by a deviation from commutativity, which is in sharp contrast to the respective result for one-point approximation, see Theorem 3 and Remark 6.

If the diffusion coefficient b satisfies the commutativity condition then, by the second part of Theorem 4, the asymptotic constant C^{det} is completely determined by the process $(b(X(t)))_{t\in[0,1]}$, which comprises the local smoothness of X in the mean square sense, see (8). In this case it turns out that L_2-approximation of X is closely related to an L_2-reconstruction problem for W with the random weight $b(X(t))$. Here we only illustrate this fact by the simple two-dimensional system

$$dX_1(t) = dt, \qquad dX_2(t) = \sigma(X_1(t))\, dW_2(t),$$

with initial value $x_0 \in \mathbb{R}^2$ and $\sigma \in C^2(\mathbb{R})$ with bounded derivatives σ' and σ''. Hence $d = m = 2$,

$$a(x) = \begin{pmatrix} 1 \\ 0 \end{pmatrix} \qquad \text{and} \qquad b(x) = \begin{pmatrix} 0 & 0 \\ 0 & \sigma(x_1) \end{pmatrix}$$

for $x \in \mathbb{R}^2$, and L_2-approximation of X is equivalent to L_2-reconstruction of the process

$$b(X(t)) \cdot W(t) = \sigma(t) \cdot W_2(t).$$

We add that reconstruction problems for stochastic processes are well-studied in the case of deterministic weight functions, see [Rit00] for results and references. Basic principles can be utilized also for the case of random weights, which leads to algorithms for L_2-approximation that are easy to implement and achieve the upper bound in Theorem 4, see Section 3.3

Remark 9. Results on minimal errors are also available for L_p-approximation with $p \neq 2$. We briefly discuss the case $p = \infty$, for which the issue of non-commutativity turns out to be irrelevant.

The respective class of algorithms $\mathcal{A}^{\text{det}} = \mathcal{A}^{\text{det}}(\mathfrak{X}_0)$ is defined as for L_2-approximation, the only change being that we take $G = L_\infty([0,1], \mathbb{R}^d)$ equipped with the norm $\|\cdot\|_{L_\infty}$, where the maximum norm on $\mathbb{R}^d$ is used as well. Furthermore, we define the error of an algorithm $A \in \mathcal{A}^{\text{det}}$ by

$$e(A) = \left(E\|X - A(W)\|_{L_\infty}^2\right)^{1/2}.$$

Recall that the local mean square smoothness of the components X_i of the solution X is determined by the respective processes $(|b_i|(X(t)))_{t\in[0,1]}$, and consider the process

$$Y_4(t) = \max_{i=1,\ldots,d} |b_i|(X(t))$$

Then the minimal errors satisfy

$$e_N(\mathcal{A}^{\text{det}}) \approx C_4 \cdot (N/\ln N)^{-1/2}$$

with

$$C_4 = \frac{1}{\sqrt{2}} \cdot E\Big(\int_0^1 Y_4^2(t)\,dt\Big)^{1/2}.$$

For the subclass $\mathcal{A}^{\text{equi}}$ we obtain

$$e_N(\mathcal{A}^{\text{equi}}) \approx C_4^{\text{equi}} \cdot (N/\ln N)^{-1/2}$$

with

$$C_4^{\text{equi}} = \frac{1}{\sqrt{2}} \cdot \Big(E \sup_{t\in[0,1]} Y_4^2(t)\Big)^{1/2}.$$

We illustrate the superiority of $\mathcal{A}^{\text{det}}$ over $\mathcal{A}^{\text{equi}}$ by comparing the respective asymptotic constants for the particular scalar equation

$$dX(t) = \sigma \cdot X(t)\,dW(t)$$

with $\sigma > 0$ and initial value $x_0 = 1$. Here we obtain

$$C_4 \leq 1/\sqrt{2} \cdot \sigma \cdot (\sigma + 2),$$

and

$$C_4^{\text{equi}} = \sigma \cdot \big(3\exp(\sigma^2) \cdot \Phi(3\sigma/2) - \Phi(\sigma/2)\big),$$

where Φ denotes the standard normal distribution function.

The general framework introduced in Section 2 can easily be extended to cover algorithms for strong approximation that use multiple Itô integrals, additionally to point evaluations of the driving Brownian motion W. For instance, Itô-Taylor schemes of higher order are of this type. In this case we take a suitable space Λ of functionals

$$\lambda : C([0,1], \mathbb{R}^m) \to \mathbb{R}$$

in the definition of the class $\mathcal{A}^{\text{det}}(\Lambda)$ instead of Dirac functionals, which correspond to $\mathcal{A}^{\text{det}}(\mathfrak{X}_0)$ for $\mathfrak{X}_0 =]0,1]$. Accordingly, the sequence of mappings

$$\psi_n : \mathbb{R}^{(n-1)\cdot m} \to \Lambda, \qquad n \geq 2,$$

associated with an algorithm $A \in \mathcal{A}^{\text{det}}(\Lambda)$ determines the functionals $\lambda \in \Lambda$ that are applied by A to a trajectory of W, and the expected number of functional evaluations of W is used to define the cost of A.

Results on minimal errors for such classes of algorithms are only available for the problem of L_2-approximation in case of a scalar equation (1), see [HMG04, HMGR02].

Simulation of the joint distribution of multiple Itô integrals is an open problem, in general, and therefore it is sometimes suggested to use suitable approximations in practice. The latter are usually based on point evaluations of W. Then, however, one actually uses an algorithm from the class $\mathcal{A}^{\text{det}}(\mathfrak{X}_0)$, and the lower bounds for the minimal errors from Theorems 2 to 4 apply.

3.3 Optimal Algorithms

The analysis of minimal errors in the Sections 3.1 and 3.2 can be used to construct algorithms that are easy to implement and achieve the corresponding upper bounds for the minimal errors.

Recall that each of the asymptotic constants $C = C_i$ from Theorems 2 to 4 and Remark 9 is determined by a space-time average of the respective process $Y = Y_i$, which suggests that, in the case $C > 0$, the number and the location of evaluation sites for a trajectory of W should be adjusted to the size of the corresponding trajectory of Y. This idea can in fact be employed to obtain algorithms $A_N \in \mathcal{A}^{\mathrm{det}}$ that satisfy

$$c(A_N) \approx N \qquad \text{and} \qquad e(A_N) \lesssim C \cdot \alpha_N,$$

where α_N specifies the corresponding order of convergence of the minimal errors $e_N(\mathcal{A}^{\mathrm{det}})$. We thus have strong asymptotic optimality, modulo the constant $\sqrt{3}$ in the case of Theorems 3 and 4.

The algorithms A_N use a step-size control and basically work as follows, see [HMGR01, MG02a, MG02b, MG04] for the details. First, we evaluate the trajectory w of W at a coarse grid, and we compute a discrete approximation y to the respective trajectory of the process Y. The approximation y determines the number and the location of the additional evaluation sites for w such that, roughly speaking, the step-size is proportional to $y^{-\beta}$, where $\beta > 0$ depends on the respective approximation problem. The resulting observations are then used to compute a discrete approximation to the corresponding trajectory of the solution X, which, in case of global approximation, is extended to a function on $[0, 1]$ by piecewise linear interpolation.

To be more precise, for one-point approximation of scalar equations we choose $\beta = 2/3$ and we apply a suitably modified Wagner-Platen scheme. In the case of one-point approximation or L_2-approximation of systems of equations we use $\beta = 1$ and we employ a modified Milstein scheme. Finally, for L_∞-approximation we take $\beta = 2$ and we use a modified Euler scheme.

We stress that for the algorithms A_N, the number of evaluations of the drift and diffusion coefficients a and b and their partial derivatives as well as the total number of all further arithmetic operations is proportional to N with a small multiplicative constant. Hence the conclusions from Remark 2 are valid, too, for $\mathcal{A}^{\mathrm{det}}$ instead of $\mathcal{A}^{\mathrm{equi}}$.

4 Weak Approximation

In this section we consider autonomous systems (1) of stochastic differential equations with drift and diffusion coefficients

$$a : \mathbb{R}^d \to \mathbb{R}^d \qquad \text{and} \qquad b : \mathbb{R}^d \to \mathbb{R}^{d \times d}$$

and initial value $x_0 \in \mathbb{R}^d$. Accordingly, the Brownian motion W as well as the solution process X both take values in $\mathbb{R}^d$.

As an important instance of weak approximation we assume that expectations $E(h(X))$ of functionals

$$h : C([0,1], \mathbb{R}^d) \to \mathbb{R}$$

w.r.t. the distribution of X have to be computed. In contrast to strong approximation we thus have to approximate deterministic quantities. To this end, however, randomness is frequently used as a computational tool. For instance, in a straightforward approach one uses a Monte Carlo simulation of a suitable approximation to the process X, which constitutes a link to strong approximation. On the other hand, deterministic algorithms are used for weak approximation as well, e.g., for solving an equivalent parabolic equation. We stress that minimal errors enable a comparison of deterministic and randomized algorithms.

We study two variants of the weak approximation problem. For the variable functional problem we only know that h belongs to some class F of functionals and we may evaluate the functionals $h \in F$ at suitable elements from the path space, while a, b, and x_0 and, as a consequence the distribution of X, are considered to be fixed. Here the mapping $S : F \to G$ with $G = \mathbb{R}$ is given by

$$S(h) = E(h(X)).$$

For the variable drift problem the functional h, b, and x_0 are considered to be fixed, while a is only known to belong to some class F of functions and partial information about the drift coefficient consists in a finite number of function values. Of course, the latter does not determine the distribution of X exactly. To stress its dependence on a we sometimes denote X by X^a. The mapping $S : F \to G$ with $G = \mathbb{R}$ is given by

$$S(a) = E(h(X^a)).$$

For both variants we present a worst-case analysis on classes F of drift coefficients $f = a$ or functionals $f = h$, respectively. Such classes are typically defined by smoothness properties and growth conditions.

The (maximal) error and the (maximal) cost of any algorithm $A \in \mathcal{A}^{\det}(\mathfrak{X}_0)$ are defined by

$$e(A) = \sup_{f \in F} |S(f) - A(f)|$$

and

$$c(A) = \sup_{f \in F} c(A, f),$$

see (6). For randomized algorithms $A \in \mathcal{A}^{\mathrm{ran}}$ on any underlying probability space $(\Omega, \mathfrak{A}, P)$ we define the (maximal) error and the (maximal) cost by

$$e(A) = \sup_{f \in F} \left(\int_{\Omega} |S(f) - A(\omega, f)|^2 \, dP(\omega) \right)^{1/2}$$

and

$$c(A) = \sup_{f \in F} \int_{\Omega} c(A(\omega, \cdot), f) \, dP(\omega).$$

See, e.g., [Nov88, TWW88, Was89].

4.1 The Variable Functional Problem

In the definition of the classes $\mathcal{A}^{\mathrm{det}}(\mathfrak{X}_0)$ and $\mathcal{A}^{\mathrm{ran}}(\mathfrak{X}_0)$ for this problem we take

$$\mathfrak{X} = C([0,1], \mathbb{R}^d) \qquad \text{and} \qquad \mathfrak{Y} = \mathbb{R},$$

and we allow $\mathfrak{X}_0$ to be any finite-dimensional subspace of $\mathfrak{X}$. Hence every algorithm $A \in \mathcal{A}^{\mathrm{ran}}(\mathfrak{X}_0)$ for approximation of $E(h(X))$ may evaluate the functionals $h : \mathfrak{X} \to \mathfrak{Y}$ at points $x \in \mathfrak{X}_0$, where $\mathfrak{X}_0$ may be chosen arbitrarily, but it is fixed for a specific algorithm. By definition the cost for each evaluation of any functional h equals $\dim(\mathfrak{X}_0)$ for every choice of $\mathfrak{X}_0$, i.e., we take

$$s(\mathfrak{X}_0) = \dim(\mathfrak{X}_0)$$

in (6). We study minimal errors on the classes

$$\mathcal{A}^{\mathrm{det}} = \bigcup_{\dim(\mathfrak{X}_0) < \infty} \mathcal{A}^{\mathrm{det}}(\mathfrak{X}_0)$$

and

$$\mathcal{A}^{\mathrm{ran}} = \bigcup_{\dim(\mathfrak{X}_0) < \infty} \mathcal{A}^{\mathrm{ran}}(\mathfrak{X}_0).$$

For a, b, and x_0 being fixed and any given class F of integrable functionals h the computation of $E(h(X))$ is a quadrature problem w.r.t. the distribution of X on the space $\mathfrak{X}$. A simple deterministic algorithm $A \in \mathcal{A}^{\mathrm{det}}$ is given by a quadrature formula

$$A(h) = \sum_{i=1}^{n} \beta_i \cdot h(x_i)$$

with a fixed choice of knots $x_i \in \mathfrak{X}$ and coefficients $\beta_i \in \mathbb{R}$, in which case $\mathfrak{X}_0 = \mathrm{span}\{x_1, \ldots, x_n\}$. A simple randomized algorithm is given by the classical Monte Carlo method

$$A(\omega, h) = \frac{1}{n} \cdot \sum_{i=1}^{n} h(X_i(\omega))$$

with i.i.d. random elements X_i that take values in a finite-dimensional subspace of $\mathfrak{X}$. In general we cannot take independent copies X_i of X, since the latter does not have a finite-dimensional support, except for trivial cases. Instead, one may take independent copies of a weak Itô-Taylor scheme with constant step-size $1/k$ and piecewise linear interpolation, see, e.g., [KP99]. In this case $\mathfrak{X}_0$ consists of piecewise linear functions $[0,1] \to \mathbb{R}^d$ with breakpoints ℓ/k, and $\dim(\mathfrak{X}_0) = d \cdot (k+1)$.

Specifically, we consider the class $F = \mathrm{Lip}(1)$ of Lipschitz continuous functionals $h : \mathfrak{X} \to \mathfrak{Y}$ with Lipschitz constant at most one. Thus $h \in \mathrm{Lip}(1)$ iff

$$|h(x) - h(y)| \leq \|x - y\|_{\mathfrak{X}}$$

holds for all $x, y \in \mathfrak{X}$, where $\|\cdot\|_{\mathfrak{X}}$ denotes the supremum norm. For the corresponding minimal errors we have the following asymptotic bounds.

Theorem 5 ([DMGR06]). *Suppose that for equation* (1) *the drift coefficient* a *is Lipschitz continuous and the diffusion coefficient* b *has bounded first and second order derivatives. Furthermore, assume that* b *is of class* C^∞ *in some neighborhood of the initial value* x_0 *and* $\det b(x_0) \neq 0$. *Then*

$$e_N(\mathcal{A}^{\mathrm{det}}) \succeq (\ln N)^{-1/2}$$

and[4]

$$N^{-1/4} \cdot (\ln N)^{-3/4} \preceq e_N(\mathcal{A}^{\mathrm{ran}}) \preceq N^{-1/4} \cdot (\ln N)^{1/4}$$

for $F = \mathrm{Lip}(1)$.

See [DMGR06] for details concerning the following remarks.

Remark 10. The minimal errors for the quadrature problem on $\mathrm{Lip}(1)$ are closely related to the average Kolmogorov widths and the quantization numbers for the stochastic process X. The n-th quantization number

$$q_n = \inf_{x_1,\ldots,x_n \in \mathfrak{X}} E\left(\min_{\ell=1,\ldots,n} \|X - x_\ell\|_{\mathfrak{X}}\right)$$

and the k-th average Kolmogorov width

$$d_k = \inf_{\dim(\mathfrak{X}_0)=k} E\left(\min_{\tilde{x} \in \mathfrak{X}_0} \|X - \tilde{x}\|_{\mathfrak{X}}\right)$$

are minimal average errors of best approximation, either from n-point sets or from k-dimensional subspaces. It turns out that

$$e_N(\mathcal{A}^{\mathrm{det}}) \succeq \min_{n \cdot k \leq N} \max(q_n, d_k) \tag{9}$$

4 By definition, $x_N \preceq y_N$ for sequences of positive real numbers x_N and y_N, if $x_N \leq \gamma \cdot y_N$ holds for every $N \in \mathbb{N}$ with a constant $\gamma > 0$.

and

$$e_N(\mathcal{A}^{\mathrm{ran}}) \succeq \min_{n \cdot k \le N} \max \left(n^{1/2} \sup_{m \ge 4n} (q_{m-1} - q_m), d_k \right). \tag{10}$$

The lower bound (9) is easily verified as follows. Let $N \in \mathbb{N}$ and consider any algorithm $A \in \mathcal{A}^{\mathrm{det}}(\mathfrak{X}_0)$ such that $c(A) \le N$. Put $k = \dim(\mathfrak{X}_0)$, and note that A uses at most $n = \lfloor N/k \rfloor$ evaluations for every functional $h \in F$. We may assume for simplicity that n actually is the number of evaluations for every h. Put $x_1 = \psi_1$ and $x_\ell = \psi_\ell(0, \ldots, 0)$ for $\ell = 2, \ldots, n$, where $\psi_1, \ldots, \psi_n$ specify the sequential selection of evaluation sites by A, see Section 2, and consider the functional

$$h(x) = \min_{\ell=1,\ldots,n} \|x - x_\ell\|_{\mathfrak{X}}.$$

Observe that $\pm h \in \mathrm{Lip}(1)$ and $A(h) = A(-h)$, since every evaluation of $\pm h$ performed by the algorithm A yields the value zero. It follows that

$$e(A) \ge S(h) \ge q_n.$$

For the functional

$$h(x) = \min_{\widetilde{x} \in \mathfrak{X}_0} \|x - \widetilde{x}\|_{\mathfrak{X}}$$

we analogously get

$$e(A) \ge S(h) \ge d_k.$$

We conclude that $e(A) \ge \max(q_n, d_k)$ for some $n, k \in \mathbb{N}$ such that $n \cdot k \le N$.

For the same reason average Kolmogorov widths also appear in the lower bound (10). Here the first term, which involves consecutive differences of quantization numbers, is obtained by means of Bakhvalov's Theorem.

Let us stress that the lower bounds (9) and (10) for the minimal errors of deterministic and randomized algorithms are valid for every random element with values in a Banach space $\mathfrak{X}$. See [DMGR06] for applications of this fact to the quadrature problem for $F = \mathrm{Lip}(1)$ and Gaussian random elements X. In the latter case it is known that the asymptotic behaviour of the Kolmogorov widths and quantization numbers is closely related to the asymptotic behaviour of the small ball function

$$\phi(\epsilon) = P(\|X\|_{\mathfrak{X}} \le \epsilon)$$

for ϵ tending to zero. In view of (9) and (10) the study of small ball functions therefore leads to lower bounds for the quadrature problem.

Remark 11. Quantization of random vectors X that take values in finite-dimensional spaces $\mathfrak{X}$ has been studied since the late 1940's, and we refer to the monograph [GL00] for an up-to-date account. For stochastic processes, i.e., for random elements X taking values in infinite-dimensional spaces $\mathfrak{X}$, quantization is studied since about ten years. Results are known for Gaussian processes, see, e.g., [Der03, DFMS03, DS06, LP02, LP04], and for diffusion processes, see [Der04, DMGR06, LP06]. In particular, under the assumptions from Theorem 5,

$$q_n \asymp (\ln n)^{-1/2}. \tag{11}$$

Average Kolmogorov widths and their relation to further scales of approximation quantities for stochastic processes are studied in, e.g., [Cre02, MW96, Mat90, Sun92]. In particular, for the Brownian motion W the weak asymptotic behaviour of d_k is determined in [Mai93], and the same asymptotics

$$d_k \asymp k^{-1/2} \tag{12}$$

holds for diffusion processes under the assumptions from Theorem 5.

The bound for $e_N(\mathcal{A}^{\text{det}})$ from Theorem 5 thus follows from (9) and (11). For $e_N(\mathcal{A}^{\text{ran}})$ the lower bound from Theorem 5 essentially follows from (10), (11), and (12).

Remark 12. The upper bound for $e_N(\mathcal{A}^{\text{ran}})$ from Theorem 5 is achieved by a Monte Carlo Euler scheme with suitably chosen numbers k of equidistant time-steps and n of replications. Consider independent random vectors $Z^{(1)}, \dots, Z^{(k)}$ that are standard-normally distributed on $\mathbb{R}^d$. Furthermore, put $X_k^{(0)}(\omega) = x_0$, and let $X_k(\omega)$ denote the piecewise linear interpolation of the values

$$X_k^{(\ell)}(\omega) = X_k^{(\ell-1)}(\omega) + a\left(X_k^{(\ell-1)}(\omega)\right)/k + b\left(X_k^{(\ell-1)}(\omega)\right)\cdot Z^{(\ell)}(\omega)/\sqrt{k}$$

at the breakpoints ℓ/k. With i.i.d. copies $X_{k,1}, \dots, X_{k,n}$ we define

$$A_{k,n}^{\text{ran}}(\omega, h) = \frac{1}{n}\cdot\sum_{i=1}^{n} h(X_{k,i}(\omega)). \tag{13}$$

Given $N \in \mathbb{N}$ we take

$$k = \lfloor N^{1/2}\cdot(\ln N)^{1/2}\rfloor, \qquad n = \lfloor N^{1/2}\cdot(\ln N)^{-1/2}\rfloor,$$

and we put $A_N^{\text{ran}} = A_{k,n}^{\text{ran}}$. For every fixed value of d we clearly have

$$c(A_N^{\text{ran}}) \preceq N,$$

and by exploiting the link to strong approximation one obtains

$$e(A_N^{\text{ran}}) \preceq N^{-1/4}\cdot(\ln N)^{1/4},$$

see Remark 9. In view of the corresponding lower bound we conclude that the sequence of randomized algorithms A_N^{ran} is optimal, up to at most a factor $\gamma\cdot\ln N$ with $\gamma = \gamma(a, b, x_0) > 0$.

For this algorithm A_N^{ran}, the number of evaluations of the drift and diffusion coefficients a and b, the number of calls of a random number generator for the standard normal distribution, and the total number of all further arithmetical operations is proportional to $c\cdot d^2\cdot N$ with a small constant $c > 0$. Hence the conclusions from Remark 2 are valid, too, for the variable functional problem of weak approximation.

Remark 13. The analysis of the variable functional problem, which we have presented so far, is non-uniform w.r.t. to a, b, and x_0, since algorithms may be particularly tuned to a specific equation and constants may depend on a, b, and x_0, too. However, the upper bound for the error $e(A_N^{\mathrm{ran}})$ of the Monte Carlo Euler algorithm does not change if the supremum over the class

$$F = \mathrm{Lip}(1) \times F_a \times F_b \times [-1,1]$$

with F_a and F_b being the classes of Lipschitz continuous mappings $a : \mathbb{R}^d \to \mathbb{R}^d$ and $b : \mathbb{R}^d \to \mathbb{R}^{d\times d}$ with Lipschitz constant at most one is considered in the definition of the error $e(A)$ of randomized algorithms. The more restrictive assumptions from Theorem 5 concerning a and b are only needed in the proof of the lower bounds, and here a non-uniform result is even stronger than a uniform one.

Remark 14. According to Theorem 5, randomized algorithms are far superior to deterministic ones for weak approximation in the worst-case on Lip(1). Moreover, deterministic algorithms seem to be not suited at all, since $(\ln N)^{-1/2}$ tends to zero too slowly. However, deterministic algorithms are successfully used in practice for computational problems with certain functionals from Lip(1). We refer in particular to [PP05] for applications of quantization and corresponding deterministic algorithms to the valuation of path-dependent options in mathematical finance. It thus seems interesting to identify classes of functionals $F \subsetneq \mathrm{Lip}(1)$ that on the one hand contain sufficiently many practically relevant functionals but on the other hand lead to substantially smaller upper bounds for suitable deterministic algorithms.

One such instance of a class F consists of the functionals

$$h(x) = h_1(x(1)) \cdot \exp\left(\int_0^1 h_2(x(s))\, ds \right), \tag{14}$$

which appear in the Feynman-Kac formula and only have a mild dependence on the whole path via the integral term. Here it is reasonable to study algorithms that may separately evaluate the functions $h_i : \mathbb{R}^d \to \mathbb{R}$. In the Brownian motion case, i.e., for $a = 0$ and $b = \mathrm{Id}_d$, corresponding minimal errors for deterministic and randomized algorithms have been determined in [Kwa05, KL03, PWW00].

Suppose, for instance, that F consists of all functionals h of the form (14) with h_i being Lipschitz continuous with Lipschitz constant at most one and vanishing outside $[-1,1]^d$, then the minimal errors are of order $N^{-1/d}$ for deterministic algorithms and of order $N^{-1/d-1/2}$ for randomized algorithms. In terms of orders of minimal errors we see that randomized algorithms are far superior to deterministic ones for large dimensions d. We refer to [Kwa05, KL03, PWW00] for further results dealing with different scales of smoothness classes. See also Remark 16.

Clearly, the quadrature problem for Lip(1) is a linear problem in the sense of [TWW88, Sec. 4.5], and general results for linear problems apply in the analysis of minimal errors. We stress that we no longer have a linear problem, if functionals of the form (14) are considered, since in this case $S(h)$ depends non-linearly on h_2.

4.2 The Variable Drift Problem

In the variable drift problem we consider a system (1) with $m = d$, a fixed initial value $x_0 \in \mathbb{R}^d$ and the fixed diffusion coefficient

$$b = \sqrt{2} \cdot \mathrm{Id}_d,$$

but with a drift coefficient a that is only known to belong to some class F of functions from $\mathbb{R}^d$ to $\mathbb{R}^d$. Furthermore, we assume that the functional $h : C([0,1], \mathbb{R}^d) \to \mathbb{R}$ is fixed and actually only depends on the value of X^a at $t = 1$, i.e.,

$$h(X) = g(X^a(1))$$

for some fixed mapping $g : \mathbb{R}^d \to \mathbb{R}$. Observe that $S(a)$ depends non-linearly on a.

To model the sequential evaluation of a we take

$$\mathfrak{X}_0 = \mathfrak{X} = \mathfrak{Y} = \mathbb{R}^d,$$

and we study minimal errors on the corresponding classes $\mathcal{A}^{\mathrm{det}} = \mathcal{A}^{\mathrm{det}}(\mathfrak{X}_0)$ and $\mathcal{A}^{\mathrm{ran}} = \mathcal{A}^{\mathrm{ran}}(\mathfrak{X}_0)$ with $s(\mathfrak{X}_0) = 1$ in the definition of the cost, see (6). Lower bounds for the minimal errors reflect the intrinsic uncertainty about the distribution of $X^a(1)$ that is due to the partial information about a.

According to the Feynman-Kac formula the computation of $E(g(X^a(1)))$ is equivalent to solving the linear parabolic initial value problem

$$\Delta u + \sum_{j=1}^{d} a_j \cdot \frac{\partial u}{\partial x_j} = \frac{\partial u}{\partial t},$$
$$u(0, \cdot) = g$$

at the single point $(1, x_0)$, i.e.,

$$S(a) = u(1, x_0).$$

This fact immediately suggests numerous promising deterministic algorithms $A \in \mathcal{A}^{\mathrm{det}}$, e.g., finite difference schemes.

As an elementary randomized algorithm $A_{k,n} \in \mathcal{A}^{\mathrm{ran}}$ we mention the Monte Carlo Euler scheme

$$A_{k,n}^{\mathrm{ran}}(\omega, a) = \frac{1}{n} \cdot \sum_{i=1}^{n} g(X_{k,i}^{(k)}(\omega)), \tag{15}$$

where n is the number of replications and k denotes the number of time-steps, see (13).

For $r \in \mathbb{N}_0$ and $0 < \alpha \leq 1$ we let $C^{r,\alpha}$ denote the Hölder class of functions $a : \mathbb{R}^d \to \mathbb{R}^d$ whose r-th order partial derivatives $\widetilde{a}$ satisfy

$$|\widetilde{a}(x) - \widetilde{a}(y)| \leq |x - y|$$

for all $x, y \in \mathbb{R}^d$. For the corresponding minimal errors on the classes

$$F^{r,\alpha} = \{a \in C^{r,\alpha} : \operatorname{supp} a \subseteq [-1,1]^d\}$$

we have the following asymptotic bounds.

Theorem 6 ([PR06]). *Suppose that the function $g : \mathbb{R}^d \to \mathbb{R}$ is continuous, non-constant, and satisfies the growth condition*

$$\sup_{x \in \mathbb{R}^d} |g(x)| \cdot \exp(-\beta \cdot |x|^2) < \infty$$

for every $\beta > 0$. Then, for every $\varepsilon > 0$,

$$N^{-(r+\alpha)/d} \preceq e_N(\mathcal{A}^{\mathrm{det}}) \preceq N^{-(r+\alpha)/d+\varepsilon}$$

and

$$N^{-(r+\alpha)/d-1/2} \preceq e_N(\mathcal{A}^{\mathrm{ran}}) \preceq N^{-(r+\alpha)/d-1/2+\varepsilon}.$$

See [PR06] for details concerning the following remarks.

Remark 15. Note that the lower bounds from Theorem 6 coincide with the asymptotic behaviour of the minimal errors for the integration problem on the class $F^{r,\alpha}$, see [Nov88, Prop. 1.3.9 and 2.2.9]. Actually, the proof of the lower bounds for the variable drift problem relies on the fact that $S(a)$ may be represented as a rapidly convergent series of weighted integrals of increasing dimension, where the integrands are tensor products of the components of a. This enables, in particular, the use of Bakhvalov's Theorem to derive lower bounds for $e_N(\mathcal{A}^{\mathrm{ran}})$ in the variable drift problem, although the latter is a non-linear problem. In this way one obtains a general theorem that relates minimal errors for the integration problem and the variable drift problem under general assumptions on the underlying function class F.

The lower bounds are also valid for algorithms that use partial derivatives of a up to the order r. This is of interest, since higher-order weak Itô-Taylor schemes need to evaluate partial derivatives of the drift and the diffusion coefficients.

Remark 16. To provide upper bounds for the minimal errors and to construct corresponding algorithms one truncates the series representation for $S(a)$ and approximates the remaining tensor products of components of a. The latter problems altogether are almost as easy as approximation of a single component, if Smolyak formulas are used. It thus turns out that solving the variable coefficient problem is almost as easy as L_∞-approximation of the single components a_j. Furthermore, the randomized algorithm uses the deterministic one for variance reduction. For this approach we also refer to the analysis of the variable functional problem in [Kwa05, KL03, PWW00], see Remark 14.

We add that optimality, up to the factor N^ε, can also be achieved in this way for classes of functions with unbounded support, provided that certain growth properties hold for a or its local Hölder constants.

However, implementation of any of these almost optimal algorithms would require extensive pre-computing. A straight-forward approach leads to more than N quadrature problems, which do not depend on the components a_j and must be solved in advance.

Remark 17. In computational practice randomized algorithms are often preferred to deterministic ones, unless the dimension d is small. Large values of d naturally arise, e.g., in computational finance, when (1) is used to model the risk-neutral dynamics of the prices of d assets and g denotes the discounted payoff of a European option with maturity $t = 1$. In this case $S(a)$ is the value of the option at time $t = 0$.

We present a simple consequence of the lower bound for the minimal error $e_N(\mathcal{A}^{\text{det}})$ on the classes $F^{r,\alpha}$. Consider the Monte Carlo Euler scheme $A^{\text{ran}}_{k,n}$, see (15). Under moderate assumptions on the smoothness and the growth of the coefficients a and b as well as of g the bias of this algorithm is proportional to the step-size $1/k$, see, e.g., [KP99]. Relating the step-size $1/k$ and the number of replications n in an optimal way, we get a randomized algorithm A^{ran}_N with error

$$e(A^{\text{ran}}_N) \leq \gamma_1 \cdot N^{-1/3} \qquad (16)$$

for some constant $\gamma_1 > 0$ and with computational cost proportional to N. For $b = \sqrt{2} \cdot \text{Id}_d$ this holds true on classes $F^{r,\alpha}$ at least if $r + \alpha > 2$. On the other hand, we have the lower bound

$$e_N(\mathcal{A}^{\text{det}}) \geq \gamma_2 \cdot n^{-(r+\alpha)/d}$$

with some constant $\gamma_2 > 0$ according to Theorem 6. We thus conclude that asymptotically the simple and easily implementable algorithm A^{ran}_N is preferable to every deterministic algorithm of the same computational cost, if

$$d > 3\,(r + \alpha).$$

For instance, if $r + \alpha$ is close to 2, this superiority already holds for $d \geq 7$.

We add that multi-level Monte Carlo Euler schemes are introduced in [Gil06], which significantly improve the upper bound (16). Under suitable assumptions on the drift and diffusion coefficients a and b as well as on the function g these algorithms achieve errors of order $\ln N \cdot N^{-1/2}$ at a cost proportional to N.

References

[BB00] C. T. H. Baker and E. Buckwar. Numerical analysis of explicit one-step methods for stochastic delay differential equations. LMS J. Comput. Math. **3**, 315–335 (2000).

[BS05] E. Buckwar and T. Shardlow. Weak approximation of stochastic differential delay equations. IMA J. Numer. Anal. **25**, 57–86 (2005).

[BT05] B. Boufoussi and C. A. Tudor. Kramers-Smoluchowski approximation for stochastic evolution equations with FBM. Rev. Roumaine Math. Pures Appl. **50**, 125–136 (2005).

[CH96] S. Cambanis and Y. Hu. Exact convergence rate of the Euler-Maruyama scheme, with application to sampling design. Stochastics Stochastics Rep. **59**, 211–240 (1996).

[CC80] J. M. C. Clark and R. J. Cameron. The maximum rate of convergence of discrete approximations for stochastic differential equations. In: Stochastic Differential Systems (B. Grigelionis, ed.), Lect. Notes Control Inf. Sci. **25**, Springer-Verlag, Berlin (1980).

[Cre02] J. Creutzig. Approximation of Gaussian Random Vectors in Banach Spaces. Ph.D. Dissertation, Fakultät für Mathematik und Inf., Friedrich-Schiller Universität Jena (2002).

[DG01] A. M. Davie and J. Gaines. Convergence of numerical schemes for the solution of parabolic partial differential equations. Math. Comp. **70**, 121–134 (2001).

[Der03] S. Dereich. High Resolution Coding of Stochastic Processes and Small Ball Probabilities. Ph.D. Dissertation, Institut für Mathematik, TU Berlin (2003).

[Der04] S. Dereich. The quantization complexity of diffusion processes. Preprint, arXiv: math.PR/ 0411597 (2004).

[DFMS03] S. Dereich, F. Fehringer, A. Matoussi, and M. Scheutzow. On the link between small ball probabilities and the quantization problem. J. Theoret. Probab. **16**, 249–265 (2003).

[DMGR06] S. Dereich, T. Müller-Gronbach, and K. Ritter. Infinite-dimensional quadrature and quantization. Preprint, arXiv: math.PR/0601240v1 (2006).

[DS06] S. Dereich and M. Scheutzow. High-resolution quantization and entropy coding for fractional Brownian motion. Electron. J. Probab. **11**, 700–722 (2006).

[Gar04] A. Gardoń. The order of approximations for solutions of Itô-type stochastic differential equations with jumps. Stochastic Anal. Appl. **22**, 679–699 (2004).

[Gil06] M. B. Giles. Multi-level Monte Carlo path simulation. Report NA-06/03, Oxford Univ. Computing Lab. (2006).

[GM04] P. Glasserman and N. Merener. Convergence of a discretization scheme for jump-diffusion processes with state-dependent intensities. Proc. R. Soc. Lond. Ser. A Math. Phys. Eng. Sci. **460**, 111–127 (2004).

[GL00] S. Graf and H. Luschgy. Foundations of Quantization for Probability Distributions, Lect. Notes in Math. **1730**, Springer-Verlag, Berlin (2000).

[GK96] W. Grecksch and P. E. Kloeden. Time-discretised Galerkin approximations of parabolic stochastic PDEs. Bull. Austr. Math. Soc. **54**, 79–85 (1996).

[GN97] I. Gyöngy and D. Nualart. Implicit scheme for stochastic parabolic partial differential equations driven by space-time white noise. Potential Analysis **7**, 725–757 (1997).

[HSV06] M. Hairer, A. M. Stuart, and J. Voss. Analysis of SPDEs arising in path sampling. Part II: The nonlinear case. Manuscript (2006).

[HSVW05] M. Hairer, A. M. Stuart, J. Voss, and P. Wiberg. Analysis of SPDEs arising in path sampling, Part I: The Gaussian case. Commun. Math. Sci. **3**, 587–603 (2005).

[Hau03] E. Hausenblas. Approximation for semilinear stochastic evolution equations. Potential Analysis **18**, 141–186 (2003).

[HK06] D. J. Higham and P. E. Kloeden. Convergence and stability of implicit methods for jump-diffusion systems. Int. J. Numer. Anal. Model. **3**, 125–140 (2006).

[HMG04] N. Hofmann and T. Müller-Gronbach. On the global error of Itô-Taylor schemes for strong approximation of scalar stochastic differential equations. J. Complexity **20**, 732–752 (2004).

[HMG06] N. Hofmann and T. Müller-Gronbach. A modified Milstein scheme for approximation of stochastic delay differential equations with constant time lag. J. Comput. Appl. Math. **197**, 89–121 (2006).

[HMGR01] N. Hofmann, T. Müller-Gronbach, and K. Ritter. The optimal discretization of stochastic differential equations. J. Complexity **17**, 117–153 (2001).

[HMGR02] N. Hofmann, T. Müller-Gronbach, and K. Ritter. Linear vs. standard information for scalar stochastic differential equations. J. Complexity **18**, 394–414 (2002).

[HMY04] Y. Hu, S.-E. A. Mohammmed, and F. Yan. Discrete-time approximation of stochastic delay equations: the Milstein scheme. Ann. Probab. **32**, 265–314 (2004).

[KP99] P. Kloeden and E. Platen. Numerical Solution of Stochastic Differential Equations. 3rd print., Springer-Verlag, Berlin (1999).

[KP02] K. Kubilius and E. Platen. Rate of weak convergence of the Euler approximation for diffusion processes with jumps. Monte Carlo Meth. Appl. **8**, 83–96 (2002).

[KP00] U. Küchler and E. Platen. Strong discrete time approximation of stochastic differential equations with time delay. Math. Comput. Simulation **54**, 189–205 (2000).

[Kwa05] M. Kwas. An optimal Monte Carlo algorithm for multivariate Feynman-Kac integrals. J. Math. Phys. **46**, 103511.

[KL03] M. Kwas and Y. Li. Worst case complexity of multivariate Feynman-Kac path integration. J. Complexity **19**, 730–743 (2003).

[Lin95] S. J. Lin. Stochastic analysis of fractional Brownian motions. Stochastics Stochastics Rep. **55**, 121–140 (1995).

[LL00] X. Q. Liu and C. W. Li. Weak approximation and extrapolations of stochastic differential equations with jumps. SIAM J. Numer. Anal. **37**, 1747–1767 (2000).

[LP02] H. Luschgy and G. Pagès. Functional quantization of Gaussian processes. J. Funct. Anal. **196**, 486–531 (2002).

[LP04] H. Luschgy and G. Pagès. Sharp asymptotics of the functional quantization problem for Gaussian processes. Ann. Appl. Prob. **32**, 1574–1599 (2004).

[LP06] H. Luschgy and G. Pagès. Functional quantization of a class of Brownian diffusions: a constructive approach. Stochastic Processes Appl. **116**, 310–336 (2006).

[Mag98] Y. Maghsoodi. Exact solutions and doubly efficient approximations of jump-diffusion Itô equations. Stochastic Anal. Appl. **16**, 1049–1072 (1998).

[Mai93] V. Maiorov. Average n-widths of the Wiener space in the $L-\infty$-norm. J. Complexity **9**, 222–230 (1993).

[MW96] V. Maiorov and G. W. Wasilkowski. Probabilistic and average linear widths in L_∞ norm with respect to r-folded Wiener measure. J. Approx. Theory **84**, 31–40 (1996).

[Mar55] G. Maruyama. Continuous Markov processes and stochastic equations. Rend. Circ. Mat. Palermo **4**, 48–90 (1955).

[Mat90] P. Mathé. s-numbers in information-based complexity. J. Complexity **6**, 41–66 (1900).

[Mil95] G. N. Milstein. Numerical Integration of Stochastic Differential Equations. Kluwer, Dordrecht (1995).

[MG02a] T. Müller-Gronbach. Strong Approximation of Systems of Stochastic Differential Equations. Habilitationsschrift, TU Darmstadt (2001).

[MG02b] T. Müller-Gronbach. Optimal uniform approximation of systems of stochastic differential equations. Ann. Appl. Prob. **12**, 664–690 (2002).

[MG04] T. Müller-Gronbach. Optimal pointwise approximation of SDEs based on Brownian motion at discrete points. Ann. Appl. Prob. **14**, 1605–1642 (2004).

[MGR07a] T. Müller-Gronbach and K. Ritter. An implicit Euler scheme with non-uniform time discretization for heat equations with multiplicative noise. BIT **47**, 393–418 (2007).

[MGR07b] T. Müller-Gronbach and K. Ritter. Lower bounds and nonuniform time discretization for approximation of stochastic heat equations. Found. Comput. Math. **7**, 135–181 (2007).

[MGRW07] T. Müller-Gronbach, K. Ritter, and T. Wagner. Optimal pointwise approximation of a linear stochastic heat equation with additive space-time white noise. In S. Heinrich, A. Keller, and H. Niederreiter, editors, *Monte Carlo and Quasi-Monte Carlo Methods 2006*, pages 577–589. Springer-Verlag, 2007.

[NY83] A. S. Nemirovsky and D. B. Yudin. Problem Complexity and Method Efficiency in Optimization. Wiley, New York (1983).

[Neu06] A. Neuenkirch. Optimal approximation of SDEs with additive fractional noise. J. Complexity **22**, 459–474 (2006).

[NN06] A. Neuenkirch and I. Nourdin. Exact rate of convergence of some approximation schemes associated to SDEs driven by a fBm. Preprint (2006).

[Nov88] E. Novak. Deterministic and Stochastic Error Bounds in Numerical Analysis. Lect. Notes in Math. **1349**, Springer-Verlag, Berlin (1988).

[Nov95] E. Novak. The real number model in numerical analysis. J. Complexity **11**, 57–73 (1995).

[Nua06] D. Nualart. The Malliavin calculus and related topics. 2nd ed., Springer-Verlag, Berlin, (2006).

[Pag07] G. Pagès. Quadratic optimal functional quantization of stochastic processes and numerical applications. In S. Heinrich, A. Keller, and H. Niederreiter, editors, *Monte Carlo and Quasi-Monte Carlo Methods 2006*, pages 101–142. Springer-Verlag, 2007.

[Pla99] E. Platen. An introduction to numerical methods for stochastic differential equations. Acta Numer. **8**, 197–246 (1999).

[PP05] G. Pagès and J. Printems. Functional quantization for numerics with an application to option pricing. Monte Carlo Meth. Appl. **11**, 407–446 (2005).

[PR06] K. Petras and K. Ritter. On the complexity of parabolic initial value problems with variable drift. J. Complexity **22**, 118–145 (2006).

[PWW00] L. Plaskota, G. W. Wasilkowski, and H. Woźniakowski. A new algorithm and worst case complexity for Feynman-Kac path integration. J. Comput. Phys. **164**, 335–353 (2000).

[Rit00] K. Ritter. Average-Case Analysis of Numerical Problems. Lect. Notes in Math. **1733**, Springer-Verlag, Berlin (2000).

[Rum82] W. Rümelin. Numerical treatment of stochastic differential equations. SIAM J. Numer. Anal. **19**, 604–613 (1982).

[SVW04] A. M. Stuart, J. Voss, and P. Wiberg. Fast communication conditional path sampling of SDEs and the Langevin MCMC method. Commun. Math. Sci. **2**, 685–697 (2004).

[Sun92] Y. Sun. Average n-width of point set in Hilbert space. Chinese Sci. Bull. **37**, 1153–1157 (1992).

[Tal95] D. Talay. Simulation of stochastic differential systems. In: Probabilistic Methods in Applied Physics (P. Krée, W. Wedig, eds.), Lect. Notes Phys. Monogr. **451**, Springer-Verlag, Berlin (1995).

[TWW88] J. F. Traub, G. W. Wasilkowski, and H. Woźniakowski. Information-Based Complexity. Academic Press, New York (1988).

[TT87] C. Tudor and M. Tudor. On approximation of solutions for stochastic delay equations. Stud. Cercet. Mat. **39**, 265–274 (1987).

[Was89] G. W. Wasilkowski. Randomization for continuous problems. J. Complexity **5**, 195–218 (1989).

Nets, (t, s)-Sequences, and Codes

Harald Niederreiter

Department of Mathematics, National University of Singapore, 2 Science Drive 2, Singapore 117543, Republic of Singapore
nied@math.nus.edu.sg

Summary. Nets and (t, s)-sequences are standard sources of quasirandom points for quasi-Monte Carlo methods. Connections between nets and error-correcting codes have been noticed for a long time, and these links have become even more pronounced with the development of the duality theory for digital nets. In this paper, we further explore these fascinating connections. We present also a recent construction of digital (t, s)-sequences using global function fields and new general constructions of nets and (t, s)-sequences.

1 Introduction and Basic Definitions

Low-discrepancy point sets and sequences are the workhorses of quasi-Monte Carlo methods. Currently, the most powerful methods for the construction of low-discrepancy point sets and sequences are based on the theory of (t, m, s)-nets and (t, s)-sequences. This paper describes further contributions to this theory.

The concept of a (t, m, s)-net is a special case of the notion of a uniform point set introduced in [Nie03]. As usual in the area, we follow the convention that a *point set* is a "multiset" in the sense of combinatorics, i.e., a set in which multiplicities of elements are allowed and taken into account. We write $I^s = [0, 1]^s$ for the s-dimensional unit cube.

Definition 1. Let $(X, \mathcal{B}, \mu)$ be an arbitrary probability space and let $\mathcal{E}$ be a nonempty subset of $\mathcal{B}$. A point set $P = \{\mathbf{x}_1, \ldots, \mathbf{x}_N\}$ of $N \geq 1$ elements of X is called *$(\mathcal{E}, \mu)$-uniform* if

$$\frac{1}{N} \sum_{n=1}^{N} \chi_E(\mathbf{x}_n) = \mu(E) \qquad \text{for all } E \in \mathcal{E},$$

where χ_E denotes the characteristic function of E.

Definition 2. Let $s \geq 1$, $b \geq 2$, and $0 \leq t \leq m$ be integers and let λ_s be the probability measure on I^s induced by the s-dimensional Lebesgue measure. Let $\mathcal{J}_{b,m,t}^{(s)}$ be the collection of all subintervals J of I^s of the form

$$J = \prod_{i=1}^{s} [a_i b^{-d_i}, (a_i + 1) b^{-d_i})$$

with integers $d_i \geq 0$ and $0 \leq a_i < b^{d_i}$ for $1 \leq i \leq s$ and with $\lambda_s(J) = b^{t-m}$. Then a $(\mathcal{J}_{b,m,t}^{(s)}, \lambda_s)$-uniform point set consisting of b^m points in I^s is called a *(t, m, s)-net in base b*.

It is important to note that the smaller the value of t for given b, m, and s, the larger the family $\mathcal{J}_{b,m,t}^{(s)}$ of intervals in Definition 2, and so the stronger the uniform point set property in Definition 1. The number t is often called the *quality parameter* of a (t, m, s)-net in base b.

For the definition of a (t, s)-sequence, we need a few preliminaries. Given a real number $x \in [0, 1]$, let

$$x = \sum_{j=1}^{\infty} y_j \, b^{-j} \qquad \text{with all } y_j \in Z_b := \{0, 1, \ldots, b-1\}$$

be a b-adic expansion of x, where the case $y_j = b - 1$ for all but finitely many j is allowed. For any integer $m \geq 1$, we define the truncation

$$[x]_{b,m} = \sum_{j=1}^{m} y_j \, b^{-j}.$$

It should be emphasized that this truncation operates on the *expansion* of x and not on x itself, since it may yield different results depending on which b-adic expansion of x is used. If $\mathbf{x} = (x^{(1)}, \ldots, x^{(s)}) \in I^s$ and the $x^{(i)}$, $1 \leq i \leq s$, are given by prescribed b-adic expansions, then we define

$$[\mathbf{x}]_{b,m} = ([x^{(1)}]_{b,m}, \ldots, [x^{(s)}]_{b,m}).$$

Definition 3. Let $s \geq 1$, $b \geq 2$, and $t \geq 0$ be integers. A sequence $\mathbf{x}_0, \mathbf{x}_1, \ldots$ of points in I^s is a *(t, s)-sequence in base b* if for all integers $k \geq 0$ and $m > t$ the points $[\mathbf{x}_n]_{b,m}$ with $kb^m \leq n < (k+1)b^m$ form a (t, m, s)-net in base b. Here the coordinates of all points $\mathbf{x}_n$, $n = 0, 1, \ldots$, are given by prescribed b-adic expansions.

As before, we are interested in small values of t in the construction of (t, s)-sequences. We call t the *quality parameter* of a (t, s)-sequence in base b. For general background on (t, m, s)-nets and (t, s)-sequences, we refer to the monograph [Nie92] and the recent survey article [Nie05].

The rest of the paper is organized as follows. In Section 2, we recall the digital method for the construction of (t, m, s)-nets and (t, s)-sequences. Section 3 presents a review of the duality theory for digital nets and its connections with the theory of error-correcting codes. Recent constructions of digital nets using duality theory and other links with coding theory are described in Section 4. The recent construction in [MN] of digital (t, s)-sequences using differentials in global function fields is presented in Section 5, together with upper bounds on the well-known quantity $d_q(s)$. Sections 6 and 7 contain new ideas on how to generalize the digital method for the construction of (t, m, s)-nets and (t, s)-sequences, respectively.

2 Digital Nets and Digital (t, s)-Sequences

Most of the known constructions of (t, m, s)-nets and (t, s)-sequences are based on the so-called digital method introduced in [Nie87, Section 6]. In order to describe the digital method for the construction of (t, m, s)-nets in base b, we need the following ingredients. First of all, let integers $m \geq 1$, $s \geq 1$, and $b \geq 2$ be given. Then we choose the following:

(i) a commutative ring R with identity and $\mathrm{card}(R) = b$;
(ii) bijections $\eta_j^{(i)} : R \to Z_b$ for $1 \leq i \leq s$ and $1 \leq j \leq m$;
(iii) $m \times m$ matrices $C^{(1)}, \ldots, C^{(s)}$ over R.

Now let $\mathbf{r} \in R^m$ be an m-tuple of elements of R and define

$$p_j^{(i)}(\mathbf{r}) = \eta_j^{(i)}(\mathbf{c}_j^{(i)} \cdot \mathbf{r}) \in Z_b \qquad \text{for } 1 \leq i \leq s,\ 1 \leq j \leq m,$$

where $\mathbf{c}_j^{(i)}$ is the jth row of the matrix $C^{(i)}$ and $\cdot$ denotes the inner product. Next we put

$$p^{(i)}(\mathbf{r}) = \sum_{j=1}^{m} p_j^{(i)}(\mathbf{r})\, b^{-j} \in [0, 1] \qquad \text{for } 1 \leq i \leq s$$

and

$$P(\mathbf{r}) = (p^{(1)}(\mathbf{r}), \ldots, p^{(s)}(\mathbf{r})) \in I^s.$$

By letting $\mathbf{r}$ range over all b^m possibilities in R^m, we arrive at a point set P consisting of b^m points in I^s.

Definition 4. If the point set P constructed above forms a (t, m, s)-net in base b, then it is called a *digital* (t, m, s)*-net in base* b. If we want to emphasize that the construction uses the ring R, then we speak also of a *digital* (t, m, s)*-net over* R.

The quality parameter of a digital (t, m, s)-net over R depends only on the so-called *generating matrices* $C^{(1)}, \ldots, C^{(s)}$ over R. A convenient algebraic condition on the generating matrices to guarantee a certain value of t is known (see [Nie92, Theorem 4.26]), and a generalization of this condition will be given in Theorem 7 below.

For (t,s)-sequences the order of the terms is important, and so in the constructions care has to be taken that the points are obtained in a suitable order. We present the digital method for the construction of (t,s)-sequences in base b in the form given in [NX96b, Section 2] which is somewhat more general than the original version in [Nie87, Section 6]. Let integers $s \geq 1$ and $b \geq 2$ be given. Then we choose the following:

(i) a commutative ring R with identity and $\mathrm{card}(R) = b$;
(ii) bijections $\psi_r : Z_b \to R$ for $r = 0, 1, \ldots$, with $\psi_r(0) = 0$ for all sufficiently large r;
(iii) bijections $\eta_j^{(i)} : R \to Z_b$ for $1 \leq i \leq s$ and $j \geq 1$;
(iv) $\infty \times \infty$ matrices $C^{(1)}, \ldots, C^{(s)}$ over R.

For $n = 0, 1, \ldots$ let

$$n = \sum_{r=0}^{\infty} a_r(n)\, b^r \tag{1}$$

be the digit expansion of n in base b, where $a_r(n) \in Z_b$ for all $r \geq 0$ and $a_r(n) = 0$ for all sufficiently large r. We put

$$\mathbf{n} = (\psi_r(a_r(n)))_{r=0}^{\infty} \in R^{\infty}. \tag{2}$$

Next we define

$$y_{n,j}^{(i)} = \eta_j^{(i)}(\mathbf{c}_j^{(i)} \cdot \mathbf{n}) \in Z_b \qquad \text{for } n \geq 0,\ 1 \leq i \leq s, \text{ and } j \geq 1,$$

where $\mathbf{c}_j^{(i)}$ is the jth row of the matrix $C^{(i)}$. Note that the inner product $\mathbf{c}_j^{(i)} \cdot \mathbf{n}$ makes sense since $\mathbf{n}$ has only finitely many nonzero coordinates. Then we put

$$x_n^{(i)} = \sum_{j=1}^{\infty} y_{n,j}^{(i)}\, b^{-j} \qquad \text{for } n \geq 0 \text{ and } 1 \leq i \leq s.$$

Finally, we define the sequence S consisting of the points

$$\mathbf{x}_n = (x_n^{(1)}, \ldots, x_n^{(s)}) \in I^s \qquad \text{for } n = 0, 1, \ldots.$$

Definition 5. If the sequence S constructed above forms a (t,s)-sequence in base b, then it is called a *digital (t,s)-sequence in base b*. If we want to emphasize that the construction uses the ring R, then we speak also of a *digital (t,s)-sequence over R*.

As in the case of digital (t,m,s)-nets, the quality parameter of a digital (t,s)-sequence over R depends only on the *generating matrices* $C^{(1)}, \ldots, C^{(s)}$ over R. A convenient algebraic condition on the generating matrices to guarantee a certain value of t is known (see [NX96b, Lemma 7 and Remark 4]), and a generalization of this condition will be given in Theorem 8 below.

The standard low-discrepancy sequences used nowadays in quasi-Monte Carlo methods, such as the sequences of Sobol' [Sob67], Faure [Fau82], and

Niederreiter [Nie88] as well as the sequences obtained by Niederreiter and Xing using algebraic-geometry methods (see [NX01, Chapter 8] for an exposition of the latter constructions), are all digital (t, s)-sequences. There are interesting generalizations and variants of (digital) (t, s)-sequences which we will not discuss here; see for instance Dick [Dic06b], [Dic06a] and Larcher and Niederreiter [LN95].

3 Codes and Duality Theory

It is known since the first paper [Nie87] on the general theory of (t, m, s)-nets and (t, s)-sequences that there are interesting links between digital nets and error-correcting codes. Recently, these links have become more pronounced with the development of a duality theory for digital nets which puts digital nets squarely into a framework of a distinctly coding-theoretic nature.

We recall the rudiments of coding theory. We refer to MacWilliams and Sloane [MWS77] for a full treatment of coding theory and to Ling and Xing [LX04] for an introduction to the area. Let $\mathbb{F}_q$ be the finite field with q elements, where q is an arbitrary prime power. For an integer $n \geq 1$, we consider the n-dimensional vector space $\mathbb{F}_q^n$ over $\mathbb{F}_q$. The number of nonzero coordinates of $\mathbf{a} \in \mathbb{F}_q^n$ is the Hamming weight $w(\mathbf{a})$. Then $d(\mathbf{a}, \mathbf{b}) = w(\mathbf{a} - \mathbf{b})$ for $\mathbf{a}, \mathbf{b} \in \mathbb{F}_q^n$ defines the Hamming metric. The vector space $\mathbb{F}_q^n$, endowed with the Hamming metric, is the Hamming space $\mathbb{F}_q^n$. A linear code over $\mathbb{F}_q$ is a nonzero $\mathbb{F}_q$-linear subspace C of the Hamming space $\mathbb{F}_q^n$. The minimum distance $\delta(C)$ of C is defined by

$$\delta(C) = \min \{d(\mathbf{a}, \mathbf{b}) : \mathbf{a}, \mathbf{b} \in C, \ \mathbf{a} \neq \mathbf{b}\}.$$

It is easy to see that we also have

$$\delta(C) = \min_{\mathbf{a} \in C \setminus \{\mathbf{0}\}} w(\mathbf{a}).$$

One of the principal aims of coding theory is to construct linear codes C over $\mathbb{F}_q$ with a large minimum distance $\delta(C)$ for given n and $k = \dim(C)$, or with a large relative minimum distance $\frac{\delta(C)}{n}$ for a given information rate $\frac{k}{n}$.

We now describe the duality theory for digital nets developed by Niederreiter and Pirsic [NP01]. We mention in passing that a completely different application of coding theory to multidimensional numerical integration occurs in the recent paper of Kuperberg [Kup06].

We first have to generalize the definition of the Hamming space. Let $m \geq 1$ and $s \geq 1$ be integers; they will have the same meaning as m and s in a digital (t, m, s)-net over $\mathbb{F}_q$. The following weight function V_m on $\mathbb{F}_q^{ms}$ was introduced by Niederreiter [Nie86] and later used in an equivalent form in coding theory by Rosenbloom and Tsfasman [RT97]. We start by defining a weight function

v on $\mathbb{F}_q^m$. We put $v(\mathbf{a}) = 0$ if $\mathbf{a} = \mathbf{0} \in \mathbb{F}_q^m$, and for $\mathbf{a} = (a_1, \ldots, a_m) \in \mathbb{F}_q^m$ with $\mathbf{a} \neq \mathbf{0}$ we set

$$v(\mathbf{a}) = \max \{j : a_j \neq 0\}.$$

Then we extend this definition to $\mathbb{F}_q^{ms}$ by writing a vector $\mathbf{A} \in \mathbb{F}_q^{ms}$ as the concatenation of s vectors of length m, that is,

$$\mathbf{A} = (\mathbf{a}^{(1)}, \ldots, \mathbf{a}^{(s)}) \in \mathbb{F}_q^{ms} \qquad \text{with } \mathbf{a}^{(i)} \in \mathbb{F}_q^m \text{ for } 1 \leq i \leq s,$$

and putting

$$V_m(\mathbf{A}) = \sum_{i=1}^{s} v(\mathbf{a}^{(i)}).$$

Note that $d_m(\mathbf{A}, \mathbf{B}) = V_m(\mathbf{A} - \mathbf{B})$ for $\mathbf{A}, \mathbf{B} \in \mathbb{F}_q^{ms}$ defines a metric on $\mathbb{F}_q^{ms}$ which for $m = 1$ reduces to the Hamming metric on $\mathbb{F}_q^s$.

Definition 6. The *minimum distance* $\delta_m(\mathcal{N})$ of a nonzero $\mathbb{F}_q$-linear subspace $\mathcal{N}$ of $\mathbb{F}_q^{ms}$ is given by

$$\delta_m(\mathcal{N}) = \min_{\mathbf{A} \in \mathcal{N} \setminus \{\mathbf{0}\}} V_m(\mathbf{A}).$$

Now let the $m \times m$ matrices $C^{(1)}, \ldots, C^{(s)}$ over $\mathbb{F}_q$ be the generating matrices of a digital net P. Set up an $m \times ms$ matrix M as follows: for $1 \leq j \leq m$, the jth row of M is obtained by concatenating the jth columns of $C^{(1)}, \ldots, C^{(s)}$. Let $\mathcal{M} \subseteq \mathbb{F}_q^{ms}$ be the row space of M and let $\mathcal{M}^\perp$ be its dual space as in coding theory, that is,

$$\mathcal{M}^\perp = \{\mathbf{A} \in \mathbb{F}_q^{ms} : \mathbf{A} \cdot \mathbf{M} = 0 \text{ for all } \mathbf{M} \in \mathcal{M}\}.$$

Then we have the following results from [NP01].

Theorem 1. *Let $m \geq 1$ and $s \geq 2$ be integers. Then, with the notation above, the point set P is a digital (t, m, s)-net over $\mathbb{F}_q$ if and only if*

$$\delta_m(\mathcal{M}^\perp) \geq m - t + 1.$$

Corollary 1. *Let $m \geq 1$ and $s \geq 2$ be integers. Then from any $\mathbb{F}_q$-linear subspace $\mathcal{N}$ of $\mathbb{F}_q^{ms}$ with $\dim(\mathcal{N}) \geq ms - m$ we can construct a digital (t, m, s)-net over $\mathbb{F}_q$ with*

$$t = m + 1 - \delta_m(\mathcal{N}).$$

Note that $\mathcal{N}$ in Corollary 1 plays the role of $\mathcal{M}^\perp$ in Theorem 1. Since $\mathcal{M}$ as the row space of an $m \times ms$ matrix has dimension at most m, we must have

$$\dim(\mathcal{N}) = \dim(\mathcal{M}^\perp) = ms - \dim(\mathcal{M}) \geq ms - m,$$

which explains the condition on $\dim(\mathcal{N})$ in Corollary 1.

It is of interest to note that the line of research started by Rosenbloom and Tsfasman [RT97] in coding theory was continued in that area. Some of the theorems obtained in this direction can be translated into results on digital nets. Typical coding-theoretic papers on this topic are Dougherty and Skriganov [DS02] and Siap and Ozen [SO04].

4 Digital Nets Inspired by Codes

Corollary 1 is a powerful tool for the construction of digital nets. It was already used in the paper [NP01] that introduced duality theory, where it was applied to obtain an analog of the classical $(u, u+v)$ construction of codes. An improved version of the $(u, u+v)$ construction for digital nets was given by Bierbrauer, Edel, and Schmid [BES02]. A considerable generalization of this construction was obtained by Niederreiter and Özbudak [NO04] who designed an analog of the matrix-product construction of codes. This yields the following result.

Theorem 2. *Let h be an integer with $2 \leq h \leq q$. If for $k = 1, \ldots, h$ there exists a digital (t_k, m_k, s_k)-net over $\mathbb{F}_q$ and if $s_1 \leq \cdots \leq s_h$, then there exists a digital $(t, \sum_{k=1}^{h} m_k, \sum_{k=1}^{h} s_k)$-net over $\mathbb{F}_q$ with*

$$t = 1 + \sum_{k=1}^{h} m_k - \min_{1 \leq k \leq h} (h - k + 1)(m_k - t_k + 1).$$

The $(u, u+v)$ construction of digital nets is the special case $h = 2$ of Theorem 2. The matrix-product construction of codes and digital nets affords a way of combining given linear codes, respectively digital nets, to produce a new linear code, respectively digital net. Another principle of this type is obtained by the Kronecker-product construction which is well known in coding theory. Kronecker-product constructions of digital nets were proposed by Bierbrauer, Edel, and Schmid [BES02] and Niederreiter and Pirsic [NP02].

Further links between coding theory and digital nets can be established by considering special families of linear codes and searching for their analogs in the realm of digital nets. For instance, an important special type of linear code is a cyclic code, i.e., a linear code that is invariant under cyclic shifts. An analog for digital nets was introduced by Niederreiter [Nie04] who adopted the viewpoint that cyclic codes can be defined by prescribing roots of polynomials (compare with [LN94, Section 8.2]). For integers $m \geq 1$ and $s \geq 2$, consider the vector space

$$\mathcal{P} = \{f \in \mathbb{F}_{q^m}[x] : \deg(f) < s\}$$

of polynomials over the extension field $\mathbb{F}_{q^m}$ of $\mathbb{F}_q$. Note that $\dim(\mathcal{P}) = ms$ as a vector space over $\mathbb{F}_q$. We fix an element $\alpha \in \mathbb{F}_{q^m}$ and define

$$\mathcal{P}_\alpha = \{f \in \mathcal{P} : f(\alpha) = 0\}.$$

It is clear that $\mathcal{P}_\alpha$ is an $\mathbb{F}_q$-linear subspace of $\mathcal{P}$ with $\dim(\mathcal{P}_\alpha) = ms - m$ as a vector space over $\mathbb{F}_q$. For each $i = 1, \ldots, s$, we choose an ordered basis B_i of $\mathbb{F}_{q^m}$ over $\mathbb{F}_q$. Next we set up a map $\tau : \mathcal{P} \to \mathbb{F}_q^{ms}$ in the following way. Take $f \in \mathcal{P}$ and write this polynomial explicitly as

$$f(x) = \sum_{i=1}^{s} \gamma_i x^{i-1}$$

with $\gamma_i \in \mathbb{F}_{q^m}$ for $1 \leq i \leq s$. For each $i = 1, \ldots, s$, let $\mathbf{c}_i(f) \in \mathbb{F}_q^m$ be the coordinate vector of γ_i with respect to the ordered basis B_i. Then we define

$$\tau : f \in \mathcal{P} \mapsto (\mathbf{c}_1(f), \ldots, \mathbf{c}_s(f)) \in \mathbb{F}_q^{ms}.$$

It is obvious that τ is an $\mathbb{F}_q$-linear isomorphism from $\mathcal{P}$ onto $\mathbb{F}_q^{ms}$. Finally, let $\mathcal{N}_\alpha$ be the image of the subspace $\mathcal{P}_\alpha$ under τ. Since τ is an isomorphism, we have

$$\dim(\mathcal{N}_\alpha) = \dim(\mathcal{P}_\alpha) = ms - m$$

as a vector space over $\mathbb{F}_q$. Thus, we can apply Corollary 1 to the $\mathbb{F}_q$-linear subspace $\mathcal{N}_\alpha$ of $\mathbb{F}_q^{ms}$. The resulting digital net is called a *cyclic digital net* over $\mathbb{F}_q$ relative to the bases $B_1, \ldots, B_s$. A theorem guaranteeing the existence of good cyclic digital nets was recently shown by Pirsic, Dick, and Pillichshammer [PDP06].

A powerful family of linear codes is that of algebraic-geometry codes. A general framework for constructing digital nets by means of algebraic curves over finite fields, or equivalently by global function fields, was developed by Niederreiter and Özbudak [NO02]. The basic construction in [NO02] uses a global function field F with full constant field $\mathbb{F}_q$ (see Section 5 for the definition of these terms) and a divisor G of F. An $\mathbb{F}_q$-linear subspace $\mathcal{N}$ of $\mathbb{F}_q^{ms}$ is defined as the image of the Riemann-Roch space $\mathcal{L}(G)$ under an $\mathbb{F}_q$-linear map from $\mathcal{L}(G)$ to $\mathbb{F}_q^{ms}$ derived from the local expansions of elements of $\mathcal{L}(G)$ at distinct places $Q_1, \ldots, Q_s$ of F. Under suitable conditions, we can invoke Corollary 1 to arrive at a digital (t, m, s)-net over $\mathbb{F}_q$ for some t.

We end this section by describing a recent construction of digital nets due to Pirsic, Dick, and Pillichshammer [PDP06]. For an integer $m \geq 1$ consider, as earlier in this section, the extension field $\mathbb{F}_{q^m}$ of $\mathbb{F}_q$. Then, for an integer $s \geq 2$, we take the s-dimensional vector space $\mathcal{Q} := \mathbb{F}_{q^m}^s$ over $\mathbb{F}_{q^m}$ which has dimension ms as a vector space over $\mathbb{F}_q$. Now fix $\boldsymbol{\alpha} \in \mathcal{Q}$ with $\boldsymbol{\alpha} \neq \mathbf{0}$ and put

$$\mathcal{Q}_{\boldsymbol{\alpha}} = \{\boldsymbol{\gamma} \in \mathcal{Q} : \boldsymbol{\alpha} \cdot \boldsymbol{\gamma} = 0\}.$$

Clearly, $\mathcal{Q}_{\boldsymbol{\alpha}}$ is an $\mathbb{F}_{q^m}$-linear subspace of $\mathcal{Q}$ of dimension $s - 1$, and so $\mathcal{Q}_{\boldsymbol{\alpha}}$ has dimension $ms - m$ as a vector space over $\mathbb{F}_q$. Since $\mathcal{Q}$ and $\mathbb{F}_q^{ms}$ are isomorphic as vector spaces over $\mathbb{F}_q$, we get in this way an $\mathbb{F}_q$-linear subspace $\mathcal{N}_{\boldsymbol{\alpha}}$ of $\mathbb{F}_q^{ms}$ of dimension $ms - m$ as a vector space over $\mathbb{F}_q$. Thus, we can apply Corollary 1 to obtain a digital (t, m, s)-net over $\mathbb{F}_q$ for some t. A digital net produced by this construction is called a *hyperplane net*. An analysis of how hyperplane nets, cyclic digital nets, and other types of digital nets are related among each other was carried out by Pirsic [Pir05].

5 Constructing Digital (t, s)-Sequences from Differentials

There are altogether four known constructions of digital (t, s)-sequences based on general global function fields, all of them due to Niederreiter and Xing.

A systematic account of these constructions is given in Niederreiter and Xing [NX96a]. In this section, we describe the first new construction of digital (t, s)-sequences using global function fields since 1996. It is also the first construction using differentials in global function fields. This construction is due to Mayor and Niederreiter [MN].

Let F be a global function field with constant field $\mathbb{F}_q$, that is, F is a finite extension of the rational function field $\mathbb{F}_q(x)$. We assume that $\mathbb{F}_q$ is the full constant field of F, which means that $\mathbb{F}_q$ is algebraically closed in F. We refer to the book of Stichtenoth [Sti93] for general background and terminology on global function fields.

Let $\mathbf{P}_F$ be the set of places of F and Ω_F the set of differentials of F, that is,

$$\Omega_F = \{f\,dz : f \in F,\ z \text{ is a separating element for } F\}.$$

For any $\omega \in \Omega_F$ and separating element z, we can write $\omega = f\,dz$ with a unique $f \in F$. If $\omega \in \Omega_F^*$ is a nonzero differential, then for every $Q \in \mathbf{P}_F$ let $\omega = f_Q\,dt_Q$, where $t_Q \in F$ is a local parameter at Q (and hence a separating element). Then we can associate ω with the divisor

$$(\omega) := \sum_{Q \in \mathbf{P}_F} \nu_Q(f_Q)\,Q,$$

where ν_Q is the normalized valuation of F corresponding to the place Q. For any divisor G of F, we define

$$\Omega(G) = \{\omega \in \Omega_F^* : (\omega) \geq G\} \cup \{0\}.$$

Note that $\Omega(G)$ is a finite-dimensional vector space over $\mathbb{F}_q$.

Now let the dimension $s \geq 1$ in the construction of a digital (t, s)-sequence be given. We assume that F contains at least one rational place Q_∞; recall that a rational place is a place of degree 1. Choose a divisor D of F with $\deg(D) = -2$ and Q_∞ not in the support of D (such a divisor always exists). Furthermore, let $Q_1, \ldots, Q_s$ be s distinct places of F with $Q_i \neq Q_\infty$ for $1 \leq i \leq s$, and put $e_i = \deg(Q_i)$ for $1 \leq i \leq s$.

The Riemann-Roch theorem can be used to show that $\dim(\Omega(D)) = g + 1$, $\dim(\Omega(D + Q_\infty)) = g$, and $\dim(\Omega(D + (2g+1)Q_\infty)) = 0$, where g is the genus of F. Hence there exist integers $0 = n_0 < n_1 < \cdots < n_g \leq 2g$ such that

$$\dim(\Omega(D + n_u Q_\infty)) = \dim(\Omega(D + (n_u + 1)Q_\infty)) + 1 \qquad \text{for } 0 \leq u \leq g.$$

Now we choose

$$\omega_u \in \Omega(D + n_u Q_\infty) \setminus \Omega(D + (n_u + 1)Q_\infty) \qquad \text{for } 0 \leq u \leq g.$$

It is easily seen that $\{\omega_0, \omega_1, \ldots, \omega_g\}$ is a basis of $\Omega(D)$. For $i = 1, \ldots, s$, consider the chain

$$\Omega(D) \subset \Omega(D - Q_i) \subset \Omega(D - 2Q_i) \subset \ldots$$

of vector spaces over $\mathbb{F}_q$. By starting from the basis $\{\omega_0, \omega_1, \ldots, \omega_g\}$ of $\Omega(D)$ and successively adding basis vectors at each step of the chain, we obtain for each integer $n \geq 1$ a basis

$$\{\omega_0, \omega_1, \ldots, \omega_g, \omega_1^{(i)}, \omega_2^{(i)}, \ldots, \omega_{ne_i}^{(i)}\}$$

of $\Omega(D - nQ_i)$. Now let $z \in F$ be a local parameter at Q_∞. For $r = 0, 1, \ldots$ we put

$$z_r = \begin{cases} z^r\, dz & \text{if } r \notin \{n_0, n_1, \ldots, n_g\}, \\ \omega_u & \text{if } r = n_u \text{ for some } u \in \{0, 1, \ldots, g\}. \end{cases}$$

Note that $\nu_{Q_\infty}((z_r)) = r$ for all $r \geq 0$. For $1 \leq i \leq s$ and $j \geq 1$, we have $\omega_j^{(i)} \in \Omega(D - kQ_i)$ for some $k \geq 1$ and also Q_∞ not in the support of $D - kQ_i$, hence $\nu_{Q_\infty}((\omega_j^{(i)})) \geq 0$. Thus, we have local expansions at Q_∞ of the form

$$\omega_j^{(i)} = \sum_{r=0}^{\infty} a_{r,j}^{(i)} z_r \qquad \text{for } 1 \leq i \leq s \text{ and } j \geq 1,$$

where all coefficients $a_{r,j}^{(i)} \in \mathbb{F}_q$. For $1 \leq i \leq s$ and $j \geq 1$, we define the sequence of elements $c_{r,j}^{(i)} \in \mathbb{F}_q$, $r = 0, 1, \ldots$, by considering the sequence of elements $a_{r,j}^{(i)}$, $r = 0, 1, \ldots$, and then deleting the terms with $r = n_u$ for some $u \in \{0, 1, \ldots, g\}$. Then we put

$$\mathbf{c}_j^{(i)} = (c_{0,j}^{(i)}, c_{1,j}^{(i)}, \ldots) \in \mathbb{F}_q^\infty \qquad \text{for } 1 \leq i \leq s \text{ and } j \geq 1.$$

Finally, for each $i = 1, \ldots, s$, we let $C^{(i)}$ be the $\infty \times \infty$ matrix over $\mathbb{F}_q$ whose jth row is $\mathbf{c}_j^{(i)}$ for $j = 1, 2, \ldots$. We write $S_\Omega(Q_\infty, Q_1, \ldots, Q_s; D)$ for a sequence obtained from the generating matrices $C^{(1)}, \ldots, C^{(s)}$ by the digital method (compare with Section 2). The following result was shown by Mayor and Niederreiter [MN].

Theorem 3. *Let F be a global function field with full constant field $\mathbb{F}_q$ and with at least one rational place Q_∞. Let D be a divisor of F with $\deg(D) = -2$ and Q_∞ not in the support of D. Furthermore, let $Q_1, \ldots, Q_s$ be distinct places of F with $Q_i \neq Q_\infty$ for $1 \leq i \leq s$. Then $S_\Omega(Q_\infty, Q_1, \ldots, Q_s; D)$ is a digital (t, s)-sequence over $\mathbb{F}_q$ with*

$$t = g + \sum_{i=1}^{s} (e_i - 1),$$

where g is the genus of F and $e_i = \deg(Q_i)$ for $1 \leq i \leq s$.

We report now on further results from the paper [MN]. We use the standard notation $d_q(s)$ for the least value of t such that there exists a digital (t, s)-sequence over $\mathbb{F}_q$.

Example 1. Let $q = 5$ and $s = 32$. Let F be the global function field given by $F = \mathbb{F}_5(x, y_1, y_2)$ with

$$y_1^2 = x(x^2 - 2), \quad y_2^5 - y_2 = \frac{x^4 - 1}{y_1 - 1}.$$

Then F has 32 rational places $Q_\infty, Q_1, \ldots, Q_{31}$ and genus $g = 11$. Furthermore, F has at least one place Q_{32} of degree 2 lying over the place $x^2 + 2x - 2$ of $\mathbb{F}_5(x)$. We can choose $D = -2Q_1$. Now we consider the sequence $S_\Omega(Q_\infty, Q_1, \ldots, Q_{32}; D)$ and apply Theorem 3. We have $e_i = 1$ for $1 \leq i \leq 31$ and $e_{32} = 2$, therefore $t = 12$. Hence we obtain $d_5(32) \leq 12$, which is an improvement on the previously best bound $d_5(32) \leq 13$ given in [Nie05, Table 1]. This improved value has already been entered into the database at

`http://mint.sbg.ac.at`

for parameters of (t, m, s)-nets and (t, s)-sequences (see [SS06] for a description of this database).

Theorem 4. *For every odd prime p and every dimension $s \geq 1$, we have*

$$d_p(s) \leq \frac{p+3}{p-1}s + \frac{p-5}{p-1}.$$

Theorem 5. *For every odd prime p and every dimension $s \geq 1$, we have*

$$d_{p^2}(s) \leq \frac{2}{p-1}s + 1.$$

Theorem 6. *For every prime power q and every dimension $s \geq 1$, we have*

$$d_{q^3}(s) \leq \frac{q(q+2)}{2(q^2-1)}s.$$

Theorems 4, 5, and 6 are derived from Theorem 3 by using towers of global function fields that were constructed in the last few years (see [MN] for the details).

Very recently, Niederreiter and Özbudak [NO07] used differentials in global function fields and the duality theory for digital nets to give a new construction of $(\mathbf{T}, s)$-sequences in the sense of [LN95]. In various cases, this construction yields low-discrepancy sequences with better discrepancy bounds than previous constructions.

6 A General Construction of Nets

We present a method of constructing (t, m, s)-nets which generalizes the digital method in Section 2. The idea is to move away from linear algebra and to allow for nonlinearity in the construction. This is motivated by the well-known fact in coding theory that there are good parameters of nonlinear codes that cannot be achieved by linear codes (see [LX04, Section 5.6]). One would hope for a similar phenomenon for nets, namely that there are parameters of nets attainable by "nonlinear" constructions, but not by the digital method in Section 2.

As in Section 2, let integers $m \geq 1$, $s \geq 1$, and $b \geq 2$ be given. We recall that $Z_b = \{0, 1, \ldots, b-1\}$ denotes the set of digits in base b. Then we choose the following:

(i) a set R with $\mathrm{card}(R) = b$;
(ii) bijections $\eta_j^{(i)} : R \to Z_b$ for $1 \leq i \leq s$ and $1 \leq j \leq m$;
(iii) maps $\phi_j^{(i)} : R^m \to R$ for $1 \leq i \leq s$ and $1 \leq j \leq m$.

Now let $\mathbf{r} \in R^m$ and define

$$p^{(i)}(\mathbf{r}) = \sum_{j=1}^{m} \eta_j^{(i)}(\phi_j^{(i)}(\mathbf{r}))\, b^{-j} \in [0,1] \qquad \text{for } 1 \leq i \leq s$$

and

$$P(\mathbf{r}) = (p^{(1)}(\mathbf{r}), \ldots, p^{(s)}(\mathbf{r})) \in I^s.$$

By letting $\mathbf{r}$ range over all b^m possibilities in R^m, we arrive at a point set P consisting of b^m points in I^s.

Theorem 7. *The point set P constructed above forms a (t, m, s)-net in base b if and only if for any nonnegative integers $d_1, \ldots, d_s$ with $\sum_{i=1}^{s} d_i = m - t$ and any $f_j^{(i)} \in R$, $1 \leq j \leq d_i$, $1 \leq i \leq s$, the system of $m - t$ equations*

$$\phi_j^{(i)}(z_1, \ldots, z_m) = f_j^{(i)} \qquad \text{for } 1 \leq j \leq d_i,\ 1 \leq i \leq s, \tag{3}$$

in the unknowns $z_1, \ldots, z_m$ over R has exactly b^t solutions.

Proof. Assume that (3) satisfies the given condition. According to Definition 2, we have to show that every interval J of the form

$$J = \prod_{i=1}^{s} [a_i b^{-d_i}, (a_i + 1) b^{-d_i})$$

with integers $d_i \geq 0$ and $0 \leq a_i < b^{d_i}$ for $1 \leq i \leq s$ and with $\sum_{i=1}^{s} d_i = m - t$ contains exactly b^t points of the point set P. For $1 \leq i \leq s$, let

$$a_i = \sum_{j=1}^{d_i} a_{i,j}\, b^{d_i - j}$$

be the digit expansion in base b, where all $a_{i,j} \in Z_b$. For the points $P(\mathbf{r})$ of P, we have $P(\mathbf{r}) \in J$ if and only if

$$p^{(i)}(\mathbf{r}) \in [a_i b^{-d_i}, (a_i + 1) b^{-d_i}) \qquad \text{for } 1 \le i \le s.$$

This is equivalent to

$$\eta_j^{(i)}(\phi_j^{(i)}(\mathbf{r})) = a_{i,j} \qquad \text{for } 1 \le j \le d_i,\ 1 \le i \le s,$$

which is, in turn, equivalent to

$$\phi_j^{(i)}(\mathbf{r}) = (\eta_j^{(i)})^{-1}(a_{i,j}) \qquad \text{for } 1 \le j \le d_i,\ 1 \le i \le s,$$

where $(\eta_j^{(i)})^{-1}$ denotes the inverse map of $\eta_j^{(i)}$. By hypothesis, the last system of equations has exactly b^t solutions $\mathbf{r} \in R^m$, and so P forms a (t, m, s)-net in base b. This shows the sufficiency part of the theorem. The converse is proved by similar arguments. □

Remark 1. The digital method for the construction of nets described in Section 2 is the special case of the present construction where R is a commutative ring with identity and the maps $\phi_j^{(i)}$ are linear forms in m variables over R. It can be argued that the construction principle in the present section is also a digital method since the coordinates of the points of the net are obtained digit by digit. We propose to refer to the nets produced by the method in this section also as *digital* (t, m, s)*-nets in base* b or as *digital* (t, m, s)*-nets over* R. The nets in Section 2 could then be called *linear digital* (t, m, s)*-nets in base* b or *linear digital* (t, m, s)*-nets over* R, to emphasize that they are obtained by the use of linear forms $\phi_j^{(i)}$.

The construction principle described above is too general to be useful in practice, so it is meaningful to consider situations in which we can introduce some structure. If we choose for R a finite field $\mathbb{F}_q$, then each map $\phi_j^{(i)} : \mathbb{F}_q^m \to \mathbb{F}_q$ can be represented by a polynomial over $\mathbb{F}_q$ in m variables and of degree less than q in each variable (see [LN97, Section 7.5]). We assume that the maps $\phi_j^{(i)}$, $1 \le i \le s$, $1 \le j \le m$, are so represented. Then, by using the concept of an orthogonal system of polynomials in $\mathbb{F}_q$ (see [LN97, Definition 7.35]), we obtain the following consequence of Theorem 7.

Corollary 2. *Let the point set* P *be obtained by the construction in this section with* $R = \mathbb{F}_q$. *Then* P *is a* (t, m, s)*-net in base* q *if and only if for any nonnegative integers* $d_1, \ldots, d_s$ *with* $\sum_{i=1}^{s} d_i = m - t$ *the polynomials* $\phi_j^{(i)}$, $1 \le j \le d_i$, $1 \le i \le s$, *form an orthogonal system in* $\mathbb{F}_q$.

There are several useful criteria for orthogonal systems of polynomials in $\mathbb{F}_q$. One such criterion, due to Niederreiter [Nie71] and given in Proposition 1 below, is in terms of permutation polynomials over $\mathbb{F}_q$. We recall that a polynomial over $\mathbb{F}_q$ (in one or several variables) is called a permutation polynomial over $\mathbb{F}_q$ if it attains each value of $\mathbb{F}_q$ equally often (see [LN97, Chapter 7] for the theory of permutation polynomials).

Proposition 1. *Let $1 \leq h \leq m$ be integers and let $g_1, \dots, g_h \in \mathbb{F}_q[z_1, \dots, z_m]$. Then $g_1, \dots, g_h$ form an orthogonal system of polynomials in $\mathbb{F}_q$ if and only if for all $b_1, \dots, b_h \in \mathbb{F}_q$ not all 0, the polynomial $b_1 g_1 + \cdots + b_h g_h$ is a permutation polynomial over $\mathbb{F}_q$.*

It follows, in particular, that every polynomial occurring in an orthogonal system of polynomials in $\mathbb{F}_q$ is a permutation polynomial over $\mathbb{F}_q$. In view of Corollary 2, this shows that a necessary condition for the polynomials $\phi_j^{(i)}$, $1 \leq i \leq s$, $1 \leq j \leq m$, to yield a (t, m, s)-net in base q is that each polynomial $\phi_j^{(i)}$ with $1 \leq i \leq s$ and $1 \leq j \leq m - t$ is a permutation polynomial over $\mathbb{F}_q$.

Example 2. Let q be an arbitrary prime power and let $m \geq 1$ be an integer. We start from a permutation polynomial g over $\mathbb{F}_{q^m}$ in one variable, for instance, $g(z) = \gamma z^k$ with $\gamma \in \mathbb{F}_{q^m}^*$ and an integer $k \geq 1$ satisfying $\gcd(k, q^m - 1) = 1$ (see [LN97, Section 7.2]). Let $B = \{\beta_1, \dots, \beta_m\}$ be an ordered basis of $\mathbb{F}_{q^m}$ over $\mathbb{F}_q$, and for each $\alpha \in \mathbb{F}_{q^m}$ let $(c_1(\alpha), \dots, c_m(\alpha)) \in \mathbb{F}_q^m$ be the coordinate vector of α with respect to B. Then there exist polynomials $g_1, \dots, g_m \in \mathbb{F}_q[z_1, \dots, z_m]$ such that

$$g(\alpha) = \sum_{j=1}^{m} g_j(c_1(\alpha), \dots, c_m(\alpha))\beta_j \qquad \text{for all } \alpha \in \mathbb{F}_{q^m}.$$

Since g is a permutation polynomial over $\mathbb{F}_{q^m}$, it follows that $g_1, \dots, g_m$ form an orthogonal system of polynomials in $\mathbb{F}_q$. Now we put $R = \mathbb{F}_q$ and $s = 2$ in the construction in this section, and we define the polynomials

$$\begin{aligned} \phi_j^{(1)} &= g_j \qquad &&\text{for } 1 \leq j \leq m, \\ \phi_j^{(2)} &= g_{m-j+1} \qquad &&\text{for } 1 \leq j \leq m. \end{aligned}$$

Then it is clear that for any integers $d_1 \geq 0$ and $d_2 \geq 0$ with $d_1 + d_2 = m$, the polynomials $\phi_1^{(1)}, \dots, \phi_{d_1}^{(1)}, \phi_1^{(2)}, \dots, \phi_{d_2}^{(2)}$ form an orthogonal system in $\mathbb{F}_q$. Thus, by Corollary 2, we obtain a digital $(0, m, 2)$-net over $\mathbb{F}_q$ (in the sense of Remark 1). This net can be viewed as a scrambled version of the well-known two-dimensional Hammersley net in base q.

7 A General Construction of (t, s)-Sequences

In this section, we present an analog of the construction principle in Section 6 for (t, s)-sequences. Let integers $s \geq 1$ and $b \geq 2$ be given. Then we choose the following:

(i) a set R with $\mathrm{card}(R) = b$ and a distinguished element $o \in R$;
(ii) bijections $\psi_r : Z_b \to R$ for $r = 0, 1, \dots$, with $\psi_r(0) = o$ for all sufficiently large r;
(iii) bijections $\eta_j^{(i)} : R \to Z_b$ for $1 \leq i \leq s$ and $j \geq 1$;

(iv) maps $\phi_j^{(i)} : \mathcal{F} \rightarrow R$ for $1 \leq i \leq s$ and $j \geq 1$, where $\mathcal{F}$ is the set of all sequences of elements of R with only finitely many terms $\neq o$.
For $n = 0, 1, \ldots$, we define $\mathbf{n}$ by (1) and (2) and observe that $\mathbf{n} \in \mathcal{F}$. Next we define

$$y_{n,j}^{(i)} = \eta_j^{(i)}(\phi_j^{(i)}(\mathbf{n})) \in Z_b \qquad \text{for } n \geq 0,\ 1 \leq i \leq s, \text{ and } j \geq 1.$$

Then we put

$$x_n^{(i)} = \sum_{j=1}^{\infty} y_{n,j}^{(i)}\, b^{-j} \qquad \text{for } n \geq 0 \text{ and } 1 \leq i \leq s.$$

Finally, we define the sequence S consisting of the points

$$\mathbf{x}_n = (x_n^{(1)}, \ldots, x_n^{(s)}) \in I^s \qquad \text{for } n = 0, 1, \ldots.$$

Theorem 8. *The sequence S constructed above is a (t, s)-sequence in base b if and only if for any integer $m > t$, any nonnegative integers $d_1, \ldots, d_s$ with $\sum_{i=1}^{s} d_i = m - t$, and any $f_j^{(i)} \in R$, $1 \leq j \leq d_i$, $1 \leq i \leq s$, the system of $m - t$ equations*

$$\phi_j^{(i)}(z_0, z_1, \ldots) = f_j^{(i)} \qquad \text{for } 1 \leq j \leq d_i,\ 1 \leq i \leq s, \tag{4}$$

has the following property: if the values of the variables $z_m, z_{m+1}, \ldots$ are fixed in R in such a way that $z_r = o$ for all sufficiently large r, then the resulting system in the unknowns $z_0, z_1, \ldots, z_{m-1}$ over R has exactly b^t solutions.

Proof. In order to prove the sufficiency, we proceed by Definition 3. For given integers $k \geq 0$ and $m > t$, we consider the point set $P_{k,m}$ consisting of the points $[\mathbf{x}_n]_{b,m}$ with $kb^m \leq n < (k+1)b^m$. We have to show that $P_{k,m}$ is a (t, m, s)-net in base b. Let J be an interval of the form

$$J = \prod_{i=1}^{s} [a_i b^{-d_i}, (a_i + 1) b^{-d_i})$$

with integers $d_i \geq 0$ and $0 \leq a_i < b^{d_i}$ for $1 \leq i \leq s$ and with $\sum_{i=1}^{s} d_i = m - t$. Then we have to prove that J contains exactly b^t points of $P_{k,m}$. For $1 \leq i \leq s$, let

$$a_i = \sum_{j=1}^{d_i} a_{i,j}\, b^{d_i - j}$$

be the digit expansion in base b, where all $a_{i,j} \in Z_b$. For the points of $P_{k,m}$ we have $[\mathbf{x}_n]_{b,m} \in J$ if and only if

$$[x_n^{(i)}]_{b,m} \in [a_i b^{-d_i}, (a_i + 1) b^{-d_i}) \qquad \text{for } 1 \leq i \leq s.$$

This is equivalent to

$$y_{n,j}^{(i)} = a_{i,j} \qquad \text{for } 1 \le j \le d_i,\ 1 \le i \le s,$$

which is, in turn, equivalent to

$$\phi_j^{(i)}(\mathbf{n}) = (\eta_j^{(i)})^{-1}(a_{i,j}) \qquad \text{for } 1 \le j \le d_i,\ 1 \le i \le s. \tag{5}$$

Recall that the range for n is $kb^m \le n < (k+1)b^m$. In this range, the digits $a_r(n)$ of n in (1) are prescribed for $r \ge m$, whereas the $a_r(n)$ with $0 \le r \le m-1$ can vary freely over Z_b. This means that the coordinates $\psi_r(a_r(n))$ of $\mathbf{n}$ in (2) are fixed for $r \ge m$ and they can vary freely over R for $0 \le r \le m-1$. Thus, the system (5) of $m-t$ equations is of the form (4), and so by the given property, (5) has exactly b^t solutions. This means that J contains exactly b^t points of $P_{k,m}$. Hence the proof of sufficiency is complete. The converse is shown by similar arguments. □

Remark 2. The digital method for the construction of (t,s)-sequences described in Section 2 is the special case of the present construction where R is a commutative ring with identity, the distinguished element o is the zero element of R, and the maps $\phi_j^{(i)}$ are linear forms over R. In analogy with Remark 1, we propose to refer to the (t,s)-sequences produced by the method in this section also as *digital (t,s)-sequences in base b* or as *digital (t,s)-sequences over R*. The (t,s)-sequences in Section 2 could then be called *linear digital (t,s)-sequences in base b* or *linear digital (t,s)-sequences over R*.

The construction principle described above is again too general to be useful in practice, so one will have to focus on interesting special cases such as R being a finite field (see Section 6).

In Sections 6 and 7, we have not really gone much beyond the description of new construction principles for (t,m,s)-nets and (t,s)-sequences, respectively. The challenge for future research on this topic is to find choices for the maps $\phi_j^{(i)}$ in these constructions that are not all linear forms and that yield good (and maybe even record) values of the quality parameter t. A source for optimism in this quest is the analogy with coding theory (compare with the first paragraph of Section 6).

Acknowledgment

This research was partially supported by the grant R-394-000-025-422 with Temasek Laboratories in Singapore.

References

[BES02] J. Bierbrauer, Y. Edel, and W.Ch. Schmid. Coding-theoretic constructions for (t,m,s)-nets and ordered orthogonal arrays. *J. Combin. Designs* **10**, 403–418 (2002).

[Dic06a] J. Dick. Walsh spaces containing smooth functions and quasi-Monte Carlo rules of arbitrary high order. Preprint, 2006.

[Dic06b] J. Dick. Explicit constructions of quasi-Monte Carlo rules for the numerical integration of high dimensional periodic functions. Preprint, 2006.

[DS02] S.T. Dougherty and M.M. Skriganov. Maximum distance separable codes in the ρ metric over arbitrary alphabets. *J. Algebraic Combinatorics* **16**, 71–81 (2002).

[Fau82] H. Faure. Discrépance de suites associées à un système de numération (en dimension s). *Acta Arith.* **41**, 337–351 (1982).

[Kup06] G. Kuperberg. Numerical cubature using error-correcting codes. *SIAM J. Numer. Analysis* **44**, 897–907 (2006).

[LN95] G. Larcher and H. Niederreiter. Generalized (t, s)-sequences, Kronecker-type sequences, and diophantine approximations of formal Laurent series. *Trans. Amer. Math. Soc.* **347**, 2051–2073 (1995).

[LN94] R. Lidl and H. Niederreiter. *Introduction to Finite Fields and Their Applications*. Revised ed., Cambridge University Press, Cambridge, 1994.

[LN97] R. Lidl and H. Niederreiter. *Finite Fields*. Cambridge University Press, Cambridge, 1997.

[LX04] S. Ling and C.P. Xing. *Coding Theory: A First Course*. Cambridge University Press, Cambridge, 2004.

[MN] D.J.S. Mayor and H. Niederreiter. A new construction of (t, s)-sequences and some improved bounds on their quality parameter. *Acta Arith.*, to appear.

[MWS77] F.J. MacWilliams and N.J.A. Sloane. *The Theory of Error-Correcting Codes*. North-Holland, Amsterdam, 1977.

[Nie03] H. Niederreiter. Error bounds for quasi-Monte Carlo integration with uniform point sets. *J. Comput. Appl. Math.* **150**, 283–292 (2003).

[Nie04] H. Niederreiter. Digital nets and coding theory. *Coding, Cryptography and Combinatorics* (K.Q. Feng, H. Niederreiter, and C.P. Xing, eds.), pp. 247–257, Birkhäuser, Basel, 2004.

[Nie05] H. Niederreiter. Constructions of (t, m, s)-nets and (t, s)-sequences. *Finite Fields Appl.* **11**, 578–600 (2005).

[Nie71] H. Niederreiter. Orthogonal systems of polynomials in finite fields. *Proc. Amer. Math. Soc.* **28**, 415–422 (1971).

[Nie86] H. Niederreiter. Low-discrepancy point sets. *Monatsh. Math.* **102**, 155–167 (1986).

[Nie87] H. Niederreiter. Point sets and sequences with small discrepancy. *Monatsh. Math.* **104**, 273–337 (1987).

[Nie88] H. Niederreiter. Low-discrepancy and low-dispersion sequences. *J. Number Theory* **30**, 51–70 (1988).

[Nie92] H. Niederreiter. *Random Number Generation and Quasi-Monte Carlo Methods*. SIAM, Philadelphia, 1992.

[NO02] H. Niederreiter and F. Özbudak. Constructions of digital nets using global function fields. *Acta Arith.* **105**, 279–302 (2002).

[NO04] H. Niederreiter and F. Özbudak. Matrix-product constructions of digital nets. *Finite Fields Appl.* **10**, 464–479 (2004).

[NO07] H. Niederreiter and F. Özbudak. Low-discrepancy sequences using duality and global function fields. *Acta Arith.*, to appear.

[NP01] H. Niederreiter and G. Pirsic. Duality for digital nets and its applications. *Acta Arith.* **97**, 173–182 (2001).

[NP02] H. Niederreiter and G. Pirsic. A Kronecker product construction for digital nets. *Monte Carlo and Quasi-Monte Carlo Methods 2000* (K.-T. Fang, F.J. Hickernell, and H. Niederreiter, eds.), pp. 396–405, Springer, Berlin, 2002.

[NX01] H. Niederreiter and C.P. Xing. *Rational Points on Curves over Finite Fields: Theory and Applications.* Cambridge University Press, Cambridge, 2001.

[NX96a] H. Niederreiter and C.P. Xing. Quasirandom points and global function fields. *Finite Fields and Applications* (S. Cohen and H. Niederreiter, eds.), pp. 269–296, Cambridge University Press, Cambridge, 1996.

[NX96b] H. Niederreiter and C.P. Xing. Low-discrepancy sequences and global function fields with many rational places. *Finite Fields Appl.* **2**, 241–273 (1996).

[PDP06] G. Pirsic, J. Dick, and F. Pillichshammer. Cyclic digital nets, hyperplane nets, and multivariate integration in Sobolev spaces. *SIAM J. Numer. Analysis* **44**, 385–411 (2006).

[Pir05] G. Pirsic. A small taxonomy of integration node sets. *Sitzungsber. Österr. Akad. Wiss. Math.-Naturwiss. Kl. Abt. II* **214**, 133–140 (2005).

[RT97] M.Yu. Rosenbloom and M.A. Tsfasman. Codes for the m-metric. *Problems Inform. Transmission* **33**, 45–52 (1997).

[SO04] I. Siap and M. Ozen. The complete weight enumerator for codes over $\mathcal{M}_{n\times s}(R)$. *Applied Math. Letters* **17**, 65–69 (2004).

[Sob67] I.M. Sobol'. Distribution of points in a cube and approximate evaluation of integrals (Russian). *Ž. Vyčisl. Mat. i Mat. Fiz.* **7**, 784–802 (1967).

[SS06] R. Schürer and W.Ch. Schmid. MinT: A database for optimal net parameters. *Monte Carlo and Quasi-Monte Carlo Methods 2004* (H. Niederreiter and D. Talay, eds.), pp. 457–469, Springer, Berlin, 2006.

[Sti93] H. Stichtenoth. *Algebraic Function Fields and Codes.* Springer, Berlin, 1993.

Quadratic Optimal Functional Quantization of Stochastic Processes and Numerical Applications

Gilles Pagès

Laboratoire de Probabilités et Modèles aléatoires, UMR 7599, Université Paris 6, case 188, 4, pl. Jussieu, F-75252 Paris Cedex 5, France
gpa@ccr.jussieu.fr

Summary. In this paper, we present an overview of the recent developments of functional quantization of stochastic processes, with an emphasis on the quadratic case. Functional quantization is a way to approximate a process, viewed as a Hilbert -valued random variable, using a nearest neighbour projection on a finite codebook. A special emphasis is made on the computational aspects and the numerical applications, in particular the pricing of some path-dependent European options.

1 Introduction

Functional quantization is a way to discretize the path space of a stochastic process. It has been extensively investigated since the early 2000's by several authors (see among others [LP02], [LP06b], [DS06], [DFMS03], [LP04], etc). It first appeared as a natural extension of the Optimal Vector Quantization theory of (finite-dimensional) random vectors which finds its origin in the early 1950's for signal processing (see [GG92] or [GL00]).

Let us consider a Hilbertian setting. One considers a random vector X defined on a probability space $(\Omega, \mathcal{A}, \mathbf{P})$ taking its values in a separable Hilbert space $(H, (.|.)_H)$ (equipped with its natural Borel σ-algebra) and satisfying $\mathbf{E}|X|^2 < +\infty$. When H is an Euclidean space ($\mathbf{R}^d$), one speaks about *Vector Quantization.* When H is an infinite dimensional space like $L^2_T := L^2([0,T], dt)$ (endowed with the usual Hilbertian norm $|f|_{L^2_T} := (\int_0^T f^2(t)dt)^{\frac{1}{2}}$) one speaks of *functional quantization* (denoted L^2_T from now on). A (bi-measurable) stochastic process $(X_t)_{t\in[0,T]}$ defined on $(\Omega, \mathcal{A}, \mathbf{P})$ satisfying $|X(\omega)|_{L^2_T} < +\infty$ $\mathbf{P}(d\omega)$-*a.s.* can always be seen, once possibly modified on a $\mathbf{P}$-negligible set, as an L^2_T-valued random variable. Although we will focus on the Hilbertian framework, other choices are possible for H, in particular some more general Banach settings like $L^p([0,T], dt)$ or $\mathcal{C}([0,T], \mathbf{R})$ spaces.

This paper is organized as follows: in Sections 2 we introduce quadratic quantization in a Hilbertian setting. In Section 3, we focus on optimal quantization, including some extensions to non quadratic quantization. Section 4 is devoted to some quantized cubature formulae. Section 5 provides some classical background on the quantization rate in finite dimension. Section 7 deals with functional quantizations of Gaussian processes, like the Brownian motion, with a special emphasis on the numerical aspects. We present here what is, to our guess, the first large scale numerical optimization of the quadratic quantization of the Brownian motion. We compare it to the optimal product quantization, formerly investigated in [PP05b]. In section, we propose a constructive approach to the functional quantization of scalar or multidimensional diffusions (in the Stratanovich sense). In Section 9, we show how to use functional quantization to price path-dependent options like Asian options (in a heston stochastic volatility model). We conclude by some recent results showing how to derive universal (often optimal) functional quantization rate from time regularity of a process in Section 10 and by a few clues in Section 11 about the specific methods that produce some lower bounds (this important subject as many others like the connections with small deviation theory is not treated in this numerically oriented overview. As concerns statistical applications of functional quantization we refer to [TK03, TPO03].

NOTATIONS. • $a_n \approx b_n$ means $a_n = O(b_n)$ and $b_n = O(a_n)$; $a_n \sim b_n$ means $a_n = b_n + o(a_n)$.

• If $X : (\Omega, \mathcal{A}, \mathbf{P}) \to (H, |\,.\,|_H)$ (Hilbert space), then $\|X\|_2 = (\mathbf{E}|X|_H^2)^{\frac{1}{2}}$.

• $\lfloor x \rfloor$ denotes the integral part of the real x.

2 What is Quadratic Functional Quantization?

Let $(H, (\,.|.\,)_H)$ denote a separable Hilbert space. Let $X \in L^2_H(\mathbf{P})$ *i.e.* a random vector $X : (\Omega, \mathcal{A}, \mathbf{P}) \longmapsto H$ (H is endowed with its Borel σ-algebra) such that $\mathbf{E}\,|X|_H^2 < +\infty$. An *N-quantizer* (or *N-codebook*) is defined as a subset

$$\Gamma := \{x_1, \ldots, x_N\} \subset H$$

with card $\Gamma = N$. In numerical applications, Γ is also called *grid*. Then, one can *quantize* (or simply discretize) X by $q(X)$ where $q : H \mapsto \Gamma$ is a Borel function. It is straightforward that

$$\forall \omega \in \Omega, \qquad |X(\omega) - q(X(\omega))|_H \geq d(X(\omega), \Gamma) = \min_{1 \leq i \leq N} |X(\omega) - x_i|_H$$

so that the best pointwise approximation of X is provided by considering for q a nearest neighbour projection on Γ, denoted Proj_Γ. Such a projection is in one-to-one correspondence with the Voronoi partitions (or diagrams) of H induced by Γ *i.e.* the Borel partitions of H satisfying

$$C_i(\Gamma) \subset \left\{\xi \in H : |\xi - x_i|_H = \min_{1\leq j\leq N} |\xi - x_j|_H\right\} = \overline{C}_i(\Gamma), \qquad i = 1, \ldots, N,$$

where $\overline{C}_i(\Gamma)$ denotes the closure of $C_i(\Gamma)$ in H (this heavily uses the Hilbert structure). Then

$$\mathrm{Proj}_\Gamma(\xi) := \sum_{i=1}^{N} x_i \mathbf{1}_{C_i(\Gamma)}(\xi)$$

is a nearest neighbour projection on Γ. These projections only differ on the boundaries of the *Voronoi cells* $C_i(\Gamma)$, $i = 1, \ldots, N$. All Voronoi partitions have the same boundary contained in the union of the median hyperplanes defined by the pairs (x_i, x_j), $i \neq j$. Figure 1 represents the Voronoi diagram defined by a (random) 10-tuple in $\mathbf{R}^2$. Then, one defines a *Voronoi* N*-quantization* of X by setting for every $\omega \in \Omega$,

$$\widehat{X}^\Gamma(\omega) := \mathrm{Proj}_\Gamma(X(\omega)) = \sum_{i=1}^{N} x_i \mathbf{1}_{C_i(\Gamma)}(X(\omega)).$$

One clearly has, still for every $\omega \in \Omega$, that

$$|X(\omega) - \widehat{X}^\Gamma(\omega)|_H = \mathrm{dist}_H(X(\omega), \Gamma) = \min_{1\leq i\leq N} |X(\omega) - x_i|_H.$$

The mean (quadratic) quantization error is then defined by

$$e(\Gamma, X, H) = \|X - \widehat{X}^\Gamma\|_2 = \sqrt{\mathbf{E}\left(\min_{1\leq i\leq N} |X - x_i|_H^2\right)}. \tag{1}$$

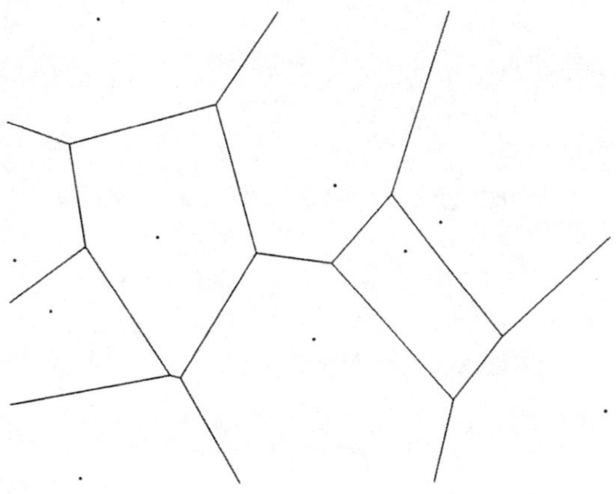

Fig. 1. *A 2-dimensional* 10*-quantizer* $\Gamma = \{x_1, \ldots, x_{10}\}$ *and its Voronoi diagram.*

The distribution of $\widehat{X}^{\Gamma}$ as a random vector is given by the N-tuple $(\mathbf{P}(X \in C_i(\Gamma)))_{1\le i\le N}$ of the Voronoi cells. This distribution clearly depends on the choice of the Voronoi partition as emphasized by the following elementary situation: if $H = \mathbf{R}$, the distribution of X is given by $\mathbf{P}_{X} = \frac{1}{3}(\delta_0 + \delta_{1/2} + \delta_1)$, $N = 2$ and $\Gamma = \{0, 1\}$ since $1/2 \in \partial C_0(\Gamma)\cap\partial C_1(\Gamma)$. However, if $\mathbf{P}_{X}$ weights no hyperplane, the distribution of $\widehat{X}^{\Gamma}$ depends only on Γ.

As concerns terminology, *Vector Quantization* is concerned with the finite dimensional case – when $\dim H < +\infty$ – and is a rather old story, going back to the early 1950's when it was designed in the field of signal processing and then mainly developed in the community of Information Theory. The term *functional quantization*, probably introduced in [Pag00, LP02], deals with the infinite dimensional case including the more general Banach-valued setting. The term "functional" comes from the fact that a typical infinite dimensional Hilbert space is the function space $H = L^2_T$. Then, any (bi-measurable) process $X : ([0, T] \times \Omega, Bor([0, T]) \otimes \mathcal{A}) \to (\mathbf{R}, Bor(\mathbf{R}))$ can be seen as a random vector taking values in the set of Borel functions on $[0, T]$. Furthermore, $((t, \omega) \mapsto X_t(\omega)) \in L^2(dt \otimes d\mathbf{P})$ if and only if $(\omega \mapsto X_.(\omega)) \in L^2_H(\mathbf{P})$ since

$$\int_{[0,T]\times\Omega} X_t^2(\omega)\, dt\, \mathbf{P}(d\omega) = \int_\Omega \mathbf{P}(d\omega) \int_0^T X_t^2(\omega)\, dt = \mathbf{E}\, |X_.|^2_{L^2_T}.$$

3 Optimal (Quadratic) Quantization

At this stage we are lead to wonder whether it is possible to design some optimally fitted grids to a given distribution $\mathbf{P}_{X}$ *i.e.* which induce the lowest possible mean quantization error among all grids of size at most N. This amounts to the following optimization problem

$$e_N(X, H) := \inf_{\Gamma\subset H, \mathrm{card}(\Gamma)\le N} e(\Gamma, X, H). \tag{2}$$

It is convenient at this stage to make a correspondence between quantizers of size at most N and N-tuples of H^N: to any N-tuple $x := (x_1, \ldots, x_N)$ corresponds a quantizer $\Gamma := \Gamma(x) = \{x_i,\ i = 1, \ldots, N\}$ (of size at most N). One introduces the quadratic distortion, denoted D^X_N, defined on H^N as a (symmetric) function by

$$D^X_N : H^N \longrightarrow \mathbf{R}_+$$
$$(x_1, \ldots, x_N) \longmapsto \mathbf{E}\left(\min_{1\le i\le N} |X - x_i|^2_H\right).$$

Note that, combining (1) and the definition of the distortion, shows that

$$D^X_N(x_1, \ldots, x_N) = \mathbf{E}\left(\min_{1\le i\le N} |X - x_i|^2_H\right) = \mathbf{E}\left(d(X, \Gamma(x))^2\right) = \|X - \widehat{X}^{\Gamma(x)}\|_2^2$$

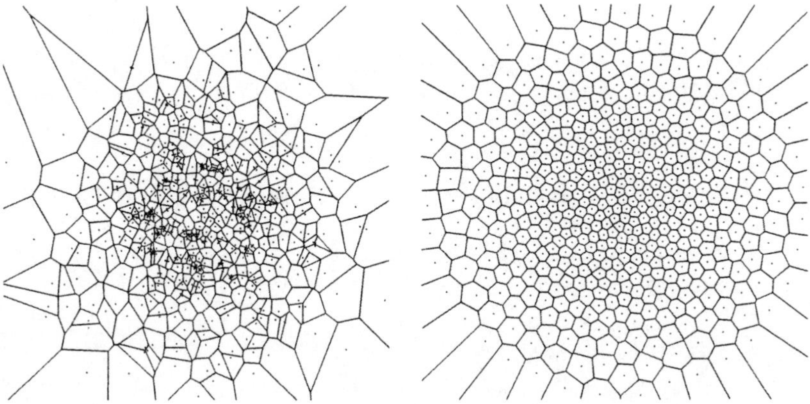

Fig. 2. *Two N-quantizers (and their Voronoi diagram) related to bi-variate normal distribution $\mathcal{N}(0; I_2)$ $(N = 500)$; which one is the best?*

so that,

$$e_N(X, H) = \inf_{(x_1,\ldots,x_N)\in H^N} \sqrt{D_N^X(x_1, \ldots, x_N)}.$$

The following proposition shows the existence of an optimal N-tuple $x^{(N,*)} \in H^N$ such that $e_N(X, H) = \sqrt{D_N^X(x^{(N,*)})}$. The corresponding optimal quantizer at level N is denoted $\Gamma^{(N,*)} := \Gamma(x^{(N,*)})$. In finite dimension we refer to [Pol82] (1982) and in infinite dimension to [CAM88] (1988) and [Par90] (1990); one may also see [Pag93], [GL00] and [LP02]. For recent developments on existence and pathwise regularity of optimal quantizer see [GLP07].

Proposition 1. *(a) The function D_N^X is lower semi-continuous for the product weak topology on H^N.*

(b) The function D_N^X reaches a minimum at a N-tuple $x^{(N,*)}$ (so that $\Gamma^{(N,*)}$ is an optimal quantizer at level *N).*

– *If* $\text{card}(\text{supp}(\mathbf{P}_X)) \geq N$, *the quantizer has full size N (i.e.* $\text{card}(\Gamma^{(N,*)}) = N$*) and $e_N(X, H) < e_{N-1}(X, H)$.*

– *If* $\text{card}(\text{supp}(\mathbf{P}_X)) \leq N$, $e_N(X, H) = 0$.

Furthermore $\lim_N e_N(X, H) = 0$.

(c) Any optimal (Voronoi) quantization at level N, $\widehat{X}^{\Gamma^{(N,*)}}$ satisfies

$$\widehat{X}^{\Gamma^{(N,*)}} = \mathbf{E}(X \mid \sigma(\widehat{X}^{\Gamma^{(N,*)}})) \tag{3}$$

where $\sigma(\widehat{X}^{\Gamma^{(N,*)}})$ denotes the σ-algebra generated by $\widehat{X}^{\Gamma^{(N,*)}}$.

(d) Any optimal (quadratic) quantization at level N is a best least square (i.e. $L^2(\mathbf{P})$) approximation of X among all H-valued random variables taking at most N values:

$$e_N(X,H)$$
$$= \|X - \widehat{X}^{\Gamma^{(N,*)}}\|_2 = \min\{\|X-Y\|_2,\ Y:(\Omega,\mathcal{A}) \to H,\ \mathrm{card}(Y(\Omega)) \le N\}.$$

Proof (sketch of): (a) The claim follows from the l.s.c. of $\xi \mapsto |\xi|_H$ for the weak topology and Fatou's Lemma.

(b) One proceeds by induction on N. If $N=1$, the optimal 1-quantizer is $x^{(N,*)} = \{\mathbf{E}\,X\}$ and $e_2(X,H) = \|X - \mathbf{E}\,X\|_2$.

Assume now that an optimal quantizer $x^{(N,*)} = (x_1^{(N,*)}, \ldots, x_N^{(N,*)})$ does exist at level N.

– If $\mathrm{card}(\mathrm{supp}(\mathbf{P})) \le N$, then the $N+1$-tuple $(x^{(N,*)}, x_N^{(N,*)})$ (among other possibilities) is also optimal at level $N+1$ and $e_{N+1}(X,H) = e_N(X,H) = 0$.

– Otherwise, $\mathrm{card}(\mathrm{supp}(\mathbf{P})) \ge N+1$, hence $x^{(N,*)}$ has pairwise distinct components and there exists $\xi_{N+1} \in \mathrm{supp}(\mathbf{P}_X) \setminus \{x_i^{(N,*)},\ i=1,\ldots,N\} \neq \emptyset$.

Then, with obvious notations,

$$D^X_{N+1}((x^{(N,*)}, \xi_{N+1})) < D^X_N(x^{(N,*)}).$$

Then, the set $F_{N+1} := \left\{x \in H^{N+1} \mid D^X_{N+1}(x) \le D^X_{N+1}((x^{(N,*)}, \xi_{N+1}))\right\}$ is non empty, weakly closed since D^X_{N+1} is l.s.c.. Furthermore, it is bounded in H^{N+1}. Otherwise there would exist a sequence $x_{(m)} \in H^{N+1}$ such that $|x_{(m),i_m}|_H = \max_i |x_{(m),i}|_H \to +\infty$ as $m \to \infty$. Then, by Fatou's Lemma, one checks that

$$\liminf_{m\to\infty} D^X_{N+1}(x_{(m)}) \ge D^X_N(x^{(N,*)}) > D^X_{N+1}((x^{(N,*)}, \xi_{N+1})).$$

Consequently F_{N+1} is weakly compact and the minimum of D^X_{N+1} on F_{N+1} is clearly its minimum over the whole space H^{N+1}. In particular

$$e_{N+1}(X,H) \le D^X_{N+1}((x^{(N,*)}, \xi_{N+1})) < e_N(X,H).$$

If $\mathrm{card}(\mathrm{supp}(\mathbf{P})) = N+1$, set $x^{(N+1,*)} = \mathrm{supp}(\mathbf{P})$ (as sets) so that t $X = \widehat{X}^{\Gamma^{(N+1,*)}}$ which implies $e_{N+1}(X,H) = 0$.

To establish that $e_N(X,H)$ goes to 0, one considers an everywhere dense sequence $(z_k)_{k\ge1}$ in the separable space H. Then, $d(\{z_1,\ldots,z_N\}, X(\omega))$ goes to 0 as $N \to \infty$ for every $\omega \in \Omega$. Furthermore, $d(\{z_1,\ldots,z_N\}, X(\omega))^2 \le |X(\omega) - z_1|_H^2 \in L^1(\mathbf{P})$. One concludes by the Lebesgue dominated convergence Theorem that $D^X_N(z_1,\ldots,z_N)$ goes to 0 as $N \to \infty$.

(c) and (d) Temporarily set $\widehat{X}^* := \widehat{X}^{\Gamma^{(N,*)}}$ for convenience. Let $Y:(\Omega,\mathcal{A}) \to H$ be a random vector taking at most N values. Set $\Gamma := Y(\Omega)$. Since $\widehat{X}^{\Gamma}$ is a Voronoi quantization of X induced by Γ,

$$|X - \widehat{X}^{\Gamma}|_{_H} = d(X, \Gamma) \le |X - Y|_{_H}$$

so that

$$\|X - \widehat{X}^{\Gamma}\|_{_2} \le \|X - Y\|_{_2}.$$

On the other hand, the optimality of $\Gamma^{(N,*)}$ implies

$$\|X - \widehat{X}^{*}\|_{_2} \le \|X - \widehat{X}^{\Gamma}\|_{_2}.$$

Consequently

$$\|X - \widehat{X}^{*}\|_{_2} \le \min\left\{\|X - Y\|_{_2},\ Y : (\Omega, \mathcal{A}) \to H,\ \mathrm{card}(Y(\Omega)) \le N\right\}.$$

The inequality holds as an equality since $\widehat{X}^{*}$ takes at most N values. Furthermore, considering random vectors of the form $Y = g(\widehat{X})$ (which take at most as many values as the size of $\Gamma^{(N,*)}$) shows, going back to the very definition of conditional expectation, that $\widehat{X}^{*} = \mathbf{E}(X \,|\, \widehat{X}^{*})$ $\mathbf{P}$-*a.s.* ◇

Item (c) introduces a very important notion in (quadratic) quantization.

Definition 1. *A quantizer $\Gamma \subset H$ is* stationary *(or* self-consistent*) if (there is a nearest neighbour projection such that $\widehat{X}^{\Gamma} = \mathrm{Proj}_{_\Gamma}(X)$ satisfying)*

$$\widehat{X}^{\Gamma} = \mathbf{E}\left(X \,|\, \widehat{X}^{\Gamma}\right). \tag{4}$$

Note in particular that any stationary quantization satisfies $\mathbf{E}X = \mathbf{E}\widehat{X}^{\Gamma}$.

As shown by Proposition 1(c) any quadratic optimal quantizer at level N is stationary. Usually, at least when $d \ge 2$, there are other stationary quantizers: indeed, the distortion function $D^X_{_N}$ is $|\,.\,|_{_H}$-differentiable at N-quantizers $x \in H^N$ with pairwise distinct components and

$$\begin{aligned}\nabla D^X_{_N}(x) &= 2\left(\int_{C_i(x)} (x_i - \xi)\mathbf{P}_{_X}(d\xi)\right)_{1 \le i \le N}\\ &= 2\left(\mathbf{E}(\widehat{X}^{\Gamma(x)} - X)\mathbf{1}_{\{\widehat{X}^{\Gamma(x)} = x_i\}}\right)_{1 \le i \le N}.\end{aligned}$$

hence, any critical points of $D^X_{_N}$ is a stationary quantizer.

Remarks and comments. • In fact (see Theorem 4.2, p. 38, [GL00]), the Voronoi partitions of $\Gamma^{(N,*)}$ always have a $\mathbf{P}_X$-negligible boundary so that (4) holds for *any* Voronoi diagram induced by Γ.

• The problem of the uniqueness of optimal quantizer (viewed as a set) is not mentioned in the above proposition. In higher dimension, this essentially never occurs. In one dimension, uniqueness of the optimal N-quantizer was first established in [Fle64] with strictly log-concave density function. This was successively extended in [Kie83] and [Tru82] and lead to the following criterion (for more general "loss" functions than the square function):

If the distribution of X is *absolutely continuous with a* log-*concave density function*, then, for every $N \geq 1$, there exists only one stationary quantizer of size N, which turns out to be the optimal quantizer at level N.

More recently, a more geometric approach to uniqueness based on the Mountain Pass Lemma first developed in [LP96] and then generalized in [Coh98]) provided a slight extension of the above criterion (in terms of loss functions).

This log-concavity assumption is satisfied by many families of probability distributions like the uniform distribution on compact intervals, the normal distributions, the gamma distributions. There are examples of distributions with a non log-concave density function having a unique optimal quantizer for every $N \geq 1$ (see *e.g.* the Pareto distribution in [FP04]). On the other hand simple examples of scalar distributions having multiple optimal quantizers at a given level can be found in [GL00].

• A stationary quantizer can be sub-optimal. This will be emphasized in Section 7 for the Brownian motion (but it is also true for finite dimensional Gaussian random vectors) where some families of sub-optimal quantizers – the *product quantizers* designed from the Karhunen-Loève basis – are stationary quantizers.

• For the uniform distribution over an interval $[a, b]$, there is a closed form for the optimal quantizer at level N given by $\Gamma^{(N,*)} = \{a + (2k-1)\frac{b-a}{N}, k = 1, \ldots, N\}$. This N-quantizer is optimal not only in the quadratic case but also for any L^r-quantization (see a definition further on). In general there is no such closed form, either in 1 or higher dimension. However, in [FP04] some semi-closed forms are obtained for several families of (scalar) distributions including the exponential and the Pareto distributions: all the optimal quantizers can be expressed using a single underlying sequence $(a_k)_{k\geq 1}$ defined by an induction $a_{k+1} = F(a_k)$.

• In one dimension, as soon as the optimal quantizer at level N is unique (as a set or as an N-tuple with increasing components), it is generally possible to compute it as the solution of the stationarity equation (3) either by a zero search (Newton-Raphson gradient descent) or a fixed point (like the specific Lloyd I procedure, see [Kie82]) procedure.

• In higher dimension, deterministic optimization methods become intractable and one uses stochastic procedures to compute optimal quantizers. The main topic of this paper being functional quantization, we postponed the short overview on these aspects to Section 7, devoted to the optimal quantization of the Brownian motion. But it is to be noticed that all efficient optimization methods rely on the so-called *splitting method* which increases progressively the quantization level N. This method is directly inspired by the induction developed in the proof of claim (b) of Proposition 1 since one designs the starting value of the optimization procedure at size $N + 1$ by "merging" the optimized N-quantizer obtained at level N with one further point of $\mathbf{R}^d$, usually randomly sampled with respect to an appropriate distribution (see [PP03] for a discussion).

• As concerns functional quantization, *e.g.* $H = L^2_T$, there is a close connection between the regularity of optimal (or even stationary) quantizers and that of $t \mapsto X_t$ form $[0,T]$ into $L^2(\mathbf{P})$. Furthermore, as concerns optimal quantizers of Gaussian processes, one shows (see [LP02]) that they belong to the reproducing space of their covariance operator, *e.g.* to the Cameron-Martin space $H^1 = \{\int_0^{\cdot} \dot{h}_s ds,\ \dot{h} \in L^2_T\}$ when $X = W$. Other properties of optimal quantization of Gaussian processes are established in [LP02].

Extensions to the $L^r(\mathbf{P})$-quantization of random variables. In this paper, we focus on the purely quadratic framework (L^2_T and $L^2(\mathbf{P})$-norms), essentially because it is a natural (and somewhat easier) framework for the computation of optimized grids for the Brownian motion and for some first applications (like the pricing of path-dependent options, see section 9). But a more general and natural framework is to consider the functional quantization of random vectors taking values in a separable Banach space $(E, |\,.\,|_E)$. Let $X : (\Omega, \mathcal{A}, \mathbf{P}) \to (E, |\ |_E)$, such that $\mathbf{E}\,|X|^r_E < +\infty$ for some $r \geq 1$ (the case $0 < r < 1$ can also be taken in consideration).

The N-level $(L^r(\mathbf{P}), |\,.\,|_E)$-quantization problem for $X \in L^r_E(\mathbf{P})$ reads

$$e_{N,r}(X,E) := \inf\left\{\|X - \widehat{X}^{\Gamma}\|_r,\ \Gamma \subset E,\ \mathrm{card}(\Gamma) \leq N\right\}.$$

The main examples for $(E, |\,.\,|_E)$ are the non-Euclidean norms on $\mathbf{R}^d$, the functional spaces $L^p_T(\mu) := L^p([0,T], \mu(dt))$, $1 \leq p \leq \infty$, equipped with its usual norm, $(E, |\,.\,|_E) = (\mathcal{C}([0,T]), \|\,.\,\|_{\sup})$, etc. As concerns, the existence of an optimal quantizer, it holds true for reflexive Banach spaces (see Pärna (90)) and $E = L^1_T$, but otherwise it may fail even when $N = 1$ (see [GLP07]). In finite dimension, the Euclidean feature is not crucial (see [GL00]). In the functional setting, many results originally obtained in a Hilbert setting have been extended to the Banach setting either for existence or regularity results (see [GLP07]) or for rates see [Der05b], [DS06], [LP04], [LP06a].

4 Cubature Formulae: Conditional Expectation and Numerical Integration

Let $F : H \longrightarrow \mathbf{R}$ be a continuous functional (with respect to the norm $|\,.\,|_H$) and let $\Gamma \subset H$ be an N-quantizer. It is natural to approximate $\mathbf{E}(F(X))$ by $\mathbf{E}(F(\widehat{X}^{\Gamma}))$. This quantity $\mathbf{E}(F(\widehat{X}^{\Gamma}))$ is simply the finite weighted sum

$$\mathbf{E}\,(F(\widehat{X}^{\Gamma})) = \sum_{i=1}^{N} F(x_i)\mathbf{P}(\widehat{X}^{\Gamma} = x_i).$$

Numerical computation of $\mathbf{E}\,(F(\widehat{X}^\Gamma))$ is possible as soon as $F(\xi)$ can be computed at any $\xi \in H$ and the distribution $(\mathbf{P}(\widehat{X} = x_i))_{1\le i\le N}$ of $\widehat{X}^\Gamma$ is known. The induced quantization error $\|X - \widehat{X}^\Gamma\|_2$ is used to control the error (see below). These quantities related to the quantizer Γ are also called *companion parameters*.

Likewise, one can consider *a priori* the $\sigma(\widehat{X}^\Gamma)$-measurable random variable $F(\widehat{X}^\Gamma)$ as a good approximation of the conditional expectation $\mathbf{E}(F(X)\,|\,\widehat{X}^\Gamma)$.

4.1 Lipschitz Functionals

Assume that the functional F is Lipschitz continuous on H. Then

$$\left|\mathbf{E}(F(X)\,|\,\widehat{X}^\Gamma) - F(\widehat{X}^\Gamma)\right| \le [F]_{\rm Lip}\mathbf{E}(|X - \widehat{X}^\Gamma|\,|\,\widehat{X}^\Gamma)$$

so that, for every real exponent $r \ge 1$,

$$\|\mathbf{E}(F(X)\,|\,\widehat{X}^\Gamma) - F(\widehat{X}^\Gamma)\|_r \le [F]_{\rm Lip}\|X - \widehat{X}^\Gamma\|_r$$

(where we applied conditional Jensen inequality to the convex function $u \mapsto u^r$). In particular, using that $\mathbf{E}\,F(X) = \mathbf{E}(\mathbf{E}(F(X)\,|\,\widehat{X}^\Gamma))$, one derives (with $r = 1$) that

$$\begin{aligned}\left|\mathbf{E}\,F(X) - \mathbf{E}\,F(\widehat{X}^\Gamma)\right| &\le \|\mathbf{E}(F(X)\,|\,\widehat{X}^\Gamma) - F(\widehat{X}^\Gamma)\|_1\\ &\le [F]_{\rm Lip}\|X - \widehat{X}^\Gamma\|_1.\end{aligned}$$

Finally, using the monotony of the $L^r(\mathbf{P})$-norms as a function of r yields

$$\left|\mathbf{E}\,F(X) - \mathbf{E}\,F(\widehat{X}^\Gamma)\right| \le [F]_{\rm Lip}\|X - \widehat{X}^\Gamma\|_1 \le [F]_{\rm Lip}\|X - \widehat{X}^\Gamma\|_2. \tag{5}$$

In fact, considering the Lipschitz functional $F(\xi) := d(\xi, \Gamma)$, shows that

$$\|X - \widehat{X}^\Gamma\|_1 = \sup_{[F]_{\rm Lip}\le 1}\left|\mathbf{E}\,F(X) - \mathbf{E}\,F(\widehat{X}^\Gamma)\right|. \tag{6}$$

The Lipschitz functionals making up a characterizing family for the weak convergence of probability measures on H, one derives that, for any sequence of N-quantizers Γ^N satisfying $\|X - \widehat{X}^{\Gamma^N}\|_1 \to 0$ as $N \to \infty$,

$$\sum_{1\le i\le N}\mathbf{P}(\widehat{X}^{\Gamma^N} = x_i^N)\,\delta_{x_i^N} \stackrel{(H)}{\Longrightarrow} \mathbf{P}_X$$

where $\stackrel{(H)}{\Longrightarrow}$ denotes the weak convergence of probability measures on $(H, |\,.\,|_H)$.

4.2 Differentiable Functionals with Lipschitz Differentials

Assume now that F is differentiable on H, with a Lipschitz continuous differential DF, and that the quantizer Γ is *stationary* (see Equation (4)).

A Taylor expansion yields

$$\left|F(X) - F(\widehat{X}^{\Gamma}) - DF(\widehat{X}^{\Gamma}).(X - \widehat{X}^{\Gamma})\right| \le [DF]_{\mathrm{Lip}}|X - \widehat{X}^{\Gamma}|^2.$$

Taking conditional expectation given $\widehat{X}^{\Gamma}$ yields

$$\left|\mathbf{E}(F(X)\,|\,\widehat{X}^{\Gamma}) - F(\widehat{X}^{\Gamma}) - \mathbf{E}\left(DF(\widehat{X}^{\Gamma}).(X - \widehat{X}^{\Gamma})\,|\,\widehat{X}^{\Gamma}\right)\right|$$
$$\le [DF]_{\mathrm{Lip}}\mathbf{E}(|X - \widehat{X}^{\Gamma}|^2\,|\,\widehat{X}^{\Gamma}).$$

Now, using that the random variable $DF(\widehat{X}^{\Gamma})$ is $\sigma(\widehat{X}^{\Gamma})$-measurable, one has

$$\mathbf{E}\left(DF(\widehat{X}^{\Gamma}).(X - \widehat{X}^{\Gamma})\right) = \mathbf{E}\left(DF(\widehat{X}^{\Gamma}).\mathbf{E}(X - \widehat{X}^{\Gamma}\,|\,\widehat{X}^{\Gamma})\right) = 0$$

so that

$$\left|\mathbf{E}(F(X)\,|\,\widehat{X}^{\Gamma}) - F(\widehat{X}^{\Gamma})\right| \le [DF]_{\mathrm{Lip}}\mathbf{E}\left(|X - \widehat{X}^{\Gamma}|^2\,|\,\widehat{X}^{\Gamma}\right).$$

Then, for every real exponent $r \ge 1$,

$$\left\|\mathbf{E}(F(X)\,|\,\widehat{X}^{\Gamma}) - F(\widehat{X}^{\Gamma})\right\|_r \le [DF]_{\mathrm{Lip}}\|X - \widehat{X}^{\Gamma}\|_{2r}^2.$$

In particular, when $r = 1$, one derives like in the former setting

$$\left|\mathbf{E}F(X) - \mathbf{E}F(\widehat{X}^{\Gamma})\right| \le [DF]_{\mathrm{Lip}}\|X - \widehat{X}^{\Gamma}\|_2^2. \tag{7}$$

In fact, the above inequality holds provided F is $\mathcal{C}^1$ with Lipschitz differential on every Voronoi cell $C_i(\Gamma)$. A similar characterization to (6) based on these functionals could be established.

Some variant of these cubature formulae can be found in [PP03] or [GLP06] for functions or functionals F having only some local Lipschitz regularity.

4.3 Quantized Approximation of $\mathbf{E}(F(X)\,|\,Y)$

Let X and Y be two H-valued random vector defined on the same probability space $(\Omega, \mathcal{A}, \mathbf{P})$ and $F : H \to \mathbf{R}$ be a Borel functional. The natural idea is to approximate $\mathbf{E}(F(X)\,|\,Y)$ by the quantized conditional expectation $\mathbf{E}(F(\widehat{X})\,|\,\widehat{Y})$ where $\widehat{X}$ and $\widehat{Y}$ are quantizations of X and Y respectively.

Let $\varphi_F : H \to \mathbf{R}$ be a (Borel) version of the conditional expectation *i.e.* satisfying

$$\mathbf{E}(F(X)\,|\,Y) = \varphi_F(Y).$$

Usually, no closed form is available for the function φ_F but some regularity property can be established, especially in a (Feller) Markovian framework. Thus assume that both F and φ_F are Lipschitz continuous with Lipschitz coefficients $[F]_{\mathrm{Lip}}$ and $[\varphi_F]_{\mathrm{Lip}}$. Then

$$\mathbf{E}(F(X)|Y)-\mathbf{E}(F(\widehat{X})|\widehat{Y})=\mathbf{E}(F(X)|Y)-\mathbf{E}(F(X)|\widehat{Y})+\mathbf{E}(F(X)-F(\widehat{X})|\widehat{Y}).$$

Hence, using that $\widehat{Y}$ is $\sigma(Y)$-measurable and that conditional expectation is an L^2-contraction,

$$\begin{aligned}\|\mathbf{E}(F(X)\,|\,Y)-\mathbf{E}(F(X)\,|\,\widehat{Y})\|_2 &= \|\mathbf{E}(F(X)|Y)-\mathbf{E}(\mathbf{E}(F(\widehat{X})|Y)|\widehat{Y})\|_2\\ &\le \|\varphi_F(Y)-\mathbf{E}(F(X)|\widehat{Y})\|_2\\ &= \|\varphi_F(Y)-\mathbf{E}(\varphi_F(Y)|\widehat{Y})\|_2\\ &\le \|\varphi_F(Y)-\varphi_F(\widehat{Y})\|_2.\end{aligned}$$

The last inequality follows form the definition of conditional expectation given $\widehat{Y}$ as the best quadratic approximation among $\sigma(\widehat{Y})$-measurable random variables. On the other hand, still using that $\mathbf{E}(\,.\,|\sigma(\widehat{Y}))$ is an L^2-contraction and this time that F is Lipschitz continuous yields

$$\|\mathbf{E}(F(X)-F(\widehat{X})\,|\,\widehat{Y})\|_2 \le \|F(X)-F(\widehat{X})\|_2 \le [F]_{\mathrm{Lip}}\|X-\widehat{X}\|_2.$$

Finally,

$$\|\mathbf{E}(F(X)\,|\,Y)-\mathbf{E}(F(\widehat{X})\,|\,\widehat{Y})\|_2 \le [F]_{\mathrm{Lip}}\|X-\widehat{X}\|_2+[\varphi_F]_{\mathrm{Lip}}\|Y-\widehat{Y}\|_2.$$

In the non-quadratic case the above inequality remains valid provided $[\varphi_F]_{\mathrm{Lip}}$ is replaced by $2[\varphi_F]_{\mathrm{Lip}}$.

5 Vector Quantization Rate ($H=\mathbf{R}^d$)

The fact that $e_N(X,\mathbf{R}^d)$ is a non-increasing sequence that goes to 0 as N goes to ∞ is a rather simple result established in Proposition 1. Its rate of convergence to 0 is a much more challenging problem. An answer is provided by the so-called Zador Theorem stated below.

This theorem was first stated and established for distributions with compact supports by Zador (see [Zad63, Zad82]). Then a first extension to general probability distributions on $\mathbf{R}^d$ is developed in [BW82]. The first mathematically rigorous proof can be found in [GL00], and relies on a random quantization argument (Pierce Lemma).

Theorem 1. (*a*) SHARP RATE. *Let $r>0$ and $X\in L^{r+\eta}(\mathbf{P})$ for some $\eta>0$. Let $\mathbf{P}_X(d\xi)=\varphi(\xi)\,d\xi\overset{\perp}{+}\nu(d\xi)$ be the canonical decomposition of the distribution of X (ν and the Lebesgue measure are singular). Then (if $\varphi\not\equiv 0$),*

$$e_{N,r}(X,\mathbf{R}^d) \sim \widetilde{J}_{r,d} \times \left(\int_{\mathbf{R}^d} \varphi^{\frac{d}{d+r}}(u)\,du\right)^{\frac{1}{d}+\frac{1}{r}} \times N^{-\frac{1}{d}} \quad as \quad N \to +\infty. \tag{8}$$

where $\widetilde{J}_{r,d} \in (0,\infty)$.

(*b*) NON ASYMPTOTIC UPPER BOUND *(see e.g. [LP06a]). Let* $d \geq 1$*. There exists* $C_{d,r,\eta} \in (0,\infty)$ *such that, for every* $\mathbf{R}^d$*-valued random vector* X,

$$\forall N \geq 1, \qquad e_{N,r}(X,\mathbf{R}^d) \leq C_{d,r,\eta}\|X\|_{r+\eta} N^{-\frac{1}{d}}.$$

Remarks. • The real constant $\widetilde{J}_{r,d}$ clearly corresponds to the case of the uniform distribution over the unit hypercube $[0,1]^d$ for which the slightly more precise statement holds

$$\lim_N N^{\frac{1}{d}} e_{N,r}(X,\mathbf{R}^d) = \inf_N N^{\frac{1}{d}} e_{N,r}(X,\mathbf{R}^d) = \widetilde{J}_{r,d}.$$

The proof is based on a self-similarity argument. The value of $\widetilde{J}_{r,d}$ depends on the reference norm on $\mathbf{R}^d$. When $d=1$, elementary computations show that $\widetilde{J}_{r,1} = (r+1)^{-\frac{1}{r}}/2$. When $d=2$, with the canonical Euclidean norm, one shows (see [New82] for a proof, see also [GL00]) that $\widetilde{J}_{2,d} = \sqrt{\frac{5}{18\sqrt{3}}}$. Its exact value is unknown for $d \geq 3$ but, still for the canonical Euclidean norm, one has (see [GL00]) using some random quantization arguments,

$$\widetilde{J}_{2,d} \sim \sqrt{\frac{d}{2\pi e}} \approx \sqrt{\frac{d}{17,08}} \quad \text{as} \quad d \to +\infty.$$

• When $\varphi \equiv 0$ the distribution of X is purely singular. The rate (8) still holds in the sense that $\lim_N N^{\frac{1}{d}} e_{r,N}(X,\mathbf{R}^d) = 0$. Consequently, this is not the right asymptotics. The quantization problem for singular measures (like uniform distribution on fractal compact sets) has been extensively investigated by several authors, leading to the definition of a quantization dimension in connection with the rate of convergence of the quantization error on these sets. For more details we refer to [GL00, GL05] and the references therein.

• A more naive way to quantize the uniform distribution on the unit hypercube is to proceed by *product quantization i.e.* by quantizing the marginals of the uniform distribution. If $N = m^d$, $m \geq 1$, one easily proves that the best quadratic product quantizer (for the canonical Euclidean norm on $\mathbf{R}^d$) is the "midpoint square grid"

$$\Gamma^{sq,N} = \left(\frac{2i_1-1}{2m},\ldots,\frac{2i_d-1}{2m}\right)_{1\leq i_1,\ldots,i_d\leq m}$$

which induces a quadratic quantization error equal to

$$\sqrt{\frac{d}{12}} \times N^{-\frac{1}{d}}.$$

Consequently, product quantizers are still *rate optimal* in every dimension d. Moreover, note that the ratio of these two rates remains bounded as $d \uparrow \infty$.

6 Optimal Quantization and QMC

The principle of Quasi-Monte Carlo method (QMC) is to approximate the integral of a function $f : [0,1]^d \to \mathbf{R}$ with respect to the uniform distribution on $[0,1]^d$, *i.e.* $\int_{[0,1]^d} f\,d\lambda_d = \int_{[0,1]^d} f(\xi^1,\ldots,\xi^d)d\xi^1\cdots d\xi^d$ (λ_d denotes the Lebesgue measure on $[0,1]^d$), by the uniformly weighted sum

$$\frac{1}{N}\sum_{k=1}^{N} f(x_k)$$

of values of f at the points of a so-called low discrepancy N-tuple $(x_1,\ldots,x_N)$ (or set). This N-tuple can the first N terms of an infinite sequence.

If f has finite variations denoted $V(f)$ – either in the measure sense (see [BL94, PX88]) or in the Hardy and Krause sense (see [Nie92] p.19) – the Koksma-Hlawka inequality provides an upper bound for the integration error induced by this method, namely

$$\left|\frac{1}{N}\sum_{k=1}^{N} f(x_k) - \int_{[0,1]^d} f\,d\lambda_d\right| \le V(f)Disc^*_N(x_1,\ldots,x_N)$$

where

$$Disc^*_N(x_1,\ldots,x_N) := \sup_{y\in[0,1]^d}\left|\frac{1}{N}\sum_{k=1}^{N}\mathbf{1}_{\{x_k\in[\![0,y]\!]\}} - \lambda_d([\![0,y]\!])\right|$$

(with $[\![0,y]\!]=\prod_{k=1}^d[0,y^i]$, $y=(y^1,\ldots,y^d)\in[0,1]^d$).

The error modulus $Disc^*_N(x_1,\ldots,x_N)$ denotes *the discrepancy at the origin* of the N-tuple $(x_1,\ldots,x_N)$. For every $N\ge 1$, there exists $[0,1]^d$-valued N-tuples $x^{(N)}$ such that

$$Disc^*_N(x^{(N)}) \le C_d\frac{(\log N)^{d-1}}{N}, \tag{9}$$

where $C_d\in(0,\infty)$ is a real constant only depending on d. This result can be proved using the so-called Hammersely procedure (see *e.g.* [Nie92], p. 31). When $x^{(N)} = (x_1,\ldots,x_N)$ is made of the first N terms of a $[0,1]^d$-valued sequence $(x_k)_{k\ge1}$, then the above upper bound has be replaced by $C'_d\frac{(\log N)^d}{N}$ ($C'_d\in(0,\infty)$). Such a sequence $x=(x_k)_{k\ge1}$ is said to be a *sequence with low discrepancy* (see [Nie92] an the references therein for a comprehensive theoretical overview, but also [BL94, PX88] for examples supported by numerical tests). When one only has $Disc^*_N(x_1,\ldots,x_N)\to 0$ as $N\to\infty$, the sequence is said to be *uniformly distributed in* $[0,1]^d$.

It is widely shared by QMC specialists that these rates are (in some sense) optimal although this remains a conjecture except when $d=1$. To be precise what is known and what is conjectured is the following:

– *Any* $[0,1]^d$-valued N-tuple $x^{(N)}$ satisfies $D_N^*(x^{(N)}) \geq B_d N^{-1}(\log N)^{\beta(d)}$ where $\beta(d) = \frac{d-1}{2}$ if $d \geq 2$ (see [Rot54] and also [Nie92] and the references therein), $\beta(1) = 0$ and $B_d > 0$ is a real constant only depending on d; the conjecture is that $\beta(d) = d - 1$.

– *Any* $[0,1]^d$-valued sequence $(x_k)_{k\geq 1}$ satisfies $D_N^*(x^{(N)}) \geq B_d N^{-1}(\log N)^{\beta'(d)}$ for infinitely many N, where $\beta'(d) = \frac{d}{2}$ if $d \geq 2$ and $\beta'(1) = 1$ and $B'_d > 0$ is a real constant only depending on d; the conjecture is that $\beta(d) = d$. This follows from the result for N-tuple by the Hammersley procedure (see *e.g.* [BL94]).

Furthermore, as concerns the use of Koksma-Hlawka inequality as an error bound for QMC numerical integration, the different notions of finite variation (which are closely connected) all become more and more restrictive – and subsequently less and less "natural" as a regularity property of functions – when the dimension d increases. Thus the Lipschitz continuous function f defined by $f(\xi^1, \xi^2, \xi^3) := (\xi^1 + \xi^2 + \xi^3) \wedge 1$ has infinite variation on $[0,1]^3$.

When applying Quasi-Monte Carlo approximation of integrals with "standard" continuous functions on $[0,1]^d$, the best known error bound, due to Proinov, is given by the following theorem.

Theorem 2. *(Proinov [Pro88]) (a) Assume* $\mathbf{R}^d$ *is equipped with the* ℓ^∞*-norm* $|(u^1, \ldots, u^d)|_\infty := \max_{1\leq i\leq d} |u^i|$. *Let* $(x_1, \ldots, x_N) \in ([0,1]^d)^N$. *For every continuous function* $f : [0,1]^d \to \mathbf{R}$.

$$\left| \int_{[0,1]^d} f(u)du - \frac{1}{N}\sum_{k=1}^{N} f(x_k) \right| \leq K_d\, \omega_f((Disc^*_N(x_1, \ldots, x_N))^{\frac{1}{d}})$$

where $\omega_f(\delta) := \sup_{x,y\in[0,1]^d, |x-y|_\infty \leq \delta} |f(x) - f(y)|$, $\delta \in (0,1)$, *is the uniform continuity modulus of* f *(with respect to the* ℓ_∞*-norm) and* $C_d \in (0,\infty)$ *is a universal constant only depending on d.*

(b) If $d = 1$, $K_d = 1$ *and if* $d \geq 2$, $K_d \in [1,4]$.

Remark. Note that if f is Lipschitz continuous, then $\omega_f(\delta) = [f]_{\mathrm{Lip}}\delta$ where $[f]_{\mathrm{Lip}}$ denotes the Lipschitz coefficient of f (with respect to the ℓ_∞-norm).

First, this result emphasizes that low discrepancy sequences or sets do suffer from the *curse of dimensionality* when a QMC approximation is implemented on functions having a "natural" regularity like Lipschitz continuity.

One also derives from this theorem an inequality between $(L^1(\mathbf{P}), \ell_\infty)$-quantization error of the uniform distribution $U([0,1]^d)$ and the discrepancy at the origin of a N-tuple $(x_1, \ldots, x_N)$, namely

$$\| \, |U - \widehat{U}^{\{x_1,\ldots,x_N\}}|_{\ell^\infty} \|_1 \leq K_d (Disc^*_N(x_1, \ldots, x_N))^{\frac{1}{d}}$$

since the function $\xi \mapsto \min_{1\le k\le N} |x_k - \xi|_\infty$ is clearly ℓ_∞-Lipschitz continuous with Lipschitz coefficient 1. The inequality also follows from the characterization established in (6) (which is clearly still true for non Euclidean norms). Then one may derive some bounds for Euclidean norms (and in fact any norms) on $\mathbf{R}^d$ (probably not sharp in terms of constant) since all the norms are strongly equivalent. However the bounds for optimal quantization error derived from Zador's Theorem ($O(N^{-\frac{1}{d}})$) and those for low discrepancy sets (see (9)) suggest that overall, optimal quantization provides lower error bounds for numerical integration of Lipschitz functions than low discrepancy sets, at least for for generic values of N. (However, standard computations show that for midpoint square grids (with $N = m^d$ points) both quantization errors and discrepancy behave like $\frac{1}{m} = N^{-\frac{1}{d}}$).

7 Optimal Quadratic Functional Quantization of Gaussian Processes

Optimal quadratic functional quantization of Gaussian processes is closely related to their so-called Karhunen-Loève expansion which can be seen in some sense as some infinite dimensional Principal Component Analysis (PCA) of a (Gaussian) process. Before stating a general result for Gaussian processes, we start by the standard Brownian motion: it is the most important example in view of (numerical) applications and for this process, everything can be made explicit.

7.1 Brownian Motion

One considers the Hilbert space $H = L^2_T := L^2([0,T], dt)$, $(f|g)_2 = \int_0^T f(t)g(t)dt$, $|f|_{L^2_T} = \sqrt{(f|f)_2}$. The covariance operator C_W of the Brownian motion $W = (W_t)_{t\in[0,T]}$ is defined on L^2_T by

$$C_W(f) := \mathbf{E}\left((f,W)_2 W\right) = \left(t \mapsto \int_0^T (s\wedge t) f(s)ds\right).$$

It is a symmetric positive trace class operator which can be diagonalized in the so-called Karhunen-Loève (K-L) orthonormal basis $(e_n^W)_{n\ge1}$ of L^2_T, with eigenvalues $(\lambda_n)_{n\ge1}$, given by

$$e_n^W(t) = \sqrt{\frac{2}{T}}\sin\left(\pi(n-\frac{1}{2})\frac{t}{T}\right), \quad \lambda_n = \left(\frac{T}{\pi(n-\frac{1}{2})}\right)^2, \quad n \ge 1.$$

This classical result can be established as a simple exercise by solving the functional equation $C_W(f) = \lambda f$. In particular, one can expand W itself on this basis so that

$$W \stackrel{L^2_T}{=} \sum_{n\geq 1} (W|e_n^W)_2\, e_n^W.$$

Now, the orthonormality of the $(K\text{-}L)$ basis implies, using Fubini's Theroem,

$$\mathbf{E}((W|e_k^W)_2 (W|e_\ell^W)_2) = (e_k^W|C_W(e_\ell^W))_2 = \lambda_\ell \delta_{k\ell}$$

where $\delta_{k\ell}$ denotes the Kronecker symbol. Hence the Gaussian sequence $((W|e_n^W)_2)_{n\geq 1}$ is pairwise non-correlated which implies that these random variables are independent. The above identity also implies that $\mathrm{Var}((W|e_n^W)_2) = \lambda_n$. Finally this shows that

$$W \stackrel{L^2_T}{=} \sum_{n\geq 1} \sqrt{\lambda_n}\, \xi_n\, e_n^W \tag{10}$$

where $\xi_n := (W|e_n^W)_2/\sqrt{\lambda_n}$, $n \geq 1$, is an i.i.d. sequence of $\mathcal{N}(0;1)$-distributed random variables. Furthermore, this K-L expansion converges in a much stronger sense since $\sup_{t\in[0,T]} |W_t - \sum_{k=1}^n \sqrt{\lambda_k}\xi_k e_k^W(t)| \to 0$ $\mathbf{P}$-$a.s.$ and

$$\|\sup_{[0,T]} |W_t - \sum_{1\leq k\leq n} \sqrt{\lambda_k}\xi_k e_k^W(t)|\|_2 = O\left(\sqrt{\log n/n}\right)$$

(see $e.g.$ [LP05]). Similar results (with various rates) hold true for a wide class of Gaussian processes expanded on "admissible" basis (see $e.g.$ [LP07]).

Theorem 3. *([LP02] (2002) and [LP04] (2003)) Let Γ^N, $N \geq 1$, be a sequence of optimal N-quantizers for W.*
(a) For every $N \geq 1$, $\mathrm{span}(\Gamma^N) = \mathrm{span}\{e_1^W, \ldots, e_{d(N)}^W\}$ with $d(N) = \Omega(\log N)$. Furthermore $\widehat{W}^{\Gamma^N}$ and $W - \widehat{W}^{\Gamma^N}$ are independent.

(b) $e_N(W, L^2_T) = \|W - \widehat{W}^{\Gamma^N}\|_2 \sim \dfrac{T\sqrt{2}}{\pi}\dfrac{1}{\sqrt{\log N}}$ *as $N \to \infty$.*

Remark. • The fact, confirmed by numerical experiments (see Section 7.3 Figure 6), that $d(N) \sim \log N$ holds as a conjecture.
• Denoting Π_d the orthogonal projection on $\mathrm{span}\{e_1^W, \ldots, e_d^W\}$, one derives from (a) that $\widehat{W}^{\Gamma^N} = \widehat{\Pi_{d(N)}(W)}^{\Gamma_N}$ (optimal quantization at level N) and

$$\begin{aligned}\|W - \widehat{W}^{\Gamma^N}\|_2^2 &= \|\Pi_{d(N)}(W) - \widehat{\Pi_{d(N)}(W)}^{\Gamma_N}\|_2^2 + \|W - \Pi_{d(N)}(W)\|_2^2\\ &= e_N\left(Z_{d(N)}, \mathbf{R}^{d(N)}\right)^2 + \sum_{n\geq d(N)+1} \lambda_n\end{aligned}$$

where $Z_{d(N)} \stackrel{d}{=} \Pi_{d(N)}(W) \sim \bigotimes_{k=1}^{d(N)} \mathcal{N}(0;\lambda_k)$.

7.2 Centered Gaussian Processes

The above Theorem 3 devoted to the standard Brownian motion is a particular case of a more general theorem which holds for a wide class of Gaussian processes

Theorem 4. *([LP02] (2002) and [LP04] (2004)) Let* $X = (X_t)_{t\in[0,T]}$ *be a Gaussian process with K-L eigensystem* $(\lambda_n^X, e_n^X)_{n\geq 1}$ *(with* $\lambda_1 \geq \lambda_2 \geq \dots$ *is non-increasing). Let* Γ^N, $N \geq 1$, *be a sequence of quadratic optimal N-quantizers for X. Assume*

$$\lambda_n^X \sim \frac{\kappa}{n^b} \qquad \textit{as } n \to \infty \qquad (b > 1).$$

(a) $\mathrm{span}(\Gamma^N) = \mathrm{span}\{e_1^X, \dots, e_{d^X(N)}^X\}$ *and* $d^X(N) = \Omega(\log N)$.
(b) $e_N(X, L_T^2) = \|X - \widehat{X}^{\Gamma^N}\|_2 \sim \sqrt{\kappa}\sqrt{b^b(b-1)^{-1}}\,(2\log N)^{-\frac{b-1}{2}}$.

Remarks. • The above result admits an extension to the case $\lambda_n^X \sim \varphi(n)$ as $n \to \infty$ with φ regularly varying, index $-b \leq -1$ (see [LP04]). In [LP02], upper or lower bounds are also established when

$$(\lambda_n^X \leq \varphi(n), \quad n \geq 1) \quad \text{or} \quad (\lambda_n^X \geq \varphi(n), \quad n \geq 1).$$

• The sharp asymptotics $d^X(N) \sim \frac{2}{b}\log N$ holds as a conjecture.

Applications to classical (centered) Gaussian processes.
• Brownian bridge: $X_t := W_t - \frac{t}{T}W_T$, $t \in [0,T]$ and $e_n^X(t) = \sqrt{2/T}\sin\left(\pi n\frac{t}{T}\right)$, $\lambda_n = \left(\frac{T}{\pi n}\right)^2$, so that $e_N(X, L_T^2) \sim T\frac{\sqrt{2}}{\pi}(\log N)^{-\frac{1}{2}}$.

• Fractional Brownian motion with Hurst constant $H \in (0,1)$

$$e_N(W^H, L_T^2) \sim T^{H+\frac{1}{2}}c(H)(\log N)^{-H}$$

where $c(H) = \left(\frac{\Gamma(2H)\sin(\pi H)(1+2H)}{\pi}\right)^{\frac{1}{2}}\left(\frac{1+2H}{2\pi}\right)^H$ and $\Gamma(t)$ denotes the Gamma function at $t > 0$.

• Some further explicit sharp rates can be derived from the above theorem for other classes of Gaussian stochastic processes (see [LP04], 2004) like the fractional Ornstein-Uhlenbeck processes, the Gaussian diffusions, a wide class Gaussian stationary processes (the quantization rate is derived from the high frequency asymptotics of its spectral density, assumed to be square integrable on the real line), for the m-folded integrated Brownian motion, the fractional Brownian sheet, etc.

• Of course some upper bounds can be derived for some even wider classes of processes, based on the above first remark (see *e.g.* [LP02], 2002).

Extensions to $r, p \neq 2$ When the processes have some self-similarity properties, it is possible to obtain some sharp rates in the non purely quadratic case: this has been done for fractional Brownian motion in [DS06] using some quite different techniques in which self-similarity properties plays there a crucial role. It leads to the following sharp rates, for $p \in [1, +\infty]$ and $r \in (0, \infty)$

$$e_{N,r}(W^H, L^p_T) \sim T^{H+\frac{1}{2}} c(r, H)(\log N)^{-H}, \quad c(r, H) \in (0, +\infty).$$

7.3 Numerical Optimization of Quadratic Functional Quantization

Thanks to the scaling property of Brownian motion, one may focus on the normalized case $T = 1$. The numerical approach to optimal quantization of the Brownian motion is essentially based on Theorem 3 and the remark that follows: indeed these results show that quadratic optimal functional quantization of a centered Gaussian process reduces to a finite dimensional optimal quantization problem for a Gaussian distribution with a diagonal covariance structure. Namely the optimization problem at level N reads

$$(\mathcal{O}_N) \equiv \begin{cases} e_N(W, L^2_T)^2 := e_N(Z_{d(N)}, \mathbf{R}^{d(N)})^2 + \displaystyle\sum_{k \geq d(N)+1} \lambda_k \\ \text{where} \quad Z_{d(N)} \stackrel{d}{=} \displaystyle\bigotimes_{k=1}^{d(N)} \mathcal{N}(0, \lambda_k). \end{cases}$$

Moreover, if $\beta^N := \{\beta^N_1, \ldots, \beta^N_N\}$ denotes an optimal N-quantizer of $Z_{d(N)}$, then, the optimal N-quantizer Γ^N of W reads $\Gamma^N = \{x^N_1, \ldots, x^N_N\}$ with

$$x^N_i(t) = \sum_{1 \leq \ell \leq d(N)} (\beta^N_i)^\ell e^W_\ell(t), \quad i = 1, \ldots, N. \tag{11}$$

The good news is that $(\mathcal{O}_N)$ is in fact a *finite dimensional quantization optimization problem* for each $N \geq 1$. The bad news is that the problem is somewhat ill conditioned since the decrease of the eigenvalues of W is very steep for small values of n: $\lambda_1 = 0.40528\ldots$, $\lambda_2 = 0.04503 \cdots \approx \lambda_1/10$. This is probably one reason for which former attempts to produce good quantization of the Brownian motion first focused on other kinds of quantizers like *scalar product quantizers* (see [PP05b] and Section 7.4 below) or d-dimensional block product quantizations (see [Wil05] and [LPW07]).

Optimization of the (quadratic) quantization of $\mathbf{R}^d$-valued random vector has been extensively investigated since the early 1950's, first in 1-dimension, then in higher dimension when the cost of numerical Monte Carlo simulation was drastically cut down (see [GG92]). Recent application of optimal vector quantization to numerics turned out to be much more demanding in terms of accuracy. In that direction, one may cite [PP03], [MBH06] (mainly focused on numerical optimization of the quadratic quantization of normal distributions). To apply the methods developed in these papers, it is more convenient to rewrite our optimization problem with respect to the standard d-dimensional distribution $\mathcal{N}(0; I_d)$ by simply considering the Euclidean norm derived from the covariance matrix $\mathrm{Diag}(\lambda_1, \dots, \lambda_{d(N)})$ *i.e.*

$$(\mathcal{O}_N) \Leftrightarrow \begin{cases} N\text{-optimal quantization of } \bigotimes_{k=1}^{d(N)} \mathcal{N}(0,1) \\ \text{for the covariance norm } |(z_1, \dots, z_{d(N)})|^2 = \sum_{k=1}^{d(N)} \lambda_k z_k^2. \end{cases}$$

The main point is of course that the dimension $d(N)$ is unknown. However (see Figure 6), one clearly verifies on small values of N that the conjecture $(d(N) \sim \log N)$ is most likely true. Then for higher values of N one relies on it to shift from one dimension to another following the rule $d(N) = d$, $N \in \{e^d, \dots, e^{d+1} - 1\}$.

A Toolbox for Quantization Optimization: A short Overview

Here is a short overview of stochastic optimization methods to compute optimal or at least locally optimal quantizers in finite dimension. For more details we refer to [PP03] and the references therein. Let $Z \stackrel{d}{=} \mathcal{N}(0; I_d)$.

Competitive Learning Vector Quantization $(CLVQ)$. This procedure is a recursive stochastic approximation gradient descent based on the integral representation of the gradient $\nabla D_N^Z(x)$, $x \in H^n$ (temporarily coming back to N-tuple notation) of the distortion as the expectation of a *local gradient i.e.*

$$\forall x^N \in H^N, \quad \nabla D_N^Z(x^N) = \mathbf{E}(\nabla D_N^Z(x^N, \zeta)), \; \zeta_k \;\; i.i.d., \; \zeta_1 \stackrel{d}{=} \mathcal{N}(0, I_d)$$

so that, starting from $x^N(0) \in (\mathbf{R}^d)^N$, one sets

$$\forall k \geq 0, \quad x^N(k+1) = x^N(k) - \frac{c}{k+1} \nabla D_N^Z(x^N(k), \zeta_{k+1})$$

where $c \in (0, 1]$ is a real constant to be tuned. As set, this looks quite formal but the operating $CLVQ$ procedure consists of two phases at each iteration:

(*i*) *Competitive Phase:* Search of the nearest neighbor $x^N(k)_{i*(k+1)}$ of ζ_{k+1} among the components of $x^N(k)_i$, $i = 1, \dots, N$ (using a "winning convention" in case of conflict on the boundary of the Voronoi cells).

(*ii*) *Cooperative Phase:* One moves the winning component toward ζ_{k+1} using a dilatation *i.e.* $x^N(k+1)_{i^*(k+1)} = \mathrm{Dilatation}_{\zeta_{k+1}, 1 - \frac{c}{k+1}}(x^N(k)_{i^*(k+1)})$.

This procedure is useful for small or medium values of N. For an extensive study of this procedure, which turns out to be singular in the world of recursive stochastic approximation algorithms, we refer to [Pag97]. For general background on stochastic approximation, we refer to [KY03, BMP90].

The randomized "Lloyd I procedure". This is the randomization of the stationarity based fixed point procedure since any optimal quantizer satisfies (4):

$$\widehat{Z}^{x^N(k+1)} = \mathbf{E}(Z \,|\, \widehat{Z}^{x^N(k)}), \qquad x^N(0) \subset \mathbf{R}^d.$$

At every iteration the conditional expectation $\mathbf{E}(Z \,|\, \widehat{Z}^{x^N(k)})$ is computed using a Monte Carlo simulation. For more details about practical aspects of Lloyd I procedure we refer to [PP03]. In [MBH06], an approach based on genetic evolutionary algorithms is developed.

For both procedures, one may substitute a sequence of quasi-random numbers to the usual pseudo-random sequence. This often speeds up the rate of convergence of the method, although this can only be proved (see [LSP90]) for a very specific class of stochastic algorithm (to which *CLVQ* does not belong).

The most important step to preserve the accuracy of the quantization as N (and $d(N)$) increase is to use the so-called *splitting method* which finds its origin in the proof of the existence of an optimal N-quantizer: once the optimization of a quantization grid of size N is achieved, one specifies the starting grid for the size $N+1$ or more generally $N+\nu$, $\nu \geq 1$, by merging the optimized grid of size N resulting from the former procedure with ν points sampled independently from the normal distribution with probability density proportional to $\varphi^{\frac{d}{d+2}}$ where φ denotes the p.d.f. of $\mathcal{N}(0; I_d)$. This rather unexpected choice is motivated by the fact that this distribution provides the lowest *in average* random quantization error (see [Coh98]).

As a result, to be downloaded on the website [PP05a] devoted to quantization:

`www.quantize.maths-fi.com`

∘ *Optimized stationary codebooks for W*: in practice, the N-quantizers β^N of the distribution $\otimes_{k=1}^{d(N)} \mathcal{N}(0; \lambda_k)$, $N = 1$ up to 10 000 ($d(N)$ runs from 1 up to 9).

∘ *Companion parameters*:

– distribution of $\widehat{W}^{\Gamma^N}$: $\mathbf{P}(\widehat{W}^{\Gamma^N} = x_i^N) = \mathbf{P}(\widehat{Z}_{d(N)}^{\beta^N} = \beta_i^N)$ ($\leftarrow$ in $\mathbf{R}^{d(N)}$).

– The quadratic quantization error: $\|W - \widehat{W}^{\Gamma^N}\|_2$.

7.4 An Alternative: Product Functional Quantization

Scalar Product functional quantization is a quantization method which produces rate optimal sub-optimal quantizers. They were used *e.g.* in [LP02] to provide exact rate (although not sharp) for a very large class of processes. The first attempts to use functional quantization for numerical computation with the Brownian motion was achieved with these quantizers (see [PP05b]). We will see further on their assets. What follows is presented for the Brownian motion but would work for a large class of centered Gaussian processes.

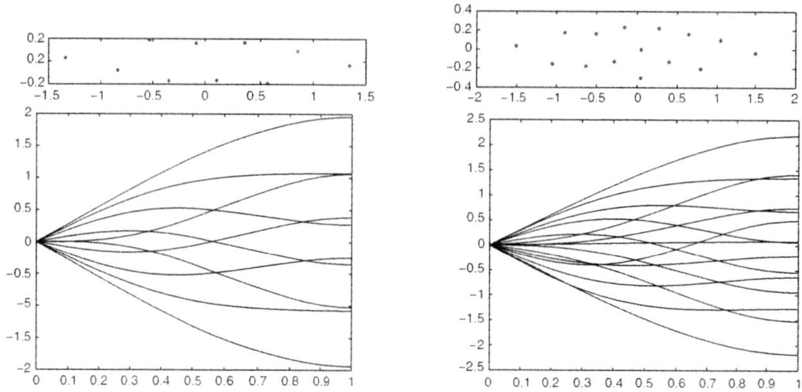

Fig. 3. *Optimized functional quantization of the Brownian motion W for $N = 10, 15$ $(d(N) = 2)$. Top: β^N depicted in $\mathbf{R}^2$. Bottom: the optimized N-quantizer Γ^N.*

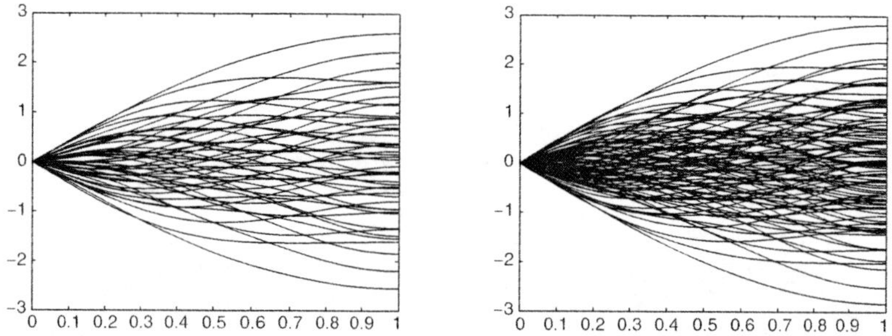

Fig. 4. *Optimized functional quantization of the Brownian motion W. The N-quantizers Γ^N. Left: $N = 48$ $(d(N) = 3)$. Right: $N = 96$, $d(96) = 4$.*

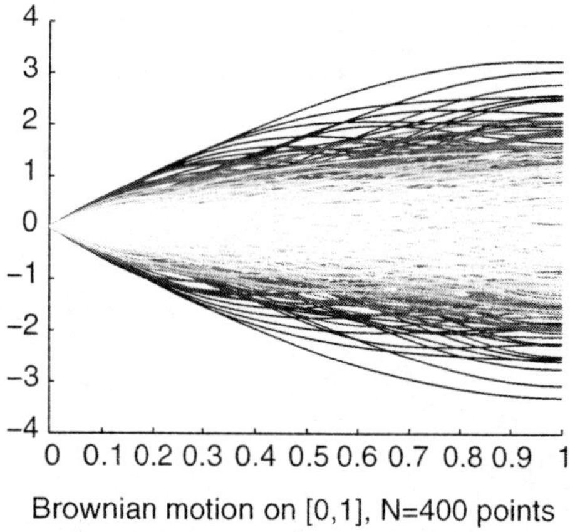

Fig. 5. *Optimized N-quantizer Γ^N of the Brownian motion W with $N = 400$. The grey level of the paths codes their weights.*

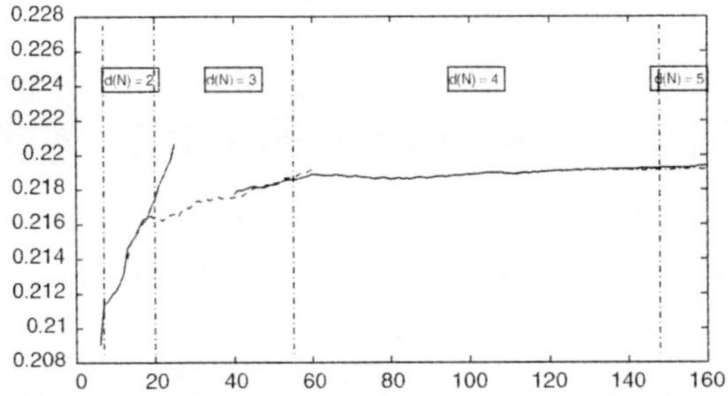

Fig. 6. *Optimal functional quantization of the Brownian motion.* $N \mapsto \log N\,(e_N(W, L_T^2))^2$, $N \in \{6, \ldots, 160\}$. *Vertical dashed lines: critical dimensions for* $d(N)$, $e^2 \approx 7$, $e^3 \approx 20$, $e^4 \approx 55$, $e^5 \approx 148$.

Let us consider again the expansion of W in its K-L basis:

$$W \overset{L_T^2}{=} \sum_{n\geq 1} \sqrt{\lambda_n}\,\xi_n\, e_n^W$$

where $(\xi_n)_{n\geq 1}$ is an i.i.d. sequence $\mathcal{N}(0;1)$-distributed random variables (keep in mind this convergence also holds $a.s.$ uniformly in $t \in [0,T]$). The idea is simply to quantize these (normalized) random coordinates ξ_n: for every $n \geq 1$, one considers an optimal N_n-quantization of ξ_n, denoted $\widehat{\xi}_n^{(N_n)}$ ($N_n \geq 1$). For $n > m$, set $N_n = 1$ and $\widehat{\xi}_n^{(N_n)} = 0$ (which is the optimal 1-quantization). The integer m is called the *length* of the product quantization. Then, one sets

$$\widehat{W}_t^{(N_1,\ldots,N_m,\,prod)} := \sum_{n\geq 1} \sqrt{\lambda_n}\;\widehat{\xi}_n^{(N_n)}\, e_n^W(t) = \sum_{n=1}^{m} \sqrt{\lambda_n}\;\widehat{\xi}_n^{(N_n)}\, e_n^W(t).$$

Such a quantizer takes $\prod_{n=1}^m N_n \leq N$ values.

If one denotes by $\alpha^M = \{\alpha_1^M, \ldots, \alpha_M^M\}$ the (unique) optimal quadratic M-quantizer of the $\mathcal{N}(0;1)$-distribution, the underlying quantizer of the above quantization $\widehat{W}^{(N_1,\ldots,N_m,\,prod)}$ can be expressed as follows (if one introduces the appropriate multi-indexation): for every multi-index $\underline{i} := (i_1, \ldots, i_m) \in \prod_{n=1}^m \{1, \ldots, N_n\}$, set

$$x_{\underline{i}}^{(N)}(t) := \sum_{n=1}^{m} \sqrt{\lambda_n}\,\alpha_{i_n}^{(N_n)} e_n^W(t) \text{ and } \Gamma^{N_1,\ldots,N_m,prod} := \left\{x_{\underline{i}}^{(N)}, \underline{i} \in \prod_{n=1}^{m}\{1,\ldots,N_n\}\right\}.$$

Then the product quantization $\widehat{W}^{(N_1,\ldots,N_m,\,prod)}$ can be rewritten as

$$\widehat{W}_t^{(N_1,\ldots,N_m,\,prod)} = \sum_{\underline{i}} \mathbf{1}_{\{W \in C_{\underline{i}}(\Gamma^{N_1,\ldots,N_m,prod})\}}\, x_{\underline{i}}^{(N)}(t).$$

where the Voronoi cell of $x_{\underline{i}}^{(N)}$ is given by

$$C_{\underline{i}}(\Gamma^{N_1,\ldots,N_m,prod}) = \prod_{n=1}^{m} (\alpha_{i_n-\frac{1}{2}}^{(N_n)}, \alpha_{i_n+\frac{1}{2}}^{(N_n)})$$

with $\alpha_{i\pm\frac{1}{2}}^{(M)} := \frac{\alpha_i^{(M)} + \alpha_{i\pm 1}^{(M)}}{2}$, $\alpha_0 = -\infty$, $\alpha_{M+1} = +\infty$.

Quantization Rate by Product Quantizers

It is clear that the optimal product quantizer is the solution to the optimal integral bit allocation

$$\min\left\{\|W-\widehat{W}^{(N_1,\ldots,N_m,\,prod)}\|_2, N_1,\ldots,N_m \geq 1, N_1\times\cdots\times N_m \leq N, m\geq 1\right\}. \tag{12}$$

Expanding $\|W - \widehat{W}^{(N_1,\ldots,N_m,\,prod)}\|_2^2 = \||W - \widehat{W}^{(N_1,\ldots,N_m,\,prod)}|_{L^2_T}\|_2^2$ yields

$$\|W - \widehat{W}^{(N_1,\ldots,N_m,\,prod)}\|_2^2 = \sum_{n\geq 1} \lambda_n \|\widehat{\xi}_n^{(N_n)} - \xi_n\|_2^2 \tag{13}$$

$$= \sum_{n=1}^{m} \lambda_n (e^2_{N_n}(\mathcal{N}(0;1), \mathbf{R}) - 1) + \frac{T^2}{2} \tag{14}$$

since $\displaystyle \sum_{n\geq 1} \lambda_n = \mathbf{E} \sum_{n\geq 1} (W \,|\, e_n^W)_2^2 = \mathbf{E} \int_0^T W_t^2 dt = \int_0^T t\, dt = \frac{T^2}{2}.$

Theorem 5. *(see [LP02]) For every $N \geq 1$, there exists an optimal scalar product quantizer of size at most N (or at level N), denoted $\widehat{W}^{(N,\,prod)}$, of the Brownian motion defined as the solution to the minimization problem (12). Furthermore these optimal product quantizers make up a rate optimal sequence: there exists a real constant $c_W > 0$ such that*

$$\|W - \widehat{W}^{(N,\,prod)}\|_2 \leq \frac{c_W T}{(\log N)^{\frac{1}{2}}}.$$

Proof (sketch of). By scaling one may assume without loss of generality that $T = 1$. Combining (13) and Zador's Theorem shows

$$\|W - \widehat{W}^{(N_1,\ldots,N_m,\,prod)}\|_2^2 \leq C\left(\sum_{n=1}^{m} \frac{1}{n^2 N_n^2}\right) + \sum_{n\geq m+1} \lambda_n$$
$$\leq C'\left(\sum_{n=1}^{m} \frac{1}{n^2 N_n^2} + \frac{1}{m}\right)$$

with $\prod_n N_n \leq N$. Setting $m := m(N) = [\log N]$ and $N_k = \left[\frac{(m!N)^{\frac{1}{m}}}{k}\right] \geq 1$, $k = 1, \ldots, m$, yields the announced upper-bound. $\diamond$

Remarks. • One can show that the length $m(N)$ of the optimal quadratic product quantizer satisfies

$$m(N) \sim \log N \qquad \text{as} \qquad N \to +\infty.$$

• The most striking fact is that very few ingredients are necessary to make the proof work as far as the quantization rate is concerned. We only need the basis of L^2_T on which W is expanded to be orthonormal *or* the random coordinates to be orthogonal in $L^2(\mathbf{P})$. This robustness of the proof has been used to obtain some upper bounds for very wide classes of Gaussian processes by considering alternative orthonormal basis of L^2_T like the Haar basis for processes having self-similarity properties (see [LP02]), or trigonometric basis for stationary processes (see [LP02]). More recently, combined with the non asymptotic Zador's Theorem, it was used to provide some connections between mean regularity of stochastic processes and quantization rate (see Section 10 and [LP06a]).

• Block quantizers combined with large deviations estimates were used to provide the sharp rate obtained in Theorem 3 in [LP04].

• d-dimensional block quantization is also possible, possibly with varying block size, providing a constructive approach to sharp rate, see [Wil05] and [LPW07].

• A similar approach can also provide some $L^r(\mathbf{P})$-rates for product quantization with respect to the sup-norm over $[0, T]$, see [LP05].

How to use Product Quantizers for Numerical Computations?

For numerics one can assume by a scaling argument that $T = 1$. To use product quantizers for numerics we need to have access to the quantizers (or grid) at a given level N, their weights (and the quantization error). All these quantities are available with product quantizers. In fact the first attempts to use functional quantization for numerics (path dependent option pricing) were carried out with product quantizers (see [PP05b]).

• The optimal product quantizers (denoted $\Gamma^{(N,prod)}$) at level N are explicit, given the optimal quantizers of the scalar normal distribution $\mathcal{N}(0;1)$. In fact the optimal allocation of the size N_i of each marginal has been already achieved up to very high values of N. Some typical optimal allocation (and the resulting quadratic quantization error) are reported in the table below.

N	N_{rec}	Quant. Error	Opti. Alloc.
1	1	0.7071	1
10	10	0.3138	5-2
100	96	0.2264	12-4-2
1 000	966	0.1881	23-7-3-2
10 000	9 984	0.1626	26-8-4-3-2-2
100 000	97 920	0.1461	34 − 10 − 6 − 4 − 3 − 2 − 2

• The weights $\mathbf{P}(\widehat{W}^{(N,prod)} = x_{\underline{i}})$ are explicit too: the normalized coordinates ξ_n of W in its K-L basis are independent, consequently

$$\begin{aligned}\mathbf{P}(\widehat{W}^{(N,prod)} = x_{\underline{i}}) &= \mathbf{P}(\widehat{\xi}_n^{(N_n)} = \alpha_{i_n}^{(N_n)}, n = 1, \ldots, m(N)) \\ &= \prod_{n=1}^{m(N)} \underbrace{\mathbf{P}(\widehat{\xi}_n^{(N_n)} = \alpha_{i_n}^{(N_n)})}_{1D\ (tabulated)\ weights}.\end{aligned}$$

• Equation (14) shows that the (squared) quantization error of a product quantizer can be straightforwardly computed as soon as one knows the eigenvalues and the (squared) quantization error of the normal distributions for the N_i's.

The optimal allocations up to $N = 12\,000$ can be downloaded on the website [PP05a] as well as the necessary 1-dimensional optimal quantizers (including the weights and the quantization error) of the scalar normal distribution (up to a size of 500 which quite enough for this purpose).

For numerical purpose we are also interested in the stationarity property since such quantizers produce lower (weak) errors in cubature formulas.

Proposition 2. *(see [PP05b]) The product quantizers obtained from the K-L basis are stationary quantizers (although sub-optimal).*

Proof. Firstly, note that

$$\widehat{W}^{N,prod} = \sum_{n\geq 1} \sqrt{\lambda_n}\, \widehat{\xi}_n^{(N_n)} e_n(t)$$

so that $\sigma(\widehat{W}^{N,prod}) = \sigma(\widehat{\xi}_k^{(N_k)},\ k \geq 1)$. Consequently

$$\begin{aligned}\mathbf{E}(W \,|\, \widehat{W}^{N,prod}) &= \mathbf{E}(W \,|\, \sigma(\widehat{\xi}_k^{(N_k)}, , , k \geq 1)) \\ \mathbf{E}(W \,|\, \widehat{W}^{N,prod}) &= \sum_{n\geq 1} \sqrt{\lambda_n}\, \mathbf{E}\left(\xi_n \,|\, \sigma(\widehat{\xi}_k^{(N_k)},\ k \geq 1)\right) e_n^W \\ &\overset{i.i.d.}{=} \sum_{n\geq 1} \sqrt{\lambda_n}\, \mathbf{E}\left(\xi_n \,|\, \widehat{\xi}_n^{(N_n)}\right) e_n^W \\ &= \sum_{n\geq 1} \sqrt{\lambda_n}\, \widehat{\xi}_n^{(N_n)} e_n^W = \widehat{W}. \qquad \diamond\end{aligned}$$

Remarks. • This result is no longer true for product quantizers based on other orthonormal basis.
• This shows the existence of non optimal stationary quantizers.

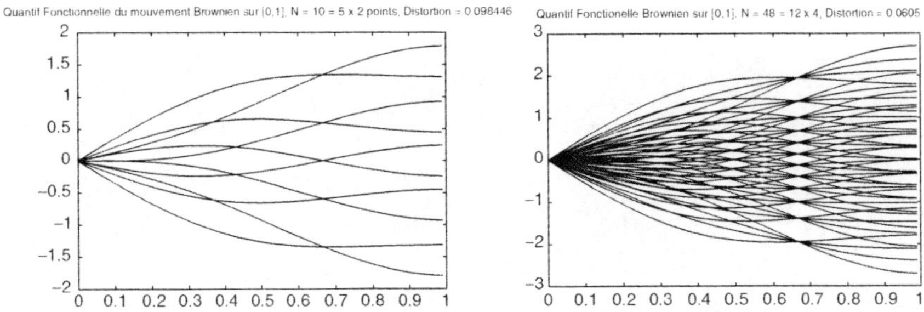

Fig. 7. *Product quantization of the Brownian motion: the* $N_{\rm rec}$*-quantizer* $\Gamma^{(N,\,prod)}$. $N = 10$*:* $N_{\rm rec} = 10$ *and* $N = 50$*:* $N_{\rm rec} = 12 \times 4 = 48$.

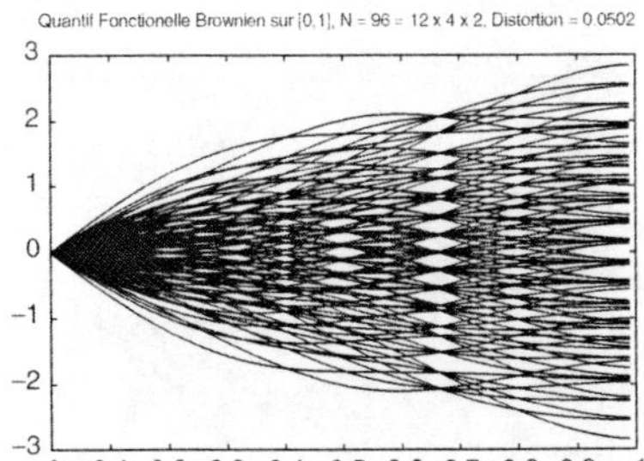

Fig. 8. *Product quantization of the Brownian motion: the* $N_{\rm rec}$*-quantizer* $\Gamma^{(N,\,prod)}$. $N = 100$*:* $N_{\rm rec} = 12 \times 4 \times 2 = 96$.

7.5 Optimal *vs* Product Quadratic Functional Quantization ($T = 1$)

∘ (Numerical) Optimized Quantization: By scaling, we can assume without loss of generality that $T = 1$. We carried out a huge optimization task in order to produce some *optimized* quantization grids for the Brownian motion by solving numerically $(\mathcal{O}_N)$ for $N = 1$ up to $N = 10\,000$.

$$e_N(W, L^2_T)^2 \approx \frac{0.2195}{\log N}, \qquad N = 1, \ldots, 10\,000.$$

This value (see Figure 9 (left)) is significantly greater than the theoretical (asymptotic) bound given by Theorem 3 which is

$$\lim_N \log N e_N(W, L^2_T)^2 = \frac{2}{\pi^2} = 0.2026...$$

Our guess, supported by our numerical experiments, is that in fact $N \mapsto \log N e_N(W, L^2_T)^2$ is possibly not monotonous but unimodal.

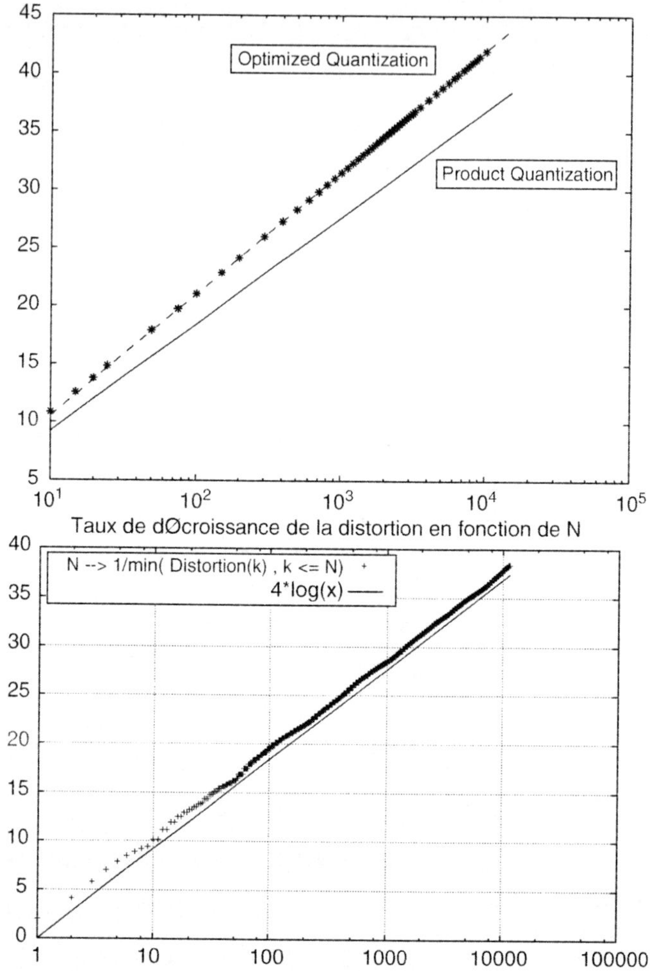

Fig. 9. *Numerical quantization rates. Top (Optimal quantization). Line +++:* $\log N \mapsto (\|W - \widehat{W}^N\|_2)^{-2}$. *Dashed line:* $\log N \mapsto \log N/0.2194$. *Solid line:* $\log N \mapsto \log N/0.25$. *Bottom (Product quantization). Line +++:* $\log N \mapsto (\min\limits_{1\le k\le N} \|W - \widehat{W}^{k,prod}\|_2^2)^{-1}$. *Solid line:* $\log N \mapsto \log N/0.25$.

∘ Optimal Product quantization: as displayed on Figure 9 (right), one has approximately

$$\min\left\{\|\,|W - \widehat{W}|_{L^2_T}\|_2^2,\ 1 \le N_1 \cdots N_m \le N,\ m \ge 1\right\}$$
$$= \|W - \widehat{W}^{(N,prod)}\|_2^2 \approx \frac{0.245}{\log N}$$

∘ OPTIMAL d-DIMENSIONAL BLOCK PRODUCT QUANTIZATION: let us briefly mention this approach developed in [Wil05] in which product quantization is achieved by quantizing some marginal blocks of size 1, 2 or 3. By this approach, the corresponding constant is approximately 0.23, *i.e.* roughly in between scalar product quantization and optimized numeric quantization.

The conclusion, confirmed by our numerical experiments on option pricing (see Section 9), is that
– Optimal quantization is significantly more accurate on numerical experiments but is much more demanding since it needs to keep off line or at least to handle large files (say 1 GB for $N = 10\,000$).
– Both approaches are included in the option pricer PREMIA (MATHFI Project, Inria). An online benchmark is available on the website [PP05a].

8 Constructive Functional Quantization of Diffusions

8.1 Rate Optimality for Scalar Brownian Diffusions

One considers on a probability space $(\Omega, \mathcal{A}, \mathbf{P})$ an homogenous Brownian diffusion process:

$$dX_t = b(X_t)dt + \vartheta(X_t)\,dW_t, \quad X_0 = x_0 \in \mathbf{R},$$

where b and ϑ are continuous on $\mathbf{R}$ with at most linear growth (*i.e.* $|b(x)| + |\sigma(x)| \leq C(1 + |x|)$) so that at least a weak solution to the equation exists.

To devise a constructive way to quantize the diffusion X, it seems natural to start from a rate optimal quantization of the Brownian motion and to obtain some "good" (but how good?) quantizers for the diffusion by solving an appropriate ODE. So let $\Gamma^N = (w_1^N, \cdots, w_N^N)$, $N \geq 1$, be a sequence of stationary rate optimal N-quantizers of W. One considers the following (non-coupled) Integral Equations:

$$dx_i^{(N)}(t) = \left(b(x_i^{(N)}(t)) - \frac{1}{2}\vartheta\theta'(x_i^{(N)}(t))\right)dt + \vartheta(t, x_i^{(N)}(t))\,dw_i^N(t). \qquad (15)$$

Set

$$\widetilde{X}_t^{x^{(N)}} = \sum_{k=1}^{N} x_i^{(N)}(t)\mathbf{1}_{\{\widehat{W}^{\Gamma^N} = w_i^N\}}.$$

The process $\widetilde{X}^{x^{(N)}}$ is a *non-Voronoi* quantizer (since it is defined using the Voronoi diagram of W). What is interesting is that it is a *computable quantizer* (once the above integral equations have been solved) since the weights $\mathbf{P}(\widehat{W}^{\Gamma^N} = w_i^N)$ are known. The Voronoi quantization defined by $x^{(N)}$ induces a lower quantization error but we have no access to its weights for numerics. The good news is that $\widetilde{X}^{x^{(N)}}$ is already rate optimal.

Theorem 6. *([LP06b] (2006)) Assume that b is differentiable, ϑ is positive twice differentiable and that $b' - b\frac{\vartheta'}{\vartheta} - \frac{1}{2}\vartheta\vartheta''$ is bounded. Then*

$$e_N(X, L^2_T) \leq \| X - \widetilde{X}^{x^{(N)}} \|_2 = O((\log N)^{-\frac{1}{2}}).$$

If furthermore, $\vartheta \geq \varepsilon_0 > 0$, then $e_N(X, L^2_T) \approx (\log N)^{-\frac{1}{2}}$.

Remarks. • For some results in the non homogenous case, we refer to [LP06b]. Furthermore, the above estimates still hold true for the $(L^r(\mathbf{P}), L^p_T)$-quantization, $1 < r, p < +\infty$ provided $\||W - \widehat{W}^{\Gamma^N}|_{L^p_T}\|_r = O((\log N)^{-\frac{1}{2}})$.

• This result is closely connected to the Doss-Sussman approach (see *e.g.* [Dos77]) and in fact the results can be extended to some classes multi-dimensional diffusions (whose diffusion coefficient is the inverse of the gradient of a diffeomorphism) which include several standard multi-dimensional financial models (including the Black-Scholes model).

• A sharp quantization rate $e_{N,r}(X, L^p_T) \sim c(\log N)^{-\frac{1}{2}}$ for scalar elliptic diffusions is established in [Der05b, Der05a] using a non constructive approach, $1 \leq p \leq \infty$.

EXAMPLE: Rate optimal product quantization of the Ornstein-Uhlenbeck process.

$$dX_t = -kX_t dt + \vartheta dW_t, \qquad X_0 = x_0.$$

One solves the non-coupled integral (linear) system

$$x_i(t) = x_0 - k\int_0^t x_i(s)\,ds + \vartheta w_i^N(t),$$

where $\Gamma^N := \{w_1^N, \ldots, w_N^N\}, N \geq 1$ is a *rate optimal* sequence of quantizers

$$w_i^N(t) = \sqrt{\frac{2}{T}} \sum_{\ell \geq 1} \varpi_{i,\ell} \frac{T}{\pi(\ell - 1/2)} \sin\left(\pi(\ell - 1/2)\frac{t}{T}\right), \quad i \in I_N.$$

If Γ^N is optimal for W then $\varpi_{i,\ell} := (\beta_i^N)^\ell$, $i = 1, \ldots, N$, $1 \leq \ell \leq d(N)$ with the notations introduced in (11). If Γ^N is an optimal product quantizer (and $N_1, \ldots, N_\ell, \ldots$ denote the optimal size allocation), then $\varpi_{i,\ell} = \alpha_{i_\ell}^{(N_\ell)}$, where $i := (i_1, \ldots, i_\ell, \ldots) \in \prod_{\ell \geq 1}\{1, \ldots, N_\ell\}$. Elementary computations show that

$$x_i^N(t) = e^{-kt}x_0 + \vartheta \sum_{\ell \geq 1} \chi_{i_\ell}^{(N_\ell)}\, \widetilde{c}_\ell\, \varphi_\ell(t)$$

$$\text{with} \quad \widetilde{c}_\ell = \frac{T^2}{(\pi(\ell - 1/2))^2 + (kT)^2}$$

$$\text{and} \quad \varphi_\ell(t) := \sqrt{\frac{2}{T}}\left(\frac{\pi}{T}(\ell-1/2)\sin\left(\pi(\ell-1/2)\frac{t}{T}\right) + k\left(\cos\left(\pi(\ell-1/2)\frac{t}{T}\right) - e^{-kt}\right)\right).$$

8.2 Multi-Dimensional Diffusions for Stratanovich SDE's

The correcting term $-\frac{1}{2}\vartheta\vartheta'$ coming up in the integral equations suggest to consider directly some diffusion in the Stratanovich sense

$$dX_t = b(t, X_t)\, dt + \vartheta(t, X_t) \circ dW_t \qquad X_0 = x_0 \in \mathbf{R}^d, \qquad t \in [0, T].$$

(see *e.g.* [RY99] for an introduction) where $W = (W^1, \ldots, W^d)$ is a d-dimensional standard Brownian Motion.

In that framework, we need to introduce the notion of p-variation: a continuous function $x : [0, T] \to \mathbf{R}^d$ has finite p-variations if

$$Var_{p,[0,T]}(x) := \sup \left\{ \left(\sum_{i=0}^{k-1} |x(t_i) - x(t_{i+1})|^p \right)^{\frac{1}{p}}, \right.$$
$$\left. 0 \le t_0 \le t_1 \le \cdots \le t_k \le T,\ k \ge 1 \right\} < +\infty.$$

Then $d_p(x, x') = |x(0) - x'(0)| + Var_{p,[0,T]}(x - x')$ defines a distance on the set of functions with finite p-variations. It is classical background that $Var_{p,[0,T]}(W(\omega)) < +\infty$ $\mathbf{P}(d\omega)$-*a.s.* for every $p > 2$.

One way to quantize W at level (at most) N is to quantize each component W^i at level $\lfloor \sqrt[d]{N} \rfloor$. One shows (see [LP04]) that $\|W - (\widehat{W}^{1,\lfloor \sqrt[d]{N} \rfloor}, \ldots, \widehat{W}^{d,\lfloor \sqrt[d]{N} \rfloor})\|_2 = O((\log N)^{-\frac{1}{2}})$.

Let $\mathcal{C}_b^r([0, T] \times \mathbf{R}^d)$ $r > 0$, denote the set of $\lfloor r \rfloor$-times differentiable bounded functions $f : [0, T] \times \mathbf{R}^d \to \mathbf{R}^d$ with bounded partial derivatives up to order $\lfloor r \rfloor$ and whose partial derivatives of order $\lfloor r \rfloor$ are $(r - \lfloor r \rfloor)$-Hölder.

Theorem 7. *(see [PS07]) Let $b, \vartheta \in \mathcal{C}_b^{2+\alpha}([0, T] \times \mathbf{R}^d)$ $(\alpha > 0)$ and let $\Gamma^N = \{w_1^N, \ldots, w_N^N\}$, $N \ge 1$, be a sequence of N-quantizers of the standard d-dimensional Brownian motion W such that $\|W - \widehat{W}^{\Gamma^N}\|_2 \to 0$ as $N \to \infty$. Let*

$$\widetilde{X}_t^{x^{(N)}} := \sum_{i=1}^{N} x_i^{(N)}(t) \mathbf{1}_{\{\widehat{W} = w_i^N\}}$$

where, for every $i \in \{1, \ldots, N\}$, $x_i^{(N)}$ is solution to

$$ODE_i \equiv dx_i^{(N)}(t) = b(t, x_i^{(N)}(t))dt + \vartheta(t, x_i^{(N)}(t))dw_i^N(t), \quad x_i^{(N)}(0) = x.$$

Then, for every $p \in (2, \infty)$,

$$Var_{p,[0,T]}(\widetilde{X}^{x^{(N)}} - X) \xrightarrow{\mathbf{P}} 0 \quad as \quad N \to \infty.$$

Remarks. • The keys of this results are the Kolmogorov criterion, stationarity (in a slightly extended sense) and the connection with rough paths theory (see [Lej03] for an introduction to rough paths theory, convergence in p-variation, etc).
• In that general setting we have no convergence rate although we conjecture that $\widetilde{X}^{x^{(N)}}$ remains rate optimal if $\widehat{W}^{\Gamma^N}$ is.
• There are also some results about the convergence of stochastic integrals of the form $\int_0^t g(\widehat{W}_s^N)\, d\widehat{B}_s^N \to \int_0^t g(W_s) \circ dB_s$, with some rates of convergence when $W = B$ or W and B independent (depending on the regularity of the function g, see [PS07]).

9 Applications to Path-Dependent Option Pricing

The typical functionals F defined on $(L_T^2, |\,.\,|_{L_T^2})$ for which $\mathbf{E}\,(F(W))$ can be approximated by the cubature formulae (5), (7) are of the form

$$F(\omega) := \varphi\left(\int_0^T f(t, \omega(t))dt\right)\mathbf{1}_{\{\omega\in\mathcal{C}([0,T],\mathbf{R})\}}$$

where $f : [0,T]\times\mathbf{R}\to\mathbf{R}$ is *locally Lipschitz continuous* in the second variable, namely

$$\forall t\in[0,T],\ \forall u,v\in\mathbf{R},\ |f(t,u)-f(t,v)|\le C_f|u-v|(1+g(|u|)+g(|v|))$$

(with $g:\mathbf{R}_+\to\mathbf{R}_+$ is increasing, convex and $g(\sup_{t\in[0,T]}|W_t|)\in L^2(\mathbf{P})$) and $\varphi:\mathbf{R}\to\mathbf{R}$ is Lipschitz continuous. One could consider for ω some càdlàg functions as well. A classical example is the Asian payoff in a Black-Scholes model

$$F(\omega) = \exp(-rT)\left(\frac{1}{T}\int_0^T s_0\exp(\sigma\omega(t)+(r-\sigma^2/2)t)dt - K\right)_+ .$$

9.1 Numerical Integration (II): log-Romberg Extrapolation

Let $F : L_T^2 \longrightarrow \mathbf{R}$ be a 3 times $|\,.\,|_{L_T^2}$-differentiable functional with bounded differentials. Assume $\widehat{W}^{(N)}$, $N\ge 1$, is a sequence of a rate-optimal stationary quantizations of the standard Brownian motion W. Assume furthermore that

$$\mathbf{E}\left(D^2F(\widehat{W}^{(N)}).(W-\widehat{W}^{(N)})^{\otimes 2}\right)\sim\frac{c}{\log N}\quad\text{as}\quad N\to\infty \tag{16}$$

and

$$\mathbf{E}\,|W-\widehat{W}^{(N)}|^3_{L_T^2} = O\left((\log N)^{-\frac{3}{2}}\right). \tag{17}$$

Then, a higher order Taylor expansion yields

$$F(W) = F(\widehat{W}^{(N)}) + DF(\widehat{W}^{(N)}).(W - \widehat{W}^{(N)}) + \frac{1}{2}D^2F(\widehat{W}^{(N)}).(W - \widehat{W}^{(N)})^{\otimes 2} + \frac{1}{6}D^2(\zeta).(W - \widehat{W}^{(N)})^{\otimes 3}, \qquad \zeta \in (\widehat{W}^{(N)}, W),$$

$$\mathbf{E}\,F(W) = \mathbf{E}F(\widehat{W}^{(N)}) + \frac{c}{2\log N} + o\left((\log N)^{-\frac{3}{2}+\varepsilon}\right).$$

Then, one can design a log-Romberg extrapolation by considering N, N', $N < N'$ (*e.g.* $N' \approx 4\,N$), so that

$$\mathbf{E}(F(W)) = \frac{\log N' \times \mathbf{E}(F(\widehat{W}^{(N')})) - \log N' \times \mathbf{E}(F(\widehat{W}^{(N)}))}{\log N' - \log N} + o\left((\log N)^{-\frac{3}{2}+\varepsilon}\right).$$

For practical implementation, it is suggested in [Wil05] to replace $\log N$ by the more consistent "estimator" $\|W - \widehat{W}^{(N)}\|_2^{-2}$.

In fact Assumption (16) holds true for optimal product quantization when F is polynomial function F, $d^0F = 2$. Assumption (17) holds true in that case as well (see [GLP06]). As concerns optimal quantization, these statements are still conjectures.

Note that the above extrapolation or some variants can be implemented with other stochastic processes in accordance with the rate of convergence of the quantization error.

9.2 Asian Option Pricing in a Heston Stochastic Volatility Model

In this section, we will price an Asian call option in a Heston stochastic volatility model using some optimal (at least optimized) functional quantization of the two Brownian motions that drive the diffusion. This model has already been considered in [PP05b] in which functional quantization was implemented for the first time with some product quantizations of the Brownian motions. The Heston stochastic volatility model was introduced in [Hes93] to model stock price dynamics. Its popularity partly comes from the existence of semi-closed forms for vanilla European options, based on inverse Fourier transform and from its ability to reproduce some skewness shape of the implied volatility surface. We consider it under its risk-neutral probability measure.

$$dS_t = S_t(r\,dt + \sqrt{v_t}dW_t^1), \qquad S_0 = s_0 > 0, \quad \text{(risky asset)}$$
$$dv_t = k(a - v_t)dt + \vartheta\sqrt{v_t}\,dW_t^2,\ v_0 > 0 \text{ with } d\langle W^1, W^2\rangle_t = \rho\,dt,\ \rho \in [-1, 1].$$

where ϑ, k, a such that $\vartheta^2/(4ak) < 1$. We consider the Asian Call payoff with maturity T and strike K. No closed form is available for its premium

$$\mathrm{AsCall}^{Hest} = e^{-rT}\mathbf{E}\left(\frac{1}{T}\int_0^T S_s ds - K\right)^+.$$

We briefly recall how to proceed (see [PP05b] for details): first, one projects W^1 on W^2 so that $W^1 = \rho W^2 + \sqrt{1-\rho^2}\,\widetilde{W}^1$ and

$$\begin{aligned} S_t &= s_0 \exp\left((r - \frac{1}{2}\bar{v}_t)t + \rho\int_0^t \sqrt{v_s}dW_s^2\right)\exp\left(\sqrt{1-\rho^2}\int_0^t \sqrt{v_s}d\widetilde{W}_s^1\right) \\ &= s_0 \exp\left(t\left((r - \frac{\rho a k}{\vartheta}) + \bar{v}_t(\frac{\rho k}{\vartheta} - \frac{1}{2})\right) + \frac{\rho}{\vartheta}(v_t - v_0)\right) \\ &\times \exp\left(\sqrt{1-\rho^2}\int_0^t \sqrt{v_s}d\widetilde{W}_s^1\right). \end{aligned}$$

The chaining rule for conditional expectations yields

$$\mathrm{AsCall}^{Hest}(s_0, K) = e^{-rT}\mathbf{E}\left(\mathbf{E}\left(\left(\frac{1}{T}\int_0^T S_s ds - K\right)^+ |\sigma(W_t^2, 0 \le t \le T)\right)\right).$$

Combining these two expressions and using that $\widetilde{W}^1$ and W^2 are independent show that $\mathrm{AsCall}^{Hest}(s_0, K)$ is a functional of $(\widetilde{W}_t^1, v_t)$ (as concerns the squared volatility process v, only v_T and $\int_0^T v_s ds$ are involved).

Let $\Gamma^N = \{w_1^N, \dots, w_N^N\}$ be an N-quantizer of the Brownian motion. One solves for $i = 1, \dots, N$, the differential equations for (v_t)

$$dy_i(t) = k\left(a - y_i(t) - \frac{\vartheta^2}{4k}\right)dt + \vartheta\sqrt{y_i(t)}\,dw_i^N(t),\ y_i(0) = v_0, \tag{18}$$

using *e.g.* a Runge-Kuta scheme. Let $y_i^{n,N}$ denote the approximation of y_i resulting from the resolution of the above ODE_i ($1/n$ is the time discretization parameter of the scheme). Set the (non-Voronoi) N-quantization of (v_t, S_t) by

$$\widetilde{v}_t^{n,N} = \sum_i y_i^{n,N}(t)\mathbf{1}_{C_i(\Gamma^N)}(W^2) \tag{19}$$

$$\widetilde{S}_t^{n,N} = \sum_{1\le i,j\le N} s_{i,j}^{n,N}(t)\mathbf{1}_{C_i(\Gamma^N)}(\widetilde{W}^1)\mathbf{1}_{C_j(\Gamma^N)}(W^2) \tag{20}$$

$$\begin{aligned} \text{with } s_{i,j}^{n,N}(t) &= s_0\exp\left(t\left((r - \frac{\rho a k}{\vartheta}) + \overline{y}_j^{n,N}(t)(\frac{\rho k}{\vartheta} - \frac{1}{2})\right) + \frac{\rho}{\vartheta}(y_j^{n,N}(t) - v_0)\right) \\ &\times \exp\left(\sqrt{1-\rho^2}\int_0^t \sqrt{y_j^{n,N}(s)}\,dw_i^N(s)\right) \end{aligned}$$

$$\text{and}\quad \overline{y}_j^{n,N}(t) = \frac{1}{t}\int_0^t y_j^{n,N}(s)\,ds.$$

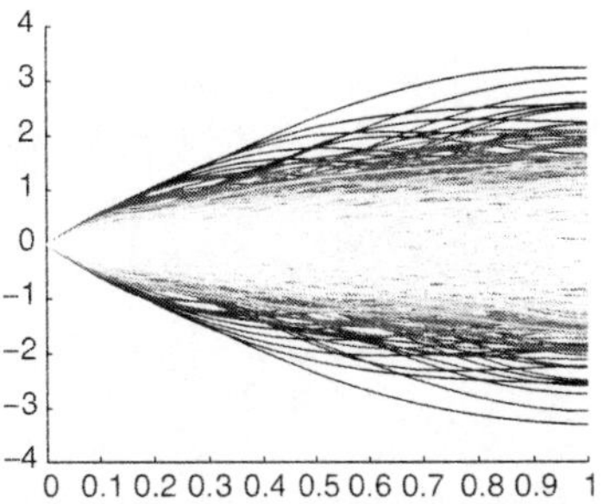

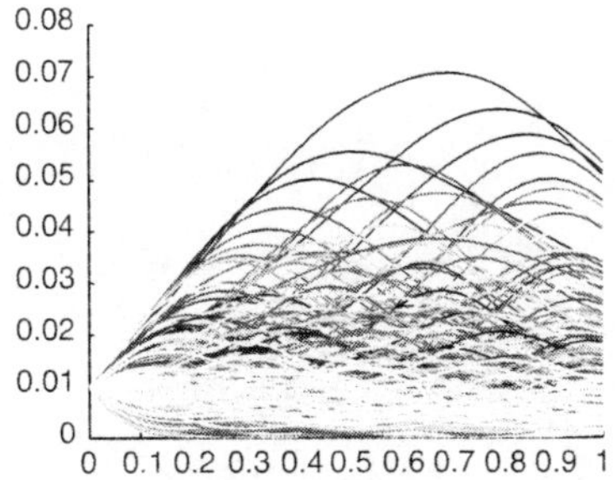

Fig. 10. *N-quantizer of the Heston squared volatility process (v_t) ($N = 400$) resulting from an (optimized) N-quantizer of W.*

Note this formula requires the computation of a quantized stochastic integral $\int_0^t \sqrt{y_j^{n,N}}(s)dw_i^N(s)$ (which corresponds to the independent case).
The weights of the product cells $\{\widetilde{W}^1 \in C_i(\Gamma^N),\, W^2 \in C_j(\Gamma^N)\}$ is given by

$$\mathbf{P}(\widetilde{W}^1 \in C_i(\Gamma^N),\, W^2 \in C_j(\Gamma^N)) = \mathbf{P}(\widetilde{W}^1 \in C_i(\Gamma^N))\mathbf{P}(W^2 \in C_j(\Gamma^N))$$

owing to the independence. For practical implementations different sizes of quantizers can be considered to quantize $\widetilde{W}^1$ and W^2.

We follow the guidelines of the methodology introduced in [PP05b]: we compute the *crude* quantized premium for two sizes N and N', then proceed a space Romberg log-extrapolation. Finally, we make a K-linear interpolation based on the (Asian) forward moneyness $s_0 e^{rT}\frac{1-e^{-rT}}{rT} \approx s_0 e^{rT}$ (like in [PP05b]) and the Asian Call-Put parity formula

$$\text{AsianCall}^{Hest}(s_0, K) = \text{AsianPut}^{Hest}(s_0, K) + s_0\frac{1-e^{-rT}}{rT} - Ke^{-rT}.$$

The *anchor strikes* $K_{\min}$ and $K_{\max}$ of the extrapolation are chosen symmetric with respect to the forward moneyness. At $K_{\max}$, the Call is deep out-of-the-money: one uses the Romberg extrapolated FQ computation; at $K_{\min}$ the Call is deep in-the-money: on computes the Call by parity. In between, one proceeds a linear interpolation in K (which yields the best results, compared to other extrapolations like the quadratic regression approach).

∘ *Parameters of the Heston model*: $s_0 = 100$, $k = 2$, $a = 0.01$, $\rho = 0.5$, $v_0 = 10\%$, $\vartheta = 20\%$.

∘ *Parameters of the option portfolio*: $T = 1$, $K = 99, \cdots, 111$ (13 strikes).

∘ *The reference price* has been computed by a 10^8 trial Monte Carlo simulation (including a time Romberg extrapolation of the Euler scheme with $2n = 256$).

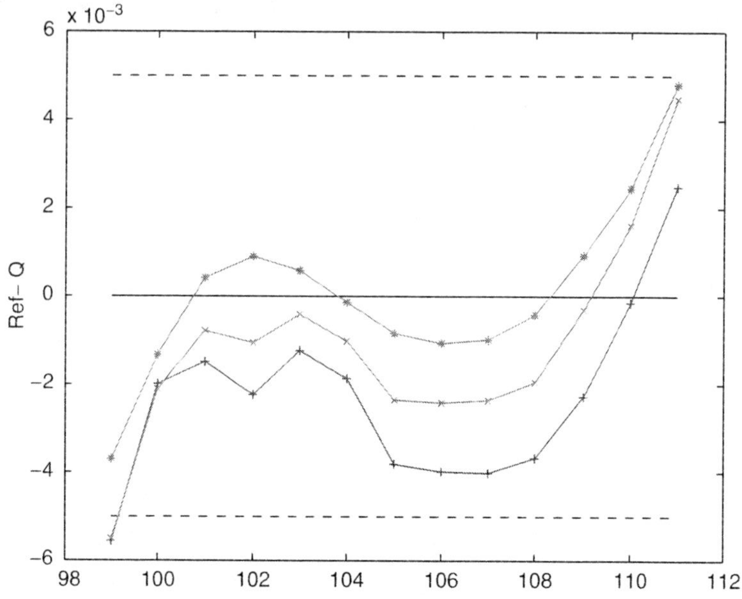

Fig. 11. *Quantized diffusions based on optimal functional quantization: Pricing by K-Interpolated-*log*-Romberg extrapolated-FQ prices as a function of K: absolute error with* $(N, M) = (400, 100)$, $(N, M) = (1000, 100)$, $(N, M) = (3200, 400)$. $T = 1$, $s_0 = 50$, $K \in \{99, \ldots, 111\}$. $k = 2$, $a = 0.01$, $\rho = 0.5$, $\vartheta = 0.1$.

∘ *The differential equations (18)* are solved with the parameters of the quantization cubature formulae $\Delta t = 1/32$, with couples of quantization levels $(N, M) = (400, 100)$, $(1000, 100)$, $(3200, 400)$.

Functional Quantization can compute a whole vector (more than 10) option premia for the Asian option in the Heston model with 1 **cent accuracy in less than** 1 **second** (implementation in C on a 2.5 GHz processor).

Further numerical tests carried out or in progress with the *B-S* model and with the *SABR* model (Asian, vanilla European options) show the same efficiency. Furthermore, recent attempt to quantize the volatility process and the asset dynamics at different level of quantizations seem very promising in two directions: reduction of the computation time and increase of the robustness of the method to parameter change.

9.3 Comparison: Optimized Quantization *vs* (Optimal) Product Quantization

The comparison is balanced and probably needs some further *in situ* experiments since it may depend on the modes of the computation. However, it seems that product quantizers (as those implemented in [PP05b]) are from 2 up to

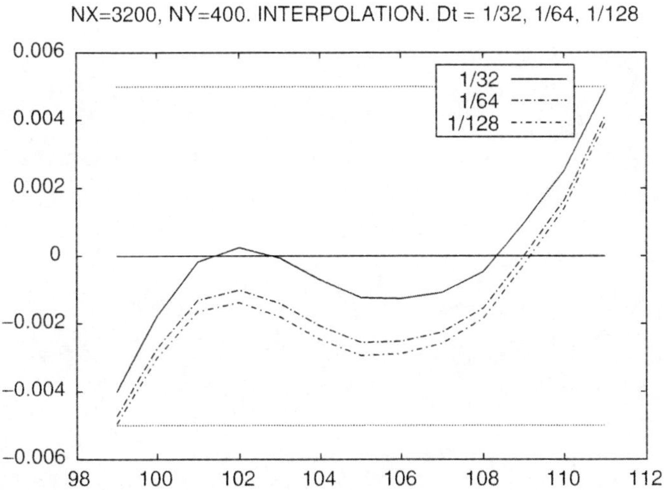

Fig. 12. *Quantized diffusions based on optimal functional quantization: Pricing by K-Interpolated-*log*-Romberg extrapolated-FQ price as a function of K: convergence as $\Delta t \to 0$ with $(N, M) = (3200, 400)$ (absolute error). $T = 1$, $s_0 = 50$, $K \in \{99, \ldots, 111\}$. $k = 2$, $a = 0.01$, $\rho = 0.5$, $\vartheta = 0.1$.*

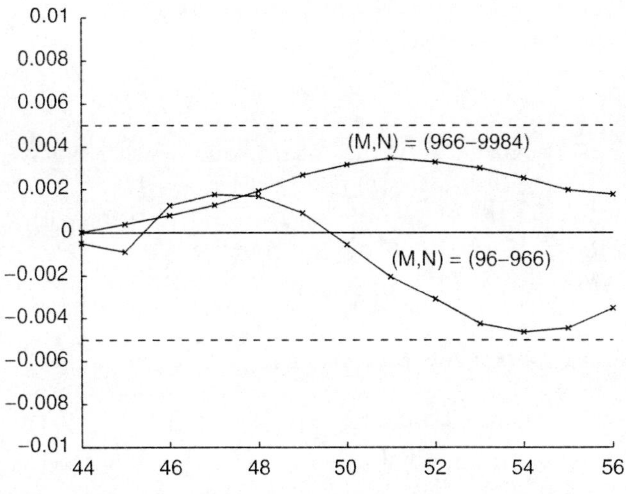

Fig. 13. *Quantized diffusions based on optimal product quantization: Pricing by K-linear interpolation of Romberg* log*-extrapolations as un function of K (absolute error) with $(M, N) = (96, 966)$, $(966, 9984)$. $T = 1$, $s_0 = 50$, $k = 2$, $a = 0.01$, $\rho = 0.5$, $\vartheta = 0.1$. $K \in \{44, \ldots, 56\}$.*

4 times less efficient than optimal quantizers within our range of application (small values of N). On the other hand, the design of product quantizer from 1-dim scalar quantizers is easy and can be made from some light elementary "bricks" (the scalar quantizer up to $N = 35$ and the optimal allocation rules).

Thus, the whole set of data needed to design all optimal product quantizers up to $N = 10\,000$ is approximately 500 KB whereas one optimal quantizer with size $10\,000 \approx 1$ MB...

10 Universal Quantization Rate and Mean Regularity

The following theorem points out the connection between functional quantization rate and mean regularity of $t \mapsto X_t$ from $[0,T]$ to $L^r(\mathbf{P})$.

Theorem 8. *([LP06a] (2005)) Let $X = (X_t)_{t\in[0,T]}$ be a stochastic process. If there is $r^* \in (0,\infty)$ and $a \in (0,1]$ such that*

$$X_0 \in L^{r^*}(\mathbf{P}), \quad \|X_t - X_s\|_{L^{r^*}(\mathbf{P})} \le C_X|t-s|^a,$$

for some positive real constant $C_X > 0$, then

$$\forall\, p, r \in (0, r^*), \quad e_{N,r}(X, L^p_T) = O((\log N)^{-a}).$$

The proof is based on a constructive approach which involves the Haar basis (instead of K-L basis), the non asymptotic version Zador Theorem and product functional quantization. Roughly speaking, we use the unconditionality of the Haar basis in every L^p_T (when $1<p<\infty$) and its wavelet feature *i.e.* its ability to "code" the path regularity of a function on the decay rate of its coordinates.

Examples (see [LP06a]): • d-dimensional Itô processes (includes d-dim diffusions with sublinear coefficients) with $a = 1/2$.
• General Lévy process X with Lévy measure ν with square integrable big jumps. If X has a Brownian component, then $a = 1/2$, otherwise if $\beta(X) > 0$ where $\beta(X) := \inf\{\theta : \int |y|^\theta \nu(dy) < +\infty\} \in (0,2)$ (Blumenthal-Getoor index of X), then $a = 1/\beta(X)$. This rate is the *exact rate i.e.*

$$e_{N,r}(X, L^p_T) \approx (\log N)^{-a}$$

for many classes of Lévy processes like symmetric stable processes, Lévy processes having a Brownian component, etc (see [LP06a] for further examples).
• When X is a compound Poisson processes, then $\beta(X) = 0$ and one shows, still with constructive methods, that

$$e_N(X) = O(e^{-(\log N)^\vartheta}), \qquad \vartheta \in (0,1),$$

which is in-between the finite and infinite dimensional settings.

11 About Lower Bounds

In this overview, we gave no clue toward lower bounds although most of the rates we mentioned are either exact ($\approx$) or sharp ($\sim$) (we tried to emphasize the numerical aspects). Several approaches can be developed to get some lower bounds. Historically, the first one was to rely on subadditivity property of the quantization error derived from self-similarity of the distribution: this works with the uniform distribution over $[0,1]^d$ but also in an infinite dimensional framework (see *e.g.* [DS06] for the fractional Brownian motion).

A second approach consists in pointing out the connection with the Shannon-Kolmogorov entropy (see *e.g.* [LP02]) using that the entropy of a random variable taking at most N values is at most $\log N$.

A third connection can be made with small deviation theory (see [DFMS03], [GLP03] and [LP06a]). Thus, in [GLP03], a connection is established between (functional) quantization and small ball deviation for Gaussian processes. In particular this approach provides a method to derive a lower bound for the quantization rate from some upper bound for the small deviation problem. A careful reading of the proof of Theorem 1.2 in [GLP03] shows that this small deviation lower bound holds for any *unimodal* (w.r.t. 0) non zero process. To be precise: assume that $\mathbf{P}_X$ is L^p_T-unimodal *i.e.* there exists a real $\varepsilon_0 > 0$ such that

$$\forall x \in L^p_T,\ \forall \varepsilon \in (0, \varepsilon_0], \qquad \mathbf{P}(|X - x|_{L^p_T} \leq \varepsilon) \leq \mathbf{P}(|X|_{L^p_T} \leq \varepsilon).$$

For centered Gaussian processes (or processes "subordinated" to Gaussian processes) this follows from the Anderson Inequality (when $p \geq 1$). If

$$G(-\log(\mathbf{P}(|X|_{L^p_T} \leq \varepsilon))) = \Omega(1/\varepsilon) \quad \text{as} \quad \varepsilon \to 0$$

for some increasing unbounded function $G : (0, \infty) \to (0, \infty)$, then

$$\forall c > 1, \quad \liminf_N G(\log(cN)) e_{N,r}(X, L^p_T) > 0, \qquad r \in (0, \infty). \tag{21}$$

This approach is efficient in the non quadratic case as emphasized in [LP06a] where several universal bounds are shown to be optimal using this approach.

Acknowledgment

I thank S. Graf, H. Luschgy J. Printems and B. Wilbertz for all the fruitful discussions and collaborations we have about functional quantization.

References

[AW82] E.F. Abaya and G.L. Wise. On the existence of optimal quantizers. *IEEE Trans. Inform. Theory*, **28**, 937–940, 1982.

[AW84] E.F. Abaya and G.L. Wise. Some remarks on the existence of optimal quantizers. *Statistics and Probab. Letters*, **2**, 349–351, 1984.

[BL94] N. Bouleau and D. Lépingle. *Numerical methods for stochastic processes*, Wiley Series in Probability and Mathematical Statistics: Applied Probability and Statistics. A Wiley-Interscience Publication. John Wiley & Sons, Inc., New York, 359 pp. ISBN: 0-471-54641-0, 1994.

[BMP90] A. Benveniste, M. Métivier, and P. Priouret. *Adaptive algorithms and stochastic approximations*, Translated from the French by Stephen S. Wilson. Applications of Mathematics **22**, Springer-Verlag, Berlin, 365 pp, 1990.

[BW82] J.A. Bucklew and G.L. Wise. Multidimensional asymptotic quantization theory with r^{th} power distortion. *IEEE Trans. Inform. Theory*, **28**(2), 239–247, 1982.

[CAM88] J.A. Cuesta-Albertos and C. Matrán. The strong law of large numbers for k-means and best possible nets of Banach valued random variables, *Probab. Theory Rel. Fields* **78**, 523–534, 1988.

[Coh98] P. Cohort. A geometric method for uniqueness of locally optimal quantizer. Pre-print LPMA-464 and Ph.D. Thesis, *Sur quelques problèmes de quantification*, 2000, Univ. Paris 6, 1998.

[Der05a] S. Dereich. The coding complexity of diffusion processes under supremum norm distortion, pre-print, 2005.

[Der05b] S. Dereich. The coding complexity of diffusion processes under $L^p[0,1]$-norm distortion, pre-print, 2005.

[DFMS03] S. Dereich, F. Fehringer, A. Matoussi, and M. Scheutzow. On the link between small ball probabilities and the quantization problem for Gaussian measures on Banach spaces, *J. Theoretical Probab.*, **16**, pp. 249–265, 2003.

[DFP04] S. Delattre, J.-C. Fort, and G. Pagès. Local distortion and μ-mass of the cells of one dimensional asymptotically optimal quantizers, *Communications in Statistics*, **33**(5), 1087–1118, 2004.

[Dos77] H. Doss. Liens entre équations différentielles stochastiques et ordinaires, *Ann. I.H.P.*, section B, **13**(2), 99–125, 1977.

[DS06] S. Dereich and M. Scheutzow. High resolution quantization and entropy coding for fractional Brownian motions, *Electron. J. Probab.*, **11**, 700–722, 2006.

[Fle64] P.E. Fleischer. Sufficient conditions for achieving minimum distortion in a quantizer. *IEEE Int. Conv. Rec.*, part I, 104–111, 1964.

[FP04] J.-C. Fort and G. Pagès. Asymptotics of optimal quantizers for some scalar distributions, *Journal of Computational and Applied Mathematics*, **146**, 253–275, 2002, 2004.

[GG92] A. Gersho and R.M. Gray. *Vector Quantization and Signal Compression.* Kluwer, Boston, 1992.

[GL00] S. Graf and H. Luschgy. *Foundations of Quantization for Probability Distributions.* Lect. Notes in Math. 1730, Springer, Berlin, 230 p, 2000.

[GL05] S. Graf and H. Luschgy. The point density measure in the quantization of self-similar probabilities. *Math. Proc. Cambridge Phil. Soc.*, **138**, 513–531, 2005.

[GLP03] S. Graf, H. Luschgy, and G. Pagès. Functional quantization and small ball probabilities for Gaussian processes, *J. Theoret. Probab.*, **16**(4), 1047–1062, 2003.

[GLP06] S. Graf, H. Luschgy, and G. Pagès. Distortion mismatch in the quantization of probability measures, to appear in *ESAIM P&S*, 2006.

[GLP07] S. Graf, H. Luschgy, and G. Pagès. Optimal quantizers for Radon random vectors in a Banach space, *J. of Approximation Theory*, **144**, 27–53, 2007.

[Hes93] S.L. Heston. A closed-form solution for options with stochastic volatility with applications to bond and currency options, *The review of Financial Studies*, **6**(2), 327–343, 1993.

[Kie82] J.C. Kieffer. Exponential rate of convergence for Lloyd's Method I, *IEEE Trans. Inform. Theory*, **28**(2), 205–210, 1982.

[Kie83] J.C. Kieffer. Uniqueness of locally optimal quantizer for log-concave density and convex error weighting functions, *IEEE Trans. Inform. Theory*, **29**, 42–47, 1983.

[KY03] H.J. Kushner and G.G. Yin. *Stochastic approximation and recursive algorithms and applications*. Second edition. Applications of Mathematics **35**. Stochastic Modelling and Applied Probability. Springer-Verlag, New York, 474 p, 2003.

[Lej03] A. Lejay. An introduction to rough paths, *Séminaire de Probabilités XXXVII*, Lecture Notes in Mathematics 1832, Stringer, Berlin, 1–59, 2003.

[LP02] H. Luschgy and G. Pagès. Functional quantization of Gaussian processes, *Journal of Functional Analysis*, **196**(2), 486–531, 2002.

[LP04] H. Luschgy and G. Pagès. Sharp asymptotics of the functional quantization problem for Gaussian processes, *The Annals of Probability*, **32**(2), 1574–1599, 2004.

[LP05] H. Luschgy and G. Pagès. High-resolution product quantization for Gaussian processes under sup-norm distortion, pre-pub LPMA-1029, forthcoming in *Bernoulli*, 2005.

[LP06a] H. Luschgy and G. Pagès. Functional Quantization Rate and mean regularity of processes with an application to Lévy Processes, pre-print LPMA-1048, 2006.

[LP06b] H. Luschgy and G. Pagès. Functional quantization of a class of Brownian diffusions: A constructive approach, *Stochastic Processes and Applications*, **116**, 310–336, 2006.

[LP07] H. Luschgy and G. Pagès. Expansion of Gaussian processes and Hilbert frames, technical report, 2007.

[LP96] D. Lamberton and G. Pagès. On the critical points of the 1-dimensional Competitive Learning Vector Quantization Algorithm. *Proceedings of the ESANN'96*, (ed. M. Verleysen), Editions D Facto, Bruxelles, 97–106, 1996.

[LPW07] H. Luschgy, G. Pagès, and B. Wilbertz. Asymptotically optimal quantization schemes for Gaussian processes, in progress, 2007.

[LSP90] B. Lapeyre, K. Sab, and G. Pagès. Sequences with low discrepancy. Generalization and application to Robbins-Monro algorithm, *Statistics*, **21**(2), 251–272, 1990.

[MBH06] M. Mrad and S. Ben Hamida. Optimal Quantization: Evolutionary Algorithm *vs* Stochastic Gradient, *Proceedings of the 9th Joint Conference on Information Sciences*, 2006.

[New82] D.J. Newman. The Hexagon Theorem. *IEEE Trans. Inform. Theory*, **28**, 137–138, 1982.

[Nie92] H. Niederreiter. *Random Number Generation and Quasi-Monte Carlo Methods*, CBMS-NSF regional conference series in Applied mathematics, SIAM, Philadelphia, 1992.

[Pag00] G. Pagès. Functional quantization: a first approach, pre-print CMP12-04-00, Univ. Paris 12, 2000.

[Pag93] G. Pagès. Voronoi tessellation, space quantization algorithm and numerical integration. *Proceedings of the ESANN'93*, M. Verleysen Ed., Editions D Facto, Bruxelles, 221–228, 1993.

[Pag97] G. Pagès. A space vector quantization method for numerical integration, *J. Computational and Applied Mathematics*, **89**, 1–38, 1997.

[Par90] K. Pärna. On the existence and weak convergence of k-centers in Banach spaces, *Tartu Ülikooli Toimetised*, **893**, 17–287, 1990.

[Pol82] D. Pollard. Quantization and the method of k-means. *IEEE Trans. Inform. Theory*, **28**(2), 199–205, 1982.

[PP03] G. Pagès and J. Printems. Optimal quadratic quantization for numerics: the Gaussian case, *Monte Carlo Methods and Appl.*, **9**(2), 135–165, 2003.

[PP05a] G. Pagès and J. Printems. Website devoted to vector and functional optimal quantization: `www.quantize.maths-fi.com`, 2005.

[PP05b] G. Pagès and J. Printems. Functional quantization for numerics with an application to option pricing, *Monte Carlo Methods and Appl.*, **11**(4), 407–446, 2005.

[PPP03] G. Pagès, H. Pham, and J. Printems. Optimal quantization methods and applications to numerical methods in finance. *Handbook of Computational and Numerical Methods in Finance*, S.T. Rachev ed., Birkhäuser, Boston, 429 p, 2003.

[Pro88] P.D. Proinov. Discrepancy and integration of continuous functions, *J. of Approx. Theory*, **52**, 121–131, 1988.

[PS07] G. Pagès and A. Sellami. Convergence of multi-dimensional quantized SDE's. In progress, 2007.

[PX88] G. Pagès and Y.J. Xiao. Sequences with low discrepancy and pseudo-random numbers: theoretical results and numerical tests, *J. of Statist. Comput. Simul.*, **56**, 163–188, 1988.

[Rot54] K.F. Roth. On irregularities of distributions, *Mathematika*, **1**, 73–79, 1954.

[RY99] D. Revuz and M. Yor. *Continuous martingales and Brownian motion*, Third edition. Grundlehren der Mathematischen Wissenschaften [Fundamental Principles of Mathematical Sciences], 293, Springer-Verlag, Berlin, 1999, 602 p, 1999.

[TK03] T. Tarpey and K.K.J. Kinateder. Clustering functional data, *J. Classification*, **20**, 93–114, 2003.

[TPO03] T. Tarpey, E. Petkova, and R.T. Ogden. Profiling Placebo responders by self-consistent partitioning of functional data, *J. Amer. Statist. Association*, **98**, 850–858, 2003.

[Tru82] A.V. Trushkin. Sufficient conditions for uniqueness of a locally optimal quantizer for a class of convex error weighting functions, *IEEE Trans. Inform. Theory*, **28**(2), 187–198, 1982.

[Wil05] B. Wilbertz. Computational aspects of functional quantization for Gaussian measures and applications, diploma thesis, Univ. Trier, 2005.

[Zad63] P.L. Zador. Development and evaluation of procedures for quantizing multivariate distributions. Ph.D. dissertation, Stanford Univ, 1963.

[Zad82] P.L. Zador. Asymptotic quantization error of continuous signals and the quantization dimension. *IEEE Trans. Inform. Theory*, **28**(2), 139–149, 1982.

Random Field Simulation and Applications

Karl Sabelfeld

Weierstrass Institute for Applied Analysis and Stochastics, Mohrenstrasse 39, 10117 Berlin, Germany, and
Institute of Computational Mathematics and Mathem. Geophysics, Russian Acad. Sci., Lavrentieva str., 6, 630090 Novosibirsk, Russia
sabelfel@wias-berlin.de

Summary. In this paper I present some new approaches to the random field simulation, and show in four different examples how this simulation technique works. The first example deals with a transport in turbulent flows, where the Lagrangian trajectories are described by a stochastic differential equation whose drift term involves the Eulerian velocity as a random field with a given spectral tensor. Studies of the second example concern with the flows in porous medium governed by the Darcy equation with random hydraulic conductivity. Elasticity system of elliptic Lamé equations with random loads is considered in the third example. Finally, in the fourth example we solve a nonlinear Smoluchowski equation which is used to model the process of crystal growth.

1 Introduction

Stochastic approach becomes more and more popular in all branches of science and technology, especially in problems where the data are highly irregular (in deterministic sense). In such problems it is very difficult and expensive to carry out measurements to extract the desired data. As important examples we mention the turbulent flow simulation [MY81]), and construction of flows through porous media [Gel93], [Dag89]. The temporal and spatial scales of the input parameters in this class of problems are varying enormously, and the behaviour is very complicated, so that there is no chance to describe it deterministically. In the stochastic approach, one needs to know a few number of parameters, like the mean and correlation tensor, whose behaviour in time and space is much more regular, so that usually, it is easier to extract them through measurements.

In most applications, it is assumed that the random fields are Gaussian, or that they can be obtained by a functional transformation of Gaussian fields. Generally, it is very difficult to construct efficient simulation methods for inhomogeneous random fields even if they are Gaussian. Therefore, the most developed methods deal with homogeneous or quasi-homogeneous random fields, i.e., the characteristic scales of the variations of the means of the field are considerably larger than the correlation scale. There are highly intensive studies and literature concerned with the simulation of homogeneous and quasi-homogeneous random fields. We have no intension to give here a detailed overview of the simulation methods even for homogeneous random fields, we just refer to the book [Sab91], and to some important papers we used [Mik83], [PS95], [Shi71], [EM94], [EM95], [SRR01], as well as to our recent paper [KS06] which includes an overview in this field.

In many practical problems (e.g., in underground hydrology, see [Dag89], [Gel93]) only data obtained through spatial averaging is at hand, for instance, statistical characteristics obtained by spatial averages, or over a family of Lagrangian trajectories generated in one fixed sample of the field (e.g., see [Dag90], [DFJ03]). If the random field is ergodic (which in practice is very often true), then the ensemble averages can be well approximated by the appropriate space averages. This is very important when a boundary value problem with random parameters is solved: then in contrast to the ensemble averaging, we have to solve the problem only once, and then make the relevant space averaging. In practical calculations, to increase the efficiency, it is sometimes reasonable to combine both the space and ensemble averaging, e.g., see [KSSV03], [KS05].

The paper is organized as follows. In section 2 we give a bit more detailed description of the four problems we mentioned above in the Abstracts. In section 3 we describe the simulation methods based on the spectral and Fourier-wavelet representations. Section 4 deals with the important case when the fluctuations are small, and the method of small perturbations can be applied. To find out the applicability limits of this method, we develop a general method which works in the general case of large fluctuations. Finally, in Section 5 we discuss a technique which we call a Double Randomization Method.

2 Four Examples of Random Field Applications

Random fields provide a useful mathematical framework for representing disordered heterogeneous media in theoretical and computational studies. Here we give four different examples where the random fields serve as natural models for the relevant processes.

Example 1: Fully developed turbulent flows. The velocity field representing the turbulent flow is modelled as a random field $\mathbf{v}(\mathbf{x},t)$ with statistics encoding important empirical features, and the temporal dynamics of the position $\mathbf{X}(t)$ and velocity $\mathbf{V}(t) = \frac{d\mathbf{X}}{dt}$ of immersed particles is then governed by

$$m\,\mathrm{d}\mathbf{V}(t) = -\gamma\Big(\mathbf{V}(t) - \mathbf{v}(\mathbf{X}(t),t)\Big)\,\mathrm{d}t + \sqrt{2k_{\mathrm{B}}T\gamma}\,\mathrm{d}\mathbf{W}(t),$$

where m is particle mass, γ is its friction coefficient, k_{B} is Boltzmann's constant, T is the absolute temperature, and $\mathbf{W}(t)$ is a random Wiener process representing molecular collisions. We mention here that it was Kolmogorov (e.g., see details in [MY81]) who has developed an elegant stochastic theory of the fully developed turbulence.

Example 2: Transport through porous media, such as groundwater aquifers, in which the hydraulic conductivity $\mathsf{K}(\mathbf{x})$ is modelled as random field reflecting the empirical variability of the porous medium. The Darcy flow rate $\mathbf{q}(\mathbf{x})$ in response to pressure applied at the boundary is governed by the Darcy equation

$$\mathbf{q}(\mathbf{x}) = -\mathsf{K}(\mathbf{x})\,\mathrm{grad}\,\phi(\mathbf{x}),$$
$$\mathrm{div}\,\mathbf{q} = 0.$$

Of course, this equation is solved under some boundary conditions, which automatically implies that the solution cannot be a homogeneous random field. But naturally assuming that the influence of the correlations decreases with the distance, it is quite plausible to expect that in 3-4 correlation lengths far from the boundary the solution can be considered as approximately homogeneous. This hypothesis is accepted by many authors, see for instance [Dag90], [Gel93], and confirmed in many calculations, see for example our recent paper [KS05].

Example 3: Elasticity problems. One has to solve the Lamé equation

$$\mu\,\Delta\mathbf{u} + (\lambda + \mu)\,\mathrm{grad}\,\mathrm{div}\,\mathbf{u} = \mathbf{f}$$

where $\mathbf{u}$ is the displacement vector, and μ and λ are the elastic Lamé coefficients, and $\mathbf{f}$ is a random vector load.

Example 4: Smoluchowski coagulation equations. Our fourth example deals with a nonlinear coagulation dynamics. Smoluchowski equation describes the size spectrum of particles which collide pairwise with frequencies proportional to the kernel of this equation, and grow due to aggregation of the colliding particles. This equation has a nice probabilistic interpretation and the relevant stochastic simulation methods are well developed (e.g., see [KS03b], [SLP07]. We are interested in the case when the kernel of the Smoluchowski equation is random. Such an example was considered in our paper [KS00] where we studied the influence of the intermittency in the process of turbulent coagulation regime.

Here we consider another example related to growth of atom islands [KPS06]. In the processes of crystal growth, the kinetics of an ensemble of atom islands that diffuse on the surface and irreversibly merge as they touch each other can be described by the set of Smoluchowski equations

$$\frac{dn_k}{dt} = \frac{1}{2}\sum_{i+j=k} K_{ij}n_i n_j - n_k \sum_{j=1}^{\infty} K_{jk}n_j.$$

Here n_j is the number of islands containing j units (atoms, or vacancies of atoms) per unit area.

The kernels describing the frequency of atom collisions depend on the sizes of colliding particles, but also on the underlying surface. We consider the so-called diffusion regime, where the atom diffusion motion has a finite correlation length. It means, for different surfaces, with different correlation lengths, the dispersion of the atoms will be different, see illustration of this situation in Figure 1 where a kind of clustering can be seen in the right panel. So we deal here with the Smoluchowski equation with random kernel.

What should be calculated in these four examples?

In the first example, important characteristics is $\langle c(x) \rangle$, the average concentration of particles which is in fact a one-particle statistics. More complicated is the fluctuation of concentration, which is a two-particle statistics related to the mean square separation $\langle \rho^2(t) \rangle$.

Take two particles initially separated by say r_0, and moving in a homogeneous gaussian random velocity field. For illustration, we show in Figure 2 a sample of the vector velocity field, when the correlation length is equal to 1 (left panel) and to 0.5 (right panel). The question is how it behaves $\langle \rho^2(t) \rangle$ as a function of time? This is an important function which is used to describe the mean square size of a diffusing cloud of particles (e.g., see [MY81], [TD05], [KS05]).

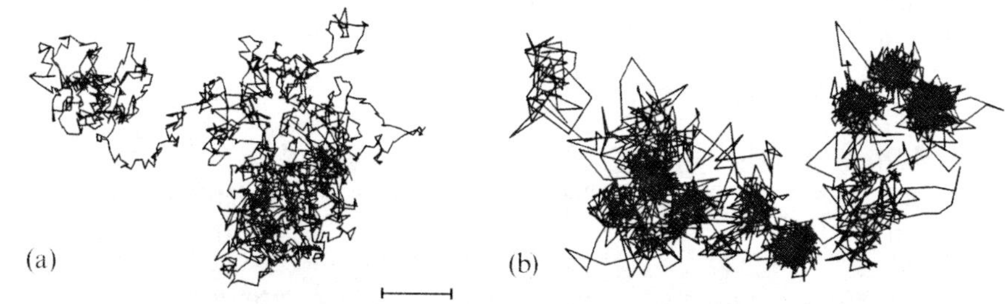

Fig. 1. Atom diffusion on different surfaces

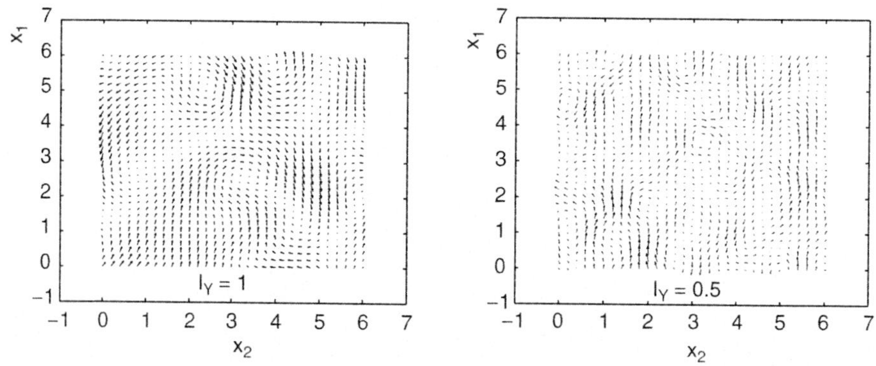

Fig. 2. Samples of an incompressible gaussian isotropic 2D random field, for a correlation length equal to 1 (left panel) and 0.5 (right panel)

In the second example, for flows in porous media: even the simplest statistical characteristics, the mean flow, is a non-trivial function to be calculated since you cannot simply average the equation with the random coefficient. More general Lagrangian statistical characteristics are necessary, to evaluate the mean concentration and its flux. For tracking the trajectories in the extremely heterogeneous velocity field, one needs a multiscale resolution through a stochastic synthesis of random fields (e.g., see [EM95], [KKS07]). In Figure 3 we show a sample of the hydraulic conductivity.

Elasticity problem. Standard statistical characteristics are the mean displacements $\langle u_i \rangle$, the second moments $\langle u_i^2 \rangle$, and the probability that the displacements exceed some critical value: $Prob(u_i > u_{cr})$.

The same statistics are evaluated for the strain and stress tensors

$$\varepsilon_{ij} = (u_{i,j} + u_{j,i})/2, \quad \tau_{ij} = 2\mu\varepsilon_{ij} + \lambda\delta_{ij}\mathrm{div}\mathbf{u}.$$

Smoluchowski coagulation equation. In the general spatially inhomogeneous case when the colliding particles are in a host flow, the governing equations are [KS03b]:

$$\begin{aligned} d\mathbf{X}(t) &= \mathbf{V}(t)dt, \\ dV_i(t) &= a_i(\mathbf{X}(t), \mathbf{V}(t))dt + \sigma_{ij}(\mathbf{X}(t), t)dB_j(t) \\ \frac{dN_l}{dt} &= \frac{1}{2}\sum_{i+j=l} K_{ij}N_iN_j - \sum_{i\geq 1} K_{li}N_lN_i. \end{aligned}$$

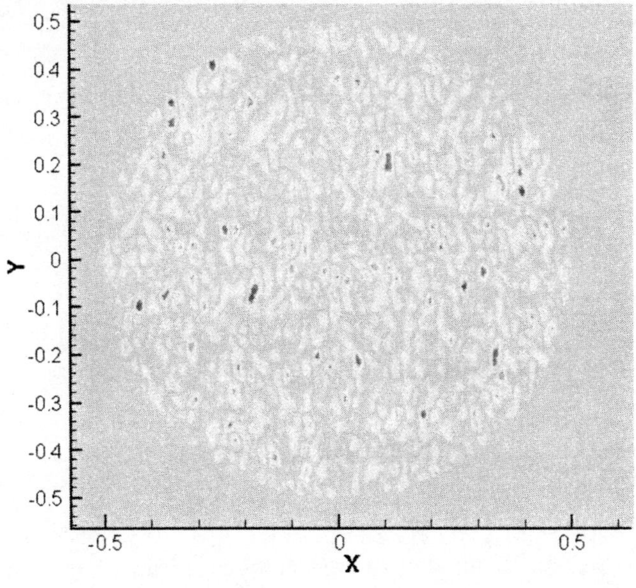

Fig. 3. A hydraulic conductivity modelled as a lognormal random field

Important quantities are the average of the mean size $\langle \bar{n}(t) \rangle$, as well as the average size spectrum $\langle n_k(t) \rangle$. More complicated functionals: what are the average and variance of the random time when the solution is exploded? (the so-called gelation phenomenon, e.g., see [Wag06]).

3 Random Field Simulation Methods

Under quite general conditions, a real-valued Gaussian homogenous random field $u(\mathbf{x})$ can be represented through a stochastic Fourier integral [MY81]:

$$u(\mathbf{x}) = \int_{R^d} \mathrm{e}^{2\pi \mathrm{i}\mathbf{k}\cdot\mathbf{x}} E^{1/2}(\mathbf{k}) \tilde{W}(\mathrm{d}\mathbf{k})$$

where $\tilde{W}(\mathrm{d}\mathbf{k})$ is a complex-valued white noise random measure on $\boldsymbol{R}^d$, with $\tilde{W}(B) = \overline{\tilde{W}(-B)}$, $\langle \tilde{W}(B) \rangle = 0$, and $\langle \tilde{W}(B)\overline{\tilde{W}(B')} \rangle = \mu(B \cap B')$ for Lebesgue measure μ and all Lebesgue-measurable sets B, B'. The spectral density $E(\mathbf{k})$ is a nonnegative even function representing the strength (energy) of the random field associated to the wavenumber $\mathbf{k}$, meaning the length scale $1/|\mathbf{k}|$ and direction $\mathbf{k}/|\mathbf{k}|$.

Multiscale random fields will have a multiscale spectral density, meaning that $E(\mathbf{k})$ will have substantial contributions over a wide range of wavenumbers $k_{min} \ll |\mathbf{k}| \ll k_{max}$, with $k_{max}/k_{min} \gg 1$. This poses a challenge for efficient simulation.

More generally, we deal with real-valued homogeneous Gaussian l-dimensional vector random fields $\mathbf{u}(\mathbf{x}) = (u_1(\mathbf{x}), \ldots, u_l(\mathbf{x}))^T$, $\mathbf{x} \in \boldsymbol{R}^d$ with a given correlation tensor $B(\mathbf{r})$:

$$B_{ij}(\mathbf{r}) = \langle u_i(\mathbf{x}+\mathbf{r})\, u_j(\mathbf{x}) \rangle, \quad i,j = 1, \ldots l,$$

or with the corresponding spectral tensor F:

$$F_{ij}(\mathbf{k}) = \int_{R^d} e^{-\mathrm{i}\,2\pi\,\mathbf{k}\cdot\mathbf{r}} B_{ij}(\mathbf{r})\, d\mathbf{r}, \quad B_{ij}(\mathbf{r}) = \int_{R^d} e^{\mathrm{i}\,2\pi\,\mathbf{r}\cdot\mathbf{k}} F_{ij}(\mathbf{k})\, d\mathbf{k}, \quad i,j = 1, \ldots l.$$

We assume that the condition $\int_{R^d} |B_{jj}(\mathbf{r})|\, d\mathbf{r} < \infty$ is satisfied which ensures that the spectral functions F_{ij} are uniformly continuous with respect to $\mathbf{k}$. Here B_{jj} is the trace of B.

Let $Q(\mathbf{k})$ be an $l \times n$-matrix defined by

$$Q(\mathbf{k})Q^*(\mathbf{k}) = F(\mathbf{k}), \quad Q(-\mathbf{k}) = \bar{Q}(\mathbf{k}).$$

Here the star stands for the complex conjugate transpose which is equivalent to taking two operations, the transpose^T, and the complex conjugation of each entry.

Then the spectral representation of the random field is written as follows

$$\mathbf{u}(\mathbf{x}) = \int\limits_{R^d} e^{\mathrm{i}\,2\pi\,\mathbf{kx}}\, Q(\mathbf{k})\, \mathbf{Z}(d\mathbf{k})$$

where the column-vector $\mathbf{Z} = (Z_1, \dots Z_n)^T$ is a complex-valued homogeneous n-dimensional white noise on $\boldsymbol{R}^d$ with a unite variance and zero mean:

$$\langle \mathbf{Z}(d\mathbf{k}) \rangle = 0, \quad \langle Z_i(d\mathbf{k}_1)\, \bar{Z}_j(d\mathbf{k}_2) \rangle = \delta_{ij}\, \delta(\mathbf{k}_1 - \mathbf{k}_2)\, d\mathbf{k}_1\, d\mathbf{k}_2$$

satisfying the condition $\mathbf{Z}(-d\mathbf{k}) = \bar{\mathbf{Z}}(d\mathbf{k})$.

Series expansions.

The random field is constructed in the form

$$\mathbf{u}(\mathbf{x}) = \sum_{\alpha \in \mathcal{A}} G_\alpha(\mathbf{x})\, \xi_\alpha$$

where $G_\alpha(\mathbf{x})$ is a system of deterministic functions (or possibly matrices), ξ_α is a family of random variables (possibly vectors) $\mathcal{A}$ is a countable (finite or not) index set.

Our purpose is to construct the system G_α and the family ξ_α so that the random field has the desired spectral tensor.

Let us choose the system of scalar functions $\varphi_\alpha(\mathbf{k})$ as a set of generally complex valued even functions ($\varphi_\alpha(-\mathbf{k}) = \bar{\varphi}_\alpha(\mathbf{k})$) which are orthonormal and complete in $L_2(\boldsymbol{R}^d)$ equipped with the scalar product $(f, g) = \int_{\boldsymbol{R}^d} f(\mathbf{k})\bar{g}(\mathbf{k})\, d\mathbf{k}$:

$$(\varphi_\alpha, \varphi_\beta) = \int_{R^d} \varphi_\alpha(\mathbf{k})\bar{\varphi}_\beta(\mathbf{k})\, d\mathbf{k} = \delta_{\alpha\beta}, \quad \alpha, \beta \in \mathcal{A},$$

where $\delta_{\alpha\beta}$ is the Kronecker symbol.

We expand $e^{\mathrm{i}\,2\pi\,\mathbf{k}\cdot\mathbf{x}}\, Q(\mathbf{k})$ as a function of $\mathbf{k}$ in the system of orthonormal functions $\varphi_\alpha(\mathbf{k})$:

$$e^{\mathrm{i}\,2\pi\,\mathbf{k}\cdot\mathbf{x}}\, Q(\mathbf{k}) = \sum_{\alpha \in \mathcal{A}} G_\alpha(\mathbf{x})\, \varphi_\alpha(\mathbf{k}), \quad G_\alpha(\mathbf{x}) = \int\limits_{R^d} e^{\mathrm{i}\,2\pi\,\mathbf{kx}} Q(\mathbf{k})\bar{\varphi}_\alpha(\mathbf{k})\, d\mathbf{k}.$$

We now substitute this into the spectral representation, and obtain the expansion with

$$\boldsymbol{\xi}_\alpha = \int\limits_{R^d} \varphi_\alpha(\mathbf{k})\, \mathbf{Z}(d\mathbf{k}), \quad \alpha \in \mathcal{A}.$$

Notice that $\boldsymbol{\xi}_\alpha$ are mutually independent standard Gaussian random vectors since

$$\langle \boldsymbol{\xi}_\alpha\, \boldsymbol{\xi}_\beta^* \rangle = \mathrm{I} \int\limits_{R^d} \varphi_\alpha(\mathbf{k})\bar{\varphi}_\beta(\mathbf{k})\, d\mathbf{k} = \mathrm{I}\, \delta_{\alpha\beta},$$

where I is a $n \times n$ identity matrix. Thus we have constructed an expansion with independent Gaussian random vectors.

Fourier-wavelet expansions for a homogeneous Gaussian vector random process $\mathbf{u}(x) = (u_1(x), \ldots, u_l(x))^T$, $x \in \boldsymbol{R}$ *with a given spectral tensor* $F(k)$.

We assume that $F = QQ^*$, where $Q(\mathbf{k})$ is $l \times n$–dimensional matrix satisfying the condition $Q(-k) = \bar{Q}(k)$. The orthonormal system of functions φ_α is constructed as follows. Let $\phi(x)$ and $\psi(x)$, $x \in \boldsymbol{R}$ be orthonormal scaling and wavelet functions, respectively, and

$$\phi_{mj}(x) = 2^{m/2}\phi(2^m x - j), \quad \psi_{mj}(x) = 2^{m/2}\psi(2^m x - j),$$

where $m, j = \ldots, -2, -1, 0, 1, 2, \ldots$. It is known (e.g., see [Chu92], [Mey90] that the system of functions

$$\{\phi_{m_0 j}\}_{j=-\infty}^{\infty}, \quad \left\{ \{\psi_{mj}\}_{j=-\infty}^{\infty}, \quad m \geq m_0 \right\}$$

is, for an arbitrary fixed integer m_0, a complete set of orthonormal functions in $L_2(\boldsymbol{R})$, and moreover, by Parseval equality, the relevant Fourier transforms of these functions

$$\{\hat{\phi}_{m_0 j}\}_{j=-\infty}^{\infty}, \quad \left\{ \{\hat{\psi}_{mj}\}_{j=-\infty}^{\infty}, \quad m \geq m_0 \right\}$$

compose also a complete set of orthonormal functions in $L_2(\boldsymbol{R})$.

Thus we choose the family φ_α as described above.

We find that

$$\mathbf{u}(x) = \sum_{j=-\infty}^{\infty} G_{m_0 j}^{(\phi)}(x)\,\boldsymbol{\xi}_j + \sum_{m=m_0}^{\infty} \sum_{j=-\infty}^{\infty} G_{mj}^{(\psi)}(x)\boldsymbol{\xi}_{mj},$$

where $\boldsymbol{\xi}_j$, $\boldsymbol{\xi}_{mj}$ is a family of mutually independent standard real valued Gaussian random vectors of dimension n, and $G_{mj}^{(\phi)}(x)$, $G_{mj}^{(\psi)}(x)$ are $l \times n$-dimensional matrices defined by

$$G_{mj}^{(\phi)}(x) = \int_{-\infty}^{\infty} e^{\mathrm{i}\,2\pi\,kx} Q(k)\,\bar{\hat{\phi}}_{mj}(k)\,dk, \quad G_{mj}^{(\psi)}(x) = \int_{-\infty}^{\infty} e^{\mathrm{i}\,2\pi\,kx} Q(k)\,\bar{\hat{\psi}}_{mj}(k)\,dk.$$

It is clear that

$$\hat{\phi}_{mj}(k) = 2^{-m/2}\,e^{-\mathrm{i}\,2\pi\,k\,j\,2^{-m}}\,\hat{\phi}(2^{-m}\,k),$$
$$\hat{\psi}_{mj}(k) = 2^{-m/2}\,e^{-\mathrm{i}\,2\pi\,k\,j\,2^{-m}}\,\hat{\psi}(2^{-m}\,k).$$

Now we can define the analog of G_α by substituting $\bar{\varphi}_\alpha(k)$

$$G^{(\phi)}_{mj}(x) = \int_{-\infty}^{\infty} e^{\mathrm{i}\,2\pi\, kx} Q(k)\,\hat{\phi}_{mj}(-k)\,dk = \int_{-\infty}^{\infty} e^{-\mathrm{i}\,2\pi\, kx}\bar{Q}(k)\,\hat{\phi}_{mj}(k)\,dk$$

$$= \int_{-\infty}^{\infty} e^{-\mathrm{i}\,2\pi\, kx}\bar{Q}(k)\,2^{-m/2}\,e^{-\mathrm{i}\,2\pi\, k\, j\, 2^{-m}}\,\hat{\phi}(2^{-m}\,k)\,dk$$

$$= \int_{-\infty}^{\infty} e^{-\mathrm{i}\,2\pi\, k'(2^m x+j)}\,2^{m/2}\bar{Q}(2^m k')\hat{\phi}(k')\,dk'.$$

Analogously,

$$G^{(\psi)}_{mj}(x) = \int_{-\infty}^{\infty} e^{-\mathrm{i}\,2\pi\, k'(2^m x+j)}\,2^{m/2}\bar{Q}(2^m k')\hat{\psi}(k')\,dk'.$$

For convenience, we define

$$\mathcal{F}^{(\phi)}_m(y) = \int_{-\infty}^{\infty} e^{-\mathrm{i}\,2\pi\, ky}\,2^{m/2}\bar{Q}(2^m k)\hat{\phi}(k)\,dk,$$

$$\mathcal{F}^{(\psi)}_m(y) = \int_{-\infty}^{\infty} e^{-\mathrm{i}\,2\pi\, ky}\,2^{m/2}\bar{Q}(2^m k)\hat{\psi}(k)\,dk,$$

hence

$$G^{(\phi)}_{m_0 j}(x) = \mathcal{F}^{(\phi)}_{m_0}(2^{m_0}x+j), \quad G^{(\psi)}_{mj}(x) = \mathcal{F}^{(\psi)}_m(2^m x+j),$$

and finally,

$$\mathbf{u}(x) = \sum_{j=-\infty}^{\infty} \mathcal{F}^{(\phi)}_{m_0}(2^{m_0}x+j)\,\boldsymbol{\xi}_j + \sum_{m=m_0}^{\infty}\sum_{j=-\infty}^{\infty} \mathcal{F}^{(\psi)}_m(2^m x+j)\,\boldsymbol{\xi}_{mj}.$$

Let us consider an l-dimensional random field $\mathbf{u}(\mathbf{x})$, $\mathbf{x} \in \boldsymbol{R}^d$ defined by the stochastic integral:

$$\mathbf{u}(\mathbf{x}) = \int_{R^d} \mathcal{H}(\mathbf{x},\mathbf{k})\,\mathbf{W}(d\mathbf{k}),$$

where (1) $\mathcal{H} : \boldsymbol{R}^d \times \boldsymbol{R}^{d_1} \to \mathcal{C}^{l\times n}$ is a matrix such that $\mathcal{H}(\mathbf{x},\cdot) \in L_2(\boldsymbol{R}^{d_1})$ for each $\mathbf{x} \in \boldsymbol{R}^d$; (2) $\mathbf{W}(\cdot) = \mathbf{W}_R(\cdot) + \mathrm{i}\,\mathbf{W}_I(\cdot)$, where $\mathbf{W}_R(\cdot)$ and $\mathbf{W}_I(\cdot)$ are two independent $n-$dimensional homogeneous Gaussian white noises on $\boldsymbol{R}^{d_1}$ with unit variance. Here $\mathcal{C}$ is the set of complex numbers.

Let us describe the randomized evaluation of the stochastic integral. Let $p : \boldsymbol{R}^{d_1} \to [0,\infty)$ be a probability density on $\boldsymbol{R}^{d_1}$: $\int p(\mathbf{k})\,d\mathbf{k} = 1$, and let

$\mathbf{k}_1, \ldots, \mathbf{k}_{n_0}$ be independent equally distributed random points in $\boldsymbol{R}^{d_1}$ with the density $p(\mathbf{k})$. Assume that $\boldsymbol{\zeta}_1, \ldots, \boldsymbol{\zeta}_{n_0}$ is a family of mutually independent standard Gaussian complex random vectors of dimension n (i.e., $\boldsymbol{\zeta}_j = \boldsymbol{\xi}_j + \mathrm{i}\,\boldsymbol{\eta}_j$ with independent, $n-$dimensional real valued standard Gaussian random vectors $\boldsymbol{\xi}_i$ and $\boldsymbol{\eta}_i$). Then the random field

$$\mathbf{u}_{n_0}(\mathbf{x}) = \frac{1}{\sqrt{n_0}} \sum_{j=1}^{n_0} \frac{1}{\sqrt{p(\mathbf{k}_j)}} \mathcal{H}(\mathbf{x}, \mathbf{k}_j) \boldsymbol{\zeta}_j$$

has the same correlation tensor as $\mathbf{u}(\mathbf{x})$, provided $p(\mathbf{k})$ satisfies the condition

$$p(\mathbf{k}) > 0, \quad if \quad \exists \mathbf{x} \in \boldsymbol{R}^d : \mathcal{H}(\mathbf{x}, \mathbf{k}) \neq 0.$$

In practical calculations, to guarantee an equal presentation of different spectral regions, one uses a stratified randomization technique. Let us describe it briefly. Let $\{\Delta_i\}_{i=1}^N$ be a subdivision of the spectral space $\boldsymbol{R}^{d_1}$: $\boldsymbol{R}^{d_1} = \cup_{i=1}^N \Delta_i$, and $\Delta_i \cap \Delta_j = \emptyset$ if $i \neq j$. This generates the representation of the random field $\mathbf{u}(\mathbf{x})$ as a sum of independent random fields:

$$\mathbf{u}(\mathbf{x}) = \sum_{i=1}^N \mathbf{u}_i(\mathbf{x}), \quad \mathbf{u}_i(\mathbf{x}) = \int_{\Delta_i} \mathcal{H}(\mathbf{x}, \mathbf{k})\, \mathbf{W}(d\mathbf{k}).$$

Let $p_i : \Delta_i \to [0, \infty)$ $(i = 1, \ldots, N)$ be a probability density on Δ_i: $\int_{\Delta_i} p(\mathbf{k})\, d\mathbf{k} = 1$ satisfying the condition $p_i(\mathbf{k}) > 0$, for $\mathbf{k} \in \Delta_i$ if $\exists \mathbf{x} \in \boldsymbol{R}^d : \mathcal{H}(\mathbf{x}, \mathbf{k}) \neq 0$. Then using the randomized representation for $\mathbf{u}_i(\mathbf{x})$ we get a stratified randomization model for $\mathbf{u}(\mathbf{x})$:

$$\mathbf{u}_{N,n_0}(\mathbf{x}) = \sum_{i=1}^N \frac{1}{\sqrt{n_0}} \sum_{j=1}^{n_0} \frac{1}{\sqrt{p_i(\mathbf{k}_{ij})}} \mathcal{H}(\mathbf{x}, \mathbf{k}_{ij}) \boldsymbol{\zeta}_{ij}.$$

Here $\{\mathbf{k}_{ij}\}_{j=1}^{n_0} \subset \Delta_i$, $i = 1, \ldots, N$ are mutually independent random points such that for fixed i the random points $\mathbf{k}_{ij}$, $j = 1, \ldots$ are all distributed with the same density $p_i(\mathbf{k})$, and $\boldsymbol{\zeta}_{ij}$, $i = 1, \ldots, N$; $j = 1, \ldots, n_0$ are mutually independent, and independent of $\{\mathbf{k}_{ij}\}_{j=1}^{n_0}$ $i = 1, \ldots, N$ family of $n-$dimensional complex valued standard Gaussian random variables.

By the construction, for any N and n_0, the random field $\mathbf{u}_{N,n_0}(\mathbf{x})$ has the same correlation tensor as that of $\mathbf{u}(\mathbf{x})$. As the Central Limit Theorem says, by increasing n_0 (N fixed) the field $\mathbf{u}_{N,n_0}(\mathbf{x})$ is convergent to a gaussian random field. So the stratified randomization model $\mathbf{u}_{N,n_0}(\mathbf{x})$ can be considered as an approximation to $\mathbf{u}(\mathbf{x})$. More details about the convergence of this type of models can be found in [Kur95].

Now we present the simulation formulae for the important case when the vector random field is isotropic.

Isotropic random fields

A homogeneous d-dimensional vector-valued random field $\mathbf{u}(x)$, $\mathbf{x} \in \boldsymbol{R}^d$ is called isotropic if the random field $U^T \mathbf{u}(U\mathbf{x})$ has the same finite-dimensional distributions as those of the random field $\mathbf{u}(\mathbf{x})$ for any rotation matrix $U \in SO(d)$ [MY81]. The spectral density tensor of an isotropic random field has the following general structure [MY81]:

$$\mathsf{F}(\mathbf{k}) = \frac{1}{2A_d\, k^{d-1}} \left\{ E_1(k)\mathsf{P}^{(1)}(\mathbf{k}) + E_2(k)\mathsf{P}^{(2)}(\mathbf{k}) \right\}$$

where $k = |\mathbf{k}|$, A_d is the area of the unit sphere in $\boldsymbol{R}^d$, E_1 and E_2 are the transverse and longitudinal radial spectra (scalar even nonnegative functions), and the projection tensors are defined componentwise as:

$$P_{ij}^{(1)}(\mathbf{k}) = \delta_{ij} - \frac{k_i k_j}{k^2}, \quad P_{ij}^{(2)}(\mathbf{k}) = \frac{k_i k_j}{k^2}, \quad i,j = 1,\ldots,d,$$

with δ_{ij} defined as the usual Kronecker delta symbol.

This representation of the random field can be used to simplify the implementation of the Randomization Method and has also been used to construct a multi-dimensional isotropic version of the Fourier-wavelet method. We describe each briefly in turn.

The isotropic spectral representation can be associated with the Helmholtz decomposition of the random field: $\mathbf{u}(\mathbf{x}) = \mathbf{u}^{(1)}(\mathbf{x}) + \mathbf{u}^{(2)}(\mathbf{x})$ where $\mathbf{u}^{(1)}$ and $\mathbf{u}^{(2)}$ are, respectively, the incompressible and potential parts of $\mathbf{u}$ with spectral density tensors

$$\mathsf{F}^{(1)}(\mathbf{k}) = \frac{1}{2A_d\, k^{d-1}} E_1(k)\mathsf{P}^{(1)}(\mathbf{k}), \quad \mathsf{F}^{(2)}(\mathbf{k}) = \frac{1}{2A_d\, k^{d-1}} E_2(k)\,\mathsf{P}^{(2)}(\mathbf{k}),$$

respectively.

Each of the random fields, $\mathbf{u}^{(1)}(\mathbf{x})$ and $\mathbf{u}^{(2)}(\mathbf{x})$, can be simulated as independent Gaussian random fields. The Cholesky factorizations

$$\mathsf{F}^{(i)}(\mathbf{k}) = p_i(\mathbf{k})\,\mathsf{Q}^{(i)}\,\mathsf{Q}^{(i)*},$$

take the special form

$$p_1(\mathbf{k}) = \sum_{i=1}^{d} F_{ii}^{(1)}(\mathbf{k}) = \frac{(d-1)E_1(k)}{2A_d\, k^{d-1}}, \quad p_2(\mathbf{k}) = \sum_{i=1}^{d} F_{ii}^{(2)}(\mathbf{k}) = \frac{E_2(k)}{2A_d\, k^{d-1}}.$$

Note in particular that $p_i(\mathbf{k}) = p_i(k)$, which generally greatly simplifies the simulation of random wavenumbers according to the probability distributions p_i.

The matrices $\mathsf{Q}^{(1)}$ and $\mathsf{Q}^{(2)}$ are to be chosen in any way such that

$$\frac{1}{d-1} P^{(1)}(\mathbf{k}) = \mathsf{Q}^{(1)}(\mathbf{k})\mathsf{Q}^{(1)*}(\mathbf{k}), \quad P^{(2)}(\mathbf{k}) = \mathsf{Q}^{(2)}(\mathbf{k})\mathsf{Q}^{(2)*}(\mathbf{k}).$$

One convenient explicit choice in three dimensions is [Sab91]

$$Q^{(1)}(\mathbf{k}) = \frac{1}{\sqrt{2}} \begin{pmatrix} 0 & \frac{k_3}{k} & -\frac{k_2}{k} \\ -\frac{k_3}{k} & 0 & \frac{k_1}{k} \\ \frac{k_2}{k} & -\frac{k_1}{k} & 0 \end{pmatrix}, \quad Q^{(2)}(\mathbf{k}) = \begin{pmatrix} \frac{k_1}{k} & 0 & 0 \\ \frac{k_2}{k} & 0 & 0 \\ \frac{k_3}{k} & 0 & 0 \end{pmatrix}.$$

Because $p_i(\mathbf{k}) = p_i(k)$ in the isotropic case, it is natural to choose the spectral subdivision $\Delta = \sum_{i=1}^{n} \Delta_i$ to be radially symmetric: $\Delta_i = \{\mathbf{k} : a_i \le |\mathbf{k}| \le b_i\}$. Using these tensors we obtain the following simulation formula for the incompressible part of an isotropic three-dimensional random vector field:

$$\mathbf{u}^{(1)}(\mathbf{x}) = \sum_{i=1}^{n} \frac{\sigma_i^{(1)}}{\sqrt{n_0}} \sum_{j=1}^{n_0} \left[\left(\boldsymbol{\Omega}_{ij}^{(1)} \times \boldsymbol{\xi}_{ij} \right) \cos(\theta_{ij}^{(1)}) + \left(\boldsymbol{\Omega}_{ij}^{(1)} \times \boldsymbol{\eta}_{ij} \right) \sin(\theta_{ij}^{(1)}) \right]$$

where $(\sigma_i^{(1)})^2 = \int_{\Delta_i} p_1(\mathbf{k})\, d\mathbf{k} = \frac{1}{2}\int_{a_i}^{b_i} E_1(k)\, dk$, $\boldsymbol{\Omega}_{ij}^{(1)}$, $i = 1, \dots, n$; $j = 1, \dots, n_0$ is a family of mutually independent random vectors distributed uniformly on the unit sphere in $\boldsymbol{R}^3$; $\boldsymbol{\xi}_{ij}$ and $\boldsymbol{\eta}_{ij}$, $i = 1, \dots, n$; $j = 1, \dots, n_0$ are mutually independent families of three-dimensional standard Gaussian random vectors; $\theta_{ij}^{(1)} = 2\pi k_{ij}^{(1)} (\boldsymbol{\Omega}_{ij}^{(1)} \cdot \mathbf{x})$; and for each $i = 1, \dots, n$, the $k_{ij}^{(1)}, j = 1, \dots, n_0$ is a sequence of independent random wavenumbers sampled from the interval (a_i, b_i) according to the probability density function proportional to $E_1(k)$.

The potential component $\mathbf{u}^{(2)}$ is simulated in three dimensions through the representation

$$\mathbf{u}^{(2)}(\mathbf{x}) = \sum_{i=1}^{n} \frac{\sigma_i^{(2)}}{\sqrt{n_0}} \sum_{j=1}^{n_0} \left[\xi_{ij}\, \boldsymbol{\Omega}_{ij}^{(2)} \cos(\theta_{ij}^{(2)}) + \eta_{ij}\, \boldsymbol{\Omega}_{ij}^{(2)} \sin(\theta_{ij}^{(2)}) \right].$$

Here, unlike in the previous simulation formula, the ξ_{ij} and η_{ij}, $i = 1, \dots, n$; $j = 1, \dots, n_0$ are families of *scalar* standard Gaussian random variables, which are all mutually independent. The remaining inputs are constructed analogously: $(\sigma_i^{(2)})^2 = \int_{\Delta_i} p_2(\mathbf{k})\, d\mathbf{k} = \frac{1}{2}\int_{a_i}^{b_i} E_2(k)\, dk$; $\boldsymbol{\Omega}_{ij}^{(2)}$ is a family of mutually independent random vectors distributed uniformly on the unit sphere in $\boldsymbol{R}^3$; $\theta_{ij}^{(2)} = 2\pi k_{ij}^{(2)} (\Omega_{ij}^{(2)} \cdot \mathbf{x})$; and for each $i = 1, \dots, n$, the $k_{ij}^{(2)}, j = 1, \dots, n_0$ is a sequence of independent random wavenumbers sampled in the interval (a_i, b_i) from the probability density function which is proportional to the function $E_2(k)$.

4 Small Perturbation Analysis

Here we show how the random field simulation can be used in the framework of small perturbation method, which works when the fluctuations are small.

4.1 Darcy Equation

We consider a steady flow through heterogeneous porous formation. For time-independent flow condition and saturated porous media the specific discharge is determined by the Darcy law:

$$\mathbf{q}(\mathbf{x}) = \theta(\mathbf{x})\mathbf{u}(\mathbf{x}) = -K(\mathbf{x})\nabla(\varphi(\mathbf{x}))$$

where $\mathbf{q}$ is the so-called Darcy's velocity, or specific discharge, $\mathbf{u}$ is the pore velocity, θ, the porosity, φ, the hydraulic potential $\varphi = \frac{p}{\rho g} + z$, p is the fluid pressure, z is the height, ρ - the density, and K - the hydraulic conductivity. The functions K and θ are key parameters of the flow. Experimental measurements show high heterogeneous behaviour of K in space with the following remarkable property [Dag90] : when considering K as a random field, its distribution is well approximated by the lognormal law. Therefore, in models, the hydraulic log-conductivity $Y = \ln K$ is commonly considered as a statistically homogeneous random field with gaussian distribution $N(m_Y, \sigma_Y)$. Here $m_Y = \langle Y \rangle$, and σ_Y is the standard deviation.

Let $C_{YY}(\mathbf{r}) = \langle Y'(\mathbf{x})Y'(\mathbf{x}+\mathbf{r})\rangle$ be the auto-correlation function, where $\mathbf{r}$ is the separation vector. We analyse the case when Y is statistically homogeneous and isotropic with the exponential auto-correlation function $C_{YY}(r) = \sigma_Y^2 exp(-r/I_Y)$ where $r = |\mathbf{r}|$, I_Y is a given correlation length.

We deal also with a random field with gaussian form of the covariance:

$$C_{YY}(r) = \sigma_Y^2 exp(-\frac{r^2}{l_Y^2}).$$

The porosity θ is also often considered in some models as a random field. However its variability is in the problems we tackle generally much smaller than that of K. We assume $\theta(x) = \theta = 1$.

Thus $\mathbf{q}$ is a random field obtained as the solution to the following diffusion equation:

$$\operatorname{div}\mathbf{q} = \operatorname{div}\,\{-K(\mathbf{x})\nabla(\varphi(\mathbf{x}))\} = \frac{\partial}{\partial x_i}\left(-K(\mathbf{x})\frac{\partial\varphi}{\partial x_i}\right) = 0.$$

Here and in what follows, we use the summation convention on repeated indices. For details, see our papers [KS03a], [KS05].

We consider two cases: (1) The fluctuations of K (say, measured via the intensity of fluctuations) are small, and (2) general case of fluctuations.

Small random perturbations about the mean values for the potential

$$\varphi = <\varphi> + \varphi' = H + h,$$

and for the specific discharge components:

$$q_i = <q_i> + q_i', \quad i = 1, 2, 3.$$

Let

$$Y'(\mathbf{x}) = \int\int\int exp(i(\mathbf{k},\mathbf{x}))dZ_Y(\mathbf{k}), \quad h(\mathbf{x}) = \int\int\int exp(i(\mathbf{k},\mathbf{x}))dZ_h(\mathbf{k}),$$

where $\mathbf{k} = (k_1, k_2, k_3)$ is the wave number vector, $\mathbf{x} = (x_1, x_2, x_3)$ is the position vector, and the integration is over three-dimensional wave number space. We use the notation $K_G = exp(\langle Y \rangle)$, and $J_i = -\partial H/\partial x_i$ for the mean hydraulic gradient in x_i-direction.

The correlation tensor $\{B_{ij}\}$ and the spectral tensor $\{S_{ij}\}$ are related through the equality

$$B_{ij}(\mathbf{r}) = \int_{R^3} S_{ij}(\mathbf{k})e^{i(\mathbf{r},\mathbf{k})}d\mathbf{k}.$$

The auto-covariance for the isotropic field has the spectrum

$$S_{YY}(k) = I_Y^3\sigma_Y^2/[\pi^2(1 + I_Y^2k^2)^2]$$

where $k = |\mathbf{k}|$.

Note that the spectrum of the field with the gaussian covariance function has also a gaussian form:

$$S_{YY}(\mathbf{k}) = \frac{\sigma_Y^2 l_Y^3}{\pi^{5/2}} exp(-\frac{l_Y^2 k^2}{4}), \quad I_Y = l_Y\sqrt{\pi}/2.$$

The following relation can be derived [Dag90], [KS05]

$$\nabla^2 h = J_i(\partial Y'/\partial x_i).$$

From this we come to the expression for the spectral tensor entries

$$S_{q_iq_j}(k) = \langle dZ_{q_i} dZ_{q_j}\rangle = K_G^2 J_m J_n(\delta_{im} - \frac{k_ik_m}{k^2})(\delta_{jn} - \frac{k_jk_n}{k^2})S_{YY}(k).$$

In Figure 4 we show trajectories of 5000 particles moving in a porous medium simulated as a random field with the given spectral tensor.

In Figure 5 we present the results of calculations of the Eulerian velocity auto-correlation functions $C_{u_iu_i}(\mathbf{r}/I_f)$. The spectrum $S_{ff}(\mathbf{k})$ is chosen in the form which corresponds to the exponential decorrelation.

Figure 5 shows the range of applicability of the small perturbation method. This can be seen by comparing the results we obtained by the small perturbation method (and spectral model) and by direct numerical solution of the Darcy equation by a SOR method. Here we plot the dimensionless functions $C_{u_1u_1}$ (left panel) and $C_{u_2u_2}$ (right panel) in longitudinal direction $r_1' = r_1/I_f$, for $\sigma_f = 0.3, 0.6$ and $\sigma_f = 1$. The left panel: as expected, the relative difference between the results is rapidly increasing with the growth of the fluctuation intensity, i.e., as σ_f increases. So, for $r_1' = 1$, this difference behaves like 4%, 18% and 62% for $\sigma_f = 0.3, 0.6$ and 1, respectively.

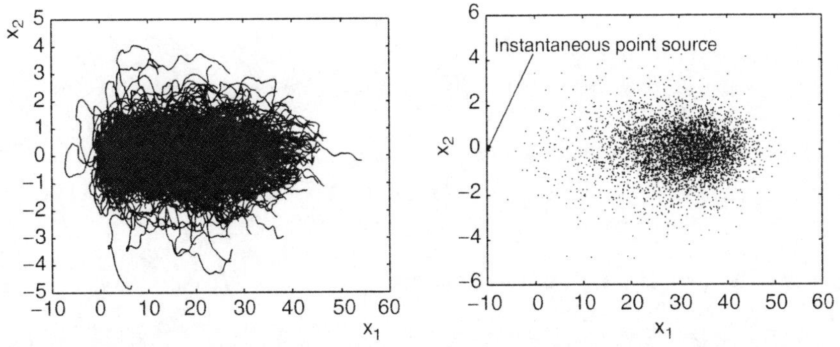

Fig. 4. Trajectories of 5000 particles started at $t' = 0$ at the origin, and finished at $t' = 30$: $\sigma_Y^2 = 1$ (left panel). In the right panel we show the resulting cloud.

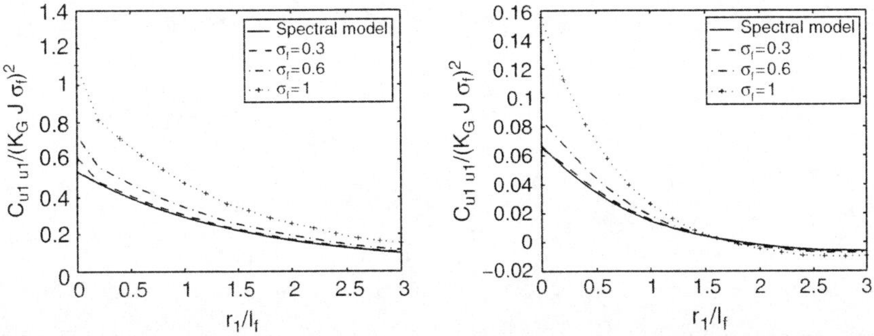

Fig. 5. The dimensionless functions $C_{u_1u_1}(\mathbf{r}/I_f)$ (left panel) and $C_{u_2u_2}(\mathbf{r}/I_f)$ (right panel) in longitudinal direction at different values σ_f in comparison against results of the small perturbation method (spectral model).

Right panel: the relative difference between the two methods (again, for $r_1' = 1$) is less than 7%, 25% and 78% for $\sigma_f = 0.3, 0.6$ and 1, respectively.

Thus the curves shown in Figure 5 present a clear picture about the region where the small perturbation approach can be applied, and how fast this approximation fails as the fluctuation intensity increases.

4.2 Lamé Equation

In this section we deal with an elasticity problem governed by an elliptic system of Lamé equations with stochastic elastic parameter and random loads. Suppose a homogeneous isotropic medium $G \subset \mathbb{R}^n$ with a boundary Γ is given, whose state in the absence of body forces is governed by the classical static equation, the Lamé equation:

$$\Delta \mathbf{u}(x) + \alpha \,\mathrm{grad}\,\mathrm{div}\, \mathbf{u}(x) = 0, \quad x \in G,$$

where $\mathbf{u}(x) = (u_1(x_1, \ldots, x_n), \ldots, u_n(x_1, \ldots, x_n))$ is a vector of displacements whose components are real-valued regular functions. The elastic constant α $\alpha = \frac{\lambda+\mu}{\mu}$ is expressed through the Lamé constants of elasticity λ and μ. It can be expressed through the Poisson ratio $\nu = \lambda/2(\lambda+\mu)$ as follows: $\alpha = 1/(1-2\nu)$. The Poisson ratio characterizes the relative amount of the change of the transverse to longitudinal displacements. It is known that due to thermodynamical reasons ν is bounded between $-1 \le \nu < 0.5$. This implies for α: $1/3 \le \alpha < \infty$. So there are materials with negative values of ν (α varies in $1/3 \le \alpha \le 1$), and materials with $\nu \approx 0.5$. The last case is very difficult for conventional deterministic methods.

In what follows, we present in this section mainly the results obtained by the small perturbation method. General case of large fluctuations is analysed in our recent paper [SSL].

We consider two different cases:

(I) The fluctuations appear in the elasticity constant α in the Lamé equation

$$\Delta\mathbf{u} + \alpha\nabla(\mathrm{div}\mathbf{u}) = 0$$

so that under the assumption of small random perturbations about mean values

$$u_i = <u_i> + u_i', \quad \alpha = <\alpha> + \alpha'.$$

The boundary conditions are deterministic: $\mathbf{u}|_\Gamma = \mathbf{u}_\gamma$.

(II) The loads $\mathbf{f}$ are random, while the constant α is fixed:

$$\Delta\mathbf{u} + \alpha\nabla(\mathrm{div}\mathbf{u}) = \mathbf{f}; \quad \alpha = const, \quad \mathbf{u}|_\Gamma = \mathbf{u}_\gamma.$$

We deal with statistically homogeneous random fields, hence we can use the Fourier-Stieltjes representations, in particular,

$$u_j'(\mathbf{x}) = \int\int exp(i(\mathbf{k},\mathbf{x}))dZ_{u_j}(\mathbf{k}),$$

$$\alpha'(\mathbf{x}) = \int\int exp(i(\mathbf{k},\mathbf{x}))dZ_\alpha(\mathbf{k})$$

where $\mathbf{k} = (k_1, k_2)$ is the wave number vector, $\mathbf{x} = (x_1, x_2)$ is the position vector, and the integration is over 2D wave number space.

Due to the small fluctuation assumption, we ignore the products of fluctuations, and from this we obtain

$$\Delta\mathbf{u}' + <\alpha> \nabla(\mathrm{div}\mathbf{u}') + \alpha'\nabla(\mathrm{div} <\mathbf{u}>) = 0.$$

We assume that $<\alpha> = A = const$ and $\nabla(div <\mathbf{u}>) = \mathbf{B} = const$, then using the above Fourier-Stiltjes representation yields

$$-k^2 dZ_{u_j} - Ak_j(k_1 dZ_{u_1} + k_2 dZ_{u_2}) + B_j dZ_\alpha = 0.$$

From this one finds

$$dZ_{u_1} = \frac{B_1k^2 + B_1Ak_2^2 - B_2Ak_1k_2}{k^4(1+A)}dZ_\alpha$$

$$dZ_{u_2} = \frac{B_2k^2 + B_2Ak_1^2 - B_1Ak_1k_2}{k^4(1+A)}dZ_\alpha,$$

which implies

$$S_{u_ju_l}(\mathbf{k})d\mathbf{k} = \langle \overline{dZ_{u_j}}dZ_{u_l}\rangle.$$

In the case of random loads, we do not need the assumption of small perturbations. Indeed,

$$\Delta\mathbf{u}' + \alpha\nabla(\mathrm{div}\mathbf{u}') = \mathbf{f}'; \quad \alpha = A = const.$$

Thus

$$-k^2dZ_{u_j} - Ak_j(k_1dZ_{u_1} + k_2dZ_{u_2}) = dZ_{f_j}.$$

$$dZ_{u_1} = \frac{Ak_1k_2dZ_{f_2} - Ak_2^2dZ_{f_1} - k^2dZ_{f_1}}{k^4(1+A)}$$

$$dZ_{u_2} = \frac{Ak_1k_2dZ_{f_1} - Ak_1^2dZ_{f_2} - k^2dZ_{f_2}}{k^4(1+A)},$$

$$S_{u_ju_l}(\mathbf{k})d\mathbf{k} = \langle \overline{dZ_{u_j}}dZ_{u_l}\rangle.$$

In Figure 6 we show two samples of the first component of the displacement vector, where in the left panel the correlation length is 5 times larger than that presented in the right panel.

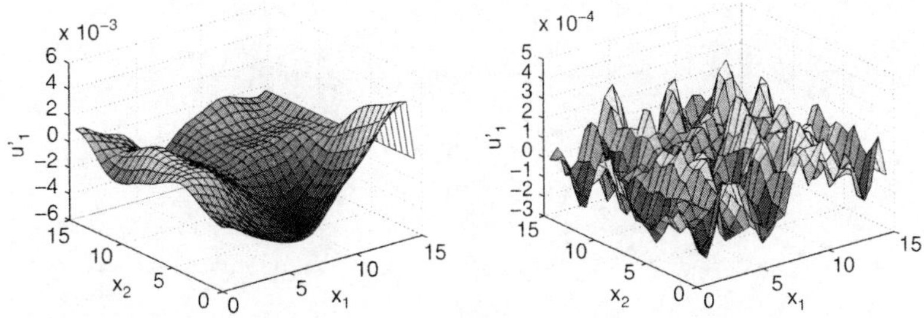

Fig. 6. Samples of displacements u_1'. Left picture: $I_\alpha = 1.5$, right picture: $I_\alpha = 0.3$. The number of harmonics $N = 100$.

5 Random Walk Methods and Double Randomization

Assume we have to solve a PDE which includes a random field σ, say in a right-hand side, in coefficients, or in the boundary conditions:

$$Lu = f, \quad u|_\Gamma = u_\gamma .$$

To solve this problem directly by constructing the ensemble of solutions via conventional numerical methods like finite elements or finite difference schemes is a hard task, which is not realistic for most practical problems. If however one of the Random Walk Methods can be applied, then a technique we call a Double Randomization Method is very useful. Let us describe it shortly.

Suppose we have constructed a stochastic method for solving this problem, for a fixed sample of σ. This implies, e.g., that an unbiased random estimator $\xi(x|\,\sigma)$ is defined so that for a fixed σ,

$$u(x,\sigma) = \langle \xi(x|\,\sigma) \rangle$$

where $\langle \cdot \rangle$ stands for averaging over the random trajectories of the stochastic method (e.g., a diffusion process, a Random Walk on Spheres, or a Random Walk on Boundary).

Let us denote by E_σ the average over the distribution of σ.

The double randomization method is based on the equality:

$$E_\sigma\, u(x,\sigma) = E_\sigma \langle \xi(x|\,\sigma) \rangle .$$

The algorithm for evaluation of $E_\sigma\, u(x,\sigma)$ then reads:

1. Choose a sample of the random field σ.

2. Construct the random walk over which the random estimator $\xi(x|\,\sigma)$ is calculated.

3. Repeat 1. and 2. N times, and take the arithmetic mean.

Suppose one needs to evaluate the covariance of the solution. Let us denote the random trajectory by ω. It is not difficult to show that

$$\langle u(x,\sigma)\, u(y,\sigma) \rangle = E_{(\omega_1,\omega_2,\sigma)}[\xi_{\omega_1}(x,\omega)\xi_{\omega_2}(y,\omega)].$$

The algorithm for calculation of $\langle u(x,\sigma)\, u(y,\sigma) \rangle$ follows from this relation:

1. Choose a sample of the random field σ.

2. Having fixed this sample, construct two conditionally independent trajectories ω_1 and ω_2, starting at x and y, respectively, and evaluate $\xi_{\omega_1}(x,\omega)\xi_{\omega_2}(y,\omega)$.

3. Repeat 1. and 2. N times, and take the arithmetic mean.

Remark. Note that for the correlation function (or tensor, in the vector case, for example, the Lamé equation), we can derive a closed equation. Indeed, assume that we have a linear equation with random right-hand side and zero boundary values

$$Lu = f, \quad x \in D, \quad u|_{\Gamma} = 0,$$

where the random field f (not necessarily homogeneous) has $B_f(x, y)$ as its correlation function (tensor).

The solution u can be represented through the Green formula

$$u(x) = \int_D G(x, y) f(y) dy$$

where $G(x, y)$ is the volume Green function for the domain D.

Under certain smoothness conditions we can derive that the correlation function (tensor) $B_u(x, y) = \langle u(x)u(y) \rangle$ satisfies the iterated equation

$$L_x \, L_y B_u(x, y) = B_f(x, y)$$

with boundary conditions $B_u|_{x\in\Gamma} = 0$, $L_x B(x, y)|_{y\in\Gamma} = 0$. Here L_x implies that the operator L acts with respect to the variable x, for fixed y.

This can be derived as follows. First, using the above Green formula, and taking the expectation, we obtain

$$B_u(x, y) = \langle u(x)u(y) \rangle = \int_D \int_D G(x, y')G(y, y'') \langle f(y')f(y'') \rangle \, dy' dy''.$$

This expression coincides obviously with the Green formula representation of the solution of the above iterated equation.

For systems of PDEs the relevant expressions are more complicated. Let us consider our system of Lamé equations. We denote the correlation tensor of the solution by $B^{(u)}(x, y) = \langle u(x)u^T(y) \rangle$, and the correlation tensor of the body forces by $B^{(f)}(x, y) = \langle f(x) \, f^T(y) \rangle$. Let $L = \Delta + \alpha \, \mathrm{grad\,div}$ be the Lamé operator. The Lamé operator L acts on a matrix W column-wise. This means, the matrix equation $LW = B$ (B is a matrix) is a pair of Lamé equations written for the relevant first and second columns of matrices W and B.

After some evaluations we arrive at

$$B^{(u)}(x, y) = \int_D \int_D G(x, y') \, B^{(f)}(y', y'') G^T(y, y'') dy' dy''. \tag{1}$$

It is also possible to write down a differential relation between the input matrix $B^{(f)}(y', y'')$ and the correlation matrix of the solution $B^{(u)}(x, y)$. Indeed, introduce a tensor $V(x, y)$, and write the following pair of coupled systems

$$L_x \, B^{(u)}(x, y) = V^T(x, y), \qquad B^{(u)}(x, y)|_{x\in\Gamma} = 0, \tag{2}$$

$$L_y \, V(x, y) = [B^{(f)}(x, y)]^T, \qquad V(x, y)|_{y\in\Gamma} = 0. \tag{3}$$

To prove that (1) solves the system (2),(3) it is enough to notice that the representation (1) can be obtained by a successive application of the Green formula representation of the solutions to (2),(3).

The system of equations (2),(3) can be written as one system of 4-th order. Indeed, using the definition $\hat{L}V = LV^T$ we apply the operator $\hat{L}_y$ to both sides of (2) . This yields

$$\hat{L}_y L_x B^{(u)}(x,y) = [B^{(f)}(x,y)]^T$$

with boundary conditions

$$B^{(u)}(x,y)|_{x\in\Gamma} = 0, \qquad L_x B^{(u)}(x,y)|_{y\in\Gamma} = 0.$$

The Double Randomization technique is often used in Monte Carlo methods also when solving deterministic PDEs. Let us show one example, - this technique is used in the "Global Random Walk" method we suggested in [Sab91].

Let us consider a boundary value problem

$$Lu(x) = f, \quad x \in D, \quad u|_{\partial D} = 0.$$

Here L is e.g., the Laplace, or the Lamé operator.

The solution can be represented as an expectation taken over random points $\tilde{y}$ distributed in G with an arbitrary probability density (such that $f(y) \neq 0$ for y where $G(x,y)f(y) \neq 0$)

$$u(x) = \int_D G(x,y)f(y)dy = E_{\tilde{y}}\left[G(x,\tilde{y})f(\tilde{y})/p(\tilde{y})\right]$$

where $G(x,y)$ is the Green function:

$$LG(x,y) = \delta(x-y), \quad x,y \in D, \quad G(x,y)|_{x\to\Gamma} = 0.$$

Thus we come to an unbiased estimator for our solution:

$$\zeta_x = G(x,\tilde{y})f(\tilde{y})/p(\tilde{y})$$

and $\tilde{y}$ is a random point distributed in D with a density $p(y)$ ($p(y) \neq 0$ for $y : G(x,y)f(y) \neq 0$.

Now, the function $G(x,\tilde{y})$ itself, is represented as an expectation taken over trajectories of a random process, say, the Random Walk on Spheres (RWS) process. Indeed, let

$$G(x,y) = \mathcal{E}(x,y) + W(x,y)$$

where $\mathcal{E}(x,y)$ is the fundamental solution (explicitly known for our operators), and hence the function $W(x,y)$ is uniquely defined by

$$Lw(x) = 0, \quad x \in D, \quad w|_{x\to\Gamma} = -\mathcal{E}(\cdot,y).$$

So we get the desired representation by using the probabilistic representation of the last problem:

$$W(x,y) = \langle -\mathcal{E}(x_\gamma, y)\rangle.$$

Thus we have the probabilistic representation $G(x,y) = \mathcal{E}(x,y) + \langle -\mathcal{E}(x_\gamma, y)\rangle$.

The direct evaluation of the double expectation gives the solution in one point x:

1. Choose a random point $\tilde{y}$ in D according to the density p.

2. Start a trajectory of RWS from the point x, and evaluate the estimator along this trajectory $\zeta_x^{(1)} = G(x,\tilde{y})f(\tilde{y})/p(\tilde{y})$.

3. Repeat N times p. 1–2, and take the arithmetic mean of $\zeta_x^{(i)}$.

It is however possible to calculate the solution simultaneously in an arbitrary set of points $x_1, \ldots, x_m$, using the symmetry property of the Green functions.

Indeed, assume, we wish to evaluate the solution in points $x_1, \ldots, x_m$. Due to the symmetry of $G(x,\tilde{y})$, we place the unit sources in these points, and the random points $\tilde{y}$ are considered now as the points where the solution should be found, i.e., the trajectories are started now from $\tilde{y}$.

The global algorithm then reads:

1. Choose random points $\tilde{y}_i$, $i = 1, \ldots, N$ in D according to the density p.

2. Start trajectories of RWS from the sampled points $\tilde{y}_i$, and evaluate $\zeta_j^{(i)} = G(\tilde{y}_i, x_j)f(\tilde{y}_i)/p(\tilde{y}_i)$, simultaneously for all $j = 1, \ldots, m$, and take the arithmetic mean of $\zeta_j^{(i)}$, $i = 1, \ldots N$. 1

Acknowledgments

This work is supported partly by the RFBR Grant N 06-01-00498, the program of Leading Scientific Schools under Grant N 4774.2006.1, and NATO Linkage Grant ESP Nr CLG 981426.

References

[Chu92] C.K. Chui. *An Introduction to Wavelets*. Academic Press, Inc., 1992.

[Dag89] G. Dagan. Flow and Transport in Porous Formations. Springer-Verlag, Berlin-Heidelberg, Germany, 1989.

[Dag90] G. Dagan. Spatial moments, Ergodicity, and Effective Dispersion. *Water Resour. Res.*, 26(6):1281–1290, 1990.

[DFJ03] G. Dagan, A. Fiori, and I. Janković. Flow and transport in highly heterogeneous formations: 1. Conceptual frameewορk and validity of first-order approximations. *Water Resour. Res.*, 39(9):1268, 2003.

[EM94] F.W. Elliott, Jr and A.J. Majda. A wavelet Monte Carlo method for turbulent diffusion with many spatial scales. *J. Comp. Phys*, 113(1):82–111, 1994.

[EM95] F.W. Elliott, Jr and A.J. Majda. A new algorithm with plane waves and wavelets for random velocity fields with many spatial scales. *J. Comp. Phys*, 117:146–162, 1995.

[Gel93] L.W. Gelhar. *Stochastic Subsurface Hydrology*. Prentice-Hall, Englewood Cliffs, N.J., 1993.

[KKS07] P. Kramer, O. Kurbanmuradov and K. Sabelfeld. Extensions of multiscale Gaussian random field simulation algorithms. Preprint No. 1040, WIAS, Berlin. To appear in J. Comp. Physics, 2007.

[KPS06] V.M. Kaganer, K.H. Ploog, and K.K. Sabelfeld. Dynamic coalescence kinetics of facetted 2D islands. Physical Review B, 73, N 11 (2006).

[KS00] O. Kurbanmuradov and K. Sabelfeld. Coagulation of aerosol particles in intermittent turbulent flows. Monte Carlo Methods and Applications. **6** (2000), N3, 211–253.

[KS03a] D. Kolyukhin and K. Sabelfeld. Stochastic Eulerian model for the flow simulation in porous media. *Monte Carlo Methods and Applications*, **9**, No. 3, 2003, 271–290.

[KS03b] A. Kolodko and K. Sabelfeld. Stochastic particle methods for Smoluchowski coagulation equation: variance reduction and error estimations. Monte Carlo Methods and Applications. vol. 9, N4, 2003, 315–340.

[KS05] D. Kolyukhin and K. Sabelfeld. Stochastic flow simulation in 3D Porous media. *Monte Carlo Methods and Applications*, **11**, No. 1, 2005, 15–38.

[KS06] O. Kurbanmuradov and K. Sabelfeld. Stochastic spectral and Fourier-wavelet methods for vector Gaussian random field. Monte Carlo Methods and Applications, **12** (2006), N 5-6, 395–445.

[KSSV03] O. Kurbanmuradov, K. Sabelfeld, O. Smidts and H.A. Vereecken. Lagrangian stochastic model for transport in statistically homogeneous porous media. *Monte Carlo Methods and Applications*. **9**, No. 4, 2003, 341–366.

[Kur95] O. Kurbanmuradov. Weak Convergence of Approximate Models of Random Fields. *Russian Journal of Numerical Analysis and Mathematical Modelling*, **10** (1995), N6, 500–517.

[Mey90] Y. Meyer. Ondelettes et opérateurs. I: Ondelettes, Hermann, Paris, 1990. (English translation: Wavelets and operators, Cambridge University Press, 1992.)

[Mik83] G.A. Mikhailov. Approximate models of random processes and fields. Russian J. Comp. Mathem. and mathem. Physics, vol. 23 (1983), N3, 558–566. (in Russian).

[MY81] A.S. Monin and A.M. Yaglom. *Statistical Fluid Mechanics: Mechanics of Turbulence*, Volume 2. The M.I.T. Press, 1981.

[PS95] F. Poirion and C. Soize. Numerical methods and mathematical aspects for simulation of homogenous and non homogenous Gaussian vector fields. In Paul Kree and Walter Wedig, editors, *Probabilistic methods in applied physics*, volume 451 of *Lecture Notes in Physics*, pages 17–53. Springer-Verlag, Berlin, 1995.

[Sab91] K. Sabelfeld. *Monte Carlo methods in boundary value problems*, chapter 1,5, pages 31–47, 228–238. Springer Series in Computational Physics. Springer-Verlag, Berlin, 1991.

[Shi71] M. Shinozuka. Simulation of multivariate and multidimensional random processes. J. of Acoust. Soc. Am. **49** (1971), 357–368.

[SLP07] K. Sabelfeld, A. Levykin, and T. Privalova. A fast stratified sampling simulation of coagulation processes. Monte Carlo Methods and Applications, **13** (2007), N1, 63–84.

[SRR01] P.D. Spanos and V. Ravi S. Rao. Random Field Representation in a Biorthogonal Wavelet Basis. *Journal of Engineering Mechanics*, **127**, No. 2 (2001), 194–205.

[SSL] K. Sabelfeld, I. Shalimova and A.I. Levykin. Stochastic simulation method for a 2D elasticity problem with random loads. Submitted to "Probabilistic Engineering Mechanics".

[TD05] D.J. Thomson and B.J. Devenish. Particle pair in kinematic simulations. *Journal of Fluid Mechanics*, **526** (2005), 277–302.

[Wag06] W. Wagner. Post-gelation behavior of a spatial coagulation model. Electron. J. Probab., 893–933. / (WIAS preprint N 1128, 2006).

Monte Carlo and Quasi-Monte Carlo Methods for Computer Graphics

Peter Shirley, Dave Edwards, and Solomon Boulos

University of Utah

Summary. Some computer graphics applications, such as architectural design, generate visually realistic images of computer models. This is accomplished by either explicitly or implicitly solving the light transport equations. Accurate solutions involve high-dimensional equations, and Monte Carlo (MC) techniques are used with an emphasis on importance sampling rather than stratification. For many applications, approximate solutions are adequate, and the dimensionality of the problem can be reduced. In these cases, the distribution of samples is important, and quasi-Monte Carlo (QMC) methods are often used. It is still unknown what sampling schemes are best for these lower dimensional graphics problems, or what "best" even means in this case. This paper reviews the work in MC and QMC computer graphics, and poses some open problems in the field.

1 Introduction

Computer graphics researchers have long attempted to generate images with the realism of photographs. There are three steps in this process:

1. build or scan a geometric model of a scene, and associate material properties, such as spectral albedo, with each object;
2. simulate the transport of light energy to compute the amount of light hitting each sensor element;
3. generate a displayable image from the sensor element responses.

This pipeline is discussed in more detail in the overview paper by Greenberg et al. [GTS+97]. The first step is usually called "modeling" or "geometric modeling", and is a field of study in its own right. Mortenson's book is a fine up-to-date introduction to modeling [Mor07]. The last step is called "tone mapping", and is analogous to photo development. More details on tone mapping methods can be found in Reinhard et al.'s book [RWPD05]. This paper addresses the middle step: solving the light transport equations either implicitly or explicitly. Our intended audience is MC and QMC researchers who want an overview of the computational issues faced by graphics researchers, including problems in the field that are still unsolved. In Section 2 we review the light transport research in computer graphics. Section 3 is an overview of the unsimplified, high-dimensional problem, in which importance sampling is more effective than stratification. In Section 4 we discuss previous work on the simplified, low-dimensional problem, where well-distributed samples are critical. Finally, in Section 5 we list some open problems in sample generation for graphics. Readers interested in more detail on the use of MC in graphics should consult Veach's dissertation [Vea97] which also summarized much standard classic MC work, or one of the general books on rendering [PH04, DBB06]. A detailed discussion of QMC in rendering can be found in Keller's recent paper [Kel06].

2 Light Transport for Computer Graphics

Almost all graphics practitioners make several assumptions to simplify the implementation of rendering software:

- light obeys geometric optics, so interference, polarization, diffraction, and other wave effects need not be modeled;
- light travels and interacts with surfaces instantaneously, so no phosphorescence is simulated;
- all wavelengths are independent, so there is no fluorescence.

Several researchers have explored softening these assumptions: some rendering software can simulate fluorescence [Gla94, WTP01], phosphorescence [Gla94], polarization [WTP01, WTU+04], thin-film interference [SM90, GMN94], and small-scale diffraction [Sta99].

Rendering programs often use a simplified model of light transport. For example, simulated light might be allowed to reflect only once between a light source and the viewer, a constraint that removes indirect lighting. This simplification lowers the dimensionality of the problem, and makes stratification and low-discrepancy sampling more beneficial. A similar constraint takes advantage of the fact that many real-world scenes have a median surface albedo of approximately 20%. For these scenes, allowing only a few reflections along a light path can still result in an image with low visual error. The low dimensionality of such cases makes QMC techniques attractive.

However, for some scenes, such as white-painted rooms, light paths with many reflections can carry significant energy. Extreme cases include scenes with participating media, which scatter light as it propagates between surfaces. Some of these media, such as clouds, have a very high scattering albedo, possibly requiring hundreds of dimensions for an accurately rendered image. There is an ongoing debate in the graphics community over whether these difficult cases should be handled with brute-force Monte Carlo rendering, as advocated by our work [MBJ+06], or by lowering the dimensionality of the physics [JMLH01].

3 Previous Work on the Full Problem

The most straightforward way to compute the effects of light transport is to explicitly simulate the transport of photon-like particles from the light to the sensor. At each interaction between a particle and a surface, the particle may be absorbed or scattered. If a particle reaches an element on the sensor, it contributes to the final image. Sensor elements that receive more light correspond to brighter pixels in the image. Appel investigated this approach for direct lighting in the 1960s [App68], but this technique has only recently become computationally feasible for non-trivial scenes. A complete implementation was first applied in Pattanaik's dissertation [Pat93].

Ray tracing is used to determine which surface a particle hits. This method computes the intersection of a geometrical ray and a set of surfaces describing the scene, and determines which intersection is closest to the ray origin. A tree-building preprocessing step allows this intersection to be computed in time proportional to the logarithm of the number of surfaces for most scenes [SAG+05].

Although a relative error as high as 2% is acceptable for most rendering applications, simulating photon-like particles without optimizations is too slow, due to the number of photons required to produce a converged image. This problem may be alleviated by several optimizations. The most critical of these is importance sampling, which was first applied by Cook [CPC84, Coo86] and first used in the formal MC sense a few years later [KA91, SW91]. Importance sampling is used to make particles more likely to scatter toward the simulated camera lens, greatly reducing the computation time for some scenes. A number of other classic MC optimizations have been successfully applied by graphics researchers. Most of this optimization research occurred in the 1990s, and is nicely summarized in Veach's dissertation [Vea97].

Several approaches have used biased methods to reduce the number of simulated photons. These techniques store the locations where particles interact with surfaces, and then use density estimation to obtain a smooth lighting function, which can be projected into image space using ray tracing or some other technique. This approach has been applied successfully by Jensen in his photon mapping technique [Jen01], and by Walter et al. in their world-space system [WHSG97].

For some scenes, tracing particles from the light is not effective. For example, outdoor scenes involve such a vast distribution of light energy that most particles will not contribute to the final image. For such scenes, tracing light backwards from the sensor to the light sources is more effective than direct photon simulation. These backward tracing approaches require the solution of adjoint equations; details of these adjoint methods have been outlined by several researchers [Pat93, Chr03]. Our own approach to the rendering problem is based on adjoint techniques, and amounts to exchanging the light source and sensor properties in a scene, and then proceeding with a forward particle tracing algorithm as described above [MBJ+06].

For some scenes, neither particle tracing nor adjoint particle tracing work well. These include scenes such as swimming pools, where light is focused twice by the same specular surface (e.g., the water-air boundary for a pool) to produce a refracted image of a caustic pattern. In these cases, researchers have rewritten the rendering integral equation as an integral over light paths, with an implicit limit on the number of reflections along the path. Veach and Guibas [VG94], and Lafortune and Willems [LW93] applied Monte Carlo techniques to this formulation to render images of several difficult configurations. A more recent system by Kollig and Keller is the simplest complete bidirectional system yet described [KK06].

Veach and Guibas noted that most energy in path space is confined to relatively small subsets of the path domain. They developed a variant of the Metropolis method, which they dubbed Metropolis light transport [VG97]. Their method was extended by Pauly et al. to render participating media [PKK00]. Although it is a promising method, Metropolis light transport has been modified and extended by few researchers [KSKAC02, SKBS04, CTE05, Tal05]. We believe this is due to the difficulty of implementing the algorithm, rather than intrinsic shortcomings of the technique.

4 Previous Work on Simplified Problem

A common simplification to full light transport is to allow light paths with only a few reflections, which greatly reduces the dimensionality of the problem. Another simplification was introduced by Cook et al. [CPC84]; for each pixel the system computes a nine-dimensional integral over pixel area, camera lens area, time, light source location, and reflection direction. Cook [Coo86] later described a stratified Monte Carlo technique for estimating the nine-dimensional integral using four sets of two-dimensional samples and one set of one-dimensional samples. The number of samples in these sets was the same constant value for every pixel. The lower-dimensional sets were combined into one nine-dimensional set using a variety of techniques including magic squares, which are the first hint of QMC in the graphics literature. It is important to note that some of the domains Cook sampled are not square: for example, the lens samples lie within a disk. Schlick used especially constructed permutations to sample in higher dimensions in similar spirit to Cook but with a more general view [Sch91].

Shirley was the first researcher to apply classical QMC techniques to computer graphics [Shi91]. He computed the star and box discrepancy of several types of two-dimensional antialiasing patterns, and showed that patterns with lower discrepancy produced more accurate renderings. In this case, accuracy was measured using RMS and maximum differences between an ideal reference image and an image rendered using a given pixel-sampling pattern. This work also indicated that classic discrepancy might not be a useful indicator of image quality in the presence of importance sampling, depending on the transformation between the unit square and the domain of interest. Chiu and Shirley later developed MC sampling techniques with similar properties to some QMC nets [CSW94].

Mitchell also proposed a deterministic sampling method based on frequency-space properties, rather than low discrepancy [Mit91]. His method constructs a blue-noise point set by incrementally adding points to a two-dimensional pattern. In each iteration, the algorithm generates several candidate points, and then adds the point with the highest minimum distance to all points already in the pattern. Mitchell also presented a similar method for generating higher-dimensional samples, which could be used to render motion blur in dynamic scenes. Mitchell's method for incremental point set generation is also known as best-candidate sampling [PH04].

Heinrich and Keller were the first to use classic QMC points, specifically the Halton and Hammersley sequences, for rendering [HK94a, HK94b]. Like Shirley, they found that discrepancy is an indicator of rendering quality. Heinrich and Keller presented a method for antialiasing using low-discrepancy patterns, rather than samples on a regular grid. They found that QMC sampling often converges faster to a reference image than pseudorandom sampling [HK94b], and that the Halton sequence is very useful for adaptive sampling techniques. These classic sequences proved much more useful for graphics by adding Faure's permutations [Fau92, Kel98].

Mitchell and Dobkin published three papers, one also with Eppstein, on a variant of discrepancy based on arbitrarily oriented edges through the square [Mit92, DM93, DEM96]. They claimed that this isotropic discrepancy measure is useful for computer graphics, since it corresponds to the common case in which object silhouettes pass through a pixel. Their articles include several tables of different discrepancy values for various sampling patterns. They also presented an incremental method for generating samples, which is similar to the best candidate algorithm, but optimizes based on discrepancy, rather than minimum distance. Mitchell summarized some of this work, and ran some simple experiments, in a SIGGRAPH 2001 course [Mit01]. He showed that points optimized with respect to arbitrary edge discrepancy resulted in less RMS error than Hammersley points for a zone plate image. The generality of these results remains unknown.

Ohbuchi and Aono performed a practical comparison of QMC sampling [OA96]. They compared several methods for sampling two-dimensional area light sources for rendering, including pseudorandom points, stratified samples, and the Sobol and Halton sequences. The Sobol and Halton sequences were chosen because they allow incremental adaptive sampling. The QMC-based samples yielded better qualitative and quantitative results than pseudorandom sampling, although stratified random samples were almost as effective. They also presented an algorithm for adaptive sampling by iteratively adding light source samples until the difference in pixel color between two iterations is less than a threshold value. In their experiments, adaptive sampling with the Halton sequence produced low-error images faster than any other method.

Szirmay-Kalos and Purgathofer performed an approximate analysis of Monte Carlo and quasi-Monte Carlo rendering that they tested experimentally [SKP98]. The authors use this analysis to show that, QMC sampling is no worse than pseudorandom sampling, and is often much more accurate for rendering, since the fraction of strata containing discontinuities is often small. However, in higher dimensions, the benefits of QMC sampling are not as pronounced. They tested their error analysis using MC and QMC integration on discontinuous two- and three-dimensional functions, and found that QMC estimates were more accurate than MC estimates, especially at higher numbers of samples. Finally, they used QMC and MC sampling to render a scene for which the value of the rendering integral is known. For light paths with one or two reflections, the QMC estimate was more accurate than the MC estimate. However, when more reflections were allowed the two sampling methods exhibit about the same amount of error. These results agree with the authors' analysis that MC and QMC sampling will exhibit approximately the same amount of error for higher-dimensional problems.

Kollig and Keller developed one of the first complete rendering systems with QMC sampling [KK02]. Their software combines QMC techniques, bidirectional path tracing, and multiple importance sampling to efficiently render scenes with global illumination effects. The authors claim that QMC point sets are well-suited to rendering, since lower dimensions of QMC point sets exhibit better distribution, and lower dimensions of the particle simulation tend to have a higher visual impact. Kollig and Keller also describe several methods for randomizing QMC samples, allowing unbiased estimates while maintaining the advantages of low discrepancy sampling. Finally, they mention techniques for creating high-dimensional sample points by concatenating randomized instances of low-dimensional samples, a technique the authors call "padded replications sampling". Padded samples are not guaranteed to have low discrepancy, but they still produce high-quality images in practice.

Keller's SIGGRAPH course notes are probably the best comprehensive description of a QMC-based rendering system [Kel03]. Although most of the information in the notes can be found in his other papers, these notes bring many important concepts together in one place. They also provide a detailed description of trajectory splitting in particle simulations, which is a useful technique for creating efficient sample sets when branching is allowed along the light paths. In more recent papers, Keller suggests that the most effective sampling methods for rendering are those that combine low discrepancy and blue noise properties [Kel04, Kel06]. He concludes that rank-1 lattice points can offer these advantages, and are simple to implement as well.

5 Open Problems in Sample Generation for Graphics

In this section we review the most important open problems in generating samples for the simplified light transport problem. Effective sampling is increasingly important, especially since improvements in hardware have made interactive ray tracing feasible, and good sampling techniques can provide higher-quality images without requiring additional rendering time.

Most computer graphics applications are designed with a sampling module that generates points on the hypercube. It is usually straightforward to test different sampling strategies by viewing hypercube sampling as a "black box" and exchanging different sampling strategies (e.g., MC or QMC) without changing any of the rendering engine code [SSB91]. Usually, the number of samples per pixel is known in advance, and ranges from around 9 to 400.

Most programs require the ability to sample non-square domains and/or non-uniform densities. This is usually handled by transforming points on the hypercube using a bijective function that produces points with the appropriate density on the desired domain. Usually the domain is a two-dimensional manifold with simple boundaries. A detailed description of this process is in Arvo's notes [Arv01], which follow the classic multidimensional inversion method of MC with examples from specific domains that arise in graphics.

For programs that simulate a camera lens, a disk must be sampled. For this two approaches are popular. The first is to transform uniformly distributed points $(\xi_0, \xi_1) \in [0,1]^2$ to polar coordinates on the disk via the mapping $\theta = 2\pi\xi_0$ and $r = R\sqrt{\xi_1}$. The Jacobian of this function has a constant determinant, and thus the method produces uniformly distributed points on the disk. A potential problem with this mapping is that it may decrease or remove some of the spatial properties of the distribution on the unit square, due to stretching. There are infinitely many constant Jacobian mappings from square to disk, and some may be more effective than the one described above. Shirley and Chiu proposed one such mapping [SC97] that has been empirically shown to reduce error [KMH95]. Little theory exists that addresses the effect of different mappings on error.

There are several basic questions that remain open:

1. Should samples on non-square domains be generated on the hypercube and transformed, or should they be generated directly on their native domain?
2. For pixel sampling, what is the appropriate measure of sample set quality? For example, is edge discrepancy more predictive than other discrepancy measures?
3. How should the human perceptual system be factored into sample set design for graphics? Perceptual factors are often used to justify minimum-distance or blue noise sampling.
4. Should numeric optimization be used to generate sample sets?
5. Is there much to be gained from better sampling, or are we already in the diminishing return stage?
6. How should the sample sets of neighboring pixels relate to each other?

Unfortunately, these questions are all fairly hard to answer. Some formal questions that are also currently unanswered include:

1. How does one generate N samples on a disk with minimal edge-discrepancy?
2. How does one generate N samples on the 3D cube with minimal 3D edge-discrepancy?
3. How does one take M^2N "good" samples on the square, and divide them into M^2 sets of N "good" samples?

The answer to the first question would be useful for generating samples on camera lenses, and the answer to the second question would be useful for images of moving objects. The answer to the third question would be helpful in tile-based sampling as advocated by Keller and Heidrich [KH01]. We use a similar tile-based architecture to avoid the costs of runtime sample generation, and agree with Keller and Heidrich's claim that it works well [BEL+07].

6 Summary

Some graphics problems are high-dimensional, and in these cases, importance sampling is the only important optimization. Other problems are low-dimensional, and sample distribution can greatly influence performance. Such graphics problems have four characteristics that are noteworthy for QMC researchers attempting to design appropriate sampling schemes: first, a high relative error is tolerable, and thus only a small number of samples is needed for each pixel; second, a separate integral estimation is performed for each pixel, and the perceptual nature of the error is important; third, non-square domains are often sampled; finally, the domain of integration is often at least seven-dimensional (two dimensions for screen, lens, and light samples, and one for time), but significant variation usually occurs in only two or three of these dimensions for a given pixel. Finding the best method for generating samples for such applications remains an open problem.

Acknowledgments

This work was supported by NSF grant 03-06151. The authors appreciate discussions with Alex Keller, Harald Niederreiter, and Kris Sikorski that influenced our thinking on several issues.

References

[App68] A. Appel. Some techniques for shading machine rendering of solids. In *Proceedings of the AFIPS 1968 Spring Joint Computer Conference*, volume 32, pages 37–45, 1968.

[Arv01] J. Arvo. Stratified sampling of 2-manifolds. In *SIGGRAPH Course: State of the Art in Monte Carlo Ray Tracing for Realistic Image Synthesis*, 2001.

[BEL+07] S. Boulos, D. Edwards, J. D. Lacewell, J. Kautz, I. Wald, and P. Shirley. Packet-based whitted and distribution ray tracing. In *Proceedings of Graphics Interface*, 2007.

[Chr03] P. Christensen. Adjoints and importance in rendering: an overview. *IEEE Transactions on Visualization and Computer Graphics*, 9(3):329–340, 2003.

[Coo86] R. Cook. Stochastic sampling in computer graphics. *ACM Transactions on Graphics*, 5(1):51–72, 1986.

[CPC84] R. Cook, T. Porter, and L. Carpenter. Distributed ray tracing. In *Proceedings of SIGGRAPH*, pages 165–174, 1984.

[CSW94] K. Chiu, P. Shirley, and C. Wang. Multi-jittered sampling. In *Graphics gems IV*, pages 370–374. Academic Press Professional, Inc., San Diego, CA, USA, 1994.

[CTE05] D. Cline, J. Talbot, and P. Egbert. Energy redistribution path tracing. In *Proceedings of SIGGRAPH*, pages 1186–1195, 2005.

[DBB06] P. Dutre, K. Bala, and P. Bekaert. *Advanced Global Illumination*. A. K. Peters, Ltd., Natick, Massachusetts, USA, second edition, 2006.

[DEM96] D. Dobkin, D. Eppstein, and D. Mitchell. Computing the discrepancy with applications to supersampling patterns. *ACM Transactions on Graphics*, 15(4):354–376, October 1996.

[DM93] D. Dobkin and D. Mitchell. Random-edge discrepancy of supersampling patterns. In *Proceedings of Graphics Interface*, pages 62–69, May 1993.

[Fau92] H. Faure. Good permutations for extreme discrepancy. *J. Number Theory*, 42:47–56, 1992.

[Gla94] A. Glassner. A model for fluorescence and phospherescence. In *Proceedings of the EUROGRAPHICS rendering workshop*, pages 57–68, 1994.

[GMN94] J. Gondek, G. Meyer, and J. Newman. Wavelength dependent reflectance functions. In *Proceedings of SIGGRAPH*, pages 213–220, 1994.

[GTS+97] D. Greenberg, K. Torrance, P. Shirley, J. Arvo, J. Ferwerda, S. Pattanaik, E. Lafortune, B. Walter, S. Foo, and B. Trumbore. A framework for realistic image synthesis. In *Proceedings of SIGGRAPH*, pages 477–494, 1997.

[HK94a] S. Heinrich and A. Keller. Quasi-Monte Carlo methods in computer graphics, Part I: The QMC-Buffer. Technical Report 242/94, University of Kaiserslautern, Kaiserslautern, Germany, 1994.

[HK94b] S. Heinrich and A. Keller. Quasi-Monte Carlo methods in computer graphics, Part II: The radiance equation. Technical Report 243/94, University of Kaiserslautern, Kaiserslautern, Germany, 1994.

[Jen01] H. Wann Jensen. *Realistic Image Synthesis Using Photon Mapping*. A. K. Peters, Ltd., Natick, Massachusetts, USA, 2001.

[JMLH01] H. Wann Jensen, S. Marschner, M. Levoy, and P. Hanrahan. A practical model for subsurface light transport. In *Proceedings of SIGGRAPH*, pages 511–518, 2001.

[KA91] D. Kirk and J. Arvo. Unbiased variance reduction for global illumination. In *Proceedings of the EUROGRAPHICS rendering workshop*, 1991.

[Kel98] A. Keller. *Quasi-Monte Carlo Methods for Photorealistic Image Synthesis*. Ph.D. thesis, Shaker Verlag Aachen, 1998.

[Kel03] A. Keller. Strictly deterministic sampling methods in computer graphics. In *SIGGRAPH Course: Monte Carlo Ray Tracing*, 2003.

[Kel04] A. Keller. Stratification by rank-1 lattices. In *Monte Carlo and Quasi-Monte Carlo Methods 2002*, pages 299–313, 2004.

[Kel06] A. Keller. Myths of computer graphics. In *Monte Carlo and Quasi-Monte Carlo Methods 2004*, pages 217–244, 2006.

[KH01] A. Keller and W. Heidrich. Interleaved sampling. In *Proceedings of the EUROGRAPHICS rendering workshop*, 2001.

[KK02] T. Kollig and A. Keller. Efficient bidirectional path tracing by randomized Quasi-Monte Carlo integration. In *Monte Carlo and Quasi-Monte Carlo Methods 2000*, pages 290–305, 2002.

[KK06] T. Kollig and A. Keller. Illumination in the presence of weak singularities. In *Monte Carlo and Quasi-Monte Carlo Methods 2004*, pages 245–257, 2006.

[KMH95] C. Kolb, D. Mitchell, and P. Hanrahan. A realistic camera model for computer graphics. In *Proceedings of SIGGRAPH*, pages 317–324, 1995.

[KSKAC02] C. Kelemen, L. Szirmay-Kalos, G. Antal, and F. Csonka. A simple and robust mutation strategy for the Metropolis light transport algorithm. *Computer Graphics Forum*, 21(3):1–10, 2002.

[LW93] E. Lafortune and Y. Willems. Bi-directional path tracing. In *Proceedings of Compugraphics*, pages 145–153, 1993.

[MBJ+06] R. Morley, S. Boulos, J. Johnson, D. Edwards, P. Shirley, M. Ashikhmin, and S. Premože. Image synthesis using adjoint photons. In *Proceedings of Graphics Interface*, pages 179–186, 2006.

[Mit91] D. Mitchell. Spectrally optimal sampling for distribution ray tracing. In *Proceedings of SIGGRAPH*, pages 157–164, 1991.

[Mit92] D. Mitchell. Ray tracing and irregularities of distribution. In *Proceedings of the EUROGRAPHICS rendering workshop*, pages 61–70, 1992.

[Mit01] D. Mitchell. Quasirandom techniques. In *SIGGRAPH Course: State of the art in Monte Carlo ray tracing for realistic image synthesis*, 2001.

[Mor07] M. Mortenson. *Geometric modeling*. Industrial Press, New York, NY, USA, third edition, 2007.

[OA96] R. Ohbuchi and M. Aono. Quasi-Monte Carlo rendering with adaptive sampling. Technical Report RT0167, IBM Tokyo Research Laboratory, November 1996.

[Pat93] S. Pattanaik. *Computational Methods for Global Illumination and Visualisation of Complex 3D Environments*. PhD thesis, Birla Inst. of Technology & Science, Pilani, India, 1993.

[PH04] M. Pharr and G. Humphreys. *Physically Based Rendering: From Theory to Implementation*. Morgan Kaufmann Publishers, San Fransisco, California, USA, 2004.

[PKK00] M. Pauly, T. Kollig, and A. Keller. Metropolis light transport for participating media. In *Proceedings of the EUROGRAPHICS rendering workshop*, 2000.

[RWPD05] E. Reinhard, G. Ward, S. Pattanaik, and P. Debevec. *High Dynamic Range Imaging: Acquisition, Display and Image-Based Lighting*. Morgan Kaufmann Publishers, San Francisco, CA, USA, 2005.

[SAG+05] P. Shirley, M. Ashikhmin, M. Gleicher, S. Marschner, E. Reinhard, K. Sung, W. Thompson, and P. Willemsen. *Fundamentals of Computer Graphics*. A. K. Peters, Ltd., Natick, Massachusetts, USA, second edition, 2005.

[SC97] P. Shirley and K. Chiu. A low distortion map between disk and square. *journal of graphics tools*, 2(3):45–52, 1997.

[Sch91] C. Schlick. An adaptive sampling technique for multidimensional ray tracing. In *Proceedings of the EUROGRAPHICS rendering workshop*, pages 48–56, 1991.

[Shi91] P. Shirley. Discrepancy as a quality measure for sample distributions. In *Proceedings of EUROGRAPHICS*, pages 183–194, 1991.

[SKBS04] L. Szirmay-Kalos, B. Balazs, and M. Sbert. Metropolis iteration. In *Proceedings of WSSG*, 2004.

[SKP98] L. Szirmay-Kalos and W. Purgathofer. Analysis of the Quasi-Monte Carlo integration of the rendering equation. Technical Report TR-186-2-98-22, Vienna University of Technology, Vienna, Austria, 1998.

[SM90] B. Smits and G. Meyer. Newton's colors: simulating interference phenomena in realistic image synthesis. In *Proceedings of the EUROGRAPHICS rendering workshop*, pages 185–194, 1990.

[SSB91] P. Shirley, K. Sung, and W. Brown. A ray tracing framework for global illumination systems. In *Proceedings of Graphics Interface*, pages 117–128, 1991.

[Sta99] J. Stam. Diffraction shaders. In *Proceedings of SIGGRAPH*, pages 101–110, 1999.

[SW91] P. Shirley and C. Wang. Direct lighting calculation by Monte Carlo integration. In *Proceedings of the EUROGRAPHICS rendering workshop*, 1991.

[Tal05] J. Talbot. Importance resampling for global illumination. Master's thesis, Brigham Young University, Provo, Utah, USA, 2005.

[Vea97] E. Veach. *Robust Monte Carlo Methods for Light Transport Simulation*. PhD thesis, Stanford University, Stanford, California, USA, 1997.

[VG94] E. Veach and L. Guibas. Bidirectional estimators for light transport. In *Proceedings of the EUROGRAPHICS Rendering Workshop*, pages 147–162, 1994.

[VG97] E. Veach and L. Guibas. Metropolis light transport. In *Proceedings of SIGGRAPH*, pages 65–76, 1997.

[WHSG97] B. Walter, P. Hubbard, P. Shirley, and D. Greenberg. Global illumination using local linear density estimation. *ACM Transactions on Graphics*, 16(3):217–259, 1997.

[WTP01] A. Wilkie, R. Tobler, and W. Purgathofer. Combined rendering of polarization and fluorescence effects. In *Proceedings of the EUROGRAPHICS rendering workshop*, pages 197–204, 2001.

[WTU+04] A. Wilkie, R. Tobler, C. Ulbricht, G. Zotti, and W. Purgathofer. An analytical model for skylight polarisation. In *Proceedings of the EUROGRAPHICS Symposium on Rendering*, pages 387–399, 2004.

Part II

Contributed Articles

Random Walk Algorithm for Estimating the Derivatives of Solution to the Elliptic BVP

Alexander Burmistrov

Institute of Computational Mathematics and Mathematical Geophysics (Siberian Branch of the Russian Academy of Sciences), prospect Akademika Lavrentjeva, 6, Novosibirsk, 630090, Russia, and
Novosibirsk State University, Pirogova, 2, Novosibirsk, 630090, Russia
burm@osmf.sscc.ru

Summary. Elliptic boundary value problem (BVP) for the stationary diffusion equation is considered. Within [BM03], we estimate the solution and its spatial derivatives by solving a system of local integral equations. We propose to use the Poisson-Boltzmann Green function instead of the Laplacian one. This enables us to obtain a convergent Neumann series for a wider class of equations.

1 Introduction

In this paper, we consider one of the classical problems of mathematical physics which often arises when studying potential theory, heat and electric conductivity, fluid dynamics, elasticity theory, geophysics, etc. Our main objective is to estimate the solution to the elliptic BVP as well as its spatial derivatives (Section 1).

For local estimation of the desired values we construct statistical algorithms, which are more suitable as compared to deterministic ones, especially, for the problems with complex geometries. In addition, statistical methods are well adapted to the up-to-date computing technique with a high degree of parallelization.

There are several statistical approaches to solving the problem: random walk on boundaries, simulation of diffusion trajectories by means of a system of stochastic differential equations, random walk inside the domain using the Green functions. We used the latter approach and constructed new statistical algorithms with the help of the central and the non-central Laplacian Green functions for the ball in [BM03]. However, algorithms proposed have some disadvantages. For example they fail to solve problems with large in absolute value negative coefficient $c(r)$.

This restriction does not have any physical interpretation but is required for convergence of the corresponding Neumann series. On the other hand, the algorithms, which use the central Poisson-Boltzmann Green function and its normal derivative are well-known. We offer to combine two approaches and to use the non-central Poisson-Boltzmann Green function for constructing a system of local integral equations (Section 2). As a result, we managed to extend a class of equations that can be solved by this algorithm. We give a detailed description of the algorithm in Section 3.

It is shown that under certain conditions the integral problem is equivalent to the differential one. We obtain the deterministic error order and investigate the estimator variance in Section 4.

We give some additional remarks concerning the algorithm in Section 5 and present numerical results in Section 6.

2 Differential Problem and Notations

Let us consider the Dirichlet problem

$$\Delta u(r) - c(r)u(r) + \big(v(r), \nabla u(r)\big) = -g(r), \quad r \in \Omega \subset \mathbb{R}^3 \tag{1}$$

$$u(s) = \psi(s), \quad s \in \Gamma = \partial\Omega \tag{2}$$

in the domain Ω with simply connected and piecewise smooth boundary Γ. Suppose that the functions $v(\cdot), c(\cdot) \in C^\delta(\mathbb{R}^3)$, i.e. satisfy the Hölder condition in $\mathbb{R}^3$ with exponent δ, $c(\cdot) > 0$, $g(\cdot) \in C^\delta(\overline{\Omega})$, and $\psi(\cdot)$ is a continuous function on Γ. Under the conditions stated above there exists a unique and smooth solution $u \in C^{2+\delta}_{loc}(\Omega) \cap C(\overline{\Omega})$ to problem (1) – (2) (see, e.g., [Kry96]).

We are interested in estimating the solution $u(r)$ and its gradient $\nabla u(r)$ at some point $r \in \Omega$.

Hereinafter the following notation is used:

$\mathcal{D}_\kappa = \Delta - \kappa^2$ – a stationary diffusion operator. We suggest to rewrite equation (1) isolating this operator on the left-hand side:

$$\mathcal{D}_\kappa u(r) \equiv \Delta u(r) - \kappa^2 u(r) = -\big(\kappa^2 - c(r)\big)u(r) - \big(v(r), \nabla u(r)\big) - g(r) \tag{3}$$

$B(r_0, R) = \{r' \in \mathbb{R}^3 : |r_0 - r'| \leqslant R(r_0) = \text{const}\}$ – the largest ball centered at r_0 and contained in $\overline{\Omega}$;

$R = R(r_0) = \text{dist}(r_0, \Gamma)$ – the radius of this ball;

$S(r_0, R) = \partial B(r_0, R) = \{r' \in \mathbb{R}^3 : |r_0 - r'| = R(r_0)\}$ – the corresponding sphere;

$R_{\max} = \max\limits_{r \in \Omega} R(r)$; c_0 – a constant such that $c_0 R^2_{\max} < 6$;

$p(r) = \left(1 - \dfrac{c_0 R^2(r)}{6}\right)$; $\varrho = \dfrac{R}{|r' - r_0|}$; $V = r' - r$; $W = \varrho(r' - r_0) - \dfrac{r - r_0}{\varrho}$;

$\Gamma_\varepsilon = \{r' \in \Omega : \exists r \in \Gamma, |r - r'| \leqslant \varepsilon\}$ – ε-strip of the boundary Γ;

ω – the unit vector corresponding to the vector function v, i.e. $\big(v(r), \nabla u(r)\big) = |v(r)| \dfrac{\partial u}{\partial \omega}(r)$ and $v(r) = |v(r)| \cdot \omega(r)$;

$a_\omega(r,r') = \cos(\widehat{V,\omega(r)})$ – the cosine of the angle between ω and $V = r' - r$;
$-c^*$ is a minimum eigenvalue of the Laplace operator in Ω;
$\mathcal{G}^\kappa_{r_0}(r,r')$ – the non-central Green function for the operator $\mathcal{D}_\kappa$ in the ball $B(r_0,R)$, which is the solution to the Dirichlet problem

$$\mathcal{D}_\kappa \mathcal{G}^\kappa_{r_0}(r,r') = \delta(r'-r), \;\; \mathcal{G}^\kappa_{r_0}(r,r')\big|_{r\in S(r_0,R)} = 0,$$

where $\delta(\cdot)$ is the Dirac delta function. The explicit form of the function $\mathcal{G}^\kappa_{r_0}$ and its derivatives, used for estimating the solution to (1) – (2), is presented below.

3 System of Integral Equations and Algorithm for its Solution

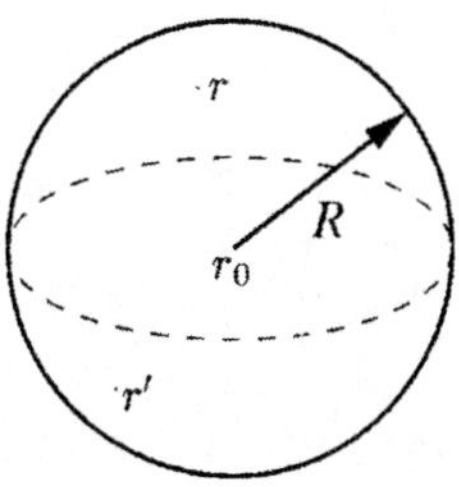

Integral equations for the function $u(r)$ from (3) and its spatial derivative $\dfrac{\partial u}{\partial \omega}(r)$ can be obtained using the mean-value theorem. We propose to use the non-central Green functions $\mathcal{G}^\kappa_{r_0}(r,r')$ for the stationary diffusion operator $\mathcal{D}_\kappa$ in the ball $B(r_0,R)$. As a result, we obtain the following equations (similar to [BM03, Mikh93]) for $r,\ r_0 \in \Omega \setminus \Gamma_\varepsilon$:

$$u_1(r) = -\int\limits_{S(r_0,R)} \frac{\partial \mathcal{G}^\kappa_{r_0}}{\partial \mathrm{n}_{r'}}(r,r')u_1(r')\mathrm{d}S_{r'} + \int\limits_{B(r_0,R)} \mathcal{G}^\kappa_{r_0}(r,r')g(r')\mathrm{d}r'$$
$$+ \int\limits_{B(r_0,R)} \mathcal{G}^\kappa_{r_0}(r,r')\left[|v(r')|\frac{\partial u_1}{\partial \omega}(r') + (\kappa^2 - c(r'))u_1(r')\right]\mathrm{d}r', \quad (4)$$

$$\frac{\partial u_1}{\partial \omega}(r) = -\int\limits_{S(r_0,R)} \frac{\partial}{\partial \omega}\left(\frac{\partial \mathcal{G}^\kappa_{r_0}}{\partial \mathrm{n}_{r'}}\right)(r,r')u_1(r')\mathrm{d}S_{r'} + \int\limits_{B(r_0,R)} \frac{\partial \mathcal{G}^\kappa_{r_0}}{\partial \omega}(r,r')g(r')\mathrm{d}r'$$
$$+ \int\limits_{B(r_0,R)} \frac{\partial \mathcal{G}^\kappa_{r_0}}{\partial \omega}(r,r')\left[|v(r')|\frac{\partial u_1}{\partial \omega}(r') + (\kappa^2 - c(r'))u_1(r')\right]\mathrm{d}r'. \quad (5)$$

Here $\mathrm{n}_{r'}$ is the outer normal to the sphere $S(r_0,R)$. We set $u_1 \equiv u$ for $r \in \Gamma_\varepsilon$.

Now let us write down the functions used in equations (4), (5) and specify the probability density functions proportional to them.

We can obtain the non-central Green function $\mathcal{G}^\kappa_{r_0}(r,r')$ by the method of images [FLS64]:

$$\mathcal{G}^\kappa_{r_0}(r,r') = \frac{1}{4\pi}\left[\frac{\sinh\{\kappa(R-|V|)\}}{\sinh\{\kappa R\}\,|V|} - \frac{\sinh\{\kappa(R-|W|)\}}{\sinh\{\kappa R\}\,|W|}\right]. \quad (6)$$

The normal derivative $\frac{\partial \mathcal{G}^{\kappa}_{r_0}}{\partial \mathrm{n}_{r'}}$ for $r' \in S(r_0, R)$, i.e. for $|r' - r_0| = R$, has the following form:

$$\frac{\partial \mathcal{G}^{\kappa}_{r_0}}{\partial \mathrm{n}_{r'}}(r, r')\bigg|_{|r'-r_0|=R} = \frac{1}{4\pi} \frac{|r_0 - r|^2 - R^2}{R \sinh\{\kappa R\}} \left[\frac{\kappa \cosh\{\kappa(R - |V|)\}}{|V|^2} + \frac{\sinh\{\kappa(R - |V|)\}}{|V|^3} \right]. \tag{7}$$

The non-central spatial derivative $\frac{\partial \mathcal{G}^{\kappa}_{r_0}}{\partial \omega}$ of the Green function has the following form:

$$\begin{aligned} \frac{\partial \mathcal{G}^{\kappa}_{r_0}}{\partial \omega}(r, r') = {} & \frac{\cos(\widehat{V, \omega})}{4\pi} \left[\frac{\kappa |V| \cosh\{\kappa(R - |V|)\} + \sinh\{\kappa(R - |V|)\}}{\sinh\{\kappa R\} |V|^2} \right] \\ & - \frac{\cos(\widehat{W, \omega})}{4\pi \varrho} \left[\frac{\kappa |W| \cosh\{\kappa(R - |W|)\} + \sinh\{\kappa(R - |W|)\}}{\sinh\{\kappa R\} |W|^2} \right], \end{aligned} \tag{8}$$

Further, we use the central Green function for the operator $\mathcal{D}_\kappa$, i.e. values of function u_1 (4) and its gradient (5) at an arbitrary point r are represented as a sum of integrals over the sphere and the ball centered at the same point ($r_0 = r$). Note that it is necessary to use non-central Green function (6) for obtaining the derivatives mentioned above. The forms of the central (i.e. for $r \to r_0$) functions (6), (7), (8) and the function $\frac{\partial}{\partial \omega}\left(\frac{\partial \mathcal{G}^{\kappa}_{r_0}}{\partial \mathrm{n}_{r'}}\right)$ are the following:

$$\mathcal{G}^{\kappa}_{r}(r, r') = \frac{1}{4\pi} \frac{\sinh\{\kappa R_V\}}{\sinh\{\kappa R\} |V|}, \tag{9}$$

$$\frac{\partial \mathcal{G}^{\kappa}_{r}}{\partial \mathrm{n}_{r'}}(r, r')\bigg|_{|V|=R} = -\frac{1}{4\pi R^2} \left(\frac{\kappa R}{\sinh\{\kappa R\}} \right), \tag{10}$$

$$\frac{\partial \mathcal{G}^{\kappa}_{r}}{\partial \omega}(r, r') = \frac{a_\omega(r, r')}{4\pi \sinh\{\kappa R\}} \left[\frac{\kappa \cosh\{\kappa R_V\}}{|V|} + \frac{\sinh\{\kappa R_V\}}{|V|^2} - \frac{\kappa |V|}{R^2} \right], \tag{11}$$

where $R_V = R - |V|$,

$$\frac{\partial}{\partial \omega}\left(\frac{\partial \mathcal{G}^{\kappa}_{r}}{\partial \mathrm{n}_{r'}}\right)(r, r')\bigg|_{|V|=R} = -\frac{a_\omega(r, r')}{4\pi R^2} \frac{\kappa R}{\sinh\{\kappa R\}} \cdot \frac{3}{R}. \tag{12}$$

Note, as the operator $\mathcal{D}_\kappa$ tends to the Laplace operator Δ when $\kappa \to 0$, functions (9) – (12) converge to the corresponding Green functions for the Laplace operator:

$$\mathcal{G}^{0}_{r}(r, r') = \frac{1}{4\pi} \left[\frac{1}{|r' - r|} - \frac{1}{R} \right],$$

$$\frac{\partial \mathcal{G}^{0}_{r}}{\partial \mathrm{n}_{r'}}(r, r')\bigg|_{|r'-r|=R} = -\frac{1}{4\pi R^2},$$

$$\frac{\partial \mathcal{G}_r^0}{\partial \omega}(r, r') = \frac{a_\omega(r, r')}{4\pi}\left[\frac{1}{|r' - r|^2} - \frac{|r' - r|}{R^3}\right],$$

$$\frac{\partial}{\partial \omega}\left(\frac{\partial \mathcal{G}_r^0}{\partial \mathrm{n}_{r'}}\right)(r, r')\bigg|_{|r'-r|=R} = -\frac{3a_\omega(r, r')}{4\pi R^3}.$$

Functions (10), (12) are proportional to the probability density F_S on the sphere and functions (9), (11) are proportional to the probability densities F_0 and F_1 on the ball:

$$\mathcal{G}_r^\kappa(r, r') = \frac{R^2}{6} \cdot C_{01}(\kappa R) \cdot F_0(r, r'), \tag{13}$$

$$-\frac{\partial \mathcal{G}_r^\kappa}{\partial \mathrm{n}_{r'}}(r, r') = C_{00}(\kappa R) \cdot F_S(r, r'), \tag{14}$$

$$\frac{\partial \mathcal{G}_r^\kappa}{\partial \omega}(r, r') = a_\omega(r, r')\frac{3R}{4} \cdot C_{11}(\kappa R) \cdot F_1(r, r'), \tag{15}$$

$$-\frac{\partial}{\partial \omega}\left(\frac{\partial \mathcal{G}_r^\kappa}{\partial \mathrm{n}_{r'}}\right)(r, r') = a_\omega(r, r')\frac{3}{R} \cdot C_{10}(\kappa R) \cdot F_S(r, r'). \tag{16}$$

Here $R = R(r)$, and the functions C_{kl} $(k, l \in \{0, 1\})$ are defined on the positive semi-axis $[0, +\infty)$ and monotonically decrease from 1 to 0; we give their explicit form in Subsection 4.2. The function F_S is the probability density of uniform distribution on the sphere:

$$F_S(\varphi, \theta)\mathrm{d}S = \frac{\mathrm{d}\varphi}{2\pi} \cdot \frac{\sin(\theta)\mathrm{d}\theta}{2},$$

where $\theta \in (0, \pi)$, $\varphi \in (0, 2\pi)$, $\rho = |r' - r| \in (0, R)$ are coordinates of the local (with the origin at the point r) spherical coordinate system with the differential of volume $\mathrm{d}r' = \sin(\theta)\rho^2\mathrm{d}\theta\mathrm{d}\varphi\mathrm{d}\rho$. The probability densities F_0 and F_1 are factorable in this coordinate system:

$$F_j(r, r')\mathrm{d}r' = F_S(\varphi, \theta)\mathrm{d}\varphi\mathrm{d}\theta \cdot F_j^\rho(\rho)\mathrm{d}\rho, \quad j = 0, 1;$$

and the factors $F_j^\rho(\rho)$ outside the angular density have the following form:

$$F_0^\rho(\rho) = \frac{6C_{01}^{-1}(\kappa R)}{R^2 \sinh\{\kappa R\}}\rho \sinh\{\kappa(R - \rho)\}, \tag{17}$$

$$F_1^\rho(\rho) = \frac{4C_{11}^{-1}(\kappa R)}{R^2 \sinh\{\kappa R\}}\left[\kappa\rho\cosh\{\kappa(R - \rho)\} + \sinh\{\kappa(R - \rho)\} - \frac{\kappa\rho^3}{R^2}\right]. \tag{18}$$

The simulation of the random variables with probability density functions (17) and (18) is described in Subsection 4.3.

To construct statistical algorithms, equations (4) and (5) are combined below into a unified integro-algebraic equation with allowance for (13) – (16). We introduce a special extension of a set of phase coordinates by the discrete variable j which can take only two values: $j = 0$ or $j = 1$. Moreover, it is reasonable to consider the variable $\frac{R(r)}{3}\frac{\partial u_1}{\partial \omega}(r)$ instead of $\frac{\partial u_1}{\partial \omega}(r)$. Let $\mathbf{w} = (r, j) \in \mathbb{R}^3 \times \{0, 1\}$ be a point of a new phase space. We define the following functions:

$$U(\mathbf{w}) \equiv U(r, j) = \begin{cases} u_1(r), & j = 0; \\ \frac{R(r)}{3}\frac{\partial u_1}{\partial \omega}(r), & j = 1. \end{cases}$$

If we choose the parameter c_0 such that $c_0 R^2_{\max} < 6$, then the variable $p(r) = \left(1 - \frac{c_0 R^2(r)}{6}\right)$ has a probability interpretation and we can use it below for the randomization of the unified equation:

$$U(r, j) = p(r) \int\limits_{S(r,R)} F_S(r, r')U(r', 0)Q_{j0}(r, r')\mathrm{d}S_{r'} + G(r, j) \tag{19}$$
$$+ (1 - p(r)) \int\limits_{B(r,R)} F_j(r, r') \left[p_1^c Q_{j1}^c(r, r')U(r', 1) + p_0^c Q_{j0}^c(r, r')U(r', 0)\right] \mathrm{d}r',$$

where $p_0^c = p_0^c(r, r')$ and $p_1^c = p_1^c(r, r')$ are, also, some probabilities: $p_0^c + p_1^c = 1$. We explain the way of defining them in Subsection 4.2. In addition, we give the explicit form of the weights Q_{k0}, Q_{kl}^c $(k, l \in \{0, 1\})$ and the form of the function $G(\mathbf{w})$ in Subsection 4.2 as well.

We construct the statistical estimator $\zeta(\mathbf{w}_0)$ for $U(\mathbf{w}_0)$ according to (19) as follows.

Algorithm. [The initial weight $Q^0 = 1$, the initial point $\mathbf{w}_0 = (r_0, j_0)$]

1 $Q^n G(\mathbf{w}_n)$ is added to the counter, $n = 0, 1, \ldots$

2a With probability $p(r_n)$, a uniformly distributed (i.e. with density $F_S(r_n, r')$) point r_{n+1} is generated on the sphere $S(r_n)$

  2a$_1$ j_{n+1} takes value 0

  2a$_2$ the weight Q^n is multiplied by Q_{j_n0}: $Q^{n+1} = Q^n \cdot Q_{j_n0}$

2b With complementary probability $1 - p(r_n)$, the next point r_{n+1} is generated in the ball $B(r_n)$ with the density $F_{j_n}(r_n, r')$

  2b$_1$ j_{n+1} takes value 1 with probability $p_1^c(r_n, r_{n+1})$, and value 0 with probability $p_0^c(r_n, r_{n+1})$

  2b$_2$ the weight Q^n is multiplied by $Q^c_{j_nj_{n+1}}$: $Q^{n+1} = Q^n \cdot Q^c_{j_nj_{n+1}}$

3 Check the termination condition, written below:

When simulating the introduced random walk on spheres and balls, with the path falling into Γ_ε at a step with a random number N, the chain terminates and the estimator for $U(\mathbf{w}_N)$ (see Subsection 4.1) multiplied by the weight

Q^N is added to the counter. As a result, we obtain the following estimator for $U(\mathbf{w}_0)$:

$$\zeta(\mathbf{w}_0) = \sum_{n=0}^{N-1} Q^n G(\mathbf{w}_n) + Q^N U(\mathbf{w}_N).$$

4 Realization of the Algorithm

First of all, let us notice that calculation of the distance $R(r)$ on every step could be fairly time-consuming in the domains with complex boundaries. Since representations (4) and (5) hold for an arbitrary ball contained in $\overline{\Omega}$, then it is possible to use not maximal balls but those with easy-to-compute radii $d(r)$ inside the domain. So, the radii $R(r)$ have to be used only in the immediate neighborhood of the boundary. This procedure will increase the number of transitions in the chain but could decrease the computer costs of the algorithm.

4.1 Estimation of the Solution $U(\mathbf{w})$ in Γ_ε

Since the values of $U(\mathbf{w})$ in Γ_ε are unknown, the corresponding estimators are obtained as follows. For $j_N = 0$, one can set

$$U(r_N, 0) = u_1(r_N) = \psi(r_N^*), \text{ where } r_N^* \in \Gamma,\ \ r_N \in \Gamma_\varepsilon,\ \ |r_N - r_N^*| = R(r_N),$$

i.e. the point r_N^* is the closest one on Γ to the point r_N in Γ_ε.

Assuming the first derivatives of the solution to be finite in Ω, and therefore

$$\frac{R(r)}{3}\frac{\partial u_1}{\partial \omega}(r) = \mathcal{O}(\varepsilon) \text{ for } r \in \Gamma_\varepsilon,$$

one can approximately set $U(r_N, 1) = 0$ for $j_N = 1$. As a result, we obtain the feasible but biased estimator $\zeta_\varepsilon(\mathbf{w}_0)$ for $U(\mathbf{w}_0)$.

4.2 Coefficients of Equation (19)

The weights Q_{k0}, Q_{kl}^c from equation (19) have the following form:

$$Q_{00}(r, r') = \frac{C_{00}(\kappa R(r))}{p(r)}, \quad Q_{10}(r, r') = \frac{a_\omega C_{10}(\kappa R(r))}{p(r)},$$

$$Q_{01}^c(r, r') = \frac{3|v(r')|C_{01}(\kappa R(r))}{p_1^c c_0 R(r')}, \quad Q_{11}^c(r, r') = \frac{9a_\omega|v(r')|C_{11}(\kappa R(r))}{2p_1^c c_0 R(r')},$$

$$Q_{00}^c(r, r') = \frac{C_{01}(\kappa R(r))(\kappa^2 - c(r'))}{p_0^c c_0}, \quad Q_{10}^c(r, r') = \frac{3a_\omega C_{11}(\kappa R(r))(\kappa^2 - c(r'))}{2p_0^c c_0}.$$

One can set $p_1^c = p_0^c = 0.5$, but it is more efficient to set these probabilities proportional to the functions $|v|$ and $|\kappa^2 - c|$:

$$p_1^c = \frac{3|v(r')|}{R(r')p_{cv}}, \quad p_0^c = \frac{|\kappa^2 - c(r')|}{p_{cv}}, \quad \text{where } p_{cv} = \frac{3|v(r')|}{R(r')} + |\kappa^2 - c(r')|.$$

The weights Q_{k0}, Q_{kl}^c differ from those introduced for the analogous method, which was proposed in [BM03] for the Laplace operator ($\kappa \equiv 0$), in factors C_{kl}, which have the following form:

$$C_{00}(x) = C_{10}(x) = \frac{x}{\sinh(x)},$$

$$C_{01}(x) = \frac{6}{x^2}\left(1 - C_{00}(x)\right),$$

$$C_{11}(x) = \frac{4}{3}\left(\frac{\tanh(x/2)}{x/2} - \frac{C_{00}(x)}{4}\right).$$

The functions $C_{00} = C_{10}, C_{01}, C_{11}$ monotonically decrease from 1 to 0.

The functions $G(\mathbf{w})$ should be estimated on every step during random walk on spheres and balls inside $\Omega \setminus \Gamma_\varepsilon$. These functions have the following forms (depending on the coordinate j):

$$G(r,0) = C_{01}(\kappa R(r))\frac{R^2(r)}{6} \int\limits_{B(r,R)} F_0(r,r')g(r')\mathrm{d}r',$$

$$G(r,1) = C_{11}(\kappa R(r))\frac{R^2(r)}{4} \int\limits_{B(r,R)} F_1(r,r')g(r')a_\omega(r,r')\mathrm{d}r'.$$

We can estimate the functions $G(r,0)$ and $G(r,1)$ by "single random sample" method [EM82], i.e. using only one sample coordinate $r_1 = (\varphi_1, \theta_1, \rho_1)$ for estimating the whole integral on every step. But it seems reasonable to use several points obtained from r_1 by the deterministic method for variance reduction. Note that if α is a uniform random variable on $(0,1)$, then the random variable $\varphi_1 = 2\pi\alpha$ has the same distribution as $2\pi\alpha + \phi \pmod{2\pi}$ for some angle ϕ. Moreover, a random variable $\mu_1 = \cos(\theta_1) = 2\alpha - 1$ has the same distribution as $-\mu_1$. Therefore, using four angles $\phi = 0, \pi/4, \pi/2, 3\pi/4$ and two variables $\pm\mu_1$, we obtain eight points instead of a single point r_1, and averaging over these points results in a lesser variance.

4.3 Sampling from Probability Density Functions $F_j^\rho(\rho)$

We can use the von Neumann rejection method [Neu51, EM82] for sampling probability densities (17) and (18). We can select a majorant function for (17) in two ways:

$$g_0(\rho) \equiv \rho\frac{\sinh\{\kappa(R-\rho)\}}{\sinh\{\kappa R\}} \leqslant \rho\exp(-\kappa\rho) \equiv g_1(\rho) \leqslant R\exp(-\kappa\rho) \equiv g_2(\rho).$$

If the function g_1, which is proportional to the gamma distribution with parameters $(2, \kappa)$, is also sampled by the rejection method (i.e. rejecting the values of ρ which are greater than R), then the ratio of the corresponding computational costs is the following:

$$\frac{S_1}{S_2} = \frac{\int_0^R g_1(\rho)\mathrm{d}\rho}{\int_0^R g_0(\rho)\mathrm{d}\rho} \cdot \frac{\int_0^\infty g_1(\rho)\mathrm{d}\rho}{\int_0^R g_1(\rho)\mathrm{d}\rho} \Bigg/ \frac{\int_0^R g_2(\rho)\mathrm{d}\rho}{\int_0^R g_0(\rho)\mathrm{d}\rho} = \frac{1}{\kappa R(1 - e^{-\kappa R})}.$$

Therefore, if $S_2 < S_1$ (it holds when $\kappa R < 1.3499764854$), then we use the majorant function g_2, otherwise g_1.

For probability density (18) we have

$$\kappa\rho \cosh\{\kappa(R-\rho)\} + \sinh\{\kappa(R-\rho)\} - \frac{\kappa\rho^3}{R^2} \leqslant (\kappa R + 1)\cosh\{\kappa(R-\rho)\}.$$

If α is a uniform random variable on $(0, 1)$ and a variable η is the solution to the equation $\sinh\{\kappa(R-\eta)\} = \alpha \sinh\{\kappa R\}$, then η has the probability density $A_{\kappa,R} \cosh\{\kappa(R-\rho)\}$ on $(0, R)$.

5 Theorems

All the theorems in this section follow from the respective statements proved in [BM03], just by taking into account the fact that the weights in the corresponding Neumann series are multiplied by the factors C_{kl} $(k, l \in \{0, 1\})$ which are less than 1.

Theorem 1. *Let for any $r \in \Omega$, the following assumptions hold:*

$$|\kappa^2 - c(r)| + 3\frac{|v(r)|}{R(r)} \leqslant \frac{2}{3}c_0, \tag{20}$$

$$c_0 < \frac{6}{\pi^2}c^* \simeq 0.6079c^*.$$

Then there exists a unique bounded solution to integral equation (19). This solution admits a representation as Neumann series and equals the solution to BVP (1) – (2):

$$U(r, 0) = u(r), \;\; U(r, 1) = \frac{R(r)}{3}\frac{\partial u}{\partial \omega}(r).$$

If $v \equiv 0$, then we can replace (20) for $|\kappa^2 - c(r)| \leqslant c_0$.

Note that there is no restriction for the coefficient $c(r)$ itself in (20), whereas an analogous assumption from [BM03] has the following form:

$$|c(r)| + 3\frac{|v(r)|}{R(r)} \leqslant \frac{2}{3}c_0.$$

So, the usage of the Green function for the operator $\mathcal{D}_\kappa$ as compared to the Green function for the Laplace operator enables us to solve problem (1) with a large (in absolute value) negative coefficient $c(r)$ that is close to some constant κ^2.

Theorem 2. *If the first derivatives of $u(r)$ are finite in Ω, then*

$$\mathbf{E}\zeta_\varepsilon(r,0) = u_\varepsilon(r) \ \text{and} \ |u(r) - u_\varepsilon(r)| = \mathcal{O}(\varepsilon), \ \ \varepsilon > 0, \ \ r \in \Omega.$$

Moreover,

$$\mathbf{E}\zeta_\varepsilon(r,1) = f_\varepsilon(r) \ \text{and} \ \left|\frac{R(r)}{3}\frac{\partial u}{\partial \omega}(r) - f_\varepsilon(r)\right| = \mathcal{O}(\varepsilon), \ \ \varepsilon > 0, \ \ r \in \Omega.$$

Theorem 3. *If $c_0 < 0.4881c^*$ and $g \equiv 0$, then the variance $\mathbf{V}\zeta_\varepsilon$ is finite for any $\varepsilon > 0$.*

To study the variance finiteness for $g \not\equiv 0$, we write down equation (19) by analogy with [BM03] in the following form:

$$U(\mathbf{w}) = p\left[\int\limits_{S(r,R)} F_S(r,r')U(r',0)Q_{j0}\mathrm{d}S_{r'} + \frac{G(\mathbf{w})}{p}\right]$$
$$+(1-p)\int\limits_{B(r,R)} F_j(r,r')[p_1^c U(r',1)Q_{j1}^c + p_0^c U(r',0)Q_{j0}^c]\mathrm{d}r'.$$

According to it the function $G(r,j)$ is calculated only for the step $(r,j) \to (r',0)$, i.e. on the sphere, but with the weight $p(r)^{-1}$. In other words, we use a new function $\tilde{G}(r,j)$ instead of $G(r,j)$:

$$\tilde{G}(r_i,j_i) = \begin{cases} 0, & \text{when } j_{i+1} = 1; \\ \left[1 - \dfrac{c_0 R^2(r_i)}{6}\right]^{-1} G(r_i,j_i), & \text{when } j_{i+1} = 0. \end{cases}$$

Thus, we obtain one more estimator $\zeta_{\varepsilon,1}$ for which $\mathbf{E}\zeta_{\varepsilon,1} = \mathbf{E}\zeta_\varepsilon$.

Theorem 4. *If $c_0 < 0.4881c^*$ and $g \not\equiv 0$, then the variance $\mathbf{V}\zeta_{\varepsilon,1}$ is finite for any $\varepsilon > 0$.*

6 Some additional Remarks

6.1 Estimating a Derivative of u along an Arbitrary Direction μ

To estimate a derivative of u along an arbitrary direction μ (which may differ from the direction of ω), we should use a representation for $\frac{R(r)}{3}\frac{\partial u_1}{\partial \mu}$ similar to (5) at the first step of the algorithm:

$$U_\mu(r) \equiv \frac{R(r)}{3}\frac{\partial u_1}{\partial \mu}(r) = p(r) \int\limits_{S(r,R)} F_S(r,r')Q_{\mu 0}(r,r')U(r',0)\mathrm{d}S_{r'} \tag{21}$$

$$+ (1-p(r)) \int\limits_{B(r,R)} F_1(r,r')\left[p_1^c Q_{\mu 1}^c U(r',1) + p_0^c Q_{\mu 0}^c U(r',0)\right]\mathrm{d}r' + G_\mu(r).$$

After the first step we apply the simulation algorithm associated with the unit vector $\omega(r)$. In particular, for the estimators $\{\zeta_\mu(r)\}_{\mu=x,y,z}$ of the solution gradient $\mathrm{grad}u(r)$, the following representation is valid

$$\zeta_\mu(r) = 3\big(G_\mu(r) + Q_\mu(\mathbf{w}_1)\zeta(\mathbf{w}_1)\big)/R(r), \quad \mu = x,y,z,$$

where $\zeta(\mathbf{w}_1)$ is the estimator for $U(\mathbf{w}_1)$ and $\mathbf{w}_1 \equiv (r_1, j_1)$ is a phase state point after the first Markov chain transition according to (21). Since the coefficients $G_\mu(r)$ and $Q_\mu(\mathbf{w}_1)$ are known after this first transition, we can estimate all three components of the gradient vector simultaneously, i.e. using the same trajectories.

6.2 Another Randomization

Another integral representation for $u(r)$ from (3) is known [ENS89] when $v \equiv 0$ and $0 \leqslant c(r) \leqslant \kappa^2$:

$$u(r) = q \int\limits_{S(r,R)} F_S(r,r')u(r')\mathrm{d}S_{r'} \tag{22}$$

$$+(1-q) \int\limits_{B(r,R)} F_0(r,r')\left[u(r')\frac{(\kappa^2 - c(r'))}{\kappa^2} + \frac{g(r')}{\kappa^2}\right]\mathrm{d}r'.$$

Here $q = C_{00}(\kappa R(r))$ is used for randomization instead of c_0 and the function $g(r')$ is calculated only for sufficiently rare transitions into the ball $B(r,R)$.

Let us rewrite (19) similar to (22):

$$U(\mathbf{w}) = \left(1 - \frac{c_0 R^2(r)}{6}\right) \int\limits_{S(r,R)} F_S(r,r')U(r',0)Q_{j0}\mathrm{d}S_{r'} \tag{23}$$

$$+ \frac{c_0 R^2(r)}{6} \int\limits_{B(r,R)} F_j(r,r')\left[p_1^c Q_{j1}^c U(r',1) + p_0^c Q_{j0}^c U(r',0) + g(r')\mathcal{Q}_j(r,r')\right]\mathrm{d}r',$$

where additional factors $\mathcal{Q}_j$ have the following form

$$\mathcal{Q}_0(r,r') = \frac{C_{01}(\kappa R(r))}{c_0}, \quad \mathcal{Q}_1(r,r') = \frac{3}{2}\frac{C_{11}(\kappa R(r))}{c_0} a_\omega(r,r').$$

Actually, to use this representation means to compute the integrals $G(r_n, j_n)$ just taking a single random sample r_{n+1}, which has been already sampled from the probability density function $F_j(r_n, r')$ in the transition $r_n \to r_{n+1}$. According to (23) we obtain one more estimator $\zeta_{\varepsilon,2}$ with $\mathbf{E}\zeta_\varepsilon = \mathbf{E}\zeta_{\varepsilon,2}$.

We can adjust the algorithm based on (22) for estimating the solution gradient $\mathrm{grad}u(r)$ by using the representation similar to (21) at the first step.

6.3 Mixed Boundary Conditions

Let us consider problem (1) with the Dirichlet boundary condition (2) given for a part of the boundary $\Gamma_1 \subset \Gamma$ only. The Neumann condition is given for the other part $\Gamma_2 = \Gamma \setminus \Gamma_1$:

$$(\nabla u(s), \gamma(s)) + \alpha(s)u(s) = \psi(s), \quad s \in \Gamma_2, \tag{24}$$

where $\forall s \in \Gamma_2$ $(\gamma(s), \mathrm{n}(s)) \geqslant \gamma > 0$, $\alpha(s) \geqslant C_\alpha > 0$, $\mathrm{n}(s)$ is the outer normal at the point s.

We suggest to use approximation of condition (24) in $\Gamma_{2\varepsilon}$ with desired accuracy (see [Mak01]). Let $d_\gamma(r)$ be the distance between r and Γ_2 along the vector field γ, and let $\pi(r)$ be the projection of the point $r \in \Gamma_{2\varepsilon}$ onto Γ_2 along the vector field γ. Then the following representation holds in $\Gamma_{2\varepsilon}$:

$$u(r) = \frac{1 + \alpha \cdot d_\gamma(r)}{1 + \alpha \cdot (d_\gamma(r) + \varepsilon)} u(r - \varepsilon\gamma) + \frac{\varepsilon}{1 + \alpha \cdot (d_\gamma(r) + \varepsilon)} \psi(\pi(r)) + \mathcal{O}(\varepsilon^2),$$

and we can randomize it for a specified probability $\tilde{p}(r)$ as follows

$$u(r) = \tilde{p}(r) \cdot \frac{z(r)}{\tilde{p}(r)} \cdot u(r - \varepsilon\gamma) + (1 - \tilde{p}(r)) \cdot 0 \cdot u(r) + \widetilde{\psi}(r).$$

According to the latter we should continue the algorithm when the trajectory gets into $\Gamma_{2\varepsilon}$ with $j_n = 0$ in the following way.

Algorithm with reflection. [The current point $\mathbf{w}_n = (r_n, 0)$, $r_n \in \Gamma_{2\varepsilon}$]

1 $Q^n \widetilde{\psi}(\pi(r_n))$ is added to the counter

2a With probability $\tilde{p}(r_n)$, the next point $r_{n+1} = r_n - \varepsilon_n \gamma_n$ is reflected back to $\Omega \setminus \Gamma_\varepsilon$

2a$_1$ j_{n+1} takes value 0

2a$_2$ the weight is changed as follows: $Q^{n+1} = Q^n \cdot \dfrac{z(r_n)}{\tilde{p}(r_n)}$

2b With probability $1 - \tilde{p}(r_n)$, the chain terminates (the trajectory is absorbed).

We can take, for example, $\tilde{p}(r) = z(r) < 1$. The considerations from Subsection 4.1 are still valid, so when the trajectory gets into Γ_ε with $j_n = 1$ it is absorbed, too.

The average number of reflections is of order $\mathcal{O}(\varepsilon^{-1})$ for this scheme [Mak01], therefore the total deterministic error remains of order $\mathcal{O}(\varepsilon)$.

7 Numerical Results

The total error δ_T of the statistical methods is equal to the sum of the deterministic part δ_d and the statistical one δ_s. Let M be the number of the simulated trajectories, then

$$\delta_T = \delta_d + \delta_s = C_d\varepsilon + C_s\frac{\sigma(\zeta)}{\sqrt{M}},$$

where $\sigma(\zeta)$ is a mean square error of the estimator ζ. We should choose $M \sim \varepsilon^{-2}$ according to this equality.

The average number of transitions in a "random walk on spheres" chain was obtained with the help of the renewal theory (see, e.g., [EM82, ENS89]) for a wide class of boundaries Γ, including those of the convex domains: $\mathbf{E}N \sim |\ln(\varepsilon)|$.

Let us demonstrate the efficiency of the proposed algorithm by estimating the solution and its gradient for test problem (1) – (2) in the domain $\Omega = [-1;1]^3$ with the known solution $u(r) \equiv u(x,y,z) = x \cdot \exp(y) \cdot \sin(\pi z/4)$ and the following coefficients:

$$c(r) = \kappa^2 + \sin\left(x - \frac{y}{z+2}\right), \quad v(r) = (R(r)/3, 0, 0), \quad \kappa = 2.12.$$

The functions $g(r)$ and $\psi(s)$ can be determined explicitly by using $u(r)$, $c(r)$ and $v(r)$. Table 1 shows the numerical results obtained at the point

Table 1. The total and statistical errors of the estimator $\zeta_{\varepsilon,2}$ from Subsection 6.2 for the solution $u(r_0) = -0.26273362$ and the gradient grad$u(r_0) = (1.05093449, -0.26273362, -0.24943491)$. The parameter $c_0 = 4.0$.

M	ε	$\delta_T \pm \frac{\sigma(\zeta)}{\sqrt{M}}$ for $u(r_0)$	$\mathbf{E}N$	M	ε	$\delta_T \pm \frac{\sigma(\zeta)}{\sqrt{M}}$ for $\frac{\partial u}{\partial x}(r_0)$	$\mathbf{E}N$
10^5	10^{-2}	$(4.973 \pm 7.887) \cdot 10^{-4}$	10.38	10^4	10^{-3}	$(2.461 \pm 0.213) \cdot 10^{-2}$	17.80
10^7	10^{-3}	$(2.295 \pm 8.006) \cdot 10^{-5}$	17.84	10^6	10^{-4}	$(2.059 \pm 0.211) \cdot 10^{-3}$	25.23
M	ε	$\delta_T \pm \frac{\sigma(\zeta)}{\sqrt{M}}$ for $\frac{\partial u}{\partial y}(r_0)$	$\mathbf{E}N$	M	ε	$\delta_T \pm \frac{\sigma(\zeta)}{\sqrt{M}}$ for $\frac{\partial u}{\partial z}(r_0)$	$\mathbf{E}N$
10^4	10^{-3}	$(3.048 \pm 0.211) \cdot 10^{-2}$	17.80	10^4	10^{-3}	$(5.770 \pm 0.203) \cdot 10^{-2}$	17.80
10^6	10^{-4}	$(2.970 \pm 0.210) \cdot 10^{-3}$	25.23	10^6	10^{-4}	$(3.020 \pm 0.205) \cdot 10^{-3}$	25.23

$r_0 = (-0.25, 0.50, 0.88)$. These numerical results confirm the predicted error order $\mathcal{O}(\varepsilon + M^{-1/2})$. Note that we estimate all the three derivatives using one set of trajectories (as described in Subsection 6.1).

In conclusion, let us point to the fact that the usage of the operator $\mathcal{D}_\kappa$ in comparison with the Laplace operator extends a class of solvable problems (1) by the ones with a large (in absolute value) negative coefficient $c(r)$ that is close to some constant value κ^2.

Acknowledgements

The author acknowledges the kind hospitality of the Ulm University and the MCQMC'2006 conference organizers.

This work was partly supported by Russian Science Support Foundation, Russian Foundation for Basic Research (grant 05-01-00268), Programme "Leading Scientific Schools" (grant SS-4774.2006.1) and Lavrentiev SB RAS Grant for Young Scientists.

The author would also like to thank Dr. Irina Shalimova for the valuable discussions.

References

[BM03] A.V. Burmistrov and G.A. Mikhailov. Monte Carlo Computation of Derivatives of a Solution to the Stationary Diffusion Equation. Comp. Math. Math. Phys., **43**, No. 10, 1459–1471 (2003)

[EM82] S.M. Ermakov and G.A. Mikhailov. Statistical modelling. Nauka, Moscow (1982) In Russian.

[ENS89] S.M. Ermakov, V.V. Nekrutkin, and A.S. Sipin. Random processes for classical equations of mathematical physics. Kluwer Academic Publishers, Dordrecht, The Netherlands (1989)

[FLS64] R.P. Feynman, R.B. Leighton, and M. Sands. The Feynman Lectures on Physics: Mainly Electromagnetism and Matter. Addison-Wesley Publishing, New York (1964)

[Kry96] N.V. Krylov. Lectures on Elliptic and Parabolic Equations in Hölder Spaces. Graduate Studies in Mathematics, **12**, 1996.

[Mak01] R.N. Makarov. Statistical modelling of solutions of stochastic differential equations with reflection of trajectories from the boundary. Russian J. Numer. Anal. Math. Modelling, **16**, No. 3, 261–278 (2001)

[Mikh93] G.A. Mikhailov. New algorithms of the Monte Carlo method for solving the Helmholtz equation. Russian Acad. Sci. Dokl. Math. **46**, No. 2, 387–391 (1993)

[Neu51] J. von Neumann. Various techniques used in connection with random digits. Monte Carlo methods. Nat. Bur. Standards AMS. **12**, 36–38 (1951)

Free-Knot Spline Approximation of Fractional Brownian Motion

Jakob Creutzig[1] and Mikhail Lifshits[2]

[1] TU Darmstadt, Dept. of Mathematics, Schloßgartenstraße 7, 64289 Darmstadt
creutzig@mathematik.tu-darmstadt.de

[2] St.Petersburg State University, Dept. of Mathematics and Mechanics, Bibliotechnaya pl., 2, 198504, Stary Peterhof, Russia
lifts@mail.rcom.ru

Summary. For a fractional Brownian motion B^H on $[0,1]$, we consider approximations of B^H by piecewise polynomial splines. Asymptotics of minimal average error rates are established and found to be of order k^{-H}, where k is the number of free knots used in the spline approximation.

1 Main Result

Let $H \in]0,1[$ and let B^H be a fractional Brownian motion (fBM) on $[0,1]$, i.e., B^H is a mean zero Gaussian process with continuous paths such that $B_0^H = 0$ and

$$\mathbb{E}\,(B_t^H - B_s^H)^2 = |t-s|^{2H}.$$

Note that for $H = 1/2$, this just boils down to classical Brownian motion. The approximation and simulation of B^H is a field of ongoing study and interest. Optimal linear approximation schemes were studied in [KL02], [AT03] and [DZ04]. We will study a certain type of nonlinear approximation; namely, we will study how well B^H can be approximated by random splines with freely chosen knots. More specifically, denote, for $k, r \in \mathbb{N}_0$, with Φ_k^r the set of all splines on $[0,1]$ with at most k free knots and polynomial degree r. For any random process X and $k, r \in \mathbb{N}_0$, $q \in]0,\infty[$, $p \in [1,\infty]$, we set

$$e_k(X,r,q,p) := \inf\{(\mathbb{E}\,\|X - X^k\|_{L_p}^q)^{1/q} \;:\; X^k \text{ has a.s. paths in } \Phi_k^r\}.$$

In other words, $e_k(X,r,q,p)$ measures the q–th moment of the smallest error (measured in L_p norm) achievable by approximating X with a random spline from Φ_k^r. (We restrict to the case of $p \geq 1$, e.g. the case where L_p is normed and not quasi–normed, since we shall rely on the triangle inequality.)

In the following, we are interested in the weak asymptotics of $e_k(B^H, r, q, p)$ as $k \to \infty$. For two sequences a_k, b_k of nonnegative real numbers we write

$a_k \preceq b_k$ iff $\overline{\lim}_k a_k/b_k$, and $a_k \asymp b_k$ iff $a_k \preceq b_k \preceq a_k$. For $H = 1/2$, it was found in [KP05], [CMGR98] that

$$e_k(B^{1/2}, r, q, p) \asymp k^{-1/2} \tag{1}$$

for all r, q, p. The aim of this paper is to generalize this result to fractional Brownian motion:

Theorem 1. *Let H, p, q, r be as above, and assume $p < \infty$. Then*

$$e_k(B^H, r, p, q) \asymp k^{-H}.$$

Since it is well-known (see, e.g., [Rit00, p. 115, Proposition 34], [KL02],) that linear approximations already achieve this rate, this shows in particular that nonlinear spline approximation of fBM is not vastly superior to linear spline approximation. This is in stark contrast e.g. to piecewise smooth processes discussed in [CA97], [CDGO02], but also to non-Gaussian Lévy processes. An intuitive explanation is that not only the increments of B^H are stationary but the local smoothness properties of its paths are rather time-homogenous, which distinguishes B^H from piecewise smooth and from non-Gaussian Lévy processes.

2 fBM and RLfBM

The main difference between BM and fBM is the independence of increments. Since this natural property is the main ingredient of the proof for (1), we need a way to control the "influence of the past" for fBM in an explicit form. The Riemann–Liouville fractional Brownian Motion, a close relative for the fBM, is ideally suited for this task. It is easily verified that that for a Brownian Motion W the stochastic integral

$$R_t^H := \int_0^t (t-s)^{H-1/2}\, \mathrm{d}W_s$$

is well–defined for every $t \geq 0$, and that the resulting process $R^H = (R_t^H)_{t \geq 0}$ is a continuous Gaussian process. R^H is called the *Riemann–Liouville fractional Brownian Motion (RLfBM)* and is H-self-similar; i.e., for any $c > 0$ we have

$$(R_{ct}^H)_{t \geq 0} \stackrel{d}{=} c^H (R_t^H)_{t \geq 0}.$$

The reason we introduce the RLfBM is that this process is at the same time convenient for analytic study and "close" to the fBM in a suitable way. Namely, we have the following connection, see [LS05, Chapter 6]:

Fact 1 *It is true that* $B_t^H \stackrel{d}{=} C_H\left(R_t^H + M_t^H\right)$, *where*

$$M_t^H = \int_0^\infty \left((t+s)^{H-1/2} - s^{H-1/2}\right) \mathrm{d}\tilde{W}_s$$

with $\tilde{W}$ *an independent copy of* W *and*

$$C_H = \frac{\left[\sin(\pi H)\Gamma(1+2H)\right]^{1/2}}{\Gamma(\frac{1}{2}+H)}.$$

Furthermore, M^H *has a.s.* C^∞ *paths.*

Thus, it is not surprising that approximation problems for fBM frequently turn out to be equivalent to the same problems for RLfBM. It follows immediately by considering e.g. piecewise constant interpolation with equidistant knots that $e_k(M^H, 0, \infty, q) \le ck^{-1}$. Note that adding or subtracting a "smoother" process doesn't really matter in approximation issues; to be more precise:

Lemma 1. *Assume that* $\nu < \sigma$, *that* X, Y, Z *are processes such that* $X = Y + Z$, *and that* $e_k(Z, r, p, q) \lesssim k^{-\sigma}$. *Then*

$$\underline{\lim}_{k\to\infty} k^\nu e_k(X, r, p, q) = \underline{\lim}_{k\to\infty} k^\nu e_k(Y, r, p, q),$$

and

$$\overline{\lim}_{k\to\infty} k^\nu e_k(X, r, p, q) = \overline{\lim}_{k\to\infty} k^\nu e_k(Y, r, p, q).$$

Proof. Fix any $\rho \in (\nu, \sigma)$. Define $k_1 = \lfloor k^{\nu/\rho} \rfloor$ and $k_2 = k - k_1$. If $q \ge 1$, then, by triangle inequality,

$$e_k(X, r, p, q) \le e_{k_2}(Y, r, p, q) + e_{k_1}(Z, r, p, q).$$

Since $k_1^{-\sigma} = o(k^{-\nu})$ and $k_2 \sim k$, this entails

$$\overline{\lim_k}\, k^\nu e_k(X, r, p, q) \le \overline{\lim_{k_1}}\, k_1^\nu e_{k_1}(Y, r, p, q).$$

If $q < 1$, then one proceeds analogously with

$$e_k(X, r, p, q)^q \le e_{k_2}(Y, r, p, q)^q + e_{k_1}(Z, r, p, q)^q$$

and derives

$$\overline{\lim_k}\, k^{q\nu} e_k(X, r, p, q)^q \le \overline{\lim_{k_1}}\, k_1^{q\nu} e_{k_1}(Y, r, p, q)^q.$$

The remaining inequalities are proved in exactly the same manner. □

We will provide now an explicit lower bound for the strong asymptotics of $e_k(B^H, r, p, q)$ in terms of RLfBM.

Let Π_r denote the space of polynomials of degree at most r. Define the stopping time

$$\tau_{H,r,p} := \inf\{t > 0 \ : \ \inf_{\pi \in \Pi_r} \|R^H - \pi\|_{L_p[0,t]} > 1\}$$

and set

$$C_{H,r,p} := (\mathbb{E}\, \tau_{H,r,p})^{-(H+1/p)}.$$

Theorem 2. *Let H, p, q, r as above, and assume $p < \infty$. Then $C_{H,r,p} > 0$ and*

$$\underline{\lim}\, k^H e_k(B^H, r, p, q) \geq 2^{-(H+1/p)} C_{H,r,p}.$$

This result yields Theorem 1 (since the proof of the upper bound follows from the known linear approximation rates [Rit00, p. 115, Proposition 34]). Moreover, it gives an asymptotical bound independent from q, vanishing as $H \to 1$ and exploding as $H \to 0$. It is thus a reasonable candidate for a possible two–sided estimate in strong asymptotics.

3 Increments and Stopping of the RLfBM

The RLfBM does not have stationary or independent increments unless $H = 1/2$; however, there is a very convenient separation of the influence of the past (as observed in [LS05]). Let first $a > 0$ be a fixed number, and define

$$R^H_{a,t} := \int_a^{a+t} (a + t - s)^{H-1/2}\, \mathrm{d}W_s = \int_0^t (t - r)^{H-1/2}\, \mathrm{d}(W_{a+r} - W_a)$$

as well as

$$S_{a,t} := R^H_{a+t} - R^H_{a,t} = \int_0^a (a + t - s)^{H-1/2}\, \mathrm{d}W_s.$$

Denote the canonical augmentation of the natural filtration of W by $\mathcal{F} = (\mathcal{F}_t)_{t\geq 0}$. Then it is easy to see that $R^H_{a,t}$ is a RLfBM independent of $\mathcal{F}_a$ and that $S_{a,t}$ is $\mathcal{F}_a$–measurable for all t (see [LS05, p. 733]).

We need a generalization of this. For a finite stopping time τ with respect to $\mathcal{F}$, we denote

$$\mathcal{F}_\tau := \{A \ : \ \forall\, t \geq 0 \ : \ A \cap \{\tau \leq t\} \in \mathcal{F}_t\}$$

the σ-algebra of events determined prior to τ. Recall also that for a family $(\xi_i)_{i\in I}$ of random variables $\sigma(\xi_i \ : \ i \in I)$ is the smallest of all σ-algebras $\mathfrak{A}$ such that all ξ_i are measurable w.r.t. $\mathfrak{A}$.

Lemma 2. *For any fixed t, the process $(R^H_{a,t})_{a\geq 0}$ is progressively $(\mathcal{F}_{a+t})_{a\geq 0}$ – measurable, and*

$$R^H_{\tau,t} = \int_0^t (t - r)^{H-1/2}\, \mathrm{d}(W_{r+\tau} - W_\tau)\,, \quad a.s.. \tag{2}$$

In particular, $\sigma(R^H_{\tau,s}: s \geq 0) \subseteq \sigma(W_{s+\tau} - W_\tau: s \geq 0)$. *Further, for any fixed* t, *the process* $(S_{a,t})_{a\geq 0}$ *is progressively* $(\mathcal{F}_a)_{a\geq 0}$*–measurable. In particular,* $S_{\tau,t}$ *is* $\mathcal{F}_\tau$*–measurable and consequently independent of* $\sigma(R^H_{\tau,s}: s \geq 0)$.

Proof. Fix $t \geq 0$. Then $S_{a,t}$ is even continuous in a, and consequently, $R^H_{a,t}$ is as well. Since $R^H_{a,t}$ is measurable w.r.t. $\mathcal{F}_{a+t}$, and $S_{a,t}$ w.r.t. $\mathcal{F}_a$, it follows that both are progressively measurable w.r.t. to $\mathcal{F}_{a+t}$, resp. $\mathcal{F}_a$ (see, e.g., [KS91, Proposition 1.1.13]). This entails immediately that $S_{\tau,t}$ is $\mathcal{F}_\tau$–measurable ([KS91, Proposition 1.2.18]). Equation (2) is easily verified if τ attains only a finite number of values, and the general case follows by standard approximation arguments. □

We recall now some notation and facts from [CMGR98]. Let $f \in C[0,\infty[$, $p \in [1,\infty]$ fixed, and $0 < u < v$. Define

$$\delta_{[u,v]}(f) := \inf_{\pi \in \Pi_r} \|f - \pi\|_{L_p[u,v]},$$

and set, for $\varepsilon > 0$, inductively $\tau_{0,\varepsilon}(f) = 0$ and

$$\tau_{j,\varepsilon}(f) = \inf\{t > \tau_{j-1,\varepsilon}(f) : \delta_{[\tau_{j-1,\varepsilon}(f),t]}(f) > \varepsilon\}.$$

Lemma 3. *We have* $0 < C_{H,r,p} < \infty$.

Proof. Note that $C_{H,r,p} = (\mathbb{E}\,\tau_{1,1}(R^H))^{-(H+1/p)}$. Let us remark that $\tau_{1,1}(R^H) > t$ entails $\delta_{[0,t]}(R^H) \leq 1$; consequently, by the H-self-similarity, we derive that

$$\mathbb{P}(\tau_{1,1}(R^H) > t) \leq \mathbb{P}(\delta_{[0,t]}(R^H) \leq 1) = \mathbb{P}(\delta_{[0,1]}(R^H) \leq t^{-(H+1/p)}). \quad (3)$$

Next, the following estimate is true (see Proposition 16 in [CMGR98]).

Fact 2 *Let* X *be a centered Gaussian random vector taking values in a normed space* $(E, \|\cdot\|)$ *and let* Π *be an* r*-dimensional subspace of* E. *Then*

$$\mathbb{P}\left(\inf_{\pi\in\Pi} \|X - \pi\| \leq \varepsilon\right) \leq (4\lambda/\varepsilon)^r \cdot \mathbb{P}(\|X\| \leq 2\varepsilon) + \mathbb{P}(\|X\| \geq \lambda - \varepsilon)$$

for all $\lambda \geq \varepsilon > 0$.

We use this fact with $X = R^H$, $E = L_p[0,1]$, and $\Pi = \Pi_r$. It is known that

$$-\log \mathbb{P}(\|R^H\|_{L_p[0,1]} \leq \varepsilon) \asymp \varepsilon^{-1/H}; \quad \text{as } \varepsilon \to 0,$$

see e.g.[LS05], and

$$-\log \mathbb{P}(\|R^H\|_{L_p[0,1]} \geq \lambda) \asymp \lambda^2; \quad \text{as } \lambda \to \infty$$

(the general rate of Gaussian large deviations, see [Lif95]). By choosing, say, $\lambda = \varepsilon^{-1/H}$ and combining three estimates one derives that

$$- \log \mathbb{P}\left(\delta_{[0,1]}(R^H) \leq \varepsilon\right) \asymp \varepsilon^{-H}.$$

From (3) we infer that

$$\mathbb{P}\left(\tau_{1,1}(R^H) > t\right) = o(\exp(-ct))$$

and hence $0 < \mathbb{E}\, \tau_{1,1}(R^H) < \infty$. □

For $k \in \mathbb{N}$ given, we now define

$$\gamma_k(f) = \inf\{\varepsilon > 0 : \tau_{k,\varepsilon}(f) \geq 1\}.$$

The intuition suggests that on each of the k intervals $[\tau_{i,\gamma_k}(f), \tau_{i+1,\gamma_k}(f)]$ dissecting $[0,1]$, f can be approximated by a polynomial with the same L_p-error γ_k, and hence $k^{1/p} \cdot \gamma_k(f)$ is a reasonable guess for a 'close-to-optimal' error estimate. Indeed, one can show ([CMGR98, Proposition 5]):

Proposition 1. *Let X be a process such that for all k,*

$$\sup_{\varepsilon>0} \tau_{k,\varepsilon}(X) = \infty, \qquad \inf_{\varepsilon>0} \tau_{k,\varepsilon}(X) = 0\,, \qquad a.s.. \tag{4}$$

Then we have for $p = \infty$ that

$$e_k(X, r, p, q) = \left(\mathbb{E}\, \gamma_k^q(X)\right)^{1/q}.$$

For $p < \infty$, we still have

$$k^{1/p}\left(\mathbb{E}\, \gamma_{2k}^q(X)\right)^{1/q} \leq e_k(X, r, p, q) \leq k^{1/p}\left(\mathbb{E}\, \gamma_k^q(X)\right)^{1/q}.$$

It is straightforward to prove that R^H satisfies (4); consequently, for proving Theorem 2 it is sufficient to determine lower bounds for the asymptotics of $\mathbb{E}\, \gamma_k(R^H)^q$ as $k \to \infty$. However,

$$\begin{aligned}\mathbb{P}\left(\gamma_k(R^H) \geq \varepsilon\right) &= \mathbb{P}\left(\tau_{k,\varepsilon}(R^H) \leq 1\right) \\ &= \mathbb{P}\left(\varepsilon^{(H+1/p)^{-1}} \tau_{k,1}(R^H) \leq 1\right) = \mathbb{P}\left((\tau_{k,1})^{-(H+1/p)}(R^H) \geq \varepsilon\right)\end{aligned}$$

by the H-self-similarity. In other words, $\gamma_k(R^H) \stackrel{d}{=} \tau_{k,1}^{-\rho}(R^H)$ with $\rho = H + 1/p$. In particular,

$$\mathbb{E}\, \gamma_k^q(R^H) = \mathbb{E}\, \tau_{k,1}^{-q\rho}(R^H). \tag{5}$$

Let us shorten $\tau_k = \tau_{k,1}(R^H)$. We are thus interested in a lower bound for

$$\mathbb{E}\, \tau_k^{-q\rho} = \int_0^\infty \mathbb{P}\left(\tau_k < \varepsilon^{-1/\rho q}\right) d\varepsilon.$$

We shall go on by using stochastic domination. Recall that a positive random variable X is said to stochastically dominate another one, Y, in short $Y \ll X$, iff $\mathbb{P}(Y \geq t) \leq \mathbb{P}(X \geq t)$ for all $t > 0$.

The following lemma is simple yet crucial for our progress.

Lemma 4. *Assume that* X, Y, Z *are positive random variables such that* $\mathbb{P}(X \geq t|Z) \geq \mathbb{P}(Y \geq t|Z)$ *a.s.. Then* $Y + Z \ll X + Z$.

Proof.

$$\begin{aligned}\mathbb{P}(Y + Z \geq t) &= \int_0^\infty \mathbb{P}(Y \geq t - s|Z = s)\,\mathrm{d}P_Z(s) \\ &\leq \int_0^\infty \mathbb{P}(X \geq t - s|Z = s)\,\mathrm{d}P_Z(s) = \mathbb{P}(X + Z \geq t).\end{aligned}$$
□

Proposition 2. *Let* $R^{(i)}$ *be a sequence of independent RLfBM and* $\eta^{(i)} = \tau_1(R^{(i)})$. *Then*

$$\tau_k(R^H) \ll \sum_{i \leq k} \eta^{(i)}.$$

In particular,

$$\mathbb{E}\,\tau_k(R^H)^{-\rho q} \geq \mathbb{E}\left(\sum_{i \leq k} \eta^{(i)}\right)^{-\rho q}.$$

Proof. We shall prove by backward induction on $j = k, \ldots, 0$ that

$$\tau_k \ll \tau_j + \sum_{j < i \leq k} \eta^{(i)}. \tag{6}$$

The induction step is done if we can show

$$\tau_j + \sum_{j < i \leq k} \eta^{(i)} \ll \tau_{j-1} + \eta^{(j)} + \sum_{j < i \leq k} \eta^{(i)}. \tag{7}$$

We wish to apply Lemma 4; set $Z = \tau_{j-1} + \sum_{j<i\leq k} \eta^{(i)}$, $Y = (\tau_j - \tau_{j-1})$ and $X = \eta^{(j)}$. Then (7) can be rewritten as $Y + Z \ll X + Z$. Thus, once we have proven that, with $\mathfrak{A} = \sigma(Z)$,

$$\mathbb{P}(X \geq t|\mathfrak{A}) \geq \mathbb{P}(Y \geq t|\mathfrak{A}) \quad a.s., \tag{8}$$

the claim (7) and thus (6) follows, and thus the Proposition is proven. So let us turn to the proof of (8). Recall that Π_r denotes the space of polynomials of degree at most r; we also denote $L_p/\Pi_r[a,b]$ the quotient space of $L_p[a,b]$ after the polynomials and Q_{Π_r} the corresponding canonical quotient mapping. We have

$$d(x, \Pi_r) := \inf_{\pi \in \Pi_r} \|x - \pi\|_{L_p[a,b]} = \left\|Q_{\Pi_r}(x)\right\|_{L_p/\Pi_r[a,b]}$$

for any $x \in L_p$. Next recall that (due to the continuity of δ in time)

$$Y < t \qquad \Leftrightarrow \qquad \left\|Q_{\Pi_r} R^H\right\|_{L_p/\Pi_r[\tau_{j-1}, \tau_{j-1}+t]} > 1.$$

However,

$$\left\|Q_{\Pi_r} R^H\right\|_{L_p/\Pi_r[\tau_{j-1},\tau_{j-1}+t]} = \inf_{\pi\in\Pi_r}\left(\int_0^t |R^H_{\tau_{j-1}+s} - \pi(s)|^p \mathrm{d}s\right)^{1/p}$$
$$= \left\|Q_{\Pi_r}(R^H_{\tau_{j-1},\cdot} + S_{\tau_{j-1},\cdot})\right\|_{L_p/\Pi_r[0,t]}.$$

In what follows, we will need a version of the Anderson inequality. Recall that this inequality ([Lif95, p. 135, Theorem 9]) yields:

Fact 3 *If U is a centered Gaussian element of a Banach space E, and V is independent of U, then*

$$\mathbb{P}\left(\|U+V\| > s\right) \geq \mathbb{P}\left(\|U\| > s\right), \qquad \forall s \geq 0.$$

We shall need a slightly generalized version:

Proposition 3. *Let U be a centered Gaussian element of a Banach space E, V independent of U, and $\mathfrak{A}$ a σ-algebra independent of U. Then*

$$\mathbb{P}\left(\|U+V\| > s|\mathfrak{A}\right) \geq \mathbb{P}\left(\|U\| > s\right), \ a.s., \qquad \forall s \geq 0.$$

Proof. Without loss of generality, we can assume that the underlying probability space is in convenient product form, i.e., $\Omega = \Omega_1 \times \Omega_2$, $\mathbb{P} = \mathbb{P}_1 \times \mathbb{P}_2$ with $U(\omega_1,\omega_2) = U(\omega_1)$, $V(\omega_1,\omega_2) = V(\omega_2)$ and $\mathfrak{A} = \Omega_1 \times \tilde{\mathfrak{A}}$. We consider some $A = \Omega_1 \times \tilde{A} \in \mathfrak{A}$ and have, using the original Anderson inequality,

$$\begin{aligned}\int_A \mathbf{1}_{\|U+V\|>s}\, \mathrm{d}\mathbb{P}(\omega) &= \int_{\tilde{A}}\left(\int_{\Omega_1} \mathbf{1}_{\|U(\omega_1)+V(\omega_2)\|>s}\, \mathrm{d}\mathbb{P}_1(\omega_1)\right) \mathrm{d}\mathbb{P}_2(\omega_2)\\ &= \int_{\tilde{A}} \mathbb{P}_1(\|U+V(\omega_2)\| > s)\, \mathrm{d}\mathbb{P}_2(\omega_2)\\ &\geq \int_{\tilde{A}} \mathbb{P}_1(\|U\| > s)\, \mathrm{d}\mathbb{P}_2(\omega_2)\\ &= \int_A \mathbf{1}_{\|U\|>s}\, \mathrm{d}\mathbb{P}(\omega).\end{aligned}$$

But this entails our claim. □

Back to the proof of Proposition 2. Recall (Lemma 2) that $Q_{\Pi_r} R^H_{\tau_{j-1},\cdot}$ is a centered Gaussian process independent of $\sigma(W_{s\wedge\tau_{j-1}} : s \leq \tau_{j-1})$ and consequently of $\mathfrak{A}$; since it is also independent of $S_{\tau_{j-1},\cdot}$, we may apply Proposition 3 to derive

$$\begin{aligned}\mathbb{P}(Y < t|\mathfrak{A}) &= \mathbb{P}\left(\left\|Q_{\Pi_r}(R^H_{\tau_{j-1},\cdot}) + Q_{\Pi_r}(S_{\tau_{j-1},\cdot})\right\|_{L_p/\Pi_r[0,t]} > 1|\mathfrak{A}\right)\\ &\geq \mathbb{P}\left(\left\|Q_{\Pi_r}(R^H_{\tau_{j-1},\cdot})\right\|_{L_p/\Pi_r[0,t]} > 1\right).\end{aligned}$$

Now we turn the table around:

$$\left\|Q_{\Pi_r}(R^H_{\tau_{j-1},\cdot})\right\|_{L_p/\Pi_r[0,t]} > 1 \quad \Leftrightarrow \quad \tau_1(R^H_{\tau_{j-1},\cdot}) < t,$$

hence

$$\mathbb{P}(Y < t|\mathfrak{A}) \geq \mathbb{P}(\tau_1(R^H_{\tau_{j-1},\cdot}) < t) = \mathbb{P}(\eta^{(j)} < t|\mathfrak{A}) = \mathbb{P}(X < t|\mathfrak{A}).$$

(Here we used that $R^H_{\tau_{j-1},\cdot}$ is again a RLfBM and the independence of $\eta^{(j)}$.) We have found that (8) is indeed correct, which was all what remained to do for proving Proposition 2. □

We now finish the proof of Theorem 2. Recall the last necessary ingredient, a result about sums of independent variables, see [CMGR98, Proposition 15].

Fact 4 *Let $(\xi_i)_{i\in\mathbb{N}}$ be an i.i.d. sequence of non-negative random variables. Put*

$$S_k = 1/k \cdot \sum_{i=1}^{k} \xi_i.$$

Then for every $\alpha > 0$,

$$\varliminf_{k\to\infty} \mathbb{E}(S_k^{-\alpha}) \geq (\mathbb{E}(\xi_1))^{-\alpha}.$$

By applying this fact with $\alpha = q\rho$ and $\xi_i = \eta^{(i)}$, we obtain

$$\varliminf_{k\to\infty} k^{q\rho} \cdot \mathbb{E}\left(\sum_{i\leq k} \eta^{(i)}\right)^{-q\rho} \geq C^q_{H,r,p}\,. \tag{9}$$

By combining Proposition 1, identity (5), Proposition 2, and inequality (9) we obtain

$$\begin{aligned}
\varliminf_{k\to\infty} k^{qH} e_k(R^H, r, p, q)^q &\geq \varliminf_{k\to\infty} k^{q\rho}\mathbb{E}\,\gamma^q_{2k} \\
&= \varliminf_{k\to\infty} k^{q\rho}\mathbb{E}\,\tau^{-q\rho}_{2k}(R^H) \\
&\geq 2^{-q\rho} \varliminf_{k\to\infty} (2k)^{q\rho} \cdot \mathbb{E}\left(\sum_{i\leq 2k} \eta^{(i)}\right)^{-q\rho} \\
&\geq 2^{-q\rho} C^q_{H,r,p}.
\end{aligned}$$

By Lemma 1, we get

$$\varliminf_{k\to\infty} k^H e_k(B^H, r, p, q) \geq 2^{-\rho} C_{H,r,p}\,,$$

as claimed in Theorem 2. □

4 Acknowledgements

The authors enjoyed many helpful discussions with Klaus Ritter. Part of this work was done while the second author was visiting TU Darmstadt. He thanks Math. Department of TU for arranging this visit and for the great hospitality. His work was also supported by grants INTAS 03-51-5018 and RFBR/DFG 04-01-04000.

References

[AT03] A. Ayache and M.S. Taqqu. Rate optimality of wavelet series approximations of fractional Brownian motion. J. Fourier Anal. Appl. **9** (2003), no. 5, 451–471.

[CA97] A. Cohen and J.-P. d'Ales. Nonlinear approximation of random functions. SIAM J. Appl. Math. **57** (1997), 518–540.

[CDGO02] A. Cohen, I. Daubechies, O.G. Guleryuz, and M.T. Orchard. On the importance of combining wavelet-based nonlinear approximation with coding strategies. IEEE Trans. Inform. Theory **48** (2002), 1895–1921.

[CMGR98] J. Creutzig, T. Müller–Gronbach, and K. Ritter. Free-knot spline approximation of stochastic processes. Preprint, arXiv number: math/0612313.

[DV98] R. DeVore. Nonlinear approximation. Acta Numer. **8** (1998), 51–150.

[DZ04] K. Dzhaparidze and H. van Zanten. A series expansion of fractional Brownian motion. Probab. Theory Related Fields **130** (2004), no. 1, 39–55.

[KL02] T. Kühn and W. Linde. Optimal series representation of fractional Brownian sheets. Bernoulli **8** (2002), no. 5, 669–696.

[KP05] M. Kon and L. Plaskota. Information-based nonlinear approximation: an average case setting. J. Complexity **21** (2005), 211–229.

[KS91] I. Karatzas and S.E. Shreve. Brownian Motion and Stochastic Calculus, 2nd Edition, Springer-Verlag, New York, 1991.

[Lif95] M. Lifshits. Gaussian Random Functions. Kluwer Academic Publishers, 1995.

[LS05] M. Lifshits and T. Simon. Small deviations for fractional processes. Ann. Inst. Henri Poincaré, Probab. Stat. **41** (2005), 725–752.

[Rit00] K. Ritter. Average-Case Analysis of Numerical Problems. Lect. Notes in Math. **1733**, Springer-Verlag, Berlin, 2000.

Simulation on Rank-1 Lattices

Holger Dammertz[1], Alexander Keller[2], and Sabrina Dammertz[3]

[1] Ulm University, Germany
holger.dammertz@uni-ulm.de
[2] Ulm University, Germany
alexander.keller@uni-ulm.de
[3] Ulm University, Germany
sabrina.dammertz@uni-ulm.de

Summary. Rank-1 lattices are available in any dimension for any number of lattice points and because their generation is so efficient, they often are used in quasi-Monte Carlo methods. Applying the Fourier transform to functions sampled on rank-1 lattice points turns out to be simple and efficient if the number of lattice points is a power of two. Considering the Voronoi diagram of a rank-1 lattice as a partition of the simulation domain and its dual, the Delauney tessellation, as a mesh for display and interpolation, rank-1 lattices are an interesting alternative to tensor product lattices. Instead of classical criteria, we investigate lattices selected by maximized minimum distance, because then the Delauney tessellation becomes as equilateral as possible. Similar arguments apply for the selection of the wave vectors. We explore the use of rank-1 lattices for the examples of stochastic field synthesis and a simple fluid solver with periodic boundary conditions.

1 Introduction

Many simulations are evaluated on Cartesian tensor product lattice structures although their sampling efficiency is not optimal [PM62]. Rank-1 lattices allow for a better sampling efficiency when selected carefully. We develop and review the basic tools like meshing, fast Fourier transform, lattice cell access, and interpolation for simulation on rank-1 lattices. In addition we give insight how to choose suitable rank-1 lattice parameters. We illustrate our new techniques for generating ocean waves as stochastic field and a simple fluid dynamics simulation. The most prominent example is the simulation of the ocean surface [AR86, Tes00] by random fields as used in the movies Titanic, Waterworld, or The Devil's Advocate [Ent]. The same principle has been applied to modeling of turbulent wind fields and various other phenomena [SF91, SF93, Sta95, Sta97].

2 Rank-1 Lattices

A discrete subset of $\mathbb{R}^s$ that contains $\mathbb{Z}^s$ and is closed under addition and subtraction is called a lattice [SJ94]. Rank-1 lattices

$$L_{n,\mathbf{g}} := \left\{ \frac{l}{n}\mathbf{g} + \Delta : \Delta \in \mathbb{Z}^s; l = 0, \ldots, n-1 \right\}$$

are defined by using only one suitable generator vector $\mathbf{g} \in \mathbb{N}^s$ for a fixed number $n \in \mathbb{N}$. Often it is more useful to consider their restriction

$$L_{n,\mathbf{g}} \cap [0,1)^s = \left\{ \mathbf{x}_l := \frac{l}{n}\mathbf{g} \bmod 1 : l = 0, \ldots, n-1 \right\}$$

to the s-dimensional unit torus $[0,1)^s$ (see the example in Figure 1). A notable advantage of rank-1 lattices over tensor product lattices is that they exist for any number n of points in any dimension s.

An $s \times s$ matrix V is called basis of the lattice $L_{n,\mathbf{g}}$, if $L_{n,\mathbf{g}} = \{\mathbf{x} = V\mathbf{l} : \mathbf{l} \in \mathbb{Z}^s\}$. Of all possible bases the Minkowski-reduced bases [AEVZ02] are the most useful for our purpose. Such a basis contains the s shortest linearly independent vectors and can be found by a computer search over all n points $\mathbf{x}_l$ using their shortest distance to the origin on the unit torus. These vectors actually describe the Delauney tessellation and can be used for accessing neighboring lattice points.

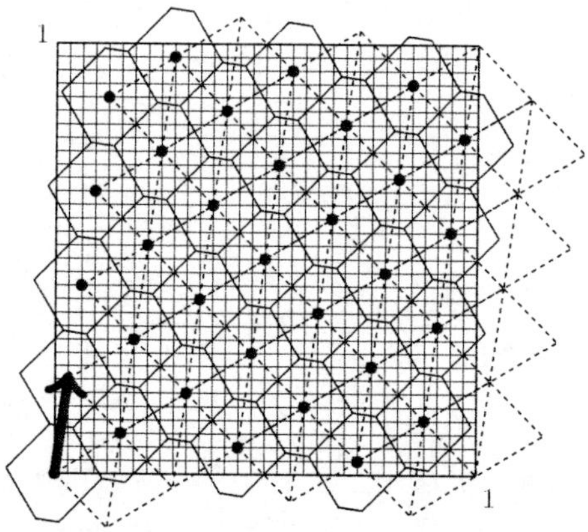

Fig. 1. Illustration of the geometry of rank-1 lattices for an example in $s = 2$ dimensions with $n = 32$ points and the generator vector $\mathbf{g} = \binom{1}{7}$ (see the solid arrow from the origin). The solid lines depict the Voronoi diagram. Each cell is generated by a lattice point and contains the points of the unit torus, which are closer to this lattice point than to any other. In addition the lattice points are the centroids of the Voronoi cells. The dashed lines represent the dual graph, which is the Delauney tessellation.

2.1 Fast Fourier Transform

The fast Fourier transform is a versatile tool in simulation. Usually the transform has to be performed for each coordinate once. Instead of the standard tensor product algorithm, rank-1 lattices in s dimensions allow for transforming the data using only the one-dimensional Fourier transform [LH03], which is simpler to implement and a little bit more efficient for the same number of lattice points.

For this purpose the set $K_n := \{\mathbf{k}_0, \ldots, \mathbf{k}_{n-1}\} \subset \mathbb{Z}^s$ of wave vectors has to be selected such that each wave vector

$$\mathbf{k}_j \in Z_j := \{\mathbf{k} \in \mathbb{Z}^s : \mathbf{k} \cdot \mathbf{g} \equiv j \pmod{n}\}, \tag{1}$$

where $\mathbf{g}$ is the generator vector of the rank-1 lattice $L_{n,\mathbf{g}}$ under consideration. Hence,

$$\mathbf{k}_j \cdot \mathbf{x}_l = \mathbf{k}_j \cdot \frac{l}{n}\mathbf{g} = (j + r_j n)\frac{l}{n} = \frac{jl}{n} + r_j l$$

for some integer $r_j \in \mathbb{Z}$. Given Fourier coefficients $\hat{\mathbf{f}}(\mathbf{k}_j)$, synthesizing a function $\mathbf{f}$ on the lattice $L_{n,\mathbf{g}}$ by

$$\mathbf{f}(\mathbf{x}_l) = \sum_{j=0}^{n-1} \hat{\mathbf{f}}(\mathbf{k}_j) e^{2\pi i \mathbf{k}_j \cdot \mathbf{x}_l} = \sum_{j=0}^{n-1} \hat{\mathbf{f}}(\mathbf{k}_j) e^{2\pi i (\frac{jl}{n} + r_j l)} = \sum_{j=0}^{n-1} \hat{\mathbf{f}}(\mathbf{k}_j) e^{2\pi i \frac{jl}{n}} \tag{2}$$

in fact turns out to be a one-dimensional finite Fourier series independent of the dimension s, because $r_j l$ is integer and therefore $e^{2\pi i r_j l} = 1$.

For n being a power of two, the fast inverse Fourier transform can synthesize the function in all lattice points most efficiently. Given a function $\mathbf{f}(\mathbf{x}_l)$ the fast Fourier transform can be used for the analysis, too:

$$\hat{\mathbf{f}}(\mathbf{k}_j) = \frac{1}{n} \sum_{l=0}^{n-1} \mathbf{f}(\mathbf{x}_l) e^{-2\pi i \frac{jl}{n}}.$$

2.2 Choosing the Wave Vectors

There are infinitly many choices of wave vectors as defined by the sets Z_j in equation (1). Understanding the connection between the wave vectors and the structure of a rank-1 lattice helps to choose the best wave vectors for a given problem and provides a way to construct these wave vectors.

The dual lattice

$$L^{\perp}_{n,\mathbf{g}} := \{\mathbf{k} \in \mathbb{Z}^s : \mathbf{k} \cdot \mathbf{g} \equiv 0 \pmod{n}\} = Z_0$$

of a rank-1 lattice $L_{n,\mathbf{g}}$ has the basis $U = (V^{-1})^T$. Now the set K_n of wave vectors is U periodic [PM62], i.e. it has the property that the sets

$$K'_n := K_n + U\mathbf{l} \qquad \mathbf{l} \in \mathbb{Z}^s$$

are valid sets of n wave vectors with $K'_n \cap K_n = \emptyset$ as well. Consequently, any tile that results in a monohedral tiling of $\mathbb{Z}^s$ (see the illustration in Figure 2) with periodicity U can be used as a set of wave vectors [LH03].

Once such a tiling is chosen all integer vectors in the interior of one cell are the wave vectors. Note that the choice of the tiling is arbitrary and the elements of a single tile are not necessarily connected. However, one can use known spectral properties of the function that should be synthesized or analyzed. If no spectral properties are known a reasonable assumption for practical problems is an isotropic spectrum (i.e. no preferred direction) and that low frequencies are most important. This results in choosing the wave vectors in the fundamental (i.e. including the origin) Voronoi cell of $L^{\perp}_{n,\mathbf{g}}$, which is illustrated in Figure 3.

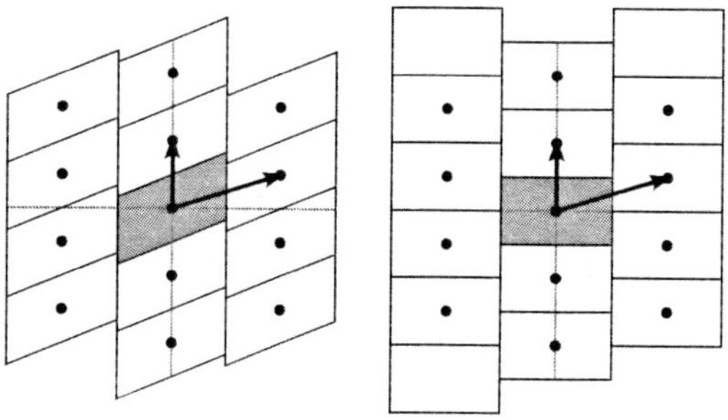

Fig. 2. Examples of monohedral tilings of the 2-dimensional plane with the dual lattice and its basis. The arrows show the Minkowski-reduced basis vectors.

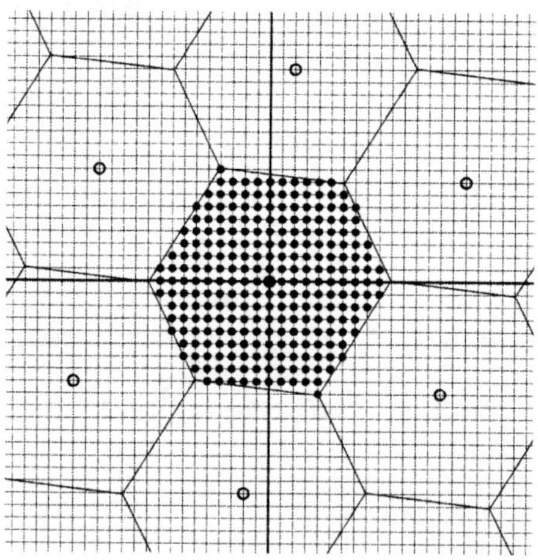

Fig. 3. Dual lattice (circles) of a rank-1 lattice with $n = 256$ and $\mathbf{g} = \binom{1}{30}$. The set of wave vectors K_n (solid disks) in the fundamental Voronoi cell is highlighted.

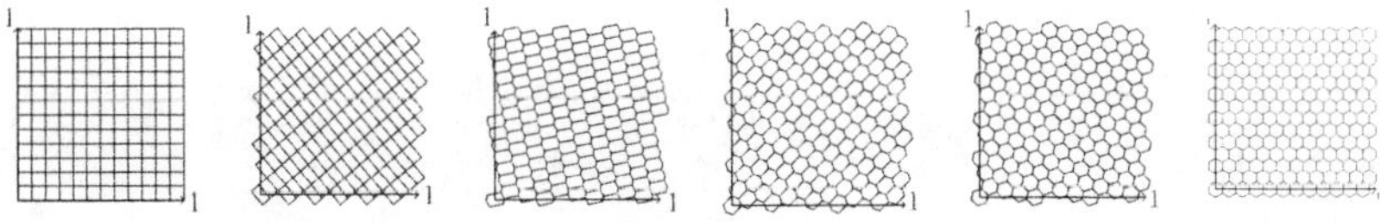

rect. lattice	$\mathbf{g} = \binom{1}{42}$	$\mathbf{g} = \binom{1}{67}$	$\mathbf{g} = \binom{1}{89}$	$\mathbf{g} = \binom{1}{33}$	hex. lattice
$\eta = 78.5\%$	$\eta = 46.3\%$	$\eta = 56.7\%$	$\eta = 69.8\%$	$\eta = 87.3\%$	$\eta = 90.7\%$

Fig. 4. Sampling efficiency η of different lattices with $n = 144$. Note that the Fibnacci lattice with $\mathbf{g} = \binom{1}{89}$ is not the best choice with respect to sampling efficiency.

The above assumptions also provide a criterion for choosing the generator vector $\mathbf{g}$ of the rank-1 lattice: Choose $\mathbf{g}$ so that the in-circle of the fundamental Voronoi cell of the dual lattice is maximized. This is equivalent to maximizing the sampling efficiency

$$\eta := \frac{R}{P}$$

as defined by Petersen [PM62], where R is the volume of the in-circle of the Voronoi region and P is the volume of the fundamental Voronoi cell. The sampling efficiency measures how many of the important frequencies are actually captured by sampling with a given lattice.

For rank-1 lattices this ratio can be maximized by choosing the generator vector such that the minimal distance between any two points of the dual lattice is maximized. Figure 4 shows the Voronoi diagrams of different rank-1 lattices, where a Cartesian tensor product and hexagonal lattice are shown for comparison. The hexagonal lattice is optimal with respect to the sampling efficiency in two dimensions and maximizing the minimum distance in a rank-1 lattice yields a good approximation. With increasing number of points the sampling efficiency of rank-1 lattices approaches the sampling efficiency of the hexagonal lattice.

Of course, if other spectral properties are known the rank-1 lattice search can be adapted for a better approximation of this kind of functions.

3 Applications in Computer Graphics

We illustrate the idea of simulation on rank-1 lattices by implementing two examples from the domain of computer graphics and animation in the new framework.

3.1 Spectral Synthesis of Ocean Waves

Along the method used by Tessendorf [Tes00], a periodic ocean tile (see Figure 5) is realized as a stochastic field using Fourier synthesis on a rank-1 lattice. Using the Fourier coefficients

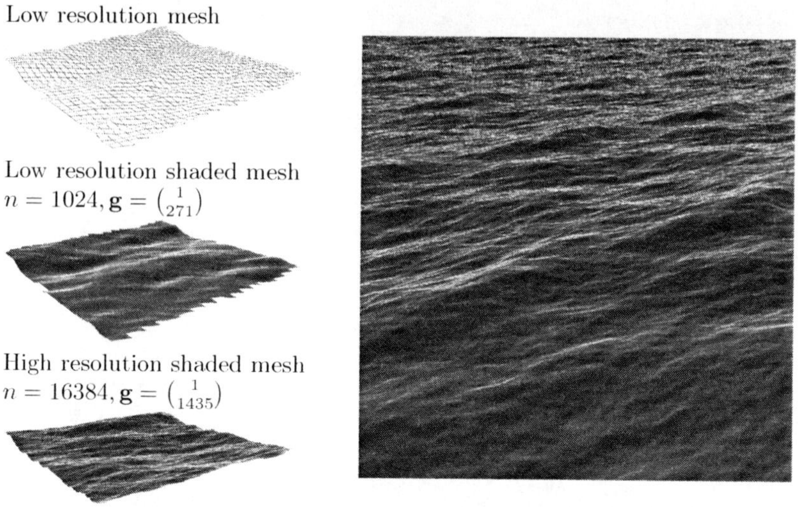

Fig. 5. Left: Synthesized periodic $50m \times 50m$ tiles in two resolutions. Right: Modeling a larger piece of the ocean by tiling the periodic patches.

$$\hat{h}(\mathbf{k}, t) := \hat{h}_0(\mathbf{k})e^{i\omega(\mathbf{k})t} + \hat{h}_0^*(-\mathbf{k})e^{-i\omega(\mathbf{k})t}$$

the height field

$$h(\mathbf{x}_l, t) := \sum_{j=0}^{n-1} \hat{h}(\mathbf{k}_j, t)e^{2\pi i \frac{jl}{n}}$$

becomes periodic in time t and real. For deep water the speed of a wave is given by the dispersion relation $\omega(\mathbf{k}) = \sqrt{g\|\mathbf{k}\|}$, where g is the gravitational constant. Based on observations from oceanography waves can be modeled statistically independent and normally distributed. Therefore, the amplitudes

$$\hat{h}_0(\mathbf{k}) := \frac{1}{\sqrt{2}}(\xi_r + i\xi_i)\sqrt{P_h(\mathbf{k})}$$

are realized using Gaussian random numbers ξ_r and ξ_i modulated by a spectrum. Out of many alternatives we chose the Phillips spectrum

$$P_h(\mathbf{k}) := A\frac{e^{-\frac{1}{(\|k\|L)^2}}}{k^4}|\mathbf{k} \cdot \mathbf{w}|^2$$

A	Phillips constant
$L = \frac{v^2}{g}$	Largest wave for windspeed v
$\mathbf{w}$	wind direction

which considers parameters like wind speed and direction. For the sake of completeness we mention that the gradient vector of the height field can be computed using the Fourier transform as well. This yields more precise normals for shading as those computed by finite differences.

Implementation

The synthesis of stochastic fields on rank-1 lattices consists of the following choices and decisions:

Number n of lattice points: Although rank-1 lattices exist for any number of points in any dimension, the fast Fourier transform is most efficient for n being a power of 2.

Generator vector g: In order to maximize the sampling efficiency we select a generator $\mathbf{g}$ that maximizes the minimum distance of the dual lattice. If the generator vector has Korobov form, i.e. $\mathbf{g} = (1, a, a^2, a^3, \ldots)$, the spectral test [Knu77] can be used to efficiently compute the minimum distance for each candidate a. While not true in general, for dimension $s = 2$ and n as a power of 2, one of the two components of the generator vector $\mathbf{g} = \binom{a_1}{a_2}$ has to be odd (w.l.o.g. $\gcd(a_1, n) = 1$). Otherwise points would coincide resulting in an obviously useless minimum distance of 0. Then for every j with $\gcd(j, n) = 1$ every vector $\mathbf{x}_j = j\mathbf{g} \bmod n$ is a generator vector of the same lattice (generator of the cyclic group), too, and there must exist an $l \in \{1, 3, 5, \ldots, n-1\}$ with $\mathbf{x}_l = \binom{1}{a}$. Thus a generator vector in Korobov form exists that obtains maximized minimum distance and the spectral test can be used. A list of all parameters for 2-dimensional maximized minimum distance rank-1 lattices is found in Table 1.

Basis V: V is determined as a Minkowski-reduced basis, which defines the Delauney triangulation that is used as the triangle mesh. Table 1 lists these basis vectors, however, multiplied by n. Given a generator vector in Korobov form, the first coordinate of each of these integer basis vectors is the increment or decrement to find the index of a neighboring lattice point.

Wave vectors K_n: We enumerate all wave vectors in a conservative bounding box of the fundamental Voronoi cell of the dual lattice and select the shortest ones. As a simple conservative convex hull we chose the axis-aligned bounding box determined by the direct lattice point neighbors of the origin. A much more involved approach is to compute the fundamental Voronoi cell in the dual lattice and rasterize it on the integer lattice.

3.2 Stable Simulation of Fluids

The stable fluids algorithm by Stam [Sta99] is a practical way of simulating incompressible fluids for animation. Note that the algorithm focuses on realtime simulation rather than on precision. The simulation is based on the Navier-Stokes equations

$$\operatorname{div} v = 0 \tag{3}$$

$$\frac{\partial v}{\partial t} = -(v \cdot \nabla)v + \nu \Delta v + f \tag{4}$$

i	$n = 2^i$	generator $\mathbf{g}$	basis vectors $V = (v_1 v_2)$
2	4	(1, 2)	(2, 0), (1, 2)
3	8	(1, 3)	(2, -2), (1, 3)
		(1, 5)	(2, 2), (1, -3)
4	16	(1, 4)	(4, 0), (1, 4)
		(1, 12)	(4, 0), (1, -4)
5	32	(1, 7)	(4, -4), (5, 3)
		(1, 9)	(4, 4), (3, -5)
		(1, 23)	(4, -4), (3, 5)
		(1, 25)	(4, 4), (5, -3)
6	64	(1, 28)	(7, 4), (2, -8)
		(1, 36)	(7, -4), (2, 8)
7	128	(1, 12)	(11, 4), (1, 12)
		(1, 116)	(11, -4), (1, -12)
8	256	(1, 30)	(9, 14), (17, -2)
		(1, 226)	(9, -14), (17, 2)
9	512	(1, 200)	(18, 16), (23, -8)
		(1, 312)	(18, -16), (23, 8)
10	1024	(1, 271)	(34, -2), (15, -31)
		(1, 495)	(2, -34), (31, -15)
		(1, 529)	(2, 34), (31, 15)
		(1, 753)	(34, 2), (15, 31)
11	2048	(1, 592)	(45, 16), (7, 48)
		(1, 1456)	(45, -16), (7, -48)
12	4096	(1, 70)	(59, 34), (58, -36)
		(1, 4026)	(59, -34), (58, 36)
13	8192	(1, 1530)	(91, -34), (75, 62)
		(1, 6662)	(91, 34), (75, -62)
14	16384	(1, 1435)	(57, -125), (137, -13)
		(1, 6291)	(125, -57), (13, -137)
		(1, 10093)	(125, 57), (13, 137)
		(1, 14949)	(57, 125), (137, 13)
15	32768	(1, 15936)	(183, -64), (146, 128)
		(1, 16832)	(183, 64), (146, -128)
16	65536	(1, 25962)	(260, -88), (53, -270)
		(1, 39574)	(260, 88), (53, 270)
17	131072	(1, 49531)	(172, -348), (217, 323)
		(1, 62899)	(348, -172), (323, 217)
		(1, 68173)	(348, 172), (323, -217)
		(1, 81541)	(172, 348), (217, -323)
18	262144	(1, 1990)	(527, 154), (395, -382)
		(1, 260154)	(527, -154), (395, 382)
19	524288	(1, 86592)	(775, -64), (442, 640)
		(1, 437696)	(775, 64), (442, -640)
20	1048576	(1, 195638)	(134, 1092), (879, -662)
		(1, 852938)	(134, -1092), (879, 662)
21	2097152	(1, 193293)	(1226, -958), (217, 1541)
		(1, 715835)	(958, 1226), (1541, -217)
		(1, 1381317)	(958, -1226), (1541, 217)
		(1, 1903859)	(1226, 958), (217, -1541)
22	4194304	(1, 1120786)	(363, -2170), (1699, 1398)
		(1, 3073518)	(363, 2170), (1699, -1398)
23	8388608	(1, 1671221)	(1807, -2533), (3097, 301)
		(1, 3288547)	(2533, 1807), (301, -3097)
		(1, 5100061)	(2533, -1807), (301, 3097)
		(1, 6717387)	(1807, 2533), (3097, -301)
24	16777216	(1, 7605516)	(2903, -3308), (1414, 4168)
		(1, 9171700)	(2903, 3308), (1414, -4168)
25	33554432	(1, 1905545)	(405, -6211), (5582, -2754)
		(1, 14462279)	(6211, 405), (2754, 5582)
		(1, 19092153)	(6211, -405), (2754, -5582)
		(1, 31648887)	(405, 6211), (5582, 2754)
26	67108864	(1, 22282116)	(6391, -6052), (8436, 2512)
		(1, 44826748)	(6391, 6052), (8436, -2512)
27	134217728	(1, 58928436)	(9147, 8444), (2740, -12144)
		(1, 75289292)	(9147, -8444), (2740, 12144)
28	268435456	(1, 86198508)	(682, 17592), (15577, 8204)
		(1, 182236948)	(682, -17592), (15577, -8204)
29	536870912	(1, 8370742)	(11737, 21958), (24885, 814)
		(1, 528500170)	(11737, -21958), (24885, -814)
30	1073741824	(1, 78999493)	(10221, -33695), (24071, 25699)
		(1, 284281075)	(33695, 10221), (25699, -24071)
		(1, 789460749)	(33695, -10221), (25699, 24071)
		(1, 994742331)	(10221, 33695), (24071, -25699)
31	2147483648	(1, 940574718)	(38453, 31638), (8176, -49120)
		(1, 1206908930)	(38453, -31638), (8176, 49120)

Table 1. Parameters of all maximized minimum distance lattices in two dimensions with $n = 2^i$ points for $i = 2, \ldots, 31$. Note that the basis vectors are given in integer precision and have to be divided by the number of lattice points.

for incompressible fluids, where v is the velocity field, ν the viscosity, and f are the external forces. The solution strategy is to simulate equation (4) and remove the divergence (3) at the end of each time step by using a projection based on the Helmholtz-Hodge decomposition $w = v + \nabla q$, which states that a vector field w can be decomposed into a divergence free part v and the gradient of a scalar field q. The velocity field for the next time step $t + \Delta t$ is computed in four steps [Sta99]:

1. The external forces are added: $v_1(\mathbf{x}_l) := v(\mathbf{x}_l) + \Delta t \cdot f(\mathbf{x}_l)$
2. The advection is computed by tracing back a particle back in time starting from point $\mathbf{x}_l$ according to the velocity field. The position $p(\mathbf{x}_l, -\Delta t)$ is computed by dividing the time step Δt into smaller time steps and performing the Euler rule for each of the small time steps (see the illustration). The velocities $v_2(\mathbf{x}_l) := v_1(p(\mathbf{x}_l, -\Delta t))$ are linearly interpolated using the closest lattice points. Due to linear interpolation the method is named stable, because the computed velocities never can exceed the old ones in magnitude.

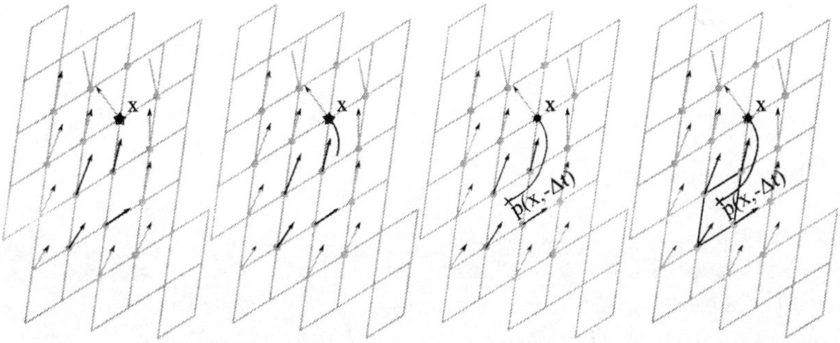

3. The diffusion by the Laplace operator is efficiently computed as low pass filter in the Fourier domain. The Fourier coefficients are computed by

$$\hat{v}_2(\mathbf{k}_j) := \sum_{l=0}^{n-1} v_2(\mathbf{x}_l) e^{-2\pi i \frac{jl}{n}}$$

   and filtered

$$\hat{v}_3(\mathbf{k}_j) := \frac{\hat{v}_2(\mathbf{k}_j)}{1 + \nu \Delta t \cdot (\mathbf{k}_j \cdot \mathbf{k}_j)}.$$

4. The divergence is removed using the projection

$$\hat{v}_4(\mathbf{k}_j) := \hat{v}_3(\mathbf{k}_j) - \frac{(\mathbf{k}_j \cdot \hat{v}_3(\mathbf{k}_j))\mathbf{k}_j}{(\mathbf{k}_j \cdot \mathbf{k}_j)}$$

   and the velocity field at time step $t + \Delta t$ is synthesized by

$$v_4(\mathbf{x}_l) := \sum_{j=0}^{n-1} \hat{v}_4(\mathbf{k}_j) e^{2\pi i \frac{jl}{n}}.$$

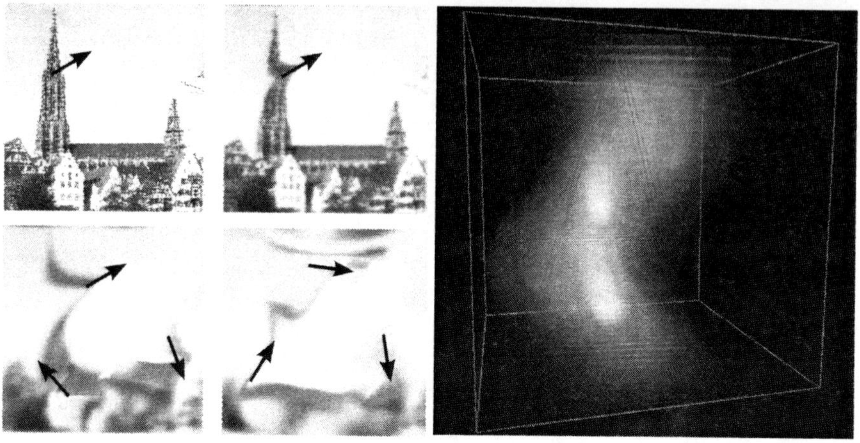

Fig. 6. Left: The four images are subsequent snapshots of the stable fluid simulation on a rank-1 lattice. The arrows indicate the external forces applied to the periodic fluid. The fluid transports the background image and the effects of advection and diffusion, i.e. blur, are clearly visible. Right: Snapshot of a wind field simulation in three dimensions for the lattice $L_{32768,\mathbf{g}}$ with $\mathbf{g} = (1, 10871, 10871^2)$, where smoke is transported in a velocity field.

Implementation

The stable fluids scheme has been implemented for two- and three-dimensional velocity fields as illustrated in Figure 6. The Fourier transformation techniques are the same as in the previous application example, except that they have to be performed for each component of the vector fields.

For linear interpolation (as required in step 2 of the algorithm) on a rank-1 lattice, the Minkowski-reduced basis V is used to access neighboring lattice points.

Scaling all lattice points by n results in integer coordinates for all $\mathbf{x}_l$ and the basis V.

Frame: Representing the velocity field v in the basis V avoids a transformation during interpolation. In this case external forces usually have to be transformed into the basis V.

Accessing lattice cells: The backtracking step requires to compute the index of a lattice cell containing a given point, which is simple if the lattice is given in Korobov form: Multiplying the basis matrix V by the integer parts of the coordinates of the point modulo n yields a lattice point in Cartesian coordinates. Obviously the first component is the lattice point or cell index. Neighboring lattice points for interpolation now are found as described in the previous example.

4 Conclusion

Spectral synthesis and simulation on rank-1 lattices can be implemented efficiently. Independent of the dimension s only a one dimensional Fourier transform is needed. Additionally the approximation can be more accurate than on a tensor product lattice [LH03, KSW04, DKKS]. It is also notable that the isotropic measure of maximized minimum distance can replace the classical measures like e.g. discrepancy (see the sampling efficiency of the Fibonacci lattice in Figure 4) often used in connection with rank-1 lattices. This measure also provides best visual quality.

In the future we like to extend our ideas to hierarchical approaches using lattice sequences and explore optimizations for anisotropic spectra. Moreover we will explore non-periodic boundaries.

Acknowledgments

The authors would like to thank mental images GmbH for support and for funding of this research. The second author is very thankful for the very fruitful communication with Li Dong, Fred Hickernell, and Peter Schröder. Note that the article entirely includes the technical report 307/01 "Random Fields on Rank-1 Lattices" by Alexander Keller, University of Kaiserslautern, that is based on the presentation "Solutions of P.D.E.s using Spectral Methods with Integration Lattices" by Li Dong and Fred Hickernell at the 4th International Conference on Monte Carlo and quasi-Monte Carlo Methods in Scientific Computing, Hong Kong Baptist University in 2000.

References

[AEVZ02] E. Agrell, T. Eriksson, A. Vardy, and K. Zeger. Closest point search in lattices. *IEEE Transactions on Information Theory*, 48(8):2201–2214, 2002.

[AR86] A. Fournier and W. Reeves. A simple model of ocean waves. In *SIGGRAPH '86: Proceedings of the 13th Annual Conference on Computer Graphics and Interactive Techniques*, pages 75–84, New York, NY, USA, 1986.

[DKKS] J. Dick, P. Kritzer, F. Kuo, and I. Sloan. Lattice-Nyström Method for Fredholm Integral Equations of the Second Kind. *Submitted to the Journal of Complexity.*

[Ent] Areté Entertainment. http://www.areteentertainment.com.

[Knu77] D. Knuth. *The Art of Computer Programming, Volume 2, Seminumerical Algorithms.* Addison Wesley, 1977.

[KSW04] F. Kuo, I. Sloan, and H. Woźniakowski. Lattice Rules for Multivariate Approximation in the Worst Case Setting. *Monte Carlo and Quasi-Monte Carlo Methods 2004*, (H. Niederreiter and D. Talay, eds.), Springer-Verlag, pages 289–330, 2006.

[LH03] D. Li and F. Hickernell. Trigonometric spectral collocation methods on lattices. *Recent Advances in Scientific Computing and Partial Differential Equations, AMS Series on Contemporary Mathematics*, pages 121–132, 2003.

[PM62] D. Petersen and D. Middleton. Sampling and reconstruction of wave-number-limited functions in N-dimensional Euclidean spaces. *Information and Control*, 5(4):279–323, 1962.

[SF91] J. Stam and E. Fiume. A multiple-scale stochastic modelling primitive. In *Proceedings of Graphics Interface '91*, pages 24–31, 1991.

[SF93] J. Stam and E. Fiume. Turbulent wind fields for gaseous phenomena. *Computer Graphics*, 27(Annual Conference Series):369–376, 1993.

[SJ94] I. Sloan and S. Joe. *Lattice Methods for Multiple Integration*. Clarendon Press Oxford, 1994.

[Sta95] J. Stam. *Multi-Scale Stochastic Modelling of Complex Natural Phenomena*. PhD thesis, University of Toronto, 1995.

[Sta97] J. Stam. Stochastic dynamics: Simulating the effects of turbulence on flexible structures. *Computer Graphics Forum*, 16(3):C159–C164, 1997.

[Sta99] J. Stam. Stable fluids. In *SIGGRAPH '99: Proceedings of the 26th Annual Conference on Computer Graphics and Interactive Techniques*, pages 121–128, 1999.

[Tes00] J. Tessendorf. Simulating ocean water. In *SIGGRAPH 2000 Course Notes: Course 25*.

Image Synthesis by Rank-1 Lattices

Sabrina Dammertz[1] and Alexander Keller[2]

[1] Ulm University, Germany
sabrina.dammertz@uni-ulm.de
[2] Ulm University, Germany
alexander.keller@uni-ulm.de

Summary. Considering uniform points for sampling, rank-1 lattices provide the simplest generation algorithm. Compared to classical tensor product lattices or random samples, their geometry allows for a higher sampling efficiency. These considerations result in a proof that for periodic Lipschitz continuous functions, rank-1 lattices with maximized minimum distance perform best. This result is then investigated in the context of image synthesis, where we study anti-aliasing by rank-1 lattices and using the geometry of rank-1 lattices for sensor and display layouts.

1 Introduction

Image synthesis can be considered as an integro-approximation problem

$$I(k,l) := \int_{I^s} f(\mathbf{x},k,l)d\mathbf{x} \approx \frac{1}{n}\sum_{i=0}^{n-1} f(\mathbf{x}_i,k,l), \tag{1}$$

where the two-dimensional image function $I(k,l)$ is given by a parametric integral. Since usually analytic solutions are hardly accessible, we are interested in efficient numerical schemes to approximate the image function. Thinking of (k,l) as pixel coordinates on the screen, the above algorithm simultaneously computes a color for each pixel on the screen by averaging samples of the integrand at positions $\mathbf{x}_i$. We consider two important aspects:

Sampling: In computer graphics the accumulation buffer [HA90] along with several extensions and improvements [Kel97, KH01, SIP06] is the most efficient implementation of integro-approximation as it can take advantage of the vast performance of rasterization hardware. In the original article [HA90] sampling points $\mathbf{x}_i$ generated by Lloyd relaxation were found to perform best. This implies that sampling points should have maximized minimum distance.

Display: Modern displays use either rectangular or hexagonal arrangements of pixels. Again, due to a larger minimum distance hexagonal arrangements expose a much better visual quality than rectangular arrangements, nevertheless, image synthesis is currently still dominated by the first.

As we will show in the following, rank-1 lattices selected by maximized minimum distance approximate hexagonal lattices. However, rank-1 lattices are simpler to generate and exist for any number of points in any dimension. We investigate the application of such lattices in two dimensions for anti-aliasing and display and sensor technology.

2 Geometry of Rank-1 Lattices

A lattice L is a discrete subset of $\mathbb{R}^s$ which is closed under addition and subtraction. Given an s-dimensional lattice basis $\{\mathbf{b}_1, \ldots, \mathbf{b}_s\}$, the lattice can be generated by all integer linear combinations

$$L(\mathbf{b}_1, \ldots, \mathbf{b}_s) := \left\{ \sum_{j=1}^{s} \lambda_j \mathbf{b}_j : \lambda_1, \ldots, \lambda_s \in \mathbb{Z} \right\}. \tag{2}$$

Of all possible bases the Minkowski-reduced bases, which contain the s shortest linearly independent vectors of L [AEVZ02], are the most useful for our purpose.

Instead of using s basis vectors the points $\mathbf{x}_i$ of a rank-1 lattice [Nie92b, SJ94]

$$L_{n,\mathbf{g}} := \left\{ \mathbf{x}_i := \frac{i}{n}\mathbf{g} \bmod 1 : i = 0, \ldots, n-1 \right\}$$

in the unit cube are easily generated by using only one suitable generator vector $\mathbf{g} \in \mathbb{N}^s$ for a fixed number $n \in \mathbb{N}$ of points. Korobov lattices $L_{n,a}$ are a special class of rank-1 lattices. Their generator vector has the form $\mathbf{g} = (1, a, a^2, \ldots, a^{s-1})$. The Fibonacci lattices are an instance of a two-dimensional Korobov lattice. Based on the Fibonacci sequence $F_k := F_{k-1} + F_{k-2}$ with $F_2 := F_1 := 1$, the number of points is set to $n = F_k, k \geq 2$ and the generator vector is defined as $\mathbf{g} = (1, F_{k-1})$. Figure 1 shows a Fibonacci lattice in the unit square with $n = F_9 = 34$ points and the generator vector $\mathbf{g} = (1, F_8) = (1, 21)$.

The generator vector $\mathbf{g}$ of a rank-1 lattice can be chosen such that the resulting point set is of low discrepancy [SJ94]. Then the elements of this point set are called good lattice points. But only very few explicit constructions for good lattice points exist. Similar to the Fibonacci lattices Niederreiter and Borosh [BN83, Nie86] showed that good two-dimensional lattice points can be explicitly constructed for n being a power of two.

Obviously the quality of rank-1 lattices is significantly influenced by their integer generator vector $\mathbf{g}$. For lattices in Korobov form the search is reduced to only one parameter a.

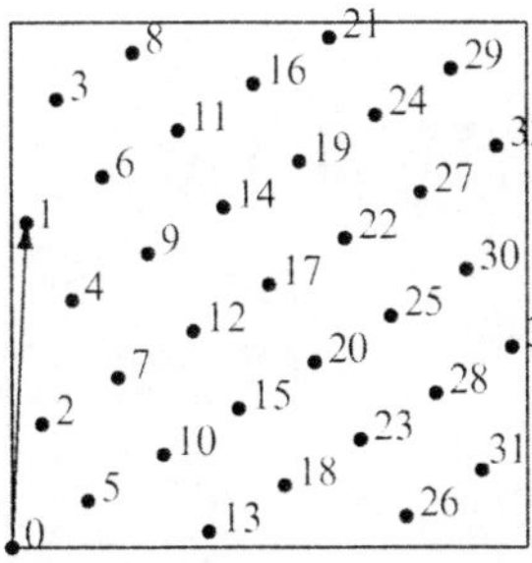

Fig. 1. The $n = 34$ points of the Fibonacci lattice $L_{34,21}$ with generator vector $\mathbf{g} = (1, 21)$.

2.1 Shifted Rank-1 Lattices

Considering shifted rank-1 lattices

$$L_{n,\mathbf{g}}^{\mathbf{\Delta}} := \left\{ \mathbf{x}_i := \frac{i}{n}\mathbf{g} + \mathbf{\Delta} \bmod 1 : \mathbf{\Delta} \in \mathbb{R}^s;\ i = 0, \ldots, n-1 \right\}$$

there exists a trivial, but nevertheless interesting connection to (t, m, s)–nets in basis b [Nie92b].

$(0, 2, 2)$-nets in base b only exhibit three kinds of elementary intervals

$$\left[\frac{i}{b}, \frac{i+1}{b}\right) \times \left[\frac{j}{b}, \frac{j+1}{b}\right) \quad \text{for } 0 \le i, j < b,$$
$$\left[\frac{i}{b^2}, \frac{i+1}{b^2}\right) \times \Big[0, 1\Big) \quad \text{for } 0 \le i < b, \text{ and}$$
$$\Big[0, 1\Big) \times \left[\frac{i}{b^2}, \frac{i+1}{b^2}\right) \quad \text{for } 0 \le i < b,$$

whereof each must contain exactly one of the $n = b^2$ points of the shifted rank-1 lattice due to $t = 0$. The latter two kinds of intervals guarantee perfect one-dimensional projections. Independent of the shift $\mathbf{\Delta}$, this is obtained by requiring $\gcd(n, g_i) = 1, i \in \{1, \ldots, s\}$. Possible shift coordinates are given by the one-dimensional projections of the lattice points. It is sufficient to search shifts in only one of the elementary intervals, such that the $t = 0$ condition is fulfilled. An illustration for $L_{25,7}^{\mathbf{\Delta}}$ with $\mathbf{\Delta} = (\frac{2}{25}, \frac{2}{25})$ is found in Figure 2 and further parameters are listed in Table 1.

Like $(0, 2, 2)$–nets the resulting lattices share both the properties of jittered grid and Latin hypercube samples but can be computed faster due to a simpler algorithm.

2.2 Maximized Minimum Distance Rank-1 Lattices

In computer graphics sampling patterns with blue noise spectral properties are used in analogy to the principle of maximized minimum distance apparent in

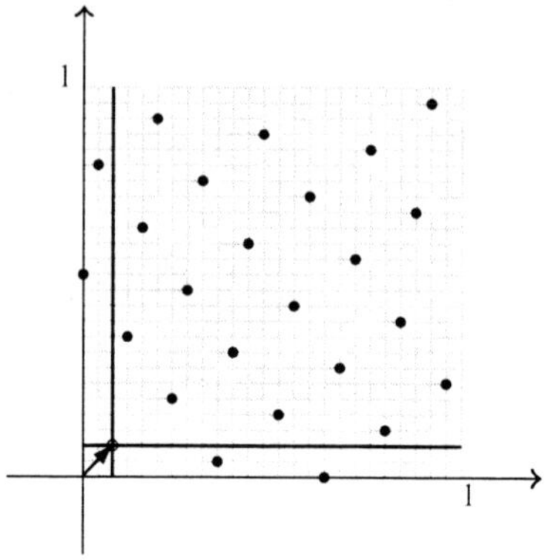

Fig. 2. Example of a shifted rank-1 lattice $L_{25,7}^{\boldsymbol{\Delta}}$ with $\boldsymbol{\Delta} = (\frac{2}{25}, \frac{2}{25})$ that is a (0, 2, 2)-net in basis $b = 5$.

b	$n = b^2$	a	$n \cdot \boldsymbol{\Delta} \in [0, b)^2]$
2	4	1	$(1, 0)$
3	9	2	$(1, 1)$
5	25	7	$(2, 2)$
7	49	6	$(5, 1)$
11	121	36	$(5, 5)$
13	169	70	$(6, 6)$
17	289	80	$(2, 4)$
19	361	100	$(14, 15)$
23	529	120	$(0, 2)$
29	841	150	$(7, 8)$
31	961	210	$(7, 9)$

Table 1. Parameters for shifted rank-1 lattices in Korobov form that are (0, 2, 2)-nets in base b. The rational shift $\boldsymbol{\Delta}$ is scaled by n for integer precision.

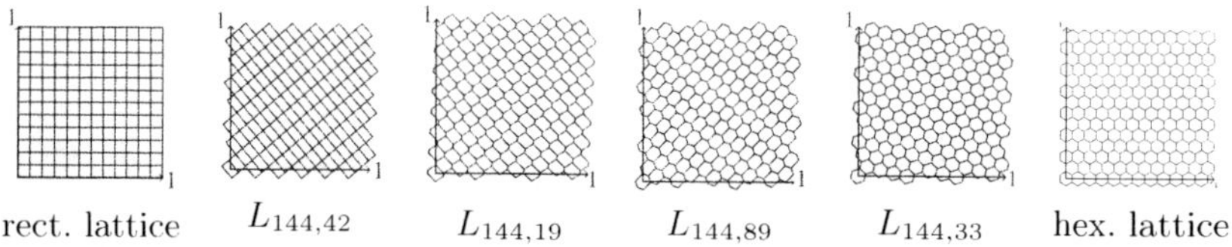

Fig. 3. In the sequence of Korobov lattices $L_{144,a}$ the minimum distance increases from left to right. For comparison the rectangular lattice is added to the left, whereas the hexagonal lattice is the rightmost of the image sequence.

nature. For example the photo receptors in the retina are distributed according to this scheme [Yel83] in order to reduce aliasing.

Similarly we can select rank-1 lattice generator vectors that maximize the minimum distance, which leads to the notion of maximized minimum distance lattices. The task of calculating the minimum distance in a lattice is a well known problem in lattice theory, namely the shortest vector problem [AEVZ02]. Since a lattice is closed under addition and subtraction the difference between two lattice points yields another point in the lattice. Therefore the minimum distance corresponds to the length of the shortest vector in the lattice. This quantity can be computed by searching the closest point to the origin, which means to consider all lattice points except $\mathbf{x}_0 = \mathbf{0}$.

For $s = 2$ the sequence of rank-1 lattices with increasing minimum distance approximates the hexagonal lattice in the limit, which is illustrated in Figure 3 for $n = 144$ points.

In [CR97] Cools and Reztsov define a family

$$L_{n,\mathbf{g}} = \left\{ \frac{i}{2F_m M_m} (M_m, F_m) \bmod 1 : 0 \leq i < 2F_m M_m \right\} \tag{3}$$

of rank-1 lattices by using the sequence of convergents

$$\left\{\frac{F_m}{M_m}\right\}_{m=1}^{\infty} = \frac{2}{1}, \frac{5}{3}, \frac{7}{4}, \frac{19}{11}, \ldots$$

of the continued fraction equal to $\sqrt{3}$. Since these lattices are constructed to exactly integrate trigonometric polynomials of a hexagonal spectrum, they actually represent maximized minimum distance lattices. As the construction only covers lattices for $n = 2F_m M_m$ points, for other n the generator vector was determined by computer search.

Computer Search

Searching for maximized minimum distance rank-1 lattices represents a computationally expensive problem, since there are $(n-1)^s$ possibilities for the generator vector $\mathbf{g} = (g_1, \ldots, g_s)$, where $g_i \in \{1, \ldots, n-1\}$. However, for $s = 2$ an exhaustive search is feasible. In order to avoid rounding errors due to floating point imprecision all computations are done in integer arithmetic allowing for exact results.

One possibility to reduce the search space is to consider only rank-1 lattices in Korobov form, which are uniquely determined by the tuple (n, a) (see Section 2). A very efficient way to search for maximized minimum distance lattices in Korobov form for $s = 2$ is given by the spectral test [Knu81], which measures the quality of linear congruential random number generators by determining the t-dimensional accuracy ν_t. It can be shown that this quantity corresponds to the length of the shortest vector in the dual lattice

$$L_{n,\mathbf{g}}^{\perp} := \{\mathbf{h} \in \mathbb{Z}^s : \mathbf{h} \cdot \mathbf{g} \equiv 0 \pmod{n}\}.$$

Since the length of the shortest vector in $L^{\perp}$ equals the length of the shortest vector in L multiplied by n, the spectral test delivers the minimum mutual distance between two lattice points for one $a \in \{1, \ldots, n-1\}$ on $[0, n)^2$. As the searching algorithm is performed on $[0, n)^2$, the two-dimensional accuracy ν_2 delivers the sought quantity, i.e. the length of the shortest vector in the lattice $L_{n,a}$.

Additionally, the search space can be restricted by demanding that n and g_i are relatively prime, i.e. $\gcd(n, g_i) = 1$. This means that the resulting lattice points will be stratified equidistantly on each coordinate axis. So the resulting rank-1 lattice is an instance of a Latin hypercube sample and the minimum distance can be bounded to mindist $\geq \frac{1}{n}$.

However, the condition $\gcd(n, g_i) = 1$ prevents to find the best lattice with regard to maximized minimum distance in some cases. This also applies to searching maximized minimum distance lattices in Korobov form. For example the maximized minimum distance of the lattices of equation 3 cannot be achieved in Korobov form.

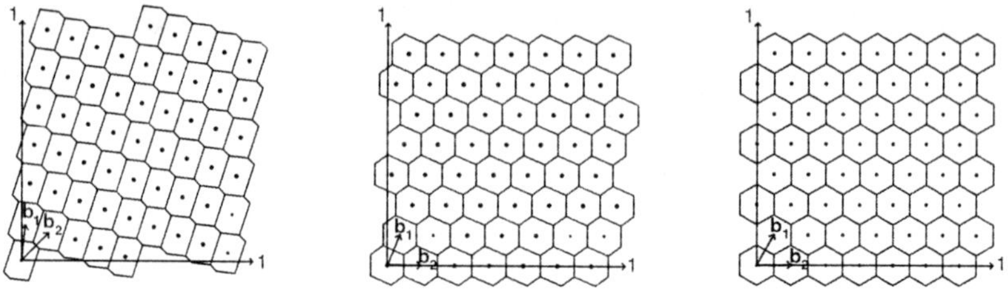

(a) $L_{56,9}$ mindist $= \frac{\sqrt{40}}{56}$ (b) $L_{56,21}$ mindist $= \frac{\sqrt{58}}{56}$ (c) $L_{56,(4,7)}$ mindist $= \frac{\sqrt{64}}{56}$

Fig. 4. Maximized minimum distance lattices for $n = 56$: (a) Rank-1 lattice searched under the restriction of $\gcd(n, a) = 1$ in Korobov form. (b) Rank-1 lattice in Korobov form. (c) Rank-1 lattice selected without restrictions.

Figure 4 compares the maximized minimum distance lattices for $n = 56$ selected in Korobov form (a), for $\gcd(n, a) = 1$ in Korobov form (b) and by using the lattice family of [CR97] (c).

3 Quasi-Monte Carlo Error Bounds

The functions in computer graphics are square integrable due to finite energy and bounded. However, they are only piecewise continuous, where the discontinuities are difficult to identify. Often the structure of the high-dimensional integrals in image synthesis comprises several $2-3$ dimensional integral operators, as it is the case for sampling the pixel area, the lens area, motion blur, depth of field, scattering events, etc. Consequently the sampling points can be padded using low-dimensional stratified patterns in a very efficient way, as Kollig and Keller have shown in [KK02].

The classical quasi-Monte Carlo error bound is given by the Koksma-Hlawka inequality [Nie92b], which bounds the integration error by the product of the discrepancy of the sampling points and the variation of the integrand in the sense of Hardy and Krause. However, the variation in the sense of Hardy and Krause already becomes infinite in the case of non-axis aligned discontinuities, thus being inapplicable to functions in computer graphics.

The error of a lattice rule [SJ94] can be formulated in terms of the Fourier coefficients of the integrand f requiring f to be periodic and to belong to the function class whose Fourier coefficients decay sufficiently fast with increasing frequency. Since these conditions usually do not hold for the setting of computer graphics, we cannot use this error bound either.

The notion of $(\mathcal{M}, \mu)$-uniformity introduced by Niederreiter [Nie03] supports partitions which are not axis aligned and relies on the stratification properties of the sampling points. The deterministic error bound based on this concept can easily be generalized to integro-approximation [Kel06]. Using the probability space $([0,1)^s, \mathcal{B}, \lambda_s)$, where $\mathcal{B}$ corresponds to the Borel-sets and λ_s

to the s-dimensional Lebesgue-measure, this bound also applies in the context of computer graphics. However, the error cannot be separated into a property of the integrand and the sampling pattern any longer.

3.1 Error Bound for Lipschitz Functions

Although the classical error bounds do not fit in the setting of computer graphics (as seen above), quasi-Monte Carlo methods achieve good results in a vast number of numerical experiments. The main reason is that the integrands are often piecewise continuous, while the discontinuities cannot be captured by the classical error bounds. Thus they cannot explain the observed convergence. We now examine an error bound for Lipschitz continuous, periodic functions with respect to parametric integration thereby completing Niederreiter's work [Nie03] for the special case of rank-1 lattices.

Given a Minkowski-reduced basis of a rank-1 lattice $L_{n,\mathbf{g}}$ the basis vectors induce the Delaunay tessellation of the lattice and its dual, the Voronoi diagram. In order to derive the error bound we need the following

Definition 1 *The radius $r(n, \mathbf{g})$ of a rank-1 lattice is the smallest circumcircle of the fundamental Voronoi cell with respect to some suitable norm.*

This quantity corresponds to the dispersion [Nie92b]

$$d_n(L_{n,\mathbf{g}}; I^s) = \sup_{x \in I^s} \min_{1 \leq i \leq n} d(\mathbf{x}, \mathbf{x}_i)$$

of a rank-1 lattice as well as the notion of the covering radius in coding theory. Figure 5 shows the Voronoi diagram along with the circumcircle of radius $r(32, (1, 7))$ of the Korobov lattice $L_{32,7}$.

Based on the results of [DFG99] and by taking advantage of the geometrical properties of rank-1 lattices the proof is very simple and resembles the proofs of the paper [Nie03].

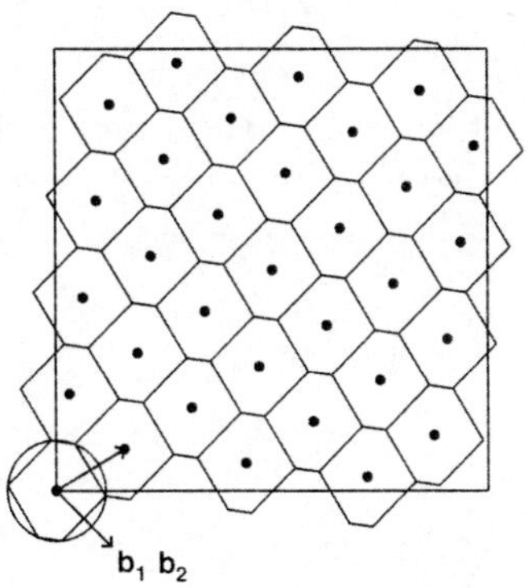

Fig. 5. Voronoi diagram of the lattice $L_{32,7}$ including the basis vectors in the sense of a Minkowski-reduced basis. The circumcircle of radius $r(n, \mathbf{g})$ encloses the fundamental Voronoi cell.

Theorem 1. *Let f be a Lipschitz function periodic on $[0,1]^{s+s'}$, with*

$$\|f(\mathbf{x}_1, \mathbf{y}) - f(\mathbf{x}_2, \mathbf{y})\| \leq L\|\mathbf{x}_1 - \mathbf{x}_2\|,$$

and Lipschitz constant L independent of $\mathbf{x}_1$, $\mathbf{x}_2$ and $\mathbf{y}$, where $\dim \mathbf{x}_1 = \dim \mathbf{x}_2 = s$ and $\dim \mathbf{y} = s'$. Further let $P_n = \{\mathbf{x}_0, \ldots, \mathbf{x}_{n-1}\}$ be a rank-1 lattice. Then

$$\left\| \int_{[0,1]^s} f(\mathbf{x}, \mathbf{y}) d\mathbf{x} - \frac{1}{n} \sum_{i=0}^{n-1} f(\mathbf{x}_i, \mathbf{y}) \right\| \leq L \cdot r(n, \mathbf{g})$$

for some suitable norm, where $r(n, \mathbf{g})$ is the radius of P_n.

Proof. Let $\mathcal{M} = \{M_0, \cdots, M_{n-1}\}$ be the partition of I^s by the Voronoi diagram of a rank-1 lattice. Then in a first step the quadrature error can be estimated similar to [DFG99]:

$$\begin{aligned}
\left\| \int_{[0,1]^s} f(\mathbf{x}, \mathbf{y}) d\mathbf{x} - \frac{1}{n} \sum_{i=0}^{n-1} f(\mathbf{x}_i, \mathbf{y}) \right\| &= \left\| \sum_{i=0}^{n-1} \int_{M_i} (f(\mathbf{x}, \mathbf{y}) - f(\mathbf{x}_i, \mathbf{y})) d\mathbf{x} \right\| \\
&\leq \sum_{i=0}^{n-1} \int_{M_i} ||f(\mathbf{x}, \mathbf{y}) - f(\mathbf{x}_i, \mathbf{y})|| d\mathbf{x} \\
&\leq L \sum_{i=0}^{n-1} \int_{M_i} ||\mathbf{x} - \mathbf{x}_i|| d\mathbf{x} \qquad (4) \\
&= L \cdot n \int_{M_0} ||\mathbf{x}|| d\mathbf{x} \qquad (5) \\
&\leq L \cdot n \cdot \lambda_s(M_0) \sup_{||\mathbf{x}|| \leq r(n,\mathbf{g})} ||\mathbf{x}|| \\
&= L \cdot n \cdot \frac{1}{n} \cdot r(n, \mathbf{g}) = L \cdot r(n, \mathbf{g}) \qquad (6)
\end{aligned}$$

Since the M_i are of identical shape and volume and due to the point symmetry of the lattice, we can choose $\mathbf{x}_i$ as $\mathbf{x}_0 = \mathbf{0}$ in equation (4) which then can further be simplified resulting in equation (5). □

Obviously the error bound results as a product of a property of the integrand and a property of the sampling pattern again. Omitting the parameter $\mathbf{y}$ in Theorem 1 yields the integration error bound.

Let $\Omega \subset \mathbb{R}^s$, $\mathcal{M} = \{M_0, \cdots, M_{n-1}\}$ an arbitrary tessellation of Ω, and $\{A_i\}_{i=0}^{n-1}$ the volumes of $\{M_i\}_{i=0}^{n-1}$. Then equation (4) represents the special case of the error estimation of [DFG99]

$$\left| \int_{\Omega} f(\mathbf{x}) d\mathbf{x} - A_i \sum_{i=0}^{n-1} f(\mathbf{x}_i) \right| \leq L \sum_{i=0}^{n-1} \int_{M_i} ||\mathbf{x} - \mathbf{x}_i|| d\mathbf{x} \qquad (7)$$

for rank-1 lattices, where $A_i = \frac{1}{n}$. It can be proved that this error bound is minimized by choosing $\{\mathbf{x}_i\}_{i=0}^{n-1}$ and $\{M_i\}_{i=0}^{n-1}$ such that the $\{M_i\}_{i=0}^{n-1}$ are the Voronoi sets for the $\mathbf{x}_i$ and the $\mathbf{x}_i$ are the mass centroids of the Voronoi sets at the same time [DFG99]. This means that rank-1 lattices are a suitable choice to minimize the integration error for Lipschitz continuous functions, since these conditions apply to these point sets due to their geometrical properties.

The theoretical rate of the new bound $\mathcal{O}(n^{-1/s})$ is already known from the field of information based complexity and approximation theory. It obviously is cursed by the dimension, which is hidden in the radius $r(n, \mathbf{g})$. However, the important issue about this theorem is not the rate as we consider $s = 2$, but that it yields a criterion for lattice search using the primal instead of the dual lattice by means of the following corollary:

Corollary 1. *Maximizing the minimum distance*

$$d_{min}(P_n) := \min_{0 \le i < n} \|\mathbf{x}_i\|$$

in a rank-1 lattice decreases the radius $r(n, \mathbf{g})$ *and thus the integration error.*

This corollary can be derived by the following observation. The minimum distance $d_{min}(P_n)$ in a rank-1 lattice corresponds to two times the radius of the in-circle of the fundamental Voronoi cell. Maximizing the minimum distance in a rank-1 lattice thus increases this radius. The Voronoi cells, being of equal size and constant volume, approximate a sphere the more, the bigger the minimum distance becomes. Consequently the gap between the radius of the circumcircle and the in-circle of the Voronoi cells decreases. This means that $r(n, \mathbf{g})$ decreases as $d_{min}(P_n)$ increases, which is stated in the corollary. Although there are similarities to sphere packings, it is important to note that this argument is not built upon them.

4 Applications

Based on the theoretical considerations of the previous sections we now investigate the effect of maximized minimum distance rank-1 lattices for integro-approximation for image synthesis and explore their geometrical properties in the context of display technology.

4.1 Anti-Aliasing by Rank-1 Lattices

A disadvantage of all current display technology that relies on regular structures to present images is that correlations between the function to be displayed and the pixel structure can be perceived as distorting artifacts. In order to avoid these so-called aliases, various sampling patterns have been investigated. It is common belief that at moderate sampling rates n random sampling points $\mathbf{x}_i$ with maximized minimum distance perform best, since aliases are mapped to noise, but low frequency details are reproduced clearly. Nevertheless, increasing the number of sample points can cause aliases to reappear as the noise vanishes. In fact these artifacts cannot be completely avoided because of the correlation of a deterministic display and a deterministic function, but they can be ameliorated by filtering parts of the image. However, using only a box filter, i.e. integrating over the pixel area by averaging samples, always causes aliasing to appear in an even converged image. This is illustrated in Figure 8 for the simple example of rendering an infinite checker board. As we are looking for the most efficient sampling patterns and as random sampling cannot always prevent aliasing, we investigate rank-1 lattices for image synthesis.

What Maximized Minimum Distance Lattices Can Do

A typical test function for anti-aliasing is given by

$$\begin{aligned} Z_2 : [0,1)^2 &\to [0,1] \\ (x,y) &\mapsto \frac{1}{2}\left(1 + \sin\left(1600 \cdot \left(x^2 + y^2\right)\right)\right). \end{aligned}$$

Figure 6 shows the results of rendering this function by stratified sampling, the Larcher-Pillichshammer points (LP) [LP01, KK02], and the maximized minimum distance lattice $L_{1048576,195638}$ in combination with a b-spline filter of degree 3. Whereas in Figure 6 (a) the aliasing is covered by noise, Moiré patterns become clearly visible in 6 (b). The best result is achieved by the maximized minimum distance lattice, which acts as a filter due to its nice Fourier properties [SJ94]. So aliasing artifacts are attenuated considerably. In fact the maximized minimum distance lattice $L_{1048576,195638}$ with approximately 4 samples per pixel achieves a similar quality to [SSA05], where 900 samples per pixel with a density proportional to a cubic b-spline filter we used to render almost the same test function.

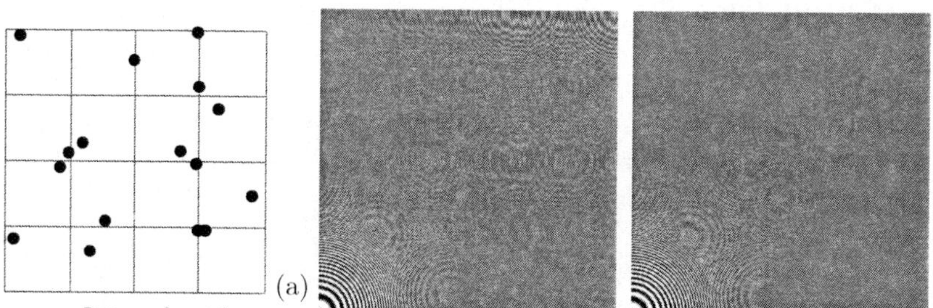

Jittered grid (a) (b) LP points (c) $L_{1048576,195638}$

Fig. 6. Rendering the test function by integro-approximation with $512 \times 512 \times 4$ samples, which means that there are about 4 samples per pixel. This figure should be viewed on screen, since otherwise the differences between the images can hardly be observed due to resampling and printing.

Speed of Convergence

In the following we compare the test patterns of Figure 7 with respect to their convergence. The sampling patterns are applied both for the integration problem (i.e. per pixel) and for the integro-approximation setting (i.e. over the whole quadratic screen). The test scene is given by the checker board of Figure 8(a).

In order to analyze the convergence of the test patterns, the L_2-norm of a converged reference image to the corresponding test image is computed for an increasing number of sampling points per pixel. Then the resulting value is diagrammed in the error graph, displaying the number of samples on the x- and the error norm on the y-axes. Both axis are scaled logarithmically. The reference image, shown in Figure 8(b), was computed by applying a jittered grid sampling pattern with 1024×1024 samples at each pixel. Obviously there are still aliasing artifacts in this image resulting from the problem of rendering a deterministic function on a deterministic display.

It is important to note that there are two different visual artifacts in image 8(a). The first one results from the case of only one edge lying in a single pixel. At low sampling rates these edges appear very jagged. Increasing the sampling rate solves this problem, though. Walking towards the horizon of the checker board, the single cells of the checker board get smaller and smaller. Therefore, many small cells, i.e. many edges fall within one pixel near the horizon as shown in Figures 8(e) and (f), yielding the aliasing structures in the converged image. Thus, we compute the error graphs only for the lower half of the checker board scene, where the convergence to the correct image is guaranteed.

Integration

We start by applying the test patterns (Figure 7) per pixel. For each pixel n rays are shot from the camera into the screen and the resulting color values

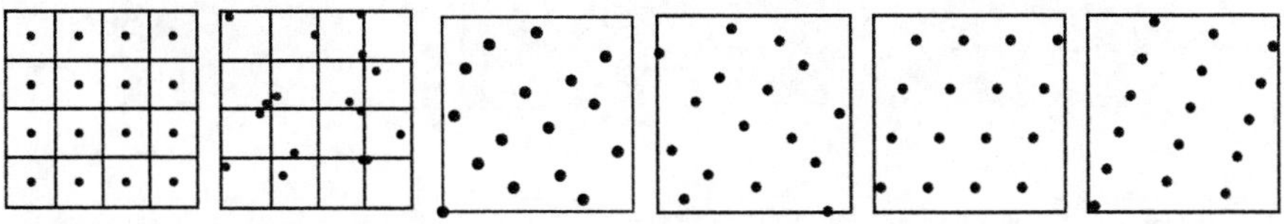

(a) Regular grid (b) Jittered grid (c) LP (d) LP rand (e) Shifted $L_{16,4}$ (f) Shifted $L_{16,3}$

Fig. 7. As sampling pattern the regular and jittered grid, the Larcher-Pillichshammer (LP) and randomized Larcher-Pillichshammer points, and rank-1 lattices are used. We analyze both the maximized minimum distance rank-1 lattices in Korobov form and the lattices resulting from the condition $\gcd(n, a) = 1$. Additionally, the lattices are shifted such that the bounding box of the lattice points is centered within the pixel.

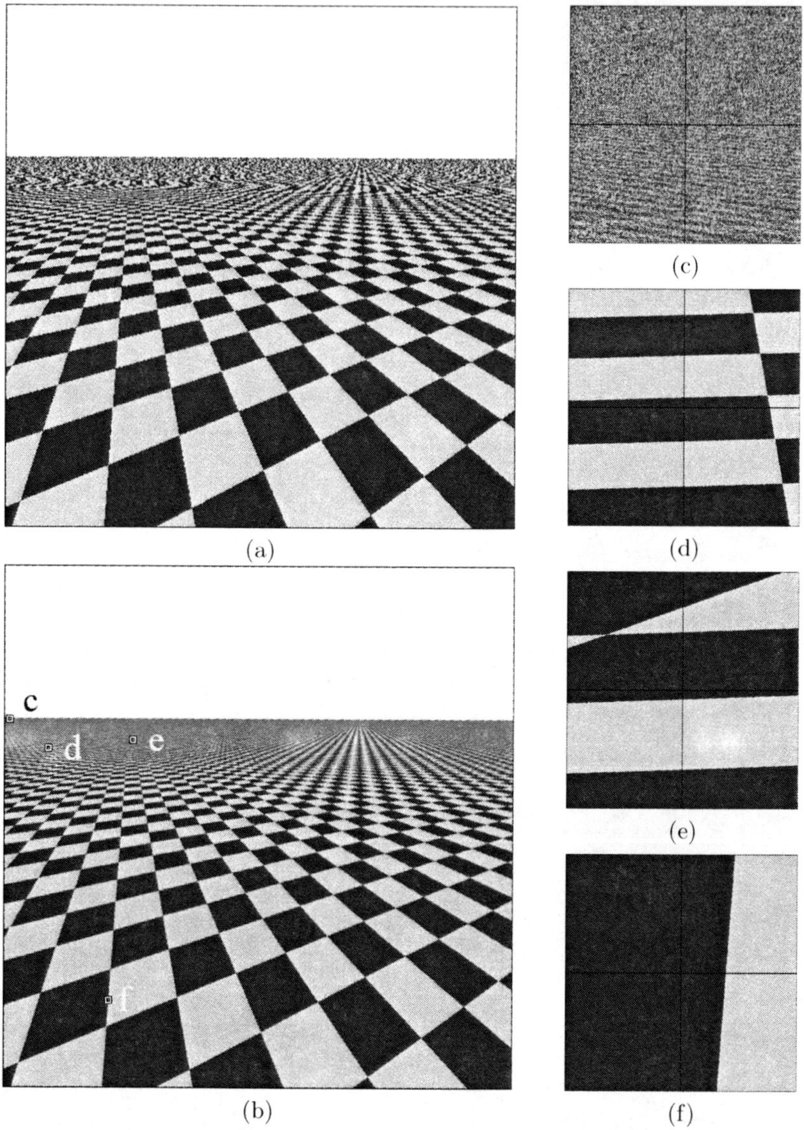

Fig. 8. (a) Aliasing due to the discrete representation of the checker board. (b) Reference image used to determine the L_2 error. (c)-(f) Magnification of the highlighted areas of (b), comprising 2×2 pixels each: (c) Roughly the same number of light and dark gray checker board cells cover this area averaging to half gray in (b). (d) The light and dark gray cells do not cover the same area in the pixels leading to aliasing. (e) The light and dark gray cells do not cover the same area in the pixels leading to aliasing. (f) Two pixels are completely covered by one dark gray cell, whereas an edge between two cells runs through the other two.

are averaged by means of the box filter. Although the regular grid possesses a worse discrepancy than the Larcher-Pillichshammer or lattice points, we notice that it performs extremely well, even surpassing them for certain n. This can be explained by the discrepancy being an anisotropic measure in fact. The large error spikes in Figure 9 arise from the factorization of n, since it is not always possible to find a good one. Jittered grid sampling turns aliasing into noise, and thus, the error proceeds on a lower level. However, this sampling pattern suffers from the same factorization problem as the regular grid.

The idea of using rank-1 lattices for computing the pixel integral is already mentioned in [Nie92a], however, not with respect to maximized minimum distance. Examining the maximized minimum distance lattices, we observe that postulating $\gcd(n, a) = 1$ sometimes severely restricts the search space; e.g. for $n = 4$ we get the pixel diagonal as sampling pattern which clearly is not a good distribution. Altogether the error curves expose a relatively strong oscillation. This is due to the structure of the lattice points featuring families of hyperplanes which are sometimes aligned to the checker board edges in such a way that these cannot be captured. As the Larcher-Pillichshammer points do not suffer from this, clearly scene dependent, problem, they offer a relatively smooth error curve.

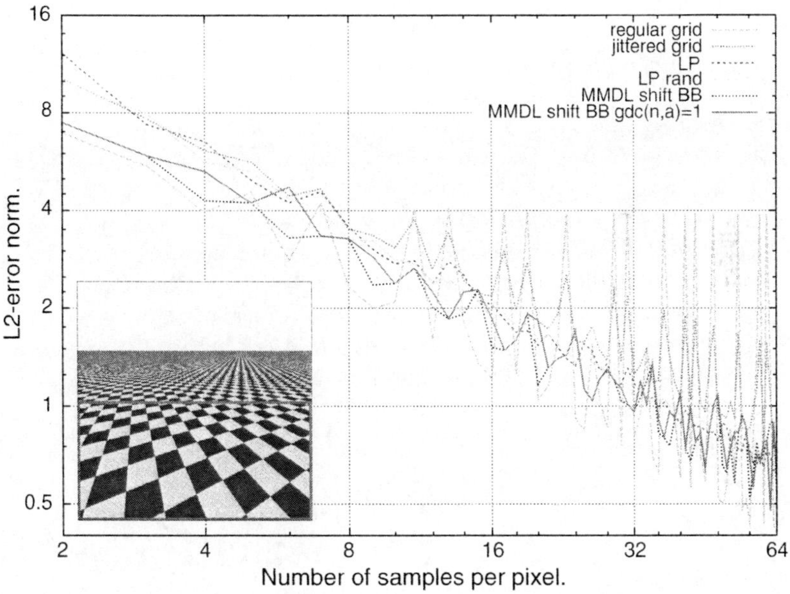

Fig. 9. Comparison of the regular and jittered grid, the Larcher-Pillichshammer points, shifted maximized minimum distance lattices and the maximized minimum distance lattices with $\gcd(n, a) = 1$.

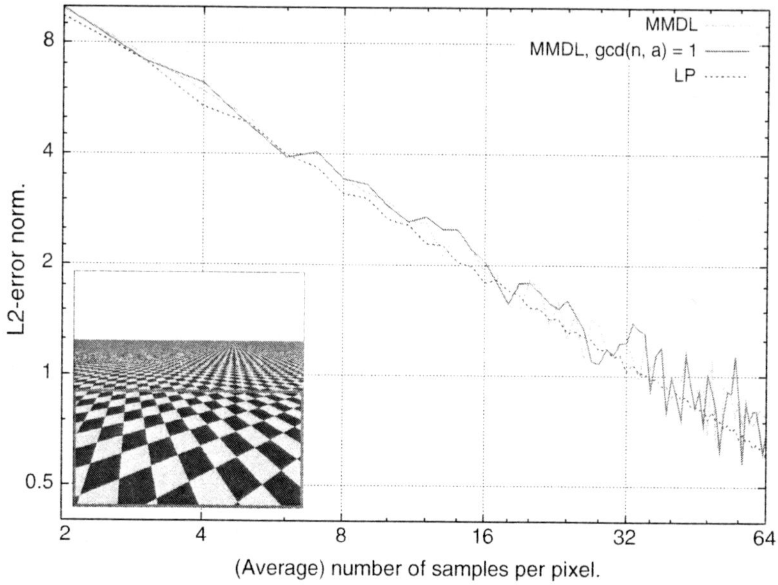

Fig. 10. Comparing the Larcher-Pillichshammer points and maximum minimum distance lattices scaled to the whole screen.

Integro-Approximation

Next we use the Larcher-Pillichshammer points and maximized minimum distance lattices over the whole quadratic screen ($xRes = yRes$), i.e. these sampling patterns are scaled from $[0, 1)^2$ to $[0, xRes) \times [0, yRes)$. To obtain a certain number n of samples per pixel, the number of sampling points has to be chosen as $xRes \cdot yRes \cdot n$. Since the Larcher-Pillichshammer pattern for $n = 2^m$ points represents a $(0, m, 2)$-net in base $b = 2$, we can guarantee a certain number of samples per pixel if the screen resolution is chosen as a power of 2. If the number of sampling points equals $xRes \cdot yRes \cdot k = 2^m$ for $k = 1$, we can take each pixel as an elementary interval of volume $b^{-m} = 2^{-m} = \frac{1}{xRes} \cdot \frac{1}{yRes}$. For $k > 1$, each pixel contains k elementary intervals of volume $\frac{1}{xRes} \cdot \frac{1}{yRes} \cdot \frac{1}{k}$. So each pixel is sampled by the same number of points, with each pixel being sampled by a different pattern at the same time. Considering the correlation between the pixels, the scaled Larcher-Pillichshammer points obtain an even smoother error curve than in the integration setting, as can be seen in Figure 12. This is even true for the case $n \neq 2^m$.

In contrast to the Larcher-Pillichshammer points, the rank-1 lattices cannot achieve the same number of sampling points per pixel and the image becomes quite noisy, especially for low sampling rates. Moreover, there are again orientations in the checker board, which fall exactly between two hyperplanes, resulting in an oscillating integration error. This also affects the comparison to the integration setting. So, in contrast to the Larcher-Pillichshammer points,

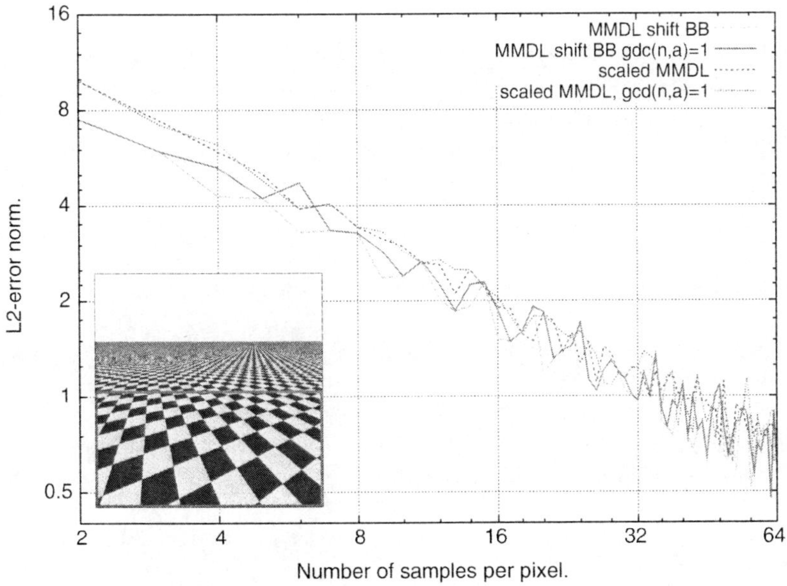

Fig. 11. Comparison integration/integro-approximation for the rank-1 lattices.

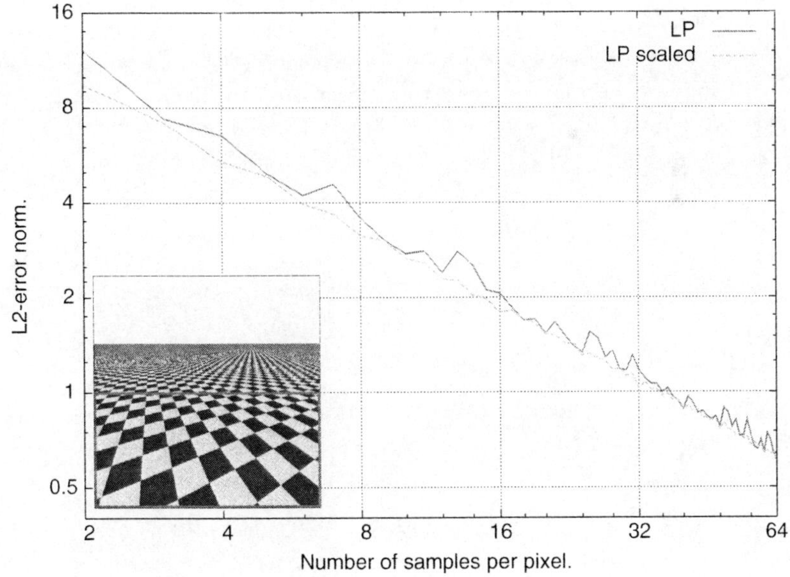

Fig. 12. Comparison integration/integro-approximation for the Larcher-Pillichshammer points.

the error graph for the integro-approximation problem is slightly worse than the one of the integration setting, which is illustrated in Figure 11.

Altogether, the test patterns converge to the reference image, which still exposes aliasing in the case of using the box filter for integration. The Larcher-Pillichshammer and maximized minimum distance sampling patterns show the fastest convergence rate, with the first outperforming the latter for this test scene.

4.2 Images on Rank-1 Lattices

A raster display is typically formed from a matrix of pixels representing the whole screen. Whereas a pixel is usually represented as a square on the integral raster being defined by the display resolution, we now structure the pixel layout by the Voronoi diagram of a maximized minimum distance rank-1 lattice, i.e. the single picture elements are represented as the cells which are induced by the Voronoi diagram of the rank-1 lattice points.

This kind of display technology has several advantages. Since rank-1 lattices are available for any number n of points, the number of pixels can be chosen freely. As seen in Section 2.2 maximizing minimum distance approximates a hexagonal grid in the limit yielding almost hexagonal picture elements. The concept of hexagonal pixels has already been studied in the context of hexagonal image processing ([MS05]) and is used in the SmartSlab LED panels (www.metropolismag.com). Moreover, this pixel layout permits optically better results than rank-2 lattices, as for example a smoother representation of curved objects. This can be seen in Figure 13, where the original image (middle) has been computed in reduced resolution, once on a traditional rank-2 lattice and once on a maximized minimum distance rank-1 lattice. At the same time, image processing algorithms are simplified in comparison to hexagonal lattices. In the same way image processing algorithms which are based upon the fast

Traditional display, 48×48 pixels   Original image, 512×512 pixels   Rank-1 lattice display, 48×48 pixels

Fig. 13. Comparison of rank-2 (left) and maximized minimum distance rank-1 lattice (right) displays at identical pixel resolution for approximating the high resolution image in the middle.

Fourier transform become simpler and can be implemented in a slightly more efficient way [DKD07].

This concept can technically be realized in a number of ways: One possibility consists in making up the display of point light sources, like for example RGB LEDs, which are arranged in the center of each Voronoi cell. Composing the display of area light sources which cover the single picture elements yields a technique for TFT and LCD displays, respectively. This may be realized by means of OLEDs for instance. Moreover, the layout of the sensors, i.e. the photosites, of a CCD (Charge-Coupled Device) camera can take the form of a rank-1 lattice cell. Further applications are given by projector technology, $3d$-Displays, etc.

2^n Display Modules, Sensors, and Images

Since rank-1 lattices can be tiled seamlessly, it is possible to compose a display of k modules each of which having the same number of lattice cells. This is illustrated in Figure 14.

The idea of 2^n display modules consists in choosing the number of picture elements for one display module as a power of 2. This has the advantage that the single cells easily can be addressed by means of a demultiplexer. If the number k of modules is set to a power of 2 as well, the single modules can be controlled the same way.

Such displays can be produced quite cheaply by fabricating small modules of the same layout and the same number of lattice cells which can easily be assembled to a display of desired resolution. More generally, the concept of 2^n displays perfectly fits all aspects of computer technology, taking advantage of memory layout, cache lines, addressing, etc.

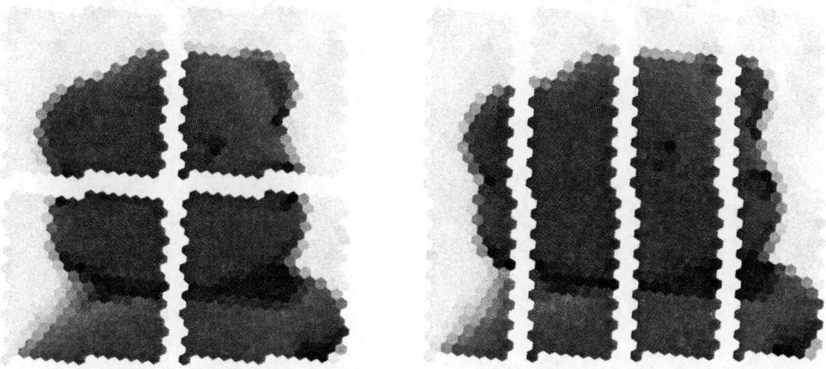

Fig. 14. The display is composed of 4 modules each of which contains 256 cells. Left: Quadratic layout of the single modules, i.e. the rank-1 lattice is searched on the unit square. Right: Rectangular layout, i.e. the rank-1 lattice is searched on a rectangular domain by means of a corresponding weighted norm.

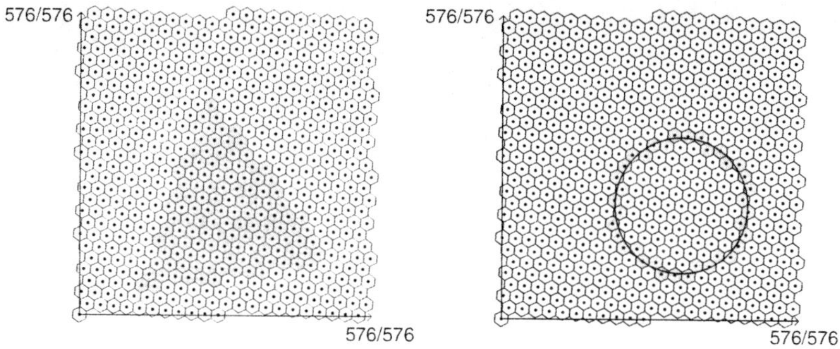

Fig. 15. Examples for rasterizing a triangle and a circle on the lattice $L_{576,155}$.

Storing images according to this scheme leads to the concept of 2^n images which equally benefit from the advantages of 2^n displays. The $\mathcal{O}(2^n)$ memory requirements ideally fit paging (memory alignment). As a further example storing a sequence of textures as $2^0 \cdots 2^n$ images naturally supports MipMapping and allows for a simple fast Fourier transforms [DKD07] processing.

Rasterization

Mathematical (ideal) primitives, such as lines, triangles, or circles areusually described in terms of 2-dimensional vertices on a Cartesian grid. In order to render them correctly a so-called rasterizer approximates them by assigning the appropriate colors to sets of pixels [FvDFH96]. The rasterizer converts the two-dimensional vertices in screen space into pixels on the display.

Changing the pixel layout by the introduction of rank-1 lattice displays also yields new algorithms for rasterization. Instead of rasterizing on a rectangular grid, the rasterization is now performed on the Voronoi cells of a maximized minimum distance rank-1 lattice, as illustrated in Figure 15.

The basic idea for converting the traditional rasterization algorithms to rank-1 lattices simply consists in changing the basis in which the rasterization is performed. This means that the rasterizer switches from the Cartesian coordinate system to that coordinate system which is formed by the basis of the corresponding rank-1 lattice. Whereas this method can be simulated on traditional raster displays by means of a software solution, it can even be performed on current graphics hardware in the following way: Since the rasterizer is only capable of operating on rectangular grids, in a first step the scene has to be transformed into the lattice basis, which in fact corresponds to a shear of the rectangular grid. After this change of frame the rasterization can be performed on the graphics hardware as usual. In order to display the rasterized scene, the resulting image has to be transformed back into the pixel basis. Performing the rasterization directly on a rank-1 lattice would have yielded the same result.

5 Conclusion

We examined maximized minimum distance rank-1 lattices in the context of integro-approximation and display technology. We derived an error bound for the class of Lipschitz continuous, periodic functions. Numerical experiments proved that these lattices perform quite well in image synthesis. However, the visual results are mixed: On the one hand extreme performance and quality gains were observed, on the other hand the convergence rate heavily depends on the function class, as the checker board example showed. Due to their algorithmical simplicity, maximized minimum distance rank-1 lattices are very promising with regard to data layout and image processing at a power of 2 pixels.

6 Acknowledgments

The authors would like to thank mental images GmbH for support and for funding of this research. The authors are very thankful for the very fruitful communication with Fred Hickernell and Ronald Cools.

References

[AEVZ02] E. Agrell, T. Eriksson, A. Vardy, and K. Zeger. Closest point search in lattices. *IEEE Transactions on Information Theory*, 48(8):2201–2214, 2002.

[BN83] I. Borosh and H. Niederreiter. Optimal multipliers for pseudo-random number generation by the linear congruential method. *BIT*, 23:65–74, 1983.

[CR97] R. Cools and A. Reztsov. Different quality indexes for lattice rules. *Journal of Complexity*, 13(2):235–258, 1997.

[DFG99] Q. Du, V. Faber, and M. Gunzburger. Centroidal Voronoi tessellations: Applications and algorithms. *SIAM Rev.*, 41(4):637–676, 1999.

[DKD07] H. Dammertz, A. Keller, and S. Dammertz. Simulation on Rank-1 Lattices. In A. Keller, S. Heinrich, and H. Niederreiter, editors, *Monte Carlo and Quasi-Monte Carlo Methods 2006*. Springer, in this volume.

[FvDFH96] J. Foley, A. van Dam, S. Feiner, and J. Hughes. *Computer Graphics, Principles and Practice, 2nd Edition in C*. Addison-Wesley, 1996.

[HA90] P. Haeberli and K. Akeley. The Accumulation Buffer: Hardware Support for High-Quality Rendering. In *Computer Graphics (SIGGRAPH 90 Conference Proceedings)*, pages 309–318, 1990.

[Kel97] A. Keller. Instant Radiosity. In *SIGGRAPH 97 Conference Proceedings*, Annual Conference Series, pages 49–56, 1997.

[Kel06] A. Keller. Myths of Computer Graphics. In H. Niederreiter and D. Talay, editors, *Monte Carlo and Quasi-Monte Carlo Methods 2004*, pages 217–243. Springer, 2006.

[KH01] A. Keller and W. Heidrich. Interleaved Sampling. In K. Myszkowski and S. Gortler, editors, *Rendering Techniques 2001 (Proc. 12th Eurographics Workshop on Rendering)*, pages 269–276. Springer, 2001.

[KK02] T. Kollig and A. Keller. Efficient Multidimensional Sampling. *Computer Graphics Forum*, 21(3):557–563, September 2002.

[Knu81] D. Knuth. *The Art of Computer Programming Vol. 2: Seminumerical Algorithms*. Addison Wesley, 1981.

[LP01] G. Larcher and F. Pillichshammer. Walsh Series Analysis of the L_2-Discrepancy of Symmetrisized Point Sets. *Monatsh. Math.*, 132:1–18, 2001.

[MS05] L. Middleton and J. Sivaswamy. *Hexagonal Image Processing: A Practical Approach (Advances in Pattern Recognition)*. Springer-Verlag, 2005.

[Nie86] H. Niederreiter. Dyadic fractions with small partial quotients. *Monatsh. Math*, 101:309–315, 1986.

[Nie92a] H. Niederreiter. Quasirandom Sampling in Computer Graphics. In *Proc. 3rd Internat. Seminar on Digital Image Processing in Medicine, Remote Sensing and Visualization of Information (Riga, Latvia)*, pages 29–34, 1992.

[Nie92b] H. Niederreiter. *Random Number Generation and Quasi-Monte Carlo Methods*. SIAM, Philadelphia, 1992.

[Nie03] H. Niederreiter. Error bounds for quasi-Monte Carlo integration with uniform point sets. *J. Comput. Appl. Math.*, 150:283–292, 2003.

[SIP06] B. Segovia, J. Iehl, and B. Péroche. Non-interleaved Deferred Shading of Interleaved Sample Patterns, September 2006. Eurographics/SIGGRAPH Workshop on Graphics Hardware '06.

[SJ94] I. Sloan and S. Joe. *Lattice Methods for Multiple Integration*. Clarendon Press, Oxford, 1994.

[SSA05] M. Stark, P. Shirley, and M. Ashikhmin. Generation of stratified samples for b-spline pixel filtering. *Journal of Graphics Tools*, 10(1):39–48, 2005.

[Yel83] J. Yellot. Spectral Consequences of Photoreceptor Sampling in the Rhesus Retina. *Science*, 221:382–385, 1983.

Continuous Runge-Kutta Methods for Stratonovich Stochastic Differential Equations

Kristian Debrabant[1] and Andreas Rößler[2]

[1] Technische Universität Darmstadt, Fachbereich Mathematik, Schlossgartenstr. 7, D-64289 Darmstadt, Germany
debrabant@mathematik.tu-darmstadt.de

[2] Technische Universität Darmstadt, Fachbereich Mathematik, Schlossgartenstr. 7, D-64289 Darmstadt, Germany
roessler@mathematik.tu-darmstadt.de

1 Introduction

Stochastic differential equations (SDEs) are applied in many disciplines like physics, biology or mathematical finance in order to describe dynamical systems disturbed by random effects. Approximation methods for the strong as well as for the weak time discrete approximation have been proposed in recent years (see, e.g., [BB00, DR06, KP99, KMS97, Mil95, MT04, New91, Roe04, Roe06b, TVA02] and the literature therein) converging with some given order at the discretization points. However, there is still a lack of higher order continuous time approximation methods guaranteeing uniform orders of convergence not only at the discretization points but also at any arbitrary time point within the approximation interval. Classical time discrete methods are inefficient in this case where the number of output points has to be very large because this forces the step size to be very small. Therefore, we develop a continuous extension of the class of stochastic Runge–Kutta (SRK) methods introduced in [Roe06c] for the weak approximation which provides continuous time approximations of the solution of Stratonovich SDE systems with uniform order two in the weak sense. Such methods are also called dense output formulas [HNW93]. The main advantage of the presented continuous extension of the SRK methods is their negligible additional computational complexity compared to the time discrete SRK methods. Especially, we are interested in continuous sample trajectories of the applied SRK method. For example, an SRK method with continuous sample trajectories allows the use of an individual discretization for each sample trajectory which needs not necessarily to contain some common discretization points for all trajectories in order to be able to calculate the expectation at these common time points. Further, in future research SRK methods with continuous sample trajectories may be applied for the numerical treatment of stochastic delay differential equations like in the deterministic setting where continuous Runge–Kutta methods are already successfully applied.

Let $(\Omega, \mathcal{F}, \mathrm{P})$ be a probability space with a filtration $(\mathcal{F}_t)_{t \geq 0}$ and let $\mathcal{I} = [t_0, T]$ for some $0 \leq t_0 < T < \infty$. We consider the solution $X = (X(t))_{t \in \mathcal{I}}$ of a d-dimensional Stratonovich stochastic differential equation system

$$dX(t) = a(t, X(t))\, dt + b(t, X(t)) \circ dW(t). \tag{1}$$

Let $X(t_0)$ be the $\mathcal{F}_{t_0}$-measurable initial condition such that for some $l \in \mathbb{N}$ holds $\mathrm{E}(\|X(t_0)\|^{2l}) < \infty$ where $\|\cdot\|$ denotes the Euclidean norm if not stated otherwise. Here, $W = ((W(t)^1, \ldots, W(t)^m))_{t \geq 0}$ is an m-dimensional Wiener process w.r.t. $(\mathcal{F}_t)_{t \geq 0}$. SDE (1) can also be written in integral form

$$X(t) = X(t_0) + \int_{t_0}^t a(s, X(s))\, ds + \sum_{j=1}^m \int_{t_0}^t b^j(s, X(s)) \circ dW(s)^j \tag{2}$$

for $t \in \mathcal{I}$ with some drift function $a : \mathcal{I} \times \mathbb{R}^d \to \mathbb{R}^d$ and a diffusion function $b \colon \mathcal{I} \times \mathbb{R}^d \to \mathbb{R}^{d \times m}$. The jth column of the $d \times m$-matrix function $b = (b^{i,j})$ is denoted by b^j for $j = 1, \ldots, m$. Further, the second integral w.r.t. the Wiener process has to be interpreted as a Stratonovich integral.

The solution $(X(t))_{t \in \mathcal{I}}$ of a Stratonovich SDE with drift a and diffusion b is also a solution of a corresponding Itô SDE and therefore also a generalized diffusion process, however with the modified drift

$$\tilde{a}^i(t, x) = a^i(t, x) + \frac{1}{2} \sum_{j=1}^d \sum_{k=1}^m b^{j,k}(t, x) \frac{\partial b^{i,k}}{\partial x^j}(t, x) \tag{3}$$

for $i = 1, \ldots, d$, provided that b is sufficiently differentiable, i.e.

$$\begin{aligned} X(t) &= X(t_0) + \int_{t_0}^t a(s, X(s))\, ds + \sum_{j=1}^m \int_{t_0}^t b^j(s, X(s)) \circ dW(s)^j && (4) \\ &= X(t_0) + \int_{t_0}^t \tilde{a}(s, X(s))\, ds + \sum_{j=1}^m \int_{t_0}^t b^j(s, X(s))\, dW(s)^j. && (5) \end{aligned}$$

We suppose that the drift $\tilde{a} \colon \mathcal{I} \times \mathbb{R}^d \to \mathbb{R}^d$ and the diffusion $b : \mathcal{I} \times \mathbb{R}^d \to \mathbb{R}^{d \times m}$ are measurable functions satisfying a linear growth and a Lipschitz condition

$$\|\tilde{a}(t, x)\| + \|b(t, x)\| \leq C\,(1 + \|x\|) \tag{6}$$

$$\|\tilde{a}(t, x) - \tilde{a}(t, y)\| + \|b(t, x) - b(t, y)\| \leq C\, \|x - y\| \tag{7}$$

for all $x, y \in \mathbb{R}^d$ and all $t \in \mathcal{I}$ with some constant $C > 0$. Then the conditions of the Existence and Uniqueness Theorem are fulfilled for the Stratonovich SDE (2) (see, e.g., Theorem 4.5.3 [KP99]).

In the following, let $C_P^l(\mathbb{R}^d, \mathbb{R})$ denote the space of all $g \in C^l(\mathbb{R}^d, \mathbb{R})$ with polynomial growth, i.e. there exist a constant $C > 0$ and $r \in \mathbb{N}$, such that

$|\partial_x^i g(x)| \leq C(1 + \|x\|^{2r})$ for all $x \in \mathbb{R}^d$ and any partial derivative of order $i \leq l$ (see [KP99], p. 153). Then g belongs to $C_P^{k,l}(\mathcal{I} \times \mathbb{R}^d, \mathbb{R})$ if $g \in C^{k,l}(\mathcal{I} \times \mathbb{R}^d, \mathbb{R})$ and $g(t, \cdot) \in C_P^l(\mathbb{R}^d, \mathbb{R})$ holds uniformly in $t \in \mathcal{I}$, i.e., the constants C and r do not depend on t.

Let $\mathcal{I}_h = \{t_0, t_1, \ldots, t_N\}$ be a discretization of the time interval $\mathcal{I} = [t_0, T]$ such that

$$0 \leq t_0 < t_1 < \ldots < t_N = T \tag{8}$$

and define $h_n = t_{n+1} - t_n$ for $n = 0, 1, \ldots, N-1$ with the maximum step size

$$h = \max_{0 \leq n \leq N-1} h_n .$$

Next to the concepts of strong or mean–square convergence which are used for good pathwise approximations of the solution process X, there is also the weak convergence which is applied if one is interested in the approximation of distributional properties of the solution process X. In the present paper, we will focus on weak convergence of some family of approximation processes $Y^h = (Y^h(t))_{t \in \mathcal{I}_h}$ with some prescribed order p to the solution X of the considered SDE and we write $Y = Y^h$ in the following.

Definition 1. *A family of approximation processes Y converges weakly with order p to the solution process X as $h \to 0$ if for each $f \in C_P^{2(p+1)}(\mathbb{R}^d, \mathbb{R})$ there exists a constant C_f, which does not depend on h, and a finite $h_0 > 0$ such that*

$$\sup_{t \in \mathcal{I}_h} |\operatorname{E}(f(X(t))) - \operatorname{E}(f(Y(t)))| \leq C_f \, h^p \tag{9}$$

holds for each $h \in \,]0, h_0[\,$.

Let $X^{t_0, X_{t_0}}$ denote the solution of the stochastic differential equation (2) in order to emphasize the initial condition. For simplicity of notation, in this section it is supposed that $\mathcal{I}_h$ denotes an equidistant discretization, i.e. $h = (t_N - t_0)/N$. In the following, we consider one-step approximations of the type

$$Y^{t,x}(t + \theta\, h) = A(t, x, h, \theta; \xi), \tag{10}$$

where ξ is a vector of random variables, with moments of sufficiently high order, $\theta \in [0, 1]$ is a parameter and A is a vector function of dimension d which is continuous in θ. We define for $t_{n+1} = t_n + h$ recursively the sequence

$$\begin{aligned} Y(t_0) &= X(t_0), \\ Y(t_{n+1}) &= A(t_n, Y(t_n), h, 1; \xi_n), \quad n = 0, 1, \ldots, N-1, \end{aligned} \tag{11}$$

where $X(t_0), \xi_0, \ldots, \xi_{N-1}$ are independent.

Since we are interested in obtaining a continuous global weak approximation $Y = (Y(t))_{t \in \mathcal{I}}$, we need an extension of the convergence theorem due to Milstein (Theorem 9.1 [Mil95], see also [MT04] p. 100) which specifies the relationship between the local and the global approximation order. The following theorem due to the authors (see [DR06]) guarantees uniform weak convergence with some prescribed order p and can be applied to any one step approximation method of type (10).

Theorem 1. *Suppose the following conditions hold:*

(i) The coefficients $\tilde{a}^i$ and $b^{i,j}$ are continuous, satisfy a Lipschitz condition (7) and belong to $C_P^{p+1,2(p+1)}(\mathcal{I} \times \mathbb{R}^d, \mathbb{R})$ for $i = 1, \ldots, d$, $j = 1, \ldots, m$.

(ii) For sufficiently large r (see, e.g., [DR06, Mil95, MT04] for details) the moments $\mathrm{E}(\|Y(t_n)\|^{2r})$ exist for $t_n \in \mathcal{I}_h$ and are uniformly bounded with respect to N and $n = 0, 1, \ldots, N$.

(iii) Assume that for all $f \in C_P^{2(p+1)}(\mathbb{R}^d, \mathbb{R})$ there exists a $K \in C_P^0(\mathbb{R}^d, \mathbb{R})$ such that the following local error estimations

$$|\,\mathrm{E}(f(X^{t,x}(t+h))) - \mathrm{E}(f(Y^{t,x}(t+h)))| \leq K(x)\, h^{p+1} \tag{12}$$

$$|\,\mathrm{E}(f(X^{t,x}(t+\theta h))) - \mathrm{E}(f(Y^{t,x}(t+\theta h)))| \leq K(x)\, h^{p} \tag{13}$$

are valid for $x \in \mathbb{R}^d$, $t, t+h \in \mathcal{I}$ and $\theta \in [0,1]$.

Then for all $t \in [t_0, T]$ the following global error estimation

$$|\,\mathrm{E}(f(X^{t_0,X(t_0)}(t))) - \mathrm{E}(f(Y^{t_0,X(t_0)}(t)))| \leq C h^p \tag{14}$$

holds for all $f \in C_P^{2(p+1)}(\mathbb{R}^d, \mathbb{R})$, where C is a constant, i.e. the method (11) has a uniform order of accuracy p in the sense of weak approximation.

Remark 1. In contrast to the original theorem due to Milstein [Mil95, MT04] now the order of convergence specified in equation (14) is not only valid in the discretization times $t \in \mathcal{I}_h$. Provided that the additional condition (13) is fulfilled, the global order of convergence (14) holds also uniformly for all $t \in [t_0, T]$.

In the following, we assume that the coefficients $\tilde{a}^i$ and $b^{i,j}$ satisfy assumption (i) of Theorem 1. Further, assumption (ii) of Theorem 1 is always fulfilled for the class of stochastic Runge–Kutta methods considered in the present paper provided that $\mathrm{E}(\|X(t_0)\|^{2r}) < \infty$ holds for sufficiently large $r \in \mathbb{N}$ (see [Roe06a, Roe06c] for details). Clearly, in the case of some deterministic initial value $X(t_0) = x_0 \in \mathbb{R}^d$ there exist all moments of $\|X(t_0)\|$.

2 Continuous Stochastic Runge–Kutta Methods

We introduce now a continuous extension of the class of stochastic Runge–Kutta methods due to Rößler [Roe06c] applicable to Stratonovich SDE systems (2). The main advantage of this class of SRK methods compared to other known SRK methods is that the number of evaluations of the diffusion functions b^j for each step does not depend on the dimension m of the driving Wiener process.

Thus, this class of SRK methods has significantly reduced computational complexity (see [Roe06c] for details). Therefore, we extend this class of SRK methods by substituting the fixed weights $\alpha_i = \bar{\alpha}_i$, $\beta_i^{(1)} = \bar{\beta}_i^{(1)}$ and $\beta_i^{(2)} = \bar{\beta}_i^{(2)}$ for some continuous weight functions

$$\alpha_i : [0,1] \to \mathbb{R}, \quad \beta_i^{(1)} : [0,1] \to \mathbb{R}, \quad \beta_i^{(2)} : [0,1] \to \mathbb{R}, \tag{15}$$

in order to obtain second order continuous SRK (CSRK) schemes for the uniform weak approximation of the solution of the Stratonovich SDE system (2). We define the d-dimensional approximation process Y by the explicit continuous SRK method of s stages with $Y(t_0) = X(t_0)$ and

$$\begin{aligned} Y(t_n + \theta\, h_n) =& Y(t_n) + \sum_{i=1}^{s} \alpha_i(\theta)\; a(t_n + c_i^{(0)} h_n, H_i^{(0)})\, h_n \\ &+ \sum_{i=1}^{s} \sum_{k=1}^{m} \beta_i^{(1)}(\theta)\; b^k(t_n + c_i^{(1)} h_n, H_i^{(k)})\, \hat{I}_{(k),n} \\ &+ \sum_{i=1}^{s} \sum_{k=1}^{m} \beta_i^{(2)}(\theta)\; b^k(t_n + c_i^{(2)} h_n, \hat{H}_i^{(k)})\, \sqrt{h_n} \end{aligned} \tag{16}$$

for $\theta \in [0,1]$ and $n = 0, 1, \ldots, N-1$ with stage values

$$\begin{aligned} H_i^{(0)} =& Y(t_n) + \sum_{j=1}^{s} A_{ij}^{(0)}\; a(t_n + c_j^{(0)} h_n, H_j^{(0)})\, h_n \\ &+ \sum_{j=1}^{s} \sum_{l=1}^{m} B_{ij}^{(0)}\; b^l(t_n + c_j^{(1)} h_n, H_j^{(l)})\, \hat{I}_{(l),n} \\ H_i^{(k)} =& Y(t_n) + \sum_{j=1}^{s} A_{ij}^{(1)}\; a(t_n + c_j^{(0)} h_n, H_j^{(0)})\, h_n \\ &+ \sum_{j=1}^{s} B_{ij}^{(1)}\, b^k(t_n + c_j^{(1)} h_n, H_j^{(k)})\, \hat{I}_{(k),n} \\ &+ \sum_{j=1}^{s} \sum_{\substack{l=1 \\ l \neq k}}^{m} B_{ij}^{(3)}\; b^l(t_n + c_j^{(1)} h_n, H_j^{(l)})\, \hat{I}_{(l),n} \\ \hat{H}_i^{(k)} =& Y(t_n) + \sum_{j=1}^{s} A_{ij}^{(2)}\; a(t_n + c_j^{(0)} h_n, H_j^{(0)})\, h_n \\ &+ \sum_{j=1}^{s} \sum_{\substack{l=1 \\ l \neq k}}^{m} B_{ij}^{(2)}\; b^l(t_n + c_j^{(1)} h_n, H_j^{(l)})\, \frac{\hat{I}_{(k,l),n}}{\sqrt{h_n}} \end{aligned}$$

for $i = 1, \ldots, s$ and $k = 1, \ldots, m$. Here, $\alpha(\theta), \beta^{(1)}(\theta), \beta^{(2)}(\theta), c^{(1)}, c^{(2)} \in \mathbb{R}^s$ for $\theta \in [0,1]$ and $A^{(q)}, B^{(r)} \in \mathbb{R}^{s \times s}$ for $q \in \{0,1,2\}$ and $r \in \{0,1,2,3\}$ with

$A_{ij}^{(q)} = B_{ij}^{(r)} = 0$ for $j \geq i$, $q \neq 2$ and $r \neq 2$ are the vectors and matrices of coefficients of the explicit CSRK method. We choose $c^{(q)} = A^{(q)} e$ with a vector $e = (1, \dots, 1)^T$ [Roe06a, Roe06c]. In the following, the product of vectors is defined component-wise.

The random variables of the CSRK method (16) are defined by

$$\hat{I}_{(k,l),n} = \begin{cases} \hat{I}_{(k),n} \, \tilde{I}_{(l),n} & \text{if } l < k \\ -\hat{I}_{(l),n} \, \tilde{I}_{(k),n} & \text{if } k < l \end{cases} \tag{17}$$

for $1 \leq k, l \leq m$ with independent random variables $\hat{I}_{(k),n}$ and $\tilde{I}_{(l),n}$ for $1 \leq k \leq m$, $1 \leq l \leq m-1$ and $n = 0, 1, \dots, N-1$. Thus, only $2m-1$ independent random variables are needed for each step. The random variables $\hat{I}_{(k),n}$ are three point distributed with $\mathrm{P}(\hat{I}_{(k),n} = \pm\sqrt{3\, h_n}) = \frac{1}{6}$ and $\mathrm{P}(\hat{I}_{(k),n} = 0) = \frac{2}{3}$ while the random variables $\tilde{I}_{(k),n}$ are defined by a two point distribution with $\mathrm{P}(\tilde{I}_{(k),n} = \pm\sqrt{h_n}) = \frac{1}{2}$ (see [Roe06c] for details).

The coefficients of the SRK method (16) can be represented by an extended Butcher array taking the form

$$\begin{array}{c|c|c|c} c^{(0)} & A^{(0)} & B^{(0)} & \\ \hline c^{(1)} & A^{(1)} & B^{(1)} & B^{(3)} \\ \hline c^{(2)} & A^{(2)} & B^{(2)} & \\ \hline & \alpha(\theta)^T & \beta^{(1)}(\theta)^T & \beta^{(2)}(\theta)^T \end{array}$$

The algorithm works as follows: First, the random variables $\hat{I}_{(k),n}$ and $\tilde{I}_{(l),n}$ have to be simulated for $1 \leq k \leq m$ and $1 \leq l \leq m-1$ w.r.t. the current step size h_n. Next, based on the approximation $Y(t_n)$ and the random variables, the stage values $H^{(0)}$, $H^{(k)}$ and $\hat{H}^{(k)}$ are calculated. Then we can determine the continuous approximation $Y(t)$ for arbitrary $t \in [t_n, t_{n+1}]$ by varying θ from 0 to 1 in formula (16). Thus, only a very small additional computational effort is needed for the calculation of the values $Y(t)$ with $t \in \mathcal{I} \setminus \mathcal{I}_h$. This is the main advantage in comparison to the application of an SRK method with very small step sizes.

In order to obtain a continuous order two CSRK approximation we firstly have to calculate order conditions for the coefficients of the CSRK method. Therefore, we apply the colored rooted tree analysis proposed in [Roe06a] to the CSRK method (16). Thus, by applying Theorem 6.4 in [Roe06a] we yield the order conditions presented in Theorem 5.1 in [Roe06c], however now we have to substitute the weights by the continuous functions (15). Then, we obtain for the colored rooted trees of order up to 1.5 the new conditions

1. $\alpha(\theta)^T e \, h = \theta \, h$
2. $(\beta^{(1)}(\theta)^T e)^2 \, h = \theta \, h$
3. $\beta^{(2)}(\theta)^T e \sqrt{h} = 0$
4. $\beta^{(1)}(\theta)^T B^{(1)} e \, h = \frac{1}{2} \theta \, h$
5. $\beta^{(2)}(\theta)^T A^{(2)} e \, h^{3/2} = 0$
6. $\beta^{(2)}(\theta)^T (B^{(2)} e)^2 \, h^{3/2} = 0$

due to conditions (12) and (13). Further, for all colored trees up to order 2.5 condition (12) has to be fulfilled for $\theta = 1$ and we can deduce these order conditions from the ones calculated in Theorem 5.1 [Roe06c].

Theorem 2. *Let $a^i \in C_P^{2,4}(\mathcal{I} \times \mathbb{R}^d, \mathbb{R})$ and $b^{i,j} \in C_P^{2,5}(\mathcal{I} \times \mathbb{R}^d, \mathbb{R})$ for $i = 1, \ldots, d$, $j = 1, \ldots, m$. If the coefficients of the continuous stochastic Runge-Kutta method (16) fulfill the equations*

1. $\alpha(\theta)^T e = \theta$ 2. $(\beta^{(1)}(\theta)^T e)^2 = \theta$ 3. $\beta^{(2)}(\theta)^T e = 0$
4. $\beta^{(1)}(\theta)^T B^{(1)} e = \frac{1}{2}\theta$ 5. $\beta^{(2)}(\theta)^T A^{(2)} e = 0$ 6. $\beta^{(2)}(\theta)^T (B^{(2)} e)^2 = 0$

for $\theta = 1$ then the method attains order 1 for the uniform weak approximation of the solution of the Stratonovich SDE (2). Further, if $a^i \in C_P^{3,6}(\mathcal{I} \times \mathbb{R}^d, \mathbb{R})$ and $b^{i,j} \in C_P^{3,7}(\mathcal{I} \times \mathbb{R}^d, \mathbb{R})$ for $1 \le i \le d$, $1 \le j \le m$, if equations 1.-6. hold for arbitrary $\theta \in [0, 1]$ and if in addition the equations

7. $\alpha(1)^T A^{(0)} e = \frac{1}{2}$ 8. $\alpha(1)^T (B^{(0)} (B^{(1)} e)) = \frac{1}{4}$
9. $\alpha(1)^T (B^{(0)} e)^2 = \frac{1}{2}$ 10. $(\beta^{(1)}(1)^T e)(\alpha(1)^T B^{(0)} e) = \frac{1}{2}$
11. $(\beta^{(1)}(1)^T e)(\beta^{(1)}(1)^T A^{(1)} e) = \frac{1}{2}$ 12. $\beta^{(1)}(1)^T (B^{(1)} (A^{(1)} e)) = \frac{1}{4}$
13. $\beta^{(1)}(1)^T ((B^{(1)} e)(A^{(1)} e)) = \frac{1}{4}$ 14. $\beta^{(1)}(1)^T B^{(3)} e = \frac{1}{2}$
15. $\beta^{(1)}(1)^T (B^{(3)} (B^{(3)} e)) = 0$ 16. $(\beta^{(2)}(1)^T B^{(2)} e)^2 = \frac{1}{4}$
17. $\beta^{(1)}(1)^T (B^{(1)} e)^3 = \frac{1}{4}$ 18. $\beta^{(1)}(1)^T (B^{(1)} (B^{(1)} e)^2) = \frac{1}{12}$
19. $\beta^{(1)}(1)^T (B^{(1)} (B^{(3)} e)^2) = \frac{1}{4}$ 20. $\beta^{(1)}(1)^T (A^{(1)} (B^{(0)} e)) = 0$
21. $\beta^{(2)}(1)^T (A^{(2)} e)^2 = 0$ 22. $\beta^{(2)}(1)^T (A^{(2)} (A^{(0)} e)) = 0$
23. $\beta^{(1)}(1)^T (B^{(1)} (B^{(1)} (B^{(1)} e))) = \frac{1}{24}$ 24. $\beta^{(2)}(1)^T (A^{(2)} (B^{(0)} e)) = 0$
25. $\beta^{(2)}(1)^T (A^{(2)} (B^{(0)} e)^2) = 0$ 26. $\beta^{(2)}(1)^T (B^{(2)} e)^4 = 0$
27. $\beta^{(2)}(1)^T (B^{(2)} (B^{(1)} e))^2 = 0$ 28. $\beta^{(2)}(1)^T (B^{(2)} (B^{(3)} e))^2 = 0$
29. $\beta^{(1)}(1)^T ((B^{(1)} e)(B^{(3)} e)^2) = \frac{1}{4}$ 30. $\beta^{(2)}(1)^T ((A^{(2)} e)(B^{(2)} e)^2) = 0$
31. $(\beta^{(1)}(1)^T e)(\beta^{(1)}(1)^T (B^{(1)} e)^2) = \frac{1}{3}$
32. $(\beta^{(1)}(1)^T e)(\beta^{(1)}(1)^T (B^{(3)} e)^2) = \frac{1}{2}$
33. $\beta^{(1)}(1)^T (B^{(1)} (B^{(3)} (B^{(1)} e))) = \frac{1}{8}$
34. $\beta^{(1)}(1)^T (B^{(3)} (B^{(3)} (B^{(3)} e))) = 0$
35. $\beta^{(1)}(1)^T (B^{(3)} (B^{(1)} (B^{(3)} e))) = 0$
36. $\beta^{(2)}(1)^T (A^{(2)} (B^{(0)} (B^{(1)} e))) = 0$
37. $(\beta^{(1)}(1)^T e)(\beta^{(1)}(1)^T ((B^{(3)} e)(B^{(1)} e))) = \frac{1}{4}$
38. $(\beta^{(1)}(1)^T e)(\beta^{(1)}(1)^T (B^{(1)} (B^{(1)} e))) = \frac{1}{6}$

39. $(\beta^{(1)}(1)^T e)(\beta^{(1)}(1)^T (B^{(3)}(B^{(1)}e))) = \frac{1}{4}$
40. $(\beta^{(1)}(1)^T e)(\beta^{(1)}(1)^T (B^{(1)}(B^{(3)}e))) = \frac{1}{4}$
41. $\beta^{(1)}(1)^T ((B^{(1)}e)(B^{(1)}(B^{(1)}e))) = \frac{1}{8}$
42. $\beta^{(1)}(1)^T ((B^{(1)}e)(B^{(3)}(B^{(1)}e))) = \frac{1}{8}$
43. $\beta^{(1)}(1)^T ((B^{(3)}e)(B^{(1)}(B^{(3)}e))) = \frac{1}{4}$
44. $\beta^{(1)}(1)^T ((B^{(3)}e)(B^{(3)}(B^{(3)}e))) = 0$
45. $\beta^{(1)}(1)^T (B^{(3)}((B^{(3)}e)(B^{(1)}e))) = 0$
46. $\beta^{(2)}(1)^T ((B^{(2)}(A^{(1)}e))(B^{(2)}e)) = 0$
47. $\beta^{(2)}(1)^T ((B^{(2)}e)(B^{(2)}(B^{(1)}e))) = 0$
48. $\beta^{(2)}(1)^T ((B^{(2)}e)(B^{(2)}(B^{(3)}e))) = 0$
49. $\beta^{(2)}(1)^T ((B^{(2)}e)(B^{(2)}((B^{(1)}e)^2))) = 0$
50. $\beta^{(2)}(1)^T ((B^{(2)}e)(B^{(2)}((B^{(3)}e)^2))) = 0$
51. $\beta^{(2)}(1)^T ((B^{(2)}e)(B^{(2)}((B^{(1)}e)(B^{(3)}e)))) = 0$
52. $\beta^{(2)}(1)^T ((B^{(2)}e)(B^{(2)}(B^{(1)}(B^{(1)}e)))) = 0$
53. $\beta^{(2)}(1)^T ((B^{(2)}e)(B^{(2)}(B^{(3)}(B^{(1)}e)))) = 0$
54. $\beta^{(2)}(1)^T ((B^{(2)}e)(B^{(2)}(B^{(3)}(B^{(3)}e)))) = 0$
55. $\beta^{(2)}(1)^T ((B^{(2)}e)(B^{(2)}(B^{(1)}(B^{(3)}e)))) = 0$

are fulfilled and if $c^{(i)} = A^{(i)}e$ *for* $i = 0, 1, 2$, *then the continuous stochastic Runge–Kutta method (16) attains order 2 for the uniform weak approximation of the solution of the Stratonovich SDE (2).*

Remark 2. Based on the order conditions presented in Theorem 2 it is of special interest to find coefficients which define SRK schemes with minimized error constants. Further, we need $s \geq 4$ for an explicit second order CSRK method (see [Roe04, Roe06c]) due to the order conditions.

Since $s = 4$ stages are needed for the Stratonovich SRK methods (16), it is possible to calculate schemes of a higher deterministic order p_D than the stochastic order of convergence p_S. Then, the SRK method converges at least with order $p = p_S$ for SDEs, however with order $p_D \geq p_S$ if it is applied to an ODE without any diffusion. In this case, the SRK method (2) reduces to a deterministic RK method where the deterministic part is represented by the coefficients $A^{(0)}$ and $\alpha(\theta)$. In the following, we will denote the order of convergence by the tuple (p_S, p_D).

In contrast to the calculation of coefficients for time discrete SRK methods considered in [Roe06c], we additionally have to look for some weight functions $\alpha_i, \beta_i^{(1)}, \beta_i^{(2)} \in C([0,1],\mathbb{R})$ fulfilling the order conditions of Theorem 2 for $1 \leq i \leq s$. However, let $\bar{\alpha}_i, \bar{\beta}_i^{(1)}, \bar{\beta}_i^{(2)} \in \mathbb{R}$ denote some constant weights of a time discrete SRK method fulfilling the order two conditions calculated in Theorem 5.1 in [Roe06c] for $1 \leq i \leq s$, which coincide with the order conditions of Theorem 2 for $\theta = 1$. Then, we simply have to look for some weight functions for the CSRK method (16) which fulfill conditions 1.-6. of Theorem 2 for all $\theta \in [0,1]$ with the boundary conditions

$$\alpha_i(0) = 0, \qquad \alpha_i(1) = \bar{\alpha}_i, \tag{18}$$

$$\beta_i^{(1)}(0) = 0, \qquad \beta_i^{(1)}(1) = \bar{\beta}_i^{(1)}, \tag{19}$$

$$\beta_i^{(2)}(0) = 0, \qquad \beta_i^{(2)}(1) = \bar{\beta}_i^{(2)}, \tag{20}$$

for $1 \leq i \leq s$. Thus, we can extend each time discrete SRK method to a continuous SRK method by replacing the weights $\bar{\alpha}_i, \bar{\beta}_i^{(1)}, \bar{\beta}_i^{(2)} \in \mathbb{R}$ by some weight functions fulfilling conditions 1.-6. of Theorem 2 and the boundary conditions (18)–(20) and by retaining all the remaining coefficients of $A^{(q)}$ and $B^{(r)}$ for $q \in \{0,1,2\}$ and $r \in \{0,1,2,3\}$.

As an example, we extend the time discrete order two SRK schemes RS1 and RS2 calculated in [Roe06c] to continuous SRK schemes. For RS1 which attains order $(2,2)$ with $s = 4$ stages we have the weights $\bar{\alpha} = [0, 0, \frac{1}{2}, \frac{1}{2}]$, $\bar{\beta}^{(1)} = [\frac{1}{8}, \frac{3}{8}, \frac{3}{8}, \frac{1}{8}]$ and $\bar{\beta}^{(2)} = [0, -\frac{1}{4}, \frac{1}{4}, 0]$, see also Table 1 for the remaining coefficients of RS1. From condition 1. of Theorem 2 follows

$$\alpha_1(\theta) = \theta - \alpha_2(\theta) - \alpha_3(\theta) - \alpha_4(\theta)$$

and the boundary condition (18) implies that $\alpha_i(0) = 0$ for $1 \leq i \leq 4$, $\alpha_1(1) = \alpha_2(1) = 0$ and $\alpha_3(1) = \alpha_4(1) = \frac{1}{2}$ has to be fulfilled. Therefore, we can

0												
0	0				0							
1	1	0			$\frac{1}{4}$	$\frac{3}{4}$						
0	0	0	0		0	0	0					
0												
0	0				$\frac{2}{3}$				0			
1	1	0			$\frac{1}{12}$	$\frac{1}{4}$			$\frac{1}{4}$	$\frac{3}{4}$		
1	1	0	0		$-\frac{5}{4}$	$\frac{1}{4}$	2		$\frac{1}{4}$	$\frac{3}{4}$	0	
0												
0	0				1							
0	0	0			-1	0						
0	0	0	0		0	0	0					
	0	0	$\frac{1}{2}\theta$	$\frac{1}{2}\theta$	$\sqrt{\theta} - \frac{7}{8}\theta$	$\frac{3}{8}\theta$	$\frac{3}{8}\theta$	$\frac{1}{8}\theta$	0	$-\frac{1}{4}\theta$	$\frac{1}{4}\theta$	0

Table 1. CSRK scheme CDRS1 with order $p_D = p_S = 2$.

choose $\alpha_1(\theta) = \alpha_2(\theta) = 0$ and $\alpha_3(\theta) = \alpha_4(\theta) = \frac{1}{2}\theta$ for $\theta \in [0,1]$. Considering $\beta^{(1)}(\theta)$ we obtain from conditions 2. and 4. that

$$\beta_1^{(1)}(\theta) = \pm\sqrt{\theta} - \beta_2^{(1)}(\theta) - \beta_3^{(1)}(\theta) - \beta_4^{(1)}(\theta)$$
$$\frac{2}{3}\beta_2^{(1)}(\theta) + \frac{1}{3}\beta_3^{(1)}(\theta) + \beta_4^{(1)}(\theta) = \frac{1}{2}\theta$$

and boundary condition (19) yields that $\beta_i^{(1)}(0) = 0$ for $1 \le i \le 4$, $\beta_1^{(1)}(1) = \beta_4^{(1)}(1) = \frac{1}{8}$ and $\beta_2^{(1)}(1) = \beta_3^{(1)}(1) = \frac{3}{8}$. Here, we can choose e.g. $\beta_1^{(1)}(\theta) = \sqrt{\theta} - \frac{7}{8}\theta$, $\beta_2^{(1)}(\theta) = \beta_3^{(1)}(\theta) = \frac{3}{8}\theta$ and $\beta_4^{(1)}(\theta) = \frac{1}{8}\theta$ for $\theta \in [0,1]$. Finally, we obtain due to conditions 3. and 6. that

$$\beta_1^{(2)}(\theta) + \beta_2^{(2)}(\theta) + \beta_3^{(2)}(\theta) + \beta_4^{(2)}(\theta) = 0$$
$$\beta_2^{(2)}(\theta) + \beta_3^{(2)}(\theta) = 0$$

and the remaining boundary condition (20) results in $\beta_i^{(2)}(0) = 0$ for $1 \le i \le 4$, $\beta_1^{(2)}(1) = \beta_4^{(2)}(1) = 0$, $\beta_2^{(2)}(1) = -\frac{1}{4}$ and $\beta_3^{(2)}(1) = \frac{1}{4}$. Then, we can choose $\beta_1^{(2)}(\theta) = \beta_4^{(2)}(\theta) = 0$, $\beta_2^{(2)}(\theta) = -\frac{1}{4}\theta$ and $\beta_3^{(2)}(\theta) = \frac{1}{4}\theta$ for $\theta \in [0,1]$. We obtain the continuous SRK scheme CDRS1 of order (2, 2) with the coefficients presented in Table 1. Here, we have to point out that there are some degrees of freedom in choosing the coefficient functions.

Analogously, we obtain for the SRK scheme RS2 of order (3, 2) with the weights $\bar{\alpha} = [\frac{1}{4}, \frac{1}{4}, \frac{1}{2}, 0]$ and with $\bar{\beta}^{(1)}$ and $\bar{\beta}^{(2)}$ equal to the weights of RS1, respectively, a continuous extension. The remaining coefficients for RS2 are presented in Table 2. We only have to determine the weight functions α_i for $1 \le i \le 4$ for the continuous SRK scheme CDRS2 if we use for $\beta_i^{(1)}$ and $\beta_i^{(2)}$ the same weight functions as calculated for CDRS1. Now, the weight functions α_i have to fulfill the condition 1. of Theorem 2 with the boundary conditions

0												
$\frac{2}{3}$	$\frac{2}{3}$				0							
$\frac{2}{3}$	$\frac{1}{6}$	$\frac{1}{2}$			$\frac{1}{4}$	$\frac{3}{4}$						
0	0	0	0		0	0	0					
0												
0	0				$\frac{2}{3}$				0			
1	1	0			$\frac{1}{12}$	$\frac{1}{4}$			$\frac{1}{4}$	$\frac{3}{4}$		
1	1	0	0		$-\frac{5}{4}$	$\frac{1}{4}$	2		$\frac{1}{4}$	$\frac{3}{4}$	0	
0												
0	0				1							
0	0	0			-1	0						
0	0	0	0		0	0	0					
	$\theta - \frac{3}{4}\theta^2$	$\frac{1}{4}\theta^2$	$\frac{1}{2}\theta^2$	0	$\sqrt{\theta} - \frac{7}{8}\theta$	$\frac{3}{8}\theta$	$\frac{3}{8}\theta$	$\frac{1}{8}\theta$	0	$-\frac{1}{4}\theta$	$\frac{1}{4}\theta$	0

Table 2. CSRK scheme CDRS2 with order $p_D = 3$ and $p_S = 2$.

$$\alpha_i(0) = 0, \qquad \alpha_i(1) = \bar{\alpha}_i, \tag{21}$$

for $1 \leq i \leq 4$. However, in order to calculate a CSRK scheme of order $p_D = 3$, condition (13) of Theorem 1 has to be fulfilled additionally with $p = 3$ in the case that the CSRK scheme is applied to an ODE. Therefore, condition 7. of Theorem 2, which is the well known deterministic order 2 condition (see, e.g., [HNW93]), has to be fulfilled not only for $\theta = 1$ but also for each $\theta \in [0, 1]$. As a result of this, the condition

$$\alpha(\theta)^T A^{(0)} e = \frac{1}{2}\theta^2 \tag{22}$$

has to be considered in addition. Then, we obtain from (22) and 1. of Theorem 2 that the conditions

$$\alpha_1(\theta) = \theta - \alpha_2(\theta) - \alpha_3(\theta) - \alpha_4(\theta) \tag{23}$$

$$\alpha_2(\theta) = \frac{3}{4}\theta^2 - \alpha_3(\theta) \tag{24}$$

and from (21) that the boundary conditions $\alpha_i(0) = 0$ for $1 \leq i \leq 4$, $\alpha_1(1) = \alpha_2(1) = \frac{1}{4}$, $\alpha_3(1) = \frac{1}{2}$ and $\alpha_4(1) = 0$ have to be fulfilled. Therefore, we can choose $\alpha_1(\theta) = \theta - \frac{3}{4}\theta^2$, $\alpha_2(\theta) = \frac{1}{4}\theta^2$, $\alpha_3(\theta) = \frac{1}{2}\theta^2$ and $\alpha_4(\theta) = 0$ for $\theta \in [0, 1]$. See also Table 2 for all coefficients of the continuous SRK scheme CDRS2.

3 Numerical Example

In the following, we approximate some moments of test equations by a Monte Carlo simulation based on the approximations of the introduced CSRK methods and analyze the empirical order of convergence. Therefore, we approximate $\mathrm{E}(f(Y(t)))$ for $t \in \mathcal{I}$ by the sample average $u_{M,h} = \frac{1}{M}\sum_{m=1}^{M} f(Y^{(m)}(t))$ of independent simulated realizations $Y^{(m)}$, $m = 1, \ldots, M$, of the considered approximation Y and we choose $M = 10^8$. Then, the mean error is given as $\hat{\mu} = u_{M,h} - \mathrm{E}(f(X(t)))$ (see, e.g., [KP99] Sec. 9.4). First, the solution $\mathrm{E}(f(X(t)))$ is considered as a mapping from $\mathcal{I}$ to $\mathbb{R}$ with $t \mapsto \mathrm{E}(f(X(t)))$. Since we are interested in a dense output, we apply the CSRK method with some fixed step size h and then calculate the intermediate points by varying the parameter $\theta \in [0, 1]$. In the following simulations, we consider the cases $\theta \in \{0.1, 0.2, \ldots, 0.9\}$.

The first considered test equation is a non–linear SDE (see (4.4.45) [KP99])

$$\mathrm{d}X(t) = \sqrt{X(t)^2 + 1}\,\mathrm{d}t + \sqrt{X(t)^2 + 1} \circ \mathrm{d}W(t), \qquad X(0) = 0. \tag{25}$$

For this equation, we choose $f(x) = p(\operatorname{arsinh}(x))$ with the polynomial $p(z) = z^3 - 6z^2 + 8z$. Since the solution is given by $X(t) = \sinh(t + W(t))$, one calculates that $\mathrm{E}(f(X(t))) = t^3 - 3t^2 + 2t$. The expectation is considered on the interval

$\mathcal{I} = [0, 2]$ as a function of the time $t \in \mathcal{I}$. So, we are not only interested in the expectation of the solution process at a certain fixed time point but on the whole trajectory on $\mathcal{I}$. For the approximation, we apply the CSRK schemes CDRS1 and CDRS2 with step size $h = 0.25$. The approximation results as well as the errors are plotted along the whole time interval $\mathcal{I}$ in the left hand side of Figure 1, respectively.

The second test equation is a non-linear SDE system for $d = m = 2$ with non-commutative noise given by

$$\begin{aligned} \mathrm{d}\begin{pmatrix} X^1(t) \\ X^2(t) \end{pmatrix} &= \begin{pmatrix} -\frac{5}{4}X^1(t) + \frac{9}{4}X^2(t) \\ \frac{9}{4}X^1(t) - \frac{5}{4}X^2(t) \end{pmatrix} \mathrm{d}t \\ &+ \begin{pmatrix} \sqrt{\frac{3}{4}(X^1(t) - X^2(t))^2 + \frac{3}{20}} \\ 0 \end{pmatrix} \circ \mathrm{d}W^1(t) \\ &+ \begin{pmatrix} -\frac{1}{2}\sqrt{(X^1(t) - X^2(t))^2 + \frac{1}{5}} \\ \sqrt{(X^1(t) - X^2(t))^2 + \frac{1}{5}} \end{pmatrix} \circ \mathrm{d}W^2(t), \end{aligned} \tag{26}$$

with initial value $X(0) = (\frac{1}{10}, \frac{1}{10})^T$. First of all, we approximate the first moment of the solution on $\mathcal{I} = [0, 1]$ with step size $h = 0.125$. The exact solution can be calculated as $\mathrm{E}(X^1(t)) = \frac{1}{10}\exp(t)$ with $f(x^1, x^2) = x^1$. The approximations calculated with the CSRK schemes CDRS1 and CDRS2 are presented in the left hand side of Figure 2 and the corresponding errors are printed below. We also approximate the second moment which can be calculated as $\mathrm{E}((X^1(t))^2) = \frac{3}{50}\exp(2t) - \frac{1}{10}\exp(-t) + \frac{1}{20}$ with $f(x^1, x^2) = (x^1)^2$ on $\mathcal{I} = [0, 1]$. The corresponding results for the CSRK schemes CDRS1 and CDRS2 are presented in the left hand side of Figure 3.

Next, SDE (25) and SDE (26) are applied for the investigation of the order of convergence. Therefore, the trajectories are simulated with step sizes $2^{-2}, \ldots, 2^{-4}$ for SDE (25) and with step sizes $2^{-1}, \ldots, 2^{-3}$ for SDE (26). As an example, we consider the error $\hat{\mu}$ at time $t = 1.4$ for SDE (25) and at $t = 0.2$

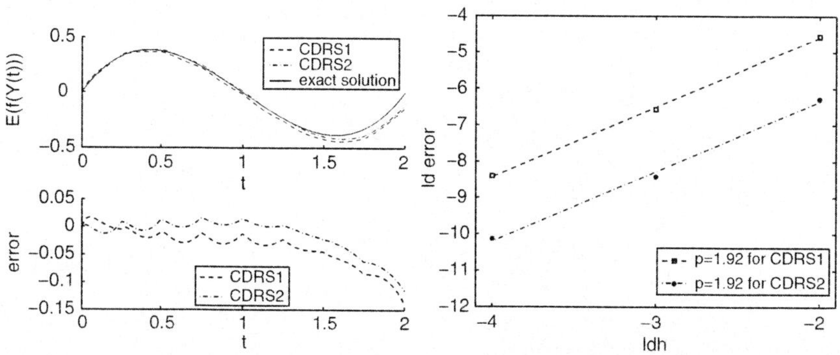

Fig. 1. Schemes CDRS1 and CDRS2 for SDE (25).

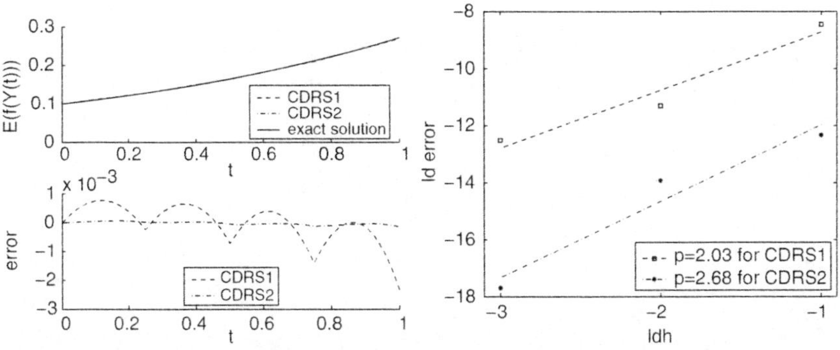

Fig. 2. Schemes CDRS1 and CDRS2 for SDE (26) w.r.t. $\mathrm{E}(X^1(t))$.

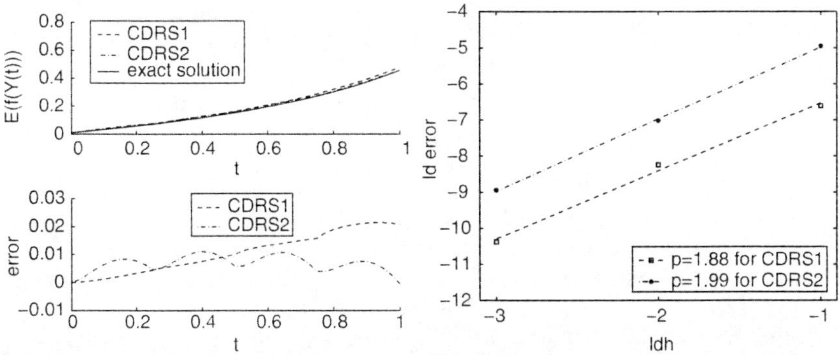

Fig. 3. Schemes CDRS1 and CDRS2 for SDE (26) w.r.t. $\mathrm{E}((X^1(t))^2)$.

for SDE (26), which are not discretization points. The results are plotted on the right hand side of Figure 1, Figure 2 and Figure 3 with double logarithmic scale w.r.t. base two. On the axis of abscissae, the step sizes are plotted against the errors on the axis of ordinates. Consequently one obtains the empirical order of convergence as the slope of the printed lines. In the case of SDE (25) we get the order $p \approx 1.92$ both for CDRS1 and for CDRS2. In the case of SDE (26) we get for the approximation of $\mathrm{E}(X^1(t))$ the order $p \approx 2.03$ for CDRS1 and $p \approx 2.68$ for CDRS2. For the approximation of $\mathrm{E}((X^1(t))^2)$ we obtain $p \approx 1.88$ for CDRS1 and $p \approx 1.99$ for CDRS2.

The good empirical orders of convergence confirm our theoretical results.

References

[BB00] K. Burrage and P. M. Burrage. Order conditions of stochastic Runge–Kutta methods by B-series. SIAM J. Numer. Anal., **38**, No. 5, 1626–1646 (2000)

[DR06] K. Debrabant and A. Rößler. Continuous weak approximation for stochastic differential equations. Journal of Computational and Applied Mathematics (2007), doi:10.1016/j.cam.2007.02.040

[HNW93] E. Hairer, S. P. Nørsett, and G. Wanner. Solving Ordinary Differential Equations I. Springer-Verlag, Berlin (1993)

[KP99] P. E. Kloeden and E. Platen. Numerical Solution of Stochastic Differential Equations. Springer-Verlag, Berlin (1999)

[KMS97] Y. Komori, T. Mitsui, and H. Sugiura. Rooted tree analysis of the order conditions of ROW-type scheme for stochastic differential equations. BIT, **37**, No. 1, 43–66 (1997)

[Mil95] G. N. Milstein. Numerical Integration of Stochastic Differential Equations. Kluwer Academic Publishers, Dordrecht (1995)

[MT04] G. N. Milstein and M. V. Tretyakov. Stochastic Numerics for Mathematical Physics. Springer-Verlag, Berlin (2004)

[New91] N. J. Newton. Asymptotically efficient Runge–Kutta methods for a class of Itô and Stratonovich equations. SIAM J. Appl. Math., **51**, No. 2, 542–567 (1991)

[Roe04] A. Rößler. Runge–Kutta methods for Stratonovich stochastic differential equation systems with commutative noise. J. Comput. Appl. Math., **164–165**, 613–627 (2004)

[Roe06a] A. Rößler. Rooted tree analysis for order conditions of stochastic Runge–Kutta methods for the weak approximation of stochastic differential equations. Stochastic Anal. Appl., **24**, No. 1, 97–134 (2006)

[Roe06b] A. Rößler. Runge–Kutta methods for Itô stochastic differential equations with scalar noise. BIT, **46**, No.1, 97–110 (2006)

[Roe06c] A. Rößler. Second order Runge–Kutta methods for Stratonovich stochastic differential equations. BIT, **47**, No. 3, 657–680 (2007)

[TVA02] A. Tocino and J. Vigo-Aguiar. Weak second order conditions for stochastic Runge–Kutta methods. SIAM J. Sci. Comput., **24**, No. 2, 507–523 (2002)

Issues on Computer Search for Large Order Multiple Recursive Generators

Lih-Yuan Deng

Department of Mathematical Sciences, University of Memphis, Memphis, TN 38152, U.S.A.
lihdeng@memphis.edu

Summary. Multiple Recursive Generators (MRGs) have become the most popular random number generators recently. They compute the next value iteratively from the previous k values using a k-th order recurrence equation which, in turn, corresponds to a k-th degree primitive polynomial under a prime modulus p. In general, when k and p are large, checking if a k-th degree polynomial is primitive under a prime modulus p is known to be a hard problem. A common approach is to check the conditions given in Alanen and Knuth [1964] and Knuth [1998]. However, as mentioned in Deng [2004], this approach has two obvious problems: (a) it requires the complete factorization of $p^k - 1$, which can be difficult; (b) it does not provide any early exit strategy for non-primitive polynomials. To avoid (a), one can consider a prime order k and prime modulus p such that $(p^k - 1)/(p - 1)$ is also a prime number as considered in L'Ecuyer [1999] and Deng [2004]. To avoid (b), one can use a more efficient iterative irreducibility test proposed in Deng [2004].

In this paper, we survey several leading probabilistic and deterministic methods for the problems of primality testing and irreducibility testing. To test primality of a large number, it is known that probabilistic methods are much faster than deterministic methods. On the other hand, a probabilistic algorithm in fact has a very tiny probability of, say, 10^{-200} to commit a false positive error in the test result. Moreover, even when such an unlikely event had happened, for a specific choice of k and p, it can be argued that such an error has a negligible effect on the successful search of a primitive polynomial. We perform a computer search for large-order DX generators proposed in Deng and Xu [2003] and present many such generators in the paper for ready implementation. An extensive empirical study shows that these large-order DX generators have passed the stringent Crush battery of the TestU01 package.

1 Introduction

Until recently, the most popular classical generators have been the Linear Congruential Generators (LCGs) which were proposed by Lehmer [1951]. LCGs are known to have several shortcomings such as a relatively short cycle by today's standard, questionable empirical performances, and lack of higher-dimensional uniformity. Since the quality of the random number generators determines the quality of any simulation study, it is important to find generators with better properties.

In Section 2, we discuss the Multiple Recursive Generators (MRGs), which may have taken the role of the classical LCG as the most popular generators. A maximum-period MRG has a nice property of equi-distribution over a high-dimensional space. A system of portable and efficient large-order MRGs with a same nonzero coefficient in the recurrence equation has been proposed by Deng and Xu [2003] and later extended by Deng [2005]. In this paper, several search algorithms for a large-order MRG are discussed and compared. The key step in these algorithms is to check if a given k-th degree polynomial is a primitive polynomial or not. In Section 3, we discuss the main issue of primitivity checking of a polynomial. We show that this problem can be converted into two easier problems: (1) irreducibility testing of a polynomial and (2) primality testing of a large integer. For both problems (1) and (2), we describe, discuss, and compare several leading methods. In Section 4, we tabulate a class of DX generators of large-order k found by the search algorithm described in Section 3. The largest order k of the DX-k generators found is $k = 10007$ with the period length approximately 10^{93384}. To evaluate the performance of these generators, we conduct an extensive empirical study. The results presented demonstrate that these large-order DX generators have passed the stringent Crush battery of the TestU01 package.

2 Multiple Recursive Generators

Multiple recursive generators (MRGs) have become one of the most commonly used random number generators in a computer simulation. MRGs are based on the k-th order linear recurrence

$$X_i = (\alpha_1 X_{i-1} + \cdots + \alpha_k X_{i-k}) \bmod p, \;\; i \geq k, \tag{1}$$

for any initial seeds $(X_0, \ldots, X_{k-1})$, not all of them being zero. Here the modulus p is a large prime number and X_i can be transformed using $U_i = X_i/p$. To avoid the possibility of obtaining 0 or 1, Deng and Xu [2003] recommended $U_i = (X_i + 0.5)/p$. It is well-known that the maximum period of an MRG is $p^k - 1$, which is reached if its characteristic polynomial

$$f(x) = x^k - \alpha_1 x^{k-1} - \cdots - \alpha_k, \tag{2}$$

is a primitive polynomial. When $k = 1$, MRG is a linear congruential generator (LCG) as proposed by Lehmer [1951]. In this case, $f(x) = x - B$ is a primitive polynomial of degree one whenever B is a primitive root in a finite field of order p.

One nice property for a maximum period MRG is that it is equidistributed up to k dimensions as stated in Lidl and Niederreiter [1994, Theorem 7.43]: every m-tuple $(1 \le m \le k)$ of integers between 0 and $p-1$ appears exactly the same number of times (p^{k-m}) over its entire period $p^k - 1$, with the exception of the all-zero tuple which appears one time less $(p^{k-m} - 1)$.

2.1 Efficient Search Algorithm

A set of necessary and sufficient conditions under which $f(x)$ as defined in (2) is a primitive polynomial has been given in Alanen and Knuth [1964] and Knuth [1998]:

AK(i) $A = (-1)^{k-1}\alpha_k$ must be a primitive element mod p. That is, $A^{(p-1)/q} \neq 1 \bmod p$ for any prime factor q of $p-1$.
AK(ii) $x^R = (-1)^{k-1}\alpha_k \bmod f(x)$, where $R = (p^k - 1)/(p-1)$.
AK(iii) For each prime factor q of R, the degree of $x^{R/q} \bmod f(x)$ is positive.

While Condition AK(i) is straightforward to check, Conditions AK(ii) and AK(iii) can be difficult to verify when k or p is large. For example, to verify Condition AK(iii), one needs to find a complete factorization of $R = (p^k - 1)/(p-1)$. Given the current technology, it is extremely hard to factor a general integer of 200 (or more) digits. To avoid the difficulty of factoring $R(k,p) = (p^k - 1)/(p-1)$, one searches for p so that $R(k,p)$ is a prime number. Clearly, k has to be an odd prime number. This idea was used first in L'Ecuyer, Blouin and Couture [1993] for $k \le 7$ and later in L'Ecuyer [1999] for $k \le 13$. Deng [2004] formally proposed a class of prime numbers of the form $R(k,p)$ and it is called *Generalized Mersenne Prime (GMP)*. This approach is based on the well-known fact that a primality check of a huge number is easier than its factorization.

When k is large, Condition AK(ii) is highly inefficient because there is no early exit strategy when $f(x)$ is not a primitive polynomial. The computing time to compute $x^R \bmod f(x)$ is constant, for any k-th degree polynomial $f(x)$. Since the chance of finding a primitive polynomial is less than $1/k$, lots of computing time is wasted. Deng [2004] proposed a much more efficient algorithm with a built-in early exit strategy:
Algorithm GMP Given a prime order k, choose a prime modulus p such that $R(k,p) = (p^k - 1)/(p-1)$ is also a prime number. Let $f(x)$ be as in (2).

(i) α_k must be a primitive element mod p. If this condition is met, then go to the next step.
(ii) Initially, let $g(x) = x$. For $i = 1, 2, 3, \ldots, \lfloor k/2 \rfloor$, do
   a) $g(x) = g(x)^p \bmod f(x)$;
   b) $d(x) = \gcd(f(x), g(x) - x)$;
   c) if $d(x) \neq 1$, then $f(x)$ cannot be a primitive polynomial.

If all the loops in Step (ii) have been passed, then $f(x)$ is a primitive polynomial.

According to our experience, an average increase of $O(k)$ folds of searching efficiency has been observed when k is large.

3 Checking Primitive Polynomials

Let $\mathbb{Z}_p = \{0, 1, 2, \ldots, p-1\}$ be a finite field of p elements. We know that if $f(x)$ is a k-th degree primitive polynomial over $\mathbb{Z}_p$, then $f(x)$ is an irreducible polynomial over $\mathbb{Z}_p$. However, the converse is not true. If $f(x)$ is a k-th degree irreducible polynomial over $\mathbb{Z}_p$, we can then use $f(x)$ to define the finite field F_{p^k} of p^k elements. Specifically, each element θ in F_{p^k} can be uniquely represented as a polynomial with degree less than k:

$$\theta = \sum_{i=0}^{k-1} g_i x^i, \quad g_i \in \mathbb{Z}_p.$$

The addition and multiplication operations over F_{p^k} are simply the polynomial addition and multiplication under the modulus $f(x)$. Furthermore, an irreducible polynomial $f(x)$ can become a k-th degree primitive polynomial if x (more precisely $x \bmod f(x)$ when $k = 1$) is a primitive element over F_{p^k}. That is, we need to show that the smallest $e > 0$ such that $x^e = 1 \bmod f(x)$ is $G = p^k - 1$. One quick way to show this is to verify $x^{G/q} \neq 1$ for any prime factor q of G and $x^G = 1 \bmod f(x)$. This observation can be useful to explain the conditions AK(i), AK(ii), and AK(iii) required. As explained earlier, one of the bottlenecks is that we need a complete factorization of $p^k - 1$ or $R = (p^k - 1)/(p - 1)$. As pointed out in L'Ecuyer [1999] and Deng [2004], one can avoid this problem by requiring R to be a prime number. Therefore, we convert the problem of checking primitive polynomial into two easier problems: the problem of checking irreducible polynomial of $f(x)$ and the problem of checking the primality of R. We discuss these two problems next.

3.1 Checking Irreducible Polynomials

There are two types of tests for checking irreducible polynomial: deterministic and probabilistic. The most popular deterministic test is the iterative irreducibility test as stated in Crandall and Pomerance [2000, Theorem 2.2.8, page 88]. The iterative irreducibility test was used in Algorithm GMP. As explained in Deng [2005], one can use the known identity

$$x^{p^i} - x = \prod_{f \text{ monic irreducible}, \deg(f) | i} f(x) \bmod p. \tag{3}$$

See Lidl and Niederreiter [1994, Theorem 3.20, page 84]. When $i = 1$, the identity in (3) becomes

$$x^p - x = \prod_{a=0}^{p-1} (x - a) \bmod p$$

which is the product of all monic linear polynomials. If $\gcd(f(x), x^p - x) \neq 1$, then $f(x)$ has at least one linear factor. Thus, $f(x)$ is not irreducible and therefore cannot be a primitive polynomial. In this case, $f(x)$ failed the first iteration of Algorithm GMP. Otherwise, we next check whether $f(x)$ has any common factor with $x^{p^2} - x$ which is the product of all monic 1st and 2nd degree irreducible polynomials. The process is similarly repeated as described in Algorithm GMP.

Using the identity in (3), Shoup [1994] gave another set of conditions for a k-th degree irreducible polynomial $f(x)$. Since the value of k discussed in this paper is a prime number, somewhat simpler conditions can be used:

(i) $\gcd(x^p - x, f(x)) = 1$ and
(ii) $x^{p^k} = x \bmod f(x)$.

Condition (i) is checking that $f(x)$ has no linear factor. Condition (ii) has a similar computational complexity as Condition AK(ii) and both do not have an efficient early exit strategy. Condition (i) is partially effective because it can provide an early exit only for those $f(x)$ with linear factors.

In addition to these deterministic tests, we consider a randomized test next.

Trace Map Test

The trace map test as considered in Gathen and Shoup [1992] is a probabilistic irreducibility test. For a review on the properties of a trace map, see Lidl and Niederreiter [1994].

Let θ be an element in the finite field F_{p^k} and let

$$Tr(\theta) = \theta + \theta^p + \theta^{p^2} + \cdots + \theta^{p^{k-1}}. \quad (4)$$

Choose a random polynomial $\theta = g(x)$, where $\deg(g) < k$, if $Tr(g)$ is in $\mathbb{Z}_p$, then declare $f(x)$ as irreducible. According to Shoup [1994], the probability of making a false positive error is less than $1/p$ which is smaller than 10^{-9} for $p = 2^{31} - c$. While $Tr(g)$ can be computed recursively and more efficiently, it is still very time-consuming when both k and p are large.

In our opinion, the trace map test is not as efficient as the iterative irreducibility test for large k and p because there is no early exit strategy for non-irreducible polynomials. In addition, it is not a deterministic test and there is a slight probability of making a false positive claim. We can further reduce this error probability by running several independent tests. Iterative irreducibility test is quite sensitive to the tiny error (either hardware or software) in the computation of $g(x) = g(x)^p \bmod f(x)$ and/or $d(x) = \gcd(f(x), g(x) - x)$ during the long iterations in the search process. Depending on the size of k, the successful search time can be up to several weeks of the computing time. Therefore, the computer hardware reliability can become an issue of concern. According to our own experience, some hardware computing errors occurred and they did caused some false positive results. Indeed, if a computing error occurred in the iterated loop of the iterative irreducibility test, it would tend to falsely pass the condition given in the iterative irreducibility test. On the other hand, to pass the trace map test, one needs to satisfy the condition $Tr(g) \in \mathbb{Z}_p$. A hardware or software error will make the condition $Tr(g) \in \mathbb{Z}_p$ less (not more) likely to be met. This is because $Tr(g)$ in the equation (4) is the summation of k terms of g^{p^i}, $i = 0, 1, \cdots k-1$. Possible computing error on individual terms tends to cause the total less likely to satisfy the condition $Tr(g) \in \mathbb{Z}_p$. Therefore, the trace map test can be very helpful to double check (to minimize the effect of the tiny possibility of computing errors) for an irreducible polynomial obtained from the deterministic iterative irreducibility test.

3.2 Primality Test for Large Integers

Like the irreducible polynomial check, there are two types of integer primality tests. The first type is a probabilistic test which can be highly efficient but there is a tiny probability of making an error. The second type is a deterministic test which can be time-consuming but the conclusion is definite. There was no general polynomial-time algorithm available until Agrawal, Kayal and Saxena announced their discovery in 2002. The algorithm is known as the AKS algorithm and its formal proof is given in Agrawal, Kayal and Saxena [2004]. However, AKS algorithm is still not yet practical for a large prime number. Next, we describe our effort to find p for a given prime order k such that $R(k,p)$ is a probable prime using the randomized test.

For each prime order k, we first find a prime modulus p for probable-prime of $R(k,p) = (p^k - 1)/(p-1)$. We then verify the primality of $R(k,p)$ using probabilistic tests via some commercial packages such as *Maple* and *MATHEMATICA*. We further perform *industrial prime test* as proposed in Damgrard, Landrock, and Pomerance [1993]. See also Algorithm 3.4.7 in Crandall and Pomerance [2000, page 126] for *industrial prime test.* They also discussed that the probability of making false positive error can be made to be much smaller than 10^{-200}. This error probability is much smaller than the computer software error or the hardware error. Therefore, according to Crandall and Pomerance [2000, page 127], it can be safely accepted as a "prime" in all but the most sensitive practical applications.

In addition to choosing p for which $(p^k - 1)/(p-1)$ is also a prime, we require that both p and $Q = (p-1)/2$ are prime numbers. Here, Q is commonly called a Sophie-Germain prime number.

The search time of p listed in Table 1 is random and it is, in general, an increasing function of k. In total, several months of CPU times were spent to search for the p as listed in Table 1.

Table 1: List of k, c and $p = 2^{31} - c$ for which $(p^k - 1)/(p - 1)$ is a prime.

k	c	$p = 2^{31} - c$	$\log_{10}(p^k - 1)$
5003	1259289	2146224359	46686
6007	9984705	2137498943	56045
7001	610089	2146873559	65332
8009	5156745	2142326903	74731
9001	7236249	2140247399	83984
10007	431745	2147051903	93384

Effect of Primality Test on Search Algorithm

Even if an unlikely mistake were made, for a specific choice of k and p, it has a negligible effect on the successful search of primitive polynomial. If $R(k,p)$ is not a prime, then we need to check condition AK(iii) as required. Let us assume that $R(k,p) = H \times Q$, where Q is the "smaller" prime factor. From some elementary number theory, one can see that the chance of satisfying condition AK(ii) but not condition AK(iii) is roughly proportional to $1/Q$. The current computer factorization programs are capable of finding a factor of 50 (or more) digits. Therefore, in the unlikely event $R(k,p)$ is not a prime number, we only have a tiny chance (say, 10^{-50} or less) to mis-classify a non-primitive polynomial. For all practical purpose, once the irreducibility test is passed and $R(k,p)$ is shown as a probable prime, we can safely assume that we have found a primitive polynomial.

4 Table of Large Order DX-k Generators

4.1 DX-k-s Generators

Deng and Lin [2000] proposed Fast MRG (FMRG) which is a maximal period MRG with minimal number terms of nonzero coefficient. FMRG is almost as efficient as the classical LCG. Deng and Xu [2003] and Deng [2005] proposed DX generators as a system of portable, efficient, and maximal period MRGs where coefficients of the nonzero multipliers are the same:

1. DX-k-1 ($\alpha_1 = 1, \alpha_k = B$).

$$X_i = X_{i-1} + BX_{i-k} \bmod p, \quad i \geq k. \tag{5}$$

2. DX-k-2 ($\alpha_1 = \alpha_k = B$).

$$X_i = B(X_{i-1} + X_{i-k}) \bmod p, \quad i \geq k. \tag{6}$$

3. DX-k-3 ($\alpha_1 = \alpha_{\lceil k/2 \rceil} = \alpha_k = B$).

$$X_i = B(X_{i-1} + X_{i-\lceil k/2 \rceil} + X_{i-k}) \bmod p, \quad i \geq k. \tag{7}$$

4. DX-k-4 ($\alpha_1 = \alpha_{\lceil k/3 \rceil} = \alpha_{\lceil 2k/3 \rceil} = \alpha_k = B$).

$$X_i = B(X_{i-1} + X_{i-\lceil k/3 \rceil} + X_{i-\lceil 2k/3 \rceil} + X_{i-k}) \bmod p, \quad i \geq k. \tag{8}$$

Here the notation $\lceil x \rceil$ is the ceiling function of a number x, returning the smallest integer $\geq x$. For the class DX-k-s, s is the number of terms with coefficient B.

4.2 List of DX-k-s Generators

Using Table 1 and Algorithm GMP, it is straightforward to find the DX-k-s generators. In Table 2, we list DX-k generators with $B < 2^{30}$.

As a simple illustration, we find DX-10007-s generator from Table 2, with $k = 10007$, and $p = 2147051903$. For $s = 2$, we have

$$X_i = 1073702542(X_{i-1} + X_{i-10007}) \bmod p.$$

For $s = 4$, we have

$$X_i = 1073730725(X_{i-1} + X_{i-3336} + X_{i-6672} + X_{i-10007}) \bmod p.$$

All of the DX-10007 generators listed in Table 2 have the same period length approximately 10^{93384}. If the generating speed is the main concern, then we recommend DX-k-2 generators. Otherwise, we recommend DX-k-4 generators for their better lattice structure over dimensions larger than k. Similarly, if the memory space is the major concern, then we recommend smaller values of k such as DX-47 proposed in Deng [2005]. Otherwise, we recommend using the largest possible value of k like the DX-10007 generators for its high-dimensional equi-distribution property and its extremely long period.

As explained in Deng [2005], such DX generators with $B < 2^{30}$ can be implemented using 64-bit data types/operations. Without using such data type, a portable implementation of MRGs can be used at the expense of slight generating inefficiency. See Deng [2005] for details.

If 64-bit data types/operations are unavailable, under IEEE double precision standard, one can use upper limits for B as considered in Deng and Xu [2003] for DX-k-s generators:

$$B < 2^e, \quad \text{where } e = 20, \text{ when } s = 1, 2; \quad e = 19, \text{ when } s = 3, 4. \tag{9}$$

Table 2: List of k, p and $B < 2^{30}$ for DX-k-s.

k	p	$s = 1$	$s = 2$	$s = 3$	$s = 4$
5003	2146224359	1073727083	1073741516	1073730698	1073740466
6007	2137498943	1073738651	1073715261	1073729141	1073738504
7001	2146873559	1073709808	1073728419	1073738188	1073735327
8009	2142326903	1073717208	1073726014	1073733016	1073719175
9001	2140247399	1073737583	1073717540	1073733156	1073732451
10007	2147051903	1073726195	1073702542	1073723329	1073730725

Table 3: List of k, p and $\min B$ and $B < 2^e$ in (9) for DX-k-s generators.

	$\min B$				$B < 2^e$			
k	$s=1$	$s=2$	$s=3$	$s=4$	$s=1$	$s=2$	$s=3$	$s=4$
5003	15851	10302	6616	8461	1041088	1039973	506762	487092
6007	932	8158	608	35	1046897	1015366	519071	519501
7001	12685	78782	171	28492	1026965	1014115	521869	506984
8009	41317	39951	35208	20374	1041446	1046062	519082	518174
9001	542	17053	5474	8057	1045508	1040383	515350	523991
10007	44166	26540	13759	9520	1042089	1042654	515671	493723

In addition, we also search for the smallest B such that DX-k-s achieving the maximum period and we list these DX generators found in Table 3.

Following a similar discussion in Deng [2005], we remark that the DX generators under "$\min B$" in Table 3 are *not* recommended but they can be useful to determine the "power" of empirical tests. So far, no general purpose empirical tests have been found to fail such DX generators with small values of B.

4.3 Empirical Evaluations

There are several well-known empirical test packages for testing a random number generator: (1) DIEHARD proposed in Marsaglia [1996], (2) NIST package and (3) TestU01 test suite which was developed by Professor L'Ecuyer with the source code from `http://www.iro.umontreal.ca/~lecuyer/`. See L'Ecuyer and Simard [2006] for more details. It is by far the most comprehensive test suite. There are three predefined test modules in TestU01:

1. *Small crush*: it has 15 tests and it takes less than 1/2 minute of computing time.
2. *Crush*: it has 144 tests and its running time is about 1.5 hours.
3. *Big crush*: it is most comprehensive with 160 tests and it may require more than 12 hours of computing time.

We apply Crush battery of tests in TestU01 with five different starting seeds: 1, 12, 123, 1234 and 12345 and we use an LCG whose multiplier is the same as B to generate the required k initial seeds. In total, there are 72 ($= 6(k) \times 3(B) \times 4(s)$) DX generators found in Table 2 and Table 3. Therefore, we obtain 51840 ($= 72 \times 144 \times 5$) p-values. The number of tests with p-values less than 0.001 are tabulated in Table 4.

Table 4: Results of Crush test on DX generators (51840 p-values).

p-value	$<10^{-3}$	$<10^{-4}$	$<10^{-5}$	$<10^{-6}$	$<10^{-7}$
counts	57	9	2	1	0
percentage	0.001033	0.000163	0.000036	0.000018	0.000000

As we can see from Table 4, none of these 51840 tests produces p-values that are smaller than 10^{-7}. The percentage of tests producing p-values which are below 10^{-3} is 0.001033 which is very close to its nominal value of 0.001. In addition, none of these 51840 tests produces p-values that are too close to 0 or 1. We believe all 72 DX generators listed in Table 2 and Table 3 have passed the Crush test.

5 Open Problem: Generalized Lucas-Lehmer Test?

The Lucas-Lehmer test is an efficient deterministic primality test for determining if a Mersenne number $M_k = 2^k - 1$ is a prime number. It is based on the sequence $S_i = S_{i-1}^2 - 4 \bmod M_k$, $i \geq 1$, $S_0 = 4$. According to Lucas-Lehmer test, M_k is a prime number if and only if $S_{k-2} = 0 \bmod M_k$. See, for example, Crandall and Pomerance [2000].

The problem is whether it is possible to find an efficient deterministic primality test (like a generalized version of the Lucas-Lehmer test) for a generalized Mersenne number $R(k,p) = (p^k - 1)/(p-1)$, where both p and k are prime numbers. Clearly, $M_k = R(k,2)$. To the best of our knowledge, it is still an open problem.

As mentioned earlier, for a general integer n, AKS algorithm is a famous polynomial-time (some polynomial of $\ln(n)$) for testing whether n is a prime number. Currently, AKS algorithm and other deterministic primality tests are not (yet) practical for a very large $R(k,p)$ where both k and p are large. The above open problem has some useful applications in the area of random number generation as described in this paper.

Acknowledgement

The author is grateful to Professor Harald Niederreiter for his careful reading of this paper. The author would like to thank an anonymous referee who made many encouraging and helpful comments.

References

[AK64] J. D. Alanen and D. E. Knuth. Tables of finite fields. *Sankhyā*, 26:305–328, 1964.

[AKS04] M. Agrawal, N. Kayal, and N. Saxena. PRIMES is in P. *Annals of Mathematics*, 160(2):781–793, 2004.

[CP00] R. Crandall and C. Pomerance. *Prime Numbers - A Computational Perspective*. Springer-Verlag, New York, N.Y., 2000.

[Den04] L. Y. Deng. Generalized mersenne prime number and its application to random number generation. In H. Niederreiter, editor, *Monte Carlo and Quasi-Monte Carlo Methods 2002*, pages 167–180. Springer-Verlag, 2004.

[Den05] L. Y. Deng. Efficient and portable multiple recursive generators of large order. *ACM Transactions on Modeling and Computer Simulation*, 15(1): 1–13, 2005.

[DL00] L. Y. Deng and D. K. J. Lin. Random number generation for the new century. *American Statistician*, pages 145–150, 2000.

[DLP93] I. Damgrard, P. Landrock, and C. Pomerance. Average case error estimates for the strong probable prime test. *Mathematics of Computation*, 61:177–194, 1993.

[DX03] L. Y. Deng and H. Xu. A system of high-dimensional, efficient, long-cycle and portable uniform random number generators. *ACM Transactions on Modeling and Computer Simulation*, 13(4):299–309, 2003.

[GS92] J. Gathen and V. Shoup. Computing frobenius maps and factoring polynomials. *Computational Complexity*, 2:187–224, 1992.

[Knu98] D. E. Knuth. *The Art of Computer Programming*, volume 2: Seminumerical Algorithms. Addison-Wesley, Reading, MA., third edition, 1998.

[LB88] P. L'Ecuyer and F. Blouin. Linear congruential generators of order $k > 1$. In *1988 Winter Simulation Conference Proceedings*, pages 432–439, 1988.

[LBC93] P. L'Ecuyer, F. Blouin, and R. Couture. A search for good multiple recursive linear random number generators. *ACM Transactions on Modeling and Computer Simulation*, 3:87–98, 1993.

[L'E99] P. L'Ecuyer. Good parameter sets for combined multiple recursive random number generators. *Operations Research*, 47:159–164, 1999.

[Leh51] D. H. Lehmer. Mathematical methods in large-scale computing units. In *Proceedings of the Second Symposium on Large Scale Digital Computing Machinery*, pages 141–146; Cambridge, MA., 1951. Harvard University Press.

[LN94] R. Lidl and H. Niederreiter. *Introduction to Finite Fields and Their Applications.* Cambridge University Press, Cambridge, MA., revised edition, 1994.

[LS06] P. L'Ecuyer and R. Simard. Testu01: A C library for empirical testing of random number generators. *ACM Transactions on Mathematical Software*, 2006. (to appear).

[Mar96] G. Marsaglia. The Marsaglia random number CDROM including the DIEHARD battery of tests of randomness. http://stat.fsu.edu/pub/diehard, 1996.

[Sho94] V. Shoup. Fast construction of irreducible polynomials over finite fields. *Journal of Symbolic Computation*, 17:371–391, 1994.

Design and Implementation of Efficient and Portable Multiple Recursive Generators with Few Zero Coefficients

Lih-Yuan Deng[1], Huajiang Li[2], Jyh-Jen Horng Shiau[3], and Gwei-Hung Tsai[4]

[1] Department of Mathematical Sciences, University of Memphis, Memphis, TN 38152, U.S.A.
lihdeng@memphis.edu

[2] Quintiles, Inc. Overland Park, KS 66211, U.S.A.
huajiang.li@quintiles.com

[3] Institute of Statistics, National Chiao-Tung University, Hsinchu, Taiwan, 30050, R.O.C.
jyhjen@stat.nctu.edu.tw

[4] Department of Applied Statistics and Information Science, Ming Chuan University, Taoyuan, Taiwan, 333, R.O.C.
herbtsai@mcu.edu.tw

Summary. DX-k, proposed by Deng and Xu [2003], is a special class of Multiple Recursive Generators (MRGs) where all nonzero coefficients of the k-th order recurrence are equal. In particular, a DX-k generator requires only up to four nonzero coefficients in its recurrence equation, hence is very efficient in computation. However, a random number generator with few nonzero coefficients has a drawback that, when the k-dimensional state vector is close to the zero vector, the subsequent numbers generated may stay within a neighborhood of zero for quite many of them before they can break away from this near-zero land, a property apparently not desirable in the sense of randomness. Consequently, two generated sequences using the same DX generator with nearly identical initial state vectors may not depart from each other quickly enough. To avoid the above potential problem, we consider MRGs with very few zero coefficients. To make such generators efficient and portable, we propose selecting the same nonzero value for all coefficients (with at most one exception) in the recurrence equation. With this feature, the proposed generators can be implemented efficiently via a higher-order recurrence of few zero coefficients. Note that the new class of generators is an opposite of the DX generators in terms of the number of nonzero coefficients. Several such generators with the maximum period have been found via computer search and presented in the paper for ready implementation.

1 Introduction

Multiple Recursive Generators (MRGs) are popular random number generators (RNGs) that each generates pseudo random numbers based on a k-th order linear recurrence equation with a large prime modulus p. When $k = 1$, MRGs reduce to Lehmer's [1951] Linear Congruential Generators (LCGs).

Deng and Xu [2003] proposed a special class of efficient and portable MRGs with modulus p and order k, called the DX-k-s generators, in which all of the s (up to 4) nonzero coefficients of the recurrence equation are equal. With the advantage that only a single multiplication is needed in computing the recurrence equation, a DX-k-s generator usually generates random numbers faster than an MRG of a general form. One potential problem with the DX-k-s generators is that the recovery from a poor state may be fairly slow. More specifically, when the k-dimensional state vector for the recurrence is close to the zero vector, the subsequent numbers generated may stay within a neighborhood of zero for quite many of them before they can break away from this near-zero land, a property apparently not desirable in the sense of randomness. This is due to the fact that only few nonzero terms are used in the recurrence. This undesirable effect can be somewhat diminished by considering more nonzero terms, such as $s = 3$ or $s = 4$. While we could have considered even larger values of s for further improvement, unfortunately it would be harder to maintain the efficiency and portability of the generators at the same time.

To overcome the above potential problem, in this paper, we propose another class of efficient and portable MRGs that have many nonzero coefficients in the recurrence, a complete opposite of the DX-k-s generators in terms of the number of nonzero coefficients. To achieve the computational efficiency and maintain the portability, we impose a special structure on the nonzero coefficients such that the generators can be efficiently implemented by a recurrence equation of order $k + 1$ in which there are only few nonzero coefficients. Using the efficient search algorithm proposed by Deng [2004], we have obtained a list of maximum-period generators with order k up to 10007.

2 MRG and DX Generators

2.1 Multiple Recursive Generator (MRG)

An MRG generates the next number, X_i, recursively based on a linear congruential combination of the components of the most recent k-dimensional state vector $(X_{i-k}, \cdots, X_{i-1})$:

$$X_i = \alpha_1 X_{i-1} + \cdots + \alpha_k X_{i-k} \bmod p, \quad i \geq k, \tag{1}$$

where the modulus p is a large prime and the multipliers $\alpha_1, \cdots, \alpha_k$ are integers between 0 and $p - 1$, inclusively. Here, k is a positive integer called the order of the MRG. The initial values $(X_0, \cdots, X_{k-1})$ are called the seeds and they can be arbitrarily chosen as long as not all of them are zero. X_i can be converted into a real value between 0 and 1 by either $U_i = X_i/p$ or, as recommended by Deng and Xu [2003], $U_i = (X_i + 0.5)/p$.

The characteristic polynomial of the MRG defined by the recurrence equation (1) is $f(x) = x^k - \alpha_1 x^{k-1} - \cdots - \alpha_k$. The largest possible period of an MRG is $p^k - 1$, which is achieved if and only if the characteristic polynomial $f(x)$ is a primitive polynomial modulo p. Alanen and Knuth [1964] and Knuth [1998] described some conditions for a polynomial to be a primitive polynomial. However, it is difficult to check their conditions directly in practice, especially when the values of k and p are large. Alternatively, Deng [2004] proposed an efficient algorithm that bypasses the difficulty of factoring a large number and provided an early exit strategy for a failed search to achieve a better efficiency. We remark that the idea of bypassing factoring a large number was first suggested by L'Ecuyer, Blouin, and Couture [1993].

The MRGs with the maximum period of $p^k - 1$ enjoy the nice property of equi-distribution up to k dimensions, i.e., every m-tuple ($m \leq k$) of integers between 0 and $p-1$ appears exactly the same number of times over its entire period $p^k - 1$ with the exception that the all-zero tuple appears one time less. See Lidl and Niederreiter [1994] for further details.

2.2 DX Generators

When the order k becomes large, generating numbers from a general MRG can be slow. To improve the efficiency of MRGs, researchers have considered the recurrence equations with only a small number of nonzero terms. Deng [2005] proposed a class of DX-k-s-t generators, where k is the order of the generator, s specifies the number of nonzero terms with the equal coefficient B, and t indicates how far back the first nonzero term is in the recurrence equation. If efficiency is a major consideration, Deng [2005] recommended the DX-k-s-2 generator with $\alpha_t = \alpha_k = B$, $1 \leq t < k$, which is

$$X_i = B(X_{i-t} + X_{i-k}) \bmod p, \quad i \geq k. \tag{2}$$

Otherwise, Deng [2005] recommended the DX-k-s-4 generator with $\alpha_t = \alpha_{\lceil k/3 \rceil} = \alpha_{\lceil 2k/3 \rceil} = \alpha_k = B$, $1 \leq t < \lceil k/3 \rceil$, which is

$$X_i = B(X_{i-t} + X_{i-\lceil k/3 \rceil} + X_{i-\lceil 2k/3 \rceil} + X_{i-k}) \bmod p, \quad i \geq k, \tag{3}$$

where the notation $\lceil x \rceil$ is the ceiling function of a number x, returning the smallest integer $\geq x$. Since it requires only one multiplication and a small number of additions to compute the recurrence equation, the DX generators are efficient.

2.3 Lattice Structure

For LCGs and low-order MRGs, it is common to evaluate the performance using the lattice structure criterion. More specifically, we study the structure of the d consecutive elements of the sequence, $\{(X_i, X_{i+1}, \cdots, X_{i+d-1}) \mid i = 0, 1, \cdots\}$, produced by a random number generator, or equivalently by its uniform(0,1) counterpart, $\{(U_i, U_{i+1}, \cdots, U_{i+d-1}) \mid i = 0, 1, \cdots\}$. If the generated sequence is indeed a realization of a sequence of truly independent uniform random variables, then these d-tuples should be uniformly distributed over the d-dimensional cube. For LCGs, Marsaglia [1968] was the first to show that successive overlapping sequences of d random numbers fall on at most $(d!\,m)^{1/d}$ hyperplanes, where m is the modulus chosen. This shortcoming may yield grossly wrong results for certain applications, such as in the Monte Carlo multiple-integration method. For an MRG, when $d > k$, all the d consecutive points lie on some parallel hyperplanes in the d-dimensional space. Therefore, the corresponding d-dimensional lattice structure can be an important property for an MRG.

One quantitative measure of the lattice structure is the spectral test corresponding to the maximum-distance between two adjacent parallel hyperplanes. Clearly, we would prefer the generator with the smaller maximum-distance because no points in between these adjacent hyperplanes can be generated. In theory, a good uniform random number generator should produce points that fill evenly the whole space. A smaller maximum-distance can avoid large slices of empty space so that the generated number sequences can be more uniformly distributed over the whole space. L'Ecuyer [1997] pointed out that a necessary but not sufficient condition for an MRG to have a good lattice structure (over dimensions larger than k) is that the sum of squares of all coefficients, $\sum_{i=1}^{k} \alpha_i^2$, is large. However, this condition will not have any effect on the equidistribution property over dimensions less than k for the maximum-period MRGs. A similar condition for a "good" RNG was given in Deng, Lin, Wang, and Yuan [1997] from a statistical justification viewpoint. Consequently, among various DX-k-s generators, we prefer the ones with a large order k and large values of s and B.

2.4 Potential Problems of DX Generators

L'Ecuyer [1997] proposed to perform spectral tests on points of subsequences taken from nonsuccessive indices. With that, L'Ecuyer and Touzin [2004] analyzed some special cases of DX-k-s generators with $s = 1$, 2. They chose some specific subsequences such as $S_k = \{(U_i, U_{i+k-1}, U_{i+k}) \mid i = 0, k+1, 2(k+1), \cdots\}$ and then performed low-dimensional spectral tests. Their empirical study showed that some DX-k-s generators have bad lattice structures especially for those with $s = 1, 2$ and a small nonzero coefficient such as $B = 23$. As pointed out in Deng [2005], for a given selection of subsequence S_k, we can avoid the bad-lattice problem by considering $t > 1$, larger values of s, and/or larger values of B in the DX-k-s-t generators. On the other hand, if the specific generator used is known, one can construct a specific subsequence with a bad lattice structure. For illustration, we give two examples. For an LCG with known prime modulus p, we can construct from the original sequence a bad subsequence of variates that are $(p-1)/q$ apart, where q is a small factor of $(p-1)$. It is easy to see that the period length of the subsequence is reduced to q. Similarly, for a maximum-period MRG of order k, we can find a bad subsequence containing variates that are $(p^k - 1)/q$ apart, where q is a factor of $p^k - 1$.

Like most other popular random number generators, DX generators can suffer from the effect of extremely bad initializations. For two distinct but extremely close initialization vectors, DX generators may require long iterations for their output sequences to become far apart. For an initial vector very close to the zero vector, the numbers in the sequence produced by DX generators may stay within a neighborhood of zero for quite many of them before getting away far enough from zero, indicating these numbers are not quite random. This effect was first observed by Panneton, L'Ecuyer, and Matsumoto [2006] for MT19937, a popular generator proposed by Matsumoto and Nishimura [1998]. MT19937 is based on a linear recurrence of order $k = 19937$ and modulo $p = 2$. It has a period of $2^{19937} - 1 \approx 10^{6001.6}$ and the equi-distribution property up to 623 dimensions.

Let $e_i = (0, 0, \cdots, 0, 1, 0, \cdots, 0)$ be the i-th unit vector in the k-dimensional Euclidean space. For the original DX-k-s generators proposed by Deng and Xu [2003] (i.e., the DX-k-s-t generators with $t = 1$), the worst initial vector is $e_{k-1} = (0, 0, \cdots, 0, 1, 0)$. When $s = 1$, the generator produces $k - 2$ zeros followed by $k - 1$ values of B. If $s = 2$, it produces $k - 2$ zeros followed by $B, B^2, B^3, \cdots$. For $s = 3$ or 4, it produces a sequence starting with $k/(s-1)$ zeros. For $s = 1$ or $s = 2$, choosing a larger value of t in DX-k-s-t generators can be helpful in getting the generated numbers more quickly away from the near-zero land.

Notice that the "bad initialization effect" occurs only for certain almost-identical k-dimensional seed vectors. Unless chosen purposely, it is extremely unlikely for two streams of random sequences to have their k-dimensional state vectors almost identical at any stage. Thus, the practical significance of the bad initialization effect is still unclear due to its extremely rare occurrence. As reported in Deng [2005], DX-k-s-t generators passed several stringent empirical tests without any problems. Nevertheless, it is still desirable to find a class of generators that do not suffer from the above-described "bad initialization effect" without losing the properties of portability and efficiency.

In the next section, we extend the DX generators to a general class of efficient and portable MRGs with many nonzero terms.

3 A General Class of Efficient Generators

The DX-k-s-t generators are efficient because there are only few nonzero terms in the recurrence equations and they have the same coefficient. To construct a class of efficient generators, one can consider a special class of MRGs that have at most two different nonzero coefficients, say, A and B. Define two index sets, $S_A = \{j \mid \alpha_j = A\}$ and $S_B = \{j \mid \alpha_j = B\}$. Since $A \neq B$, $S_A \cap S_B = \emptyset$.

A generator in this general class has the following form:

$$X_i = A \sum_{j \in S_A} X_{i-j} + B \sum_{j \in S_B} X_{i-j} \bmod p. \tag{4}$$

Clearly, one simple way to make this class of generators efficient is to have only few elements in both S_A and S_B. Indeed, DX-k-s-t generators are constructed using this principle. Notice that $s = \#(S_B)$, the number of indices in the set S_B.

When the numbers of the indices in S_A and/or S_B are large (i.e., $\#(S_A)$ and/or $\#(S_B)$ are large), it is still possible to find an efficient implementation for some generators as in (4). The idea is to impose a special structure on both S_A and S_B so that the equation (4) can be rewritten as a simpler higher-order recurrence equation. Two classes of such generators are discussed next.

3.1 DL Generators

Li [2005] considered a class of DL generators as in (4) corresponding to $S_A = \{1, 2, 3, \cdots, t-1\}$ and $S_B = \{t, t+1, \cdots, k\}, 1 \leq t < k$. Specifically, a DL generator is defined as:

$$X_i = A(X_{i-1} + \cdots + X_{i-t+1}) + B(X_{i-t} + \cdots + X_{i-k}) \bmod p, \tag{5}$$

for $i > k$. Note that the t here plays a similar role as the t in DX-k-s-t. Utilizing higher-order recurrence, DL generators can be implemented efficiently as:

$$X_i = X_{i-1} + A(X_{i-1} - X_{i-t}) + B(X_{i-t} - X_{i-(k+1)}) \bmod p, \quad i \geq k+1, \tag{6}$$

where $X_0, X_1, \cdots, X_{k-1}$ are the initial seeds and X_k is computed according to equation (5). Li [2005] and Deng, Li, and Shiau [2005] considered and tabulated this general class of DL generators for the order $k \leq 1709$.

From the above equation, we can see that only two multiplications and several additions/subtractions are needed for calculating the next value. To further improve the efficiency and portability, we can take the coefficient $A = 0, -1, 1$, or $-B$ to reduce one multiplication and several addition/subtraction operations. In particular, when $A = 0$, it leads to a simpler form:

$$X_i = B(X_{i-t} + X_{i-t-1} + \cdots + X_{i-k}) \bmod p, \quad i \geq k, t \geq 1. \tag{7}$$

For simplicity, we consider and study in this paper this special case of $A = 0$ and $t = 1$ and refer to it as the DL-k generators. Such DL generators can be implemented efficiently by:

$$X_i = X_{i-1} + B(X_{i-1} - X_{i-(k+1)}) \bmod p, \quad i \geq k+1. \tag{8}$$

It is interesting to note that two generators considered in Marsaglia [1996] are special cases of DL-k generators of a very small order ($k = 3$) with $B = 2^{10}, p = 2^{32} - 5$ and $B = 2^{20}, p = 2^{32} - 209$.

As we can see from (8), the $(k+1)$-th order recurrence equation for the DL-k generator has a simple structure similar to that of the DX-k-s-2 generator. As discussed before, its lattice structure over dimensions larger than k may not be great. However, the order k under discussion in this paper is already so large that this imperfection is probably of no practical significance. Nevertheless, in the next subsection, we present another special class of generators (4) that has better lattice structure than the DL-k generators.

3.2 The New DS Generators

In this paper, in addition to the DL-k generators (8), we consider a new class of generators with many nonzero coefficients:

$$X_i = B \sum_{j=1}^{k} X_{i-j} - DX_{i-d} \bmod p. \tag{9}$$

By introducing parameters B, D for the multipliers and d for the index, we expand the search parameter space for the maximum-period generators in the new class. Furthermore, the complexity of the recurrence equation for the corresponding generators is increased as well. We refer to this class of MRGs as the DS generators. Like DL generators, DS generators can be efficiently implemented via the following $(k+1)$-th order recurrence:

$$X_i = X_{i-1} + B(X_{i-1} - X_{i-k-1}) - D(X_{i-d} - X_{i-(d+1)}) \bmod p, \quad i \geq k+1, \tag{10}$$

where $X_0, X_1, \cdots, X_{k-1}$ are the initial seeds and X_k is computed by equation (9).

There are several special cases of interest for the DS generators. When $D = 0$, the DS generator is the same as the DL generators with $A = 0$ and $t = 1$ (i.e., the DL-k generator in (8)). When $D = B$, the DS generator has exactly one zero coefficient at the d-th term:

$$X_i = B \sum_{j=1, j \neq d}^{k} X_{i-j} \bmod p, \tag{11}$$

which can be efficiently implemented as

$$X_i = X_{i-1} + B(X_{i-1} - X_{i-d} + X_{i-d-1} - X_{i-k-1}) \bmod p, \quad i \geq k+1. \tag{12}$$

The parameter d of the zero-coefficient index can be chosen arbitrarily. For simplicity, we refer to the case of $d = \lceil k/2 \rceil$ as the DS-k generators.

Comparing equations (8) and (12), we can see that the higher-order implementation of DS-k generators has a more complex recurrence than that of DL-k generators. Therefore, DS-k generators may have a better lattice structure for dimensions larger than k as described in Section 2.3. Note that both DS and DL have the "perfect" lattice structure for dimensions up to k.

3.3 DL and DS Generators of Large Order

To select a prime modulus p for a prime order k, we follow the approach taken in L'Ecuyer, Blouin, and Couture [1993] and Deng [2004]. For a prime k, select a prime p such that $R(k,p) = (p^k - 1)/(p-1)$ is prime. For a 32-bit RNG, fix k and find c such that both $p = 2^{31} - c$ and $R(k,p)$ are prime. Note that some prime modulus was found in L'Ecuyer, Blouin, and Couture [1993] for $k = 7$ and in L'Ecuyer [1999] for $k = 13$. Deng [2004] listed some prime modulus p for k up to 1511. Deng [2007] found some prime modulus p for k up to 10007.

With the k and p given in Deng [2007], we can then apply the "Algorithm GMP" proposed in Deng [2004] to find the DL-k and DS-k generators. Following the approach proposed in Deng [2005], we search for the coefficient B for the corresponding DS-k or DL-k generators sequentially from the upper bound of $B < 2^{30}$. Table 1 lists some generators found.

The search time for DS-k or DL-k generators is generally an increasing function of k and the search time varies from a few hours to a month.

We use $k = 10007$ to illustrate the DL and DS generators. For the DL-10007 generator listed in Table 1, we find

$$X_i = 1073730057(X_1 + \cdots + X_{i-10007}) \bmod 2147051903,$$

which can be implemented efficiently as

$$X_i = X_{i-1} + 1073730057(X_{i-1} - X_{i-10008}) \bmod 2147051903, i \geq 10008.$$

Similarly, the DS-10007 generator can be implemented efficiently as

$$X_i = X_{i-1} + 1073668540(X_{i-1} - X_{i-5004} - X_{i-5005} - X_{i-10008}) \\ \bmod 2147051903.$$

The period length of the DS-10007 and DL-10007 generators is approximately 10^{93384}.

As discussed in Deng and Xu [2003] and Deng [2005], for maintaining portability, it is common to impose certain limits on the size of B. Thus, in addition to the largest $B < 2^{30}$ given in Table 1, we also search for the smallest B and the largest $B < 2^e$ of the maximum-period DL-k (with $e = 20$) and DS-k (with $e = 19$) generators. We list these generators in Table 2.

Table 1: List of k, c, $p = 2^{31} - c$ and B for DL/DS generators.

k	c	$p = 2^{31} - c$	$\log_{10}(p^k - 1)$	B for DL-k	B for DS-k
5003	1259289	2146224359	46686	1073741664	1073737044
6007	9984705	2137498943	56045	1073739168	1073741104
7001	610089	2146873559	65332	1073741583	1073738430
8009	5156745	2142326903	74731	1073734663	1073740201
9001	7236249	2140247399	83984	1073696126	1073727087
10007	431745	2147051903	93384	1073730057	1073668540

Table 2: List of k, p, with $\min B$ and $B < 2^e$ for DL/DS generators.

k	p	DL ($\min B$)	DS ($\min B$)	DL ($B < 2^{20}$)	DS ($B < 2^{19}$)
5003	2146224359	8724	55302	1043251	510495
6007	2137498943	1900	342	1048062	510422
7001	2146873559	15167	32335	1013185	523310
8009	2142326903	27417	9776	1047374	511773
9001	2140247399	12431	19109	1043023	499683
10007	2147051903	11507	17267	1044762	523227

The DL and DS generators under "$\min B$" in Table 2 are *not* recommended because the value of B is small. But they can be useful to determine the "power" of the empirical tests used. We remark that no empirical tests that we know of have failed these generators with a small B. The generators under "($B < 2^e$)" are recommended when 64-bit integer type is not available and the generating efficiency is a major concern. For a portability consideration, different upper bounds are used for DL and DS generators. This is because there are different numbers of terms with coefficient B in equations (8) and (12). See Deng and Xu [2003] and Deng [2005] for more explanations.

4 Summary and Conclusion

In addition to the DL generators, we propose the DS generators as a special class of MRGs and as an alternative to DX generators. DS-k generators may have a better lattice structure for dimensions higher than k than DL-k generators. Both DL and DS generators have the following nice features: (1) they enjoy many nice properties such as the equi-distribution and huge period, since they are MRGs with special structures; (2) they can be implemented efficiently because of their special forms of the common nonzero coefficients (typically, only one multiplication operation and few addition or subtraction operations are needed); (3) they have excellent empirical performances when tested with the comprehensive and stringent test package TestU01. Above all, the great feature that makes the DL/DS generators distinct is that the DL/DS generators can recover a lot more quickly from bad initializations such as near-zero initial vectors than many other popular efficient generators.

Acknowledgements

The authors are grateful to an anonymous referee who made many helpful comments and suggestions that led to a significant improvement of this paper. The authors would also like to thank Professor Harald Niederreiter for his careful reading of this paper.

References

[AK64] J. D. Alanen and D. E. Knuth. Tables of finite fields. *Sankhyā*, 26:305–328, 1964.

[Den04] L. Y. Deng. Generalized mersenne prime number and its application to random number generation. In H. Niederreiter, editor, *Monte Carlo and Quasi-Monte Carlo Methods 2002*, pages 167–180. Springer-Verlag, 2004.

[Den05] L. Y. Deng. Efficient and portable multiple recursive generators of large order. *ACM Transactions on Modeling and Computer Simulation*, 15(1):1–13, 2005.

[Den07] L. Y. Deng. Issues on computer search for large order multiple recursive generators. In S. Heinrich, A. Keller, and H. Niederreiter, editors, *Monte Carlo and Quasi-Monte Carlo Methods 2006*, pages 251–261. Springer-Verlag, 2007.

[DLS05] L. Y. Deng, H. Li, and J. J. H. Shiau. Efficient and portable multiple recursive generators of large order with many nonzero terms. preprint, 2005.

[DLWY97] L. Y. Deng, D. K. J. Lin, J. Wang, and Y. Yuan. Statistical justification of combination generators. *Statistica Sinica*, 7:993–1003, 1997.

[DX03] L. Y. Deng and H. Xu. A system of high-dimensional, efficient, long-cycle and portable uniform random number generators. *ACM Transactions on Modeling and Computer Simulation*, 13(4):299–309, 2003.

[Knu98] D. E. Knuth. *The Art of Computer Programming*, volume 2: Seminumerical Algorithms. Addison-Wesley, Reading, MA., third edition, 1998.

[LBC93] P. L'Ecuyer, F. Blouin, and R. Couture. A search for good multiple recursive linear random number generators. *ACM Transactions on Modeling and Computer Simulation*, 3:87–98, 1993.

[L'E97] P. L'Ecuyer. Bad lattice structures for vectors of non-successive values produced by some linear recurrences. *INFORMS Journal on Computing*, 9:57–60, 1997.

[L'E99] P. L'Ecuyer. Good parameter sets for combined multiple recursive random number generators. *Operations Research*, 47:159–164, 1999.

[Leh51] D. H. Lehmer. Mathematical methods in large-scale computing units. In *Proceedings of the Second Symposium on Large Scale Digital Computing Machinery*, pages 141–146, Cambridge, MA., 1951. Harvard University Press.

[Li05] H. Li. *A System of Efficient and Portable Multiple Recursive Generators of Large Order*. PhD thesis, University of Memphis, Memphis, TN., U.S.A., 2005.

[LN94] R. Lidl and H. Niederreiter. *Introduction to Finite Fields and Their Applications*. Cambridge University Press, Cambridge, MA., revised edition, 1994.

[LT04] P. L'Ecuyer and R. Touzin. On the deng–lin random number generators and related methods. *Statistical Computing*, 14:5–9, 2004.

[Mar68] G. Marsaglia. Random numbers fall mainly in the planes. In *Proceedings of the National Academy of Sciences*, volume 61, pages 25–28, 1968.

[Mar96] G. Marsaglia. The Marsaglia random number CDROM including the DIEHARD battery of tests of randomness. http://stat.fsu.edu/pub/diehard, 1996.

[MN98] M. Matsumoto and T. Nishimura. Mersenne twister: A 623-dimensionally equidistributed uniform pseudo-random number generator. *ACM Transactions on Modeling and Computer Simulation*, 8(1):3–30, 1998.

[PLM06] F. Panneton, P. L'Ecuyer, and M. Matsumoto. Improved long-period generators based on linear recurrences modulo 2. *ACM Transactions on Mathematical Software*, 32(1):1–16, 2006.

Approximation of Functions Using Digital Nets

Josef Dick[1], Peter Kritzer[2], and Frances Y. Kuo[3]

[1] School of Mathematics and Statistics, University of New South Wales, Sydney NSW 2052, Australia
josi@maths.unsw.edu.au

[2] Fachbereich Mathematik, Universität Salzburg, Hellbrunnerstraße 34, A-5020 Salzburg, Austria
peter.kritzer@sbg.ac.at

[3] School of Mathematics and Statistics, University of New South Wales, Sydney NSW 2052, Australia
f.kuo@unsw.edu.au

Summary. In analogy to a recent paper by Kuo, Sloan, and Woźniakowski, which studied lattice rule algorithms for approximation in weighted Korobov spaces, we consider the approximation problem in a weighted Hilbert space of Walsh series. Our approximation uses a truncated Walsh series with Walsh coefficients approximated by numerical integration using digital nets. We show that digital nets (or more precisely, polynomial lattices) tailored specially for the approximation problem lead to better error bounds. The error bounds can be independent of the dimension s, or depend only polynomially on s, under certain conditions on the weights defining the function space.

1 Introduction

We introduce an algorithm to approximate functions $f : [0,1]^s \to \mathbb{R}$ in certain Hilbert spaces. These spaces are in analogy to *weighted Korobov spaces* (see [SW01]), but instead of trigonometric functions we use *Walsh functions*, see Section 2. Recently, the approximation problem has been studied in [KSW06], where a function from the weighted Korobov space is approximated by a truncated Fourier series, with the remaining Fourier coefficients approximated using *lattice rules*. Here, in analogy, we want to approximate functions from a Hilbert space of *Walsh series* using *digital nets* (see [Nie92b] or Section 4 below).

More precisely, every function f in our Hilbert space H_s is given by its Walsh-series representation

$$f(\boldsymbol{x}) = \sum_{\boldsymbol{k}\in\mathbb{N}_0^s} \hat{f}(\boldsymbol{k})\mathrm{wal}_{\boldsymbol{k}}(\boldsymbol{x}), \quad \text{with} \quad \hat{f}(\boldsymbol{k}) := \int_{[0,1]^s} f(\boldsymbol{x})\overline{\mathrm{wal}_{\boldsymbol{k}}(\boldsymbol{x})}\,\mathrm{d}\boldsymbol{x}, \qquad (1)$$

where $\mathbb{N}_0 = \{0, 1, 2, \ldots\}$ denotes the set of nonnegative integers, and $\hat{f}(\boldsymbol{k})$ are the *Walsh coefficients* associated with the Walsh functions $\mathrm{wal}_{\boldsymbol{k}}(\boldsymbol{x})$ (see (3) and (4) below). For functions $f \in H_s$, the values of $|\hat{f}(\boldsymbol{k})|$ are larger for $\boldsymbol{k}$ "closer" to $\boldsymbol{0}$. We introduce a set $\mathcal{A}_s$ of vectors $\boldsymbol{k} \in \mathbb{N}_0^s$ that are close to $\boldsymbol{0}$ in some sense, and we approximate f by the Walsh polynomial

$$F(\boldsymbol{x}) := \sum_{\boldsymbol{k} \in \mathcal{A}_s} \hat{F}(\boldsymbol{k}) \mathrm{wal}_{\boldsymbol{k}}(\boldsymbol{x}), \quad \text{with} \quad \hat{F}(\boldsymbol{k}) := \frac{1}{N} \sum_{n=0}^{N-1} f(\boldsymbol{x}_n) \overline{\mathrm{wal}_{\boldsymbol{k}}(\boldsymbol{x}_n)}, \tag{2}$$

where $\{\boldsymbol{x}_0, \ldots, \boldsymbol{x}_{N-1}\} \subseteq [0, 1)^s$ is a digital net. A similar algorithm for lattice rules was proposed in [Kor63] and has recently been studied in [KSW06] (see also [KSW07, LH03, ZLH06]).

It is natural to use digital nets for the approximation of the integrals arising from the Walsh coefficients, since Walsh functions are characters over the group formed by digital nets (see [DP05b] or (13) below), which implies that the Walsh coefficients are aliased via the so-called *dual net* $\mathcal{D}$ (see [DP05b] or (12) below), i.e., it can be shown that

$$\hat{F}(\boldsymbol{k}) = \hat{f}(\boldsymbol{k}) + \sum_{\boldsymbol{h} \in \mathcal{D}} \hat{f}(\boldsymbol{h} \oplus \boldsymbol{k}),$$

where $\oplus$ denotes digit-wise addition modulo b, and it is to act on the vectors component-wise. If the dual net $\mathcal{D}$ contains only elements $\boldsymbol{k}$ which are in some sense large and $\hat{f}(\boldsymbol{k})$ is small for large $\boldsymbol{k}$, then $\hat{F}(\boldsymbol{k})$ will be a good approximation of $\hat{f}(\boldsymbol{k})$, as $\sum_{\boldsymbol{h} \in \mathcal{D}} \hat{f}(\boldsymbol{h} \oplus \boldsymbol{k})$ is small in this case compared to $\hat{f}(\boldsymbol{k})$. Hence the Walsh polynomial $F(\boldsymbol{x})$ will give a good approximation to $f(\boldsymbol{x})$.

There are several ways of finding suitable digital nets. One choice is to construct *polynomial lattices* which are suitable for integration in the space H_s (see [DKPS05]). This way one can make use of the *weights* (see [SW98]), which are introduced to moderate the importance of successive variables. Another way is to use existing digital nets, say, obtained from the Sobol$'$ sequence or the Niederreiter sequence. The third method is to construct polynomial lattices for approximation directly. This construction is similar to the one considered in [DKPS05], but with a different quality measure which appears in the upper bound on the approximation error and, at least theoretically, yields a better approximation algorithm. This is also in analogy to the results for lattice rule algorithms in [KSW06] for approximation in weighted Korobov spaces.

We also study *tractability* and *strong tractability* of the approximation problem in H_s. Strong tractability means that the error converges to zero with increasing N independently of the dimension s whereas tractability means that the error converges with N with at most a polynomial dependence on s. We show that our approximation algorithms based on digital nets achieve tractability or strong tractability error bounds under certain conditions on the weights. These results are again analogous to the results in [KSW06].

This paper is organized as follows. We introduce the weighted Hilbert space of Walsh series in Section 2, and we discuss the approximation problem in Section 3. In Section 4 we review and develop results on digital nets for the integration problem that are relevant to the approximation problem. The final section, Section 5, contains the main results of this paper as discussed above.

2 Weighted Hilbert Spaces of Walsh Series

Let $b \geq 2$ be an integer – the *base*. (Later we will restrict ourselves to a *prime* base b for simplicity.) Let $\mathbb{N}_0$ denote the set of nonnegative integers.

Each $k \in \mathbb{N}_0$ has a b-adic representation $k = \sum_{i=0}^{\infty} \kappa_i b^i$, $\kappa_i \in \{0, \ldots, b-1\}$. Each $x \in [0,1)$ has a b-adic representation $x = \sum_{i=1}^{\infty} \chi_i b^{-i}$, $\chi_i \in \{0, \ldots, b-1\}$, which is unique in the sense that infinitely many of the χ_i must differ from $b-1$. If $\kappa_a \neq 0$ is the highest nonzero digit of k, we define the *Walsh function* $\mathrm{wal}_k : [0,1) \longrightarrow \mathbb{C}$ by

$$\mathrm{wal}_k(x) := e^{2\pi \mathrm{i}(\chi_1\kappa_0 + \cdots + \chi_{a+1}\kappa_a)/b}. \tag{3}$$

For dimension $s \geq 2$ and vectors $\boldsymbol{k} = (k_1, \ldots, k_s) \in \mathbb{N}_0^s$ and $\boldsymbol{x} = (x_1, \ldots, x_s) \in [0,1)^s$ we define $\mathrm{wal}_{\boldsymbol{k}} : [0,1)^s \longrightarrow \mathbb{C}$ by

$$\mathrm{wal}_{\boldsymbol{k}}(\boldsymbol{x}) := \prod_{j=1}^{s} \mathrm{wal}_{k_j}(x_j). \tag{4}$$

It follows from the definition above that Walsh functions are piecewise constant functions. For more information on Walsh functions, see, e.g., [Chr55, Wal23].

We consider functions in a weighted Hilbert space of Walsh series. This function space was considered in [DKPS05, DP05a, DP05b]; the notion of weights was first introduced in [SW98].

Let $\alpha > 1$, $s \geq 1$, and $b \geq 2$ be fixed. Let $\boldsymbol{\gamma} = (\gamma_j)_{j=1}^{\infty}$ be a sequence of non-increasing weights, with $0 < \gamma_j \leq 1$ for all j. The weighted Hilbert space $H_s = H_{\mathrm{wal},b,s,\alpha,\boldsymbol{\gamma}}$ is a tensor product of s one-dimensional Hilbert spaces of univariate functions, each with weight γ_j. Every function f in H_s can be written in a Walsh-series representation (1).

The inner product and norm in H_s are defined by

$$\langle f, g \rangle_{H_s} := \sum_{\boldsymbol{k} \in \mathbb{N}_0^s} r(\alpha, \boldsymbol{\gamma}, \boldsymbol{k})^{-1} \hat{f}(\boldsymbol{k}) \overline{\hat{g}(\boldsymbol{k})},$$

and $\|f\|_{H_s} := \langle f, f \rangle_{H_s}^{1/2}$, where $r(\alpha, \boldsymbol{\gamma}, \boldsymbol{k}) := \prod_{j=1}^{s} r(\alpha, \gamma_j, k_j)$, with

$$r(\alpha, \gamma, k) := \begin{cases} 1 & \text{if } k = 0, \\ \gamma\, b^{-\alpha \psi_b(k)} & \text{if } k \neq 0, \end{cases} \qquad \text{and} \qquad \psi_b(k) := \lfloor \log_b(k) \rfloor. \tag{5}$$

(Equivalently, $\psi_b(k) = a$ iff $\kappa_a \neq 0$ is the highest nonzero digit in the b-adic representation of $k = \sum_{i=0}^{\infty} \kappa_i b^i$.) For $x > 1$ we define

$$\mu(x) := \sum_{k=1}^{\infty} b^{-x\psi_b(k)} = (b-1)\sum_{a=0}^{\infty} b^{-(x-1)a} = \frac{b^x(b-1)}{b^x - b}. \tag{6}$$

(The equalities hold since for any $a \geq 0$ there are $b^a(b-1)$ values of $k \geq 1$ for which $\psi_b(k) = a$.)

The space H_s is a Hilbert space with the reproducing kernel (see [Aro50, DP05b])

$$K(\boldsymbol{x}, \boldsymbol{y}) = \sum_{\boldsymbol{k} \in \mathbb{N}_0^s} r(\alpha, \boldsymbol{\gamma}, \boldsymbol{k}) \, \mathrm{wal}_{\boldsymbol{k}}(\boldsymbol{x}) \, \overline{\mathrm{wal}_{\boldsymbol{k}}(\boldsymbol{y})}.$$

The kernel satisfies the reproducing property $\langle f, K(\cdot, \boldsymbol{y})\rangle_{H_s} = f(\boldsymbol{y})$ for all $f \in H_s$ and all $\boldsymbol{y} \in [0,1)^s$.

As we have said in the introduction, we approximate functions from H_s by truncated Walsh series, see (2). Now we define precisely the set of Walsh terms to remain in the truncated Walsh series. In analogy to [KSW06], let $M > 0$ and define

$$\mathcal{A}_s(M) := \{\boldsymbol{k} \in \mathbb{N}_0^s : r(\alpha, \boldsymbol{\gamma}, \boldsymbol{k})^{-1} \leq M\}. \tag{7}$$

Following [KSW06, Lemma 1] and its proof, we can derive a number of properties for our set $\mathcal{A}_s(M)$ here; the most important one is an upper bound on the cardinality of the set, which we state as a lemma below.

Lemma 1. (cf. [KSW06, Lemma 1(d)]) *For any $M > 0$ we have*

$$|\mathcal{A}_s(M)| \leq M^q \prod_{j=1}^{s} \left(1 + \mu(\alpha q)\gamma_j^q\right)$$

for all $q > 1/\alpha$, where the function μ is defined in (6).

We end this section with a useful property that will be needed later. For $k, h \in \mathbb{N}_0$ with b-adic representations $k = \sum_{i=0}^{\infty} \kappa_i b^i$ and $h = \sum_{i=0}^{\infty} \hbar_i b^i$, let $\oplus$ and $\ominus$ denote digit-wise addition and subtraction modulo b, i.e.,

$$k \oplus h := \sum_{i=0}^{\infty} ((\kappa_i + \hbar_i) \bmod b)\, b^i \quad \text{and} \quad k \ominus h := \sum_{i=0}^{\infty} ((\kappa_i - \hbar_i) \bmod b)\, b^i.$$

For vectors $\boldsymbol{h}, \boldsymbol{k} \in \mathbb{N}_0^s$, the operations are defined component-wise.

Lemma 2. (cf. [NSW04, Formula (23)]) *For any $\boldsymbol{h}, \boldsymbol{k} \in \mathbb{N}_0^s$, we have*

$$r(\alpha, \boldsymbol{\gamma}, \boldsymbol{h} \oplus \boldsymbol{k}) \leq r(\alpha, \boldsymbol{\gamma}, \boldsymbol{k})\, r(\alpha, \boldsymbol{\gamma}, \boldsymbol{h})^{-1}.$$

Proof. It is sufficient to prove the result in one dimension, i.e., $r(\alpha,\gamma,h\oplus k) \le r(\alpha,\gamma,k)\, r(\alpha,\gamma,h)^{-1}$. Clearly this holds when $h = 0$ or $k = 0$. When $h \neq 0$ and $k \neq 0$, we have

$$r(\alpha,\gamma,h\oplus k)\frac{r(\alpha,\gamma,h)}{r(\alpha,\gamma,k)} = \gamma\left(\frac{b^{\psi_b(k)-\psi_b(h)}}{b^{\psi_b(k\oplus h)}}\right)^{\alpha} \le 1,$$

because $\psi_b(k) - \psi_b(h) \le \psi_b(k \oplus h)$. This completes the proof.

3 Approximation in the Weighted Hilbert Space H_s

We now discuss the approximation problem in the weighted Hilbert space H_s following closely the discussions from [KSW06, NSW04] for the weighted Korobov space, see also [TWW88, WW99, WW01] for general results.

Without loss of generality (see, e.g., [TWW88]), we approximate f by a linear algorithm of the form

$$A_{N,s}(f) = \sum_{n=0}^{N-1} a_n L_n(f),$$

where each a_n is a function from $L_2([0,1]^s)$ and each L_n is a continuous linear functional defined on H_s from a permissible class Λ of information. We consider two classes: $\Lambda^{\rm all}$ is the class of all continuous linear functionals, while $\Lambda^{\rm std}$ is the class of standard information consisting only of function evaluations. In other words, $L_n \in \Lambda^{\rm std}$ iff there exists $\boldsymbol{x}_n \in [0,1]^s$ such that $L_n(f) = f(\boldsymbol{x}_n)$ for all $f \in H_s$. (The approximation (2) in the introduction is of the linear form above and uses standard information from $\Lambda^{\rm std}$.)

The *worst case error* of the algorithm $A_{N,s}$ is defined as

$$e^{\rm wor-app}_{N,s}(A_{N,s}) := \sup_{\|f\|_{H_s}\le 1} \|f - A_{N,s}(f)\|_{L_2([0,1]^s)}.$$

The *initial error* associated with $A_{0,s} \equiv 0$ is

$$e^{\rm wor-app}_{0,s} := \sup_{\|f\|_{H_s}\le 1} \|f\|_{L_2([0,1]^s)} = 1,$$

where the exact value 1 is obtained by considering $f \equiv 1$.

For $\varepsilon \in (0,1)$, $s \ge 1$, and $\Lambda \in \{\Lambda^{\rm all}, \Lambda^{\rm std}\}$, we define

$$N^{\rm wor}(\varepsilon, s, \Lambda) := \min\left\{N : \exists A_{N,s} \text{ with } L_n \in \Lambda \text{ so that } e^{\rm wor-app}_{N,s}(A_{N,s}) \le \varepsilon\right\}.$$

(Note that in the definition above we actually require $e^{\rm wor-app}_{N,s}(A_{N,s}) \le \varepsilon\, e^{\rm wor-app}_{0,s}$, but since the initial error is conveniently 1, from this point on we omit the initial error from our discussion.)

We say that the approximation problem for the space H_s is *tractable* in the class Λ iff there are nonnegative numbers C, p, and a such that

$$N^{\mathrm{wor}}(\varepsilon, s, \Lambda) \leq C\varepsilon^{-p}s^{a} \qquad \forall \varepsilon \in (0,1) \quad \text{and} \quad \forall s \geq 1. \tag{8}$$

The approximation problem is *strongly tractable* in the class Λ iff (8) holds with $a = 0$. In this case, the infimum of the numbers p is called the *exponent of strong tractability*, and is denoted by $p^{\mathrm{wor-app}}(\Lambda)$.

It is known from classical results (see, e.g., [TWW88]) that the optimal algorithm in the class Λ^{all} is the truncated Walsh series

$$A_{N,s}^{(\mathrm{opt})}(f)(\boldsymbol{x}) := \sum_{\boldsymbol{k} \in \mathcal{A}_s(\varepsilon^{-2})} \hat{f}(\boldsymbol{k}) \,\mathrm{wal}_{\boldsymbol{k}}(\boldsymbol{x}),$$

where we have taken $M = \varepsilon^{-2}$ in (7) and $N = |\mathcal{A}_s(\varepsilon^{-2})|$, which ensures that the worst case error satisfies $e_{N,s}^{\mathrm{wor-app}}(A_{N,s}^{(\mathrm{opt})}) \leq \varepsilon$. In fact, it is known from the general result in [WW99] that strong tractability and tractability in the class Λ^{all} are equivalent, and they hold iff $s_{\boldsymbol{\gamma}} < \infty$, where

$$s_{\boldsymbol{\gamma}} := \inf \left\{ \lambda > 0 : \sum_{j=1}^{s} \gamma_j^{\lambda} < \infty \right\} \tag{9}$$

is known as the *sum exponent* of the weights $\boldsymbol{\gamma} = (\gamma_j)_{j=1}^{\infty}$. Furthermore, the exponent of strong tractability is $p^{\mathrm{wor-app}}(\Lambda^{\mathrm{all}}) = 2\max(1/\alpha, s_{\boldsymbol{\gamma}})$.

For the class Λ^{std}, which is the focus of this paper, a lower bound on the worst case error for any algorithm $A_{N,s}(f) = \sum_{n=0}^{N-1} a_n f(\boldsymbol{x}_n)$ can be obtained following the argument in [NSW04], i.e.,

$$e_{N,s}^{\mathrm{wor-app}}(A_{N,s}) \geq \sup_{\|f\|_{H_s} \leq 1} \left| \int_{[0,1]^s} f(\boldsymbol{x}) \,\mathrm{d}\boldsymbol{x} - \sum_{n=0}^{N-1} b_n f(\boldsymbol{x}_n) \right|,$$

where $b_n := \int_{[0,1]^s} a_n(\boldsymbol{x}) \,\mathrm{d}\boldsymbol{x}$. This lower bound is exactly the worst case integration error in H_s for the linear integration rule $\sum_{n=0}^{N-1} b_n f(\boldsymbol{x}_n)$. Hence the approximation problem is no easier than the integration problem in H_s, and thus the necessary condition for (strong) tractability for the integration problem in H_s is also necessary for the approximation problem.

(Strong) tractability in the weighted Hilbert space H_s for the family of equal-weight integration rules have been analyzed in [DP05b], where it is shown that strong tractability holds iff $\sum_{j=1}^{\infty} \gamma_j < \infty$, and tractability holds iff

$$\limsup_{s \to \infty} \sum_{j=1}^{s} \gamma_j / \ln(s+1) < \infty.$$

The same conditions can be obtained for the family of linear integration rules following the argument used in [SW01] for the weighted Korobov space. Hence, the same conditions are necessary for (strong) tractability of the approximation problem in the class $\Lambda^{\rm std}$. It follows from [WW01] that these conditions are also sufficient for (strong) tractability of approximation. Moreover, if $\sum_{j=1}^{\infty} \gamma_j < \infty$ then the exponent of strong tractability satisfies $p^{\rm wor-app}(\Lambda^{\rm std}) \in [p^{\rm wor-app}(\Lambda^{\rm all}), p^{\rm wor-app}(\Lambda^{\rm all}) + 2]$, see [WW01, Corollary 2(i)].

We summarize this discussion in the following theorem.

Theorem 1. *Consider the approximation problem in the worst case setting in the weighted Hilbert space H_s.*

- *Strong tractability and tractability in the class $\Lambda^{\rm all}$ are equivalent, and they hold iff $s_\gamma < \infty$, where s_γ is defined in (9). When this holds, the exponent of strong tractability is*

$$p^{\rm wor-app}(\Lambda^{\rm all}) \;=\; 2 \max\left(\tfrac{1}{\alpha}, s_\gamma\right).$$

- *The problem is strongly tractable in the class $\Lambda^{\rm std}$ iff*

$$\sum_{j=1}^{\infty} \gamma_j \;<\; \infty. \qquad (10)$$

When this holds, the exponent of strong tractability satisfies

$$p^{\rm wor-app}(\Lambda^{\rm std}) \;\in\; \left[p^{\rm wor-app}(\Lambda^{\rm all}), p^{\rm wor-app}(\Lambda^{\rm all}) + 2\right].$$

- *The problem is tractable in the class $\Lambda^{\rm std}$ iff*

$$\ell \;:=\; \limsup_{s\to\infty} \frac{\sum_{j=1}^{s} \gamma_j}{\ln(s+1)} \;<\; \infty. \qquad (11)$$

Note that when (10) holds, we have $s_\gamma \le 1$. When (10) does not hold but (11) holds, we have $s_\gamma = 1$.

The known results for the class $\Lambda^{\rm std}$ are non-constructive. In this paper we obtain constructive algorithms based on digital nets, and we reduce the upper bound on the exponent of strong tractability to $p^{\rm wor-app}(\Lambda^{\rm std}) \le 2\,p^{\rm wor-app}(\Lambda^{\rm all})$.

4 Integration Using Digital Nets

In this section we introduce nets and review results on numerical integration rules using those point sets.

A detailed theory of (t,m,s)-nets and (t,s)-sequences was developed in [Nie87] (see also [Nie92b, Chapter 4] and [Nie05] for a recent survey). The

(t, m, s)-nets in base b provide sets of b^m points in the s-dimensional unit cube $[0, 1)^s$ which are well distributed if the quality parameter t is small.

Definition 1. *Let $b \geq 2$, $s \geq 1$ and $0 \leq t \leq m$ be integers. A point set P consisting of b^m points in $[0, 1)^s$ forms a (t, m, s)-net in base b if every subinterval $J = \prod_{j=1}^{s} [a_j b^{-d_j}, (a_j + 1)b^{-d_j}) \subseteq [0, 1)^s$ of volume b^{t-m}, with integers $d_j \geq 0$ and integers $0 \leq a_j < b^{d_j}$ for $1 \leq j \leq s$, contains exactly b^t points of P.*

In practice, all concrete constructions of (t, m, s)-nets are based on the general construction scheme of *digital nets*. To avoid too many technical notions we restrict ourselves to digital point sets defined over the finite field $\mathbb{Z}_b = \{0, 1, \ldots, b-1\}$ with b *prime*. For a more general definition, see, e.g., [Lar98, LNS96, Nie92b]. Throughout the paper, $\top$ means the transpose of a vector or matrix.

Definition 2. *Let b be a prime and let $s \geq 1$ and $m \geq 1$ be integers. Let $C_1, \ldots, C_s$ be $m \times m$ matrices over the finite field $\mathbb{Z}_b$. For each $0 \leq n < b^m$ with b-adic representation $n = \sum_{i=0}^{m-1} \eta_i b^i$, and each $1 \leq j \leq s$, we multiply the matrix C_j by the vector $(\eta_0, \ldots, \eta_{m-1})^\top \in \mathbb{Z}_b^m$, i.e.,*

$$C_j \left(\eta_0, \ldots, \eta_{m-1}\right)^\top =: \left(\chi_{n,j,1}, \ldots, \chi_{n,j,m}\right)^\top \in \mathbb{Z}_b^m,$$

and set

$$x_{n,j} := \frac{\chi_{n,j,1}}{b} + \cdots + \frac{\chi_{n,j,m}}{b^m}.$$

If the point set $\{\boldsymbol{x}_n = (x_{n,1}, \ldots, x_{n,s}) : 0 \leq n < b^m\}$ is a (t, m, s)-net in base b for some integer t with $0 \leq t \leq m$, then it is called a digital (t, m, s)-net over $\mathbb{Z}_b$.

See [Nie92b, Theorem 4.28] and [PS01] for results concerning the determination of the quality parameter t of digital nets.

Niederreiter introduced in [Nie92a] (see also [Nie92b, Section 4.4]) a special family of digital nets known now as *polynomial lattices*. In the following, let $\mathbb{Z}_b((x^{-1}))$ be the field of formal Laurent series over $\mathbb{Z}_b$, $\sum_{l=w}^{\infty} t_l x^{-l}$, where w is an arbitrary integer and all $t_l \in \mathbb{Z}_b$. Further, let $\mathbb{Z}_b[x]$ be the set of all polynomials over $\mathbb{Z}_b$, and let

$$R_{b,m} := \{q \in \mathbb{Z}_b[x] : \deg(q) < m \text{ and } q \neq 0\}.$$

Definition 3. *Let b be a prime and let $s \geq 1$ and $m \geq 1$ be integers. Let v_m be the map from $\mathbb{Z}_b((x^{-1}))$ to the interval $[0, 1)$ defined by*

$$v_m \left(\sum_{l=w}^{\infty} t_l x^{-l} \right) := \sum_{l=\max(1,w)}^{m} t_l b^{-l}.$$

Choose polynomials $p \in \mathbb{Z}_b[x]$ *with* $\deg(p) = m$ *and* $\boldsymbol{q} := (q_1, \ldots, q_s) \in R_{b,m}^s$. *For each* $0 \le n < b^m$ *with b-adic representation* $n = \sum_{i=0}^{m-1} \eta_i b^i$, *we associate* n *with the polynomial* $n(x) = \sum_{i=0}^{m-1} \eta_i x^i \in \mathbb{Z}_b[x]$. *Then the point set*

$$P_{\mathrm{PL}} := \left\{ \boldsymbol{x}_n = \left(v_m\left(\frac{n(x)q_1(x)}{p(x)}\right), \ldots, v_m\left(\frac{n(x)q_s(x)}{p(x)}\right) \right) : 0 \le n < b^m \right\}$$

is a polynomial lattice.

We are ready to review known results on digital nets for integration. Let $P = \{\boldsymbol{x}_0, \ldots, \boldsymbol{x}_{N-1}\}$ denote a digital (t, m, s)-net over $\mathbb{Z}_b$ consisting of $N = b^m$ points. For $f \in H_s$, we approximate the integral of f by an equal-weight integration rule using the point set P. The *worst case error* of the point set P (or more precisely, of the equal-weight integration rule using the point set) for integration in the space H_s is defined by

$$e_{N,s}^{\mathrm{wor-int}}(P) := \sup_{\|f\|_{H_s} \le 1} \left| \int_{[0,1]^s} f(\boldsymbol{x}) \,\mathrm{d}\boldsymbol{x} - \frac{1}{N} \sum_{n=0}^{N-1} f(\boldsymbol{x}_n) \right|.$$

First we discuss the results from [DP05b]. For $k \in \mathbb{N}_0$ with b-adic representation $k = \sum_{i=0}^{\infty} \kappa_i b^i$, we write

$$\mathrm{tr}_m(k) := (\kappa_0, \ldots, \kappa_{m-1})^\top \in \mathbb{Z}_b^m$$

to denote the truncated digit vector of k. For a digital net P over $\mathbb{Z}_b$ generated by matrices $C_1, \ldots, C_s$, we define the *dual net* $\mathcal{D}$ by

$$\mathcal{D} := \{\boldsymbol{k} \in \mathbb{N}_0^s \setminus \{\boldsymbol{0}\} : C_1^\top \mathrm{tr}_m(k_1) + \cdots + C_s^\top \mathrm{tr}_m(k_s) = \boldsymbol{0}\}, \tag{12}$$

where the matrix-vector multiplications and vector additions are to be carried out in $\mathbb{Z}_b$. It is well known that Walsh functions are characters over the group formed by digital nets (see, e.g., [DP05b]), i.e.,

$$\frac{1}{N} \sum_{n=0}^{N-1} \mathrm{wal}_{\boldsymbol{k}}(\boldsymbol{x}_n) = \begin{cases} 1 & \text{if } \boldsymbol{k} \in \mathcal{D} \cup \{\boldsymbol{0}\}, \\ 0 & \text{otherwise.} \end{cases} \tag{13}$$

It follows from the character property that for any $f \in H_s$,

$$\int_{[0,1]^s} f(\boldsymbol{x}) \,\mathrm{d}\boldsymbol{x} - \frac{1}{N} \sum_{n=0}^{N-1} f(\boldsymbol{x}_n) = - \sum_{\boldsymbol{k} \in \mathcal{D}} \hat{f}(\boldsymbol{k}), \tag{14}$$

and hence (see [DP05b])

$$[e_{N,s}^{\mathrm{wor-int}}(P)]^2 = \sum_{\boldsymbol{k} \in \mathcal{D}} r(\alpha, \boldsymbol{\gamma}, \boldsymbol{k}). \tag{15}$$

4.1 Results for Polynomial Lattices

Now we discuss the results from [DKPS05] concerning polynomial lattices. We need some further notation: for every nonnegative integer $k = \sum_{i=0}^{\infty} \kappa_i b^i$ we define the polynomial

$$\widetilde{\mathrm{tr}}_m(k)(x) := \kappa_0 + \kappa_1 x + \cdots + \kappa_{m-1} x^{m-1} \in \mathbb{Z}_b[x],$$

and for the vector $\boldsymbol{k} = (k_1, \ldots, k_s) \in \mathbb{N}_0^s$ we consider

$$\widetilde{\mathrm{tr}}_m(\boldsymbol{k}) := (\widetilde{\mathrm{tr}}_m(k_1), \ldots, \widetilde{\mathrm{tr}}_m(k_s))^\top \in \mathbb{Z}_b[x]^s$$

to be a vector of polynomials. It is shown in [DKPS05] that the dual net for the polynomial lattice P_{PL}, with polynomials $p \in \mathbb{Z}_b[x]$ and $\boldsymbol{q} = (q_1, \ldots, q_s) \in R_{b,m}^s$, can be expressed as

$$\mathcal{D}_{\mathrm{PL}} := \left\{ \boldsymbol{k} \in \mathbb{N}_0^s \setminus \{\boldsymbol{0}\} : \widetilde{\mathrm{tr}}_m(\boldsymbol{k}) \cdot \boldsymbol{q} \equiv 0 \pmod{p} \right\}, \tag{16}$$

where $\widetilde{\mathrm{tr}}_m(\boldsymbol{k}) \cdot \boldsymbol{q} \equiv 0 \pmod{p}$ means that the polynomial p divides the polynomial

$$\widetilde{\mathrm{tr}}_m(\boldsymbol{k}) \cdot \boldsymbol{q} := \sum_{j=1}^{s} \widetilde{\mathrm{tr}}_m(k_j)\, q_j \in \mathbb{Z}_b[x].$$

The main result of [DKPS05] is summarized in the following lemma.

Lemma 3. (cf. [DKPS05, Algorithm 4.3 and Theorem 4.4]) *Given prime $b \geq 2$, positive integer m, and irreducible polynomial $p \in \mathbb{Z}_b[x]$, a vector of polynomials $\boldsymbol{q} = (q_1, \ldots, q_s) \in R_{b,m}^s$ for a polynomial lattice P_{PL} with $N = b^m$ points can be constructed by a component-by-component algorithm such that*

$$[e_{N,s}^{\mathrm{wor-int}}(P_{\mathrm{PL}})]^2 \leq (b^m - 1)^{-1/\lambda} \prod_{j=1}^{s} \left(1 + \mu(\alpha\lambda)\gamma_j^\lambda\right)^{1/\lambda}$$

for all $\lambda \in (1/\alpha, 1]$, where the function μ is defined in (6).

Using the property $\prod_{j=1}^{s}(1 + x_j) = \exp(\sum_{j=1}^{s} \ln(1 + x_j)) \leq \exp(\sum_{j=1}^{s} x_j)$ for all nonnegative x_j, we see from Lemma 3 that if $s_\gamma \leq 1/\alpha$ then

$$e_{N,s}^{\mathrm{wor-int}}(P_{\mathrm{PL}}) = \mathcal{O}(N^{-\alpha/2+\delta}), \quad \delta > 0,$$

with the implied factor in the big-$\mathcal{O}$ notation is independent of N and s. This is the optimal rate of convergence for integration in H_s.

4.2 Results for General Digital Nets

For any digital (t, m, s)-net with regular generating matrices, we can obtain a worst case error bound in terms of its t-value. This is in analogy to results obtained in [CDP06, DP05a, DP05c].

Lemma 4. *Let P be a digital (t,m,s)-net over $\mathbb{Z}_b$ with non-singular generating matrices. For each $\emptyset \neq \mathfrak{u} \subseteq \{1,\ldots,s\}$, suppose that the projection of P onto the coordinates in $\mathfrak{u}$ is a $(t_{\mathfrak{u}}, m, |\mathfrak{u}|)$-net. Then we have*

$$[e_{N,s}^{\mathrm{wor-int}}(P)]^2 \leq \frac{1}{b^{\alpha m}}\left(1 + \sum_{\emptyset \neq \mathfrak{u} \subseteq \{1,\ldots,s\}} b^{\alpha t_{\mathfrak{u}}} \prod_{j \in \mathfrak{u}} \left(b^{\alpha+1}(m+2)\mu(\alpha)\gamma_j\right)\right).$$

Proof. We start with (15) and consider all vectors $\boldsymbol{k}$ in the dual net $\mathcal{D}$ given by (12). If $\boldsymbol{k} = b^m \boldsymbol{l}$ with $\boldsymbol{l} \in \mathbb{N}_0^s \setminus \{\boldsymbol{0}\}$, then $\mathrm{tr}_m(k_j) = \boldsymbol{0}$ for $1 \leq j \leq s$. Otherwise we can write $\boldsymbol{k} = \boldsymbol{k}^* + b^m \boldsymbol{l}$ with $\boldsymbol{l} \in \mathbb{N}_0^s$, $\boldsymbol{k}^* = (k_1^*, \ldots, k_s^*) \neq \boldsymbol{0}$ and $0 \leq k_j^* < b^m$ for all $1 \leq j \leq s$. In the latter case we have $\mathrm{tr}_m(k_j) = \mathrm{tr}_m(k_j^*)$ for all $1 \leq j \leq s$. Thus we have (after renaming $\boldsymbol{k}^*$ to $\boldsymbol{k}$)

$$[e_{N,s}^{\mathrm{wor-int}}(P)]^2 = \sum_{\boldsymbol{l} \in \mathbb{N}_0^s \setminus \{\boldsymbol{0}\}} r(\alpha, \boldsymbol{\gamma}, b^m \boldsymbol{l}) + \sum_{\boldsymbol{k} \in \mathcal{D}^*} \sum_{\boldsymbol{l} \in \mathbb{N}_0^s} r(\alpha, \boldsymbol{\gamma}, \boldsymbol{k} + b^m \boldsymbol{l}) =: \Sigma_1^* + \Sigma_2^*,$$

where

$$\mathcal{D}^* := \left\{\boldsymbol{k} \in \{0,\ldots,b^m-1\}^s \setminus \{\boldsymbol{0}\} : C_1^\top \mathrm{tr}_m(k_1) + \cdots + C_s^\top \mathrm{tr}_m(k_s) = \boldsymbol{0}\right\}.$$

It follows from the definition (5) that for $0 \leq k_j < b^m$ we have

$$\begin{aligned}\sum_{l=0}^{\infty} r(\alpha, \gamma_j, k_j + b^m l) &= r(\alpha, \gamma_j, k_j) + \sum_{l=1}^{\infty} r(\alpha, \gamma_j, b^m l) \\ &= r(\alpha, \gamma_j, k_j) + \frac{\mu(\alpha)}{b^{m\alpha}} \gamma_j.\end{aligned}$$

Thus

$$\Sigma_1^* = \prod_{j=1}^{s} \left(1 + \frac{\mu(\alpha)}{b^{m\alpha}} \gamma_j\right) - 1$$

and

$$\begin{aligned}\Sigma_2^* &= \sum_{\boldsymbol{k} \in \mathcal{D}^*} \prod_{j=1}^{s} \left(r(\alpha, \gamma_j, k_j) + \frac{\mu(\alpha)}{b^{m\alpha}} \gamma_j\right) \\ &= \sum_{\boldsymbol{k} \in \mathcal{D}^*} r(\alpha, \boldsymbol{\gamma}, \boldsymbol{k}) + \sum_{\mathfrak{u} \subsetneq \{1,\ldots,s\}} \left[\left(\sum_{\boldsymbol{k} \in \mathcal{D}^*} \prod_{j \in \mathfrak{u}} r(\alpha, \gamma_j, k_j)\right) \prod_{j \notin \mathfrak{u}} \left(\frac{\mu(\alpha)\gamma_j}{b^{m\alpha}}\right)\right]. \quad (17)\end{aligned}$$

First we investigate the sum $\sum_{\boldsymbol{k} \in \mathcal{D}^*} \prod_{j \in \mathfrak{u}} r(\alpha, \gamma_j, k_j)$ where $\mathfrak{u}$ is a proper subset of $\{1,\ldots,s\}$. Let $\boldsymbol{k} = (k_1, \ldots, k_s) \in \{0,\ldots,b^m-1\}^s \setminus \{\boldsymbol{0}\}$ and $j_0 \notin \mathfrak{u}$. Since the generating matrices $C_1, \ldots, C_s$ are non-singular, for any combination of the $s-1$ components $k_j \in \{0,\dots,b^m-1\}$ with $j \neq j_0$, there is exactly one value of $k_{j_0} \in \{0,\dots,b^m-1\}$ which ensures that $\boldsymbol{k} \in \mathcal{D}^*$. Hence we have

$$\sum_{\boldsymbol{k}\in\mathcal{D}^*} \prod_{j\in\mathfrak{u}} r(\alpha,\gamma_j,k_j) = b^{m(s-|\mathfrak{u}|-1)} \prod_{j\in\mathfrak{u}} \left(\sum_{k=0}^{b^m-1} r(\alpha,\gamma_j,k) \right) - 1$$
$$\leq b^{m(s-|\mathfrak{u}|-1)} \prod_{j\in\mathfrak{u}} (1+\mu(\alpha)\gamma_j) - 1,$$

from which we can show that the second term in (17) is bounded by

$$\frac{1}{b^{\alpha m}} \prod_{j=1}^{s} (1+2\mu(\alpha)\gamma_j) - \Sigma_1^*.$$

It remains to obtain a bound on the first term in (17). Here we only outline the most important steps; the details follow closely the proofs of [CDP06, Lemma 7] and [DP05c, Lemma 7], see also [DP05a, Lemma 3].

We have

$$\sum_{\boldsymbol{k}\in\mathcal{D}^*} r(\alpha,\boldsymbol{\gamma},\boldsymbol{k}) = \sum_{\substack{\emptyset\neq\mathfrak{u}\subseteq\{1,\dots,s\} \\ \mathfrak{u}=\{u_1,\dots,u_e\}}} \sum_{\substack{k_{u_1},\dots,k_{u_e}=1 \\ C_{u_1}^\top \mathrm{tr}_m(k_{u_1})+\cdots+C_{u_e}^\top \mathrm{tr}_m(k_{u_e})=\boldsymbol{0}}}^{b^m-1} \prod_{j\in\mathfrak{u}} r(\alpha,\gamma_j,k_j). \quad (18)$$

The $\mathfrak{u}=\{1,\dots,s\}$ term in (18) is

$$\sum_{\substack{k_1,\dots,k_s=1 \\ C_1^\top \mathrm{tr}_m(k_1)+\cdots+C_s^\top \mathrm{tr}_m(k_s)=\boldsymbol{0}}}^{b^m-1} \prod_{j=1}^{s} r(\alpha,\gamma_j,k_j)$$
$$= \sum_{v_1,\dots,v_s=0}^{m-1} \frac{\prod_{j=1}^s \gamma_j}{b^{\alpha(v_1+\cdots+v_s)}} \sum_{l_1,\dots,l_s=1}^{b-1} \underbrace{\sum_{k_1=l_1 b^{v_1}}^{(l_1+1)b^{v_1}-1} \cdots \sum_{k_s=l_s b^{v_s}}^{(l_s+1)b^{v_s}-1}}_{C_1^\top \mathrm{tr}_m(k_1)+\cdots+C_s^\top \mathrm{tr}_m(k_s)=\boldsymbol{0}} 1.$$

Using the fact that P is a digital (t,m,s)-net, it can be shown that

$$\underbrace{\sum_{k_1=l_1 b^{v_1}}^{(l_1+1)b^{v_1}-1} \cdots \sum_{k_s=l_s b^{v_s}}^{(l_s+1)b^{v_s}-1}}_{C_1^\top \mathrm{tr}_m(k_1)+\cdots+C_s^\top \mathrm{tr}_m(k_s)=\boldsymbol{0}} 1 \leq (b-1)^s \sum_{\substack{v_1,\dots,v_s=0 \\ m-t-s+1\leq v_1+\cdots+v_s\leq m-t}}^{m-1} \frac{\prod_{j=1}^s \gamma_j}{b^{\alpha(v_1+\cdots+v_s)}}$$
$$+ (b-1)^s \sum_{\substack{v_1,\dots,v_s=0 \\ v_1+\cdots+v_s>m-t}}^{m-1} \frac{\prod_{j=1}^s \gamma_j}{b^{\alpha(v_1+\cdots+v_s)}} b^{v_1+\cdots+v_s-m+t}.$$

The $\emptyset \neq \mathfrak{u} \subsetneq \{1,\dots,s\}$ terms in (18) can be estimated in a similar way by making use of the fact that the projection of P onto the coordinates in $\mathfrak{u}$ is a digital $(t_{\mathfrak{u}},m,|\mathfrak{u}|)$-net. Combining all the terms together, we finally obtain

$$\sum_{\boldsymbol{k}\in\mathcal{D}^*} r(\alpha,\boldsymbol{\gamma},\boldsymbol{k}) \leq \sum_{\emptyset\neq\mathfrak{u}\subseteq\{1,\ldots,s\}} \left(\frac{b-1}{b^{\alpha-1}-1}\right)^{|\mathfrak{u}|} \frac{2(m-t_{\mathfrak{u}}+2)^{|\mathfrak{u}|-1}}{b^{\alpha(m-t_{\mathfrak{u}}+1-2|\mathfrak{u}|)}} \prod_{j\in\mathfrak{u}} \gamma_j$$
$$\leq \frac{1}{b^{\alpha(m+1)}} \sum_{\emptyset\neq\mathfrak{u}\subseteq\{1,\ldots,s\}} b^{\alpha t_{\mathfrak{u}}} \prod_{j\in\mathfrak{u}} \left(b^{\alpha+1}(m+2)\mu(\alpha)\gamma_j\right),$$

from which the result can be derived.

We now give two examples of digital nets for which explicit bounds on the values of $t_{\mathfrak{u}}$ are known. Let P_{Sob} and P_{Nie} denote the digital net generated by the modified (as discussed below) left upper $m \times m$ sub-matrices of the generating matrices of the *Sobol′ sequence* and the *Niederreiter sequence* (which are examples of *digital* (t,s)*-sequences*, see, e.g., [Nie92a]), respectively. We need to modify the generating matrices to make them regular; this can be achieved by changing the least significant rows of the matrices without influencing the digital net property nor the quality parameter of the net and its projections.

Lemma 5. *Let $P \in \{P_{\text{Sob}}, P_{\text{Nie}}\}$ be a digital (t,m,s)-net over $\mathbb{Z}_b$ obtained from either the Sobol′ sequence ($b=2$) or the Niederreiter sequence. We have*

$$[e_{N,s}^{\text{wor-int}}(P_{\text{Sob}})]^2$$
$$\leq \frac{1}{2^{\alpha m}} \prod_{j=1}^{s} \left(2^{\alpha c+1}\left(j\log_2(j+1)\log_2\log_2(j+3)\right)^{\alpha}(m+2)\mu(\alpha)\gamma_j\right),$$

where c is some constant independent of all parameters, and

$$[e_{N,s}^{\text{wor-int}}(P_{\text{Nie}})]^2 \leq \frac{1}{b^{\alpha m}} \prod_{j=1}^{s} \left(b^{2\alpha+1}\left(j\log_b(j+b)\right)^{\alpha}(m+2)\mu(\alpha)\gamma_j\right).$$

If

$$\begin{cases} \sum_{j=1}^{\infty}(j\ln j\,\ln\ln j)^{\alpha}\gamma_j < \infty & \text{when } P = P_{\text{Sob}}, \\ \sum_{j=1}^{\infty}(j\ln j)^{\alpha}\gamma_j < \infty & \text{when } P = P_{\text{Nie}}, \end{cases} \tag{19}$$

then

$$[e_{N,s}^{\text{wor-int}}(P)]^2 \leq C_{\delta}\,N^{-\alpha+\delta}, \quad C_{\delta} \in \{C_{\text{Sob},\delta}, C_{\text{Nie},\delta}\}, \quad \delta > 0,$$

where $C_{\text{Sob},\delta}$ and $C_{\text{Nie},\delta}$ are independent of m and s but depend on δ, b, α, and $\boldsymbol{\gamma}$.

Proof. The construction of the Sobol′ sequence makes use of primitive polynomials in base $b=2$, one polynomial p_j for each dimension j, with non-decreasing degrees as the dimension increases. It is known (see, e.g., [Wan02]) that $t_{\mathfrak{u}} = \sum_{j\in\mathfrak{u}}(\deg(p_j)-1)$ and $\deg(p_j) \leq \log_2 j + \log_2\log_2(j+1) + \log_2\log_2\log_2(j+3) + c$, where c is a constant independent of j. (Note that the above formula for $t_{\mathfrak{u}}$ is associated with the whole sequence; for a net of b^m points $t_{\mathfrak{u}}$ is bounded by the minimum of m and the given formula.) Thus we have

$$b^{t_{\mathfrak{u}}} = 2^{t_{\mathfrak{u}}} \leq \prod_{j\in\mathfrak{u}} \left(2^{c-1} j \log_2(j+1) \log_2 \log_2(j+3)\right).$$

On the other hand, the construction of the Niederreiter sequence makes use of monic irreducible polynomials, and it is known (see [Wan02, Lemma 2]) that $\deg(p_j) \leq \log_b j + \log_b \log_b(j+b) + 2$. Thus in this case

$$b^{t_{\mathfrak{u}}} \leq \prod_{j\in\mathfrak{u}} \left(bj \log_b(j+b)\right).$$

Substituting these bounds on $b^{t_{\mathfrak{u}}}$ into Lemma 4 proves the first part of this lemma.

To prove the second part of this lemma, we follow closely the proof of [HN03, Lemma 3]. Consider first the Niederreiter sequence and suppose that $\sum_{j=1}^{\infty} (j \ln j)^{\alpha} \gamma_j < \infty$. For $k \geq 0$, define $\sigma_k := b^{2\alpha+1} \mu(\alpha) \sum_{j=k+1}^{\infty} (j \log_b(j+b))^{\alpha} \gamma_j$. Then we have

$$\prod_{j=1}^{s} \left(b^{2\alpha+1} (j \log_b(j+b))^{\alpha} (m+2) \mu(\alpha) \gamma_j\right) \leq (1+\sigma_k^{-1})^k \, b^{(m+2)\sigma_k(\sigma_0+1)}.$$

Let $\delta > 0$ and choose k_δ such that $\sigma_{k_\delta}(\sigma_0+1) \leq \delta$. The desired result is obtained with $C_{\mathrm{Nie},\delta} := b^{2\delta}(1+\sigma_{k_\delta}^{-1})^{k_\delta}$. The result for the Sobol′ sequence can be obtained in the same way.

5 Approximation Using Digital Nets

Now we formalize the approximation algorithm (2). For $M > 0$ and $P = \{\boldsymbol{x}_0, \ldots, \boldsymbol{x}_{N-1}\}$ a digital net with $N = b^m$ points, we define

$$A_{N,s,M}(f) := \sum_{\boldsymbol{h} \in \mathcal{A}_s(M)} \left(\frac{1}{N} \sum_{n=0}^{N-1} f(\boldsymbol{x}_n) \overline{\mathrm{wal}_{\boldsymbol{h}}(\boldsymbol{x}_n)} \right) \mathrm{wal}_{\boldsymbol{h}}(\boldsymbol{x}).$$

Recall that the *worst case error* for the algorithm $A_{N,s,M}$ using the point set P is defined by

$$\mathrm{e}^{\mathrm{wor-app}}_{N,s,M}(P) = \mathrm{e}^{\mathrm{wor-app}}_{N,s,M}(A_{N,s,M}) := \sup_{\|f\|_{H_s} \leq 1} \|f - A_{N,s,M}(f)\|_{L_2([0,1]^s)}.$$

We have

$$\begin{aligned}
&\|f - A_{N,s,M}(f)\|^2_{L_2([0,1]^s)} \\
&= \sum_{\boldsymbol{h} \notin \mathcal{A}_s(M)} |\hat{f}(\boldsymbol{h})|^2 + \sum_{\boldsymbol{h} \in \mathcal{A}_s(M)} \left| \int_{[0,1]^s} f(\boldsymbol{x}) \overline{\mathrm{wal}_{\boldsymbol{h}}(\boldsymbol{x})} \,\mathrm{d}\boldsymbol{x} - \frac{1}{N} \sum_{n=0}^{N-1} f(\boldsymbol{x}_n) \overline{\mathrm{wal}_{\boldsymbol{h}}(\boldsymbol{x}_n)} \right|^2 \\
&\le \frac{1}{M} \|f\|^2_{H_s} + \sum_{\boldsymbol{h} \in \mathcal{A}_s(M)} |\langle f, \tau_{\boldsymbol{h}} \rangle_{H_s}|^2 ,
\end{aligned}$$

where

$$\tau_{\boldsymbol{h}}(\boldsymbol{t}) := \int_{[0,1]^s} K(\boldsymbol{t}, \boldsymbol{x}) \mathrm{wal}_{\boldsymbol{h}}(\boldsymbol{x}) \,\mathrm{d}\boldsymbol{x} - \frac{1}{N} \sum_{n=0}^{N-1} K(\boldsymbol{t}, \boldsymbol{x}_n) \mathrm{wal}_{\boldsymbol{h}}(\boldsymbol{x}_n).$$

Hence

$$\mathrm{e}^{\mathrm{wor-app}}_{N,s,M}(P) = \left(\frac{\beta}{M} + \sup_{\|f\|_{H_s} \le 1} \sum_{\boldsymbol{h} \in \mathcal{A}_s(M)} |\langle f, \tau_{\boldsymbol{h}} \rangle_{H_s}|^2 \right)^{1/2}$$

for some $\beta \in [0,1]$. Moreover, it can be shown that the second term involving the supremum is essentially the spectral radius ρ of the matrix T_P whose entries are given by $\langle \tau_{\boldsymbol{h}}, \tau_{\boldsymbol{p}} \rangle_{H_s}$.

Using (14), it can be shown that

$$\begin{aligned}
\tau_{\boldsymbol{h}}(\boldsymbol{t}) &= - \sum_{\boldsymbol{k} \in \mathcal{D}} r(\alpha, \boldsymbol{\gamma}, \boldsymbol{h} \ominus \boldsymbol{k}) \,\mathrm{wal}_{\boldsymbol{h} \ominus \boldsymbol{k}}(\boldsymbol{t}) \\
&= - \sum_{\substack{\boldsymbol{q} \in \mathbb{N}_0^s \setminus \{\boldsymbol{h}\} \\ C_1^\top \mathrm{tr}_m(h_1 \ominus q_1) + \cdots + C_s^\top \mathrm{tr}_m(h_s \ominus q_s) = \boldsymbol{0}}} r(\alpha, \boldsymbol{\gamma}, \boldsymbol{q}) \,\mathrm{wal}_{\boldsymbol{q}}(\boldsymbol{t}).
\end{aligned}$$

Consequently,

$$\langle \tau_{\boldsymbol{h}}, \tau_{\boldsymbol{p}} \rangle_{H_s} = \begin{cases} 0 & \text{if } C_1^\top \mathrm{tr}_m(h_1 \ominus p_1) + \cdots + C_s^\top \mathrm{tr}_m(h_s \ominus p_s) \neq \boldsymbol{0}, \\ \displaystyle\sum_{\substack{\boldsymbol{k} \in \mathbb{N}_0^s \setminus \{\boldsymbol{0}, \boldsymbol{p} \ominus \boldsymbol{h}\} \\ C_1^\top \mathrm{tr}_m(k_1) + \cdots + C_s^\top \mathrm{tr}_m(k_s) = \boldsymbol{0}}} r(\alpha, \boldsymbol{\gamma}, \boldsymbol{h} \oplus \boldsymbol{k}) & \text{otherwise.} \end{cases} \tag{20}$$

We state the result in the following lemma.

Lemma 6. (cf. [KSW06, Lemma 2]) *The worst case error for the approximation algorithm $A_{N,s,M}$ using a digital net P satisfies*

$$e^{\mathrm{wor-app}}_{N,s,M}(P) = \left(\frac{\beta}{M} + \rho(T_P) \right)^{1/2} \quad \textit{for some} \quad \beta \in [0,1],$$

where T_P is a nonnegative-definite symmetric $|\mathcal{A}_s(M)| \times |\mathcal{A}_s(M)|$ matrix with entries given by $\langle \tau_{\boldsymbol{h}}, \tau_{\boldsymbol{p}} \rangle_{H_s}$ in (20) for $\boldsymbol{h}, \boldsymbol{p} \in \mathcal{A}_s(M)$.

Unfortunately we do not have a computable expression for the spectral radius $\rho(T_P)$. Therefore we consider its upper bound, the trace of T_P,

$$\rho(T_P) \le \mathrm{trace}(T_P) = \sum_{\boldsymbol{h}\in\mathcal{A}_s(M)} \sum_{\boldsymbol{k}\in\mathcal{D}} r(\alpha,\boldsymbol{\gamma},\boldsymbol{h}\oplus\boldsymbol{k}). \tag{21}$$

5.1 Nets Constructed for Integration

A natural question to ask is: how good are the nets constructed for integration when they are used for approximation? To relate the worst case error for approximation $e_{N,s}^{\mathrm{wor-app}}(P)$ to the worst case error for integration $e_{N,s}^{\mathrm{wor-int}}(P)$, we apply Lemma 2 to (21) and obtain

$$\rho(T_P) \le \sum_{\boldsymbol{h}\in\mathcal{A}_s(M)} \frac{1}{r(\alpha,\boldsymbol{\gamma},\boldsymbol{h})} \sum_{\boldsymbol{k}\in\mathcal{D}} r(\alpha,\boldsymbol{\gamma},\boldsymbol{k}) \le M|\mathcal{A}_s(M)|[e_{N,s}^{\mathrm{wor-int}}(P)]^2.$$

Hence it follows from Lemma 6 that

$$e_{N,s,M}^{\mathrm{wor-app}}(P) \le \left(\frac{1}{M} + M|\mathcal{A}_s(M)|[e_{N,s}^{\mathrm{wor-int}}(P)]^2\right)^{1/2}. \tag{22}$$

Applying Lemmas 1 and 3 to (22), we obtain the following result for polynomial lattices constructed for the integration problem.

Lemma 7. (cf. [KSW06, Lemma 3]) *Let P_{PL} be a polynomial lattice constructed component-by-component for integration. Then the worst case error for the approximation algorithm $A_{N,s,M}$ using P_{PL} satisfies*

$$e_{N,s,M}^{\mathrm{wor-app}}(P_{\mathrm{PL}}) \le \left(\frac{1}{M} + \frac{C_{s,q,\lambda}M^{q+1}}{(N-1)^{1/\lambda}}\right)^{1/2}$$

for all $q > 1/\alpha$ and $\lambda \in (1/\alpha, 1]$, where

$$C_{s,q,\lambda} := \prod_{j=1}^{s} \left(1+\mu(\alpha\lambda)\gamma_j^{\lambda}\right)^{1/\lambda} \left(1+\mu(\alpha q)\gamma_j^{q}\right).$$

Given $\varepsilon \in (0,1)$, we want to find small M and $N = b^m$ for which the error bound in Lemma 7 is at most ε. To ensure that the two terms in the error bound are of the same order, we first choose $M = 2\varepsilon^{-2}$, and then choose N such that the second term is no more than the first term. Hence it is sufficient that we take $N = b^m$ with

$$m = \left\lceil \log_b\left(\left(C_{s,q,\lambda}M^{q+2}\right)^{\lambda} + 1\right)\right\rceil. \tag{23}$$

Using the property $\prod_{j=1}^{s}(1+x_j) = \exp(\sum_{j=1}^{s}\log(1+x_j)) \le \exp(\sum_{j=1}^{s} x_j)$ for nonnegative x_j, we can write

$$C_{s,q,\lambda} \le \exp\left(\frac{\mu(\alpha\lambda)}{\lambda}\sum_{j=1}^{s}\gamma_j^{\lambda} + \mu(\alpha q)\sum_{j=1}^{s}\gamma_j^{q}\right) \tag{24}$$

$$= (s+1)^{\mu(\alpha\lambda)\lambda^{-1}\sum_{j=1}^{s}\gamma_j^{\lambda}/\ln(s+1) + \mu(\alpha q)\sum_{j=1}^{s}\gamma_j^{q}/\ln(s+1)}. \tag{25}$$

Let $p^* = 2\max(1/\alpha, s_\gamma)$. When (10) holds but $s_\gamma = 1$, we have $p^* = 2$ and we take $q = \lambda = p^*/2 = 1$. Then we see from (24) that $\sup_{s\ge1} C_{s,q,\lambda} < \infty$. When (10) holds and $s_\gamma < 1$, we have $p^* < 2$ and we choose $q = \lambda = p^*/2 + \delta$ for some $\delta > 0$. Then $\mu(\alpha\lambda) < \infty$ and $\sum_{j=1}^{\infty}\gamma_j^{\lambda} < \infty$, and once again we see from (24) that $\sup_{s\ge1} C_{s,q,\lambda} < \infty$. In both cases, we see from (23) that $N = \mathcal{O}(\varepsilon^{-p})$, with p equal to or arbitrarily close to $2p^* + {p^*}^2/2$ as δ goes to 0.

When (11) holds but not (10), we have $s_\gamma = 1$. We take $q = \lambda = 1$ and it follows from (23) and (25) that $N = \mathcal{O}(\varepsilon^{-6})$ and $C_{s,q,\lambda} = \mathcal{O}(s^a)$, with a arbitrarily close to $2\mu(\alpha)\ell$.

We summarize the analysis in the following theorem.

Theorem 2. (cf. [KSW06, Theorem 1]) *Let P_{PL} be a polynomial lattice constructed component-by-component for integration. For $\varepsilon \in (0,1)$ set $M = 2\varepsilon^{-2}$. If (10) holds, then the approximation algorithm $A_{N,s,M}$ using P_{PL} achieves the worst case error bound $e^{\mathrm{wor-app}}_{N,s,M}(P_{\mathrm{PL}}) \le \varepsilon$ using $N = \mathcal{O}(\varepsilon^{-p})$ function values, with p equal to or arbitrarily close to*

$$2\,p^{\mathrm{wor-app}}(\Lambda^{\mathrm{all}}) + \frac{[p^{\mathrm{wor-app}}(\Lambda^{\mathrm{all}})]^2}{2}.$$

If (10) does not hold but (11) holds, then the error bound $e^{\mathrm{wor-app}}_{N,s,M}(P_{\mathrm{PL}}) \le \varepsilon$ is achieved using $N = \mathcal{O}(s^a\varepsilon^{-6})$ function values, with a arbitrarily close to $2\mu(\alpha)\ell$. The implied factors in the big $\mathcal{O}$-notations are independent of ε and s.

Now we use Lemmas 1 and 5 in (22) to derive results for digital nets obtained from the Sobol′ sequence or the Niederreiter sequence.

Lemma 8. *Let $P \in \{P_{\mathrm{Sob}}, P_{\mathrm{Nie}}\}$ be a digital net obtained from either the Sobol′ sequence or the Niederreiter sequence. If (19) holds, then the worst case error for the approximation algorithm $A_{N,s,M}$ using P satisfies*

$$e^{\mathrm{wor-app}}_{N,s,M}(P) \le \left(\frac{1}{M} + \frac{\bar{C}_{s,q,\delta}M^{q+1}}{N^{\alpha-\delta}}\right)^{1/2}, \quad \bar{C}_{s,q,\delta} := C_\delta\prod_{j=1}^{s}\left(1+\mu(\alpha q)\gamma_j^q\right),$$

for all $q > 1/\alpha$ and $\delta > 0$, with $C_\delta \in \{C_{\mathrm{Sob},\delta}, C_{\mathrm{Nie},\delta}\}$ given in Lemma 5.

Note that both conditions on the weights in (19) imply (10) as well as $s_\gamma \le 1/\alpha$. For $\varepsilon \in (0,1)$ we take $q = 1/\alpha + \delta$, $M = 2\varepsilon^{-2}$ and $N = b^m$ with

$$m = \left\lceil \log_b\left(\left(\bar{C}_{s,q,\delta}M^{q+2}\right)^{1/(\alpha-\delta)}\right)\right\rceil.$$

Then we have $\sup_{s\geq 1} \bar{C}_{s,q,\delta} < \infty$ and $N = \mathcal{O}(\varepsilon^{-p})$, with p arbitrarily close to $4/\alpha + 2/\alpha^2$ as δ goes to 0. This is summarized in the theorem below.

Theorem 3. *Let $P \in \{P_{\mathrm{Sob}}, P_{\mathrm{Nie}}\}$ be a digital net obtained from either the Sobol′ sequence or the Niederreiter sequence. For $\varepsilon \in (0,1)$ set $M = 2\varepsilon^{-2}$. If* (19) *holds, then the approximation algorithm $A_{N,s,M}$ using P achieves the worst case error bound $e^{\mathrm{wor-app}}_{N,s,M}(P) \leq \varepsilon$ using $N = \mathcal{O}(\varepsilon^{-p})$ function values, with p arbitrarily close to*

$$2\, p^{\mathrm{wor-app}}(\Lambda^{\mathrm{all}}) + \frac{[p^{\mathrm{wor-app}}(\Lambda^{\mathrm{all}})]^2}{2}.$$

The implied factor in the big $\mathcal{O}$-notation is independent of ε and s.

5.2 Polynomial Lattices Constructed for Approximation

In this section we study polynomial lattices with the generating polynomials specially constructed for the approximation problem. It is perhaps not surprising that such polynomial lattices yield smaller error bounds than those studied in the previous subsection.

Since $M\, r(\alpha, \boldsymbol{\gamma}, \boldsymbol{h}) \geq 1$ for all $\boldsymbol{h} \in \mathcal{A}_s(M)$, we have from (21) that $\rho(T_P) \leq M\, S_{N,s}(P)$, where

$$S_{N,s}(P) := \sum_{\boldsymbol{h}\in\mathbb{N}_0^s} \sum_{\boldsymbol{k}\in\mathcal{D}} r(\alpha,\boldsymbol{\gamma},\boldsymbol{h})\, r(\alpha,\boldsymbol{\gamma},\boldsymbol{k}\oplus\boldsymbol{h}). \tag{26}$$

Thus it follows from Lemma 6 that

$$e^{\mathrm{wor-app}}_{N,s,M}(P) \leq \left(\frac{1}{M} + M\, S_{N,s}(P)\right)^{1/2}. \tag{27}$$

An analogous expression to $S_{N,s}(P)$ for lattice rule algorithms in weighted Korobov spaces was considered in [DKKS07] for some integral equation problem. (It is advocated that the expression $S_{n,d}(\boldsymbol{z})$ in [DKKS07] should be considered instead of the quantity $E_{n,d}(\boldsymbol{z})$ in [KSW06] for the approximation problem.) Observe that the quantity $S_{N,s}(P)$ depends only on the digital net $P = \{\boldsymbol{x}_0, \ldots, \boldsymbol{x}_{N-1}\}$ and does not depend on the value of M nor the set $\mathcal{A}_s(M)$. Following [DP05b, Proof of Theorem 2]), we can rewrite $S_{N,s}(P)$ in an easily computable form

$$\begin{aligned} S_{N,s}(P) &= -\prod_{j=1}^{s}\left(1+\mu(2\alpha)\gamma_j^2\right) + \frac{1}{N}\prod_{j=1}^{s}\left(1+\mu(\alpha)\gamma_j\right)^2 \\ &\quad + \frac{1}{N}\sum_{n=1}^{N-1}\prod_{j=1}^{s}\left(1+\omega(x_{n,j})\gamma_j\right)^2, \end{aligned}$$

where $\omega(0) = \mu(\alpha)$, and $\omega(x) = \mu(\alpha) - b^{(a-1)(1-\alpha)}(\mu(\alpha)+1)$ if $x \neq 0$ and $\chi_a \neq 0$ is the first nonzero digit in the b-adic representation $x = \sum_{i=0}^{\infty} \chi_i b^{-i}$.

Let p be an irreducible polynomial of degree m. We wish to construct a vector of polynomials $\boldsymbol{q} = (q_1, \ldots, q_s)$ for a polynomial lattice P_{PL}, one polynomial at a time, such that the quantity $S_{N,s}(\boldsymbol{q}) = S_{N,s}(q_1, \ldots, q_s) = S_{N,s}(P_{\mathrm{PL}})$ is as small as possible.

Algorithm 1 *Let $m \geq 1$ and $N = b^m$. Let p be an irreducible polynomial in $\mathbb{Z}_b[x]$ with $\deg(p) = m$.*

1. *Set $q_1 = 1$.*
2. *For $d = 2, 3, \ldots, s$ find q_d in $R_{b,m}$ to minimize $S_{N,s}(q_1, \ldots, q_{d-1}, q_d)$.*

Lemma 9. (cf. [KSW06, Lemma 6]) *Let P_{PL}^* denote the polynomial lattice constructed by Algorithm 1. Then the worst case error for the approximation algorithm $A_{N,s,M}$ using P_{PL}^* satisfies*

$$e_{N,s,M}^{\mathrm{wor-app}}(P_{\mathrm{PL}}^*) \leq \left(\frac{1}{M} + \frac{\widetilde{C}_{s,\lambda,\delta} M}{N^{1/\lambda}} \right)^{1/2}$$

for all $\lambda \in (1/\alpha, 1]$ and $\delta > 0$, where

$$\widetilde{C}_{s,\lambda,\delta} := \frac{1}{\delta} \prod_{j=1}^{s} \left(1 + (1 + \delta^\lambda)\mu(\alpha\lambda)\gamma_j^\lambda\right)^{2/\lambda}.$$

Proof. We prove by induction that the polynomials $q_1^*, \ldots, q_s^*$ for a polynomial lattice P_{PL}^* constructed by Algorithm 1 satisfy, for each $d = 1, \ldots, s$,

$$S_{N,d}(q_1^*, \ldots, q_d^*) \leq \widetilde{C}_{d,\lambda,\delta} N^{-1/\lambda} \tag{28}$$

for all $\lambda \in (1/\alpha, 1]$ and $\delta > 0$. Our proof follows the argument used in the proofs of [DKKS07, Lemma 4] and [KSW06, Lemma 6]. We present here only a skeleton proof; the technical details can be verified in analogy to [DKKS07, KSW06].

The $d = 1$ case can easily be verified. Suppose now that $\boldsymbol{q}^* = (q_1^*, \ldots, q_d^*) \in R_{b,m}^d$ is chosen according to Algorithm 1 and that $S_{N,d}(\boldsymbol{q}^*)$ satisfies (28) for all $\lambda \in (1/\alpha, 1]$ and $\delta > 0$. By separating the $k_{d+1} = 0$ and $k_{d+1} \neq 0$ terms in (26) (with the dual net $\mathcal{D}$ replaced by $\mathcal{D}_{\mathrm{PL}}$ in (16)), we can write

$$S_{N,d+1}(\boldsymbol{q}^*, q_{d+1}) = \phi(\boldsymbol{q}^*) + \theta(\boldsymbol{q}^*, q_{d+1}),$$

where

$$\begin{aligned}\phi(\boldsymbol{q}^*) &= \sum_{h_{d+1} \in \mathbb{N}_0} r^2(\alpha, \gamma_{d+1}, h_{d+1}) \sum_{\boldsymbol{h} \in \mathbb{N}_0^d} \sum_{\substack{\boldsymbol{k} \in \mathbb{N}_0^d \setminus \{\boldsymbol{0}\} \\ \mathrm{tr}_m(\boldsymbol{k}) \cdot \boldsymbol{q}^* \equiv 0 \pmod{p}}} r(\alpha, \boldsymbol{\gamma}, \boldsymbol{h}) r(\alpha, \boldsymbol{\gamma}, \boldsymbol{k} \oplus \boldsymbol{h}) \\ &= \left(1 + \mu(2\alpha)\gamma_{d+1}^2\right) S_{N,d}(\boldsymbol{q}^*),\end{aligned}$$

and

$$\theta(\boldsymbol{q}^*, q_{d+1}) = \sum_{(\boldsymbol{h},h_{d+1})\in\mathbb{N}_0^{d+1}} \sum_{k_{d+1}=1}^{\infty} \sum_{\substack{\boldsymbol{k}\in\mathbb{N}_0^d \\ \widetilde{\mathrm{tr}}_m(\boldsymbol{k})\cdot\boldsymbol{q}^* \equiv -\widetilde{\mathrm{tr}}_m(k_{d+1})\cdot q_{d+1} \pmod p}} r(\alpha,\boldsymbol{\gamma},\boldsymbol{h}) r(\alpha,\gamma_{d+1},h_{d+1})$$
$$\times\, r(\alpha,\gamma_{d+1},k_{d+1}\oplus h_{d+1}) r(\alpha,\boldsymbol{\gamma},\boldsymbol{k}\oplus\boldsymbol{h}).$$

We choose q_{d+1}^* to minimize $S_{N,d+1}(\boldsymbol{q}^*, q_{d+1})$. Then for any $\lambda \in (1/\alpha, 1]$ we have

$$\theta(\boldsymbol{q}^*, q_{d+1}^*) \leq \left(\frac{1}{N-1} \sum_{q_{d+1}\in R_{b,m}} (\theta(\boldsymbol{q}^*, q_{d+1}))^\lambda \right)^{1/\lambda}.$$

After some very long and tedious calculations to estimate this average on the right hand side, with the aid of Jensen's inequality and the property $[r(\alpha,\gamma,h_j)]^\lambda = r(\alpha\lambda,\gamma^\lambda,h_j)$, we finally obtain

$$\theta(\boldsymbol{q}^*, q_{d+1}^*) \leq \left(2\mu(\alpha\lambda)\gamma_{d+1}^\lambda + 4(\mu(\alpha\lambda))^2\gamma_{d+1}^{2\lambda}\right)^{1/\lambda} N^{-1/\lambda} \prod_{j=1}^{d} \left(1+\mu(\alpha\lambda)\gamma_j^\lambda\right)^{2/\lambda}.$$

Hence it follows from the induction hypothesis that

$$S_{N,d+1}(\boldsymbol{q}^*, q_{d+1}^*) \leq \left((1+\mu(2\alpha)\gamma_{d+1}^2) + \delta\left(2\mu(\alpha\lambda)\gamma_{d+1}^\lambda + 4(\mu(\alpha\lambda))^2\gamma_{d+1}^{2\lambda}\right)^{1/\lambda} \right)$$
$$\times\, \delta^{-1} N^{-1/\lambda} \prod_{j=1}^{d} \left(1+(1+\delta^\lambda)\mu(\alpha\lambda)\gamma_j^\lambda\right)^{2/\lambda}.$$

With some elementary inequalities we can show that the multiplying factor in the expression above is bounded by $(1+(1+\delta^\lambda)\mu(\alpha\lambda)\gamma_{d+1}^\lambda)^{2/\lambda}$. This completes the proof.

For $\varepsilon \in (0,1)$, we choose $M = 2\varepsilon^{-2}$ and $N = b^m$ with

$$m = \left\lceil \log_b \left(\widetilde{C}_{s,\lambda,\delta} M^2\right)^\lambda \right\rceil.$$

We have

$$\widetilde{C}_{s,\lambda,\delta} \leq \frac{1}{\delta} \exp\left(\frac{2(1+\delta^\lambda)\mu(\alpha\lambda)}{\lambda} \sum_{j=1}^{s} \gamma_j^\lambda \right)$$
$$= \delta^{-1} (s+1)^{2(1+\delta^\lambda)\mu(\alpha\lambda)\lambda^{-1}\sum_{j=1}^{s}\gamma_j^\lambda/\ln(s+1)}.$$

Let $p^* = 2\max(1/\alpha, s_\gamma)$. When (10) holds we take $\lambda = p^*/2 = 1$ if $s_\gamma = 1$, and $\lambda = p^*/2 + \delta$ if $s_\gamma < 1$. In both cases we have $\sup_{s\geq 1} \widetilde{C}_{s,\lambda,\delta} < \infty$ and $N = \mathcal{O}(\varepsilon^{-p})$, with p equal to or arbitrarily close to $2p^*$ as δ goes to 0. When (11) holds but not (10), we have $s_\gamma = 1$ and we take $\lambda = 1$. Then $N = \mathcal{O}(\varepsilon^{-4})$ and $\widetilde{C}_{s,\lambda,\delta} = \mathcal{O}(s^a)$, with a arbitrarily close to $2\mu(\alpha)\ell$ as δ goes to 0. We summarize the analysis in the following theorem.

Theorem 4. (cf. [KSW06, Theorem 3]) *Let P_{PL}^* be a polynomial lattice constructed component-by-component by Algorithm 1. For $\varepsilon \in (0,1)$ set $M = 2\varepsilon^{-2}$. If* (10) *holds, then the approximation algorithm $A_{N,s,M}$ using P_{PL}^* achieves the worst case error bound $e_{N,s,M}^{\mathrm{wor-app}}(P_{\mathrm{PL}}^*) \leq \varepsilon$ using $N = \mathcal{O}(\varepsilon^{-p})$ function values, with p equal to or arbitrarily close to*

$$2\, p^{\mathrm{wor-app}}(\Lambda^{\mathrm{all}}).$$

If (10) *does not hold but* (11) *holds, then the error bound $e_{N,s,M}^{\mathrm{wor-app}}(P_{\mathrm{PL}}^*) \leq \varepsilon$ is achieved using $N = \mathcal{O}(s^a \varepsilon^{-4})$ function values, with a arbitrarily close to $2\mu(\alpha)\ell$. The implied factors in the big $\mathcal{O}$-notations are independent of ε and s.*

Observe that when (10) holds we have $p^{\mathrm{wor-app}}(\Lambda^{\mathrm{all}}) \leq 2$. Therefore

$$2\, p^{\mathrm{wor-app}}(\Lambda^{\mathrm{all}}) \leq p^{\mathrm{wor-app}}(\Lambda^{\mathrm{all}}) + 2,$$

and we have improved the result in Theorem 1 using a fully constructive argument.

Remark. (cf. Theorem 1) *The exponent of strong tractability in the class Λ^{std} satisfies*

$$p^{\mathrm{wor-app}}(\Lambda^{\mathrm{std}}) \in [p^{\mathrm{wor-app}}(\Lambda^{\mathrm{all}}), 2\, p^{\mathrm{wor-app}}(\Lambda^{\mathrm{all}})].$$

Acknowledgments

The second author is supported by the Austrian Science Foundation (FWF), Project S9609, which is part of the Austrian National Research Network "Analytic Combinatorics and Probabilistic Number Theory". The first and third authors were supported by University of New South Wales Vice-chancellor's Postdoctoral Research Fellowships.

References

[Aro50] N. Aronszajn. Theory of reproducing kernels. *Trans. Amer. Math. Soc.*, 68:337–404, 1950.

[CDP06] L. L. Cristea, J. Dick, and F. Pillichshammer. On the mean square weighted $\mathcal{L}_2$ discrepancy of randomized digital nets in prime base. *J. Complexity*, 22:605–629, 2006.

[Chr55] H.E. Chrestenson. A class of generalized Walsh functions. *Pacific J. Math.*, 5:17–31, 1955.

[DKKS07] J. Dick, P. Kritzer, F. Y. Kuo, and I. H. Sloan. Lattice-Nyström method for Fredholm integral equations of the second kind with convolution type kernels. *J. Complexity*, in press, doi:10.1016/j.jco.2007.03.004.

[DKPS05] J. Dick, F. Y. Kuo, F. Pillichshammer, and I. H. Sloan. Construction algorithms for polynomial lattice rules for multivariate integration. *Math. Comp.*, 74:1895–1921, 2005.

[DP05a] J. Dick and F. Pillichshammer. Dyadic diaphony of digital nets over $\mathbb{Z}_2$. *Monatsh. Math.*, 145:285–299, 2005.

[DP05b] J. Dick and F. Pillichshammer. Multivariate integration in weighted Hilbert spaces based on Walsh functions and weighted Sobolev spaces. *J. Complexity*, 21:149–195, 2005.

[DP05c] J. Dick and F. Pillichshammer. On the mean square weighted L_2 discrepancy of randomized digital (t, m, s)-nets over $\mathbb{Z}_2$. *Acta Arith.*, 117:371–403, 2005.

[HN03] F. J. Hickernell and H. Niederreiter. The existence of good extensible rank-1 lattices. *J. Complexity*, 19:286–300, 2003.

[Kor63] N. M. Korobov. *Number-theoretic Methods in Approximate Analysis*. Moscow: Fizmatgiz, 1963.

[KSW06] F. Y. Kuo, I. H. Sloan, and H. Woźniakowski. Lattice rules for multivariate approximation in the worst case setting. In H. Niederreiter and D. Talay, editors, *Monte Carlo and Quasi-Monte Carlo Methods 2004*, pages 289–330. Berlin: Springer, 2006.

[KSW07] F. Y. Kuo, I. H. Sloan, and H. Woźniakowski. Lattice rule algorithms for multivariate approximation in the average case setting. *J. Complexity*, in press, doi:10.1016/j.jco.2006.10.006.

[Lar98] G. Larcher. Digital point sets: Analysis and application. In P. Hellekalek and G. Larcher, editors, *Random and Quasi-Random Point Sets*, volume 138 of *Lecture Notes in Statistics*, pages 167–222. New York: Springer, 1998.

[LH03] D. Li and F. J. Hickernell. Trigonometric spectral collocation methods on lattices.In S. Y. Cheng, C.-W. Shu, and T. Tang, editors, *Recent Advances in Scientific Computing and Partial Differential Equations*, volume 330 of *AMS Series in Contemporary Mathematics*, pages 121–132. Providence: American Mathematical Society, 2003.

[LNS96] G. Larcher, H. Niederreiter, and W. Ch. Schmid. Digital nets and sequences constructed over finite rings and their application to quasi-Monte Carlo integration.*Monatsh. Math.*, 121:231–253, 1996.

[Nie87] H. Niederreiter. Point sets and sequences with small discrepancy. *Monatsh. Math.*, 104:273–337, 1987.

[Nie88] H. Niederreiter. Low-discrepancy and low-dispersion sequences. *J. Number Theory*, 30:51–70, 1988.

[Nie92a] H. Niederreiter. Low-discrepancy point sets obtained by digital constructions over finite fields. *Czechoslovak Math. J.*, 42:143–166, 1992.

[Nie92b] H. Niederreiter. *Random Number Generation and Quasi-Monte Carlo Methods*, volume 63 of *CBMS-NSF Series in Applied Mathematics*. Philadelphia: SIAM, 1992.

[Nie05] H. Niederreiter. Constructions of (t, m, s)-nets and (t, s)-sequences. *Finite Fields Appl.*, 11:578–600, 2005.

[NSW04] E. Novak, I. H. Sloan, and H. Woźniakowski. Tractability of approximation for weighted Korobov spaces on classical and quantum computers. *Found. Comput. Math.*, 4:121–156, 2004.

[PS01] G. Pirsic and W. Ch. Schmid. Calculation of the quality parameter of digital nets and application to their construction. *J. Complexity*, 17:827–839, 2001.

[SW98] I. H. Sloan and H. Woźniakowski. When are quasi-Monte Carlo algorithms efficient for high dimensional integrals?. *J. Complexity*, 14:1–33, 1998.

[SW01] I. H. Sloan and H. Woźniakowski. Tractability of multivariate integration for weighted Korobov classes. *J. Complexity*, 17:697–721, 2001.

[TWW88] J. F. Traub, G. W. Wasilkowski, and H. Woźniakowski. *Information-Based Complexity*. New York: Academic Press, 1988.

[Wal23] J. L. Walsh. A closed set of normal orthogonal functions. *Amer. J. Math.*, 45:5–24, 1923.

[Wan02] X. Wang. Strong tractability of multivariate integration using quasi-Monte Carlo algorithms. *Math. Comp.*, 72:823–838, 2002.

[WW99] G. W. Wasilkowski and H. Woźniakowski. Weighted tensor product algorithms for linear multivariate problems. *J. Complexity*, 15:402–447, 1999.

[WW01] G. W. Wasilkowski and H. Woźniakowski. On the power of standard information for weighted approximation. *Found. Comput. Math.*, 1:417–434, 2001.

[ZLH06] X. Y. Zeng, K. T. Leung, and F. J. Hickernell. Error analysis of splines for periodic problems using lattice designs. In H. Niederreiter and D. Talay, editors, *Monte Carlo and Quasi-Monte Carlo Methods 2004*, pages 501–514. Berlin: Springer, 2006.

Construction of Low-Discrepancy Point Sets of Small Size by Bracketing Covers and Dependent Randomized Rounding

Benjamin Doerr[1] and Michael Gnewuch[2]

[1] Max-Planck-Institut für Informatik, Stuhlsatzenhausweg 85, 66123 Saarbrücken, Germany
doerr@mpi-sb.mpg.de

[2] Institut für Informatik, Christian-Albrechts-Universität zu Kiel, Christian-Albrechts-Platz 4, 24098 Kiel, Germany
mig@informatik.uni-kiel.de

In memory of our friend, colleague and former fellow student Manfred Schocker

Summary. We provide a deterministic algorithm that constructs small point sets exhibiting a low star discrepancy. The algorithm is based on bracketing and on recent results on randomized roundings respecting hard constraints. It is structurally much simpler than the previous algorithm presented for this problem in [B. Doerr, M. Gnewuch, A. Srivastav. Bounds and constructions for the star discrepancy via δ-covers. J. Complexity, 21:691–709, 2005]. Besides leading to better theoretical run time bounds, our approach also can be implemented with reasonable effort.

1 Introduction

The L^∞*-star discrepancy* or, more shortly, *star discrepancy* of an n-point set T in the d-dimensional unit cube $[0,1]^d$ is given by

$$d_\infty^*(T) := \sup_{x\in[0,1]^d} \left| \frac{1}{n}|T \cap [0,x[| - \mathrm{vol}([0,x[) \right| ,$$

where $[0,x[$ is the d-dimensional anchored half-open box $[0,x_1[\times \ldots \times [0,x_d[$. Here, as in the whole article, the cardinality of a finite set S is denoted by $|S|$ and the ith component of a vector x by x_i. The smallest possible discrepancy of any n-point configuration in $[0,1]^d$ is

$$d_\infty^*(n,d) := \inf_{T\subset[0,1]^d\,;\,|T|=n} d_\infty^*(T).$$

The *inverse of the star discrepancy* is given by

$$n^*_\infty(\varepsilon, d) := \min\{n \in \mathbb{N} \,|\, d^*_\infty(n,d) \le \varepsilon\}\,.$$

The star discrepancy is related to the worst case error of multivariate integration of a certain class of functions by the Koksma-Hlawka inequality (see, e.g., [DT97, HSW04, Nie92]). The inequality shows that points with small star discrepancy induce quasi-Monte Carlo algorithms with small worst case errors. Since the number of sample points is roughly proportional to the costs of those algorithms, it is of interest to find n-point configurations with small discrepancy and n not too large. In particular, n should not depend exponentially on d.

For fixed dimension d the asymptotically best upper bounds for $d^*_\infty(n,d)$ that have been proved so far are of the form

$$d^*_\infty(n,d) \le C_d \ln(n)^{d-1} n^{-1}\,, \quad n \ge 2\,. \tag{1}$$

These bounds give us no helpful information for moderate values of n, since $\ln(n)^{d-1}n^{-1}$ is an increasing function for $n \le e^{d-1}$. Additionally, point configurations satisfying (1) will in general lead to constants C_d that depend critically on d. (Actually, it is known for some constructions that the constant C'_d in the representation

$$d^*_\infty(n,d) \le \left(C'_d \ln(n)^{d-1} + o(\ln(n)^{d-1})\right) n^{-1}$$

of (1) tends to zero as d approaches infinity, see, e.g., [Nie92, NX96, Ata04]. But as far as we know, no good bounds have been published for the implicit constant of the o-notation or, respectively, the "whole" constant C_d in (1).)

A bound more suitable for high-dimensional integration was established by Heinrich, Novak, Wasilkowski and Woźniakowski [HNWW01], who proved

$$d^*_\infty(n,d) \le c d^{1/2} n^{-1/2} \quad \text{and} \quad n^*_\infty(d,\varepsilon) \le \lceil c^2 d \varepsilon^{-2} \rceil\,, \tag{2}$$

where c does not depend on d, n or ε. Here the dependence of the inverse of the star discrepancy on d is optimal. This was also established in [HNWW01] by a lower bound for $n^*_\infty(d,\varepsilon)$, which was later improved by Hinrichs [Hin04] to $n^*_\infty(d,\varepsilon) \ge c_0 d \varepsilon^{-1}$ for $0 < \varepsilon < \varepsilon_0$, where $c_0, \varepsilon_0 > 0$ are constants. The proof of (2) is not constructive but probabilistic, and the proof approach does not provide an estimate for the value of c. (A. Hinrichs presented a more direct approach to prove (2) with $c = 10$ at the Dagstuhl Seminar 04401 "Algorithms and Complexity for Continuous Problems" in 2004.)

In the same paper the authors proved a slightly weaker bound with an explicitly known small constant k:

$$d^*_\infty(n,d) \le k d^{1/2} n^{-1/2} \big(\ln(d) + \ln(n)\big)^{1/2}\,. \tag{3}$$

The proof is again probabilistic and uses Hoeffding's inequality. (A similar probabilistic approach was already used by Beck in [Bec84] to prove upper bounds for other kinds of geometric discrepancy.) For the sake of explicit constants the proof technique has been adapted in subsequent papers on high-dimensional integration of certain function classes [HSW04, Mha04]. In [DGS05] Srivastav and the authors were able to improve (3) to

$$d^*_\infty(n,d) \leq k' d^{1/2} n^{-1/2} \ln(n)^{1/2}\,, \tag{4}$$

where k' is smaller than k. (A slightly better bound for the star discrepancy and a corresponding bound for the so-called extreme discrepancy can be found in [Gne07].) Of course the estimate (4) is asymptotically not as good as (2). But the constant k' is small—essentially we have $k' = \sqrt{2}$. If, e.g., $c = 10$, then (4) is superior to (2) for all n that are roughly smaller than e^{50}, i.e., for all values of n of practical interest. By derandomizing the probabilistic argument used in the proof of the inequality (4), the authors and Srivastav additionally gave a deterministic algorithm constructing point sets satisfying (4). The algorithm is based on a quite general derandomization approach of Srivastav and Stangier [SS96] and essentially a point-by-point construction using the method of conditional probabilities and so-called pessimistic estimators.

Our Results

In this paper, we use a novel approach to randomized rounding presented in [Doe06]. Contrary to the classical one, it allows to generate randomized roundings that respect certain hard constraints. This enables us to use a construction that needs significantly fewer random variables, which in turn speeds up the randomized construction.

A second speed-up and considerable simplification from the implementational point of view stems from the fact that the general approach in [Doe06] may be derandomized via the more restricted approach of Raghavan [Rag88]. This runs in time $O(mn)$, where n is the number of (random) variables and m the number of constraints.

It thus avoids the general, but more costly solution by Srivastav and Stangier [SS96]. The latter was a break-through from the theoretical point of view as it showed that randomized rounding for arbitrary linear constraints can be derandomized. From the practical point of view, it suffers from a higher run-time of $O(mn^2 \log(mn))$ and its extremely high technical demands. To the best of our knowledge, the algorithm implicit in the 30 pages proof has never been implemented.

We show the following result. For a given $n \in \mathbb{N}$ the algorithm computes an n-point set T with discrepancy

$$d^*_\infty(T) \leq (4+\sqrt{3})\sqrt{n^{-1}\left(\tfrac{1}{2}d\ln(\sigma n) + \ln 2\right)} + 2^{-d\ln(dn)-1}n^{-1}$$

in time $O(d(\sigma n)^d \log(dn))$. Here $\sigma = \sigma(d)$ is less than one and converges to zero as d tends to infinity. In [DGS05] the running time for constructing an n-point set with the same discrepancy order was $O(C^d n^{d+2} \log(d)^d / \log(n)^{d-1})$, C some constant. That the running times of our deterministic algorithms are exponential in d may not be too surprising. Already any deterministic algorithm known so far that approximates the L^∞-star discrepancy of arbitrary given n-point sets has running time exponential in d (see [Thi01], the literature mentioned therein, and the discussion in [Gne07]). A comparison with other deterministic algorithms for constructing low-discrepancy sets of small size can be found in [DGS05].

Let us stress that the main advance in this paper is providing a simple solution (even though we are also faster than the previous one). We feel that our solution can be implemented with reasonable effort (and our future research will include this implementation project). Since often the quality of computed solutions is much better than what is guaranteed by theoretical worst-case error bounds, our solution presented in this paper opens an interesting line of research.

2 Randomized Construction

We start by introducing some useful notation: For arbitrary $n \in \mathbb{N}$ put $[n] := \{1, \ldots, n\}$. If $x, y \in [0,1]^d$, we write $x \leq y$ if $x_i \leq y_i$ holds for all $i \in [d]$. We write $[x, y] = \prod_{i \in [d]} [x_i, y_i]$ and use corresponding notation for open and half-open intervals. For a point $x \in [0,1]^d$ we denote by V_x the volume of the box $[0, x]$. Similarly, we denote the volume of a Lebesgue measurable subset S of $[0,1]^d$ by V_S.

2.1 Grids and Covers

Let $0 = q_0 < q_1 < \ldots < q_k = 1$ and $G := \{q_i \,|\, 1 \leq i \leq k\}^d$. G is a (not necessarily equidistant) grid in the d-dimensional unit cube $[0,1]^d$. Let $\delta = \delta(G)$ be the smallest real number such that for all $y \in [0,1]^d$ there are $x, z \in G \cup \{0\}$ with $x \leq y \leq z$ and $V_z - V_x \leq \delta$. In the language of [DGS05] δ is minimal such that G is a δ–cover. Let us restate the definition from [DGS05]:

A finite set $\Gamma \subset [0,1]^d$ is a *δ-cover* of $[0,1]^d$ if for every $y \in [0,1]^d$ there exist $x, z \in \Gamma \cup \{0\}$ with $V_z - V_x \leq \delta$ and $x \leq y \leq z$. Essentially the same concept is known in the literature of empirical processes as *bracketing*, see also [Gne07].

The helpfulness of δ-covers in discrepancy theory lies in the fact that one can use them to discretize discrepancy while controlling the discretization error:

Lemma 1. *Let Γ be a δ-cover of $[0,1]^d$. Then for all n-point sets $T \subset [0,1]^d$*

$$d^*_\infty(T) \leq d^*_\Gamma(T) + \delta\,, \quad \textit{where} \quad d^*_\Gamma(T) := \max_{x \in \Gamma} \left| \frac{1}{n} |T \cap [0, x[| - V_x \right|. \tag{5}$$

The proof is straightforward and can, e.g., be found in [DGS05].

Let now $\mathcal{I} := \{[q_{i-1}, q_i[\mid 1 \leq i \leq k\}$ and $\mathcal{B} := \{\prod_{i=1}^d I_i \mid I_1, \ldots, I_d \in \mathcal{I}\}$. Note that $\mathcal{B}$ is a partition of $[0,1[^d$ into axis-parallel boxes with upper right corners in G. Let $\mathcal{C}_0 := \{[0, g[\mid g \in G\}$. $\mathcal{C}_0$ is a subset of the set $\mathcal{C}$ of all axis-parallel boxes that are anchored in 0 (these boxes are sometimes called *corners*). If $g \in G$, let $B(g)$ be the uniquely determined $B \in \mathcal{B}$ and $C(g)$ the uniquely determined $C \in \mathcal{C}_0$ whose upper right corners are g. Furthermore, let $\mathcal{B}(g) := \{B \in \mathcal{B} \mid B \subseteq [0, g[\}$. If $B = B(g)$ or $C = C(g)$, we denote $\mathcal{B}(g)$ also by $\mathcal{B}(B)$ or $\mathcal{B}(C)$ respectively. Note that $(G, \leq)$ is a partially ordered set and that via the identification of elements from $\mathcal{B}$ and $\mathcal{C}_0$ with their upper right corners we get induced partial orderings on $\mathcal{B}$ and $\mathcal{C}_0$ respectively. Let us denote all these orderings simply by $\leq$. Finally, denote for a given $B \in \mathcal{B}$ the smallest set in $\mathcal{C}_0$ that contains B by $C(B)$. That is, B and $C(B)$ share the same upper right corner.

2.2 Reducing the Binary Length

Our aim is to construct n points in the unit cube exhibiting a fairly good discrepancy. We proceed as follows. For $B \in \mathcal{B}$, let $x_B := n \operatorname{vol}(B)$ be the fair number of points to lie in B.

We first round the x_B, $B \in \mathcal{B}$, to non-negative numbers having a finite binary expansion. Having numbers with finite binary expansion is necessary in the subsequent rounding step, but also carries the advantages of allowing (from this point on) efficient and exact computations.

Using a simple rounding approach, we could obtain $(\tilde{x}_B)$ such that $2^\ell \tilde{x}_B \in \mathbb{Z}$, $|x_B - \tilde{x}_B| < 2^{-\ell}$ and $\sum_{B \in \mathcal{B}} \tilde{x}_B = \sum_{B \in \mathcal{B}} x_B = n$. This yields a rounding error of $|\sum_{B \in \mathcal{B}(C)} (x_B - \tilde{x}_B)| < 2^{-\ell} |\mathcal{B}(C)|$ in all corners $C \in \mathcal{C}_0$.

However, we can achieve much smaller rounding errors by using a very recent result of Güntürk, Yılmaz and the first author [DGY06]. Let us remark first that there are higher-dimensional matrices $A \in [0,1]^{[k]^d}$ such that for any roundings $B \in \{0,1\}^{[k]^d}$ there is an $x \in [k]^d$ such that the rounding error

$$\left| \sum_{i_1=1}^{x_1} \cdots \sum_{i_d=1}^{x_d} (a_{i_1,\ldots,i_d} - b_{i_1,\ldots,i_d}) \right|$$

is of order $\Omega((\log k)^{(d-1)/2})$. Hence if we round the x_B to multiples of $2^{-\ell}$, we may get rounding errors $|\sum_{B \in \mathcal{B}(C)} (x_B - \tilde{x}_B)|$ of order $\Omega(2^{-\ell} (\log k)^{(d-1)/2})$. This follows from a result of Beck [Bec81], which in turn relies heavily on lower bounds for geometric discrepancies (Roth [Rot64], Schmidt [Sch72]). We refer to [Doe07] for a more extensive discussion of these connections.

The surprising result of [DGY06] is that by allowing larger deviations in the variables we can guarantee much better errors in the subarrays. In particular, we are able to remove any dependence on the grid size k. To ease reading, we reformulate and prove their result in our language.

Lemma 2. *Let $\ell \in \mathbb{N}$. There is a simple $O(|\mathcal{B}|)$ time algorithm computing $(\tilde{x}_B)$ such that*

(i) $2^\ell \tilde{x}_B \in \mathbb{Z}$ *for all* $B \in \mathcal{B}$;
(ii) $|x_B - \tilde{x}_B| \le 2^{-\ell-1+d}$ *for all* $B \in \mathcal{B}$;
(iii) $\sum_{B\in\mathcal{B}} \tilde{x}_B = \sum_{B\in\mathcal{B}} x_B = n$;
(iv) $|\sum_{B\in\mathcal{B}(C)}(x_B - \tilde{x}_B)| \le 2^{-\ell-1}$ *for all* $C \in \mathcal{C}_0$.

To assure that all quantities $\tilde{x}_B$, $B \in \mathcal{B}$, are non-negative, we will later choose $\ell \ge d - 1 + \log_2(\max_{B\in\mathcal{B}} x_B^{-1})$.

Proof. We may sort $\mathcal{B}$ in a way that B_1 is prior to B_2 if $B_1 \le B_2$. In this order, we traverse (x_B) and choose an $\tilde{x}_B$ satisfying $2^\ell \tilde{x}_B \in \mathbb{Z}$ and

$$-2^{-\ell-1} < \sum_{B'\in\mathcal{B}(B)} (\tilde{x}_{B'} - x_{B'}) \le 2^{-\ell-1}\,. \tag{6}$$

That such an $\tilde{x}_B$ exists follows easily: If we have to choose $\tilde{x}_B$ for some $B \in \mathcal{B}$, then, due to our sorting, for all $B' \in \mathcal{B}(B)$, $B' \neq B$, the value of $\tilde{x}_{B'}$ has already been fixed. Now take the uniquely determined integer y_B satisfying

$$-\frac{1}{2} < y_B - \left(2^\ell x_B - 2^\ell \sum_{B\neq B'\in\mathcal{B}(B)} (\tilde{x}_B - x_{B'})\right) \le \frac{1}{2}\,.$$

Then $\tilde{x}_B := 2^{-\ell} y_B$ satisfies (6) (and is actually uniquely determined). We may express any B by unions and differences of at most 2^d corners $C(\tilde{B})$. Hence we can write $\tilde{x}_B - x_B$ as sum and difference of at most 2^d terms $\sum_{B'\in\mathcal{B}(\tilde{B})}(\tilde{x}_{B'} - x_{B'})$, resulting in $|\tilde{x}_B - x_B| \le 2^{-\ell-1+d}$.

Finally, from $\sum_{B\in\mathcal{B}} \tilde{x}_B \in 2^{-\ell}\mathbb{Z}$, $\sum_{B\in\mathcal{B}} x_B = n \in \mathbb{Z}$ and $|\sum_{B\in\mathcal{B}}(x_B - \tilde{x}_B)| \le 2^{-\ell-1}$, we conclude $\sum_{B\in\mathcal{B}} \tilde{x}_B = \sum_{B\in\mathcal{B}} x_B = n$.

2.3 Randomized Rounding with Cardinality Constraint

We now randomly round $(\tilde{x}_B)$ to integers (y_B) and then choose our point set in a way that it has exactly y_B points in the box B. This rounding is done via a recent extension of the classical randomized rounding method due to Raghavan [Rag88]. We briefly review the basics.

Randomized Rounding

For a number r we write $\lfloor r \rfloor = \max\{z \in \mathbb{Z} \,|\, z \le r\}$, $\lceil r \rceil = \min\{z \in \mathbb{Z} \,|\, z \ge r\}$ and $\{r\} = r - \lfloor r \rfloor$. Let $\xi \in \mathbb{R}$. An integer-valued random variable y is called *randomized rounding of* ξ if

$$\Pr(y = \lfloor \xi \rfloor + 1) = \{\xi\},$$
$$\Pr(y = \lfloor \xi \rfloor) = 1 - \{\xi\}.$$

Since only the fractional part of ξ is relevant, we often may ignore the integer part and then have $\xi \in [0,1]$. In this case, a randomized rounding y of ξ satisfies

$$\Pr(y = 1) = \xi,$$
$$\Pr(y = 0) = 1 - \xi.$$

For $\xi \in \mathbb{R}^n$, we call $y = (y_1, \ldots, y_n)$ *randomized rounding of* ξ if y_j is a randomized rounding of ξ_j for all $j \in [n]$. We call y *independent randomized rounding* of ξ, if the y_i are mutually independent random variables.

Independent randomized rounding was introduced by Raghavan [Rag88] and since has found numerous applications. It takes its strength from the fact that sums of independent random variables are strongly concentrated around their mean. This allows to bound the deviation of a weighted sum of the ξ_i from the corresponding sum of the y_i (this is done via so-called Chernoff bounds).

Independent randomized rounding can be derandomized. That is, one can transform the above sketched approach into a deterministic rounding algorithm (at the price of a slightly higher run-time) that guarantees large deviation bounds comparable to those that randomized rounding satisfies with high probability.

For our purposes, independent randomized rounding is not fully satisfactory since we would like to construct exactly n points. In other words, we prefer to have $\sum_{B\in\mathcal{B}} y_B = \sum_{B\in\mathcal{B}} \tilde{x}_B = n$ without any deviation. Fortunately, this can be achieved relatively easy with the randomized rounding method proposed by the first author in [Doe06]. It allows to generate randomized roundings that always fulfill constraints like $\sum_{B\in\mathcal{B}} y_B = \sum_{B\in\mathcal{B}} \tilde{x}_B = n$. In consequence, this is not independent randomized rounding. However, though not being independent, these roundings still satisfy Chernoff bounds and can be derandomized. The following makes this precise.

Theorem 1 ([Doe06]). *Let $\xi \in \mathbb{R}^N$ such that all ξ_i have binary length at most ℓ and $\sum_{i=1}^N \xi_i \in \mathbb{N}$. Then in time $O(\ell N)$ a randomized rounding y of ξ can be generated such that* $\Pr(\sum_{i=1}^N y_i = \sum_{i=1}^N \xi_i) = 1$ *and for all $a \in [0,1]^N$, $Y := \sum_{i=1}^N a_i y_i$, $\mu := E(Y) = \sum_{i=1}^N a_i \xi_i$ and all $\delta \in [0,1]$,*

$$\Pr(Y \geq (1+\delta)\mu) \leq \exp(-\tfrac{1}{3}\mu\delta^2),$$
$$\Pr(Y \leq (1-\delta)\mu) \leq \exp(-\tfrac{1}{2}\mu\delta^2).$$

The Chernoff bounds given above are not strongest possible. Bounds like $\Pr(Y \geq (1+\delta)\mu) \leq \left(\frac{e^\delta}{(1+\delta)^{(1+\delta)}}\right)^\mu$ would also hold, but are often not practical to work with. Since our roundings do not change the sum of all variables, we also have the following bound, which often is easier to handle.

Lemma 3. *In the setting of Theorem 1, assume further that ξ is non-negative and $n := \sum_{i=1}^{N} \xi_i$. Then for all $\lambda \geq 0$, we have*

$$\Pr(|Y - \mu| \geq \lambda) \leq 2\exp(-\tfrac{1}{3}\lambda^2/n).$$

Proof. We may assume $\lambda \leq n$, as Y never exceeds n by non-negativity of ξ. Let $\xi_{N+1} = \lceil n - \mu \rceil$ and $a_{N+1} = (n-\mu)/\lceil n - \mu \rceil$. Note that since ξ_{N+1} is integral, any randomized rounding of $\xi_1, \ldots, \xi_{N+1}$ as in Theorem 1 yields a randomized rounding for $\xi_1, \ldots, \xi_N$ as in Theorem 1 (by just forgetting the $(N+1)$-st variable) and vice versa (by taking $y_{N+1} = \xi_{N+1}$ with probability one). Hence we need not to distinguish between the two.

Let $\tilde{Y} = \sum_{i=1}^{N+1} a_i y_i$ and $\tilde{\mu} = E(\tilde{Y})$. Note that by construction, $\tilde{\mu} = n$. Hence with $\delta = \lambda/n$, the first bound of Theorem 1 yields

$$\begin{aligned}
\Pr(Y - \mu \geq \lambda) &= \Pr(\tilde{Y} - \tilde{\mu} \geq \lambda) \\
&= \Pr(\tilde{Y} \geq (1+\delta)\tilde{\mu}) \\
&\leq \exp(-\tfrac{1}{3}n\delta^2) \;=\; \exp(-\tfrac{1}{3}\lambda^2/n).
\end{aligned}$$

The second bound of Theorem 1 analogously yields $\Pr(-(Y-\mu) \geq \lambda) \leq \exp(-\frac{1}{3}\lambda^2/n)$. Both estimates give this lemma.

Construction of the Point Set

We use the theorem above to generate random variables (y_B) as randomized roundings of $(\tilde{x}_B)$. Since we required the $\tilde{x}_B$, $B \in \mathcal{B}$, to be non-negative, the y_B, $B \in \mathcal{B}$, are non-negative integers. Let T be an n-point set in the unit cube such that for all $B \in \mathcal{B}$, T contains exactly y_B points in B.

Lemma 4. *Let $C \in \mathcal{C}_0$. Then for all non-negative λ we have*

$$\Pr\left(\big||C \cap T| - nV_C\big| > \lambda + 2^{-\ell-1}\right) \leq 2\exp\left(-\frac{\lambda^2}{3n}\right).$$

Proof. By construction, we have $|C \cap T| = \sum_{B \in \mathcal{B}(C)} y_B$ and $nV_C = \sum_{B \in \mathcal{B}(C)} x_B$. From Lemma 2 we get

$$\big||C \cap T| - nV_C\big| = \left|\sum_{B \in \mathcal{B}(C)} (y_B - x_B)\right| \leq \left|\sum_{B \in \mathcal{B}(C)} (y_B - \tilde{x}_B)\right| + 2^{-\ell-1}.$$

Put $Y := \sum_{B \in \mathcal{B}(C)} y_B$. Since the $\tilde{x}_B$ are non-negative and $n = \sum_{B \in \mathcal{B}} \tilde{x}_B$, we get from Lemma 3

$$\Pr\left(\left|\sum_{B \in \mathcal{B}(C)} (y_B - \tilde{x}_B)\right| \geq \lambda\right) \leq 2\exp\left(-\frac{\lambda^2}{3n}\right).$$

Theorem 2. *Let T be as above. Let $\delta \in]0,1]$. For all $\theta \in [1,\infty[$ we have*

$$\Pr\left(d_\infty^*(T) > \sqrt{3n^{-1}\ln(2\theta|\mathcal{B}|)} + \delta + 2^{-\ell-1}n^{-1}\right) \le \theta^{-1}. \tag{7}$$

Proof. By Lemma 1, we have $d_\infty^*(T) \le d_G^*(T) + \delta$. (Of course, G should be a grid as in Subsection 2.1.) Choosing $\lambda = \sqrt{3n\ln(2\theta|\mathcal{B}|)}$, we deduce from Lemma 4 that

$$\Pr\left(\big||C\cap T| - nV_C\big| > \lambda + 2^{-\ell-1}\right) \le (\theta|\mathcal{B}|)^{-1} \quad \text{for all } C \in \mathcal{C}_0.$$

Hence, since $|\mathcal{C}_0| = |\mathcal{B}|$,

$$\Pr(d_G^*(T) > (\lambda + 2^{-\ell-1})/n) \le \sum_{C\in\mathcal{C}_0} \Pr\left(\big||C\cap T| - nV_C\big| > \lambda + 2^{-\ell-1}\right) \le \theta^{-1}.$$

Choice of Parameters

Note that inequality (7) depends on the parameters θ, ℓ and δ (in particular, $|\mathcal{B}|$ depends on δ). In the following we make some reasonable choices for these parameters to get a version of inequality (7) that only depends on d and n.

Let $d \ge 2$. In [DGS05, Thm.2.3] a δ-cover in form of a non-equidistant grid $G = \{q_1, \ldots, q_k\}^d$ was constructed satisfying

$$k = \left\lceil \frac{d}{d-1}\,\frac{\ln(1-(1-\delta)^{1/d}) - \ln\delta}{\ln(1-\delta)} \right\rceil + 1 \le \left\lceil \frac{d}{d-1}\,\frac{\ln d}{\delta} \right\rceil + 1.$$

The explicit construction goes as follows: Put $p_0 := 1$ and $p_1 := (1-\delta)^{1/d}$. If $p_i > \delta$, then define $p_{i+1} := (p_i - \delta)p_1^{1-d}$. If $p_{i+1} \le \delta$, then put $\kappa(\delta, d) := i+1$, otherwise proceed by calculating p_{i+2}. Then $k = \kappa(\delta, d) + 1$ and $q_{k-i} = p_i$.

For this grid G and $\delta \le 1/2$ we get

$$\ln|\mathcal{B}| = \ln|G| = d\ln k \le d(\ln\delta^{-1} + \ln\ln d + \ln 4)\,.$$

Choosing

$$\delta = \left(3n^{-1}(d\,(\ln\ln d + \ln 8) + \ln 4)\right)^{1/2} \tag{8}$$

leads to

$$\ln|\mathcal{B}| \le \frac{d}{2}\ln(\sigma n)\,, \quad \text{where} \quad \sigma = \sigma(d) := \frac{16(\ln d)^2}{3(d(\ln\ln d + \ln 8) + \ln 4)}\,. \tag{9}$$

An elementary analysis shows that σ takes its maximum in $d = 6$ and therefore $\max_{d\ge 2}\sigma(d) < 0.9862$. In the table below we listed some values of σ.

d	$\sigma(d)$	d	$\sigma(d)$
2	0.5324886424	20	0.7528969387
3	0.8141209699	30	0.6139386902
4	0.9308908286	40	0.5306094834
5	0.9754256341	50	0.4702720050
6	0.9861774970	60	0.4242704800
7	0.9802221264	70	0.3878425250
8	0.9657904472	80	0.3581541672
9	0.9471133088	90	0.3334088723
10	0.9264689299	100	0.3124089382
11	0.9051224430	360	0.1331152560
12	0.8837875877	1000	0.0634092061

Let us assume that for a given dimension $d \geq 2$ the number of points n is large enough to imply $\delta \leq 1/2$. Then, for $\theta = 2$,

$$\sqrt{3n^{-1}\ln(2\theta|\mathcal{B}|)} + \delta \leq 2\sqrt{3n^{-1}\left(\frac{d}{2}\ln(\sigma n) + \ln 4\right)}\,.$$

Now let us specify the choice of ℓ. Due to our choice of the δ-cover G we may assume that

$$\min_{B\in\mathcal{B}} V_B > \frac{\delta^d}{d^d}\,. \tag{10}$$

(It is easy to see that $q_k - q_{k-1} > \delta/d$ and that $q_i - q_{i-1}$, $i = 2,\ldots,k$, is a strictly monotonic decreasing sequence. Nevertheless, (10) does not hold in the situation where q_1 is almost zero. In this case we substitute $q_1 := q_2/2$. It is easy to see that the new grid G is still a δ-cover. Then $q_1 > \delta/2$ and therefore $q_i - q_{i-1} > \delta/d$ for all $i = 1,\ldots,k$.)

To guarantee that our quantities $\tilde{x}_B$, $B \in \mathcal{B}$, from Lemma 2 are non-negative, we choose ℓ such that $2^{-\ell+d-1} \leq \delta^d/d^d$. According to our choice of δ in (8), the last inequality holds if

$$\ell \geq \left\lceil \frac{d}{2\ln 2}\ln\left(\frac{2}{3}dn\right)\right\rceil - 1\,.$$

For simplicity we choose

$$\ell = \lceil d\ln(dn)\rceil.$$

Our choices of G, δ, θ and ℓ result in the following corollary.

Corollary 1. *Let G, δ, θ and ℓ be chosen as above, and let $\sigma = \sigma(d) = \frac{16(\ln d)^2}{3(d(\ln\ln d + \ln 8) + \ln 4)}$ be as above. Then*

$$\Pr\left(d^*_\infty(T) > 2\sqrt{3n^{-1}\left(\tfrac{1}{2}d\ln(\sigma n) + \ln 4\right)} + 2^{-d\ln(dn)-1}n^{-1}\right) \leq \frac{1}{2}. \tag{11}$$

Remark 1. Note that above we were using a non-equidistant grid as δ-cover. In [DGS05, Gne07], also δ-covers were constructed that had no grid structure (by a grid, we shall always mean a point set G in $[0,1]^d$ that can be written as $G = (G_0)^d$ for some $G_0 \subset [0,1]$). These δ-covers were superior in the sense that they needed fewer points. For the approach we use in this paper, however, they cannot be applied. The reason is that in Lemma 2 and 4, we heavily use the fact that corners (elements from $\mathcal{C}_0$) are the union of all boxes (elements from $\mathcal{B}$) which they have a non-trivial intersection with.

3 Derandomized Construction

The randomized roundings of Theorem 1 and hence the whole construction above can be derandomized. Combining Theorem 4 of [Doe06] with the simple derandomization without pessimistic estimators (this is derandomization (i) in Section 3.2 of [Doe06]) yields the following.

Theorem 3. *Let $A \in \{0,1\}^{m \times n}$. Let $\xi \in \mathbb{R}^n$ such that $\sum_{i=1}^n \xi_i \in \mathbb{Z}$ and $2^\ell \xi \in \mathbb{Z}^n$. Then a rounding y of ξ such that $\sum_{i=1}^n y_i = \sum_{i=1}^n \xi_i$ and*

$$\forall i \in [m] : |(A\xi)_i - (Ay)_i| \le 13\sqrt{\max\{(A\xi)_i, \ln(4m)\}\ln(4m)}$$

can be computed in time $O(mn\ell)$.

The rounding errors we are interested in are all of the kind $\sum_{B \in \mathcal{B}(C)} (\tilde{x}_B - y_B)$ for some $C \in \mathcal{C}_0$. Hence the matrix encoding all these errors is an $|\mathcal{C}_0| \times |\mathcal{B}|$ matrix having entries 0 and 1 only. More precisely, we consider the matrix $A = (a_{C,B})_{C \in \mathcal{C}_0, B \in \mathcal{B}}$, where $a_{C,B} = 1$ if $B \subseteq C$ and $a_{C,B} = 0$ else. For each $C \in \mathcal{C}_0$ we have

$$(A\tilde{x})_C = \sum_{B \in \mathcal{B}(C)} \tilde{x}_B \le \sum_{B \in \mathcal{B}} \tilde{x}_B = n\,.$$

Thus, if $n \ge \ln(4|\mathcal{C}_0|)$, we get from Theorem 3 the bound

$$|(A\tilde{x})_C - (Ay)_C| \le 13\sqrt{n \ln(4|\mathcal{C}_0|)}\,.$$

If $n \le \ln(4|\mathcal{C}_0|)$, this bound holds trivially, since always $|(A\tilde{x})_C - (Ay)_C| \le n$. Altogether we get the following theorem.

Theorem 4. *Let $n \in \mathbb{N}$ be given. There is a deterministic algorithm that*

(i) computes a point set $T \subseteq [0,1]^d$ that has exactly n points;

(ii) $d_\infty^*(T) \le 13\sqrt{n^{-1}\ln(4|\mathcal{C}_0|)} + \delta + 2^{-\ell-1}n^{-1}$;

(iii) has run time $O(\ell|\mathcal{B}||\mathcal{C}_0|)$.

We get the following corollary.

Corollary 2. *Let $n \in \mathbb{N}$ be given. Let G, δ and ℓ be as chosen in the last section. Furthermore, let σ be as defined in (9). There is a deterministic algorithm that*

(i) computes a point set $T \subseteq [0,1]^d$ that has exactly n points;
(ii) $d_\infty^*(T) \le (13+\sqrt{3})\sqrt{n^{-1}\left(\frac{d}{2}\ln(\sigma n) + \ln 4\right)} + 2^{-d\ln(dn)-1}n^{-1}$;
(iii) has run time $O(d\ln(dn)(\sigma n)^d)$.

3.1 Improvements on the Constants

In [Doe06], having good estimates for the constant in the error term (which e.g. yield the 13 in Theorem 3) was not too important. Here, this constant has roughly a quadratic influence on the size of the point set having a fixed discrepancy. We therefore discuss a simple improvement of the result in [Doe06].

We note that the simple derandomization for $\{0, \frac{1}{2}\}$ vectors and $A \in \{0,1\}^{m\times n}$ (this is derandomization (i) in Section 3.2 of [Doe06]) works as well for $A \in \{-1,0,1\}^{m\times n}$. This saves us from separating positive and negative entries of A as in the proof of Lemma 4 in [Doe06]. Consequently, the $\ln(4m)$ terms become $\ln(2m)$ and the constant $13 \ge f(2\sqrt{\frac{1}{2}})$ becomes $f(\sqrt{\frac{1}{2}}) \le 4$ (where f is defined as in [Doe06]).

Further improvements, in particular, for certain values of the variables involved, are definitely possible (cf. also the note after Theorem 1). We feel, however, that in this paper further technicalities would rather hide the main ideas. We thus decided not to follow such lines of research in this paper.

Future Work

We provided a deterministic algorithm to construct low-discrepancy sets of small size. This algorithm can be implemented with reasonable effort, and we will concentrate on this task in the near future. It would be interesting to test the quality of the resulting point sets T. Further "fine tuning" may improve the discrepancy of T. Notice, e.g., that so far we only distributed n points in boxes $B \in \mathcal{B}$, but we have not specified where to place them inside these boxes. Indeed, this has no influence on our given analysis. But in practise it may, e.g., lead to better results if one places the points in a more sophisticated way inside the box instead of just putting them into the lower left corner. A further investigation of this topic seems to be interesting.

Acknowledgement

We would like to thank an anonymous referee for valuable suggestions and comments.

References

[Ata04] E. Atanassov. On the discrepancy of the Halton sequences. Math. Balkanica (N.S.), **18**, 15–32 (2004)

[Bec81] J. Beck. Balanced two-colorings of finite sets in the square. Combinatorica, **1**, 327–335 (1981)

[Bec84] J. Beck. Some upper bounds in the theory of irregularities of distribution. Acta Arith., **44**, 115–130 (1984)

[DGS05] B. Doerr, M. Gnewuch, and A. Srivastav. Bounds and constructions for the star discrepancy via δ-covers. J. Complexity, **21**, 691–709 (2005)

[DGY06] B. Doerr, S. Güntürk, and O. Yılmaz. Matrix Quasi-Rounding. Preprint (2006)

[Doe06] B. Doerr. Generating randomized roundings with cardinality constraints and derandomizations. In: Durand, B., Thomas, W. (eds) Proceedings of the 23rd Annual Symposium on Theoretical Aspects of Computer Science (STACS'06), Lecture Notes in Comput. Sci., **3884**, 571–583 (2006)

[Doe07] B. Doerr. Matrix approximation and Tusnády's problem. European J. Combin., **28**, 990–995 (2007)

[DT97] M. Drmota and R.F. Tichy. Sequences, Discrepancies and Applications. Lecture Notes in Math., **1651**, Springer, Berlin Heidelberg New York (1997)

[Gne07] M. Gnewuch. Bracketing numbers for d-dimensional boxes and applications to geometric discrepancy. Preprint 21/2007, Max Planck Institute for Mathematics in the Sciences, Leipzig (2007)

[Hin04] A. Hinrichs. Covering numbers, Vapnik-Červonenkis classes and bounds for the star-discrepancy. J. Complexity, **20**, 477–483 (2004)

[HNWW01] S. Heinrich, E. Novak, G.W. Wasilkowski, and H. Woźniakowski. The inverse of the star-discrepancy depends linearly on the dimension. Acta Arith., **96**, 279–302 (2001)

[HSW04] F.J. Hickernell, I.H. Sloan, and G.W. Wasilkowski. On tractability of weighted integration over bounded and unbounded regions in $\mathbb{R}^s$. Math. Comp., **73**, 1885–1901 (2004)

[Mha04] H.N. Mhaskar. On the tractability of multivariate integration and approximation by neural networks. J. Complexity, **20**, 561–590 (2004)

[Nie92] H. Niederreiter. Random Number Generation and Quasi-Monte Carlo Methods. SIAM, Philadelphia (1992)

[NX96] H. Niederreiter and C. Xing. Low-discrepancy sequences and global function fields with many rational places. Finite Fields Appl., **2**, 241–273 (1996)

[Rag88] P. Raghavan. Probabilistic construction of deterministic algorithms: Approximating packing integer programs. J. Comput. Syst. Sci., **37**, 130–143 (1988)

[Rot64] K.F. Roth. Remark concerning integer sequences. Acta Arith., **9**, 257–260 (1964)

[Sch72] W.M. Schmidt. On irregularities of distribution VII. Acta Arith., **21**, 45–50 (1972)

[SS96] A. Srivastav and P. Stangier. Algorithmic Chernoff-Hoeffding inequalities in integer programming. Random Structures Algorithms, **8**, 27–58 (1996)

[Thi01] E. Thiémard. An algorithm to compute bounds for the star discrepancy. J. Complexity, **17**, 850–880 (2001)

A Coding Theoretic Approach to Building Nets with Well-Equidistributed Projections

Yves Edel[1] and Pierre L'Ecuyer[2]

[1] Mathematisches Institut der Universität, Im Neuenheimer Feld 288, 69120 Heidelberg, Germany
y.edel@mathi.uni-heidelberg.de

[2] Département d'informatique et de recherche opérationnelle, Université de Montréal, C.P. 6128, Succ. Centre-Ville, Montréal (Québec), H3C 3J7, Canada, and
IRISA, Rennes, France
lecuyer@iro.umontreal.ca

Summary. Starting from coding-theoretic constructions, we build digital nets with good figures of merit, where the figure of merit takes into account the equidistribution of a preselected set of low-dimensional projections. This type of figure of merit turns out to be a better predictor than the t-value for the variance of randomized quasi-Monte Carlo (RQMC) estimators based on nets, for certain classes of integrals. Our construction method determines the most significant digits of the points by exploiting the equivalence between the desired equidistribution properties used in our criterion and the property of a related point set to be an orthogonal array, and using existing orthogonal array constructions. The least significant digits are then adjusted to improve the figure of merit. Known results on orthogonal arrays provide bounds on the best possible figure of merit that can be achieved. We present a concrete construction that belongs to the class of cyclic digital nets and we provide numerical illustrations of how it can reduce the variance of an RQMC estimator, compared with more standard constructions.

1 Introduction

This paper deals with the construction of finite sets of points that are more evenly distributed in the s-dimensional unit hypercube, in some sense, than a typical set of random points. The two main issues that arise in building such point sets are: (a) to define an appropriate measure of uniformity, or measure of *discrepancy* between the uniform distribution and the empirical distribution of the points; (b) to find construction methods for point sets having high uniformity, or low discrepancy, with respect to the retained definition.

A popular class of construction is that of digital nets [Nie92, Nie05], whose uniformity is usually measured by figures of merit defined in terms of the equidistribution of the points in certain families of rectangular boxes that partition the unit hypercube. A widely-used figure of merit in this context is the t-value [LL02, Nie92, Nie05, PS01, SS05]. One limitation of this measure, however, is that when the dimension of the point set is much larger than the basis of the net, the t-value is necessarily large, and it does not really take into account the quality of the low-dimensional projections. There are several applications in RQMC integration where for a given t-value, the uniformity of certain low-dimensional projections can make an important difference [LL99, LL01, PL06].

The aim of this paper is to propose digital net constructions with good t-values and high-quality low-dimensional projections and to exhibit theoretical bounds on what can be achieved in this direction. We do this by exploiting the links between digital net constructions on the one hand, and some established results on orthogonal arrays and error correcting codes on the other hand. Results from coding theory have already been exploited extensively to construct digital nets with a small t-value and to compute tables of the best known t-value for a given dimension, basis, and number of points [BE98, Nie04, Nie05, SS05]. Here we use similar techniques to define skeletons for our nets, i.e., to determine the most significant digits of the points. The construction is then refined by adjusting the least significant digits to improve our figure of merit. Known results on orthogonal arrays also provide bounds on the best possible figure of merit that can be achieved.

As a concrete example, we propose an algebraic construction of a family of cyclic digital nets with well-equidistributed projections. These nets are *cyclic* in the sense that if we shift all coordinates of any given s-dimensional point of the net by one position to the left and put the old first coordinate at the end, the resulting point is always in the net. (This definition differs from that of [Nie04].) By repeating the blocks of s successive coordinates ad infinitum, these nets provide point sets that are infinite-dimensional and dimension-stationary, in the sense of [LL99, PL06]. We present a family of cyclic (t, m, s)-nets that belong to that class; they have the same parameters t, m, and s as in [BE98] (which give the best t known so far for certain values of s and m), and improved equidistribution properties for certain projections. We give a numerical illustration showing that this type of point set can be more accurate than other well-established digital nets (such as Sobol' nets) for QMC integration, at least for certain types of integrands.

The rest of the paper is organized as follows. In Section 2, we recall and discuss various ways of measuring the uniformity of digital nets. In Section 3, we make the links between the nets that we want to construct and orthogonal arrays, whose additive versions are the duals of additive error-correcting codes. A specific class of cyclic net constructions is proposed and analyzed in Section 4. The numerical illustrations are in Section 5.

2 Digital Nets and Their Figures of Merit

QMC and RQMC.

We want to construct finite point sets of the form $P_n = \{\mathbf{u}_0, \ldots, \mathbf{u}_{n-1}\}$ in $[0,1)^s$ with *low discrepancy* (i.e., *high uniformity*) in some sense. These point sets can be used, for instance, to estimate the integral of some function f over $[0,1)^s$ by quasi-Monte Carlo (QMC):

$$\mu = \int_{[0,1)^s} f(\mathbf{u})d\mathbf{u} \approx \frac{1}{n}\sum_{i=0}^{n-1} f(\mathbf{u}_i). \tag{1}$$

Randomized QMC (RQMC) also uses the approximation (1), but after randomizing the point set P_n in a way that each individual point has the uniform distribution over $[0,1)^s$ even though the point set as a whole keeps its high uniformity [LL02, Nie92, Owe98]. It has the advantage of providing an unbiased estimator of μ, and also an unbiased variance estimator if we make several independent randomizations.

Digital nets.

The two most widely used classes of constructions for P_n are digital nets and lattice rules [Nie92, SJ94]. We focus on the former. For given integers $b \geq 2$ (usually a prime or a prime power) and $m \geq 1$, a *digital net in base* b with $n = b^m$ points is defined as follows. For $j = 1, \ldots, s$, select a $w \times m$ *generator matrix* $\mathbf{C}^{(j)}$ whose elements are either in the finite ring $\mathbb{Z}_b$ or in the finite field $\mathbb{F}_b$. (If $b = p^e$ where p is prime and $e > 1$, the operations in $\mathbb{F}_b$ and in $\mathbb{Z}_b$ are not equivalent, so one must make sure that the correct arithmetic is used, depending on how the $\mathbf{C}^{(j)}$ where constructed.) To define the ith point $\mathbf{u}_i$, for $i = 0, \ldots, b^m - 1$, we write the digital expansion of i in base b and multiply the vector of its digits by $\mathbf{C}^{(j)}$, modulo b, to obtain the digits or the expansion of $u_{i,j}$, the jth coordinate of $\mathbf{u}_i$. That is,

$$i = a_{i,0} + a_{i,1}b + \cdots + a_{i,m-1}b^{m-1},$$

$$\begin{pmatrix} u_{i,j,1} \\ u_{i,j,2} \\ \vdots \end{pmatrix} = \mathbf{C}^{(j)} \begin{pmatrix} a_{i,0} \\ a_{i,1} \\ \vdots \\ a_{i,m-1} \end{pmatrix}$$

$$u_{i,j} = \sum_{\ell=1}^{\infty} u_{i,j,\ell} b^{-\ell}, \qquad \mathbf{u}_i = (u_{i,1}, \ldots, u_{i,s}).$$

In practice, we take w and m finite, but there is no limit on their size. If the generating matrices are defined with an infinite number of columns, then we have a *digital sequence* of points. If we have an infinite sequence of generating matrices, then the points can be thought as having infinite dimension. Typically, these infinite sequences are defined via recurrences, either for the successive columns or the successive generating matrices. Well-known digital net constructions are those of Sobol', Faure, Niederreiter, and Niederreiter-Xing.

Equidistribution

Let $(q_1, \ldots, q_s)$ be a vector of nonnegative integers such that $q = q_1 + \ldots + q_s \leq m$. A $(q_1, \ldots, q_s)$-*equidissection in base* b is a partition of the unit hypercube in $b^{q_1+\cdots+q_s}$ rectangular boxes aligned with the axes, of equal volume b^{-q}, defined by dividing the interval $[0, 1)$ along the j-th coordinate into b^{q_j} equal parts, for each j. A point set P_n with $n = b^m$ is said to be $(q_1, \ldots, q_s)$-*equidistributed in base* b if every cell defined by the $(q_1, \ldots, q_s)$-equidissection contains exactly b^{m-q} points from P_n. It is easy to see that a digital net in base b is $(q_1, \ldots, q_s)$-equidistributed in base b if and only if the matrix constructed with the first q_1 rows of $\mathbf{C}^{(1)}$, the first q_2 rows of $\mathbf{C}^{(2)}$, ..., and the first q_s rows of $\mathbf{C}^{(s)}$, has full rank $q_1 + \cdots + q_s$. This is possible only if $q_1 + \cdots + q_s \leq m$.

These definitions apply more generally to lower-dimensional projections of P_n. For $I = \{i_1, \ldots, i_\eta\} \subseteq \{1, \ldots, s\}$, $P_n(I)$ denotes the η-dimensional *projection* of P_n on the coordinates determined by I. The set $P_n(I)$ is $(q_{i_1}, \ldots, q_{i_\eta})$-*equidistributed in base* b if each box of the $(q_{i_1}, \ldots, q_{i_\eta})$-equidissection has the same number of points. This is equivalent to saying that P_n is $(\tilde{q}_1, \ldots, \tilde{q}_s)$-equidistributed with $\tilde{q}_j = q_{i_h}$ if $j = i_h \in I$ and $\tilde{q}_j = 0$ otherwise. This equidistribution can thus be verified by checking the rank of a matrix as explained earlier.

The t-value.

A digital net in base b with $n = b^m$ points is a (t, m, s)-*net in base* b, also denoted $(t, m, s)_b$ net, if it is $(q_1, \ldots, q_s)$-equidistributed whenever $q_1 + \cdots + q_s \leq m - t$ [Nie92]. The smallest integer $t \geq 0$ such that this holds is called the t-*value* of the net. Ideally, we want t to be as small as possible. But $t = 0$ is possible only if $s \leq b + 1$ [Nie92]. Otherwise, the best possible t-value can be much larger than 0; the best possible t-value as a function of b and s, together with the best known t-values, can be found in the `MinT` tables of [SS05].

For example, in base $b = 2$, for $m = 14$ and $s = 23$, the best known t-value is $t = 8$. This guarantees equidistribution only for $q_1 + \cdots + q_s \leq 6$, i.e., when considering no more than 6 output bits. But why not be more demanding for low-dimensional projections? For instance, an easily achieved requirement would be that all one-dimensional projections be (m)-equidistributed. We could also ask that other low-dimensional projections have a smaller t-value; in the previous example where $t = 8$, for instance, we may ask that several of the two-dimensional projections have a t-value of 0.

Another way to compromise when the lower bound on the t-value is deemed too high is to define a figure of merit that takes the worst case over a smaller number of equidissections, i.e., fewer shapes of boxes. This is the direction we take in what follows.

Looking at square boxes only

We say that P_n or $P_n(I)$ is *η-distributed with ℓ digits of accuracy* if it is $(\ell, \ldots, \ell)$-equidistributed. This means that if we partition the hypercube into $b^{\eta\ell}$ cubic boxes or equal size, each box contains exactly $b^{m-\eta\ell}$ points. The largest ℓ for which this holds is the *η-dimensional resolution of $P_n(I)$ in base b*, denoted $\ell(I)$. One has $\ell(I) \leq \lfloor m/\eta \rfloor$. The *resolution gap of $P_n(I)$* is defined by $\delta(I) = \lfloor m/\eta \rfloor - \ell(I)$. This can be used to define a worst-case criterion based on (cubic) equidistribution [LL00, LL02]:

$$\Delta_{\mathcal{J}} = \max_{I \in \mathcal{J}} \delta(I)$$

where $\mathcal{J}$ is a selected class of sets $I \subseteq \{1, \ldots, s\}$. The choice of $\mathcal{J}$ is arbitrary. If $\mathcal{J}$ contains too many projections, typically there are inevitably some bad ones and the criterion loses its discriminatory power, because it only cares about the worst projections. A leaner $\mathcal{J}$ can concentrate on the most important projections, if it diminishes the theoretical lower bound on $\Delta_{\mathcal{J}}$. As a practical compromise, Lemieux and L'Ecuyer [LL02] suggested the form

$$\begin{aligned} \mathcal{J} = \{\{0, 1, \ldots, i\} : i < s_1\} \cup \{\{i_1, i_2\} : 0 = i_1 < i_2 < s_2\} \cup \cdots \\ \cup \{\{i_1, \ldots, i_d\} : 0 = i_1 < \ldots < i_d < s_d\} \end{aligned} \qquad (2)$$

for arbitrarily selected values of $d, s_1, \ldots, s_d$.

3 A Coding Theoretic Link: Orthogonal Arrays

An *orthogonal array* $\mathrm{OA}(n, s, q, t)$ is an array of n rows and s columns, with entries in $\{0, 1, \ldots, q-1\}$, such that in the submatrix formed by any t columns of the array, each of the q^t possibilities for a row appear equally often, namely n/q^t times each. We say that we have an orthogonal array (OA) with n *words* (or *runs*), *length* s (or s *factors*), q *levels*, and *strength* t. For further details on OAs, see [Bie04, HSS99, Owe92].

We can define a correspondence between an OA and a point set P_n in $[0, 1)^s$ simply by dividing each entry of the array by q and viewing each row of the array as representing an s-dimensional point. With a slight abuse of language, we also call this point set an OA (i.e., identify it with the OA). Note that all coordinates of all points in this point set are multiples of $1/q$. If $q = b^\ell$ for some positive integers b and ℓ, the $\mathrm{OA}(n, s, q, t)$ property means that every t-dimensional projection of P_n is t-distributed with ℓ digits of accuracy, in base b.

Let $\mathcal{J}_\eta$ denotes the class of all subsets of exactly η coordinates, i.e., of the form $I = \{i_1, \ldots, i_\eta\} \subseteq \{1, \ldots, s\}$. If P_n is a point set whose coordinates are all multiples of $b^{-\ell}$, then P_n is an $\mathrm{OA}(n, s, b^\ell, \eta)$ if and only if

$$\min_{I \in \mathcal{J}_\eta} \ell(I) \geq \ell,$$

if and only if

$$\Delta_{\mathcal{J}_\eta} \leq \lfloor m/\eta \rfloor - \ell.$$

If P_n is a *digital* net with $n = b^m$ where b is prime, then the sum (digitwise, modulo b) of two points of the net is again a point of the net; that is, the corresponding OA is an *additive* OA, which is the *dual* of an *additive error-correcting code* $(s, b^{\ell s - m}, \eta + 1)_{b^\ell}$ [Bie04, HSS99]. In fact, each additive error-correcting code gives an additive orthogonal array, and vice-versa.

Our aim here is to construct digital nets P_n in base b, such that P_n truncated to its first ℓ_η digits is an $\mathrm{OA}(b^m, s, b^{\ell_\eta}, \eta)$, *simultaneously* for $\eta = 1, 2, \ldots, d$, where each ℓ_η *is as large as possible.* So our task is more than just looking up for existing OAs or codes. A trivial upper bound for each ℓ_η is $\ell_\eta \leq \lfloor m/\eta \rfloor$. Known bounds on the largest ℓ for which there can exist an $\mathrm{OA}(b^m, s, b^\ell, \eta)$ are generally tighter than this trivial bound. In some cases, there are known constructions that match the bounds. Note that the closer η is to $s/2$, the more η-dimensional projections there are. To verify the OA property, $\eta\ell$ digits of each projection are examined, so a larger ℓ means that more digits are involved (the boxes have smaller volume) and the corresponding OA is then harder to construct.

Example 1. Take $b = 2$, $s = 65$, and $m = 12$, so $n = 2^{12}$. Table 1 gives upper bounds on the largest ℓ for which there can be an $\mathrm{OA}(2^{12}, 65, 2^\ell, \eta)$, as well as the values of ℓ achieved by known OA constructions.

The upper bounds from `MinT` [SS05] are obtained as follows. For $\eta = 3$ there is no $\mathrm{OA}(16^3, 19, 2^4, 3)$ which is a consequence of the bound on OAs with index unity [Bus52]. For $\eta = 4$, from the linear programming bound we find that there is no $\mathrm{OA}(4^6, 30, 2^2, 4)$ [SS06]. For $\eta = 5$, there is no $\mathrm{OA}(2^{12}, 65, 2, 5)$, because otherwise its truncation would provide an $\mathrm{OA}(2^{11}, 64, 2, 4)$, which would violate the sphere-packing bound [Bie04].

Table 1. Upper bounds on the values of ℓ for which there can exist an $\mathrm{OA}(2^{12}, 65, 2^\ell, \eta)$, and values for which there exist known constructions.

η	1	2	3	4	5	6	7	$\cdots$	12
$\lfloor 12/\eta \rfloor$	12	6	4	3	2	2	1	$\cdots$	1
`MinT` upper bound for OA			3	1	0	0	0	$\cdots$	0
Best known additive OA	12	6	3	1	0	0	0	$\cdots$	0
Best known net	12	6	3	1	0	0	0	$\cdots$	0

The best known additive OAs, on the other hand, can be found from the best known linear error-correcting codes $[65, 65 - 12/\ell, \eta + 1]_{2^\ell}$. For the case $\eta = 1$, there is an obvious construction. For $\eta = 2$, there is a Hamming code $[65, 65 - 2, 3]_{64}$. For $\eta = 3$, there is an ovoid code $[65, 65 - 4, 4]_8$. For $\eta = 4$, there is a binary linear code $[65, 53, 5]_2$.

Our strategy for building our nets will be to start with a good (known) $\mathrm{OA}(n, s, b^\ell, \eta)$ for some reasonably large η (and a rather small ℓ, necessarily) and fix the first ℓ digits of the net; then, in a second stage, we "optimize" the other digits, either by algebraic construction or via computer search, to obtain a point set whose $\ell(\eta')$-digit truncation is an $\mathrm{OA}(n, s, b^{\ell(\eta')}, \eta')$ for reasonably large $\ell(\eta')$, for all $\eta' \leq \eta$.

4 A Cyclic Net Construction

We call a digital net P_n *cyclic* if for any point $(u_0, \ldots, u_{s-2}, u_{s-1}) \in P_n$, we also have $(u_1, \ldots, u_{s-1}, u_0) \in P_n$. For a cyclic digital net P_n, whenever b is prime and $\gcd(b, s) = 1$, the net is a direct product of rings (principal ideal domains). These rings turn out to be linear cyclic codes, one of the favorite classes of codes of coding theorists, and their structure can be exploited for efficient computer search and algebraic constructions of good instances of these nets. The following special case illustrates this.

A cyclic net construction

The following construction gives a cyclic digital net P_n in base $b = 2$, with $n = 2^{4r}$ points in $s = 2^{2r} + 1$ dimensions, for some integer r. The dimensions of the generating matrices will be $w = m = 4r$. This net can be used to approximate integrals in $s' \leq s$ dimensions by taking only the first s' coordinates of each point. For $s' > s$, we can take advantage of the cyclic property to get as many coordinates as needed. The periodicity of the coordinates will be destroyed by the randomization (see Section 5).

The generator matrices of the net are defined as follows. Recall that

$$\mathbb{F}_2 \subset \mathbb{F}_{2^r} \subset \mathbb{F}_{2^{2r}} \subset \mathbb{F}_{2^{4r}}.$$

Let $\zeta \in \mathbb{F}_{2^{4r}}$ be a $(2^{2r} + 1)$th primitive root of unity, i.e., such that $\zeta^{2^{2r}+1} = 1$. Such a ζ exists because we know that there is an element ζ' of multiplicative order $2^{4r} - 1$, so it suffices to take $\zeta = (\zeta')^{2^{2r}-1}$, which has multiplicative order $2^{2r} + 1$. Choose a basis $1 = \alpha_1, \ldots, \alpha_r$ of $\mathbb{F}_{2^r}$ over $\mathbb{F}_2$ and choose some elements $\beta \in \mathbb{F}_{2^{2r}} \setminus \mathbb{F}_{2^r}$, and $\gamma \in \mathbb{F}_{2^{4r}} \setminus \mathbb{F}_{2^{2r}}$. Put

$$a_i = \alpha_i, \quad a_{i+r} = \beta\alpha_i, \quad a_{i+2r} = \gamma\alpha_i, \quad a_{i+3r} = \gamma\beta\alpha_i,$$

for $i = 1, \ldots, r$. Then, $a_1, \ldots, a_r$ form a basis of $\mathbb{F}_{2^r}$, $a_1, \ldots, a_{2r}$ are a basis of $\mathbb{F}_{2^{2r}}$, and $a_1, \ldots, a_{4r}$ are a basis of $\mathbb{F}_{2^{4r}}$.

To define the ith row of the matrix $\mathbf{C}^{(j)}$, we compute $a_i\zeta^j \in \mathbb{F}_{2^{4r}}$ and represent it as a vector of $4r$ elements (or coordinates) over $\mathbb{F}_2$.

Proposition 1. *For $r > 1$, the cyclic net just constructed has the following properties:*

1. *It is a digital $(4r-4, 4r, 2^{2r}+1)$-net in base 2*
2. *It is $(4r)$-equidistributed for all one-dimensional projections.*
3. *It is $(2r, 2r)$-equidistributed for all two-dimensional projections.*
4. *It is (r, r, r)-equidistributed for all three-dimensional projections.*
5. *It is $(1, 1, 1, 1)$-equidistributed for all four-dimensional projections.*
6. *It is $(1, \ldots, 1)$-equidistributed whenever $I = \{j, j+1, \ldots, j+4r-1\}$.*
7. *It is (r, r, r, r)-equidistributed whenever $I = \{j, j+1, j+2, j+3\}$ or $I = \{j, k, l, m(j, k, l)\}$, for any pairwise different j, k, l and some $2^r - 2$ different $m(j, k, l)$. Thus, the proportion of four-dimensional projections that are (r, r, r, r)-equidistributed is approximately $1/(1-1/n^2)$.*

Proof. That the net is $(4r)$-equidistributed for all one-dimensional projections is equivalent to the fact that the matrix $\mathbf{C}^{(j)}$ has full rank. This is obvious, as the $\alpha_i, \beta\alpha_i, \gamma\alpha_i, \gamma\beta\alpha_i$ are a basis of $\mathbb{F}_{2^{4r}}$ over $\mathbb{F}_2$ and $\zeta^j \neq 0$.

Now we want to show that the net is $(2r, 2r)$-equidistributed for all two-dimensional projections. We have to show that the first $2r$ rows of $\mathbf{C}^{(j)}$ and the first $2r$ rows of $\mathbf{C}^{(j')}$ have full rank. The first $2r$ rows of $\mathbf{C}^{(j)}$ are of the form $a_i\zeta^j$ and the a_i, for $1 \leq i \leq 2r$, are a basis of $\mathbb{F}_{2^{2r}}$. So we have to show that ζ^j and $\zeta^{j'}$ are linearly independent over $\mathbb{F}_{2^{2r}}$. Let $W = \{\zeta^j | 0 \leq j < 2^{2r}+1\} \subset \mathbb{F}_{4^r}$ be the group of elements of multiplicative order $2^{2r}+1$. As $\gcd(2^{2r}+1, 2^{2r}-1) = 1$, we have that $W \cap \mathbb{F}_{2^{2r}} = \{1\}$. So two different $\zeta^j, \zeta^{j'} \in W$ are linearly independent over $\mathbb{F}_{2^{2r}}$.

For the (r, r, r)-equidistribution, with the same argument, we have to show the linear independence of different powers of ζ over $\mathbb{F}_{2^r}$. It is known, that the $(2^{2r}+1)$-th roots of unity, i.e., W is an ovoid in $\mathrm{PG}(3, 2^r)$, where $\mathrm{PG}(k, q)$ denotes the k-dimensional projective geometry over the finite field $\mathbb{F}_q$ [Bie04, Hir85] (see the proof of Theorem 17 of [Bie03]). The defining property of the ovoid implies that the $\zeta^j \in W$ are a $\mathbb{F}_{2^r}$-linear OA of strength 3. This means that for any distinct indexes $\{j, j', j''\}$, $\zeta^j, \zeta^{j'}, \zeta^{j''}$ are linearly independent over $\mathbb{F}_{2^r}$. Hence the net is (r, r, r)-equidistributed for all three-dimensional projections.

For the (r, r, r, r)-equidistribution we consider again the ζ^j as *points* in $PG(3, 2^r)$ (not to be confused with the points of P_n). Consider four points $\zeta^j, \zeta^k, \zeta^l, \zeta^{m(j,k,l)}$. Since W is an ovoid and three independent points define a plane in $PG(3, 2^r)$, the points $\zeta^{m(j,k,l)}$ that are not independent from $\{\zeta^j, \zeta^k, \zeta^l\}$ are those points of the ovoid that lie in the plane generated by $\{\zeta^j, \zeta^k, \zeta^l\}$. The claimed property follows from the fact that every plane that contains more than one point of the ovoid in $PG(3, 2^r)$ contains exactly $2^r + 1$ points of the ovoid (Theorem 16.1.6.ii in [Hir85]).

That the net is (r, r, r, r)-equidistributed for $I = \{j, j+1, j+2, j+3\}$ follows from the fact that $\{1, \zeta, \zeta^2, \zeta^3\}$ are linearly independent over $\mathbb{F}_{2^r}$, because $\mathbb{F}_{2^{4r}}$ is the smallest field that contains ζ. That the net is $(1, \ldots, 1)$-equidistributed whenever $I = \{j, j+1, \ldots, j+4r-1\}$ follows from the fact that $\{1, \zeta, \ldots, \zeta^{4r-1}\}$ are linearly independent over $\mathbb{F}_2$.

The net is $(1,1,1,1)$-equidistributed for all four-dimensional projections. This follows from the fact that the binary code we obtain by restriction to the first digit has strength 4; see [BE98] for the proof.

For the (t,m,s)-net property (i), we have to show that the net is (l_1, l_2, l_3, l_4)-equidistributed whenever $l_1 + l_2 + l_3 + l_4 = 4$. The only case that is not covered by what we already have shown is the $(3,1)$-equidistribution for all two dimensional projections, for $r = 2$. But for $r = 2$ we are exactly in the same situation as in [BE98], where the corresponding (t,m,s)-net property is proved.

5 Numerical Illustrations

We report (a subset of) the results of numerical experiments where we try our cyclic nets for estimating some multivariate integrals by RQMC. We compare their performance with that of Sobol' nets when both are randomized by a random binary digital shift [LL02, Owe03]. In each case, we estimate the variance per run, defined as n times the variance of the average over the n points, and compare it with the empirical variance of standard Monte Carlo (MC). The *variance reduction factor* (VRF) reported is the ratio of the MC variance over the RQMC variance per run. The digital nets are randomized by a random digital shift (DS), which consists in generating a single point $\mathbf{u} = (u_1, \ldots, u_s)$ uniformly over $[0,1)^s$, and performing a digit-wise addition modulo b of u_j with the jth coordinate of each point of P_n, for each j. For $b = 2$, the digit-wise addition becomes a bitwise exclusive-or. This randomization preserves the equidistribution for every equidissection in base 2; in particular, it preserves the (t,m,s)-net properties. The primitive polynomials and the direction numbers for the Sobol' sequence were taken from [LCL04].

Example 2. This example is from [KW97] and [PL06]. We consider the function f defined by

$$f(u_1, \ldots, u_s) = \sqrt{\frac{2}{t(t-1)}} \sum_{j=0}^{t-1} \sum_{i=0}^{j-1} g(u_i) g(u_j),$$

where $g(x) = 27.20917094x^3 - 36.19250850x^2 + 8.983337562x + 0.7702079855$ and $s = 120$. We take n from 2^{14} to 2^{16} and we use RQMC with 100 independent digital random shifts (DS) to estimate the variance for each method. Table 2 gives the VRF for different digital nets. The $\mathbb{F}_{2^w}$ nets were proposed by Panneton and L'Ecuyer [PL06]; these authors tried 12 instances of these nets on this example and obtained VRFs ranging from 10 to 4×10^5. With the $(0,2,126)_{125}$-net, we obtain a competitive VRF. (Note that this net cannot be written as a net in base 2.) These $(0,2,q+1)_q$-nets are essentially the duals

Table 2. Variance reduction factors of RQMC compared with MC, with various digital nets.

net	n	VRF
Sobol	2^{14}	2
Sobol	2^{16}	2
$\mathbb{F}_{2^w}$-nets	$2^{14} - 2^{16}$	10 to 4×10^5
$(0, 2, 129)_{128}$-net	2^{14}	330
$(0, 2, 126)_{125}$-net	5^6	8.3×10^4
Proposition 1	2^{16}	1.8×10^6

of Hamming codes. They provide an optimal resolution for the projections in one and two dimensions, which seems to be what we need for the function f considered here. With the net from Proposition 1, with $r = 4$, we obtain a significantly larger VRF.

Example 3. This example is from [IT02] and [L'E04]. We consider a Bermudan-Asian option on c assets. For $1 \le i \le c$, the value of asset i evolves as a geometric Brownian motion (GMB) $\{S_i(t), t \ge 0\}$ with *drift parameter* μ_i and *volatility parameter* σ_i. That is,

$$S_i(t) = S_i(0) \exp\left[(\mu_i - \sigma_i 2/2)t + \sigma_i W_i(t)\right]$$

where W_i is a standard Brownian motion. The W_i's are also correlated, with Cov $[W_i(t+\delta) - W_i(t), W_j(t+\delta) - W_j(t)] = \rho_{i,j}\delta$ for all $\delta > 0$. The option has discounted payoff $e^{-rT} \max[\bar{S}^{(A)} - K, 0]$ for some constants $K > 0$ and $T > 0$, where

$$\bar{S}^{(A)} = \frac{1}{cd} \sum_{i=1}^{c} \sum_{j=1}^{d} S_i(t_j) \tag{3}$$

is the arithmetic average at the fixed observation times $t_j = jT/d$ for $j = 1, \ldots, d$. The vector $\mathbf{Y} = (W_1(t_1), \ldots, W_c(t_1), W_1(t_2), \ldots, W_c(t_2), \ldots, W_1(t_d), \ldots, W_c(t_d))^{\mathrm{t}}$, has a multivariate normal distribution with mean zero and covariance matrix $\boldsymbol{\Sigma}$ whose element $((i-1)c + j), (i'-1)c + j')$ is $\rho_{i,i'}\sigma_i\sigma_{i'}|t_{j'} - t_{j-1}|)$ for $j' \ge j$.

To generate $\mathbf{Y}$, we can decompose $\boldsymbol{\Sigma}$ as $\boldsymbol{\Sigma} = \mathbf{CC}^{\mathrm{t}}$ for some matrix $\mathbf{C}$, generate a vector $\mathbf{Z} = (Z_1, \ldots, Z_s)$ of independent $N(0,1)$ (standard normal) random variates by inversion from s independent $U(0,1)$ random variates $U_1, \ldots, U_s$, i.e., $Z_j = \Phi^{-1}(U_j)$, and return $\mathbf{Y} = \mathbf{CZ}$. There are several possibilities for the choice of factorization $\boldsymbol{\Sigma} = \mathbf{CC}^{\mathrm{t}}$. For instance, the *Cholesky factorization*, takes $\mathbf{C}$ lower triangular, whereas *principal component analysis* (PCA) selects $\mathbf{C}$ so that each Z_j accounts for the maximum amount of variance conditional on $Z_1, \ldots, Z_{j-1}$. Its combination with QMC was suggested in [ABG98].

Table 3. Empirical variance reduction factors of RQMC with respect to MC for Example 3 (in 250 Dimensions), for a Sobol' net and for the net of Proposition 1, with $n = 2^{16}$ points.

net	Cholesky	PCA
Sobol'	16	6144
Proposition 1	50	2108

We take the same parameters as in Example 2 of [L'E04]: $c = 10$, $d = 25$ (so $s = 250$), $\rho_{i,j} = 0.4$ for all $i \neq j$, $T = 1$, $\sigma_i = 0.1 + 0.4(i-1)/9$ for all i, $r = 0.04$, $S(0) = 100$, and $K = 100$. We thus have a 250-dimensional integral. Simulations with a huge number of runs told us that $\mu \approx 5.818$ and the MC variance is $\sigma^2 \approx 72.3$.

Recall that the Sobol' nets are constructed to behave well for the projections over the first successive coordinates, but not for arbitrary projections over coordinates with a large index, whereas the net of Proposition 1 has been built precisely to have good uniformity for projections over a small number of arbitrary coordinates. With the Cholesky decomposition, the variance is spread over pretty much all coordinates, whereas PCA pushes most of the variance in the first coordinates. Thus, we expect the Sobol' nets to work well when PCA is used and the new nets to be more competitive if one is forced to use the Cholesky decomposition. The simulation results reported in Table 3 agree with these expectations. We also tried with different values of r, from 0.03 to 0.07, and the VRFs were similar. The VRFs are much larger with PCA than with Cholesky, due to the fact that PCA reduces significantly the effective dimension in the truncation sense [IT02, LL02]. But PCA is not always practical for real-life problems; for instance when the dimension is very large. Then, one may have to use more traditional simulation schemes that do not reduce the effective dimension in the truncation sense. The new nets can be useful in this type of situation.

Acknowledgments

This research has been supported by NSERC-Canada grant No. ODGP0110050 and a Canada Research Chair to the second author.

References

[ABG98] P. Acworth, M. Broadie, and P. Glasserman. A comparison of some Monte Carlo and quasi-Monte Carlo techniques for option pricing. In P. Hellekalek, G. Larcher, H. Niederreiter, and P. Zinterhof, editors, *Monte Carlo and Quasi-Monte Carlo Methods 1996*, volume 127 of *Lecture Notes in Statistics*, pages 1–18. Springer-Verlag, New York, 1998.

[BE98] J. Bierbrauer and Y. Edel. Construction of digital nets from BCH-codes. In P. Hellekalek, G. Larcher, H. Niederreiter, and P. Zinterhof, editors, *Monte Carlo and Quasi-Monte Carlo Methods 1996*, volume 127 of *Lecture Notes in Statistics*, pages 221–231. Springer-Verlag, New York, 1998.

[Bie03] J. Bierbrauer. Large caps. *Journal of Geometry*, 76:16–51, 2003.

[Bie04] J. Bierbrauer. *Introduction to Coding Theory*. Chapman and Hall/CRC Press, Boca Raton, FL, USA, 2004.

[Bus52] K. A. Bush. Orthogonal arrays of index unity. *Annals of Mathematical Statistics*, 13:426–434, 1952.

[Hir85] J. W. P. Hirschfeld. *Finite Projective Spaces of Three Dimensions*. Clarendon Press, Oxford, 1985.

[HSS99] A. S. Hedayat, N. J. A. Sloane, and J. Stufken. *Orthogonal Arrays: Theory and Applications*. Springer-Verlag, New York, 1999.

[IT02] J. Imai and K. S. Tan. Enhanced quasi-Monte Carlo methods with dimension reduction. In E. Yücesan, C. H. Chen, J. L. Snowdon, and J. M. Charnes, editors, *Proceedings of the 2002 Winter Simulation Conference*, pages 1502–1510, Piscataway, New Jersey, 2002. IEEE Press.

[KW97] L. Kocis and W. J. Whiten. Computational investigations of low-discrepancy sequences. *ACM Transactions on Mathematical Software*, 23(2):266–294, June 1997.

[LCL04] C. Lemieux, M. Cieslak, and K. Luttmer. *RandQMC User's Guide: A Package for Randomized Quasi-Monte Carlo Methods in C*, 2004. Software user's guide, available from `http://www.math.uwaterloo.ca/~clemieux/`.

[L'E04] P. L'Ecuyer. Quasi-Monte Carlo methods in finance. In R. G. Ingalls, M. D. Rossetti, J. S. Smith, and B. A. Peters, editors, *Proceedings of the 2004 Winter Simulation Conference*, Piscataway, New Jersey, 2004. IEEE Press.

[LL99] P. L'Ecuyer and C. Lemieux. Quasi-Monte Carlo via linear shift-register sequences. In *Proceedings of the 1999 Winter Simulation Conference*, pages 632–639. IEEE Press, 1999.

[LL00] P. L'Ecuyer and C. Lemieux. Variance reduction via lattice rules. *Management Science*, 46(9):1214–1235, 2000.

[LL01] C. Lemieux and P. L'Ecuyer. Selection criteria for lattice rules and other low-discrepancy point sets. *Mathematics and Computers in Simulation*, 55(1–3):139–148, 2001.

[LL02] P. L'Ecuyer and C. Lemieux. Recent advances in randomized quasi-Monte Carlo methods. In M. Dror, P. L'Ecuyer, and F. Szidarovszky, editors, *Modeling Uncertainty: An Examination of Stochastic Theory, Methods, and Applications*, pages 419–474. Kluwer Academic, Boston, 2002.

[Nie92] H. Niederreiter. *Random Number Generation and Quasi-Monte Carlo Methods*, volume 63 of *SIAM CBMS-NSF Regional Conference Series in Applied Mathematics*. SIAM, Philadelphia, 1992.

[Nie04] H. Niederreiter. Digital nets and coding theory. In K. Q. Feng, H. Niederreiter, and C. P. Xing, editors, *Coding, Cryptography and Combinatorics*, volume 23 of *Progress in Computer Science and Applied Logic*, pages 247–257. Birkhäuser, Basel, 2004.

[Nie05] H. Niederreiter. Constructions of (t, m, s)-nets and (t, s)-sequences. *Finite Fields and Their Applications*, 11:578–600, 2005.

[Owe92] A. B. Owen. Orthogonal arrays for computer experiments, integration and visualization. *Statistica Sinica*, 2(2):439–452, 1992.

[Owe98] A. B. Owen. Latin supercube sampling for very high-dimensional simulations. *ACM Transactions on Modeling and Computer Simulation*, 8(1):71–102, 1998.

[Owe03] A. B. Owen. Variance with alternative scramblings of digital nets. *ACM Transactions on Modeling and Computer Simulation*, 13(4):363–378, 2003.

[PL06] F. Panneton and P. L'Ecuyer. Infinite-dimensional highly-uniform point sets defined via linear recurrences in F_{2^w}. In H. Niederreiter and D. Talay, editors, *Monte Carlo and Quasi-Monte Carlo Methods 2004*, pages 419–429, Berlin, 2006. Springer-Verlag.

[PS01] G. Pirsic and W. Ch. Schmid. Calculation of the quality parameter of digital nets and application to their construction. *Journal of Complexity*, 17(4):827–839, 2001.

[SJ94] I. H. Sloan and S. Joe. *Lattice Methods for Multiple Integration*. Clarendon Press, Oxford, 1994.

[SS05] W. Ch. Schmid and R. Schürer. MinT, the database for optimal (t, m, s)-net parameters. `http://mint.sbg.ac.at`, 2005.

[SS06] R. Schürer and W. Ch. Schmid. Linear programming bounds; in MinT, the database for optimal (t, m, s)-net parameters. `http://mint.sbg.ac.at/desc_CBoundLP.html`, version: 2006-12-20, 2006.

Improvements on Low Discrepancy One-Dimensional Sequences and Two-Dimensional Point Sets

Henri Faure

Institut de Mathématiques de Luminy, UMR 6206 CNRS, 163 Av. de Luminy, case 907, 13288 Marseille cedex 9, France
faure@iml.univ-mrs.fr

Summary. We obtain significant improvements for the star discrepancy D^* of generalized van der Corput sequences by means of linear digit scramblings (see Section 5.2 for the definition). We also find a new lower bound for the extreme discrepancy D of these sequences which permits to show that linearly-scrambled sequences are set in a good place among generalized van der Corput sequences. Finally, we derive the corresponding properties for generalized Hammersley point sets in arbitrary bases and link recent developments in base 2 by Kritzer, Larcher and Pillichshammer to former studies of Béjian and the author.

1 Introduction

For a long time, permutations and linear scramblings play a leading part in QMC methods, especially since the founding works of Owen [Owe95] and Tezuka [Tez94], [Tez95] and the clever classification they received by Matoušek [Mat98]. On the other hand, practitioners utilize systematically random shifts for computing variance estimators, in particular random digit shifts of the nets used to perform quadratures in Randomized QMC methods. As far as we know, no theoretical study justify the superiority of such shifts on scramblings by means of other classes of permutations like linear scramblings.

One of the aims of this paper is to show, in elementary situations (one-dimensional sequences or two-dimensional point sets), that shifting and linearly-scrambling the generators (here van der Corput sequences) improves their discrepancy behavior. Especially, digital shifts improve significantly the behavior of the star discrepancy (Sections 5.2 and 6.2). Such subtle distinction has been made possible thanks to our good knowledge of permuted van der Corput sequences and special $(0,1)-$sequences (Sections 3 and 4.1).

Another motivation is the question of the position of linearly-scrambled sequences in the hierarchy of low discrepancy sequences and in particular among generalized van der Corput sequences. Finding efficient lower bounds for low discrepancy sequences is a difficult task and in this prospect, we have improved a preceding lower bound which permits to sharpen the gap and confirms that many linear digit scramblings produce very good sequences (Sections 4.2–3 and 5.2).

Finally, according to the revival of interest for Hammersley two-dimensional point sets with the works in base 2 by Kritzer, Larcher and Pillichshammer [LP01], [LP03], [Kri06], [KLP06], it was important to link firmly these studies to the preceding ones of Béjian [Bej78] and the author [Fau81], [Fau86], a lot of results being in common although the approachs are –or at least presently seem– quite different (see Section 5 and end of Section 6 for a temporary conclusion).

2 Definitions

2.1 Irregularities of Distribution

The Discrepancies

For a point set $P_N = \{X_1, X_2, \ldots, X_N\}$ in $I^s = [0,1]^s$ and a subinterval J of I^s, we define the *remainder* (to ideal distribution) by

$$E(J; N) = A(J; N) - NV(J)$$

where $A(J; N) = \#\{n; 1 \leq n \leq N, X_n \in J\}$ and $V(J)$ is the volume of J.

Then, the *star discrepancy* D^* and the *discrepancy* D of P_N are defined by

$$D^*(P_N) = \sup_{J^*} |E(J^*; N)| \quad \text{and} \quad D(P_N) = \sup_J |E(J; N)|$$

where J^* (resp. J) is in the shape of $\prod_{i=1}^s [0, y_i[$ (resp. $\prod_{i=1}^s [y_i, z_i[$).

For an infinite sequence X, we denote by $D(N, X)$ and $D^*(N, X)$ respectively the discrepancy and the star discrepancy of its first N points. To emphasize that we deal with the infinite sequence X, we set also

$$E(J; N; X) = A(J; N; X) - NV(J).$$

Note that $D^* \leq D \leq 2^s D^*$.

In the following, we only deal with $s = 1$ or 2.

Relations between sequences and point sets

General principle (also valid for all dimensions), see [Nie92] Lemma 3.7 and [Fau86] Section III:

Let $X = (X_n)$ be an infinite sequence taking its values in I and let P_N be the point set

$$P_N = \left\{ \left(\frac{k-1}{N}, X(k) \right) \; ; \; 1 \leq k \leq N \right\} \subset I^2.$$

Then

$$\max_{1 \leq M \leq N} D^*(M, X) \leq D^*(P_N) \leq \max_{1 \leq M \leq N} D^*(M, X) + 1.$$

The left inequality will be useful to obtain lower bounds for Hammersley point sets with $N = b^n$ points.

2.2 The Sequences

Permuted van der Corput Sequences

Let $b \geq 2$ be an arbitrary integer and let $\Sigma = (\sigma_r)_{r \geq 0}$ be a sequence of permutations of $Z_b = \{0, 1, \ldots, b-1\}$.

For any integer $N \geq 1$, the *permuted (or generalized) van der Corput sequence S_b^Σ in base b associated with Σ* (see [Fau81]) is defined by

$$S_b^\Sigma(N) = \sum_{r=0}^{\infty} \frac{\sigma_r(a_r(N))}{b^{r+1}},$$

where $a_r(N)$ is the r-th digit of the b–adic expansion of $N - 1 = \sum_{r=0}^{\infty} a_r(N)\, b^r$.

If $\Sigma = (\sigma_r) = (\sigma)$ is constant, we write $S_b^\Sigma = S_b^\sigma$.

The *van der Corput sequence in base b*, S_b^I, is obtained with the identical permutation I.

The *original van der Corput sequence* (1935) is S_2^I.

The *permuted van der Corput sequences* are $(0,1)$–sequences (in the sense of Niederreiter–Xing, [NX96]), see [Fau07] Proposition 3.1.

NUT Digital $(0,1)$*–Sequences*

In this case, we deal only with prime bases b. Instead of permutations, we consider the action on the digits of infinite $(\mathbb{N} \times \mathbb{N})$, nonsingular, upper triangular matrices over $\mathbb{F}_b$.

Let $C = (c_k^r)_{r\geq 0, k\geq 0}$ be such a matrix and $N-1 = \sum_{r=0}^{\infty} a_r(N)\, b^r$. Then the *NUT digital* $(0,1)$*-sequence* X_b^C *in base* b *associated with* C is defined by

$$X_b^C(N) = \sum_{r=0}^{\infty} \frac{x_{N,r}}{b^{r+1}} \quad \text{in which} \quad x_{N,r} = \sum_{k=r}^{\infty} c_r^k a_k(N).$$

With the identity matrix I, we obtain $X_b^I = S_b^I$, the van der Corput sequence in base b.

Of course, the NUT digital $(0,1)-$sequences are a special case of $(t,s)-$sequences introduced by Niederreiter, see for instance [Nie92].

2.3 The Point Sets

The point sets we are concerned with here are the usual *Hammersley point sets* associated with the preceding sequences: for any integer $N \geq 1$,

$$HS_b^{\Sigma}(N) = \left\{ \left(\frac{k-1}{N}, S_b^{\Sigma}(k) \right); 1 \leq k \leq N \right\}$$

$$HX_b^C(N) = \left\{ \left(\frac{k-1}{N}, X_b^C(k) \right); 1 \leq k \leq N \right\}.$$

Remark: we write $HX(N)$ instead of HX_N to avoid too many subscripts. In almost all studies on two-dimensional Hammersley point sets, N is a power of the base b ([DeC86], [Whi75]) and moreover very often $b = 2$ ([DLP05], [HZ69], [Kri06], [KLP06], [LP01], [LP03]).

3 Exact Formulas for the One–Dimensional Sequences

3.1 Functions Related to a Pair (b, σ)

Let σ be a permutation of Z_b and set $Z_b^{\sigma} := \left(\frac{\sigma(0)}{b}, \ldots, \frac{\sigma(b-1)}{b} \right)$. For any integer h with $0 \leq h \leq b-1$, define the real function $\varphi_{b,h}^{\sigma}$ as follows:
Let k be an integer with $1 \leq k \leq b$; then for every $x \in [\frac{k-1}{b}, \frac{k}{b}[$ set:

$$\varphi_{b,h}^{\sigma}(x) = A\left(\left[0, \frac{h}{b}\right[; k; Z_b^{\sigma} \right) - hx \quad \text{if } 0 \leq h \leq \sigma(k-1) \quad \text{and}$$

$$\varphi_{b,h}^{\sigma}(x) = (b-h)x - A\left(\left[\frac{h}{b}, 1\right[; k; Z_b^{\sigma} \right) \text{ if } \sigma(k-1) < h < b.$$

Finally the function $\varphi_{b,h}^{\sigma}$ is extended to $\mathbb{R}$ by periodicity. Note that $\varphi_{b,0}^{\sigma} = 0$. These functions are linearizations of remainders associated with Z_b^{σ}.

In the special case $b = 2$, we only have two permutations which give either $\varphi_{2,1}^{\sigma} = \|\cdot\|$ if $\sigma = I$ or $\varphi_{2,1}^{\sigma} = -\|\cdot\|$ if $\sigma = (0\ 1)$, where $\|\cdot\|$ is the distance to the nearest integer (occuring in many austrian papers).

Actually, the b functions $\varphi_{b,h}^{\sigma}$ give rise to other functions, depending only on (b, σ), according to the notion of discrepancy at work: for the extreme discrepancies D^* and D, we need

$$\psi_b^{\sigma,+} = \max_{0 \le h \le b-1} (\varphi_{b,h}^{\sigma}),\ \psi_b^{\sigma,-} = \max_{0 \le h \le b-1} (-\varphi_{b,h}^{\sigma})$$

and

$$\psi_b^{\sigma} = \psi_b^{\sigma,+} + \psi_b^{\sigma,-} = \sup_{0 \le h < h' < b} |\varphi_{b,h'}^{\sigma} - \varphi_{b,h}^{\sigma}|.$$

These functions have been introduced in [Fau81]. Other functions are necessary for the study of the L_2–discrepancy and the diaphony (see [CF93]), notions we shall not consider in this paper.

3.2 The Exact Formulas for the Sequences S_b^{Σ} and X_b^C

For any *permuted van der Corput sequence* S_b^{Σ} *in base* b *associated with* Σ, we have [Fau81], for all integers n and N with $1 \le N \le b^n$,

$$D^+(N, S_b^{\Sigma}) := \sup_{0 \le \alpha \le 1} E([0, \alpha[; N; X)$$
$$= \sum_{j=1}^{n} \psi_b^{\sigma_{j-1},+}\left(\frac{N}{b^j}\right) + \frac{N}{b^n} - N \sum_{j=n+1}^{\infty} \frac{\sigma_{j-1}(0)}{b^j},$$

$$D^-(N, S_b^{\Sigma}) := \sup_{0 \le \alpha \le 1} (-E([0, \alpha[; N; X))$$
$$= \sum_{j=1}^{n} \psi_b^{\sigma_{j-1},-}\left(\frac{N}{b^j}\right) + N \sum_{j=n+1}^{\infty} \frac{\sigma_{j-1}(0)}{b^j}.$$

Recall that for any one–dimensional sequence,

$$D^*(N, X) = \max(D^+(N, X), D^-(N, X)) \text{ and}$$
$$D(N, X) = D^+(N, X) + D^-(N, X).$$

Therefore, the star discrepancy of S_b^{Σ} is obvious and its discrepancy is given by

$$D(N, S_b^{\Sigma}) = \sum_{j=1}^{n} \psi_b^{\sigma_{j-1}}\left(\frac{N}{b^j}\right) + \frac{N}{b^n}.$$

Concerning the *NUT digital* $(0,1)$–*sequences* X_b^C, we have been able to obtain the corresponding formulas for D^+, D^- and D (see [Fau05a]), but here things are more complicated and we shall restrict in the present paper to D, for which a simple formula occurs by means of the sequence of permutations $\Delta = (\delta_r)$ defined by $\delta_r(i) = c_r^r i \pmod b$ (the c_r^r are the diagonal entries of C). We obtain

$$D(N, X_b^C) = D(N, S_b^\Delta) = \sum_{j=1}^{n} \psi_b^{\delta_{j-1}}\left(\frac{N}{b^j}\right) + \frac{N}{b^n}.$$

There are analogous formulas for the L_2–discrepancy and the diaphony of S_b^Σ and X_b^C (see [CF93] and [Fau05a]).

4 Bounds and Asymptotic Behavior for $D(N, S_b^\sigma)$

The experimental research of pairs (b, Σ) giving the lowest discrepancy D shows that it is better to work with constant sequences $\Sigma = (\sigma)$. On the contrary, concerning D^*, it appears that specific sequences of permutations Σ associated with σ give the best results (even optimal with the identical permutation); this case of D^* will be considered in the next section. Note also that, according to the last formula of Section 3 above, the case of NUT digital $(0,1)$–sequences X_b^C reduces to sequences S_b^Σ for the discrepancy D.

4.1 Upper Bounds for $D(N, S_b^\sigma)$

Set $d_b^\sigma(n) := \sup_{x\in\mathbb{R}} \sum_{j=1}^n \psi_b^\sigma\left(\frac{x}{b^j}\right)$ and $\alpha_b^\sigma := \inf_{n\geq 1} \frac{d_b^\sigma(n)}{n}$. Then

$$\limsup_{N\to\infty} \frac{D(N, S_b^\sigma)}{\log N} = \frac{\alpha_b^\sigma}{\log b} \quad \text{and} \quad D(N, S_b^\sigma) \leq \frac{\alpha_b^\sigma}{\log b}\log N + \alpha_b^\sigma + 2.$$

Moreover, $\alpha_b^\sigma = \lim_{n\to\infty} \frac{d_b^\sigma(n)}{n}$ and there exists β_n, with $0 \leq \beta_n \leq 1$, such that $d_b^\sigma(n) = \alpha_b^\sigma n + \beta_n$, so that

$$0 \leq d_b^\sigma(1) - \alpha_b^\sigma \leq 1.$$

We also have

$$d_b^\sigma(1) = \sup \psi_b^\sigma = \max_{1\leq k\leq b} \max_{0\leq h'<h<b} \left| E\left(\left[\frac{h'}{b}, \frac{h}{b}\right[; k; Z_b^\sigma\right)\right| =: d_b^\sigma.$$

The quantity d_b^σ is very important since it is easy to compute and it permits to bound α_b^σ from above (and below for large bases b). We will call it the *discrete discrepancy* of the net $\left\{\left(\frac{k-1}{b}, Z_b^\sigma(k)\right); 1 \leq k \leq b\right\}$ (a slight variant of the *discrete discrepancy* introduced in [LP03] p. 399). Numerical examples can be found in Table 1 below.

We recall also the properties $D(N, S_b^\sigma) \leq D(N, S_b^I)$ and $\alpha_b^I = \dfrac{b-1}{4\log b}$ if b is odd and $\alpha_b^I = \dfrac{b^2}{4(b+1)\log b}$ if b is even. All these results come from [Fau81] (Theorem 2, Property 3.2.2, Lemma 4.2.2, Section 5.5.4 and Theorem 6).

Table 1. Examples

b	$\frac{\tilde{f}(b)}{\log b}$	$\frac{\overline{f}(b)}{\log b}$	$\frac{d_b^{\sigma_0}-1}{\log b}$	$\frac{d_b^{\sigma_0}}{\log b}$	$\frac{\alpha_b^I}{\log b}$
19	.339	.375	.107	.447	1.528
36	.318	.338	.116	.396	2.443
101	.279	.287	.293	.511	5.416
233	.251	.254	.261	.444	10.64
367	.232	.238	.331	.503	15.49
1301	.2020	.2024	.327	.467	45.32

4.2 Lower Bounds for $D(N, S_b^\sigma)$

The problem of lower bounds for the discrepancy D is still a very challenging problem. In one dimension, the famous result of Schmidt (1972, improved by Béjian [Bej82], see the first inequality below) solves the problem for the order of magnitude, but improving the constants is still an open question. We have from the general to the specific:

$$\text{for any sequence } X,\; 0.12 < \limsup_{N\to\infty} \frac{D(N,X)}{\log N},$$

$$\text{for any sequence } S_b^\sigma,\quad \frac{b-2}{(b-1)\log b} \leq \limsup_{N\to\infty} \frac{D(N,S_b^\sigma)}{\log N} \left(< \frac{b}{4\log b}\right),$$

$$\text{for a specific sequence } S_b^\sigma,\quad \frac{d_b^\sigma - 1}{\log b} < \limsup_{N\to\infty} \frac{D(N,S_b^\sigma)}{\log N} \left(< \frac{d_b^\sigma}{\log b}\right).$$

The lower bound for arbitrary sequences S_b^σ results from the property $\theta_b \leq \psi_b^\sigma$ with θ_b the 1-periodic function defined by $\theta_b(x) = (b-1)x$ if $x \in [0, \frac{1}{b}]$, $\theta_b(x) = 1 - x$ if $x \in [\frac{1}{b}, \frac{1}{2}]$ and $\theta_b(x) = \theta_b(1-x)$ if $x \in [\frac{1}{2}, 1]$ (see [Fau81] Section 4.1). The lower bound for a specific sequence S_b^σ is poor for small b (see Table 1).

4.3 A New Lower Bound for $D(N, S_b^\sigma)$

With increasing b, it is difficult to find good lower bounds for $D(N, S_b^\sigma)$. This research seems to be of the same kind as for arbitrary sequences X. But nevertheless, by optimization of the remainders with intervals containing 0 or 3 points, we have been able to improve the preceding lower bound of $\frac{b-2}{(b-1)\log b}$ slightly. The *idea* is to consider the remainders (for $2 \leq k \leq b-1$)

$$E\left(\left[\frac{l+1}{b}, \frac{l+1+h}{b}\right[; k; Z_b^\sigma\right) \quad \text{and} \quad E\left(\left[\frac{l}{b}, \frac{l+2+h}{b}\right[; k+1; Z_b^\sigma\right)$$

with h and l such that

- $Z_b^\sigma(k+1) \in \left[\frac{l+1}{b}, \frac{l+1+h}{b}\right[$,
- $\frac{l}{b}$ is the abscis of the first point $Z_b^\sigma(i)$ $(i \leq k)$ preceding $Z_b^\sigma(k+1)$,
- $\frac{l+1+h}{b}$ is the abscis of the first point $Z_b^\sigma(j)$ $(j \leq k)$ following $Z_b^\sigma(k+1)$.

In that way, $E\left(\left[\frac{l+1}{b}, \frac{l+1+h}{b}\right[; k; Z_b^\sigma\right) = h\frac{k}{b}$ (no point) and (3 points) $E\left(\left[\frac{l}{b}, \frac{l+2+h}{b}\right[; k+1; Z_b^\sigma\right) = 3 - (h+2)\frac{k+1}{b}$. Since one increases with h and the other decreases with h, we obtain the minimum for the equality $hk = 3b - (h+2)(k+1)$. Leaving aside the integer nature of h and k, we get the function $f(k) = \frac{k(3b-2k-2)}{2k+1}$ for $2 \leq k \leq b-1$ and computing the maximum of f (obtained for $k_0 = \frac{\sqrt{3b-1}-1}{2}$ with $f(k_0) = \frac{3b}{2} - \sqrt{3b-1}$), we arrive at:

Proposition 1. *For any sequence S_b^σ, we have*

$$\frac{\tilde{f}(b)}{\log b} \leq \limsup_{N\to\infty} \frac{D(N, S_b^\sigma)}{\log N} \quad \text{with} \quad \tilde{f}(b) \approx \frac{3}{2} - \frac{\sqrt{3b-1}}{b} := \overline{f}(b).$$

Of course, with specific bases b, more precise results can be obtained with the ceiling or the floor of reals involved in the computations. Table 1 gives some examples of such computations (the reals are rounded to 3 significant digits), some comments follow. The lower bound $\frac{d_b^{\sigma_0}-1}{\log b}$ is not interesting for small bases. The lower bound of Proposition 1 makes sense until bases as large as $b \approx 10^5$; after that the general lower bound 0.12 is better. The best permutations σ_0 (presently known) for large bases are linear digit scramblings without additive term (see the next section). The general lower bound 0.12 seems far from the lowest possible value for S_b^σ sequences; for very large b, our lower bounds also. The method of Proposition 1 facilitate too the research of the best permutations for small bases; moreover, in the case of $b = 36$, it was possible to compute the exact value $\alpha_{36}^{\sigma_0} = \frac{46}{35}$, so that $\frac{\alpha_{36}^{\sigma_0}}{\log 36} = 0.366\ldots$, the smallest discrepancy D presently known among all one-dimensional sequences [Fau92]. Note also that for each prime base up to 1301, we have found many digit scramblings σ with constants $\frac{d_b^\sigma}{\log b}$ around 0.5 [Fau05b], [Fau06]. Compared with the general lower bounds $\frac{\overline{f}(b)}{\log b}$ and the upper bounds $\frac{\alpha_b^I}{\log b}$, we see that a lot of linearly-scrambled sequences are really set in a good place among the family of sequences S_b^σ.

5 Formulas and Asymptotic Behavior for D^*

As announced at the beginning of Section 4, we come now to the study of the star discrepancy D^*. First, we introduce the method to obtain low star discrepancy sequences and give the main results we got with it (see [Fau81]).

Then we link our results with recent publications of Kritzer, Larcher and Pillichshammer by means of linear digit scramblings and finally, we obtain new remarkable simple sequences with such scramblings judiciously chosen.

5.1 Swapping Permutations

Let $\mathcal{E}$ be a subset of $\mathbb{Z}^+$ and σ a permutation of Z_b. Let τ be the permutation of Z_b defined by $\tau(k) = b-k-1,\ 0 \leq k \leq b-1$. Define the sequence $\Sigma_{\mathcal{E}}^{\sigma} = (\sigma_j)_{j\geq 0}$ by $\sigma_j = \sigma$ if $j \in \mathcal{E}$ and $\sigma_j = \tau \circ \sigma$ if $j \notin \mathcal{E}$. Then ([Fau81] Lemma 4.4.1)

$$D^+(N, S_b^{\Sigma_{\mathcal{E}}^{\sigma}}) = \sum_{j=1, j\in\mathcal{E}}^{\infty} \psi_b^{\sigma,+}\left(\frac{N}{b^j}\right) + \sum_{j=1, j\notin\mathcal{E}}^{\infty} \psi_b^{\sigma,-}\left(\frac{N}{b^j}\right),$$

$$D^-(N, S_b^{\Sigma_{\mathcal{E}}^{\sigma}}) = \sum_{j=1, j\in\mathcal{E}}^{\infty} \psi_b^{\sigma,-}\left(\frac{N}{b^j}\right) + \sum_{j=1, j\notin\mathcal{E}}^{\infty} \psi_b^{\sigma,+}\left(\frac{N}{b^j}\right).$$

The permutation τ swaps the functions $\psi_b^{\sigma,+}$ and $\psi_b^{\sigma,-}$ hence, to get lower $D^* = \max(D^+, D^-)$, we must find $\mathcal{E}$ so that the sums with $\psi_b^{\sigma,+}$ and $\psi_b^{\sigma,-}$ divide into two equal parts. This is achieved, among others, by the following set:

$$\mathcal{A} = \{0, 2, 3, 6, 7, 8, 12, 13, 14, 15, \ldots\}.$$

With that $\mathcal{A}$ and with the condensed notation $\overline{\sigma} = \tau \circ \sigma$, we obtain

$$\Sigma_{\mathcal{A}}^{\sigma} = (\sigma, \overline{\sigma}, \sigma, \sigma, \overline{\sigma}, \overline{\sigma}, \sigma, \sigma, \sigma, \overline{\sigma}, \overline{\sigma}, \overline{\sigma}, \ldots) \quad \text{and}$$

$$\limsup_{N\to\infty} \frac{D^*(N, S_b^{\Sigma_{\mathcal{A}}^{\sigma}})}{\log N} = \frac{\alpha_b^{\sigma,+} + \alpha_b^{\sigma,-}}{2\log b} \quad \text{in which}$$

$$\alpha_b^{\sigma,+} = \inf_{n\geq 1} \sup_{x\in[0,1]} \sum_{j=1}^{n} \psi_b^{\sigma,+}\left(\frac{x}{b^j}\right) \quad \text{and} \quad \alpha_b^{\sigma,-} = \inf_{n\geq 1} \sup_{x\in[0,1]} \sum_{j=1}^{n} \psi_b^{\sigma,-}\left(\frac{x}{b^j}\right).$$

For small b, the constants $\alpha_b^{\sigma,+}$ and $\alpha_b^{\sigma,-}$ are not difficult to compute as well as for the identical permutation in any base b (in this case $\psi_b^{I,-} = 0$). We write down *two important consequences*:

$$\limsup_{N\to\infty} \frac{D^*(N, S_b^{\Sigma_{\mathcal{A}}^{I}})}{\log N} = \frac{\alpha_b^{I,+}}{2\log b} = \frac{b-1}{8\log b} \quad \text{if } b \text{ is odd,}$$

$$\limsup_{N\to\infty} \frac{D^*(N, S_b^{\Sigma_{\mathcal{A}}^{I}})}{\log N} = \frac{\alpha_b^{I,+}}{2\log b} = \frac{b^2}{8(b+1)\log b} \quad \text{if } b \text{ is even,}$$

and in base 12, there exists a permutation σ_0 such that (this is the smallest star discrepancy currently known):

$$\limsup_{N\to\infty} \frac{D^*(N, S_{12}^{\Sigma_A^{\sigma_0}})}{\log N} = \frac{1919}{3454 \log 12} = .223\ldots$$

(see [Fau81], Theorems 3, 5, 6).
When $b = 2$, we recover the result of Béjian [Bej78], re-discovered recently in [KLP06] Section 5 after a lot of computations involving shifted Hammersley point sets (see Section 6 below):

$$\limsup_{N\to\infty} \frac{D^*(N, S_2^{\Sigma_A^I})}{\log N} = \frac{1}{6 \log 2} = .2404\ldots.$$

When $b = 3$, we obtain "almost" the best sequence, anyhow the best among the sequences S_b^I:

$$\limsup_{N\to\infty} \frac{D^*(N, S_3^{\Sigma_A^I})}{\log N} = \frac{1}{4 \log 3} = .227\ldots.$$

As to NUT digital $(0,1)$–sequences, things appear much more complicated for D^* than for D. The only result we know is from Pillichshammer [Pil04], in base 2, with the NUT matrix whose all entries are 1. The special case of diagonal matrices Δ, $X_b^\Delta = S_b^\Delta$, follows.

5.2 Linear Digit Scramblings

This denomination comes from J. Matoušek [Mat98] in his attempt to classify the very general scramblings introduced four years before by A. Owen.
A *linear digit scrambling* is a permutation of the set Z_b of the form

$$\pi(k) = fk + g \pmod b \quad (0 \le k \le b-1),$$

where $f \neq 0$ and g are given in $\mathbb{Z}_b$ (identified as a set to Z_b) with b *prime*. The definition works also for any base b, provided that the multiplication by f remains a bijection.
If $g = 0$, we obtain the so-called *multiplicative factors* f of our preceding papers on NUT digital sequences ([Fau05b], [Fau06]). The additive factor g is a translation also called *digital shift* ([Kri06], [KLP06]).
It is quite remarkable that the swapping of permutations is a linear scrambling for any base, even though it is also quite trivial since $\tau(k) = b - 1 - k = (b-1)k + b - 1 \pmod b$. But, according to the importance of this property to link our former study on D^* to recent studies on digital scramblings, we defer it to a proposition.

Proposition 2. *The permutation τ in the definition of sequences $S_b^{\Sigma_\varepsilon^\sigma}$ is the linear digit scrambling $\tau(k) = (b-1)k + b - 1$.*

The case of the base 2:

In base $b = 2$, since $\tau(k) = (b-1)k + b - 1 = k + 1$, the *digital shift* on $\mathbb{F}_2$ introduced in [Kri06] and [KLP06] is the permutation τ. In other words, *in base 2, shifting is swapping.* Typical results in these papers follow from [Fau81], [Fau86] and from the former Note of Béjian [Bej78] where, unfortunately, the proofs are only outlined. As an example, we recall Theorem 2 of [Bej78] (compare with [KLP06], end of Section 4):

$$\limsup_{N\to\infty} \frac{D^*(N, S_2^{\Sigma_{\mathcal{M}}^I})}{\log N} = \frac{1}{6\log 2} + \frac{(2^m-1)(2^m-(-1)^m)}{18m(2^{2m}-(-1)^m)\log 2}$$

with $\Sigma_{\mathcal{M}}^I = (\underbrace{I, I, \ldots, I}_{m}, \underbrace{\tau, \tau, \ldots, \tau}_{m}, \underbrace{I, I, \ldots, I}_{m}, \underbrace{\tau, \tau, \ldots, \tau}_{m}, \ldots)$.

The case of the base 3:

The 6 permutations are linear digit scramblings, in short $I, I+1, I+2, 2I, 2I+1$ and $2I + 2 = \tau$. All of them have the same asymptotic behavior and there is no improvement with regards to $S_3^{\Sigma_{\mathcal{A}}^I}$.

The case of the base 4:

The sequence $S_4^{(12)}$ ((12) means the transposition which exchanges 1 and 2) is the van der Corput sequence S_2^I and therefore $S_4^{\Sigma_{\mathcal{A}}^{(12)}} = S_2^{\Sigma_{\mathcal{A}}^I}$ with asymptotic constant $\frac{1}{6\log 2}$. No improvement with the linear digit scramblings $3I + g$, but the permutation $\sigma = 3(12) + 1$ produces the sequence $S_4^\sigma = S_2^{\Sigma_{\mathcal{M}}^I}$ (for $m = 1$) with asymptotic constant $\frac{1}{5\log 2}$, the sequence associated with the Halton-Zaremba plane point set [HZ69]. We see that the permutation $\sigma = 3(12) + 1$ in base 4 is equivalent to the sequence of permutations $\Sigma_{\mathcal{M}}^I$ (with $m = 1$) in base 2. Moreover, $S_4^{\Sigma_{\mathcal{A}}^I}$ has also the same asymptotic constant $\frac{1}{5\log 2}$, hence the only permutation σ is also equivalent to the sequence of permutations $\Sigma_{\mathcal{A}}^I$ in base 4.

The case of the bases $b > 4$:

With increasing b, things become more intricate and we cannot give an exhaustive study. Let us only remark that the considerations of Section 4.1 apply to functions $\psi_b^{\sigma,+}$ and $\psi_b^{\sigma,-}$ and permit to obtain upper bounds for the star discrepancy as well, see [Fau86]. We give in the following two examples showing the great interest of linear digit scramblings, especially with a digital shift $g \neq 0$.

In base 5, we observe the same phenomenon as in base 4, but this time with the linear digit scramblings $\pi(k) = 3k + 1$ and $\pi'(k) = 2k + 3$. The sequences S_5^π, $S_5^{\pi'}$ and $S_5^{\Sigma_A^I}$ have the same asymptotic constant $\dfrac{1}{2\log 5}$, but the first ones are much simpler. A simple bound is $\dfrac{3}{5\log 5}$; finding the exact constant needs a bit more computations.

In base 233, which is our best prime base for D up to 1301 (see [Fau05b]), the phenomenon is magnified: with the two linear digit scramblings $\rho(k) = 89k$ and $\pi(k) = 89k + 44$ we have

$$\limsup_{N\to\infty} \frac{D^*(N, S_{233}^{\Sigma_A^\rho})}{\log N} < \frac{469 + 365}{466\log 233} = .328\ldots,$$

$$\limsup_{N\to\infty} \frac{D^*(N, S_{233}^{\pi})}{\log N} < \frac{368}{233\log 233} = .289\ldots$$

$$\limsup_{N\to\infty} \frac{D^*(N, S_{233}^{\Sigma_A^\pi})}{\log N} < \frac{368}{233\log 233} = .289\ldots \text{ while}$$

$$\limsup_{N\to\infty} \frac{D^*(N, S_{233}^{\Sigma_A^I})}{\log N} = \frac{232}{8\log 233} = 5.32\ldots.$$

Here, we observe more: adding a digital shift (44) still improves the behavior of the sequence and without using the "sophisticated" sequence Σ_A. We bring this fact together with the remark of Matoušek in [Mat98] p. 537: "Introducing additive terms makes the situation much simpler and more regular".

6 Hammersley Point Sets

6.1 Foreword

The study of two-dimensional Hammersley point sets (see Section 2.3) can be approched by two ways (at least): directly, with the investigation of the two-dimensional remainder $E([0,\alpha[\times[0,\beta[; N) = A([0,\alpha[\times[0,\beta[; N) - N\alpha\beta$, or as a by-product of the study of the infinite one-dimensional sequences associated with the second coordinate of the point set.

In the first approach, systematically $N = b^n$ is a power of the base. Precise studies, with exact formulas, have been done with combinatorial tools in arbitrary bases [DeC86], [Whi75] and in base two [HZ69]. More recently, by means of Walsh series analysis in base two, important new advances were provided by the austrian team of Linz and Salzburg [LP01], [LP03], [Kri06], [KLP06], especially in the study of digitally shifted Hammersley point sets in base two.

The second approach is based on the relations between sequences and point sets (see Section 2.1). Here, there is no necessity for N to be a power of b but, since these relations are inequalities, this approach cannot claim to get exact formulas for the star discrepancy of Hammersley point sets. Nevertheless, the precise knowledge of the sequences S_b^Σ in arbitrary bases permits to obtain the best results currently known and to derive very good approximations of the exact formulas of De Clerck (see [Fau86] for details). In the same way our new lower bound (Section 4.3) and our linear digit scramblings for D^* (Section 5.2) give new results and allow investigations in very large (prime) bases for two-dimensional Hammersley point sets. We make them explicit in the next subsection.

6.2 New Results for Hammersley Point Sets

General estimates

Proposition 3. *For any pair* (b, Σ) *and any integer* $n \geq 1$ *we have*

$$\frac{n\tilde{f}(b)}{2} \leq D^*(HS_b^\Sigma(b^n)) \leq D^*(HS_b^I(b^n)) \leq \frac{b}{4}n + 3$$

in which $\tilde{f}(b) \approx \dfrac{3}{2} - \dfrac{\sqrt{3b-1}}{b}$.

Proposition 3 is also valid for the Hammersley point sets $HX_b^C(b^n)$.
The lower bound results from Proposition 1 and from the relation $D \leq 2D^*$.
The right upper bound is not optimal and is a compression of formulas for odd and even b given in [Fau86], in which also the upper bounds are in terms of N instead of b^n.
In particular, we see that the usual Hammersley point sets HS_b^I are the worst distributed among the HS_b^Σ and the HX_b^C, a property re-discovered 20 years later for $b = 2$ by Kritzer [Kri06], in the more general setting of $(0, n, 2)-$Hammersley nets over $\mathbb{Z}_2$.
As to the lower bound, we recall the preceding one obtained in [Fau86]: $D^*(HS_b^\Sigma(b^n)) \geq \dfrac{n(b-2)}{2(b-1)}$. The gain on the constant is about 0.5 to 0.75, see Section 4.3 for comments on these poor (but only known) lower bounds.

Linear digit scramblings

Very good Hammersley point sets in any given prime base can be obtained with linear digit scramblings of the original Hammersley point sets HS_b^I. They are deduced with the help of the general principle, right inequality, from the method and the results of Section 5.2. For instance, and *without* the complicated generic sequence of permutations $\Sigma_{\mathcal{A}}^\sigma$, we have:

$$\text{with } b = 5 \text{ and } \pi(k) = 3k + 1, \; D^*(HS_5^\pi(N)) < \frac{\log N}{2 \log 5} + c < .32 \log N + c,$$

with $b = 233$ and $\pi(k) = 89k + 44$, $D^*(HS^{\pi}_{233}(N)) < .29 \log N + c$,
with $b = 1301$ and $\pi(k) = 498k + 243$, $D^*(HS^{\pi}_{1301}(N)) < .31 \log N + c$

So, until now, bases 2, 3 and 12 remain the best with respectively the constants 0.240..., 0.227... and 0.223..., but with our method we need the sequence $\Sigma^{\sigma}_{\mathcal{A}}$ for these three sequences. Revisiting the proofs in our papers [Fau81] and [Fau86] will permit to obtain the same constants without using the sequence $\Sigma^{\sigma}_{\mathcal{A}}$, as did Kritzer, Larcher and Pillichshammer in base 2 with their direct approach of the remainder $E([0,\alpha[\times[0,\beta[;N) = A([0,\alpha[\times[0,\beta[;N) - N\alpha\beta$ (see [Kri05] and [KLP06]). This study, together with other comparisons between the two methods, will be considered in a forthcoming paper.

Acknowledgements

We want to thank the referee for the correction of various misprints and inaccuraties in English.

References

[Bej78] R. Béjian. Sur certaines suites présentant une faible discrépance à l'origine. C. R. Acad. Sc.,Paris, **286A**, 135–138 (1978).

[Bej82] R. Béjian. Minoration de la discrépance d'une suite quelconque sur $\mathbb{T}$. Acta Arith., **41**, 185–202 (1982).

[CF93] H. Chaix and H. Faure. Discrépance et diaphonie en dimension un. Acta Arith., **63**, 103–141 (1993).

[DeC86] L. De Clerck. A method for the exact calculation of the star-discrepancy of plane sets applied to the sequences of Hammersley. Mh. Math., **101**, 261–278 (1986).

[DLP05] M. Drmota, G. Larcher, and F. Pillichshammer. Precise distribution properties of the van der Corput sequence and related sequences. Manuscripta Math. **118**, 11–41 (2005).

[Fau81] H. Faure. Discrépance de suites associées à un système de numération (en dimension un). Bull. Soc. Math. France, **109**, 143–182 (1981).

[Fau86] H. Faure. On the star-discrepancy of generalized Hammersley sequences in two dimensions. Mh. Math., **101**, 291–300 (1986).

[Fau92] H. Faure. Good permutations for extreme discrepancy. J. Number Theory, **41**, 47–56 (1992).

[Fau05a] H. Faure. Discrepancy and diaphony of digital (0,1)–sequences in prime base. Acta Arith. **117.2**, 125–148 (2005).

[Fau05b] H. Faure. Irregularities of distribution of digital $(0,1)$–sequences in prime base. Integers, Electronic Journal of Combinatorial Number Theory **5(3) A07**, 1–12 (2005).

[Fau06] H. Faure. Selection criteria for (random) generation of digital $(0,s)$–sequences. In: Niederreiter, H., Talay, D. (Eds.) Monte Carlo and Quasi-Monte Carlo Methods 2004. Springer, Berlin Heidelberg New York, 113–126 (2006).

[Fau07] H. Faure. Van der Corput sequences towards general $(0,1)-$sequences in base b. J. Théor. Nombres Bordeaux, to appear (2007).

[HZ69] J.H. Halton and S.K. Zaremba. The extreme and L_2–discrepancies of some plane sets. Mh. Math., **73**, 316–328 (1969).

[Kri05] P. Kritzer. A new upper bound on the star discrepancy of $(0,1)$–sequences. Integers, Electronic Journal of Combinatorial Number Theory **5(3) A11**, 1–9 (2005).

[Kri06] P. Kritzer. On some remarkable properties of the two-dimensional Hammersley point set in base 2. J. Théor. Nombres Bordeaux **18**, 203–221 (2006).

[KLP06] P. Kritzer, G. Larcher, and F. Pillichshammer. A thorough analysis of the discrepancy of shifted Hammersley and van der Corput point sets. Ann. Math. Pura Appl., to appear (2006).

[LP01] G. Larcher and F. Pillichshammer. Walsh series analysis of the L_2–discrepancy of symmetrisized points sets. Mh. Math., **132**, 1–18 (2001).

[LP03] G. Larcher and F. Pillichshammer. Sums of distances to the nearest integer and the discrepancy of digital nets. Acta Arith., **106.4**, 379–408 (2003).

[Mat98] J. Matoušek. On the L_2-discrepancy for anchored boxes. J. Complexity, **14**, 527–556 (1998).

[Nie92] H. Niederreiter. *Random Number Generation and Quasi-Monte Carlo Methods*. CBMS-NSF Regional Conference Series in Applied Mathematics. SIAM, Philadelphia (1992).

[NX96] H. Niederreiter and C.P. Xing. Quasirandom points and global functions fields. In: Cohen, S. and Niederreiter, H. (eds) Finite Fields and Applications. London Math. Soc. Lectures Notes Series, **233**, 269–296 (1996).

[Owe95] A.B. Owen. Randomly permuted (t, m, s)-nets and (t, s)-sequences. In: Niederreiter, H., Shiue, P. (eds.) Monte Carlo and Quasi-Monte Carlo Methods in Scientific Computing. Springer, Lectures Notes in Statistics, **126**, 299–317 (1995).

[Pil04] F. Pillichshammer. On the discrepancy of (0,1)–sequences. J. Number Theory, **104**, 301–314 (2004).

[Tez94] S. Tezuka. A generalization of Faure sequences and its efficient implementation. Research Report IBM RT0105 (1994).

[Tez95] S. Tezuka. *Uniform Random Numbers: Theory and Practice*. Kluwer Academic Publishers, Boston (1995).

[Whi75] B.E. White. Mean-square discrepancies of the Hammersley and Zaremba sequences for arbitrary radix. Mh. Math., **80**, 219–229 (1975).

Improved Multilevel Monte Carlo Convergence using the Milstein Scheme

Mike Giles

Oxford University Computing Laboratory, Parks Road, Oxford, U.K.
Mike.Giles@comlab.ox.ac.uk

Summary. In this paper we show that the Milstein scheme can be used to improve the convergence of the multilevel Monte Carlo method for scalar stochastic differential equations. Numerical results for Asian, lookback, barrier and digital options demonstrate that the computational cost to achieve a root-mean-square error of ϵ is reduced to $O(\epsilon^{-2})$. This is achieved through a careful construction of the multilevel estimator which computes the difference in expected payoff when using different numbers of timesteps.

1 Introduction

In many financial engineering applications, one is interested in the expected value of a financial option whose payoff depends upon the solution of a stochastic differential equation. To be specific, we consider an SDE with general drift and volatility terms,

$$\mathrm{d}S(t) = a(S,t)\,\mathrm{d}t + b(S,t)\,\mathrm{d}W(t), \quad 0 < t < T, \tag{1}$$

with given initial data S_0. In the case of European and digital options, we are interested in the expected value of a function of the terminal state, $f(S(T))$, but in the case of Asian, lookback and barrier options the valuation depends on the entire path $S(t), 0 < t < T$.

Using a simple Monte Carlo method with a numerical discretisation with first order weak convergence, to achieve a root-mean-square error of $O(\epsilon)$ would require $O(\epsilon^{-2})$ independent paths, each with $O(\epsilon^{-1})$ timesteps, giving a computational complexity which is $O(\epsilon^{-3})$. We have recently introduced a new multilevel approach [Gil06] which reduces the cost to $O(\epsilon^{-2}(\log \epsilon)^2)$ when using an Euler path discretisation for a European option with a payoff with a uniform Lipschitz bound. This multilevel approach is related to the two-level method of Kebaier [Keb05], and is similar to the multi-level method proposed by Speight [Spe05] based on the quasi control variate method of Emsermann and Simon [ES02]. There are also strong similarities to Heinrich's multilevel approach for parametric integration [Hei01].

In the previous work, it was also proved that the computational cost can be further reduced to $O(\epsilon^{-2})$ for numerical discretisations with certain multilevel convergence properties. The objective of this paper is to demonstrate that this improved complexity is attainable for scalar SDEs with a variety of exotic options through using the Milstein path discretisation. For European options with a Lipschitz continuous payoff, it can be proved that this an immediate consequence of the improved strong order of convergence of the Milstein discretisation compared to the simpler Euler discretisation. However, for Asian, lookback, barrier and digital options, special numerical treatments have to be introduced, and that is the focus of the paper. Furthermore, no *a priori* convergence proofs have yet been constructed for these cases and so the paper relies on numerical demonstration of the effectiveness of the algorithms that have been developed.

The paper begins by reviewing the multilevel approach, and the theorem which describes its computational cost given certain properties of the numerical discretisation. The next section discusses the Milstein discretisation and the challenges of achieving higher order variance convergence within the multilevel method. Asian, lookback, barrier and digital options are all considered, and $O(\epsilon^{-2})$ computational cost is demonstrated for each through the use of Brownian interpolation to approximate the behaviour of paths within each timestep.

The final section indicates the direction of future research, including the need for *a priori* convergence analysis, the challenges of extending this work to multi-dimensional SDEs, and the use of quasi-Monte Carlo methods for further reduction of the computational complexity.

2 Multilevel Monte Carlo Method

Consider Monte Carlo path simulations with different timesteps $h_l = 2^{-l}\,T$, $l = 0, 1, \ldots, L$. Thus on the coarsest level, $l=0$, the simulations use just 1 timestep, while on the finest level, $l=L$, the simulations use 2^L timesteps. For a given Brownian path $W(t)$, let P denote the payoff, and let $\widehat{P}_l$ denote its approximation using a numerical discretisation with timestep h_l. Because of the linearity of the expectation operator, it is clearly true that

$$E[\widehat{P}_L] = E[\widehat{P}_0] + \sum_{l=1}^{L} E[\widehat{P}_l - \widehat{P}_{l-1}]. \qquad (2)$$

This expresses the expectation on the finest level as being equal to the expectation on the coarsest level plus a sum of corrections which give the difference in expectation between simulations using different numbers of timesteps. The idea behind the multilevel method is to independently estimate each of the expectations on the right-hand side in a way which minimises the overall variance for a given computational cost.

Let $\widehat{Y}_0$ be an estimator for $E[\widehat{P}_0]$ using N_0 samples, and let $\widehat{Y}_l$ for $l>0$ be an estimator for $E[\widehat{P}_l - \widehat{P}_{l-1}]$ using N_l paths. The simplest estimator is a mean of N_l independent samples, which for $l>0$ is

$$\widehat{Y}_l = N_l^{-1} \sum_{i=1}^{N_l} \left(\widehat{P}_l^{(i)} - \widehat{P}_{l-1}^{(i)} \right). \tag{3}$$

The key point here is that the quantity $\widehat{P}_l^{(i)} - \widehat{P}_{l-1}^{(i)}$ comes from two discrete approximations with different timesteps but the same Brownian path. The variance of this simple estimator is $V[\widehat{Y}_l] = N_l^{-1} V_l$ where V_l is the variance of a single sample. Combining this with independent estimators for each of the other levels, the variance of the combined estimator $\widehat{Y} = \sum_{l=0}^{L} \widehat{Y}_l$ is $V[\widehat{Y}] = \sum_{l=0}^{L} N_l^{-1} V_l$, while its computational cost is proportional to $\sum_{l=0}^{L} N_l\, h_l^{-1}$. Treating the N_l as continuous variables, the variance is minimised for a fixed computational cost by choosing N_l to be proportional to $\sqrt{V_l\, h_l}$.

In the particular case of an Euler discretisation, provided $a(S,t)$ and $b(S,t)$ satisfy certain conditions [BT95, KP92, TT90] there is $O(h^{1/2})$ strong convergence. From this it follows that $V[\widehat{P}_l - P] = O(h_l)$ for a European option with a Lipschitz continuous payoff. Hence for the simple estimator (3), the single sample variance V_l is $O(h_l)$, and the optimal choice for N_l is asymptotically proportional to h_l. Setting $N_l = O(\epsilon^{-2} L\, h_l)$, the variance of the combined estimator $\widehat{Y}$ is $O(\epsilon^2)$. If L is chosen such that $L = \log \epsilon^{-1} / \log 2 + O(1)$, as $\epsilon \to 0$, then $h_L = 2^{-L} = O(\epsilon)$, and so the bias error $E[\widehat{P}_L - P]$ is $O(\epsilon)$ due to standard results on weak convergence. Consequently, we obtain a Mean Square Error which is $O(\epsilon^2)$, with a computational complexity which is $O(\epsilon^{-2} L^2) = O(\epsilon^{-2} (\log \epsilon)^2)$.

This analysis is generalised in the following theorem:

Theorem 1. *Let P denote a functional of the solution of stochastic differential equation (1) for a given Brownian path $W(t)$, and let $\widehat{P}_l$ denote the corresponding approximation using a numerical discretisation with timestep $h_l = M^{-l}\, T$.*

If there exist independent estimators $\widehat{Y}_l$ based on N_l Monte Carlo samples, and positive constants $\alpha \geq \frac{1}{2}, \beta, c_1, c_2, c_3$ such that

i) $E[\widehat{P}_l - P] \leq c_1\, h_l^{\alpha}$

ii) $E[\widehat{Y}_l] = \begin{cases} E[\widehat{P}_0], & l = 0 \\ E[\widehat{P}_l - \widehat{P}_{l-1}], & l > 0 \end{cases}$

iii) $V[\widehat{Y}_l] \leq c_2\, N_l^{-1} h_l^{\beta}$

iv) C_l*, the computational complexity of* $\widehat{Y}_l$*, is bounded by*

$$C_l \leq c_3\, N_l\, h_l^{-1},$$

then there exists a positive constant c_4 *such that for any* $\epsilon < e^{-1}$ *there are values* L *and* N_l *for which the multilevel estimator*

$$\widehat{Y} = \sum_{l=0}^{L} \widehat{Y}_l,$$

has a mean-square-error with bound

$$MSE \equiv E\left[\left(\widehat{Y} - E[P]\right)^2\right] < \epsilon^2$$

with a computational complexity C *with bound*

$$C \leq \begin{cases} c_4\, \epsilon^{-2}, & \beta > 1, \\ c_4\, \epsilon^{-2} (\log \epsilon)^2, & \beta = 1, \\ c_4\, \epsilon^{-2-(1-\beta)/\alpha}, & 0 < \beta < 1. \end{cases}$$

Proof. See [Gil06].

The remainder of this paper addresses the use of the Milstein scheme [Gla04, KP92] to construct estimators with variance convergence rates $\beta > 1$, resulting in an $O(\epsilon^{-2})$ complexity bound. Provided certain conditions are satisfied [KP92], the Milstein scheme gives $O(h)$ strong convergence. In the case of a Lipschitz continuous European payoff, this immediately leads to the result that $V_l = O(h_l^2)$, corresponding to $\beta{=}2$. Numerical results which are not presented here demonstrate this convergence rate, and the associated $O(\epsilon^{-2})$ complexity

This paper addresses the tougher challenges of Asian, lookback, barrier and digital options. These cases require some ingenuity to construct estimators for which $\beta > 1$. Unfortunately, there is no accompanying theoretical analysis as yet, and so the paper relies on numerical demonstration of their effectiveness.

3 Milstein Discretisation

For a scalar SDE, the Milstein discretisation of equation (1) is

$$\widehat{S}_{n+1} = \widehat{S}_n + a\, h + b\, \Delta W_n + \tfrac{1}{2}\, \frac{\partial b}{\partial S}\, b\, (\Delta W_n)^2. \tag{4}$$

In the above equation, the subscript n is used to denote the timestep index, and a, b and $\partial b/\partial S$ are evaluated at $\widehat{S}_n, t_n$.

All of the numerical results to be presented are for the case of geometric Brownian motion for which the SDE is

$$\mathrm{d}S(t) = r\, S\, \mathrm{d}t + \sigma\, S\, \mathrm{d}W(t), \quad 0 < t < T.$$

By switching to the new variable $X = \log S$, it is possible to construct numerical approximations which are exact, but here we directly simulate the geometric Brownian motion using the Milstein method as an indication of the behaviour with more complicated models, for example those with a local volatility function $\sigma(S,t)$.

3.1 Estimator Construction

In all of the cases to be presented, we simulate the paths using the Milstein method. The refinement factor is $M=2$, so each level has twice as many timesteps as the previous level. The difference between the applications is in how we use the computed discrete path data to estimate $E[\widehat{P}_l - \widehat{P}_{l-1}]$.

In each case, the estimator for $E[\widehat{P}_l - \widehat{P}_{l-1}]$ is an average of values from N_l independent path simulations. For each Brownian input, the value which is computed is of the form $\widehat{P}_l^f - \widehat{P}_{l-1}^c$. Here $\widehat{P}_l^f$ is a fine-path estimate using timestep $h = 2^{-l}T$, and $\widehat{P}_{l-1}^c$ is the corresponding coarse-path estimate using timestep $h = 2^{-(l-1)}T$. To ensure that the identity (2) is correctly respected, to avoid the introduction of an undesired bias, we require that

$$E[\widehat{P}_l^f] = E[\widehat{P}_l^c]. \tag{5}$$

This means that the definitions of $\widehat{P}_l$ when estimating $E[\widehat{P}_l - \widehat{P}_{l-1}]$ and $E[\widehat{P}_{l+1} - \widehat{P}_l]$ must have the same expectation.

In the simplest case of a European option, this can be achieved very simply by defining $\widehat{P}_l^f$ and $\widehat{P}_l^c$ to be the same; this is the approach which was used for all applications in the previous work using the Euler discretisation [Gil06]. However, for more challenging applications such as Asian, lookback, barrier and digital options, the definition of $\widehat{P}_l^c$ will involve information from the discrete simulation of $\widehat{P}_{l+1}^f$, which is not available in computing $\widehat{P}_l^f$. The reason for doing this is to reduce the variance of the estimator, but it must be shown that equality (5) is satisfied. This will be achieved in each case through a construction based on a simple Brownian motion approximation.

3.2 Asian Option

The Asian option we consider has the discounted payoff

$$P = \exp(-rT)\ \max\left(0, \overline{S} - K\right),$$

where

$$\overline{S} = T^{-1} \int_0^T S(t)\ \mathrm{d}t.$$

The simplest approximation of $\overline{S}$, which was used in previous work [Gil06], is

$$\overline{\widehat{S}} = T^{-1} \sum_{0}^{n_T-1} \tfrac{1}{2}\, h\, (\widehat{S}_n + \widehat{S}_{n+1}),$$

where $n_T = T/h$ is the number of timesteps. This corresponds to a piecewise linear approximation to $S(t)$ but improved accuracy can be achieved by approximating the behaviour within a timestep as simple Brownian motion, with constant drift and volatility, conditional on the computed values $\widehat{S}_n$. Taking b_n to be the constant volatility within the time interval $[t_n, t_{n+1}]$, standard Brownian Bridge results (see section 3.1 in [Gla04]) give

$$\int_{t_n}^{t_{n+1}} S(t)\ \mathrm{d}t = \tfrac{1}{2} h (S(t_n) + S(t_{n+1})) + b_n \Delta I_n,$$

where ΔI_n, defined as

$$\Delta I_n = \int_{t_n}^{t_{n+1}} (W(t) - W(t_n))\ \mathrm{d}t \ - \ \tfrac{1}{2}\, h \Delta W,$$

is a $N(0, h^3/12)$ Normal random variable, independent of ΔW. Using $b_n = b(\widehat{S}_n, t_n)$, this gives the fine-path approximation

$$\overline{S} = T^{-1} \sum_{0}^{n_T-1} \left(\tfrac{1}{2}\, h\, (\widehat{S}_n + \widehat{S}_{n+1}) + b_n \Delta I_n \right).$$

The coarse path approximation is the same except that the values for ΔI_n are derived from the fine path values, noting that

$$\begin{aligned}
&\int_{t_n}^{t_n+2h} (W(t) - W(t_n))\ \mathrm{d}t - h(W(t_n+2h) - W(t_n)) \\
&= \int_{t_n}^{t_n+h} (W(t) - W(t_n))\ \mathrm{d}t - \tfrac{1}{2}\, h\, (W(t_n+h) - W(t_n)) \\
&\quad + \int_{t_n+h}^{t_n+2h} (W(t) - W(t_n+h))\ \mathrm{d}t - \tfrac{1}{2}\, h\, (W(t_n+2h) - W(t_n+h)) \\
&\quad + \tfrac{1}{2}\, h\, (W(t_n+h) - W(t_n)) - \tfrac{1}{2}\, h\, (W(t_n+2h) - W(t_n+h))\,,
\end{aligned}$$

and hence

$$\Delta I^c = \Delta I^{f1} + \Delta I^{f2} + \tfrac{1}{2}\, h (\Delta W^{f1} - \Delta W^{f2}),$$

where ΔI^c is the value for the coarse timestep, and ΔI^{f1} and ΔW^{f1} are the values for the first fine timestep, and ΔI^{f2} and ΔW^{f2} are the values for the second fine timestep.

Figure 1 shows the numerical results for parameters $S(0) = 1$, $K = 1$, $T = 1$, $r = 0.05$, $\sigma = 0.2$. The top left plot shows the behaviour of the

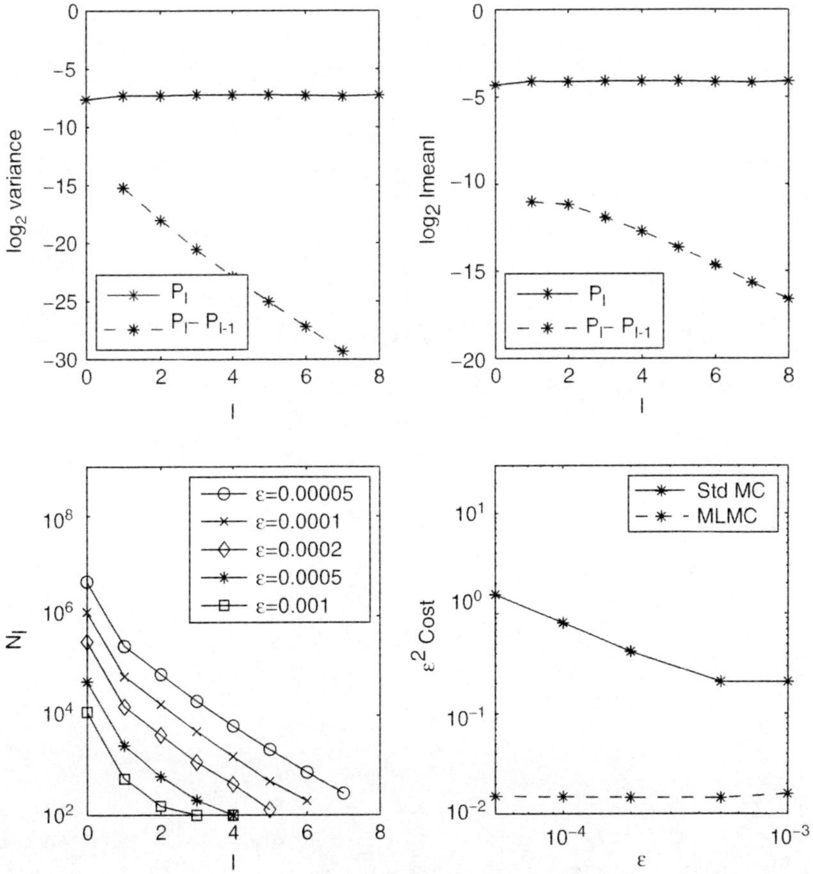

Fig. 1. Asian option

variance of both $\widehat{P}_l$ and $\widehat{P}_l - \widehat{P}_{l-1}$. The slope of the latter is approaching a value approximately equal to -2, indicating that $V_l = O(h_l^2)$, corresponding to $\beta = 2$. On level $l = 2$, which has just 4 timesteps, V_l is already more than 1000 times smaller than the variance $V[\widehat{P}_l]$ of the standard Monte Carlo method with the same timestep. The top right plot shows that $E[\widehat{P}_l - \widehat{P}_{l-1}]$ is approximately $O(h_l)$, corresponding to first order weak convergence, $\alpha = 1$. This is used to determine the number of levels that are required to reduce the bias to an acceptable level [Gil06].

The bottom two plots have results from five multilevel calculations for different values of ϵ. Each line in the bottom left plot shows the values for $N_l, l = 0, \ldots, L$, with the values decreasing with l because of the decrease in both V_l and h_l. It can also be seen that the value for L, the maximum level of timestep refinement, increases as the value for ϵ decreases, requiring a lower bias error. The bottom right plot shows the variation with ϵ of $\epsilon^2 C$ where the computational complexity C is defined as

$$C = \sum_l 2^l N_l,$$

which is the total number of fine grid timesteps on all levels. One line shows the results for the multilevel calculation and the other shows the corresponding cost of a standard Monte Carlo simulation of the same accuracy, i.e. the same bias error corresponding to the same value for L, and the same variance. It can be seen that $\epsilon^2 C$ is almost constant for the multilevel method, as expected, whereas for the standard Monte Carlo method it is approximately proportional to ϵ^{-1}. For the most accurate case, $\epsilon = 5\times 10^{-5}$, the multilevel method is more than 100 times more efficient than the standard method.

3.3 Lookback Option

The lookback option we consider has the discounted payoff

$$P = \exp(-rT)\left(S(T) - \min_{0<t<T} S(t) \right).$$

In previous work [Gil06], the minimum value of $S(t)$ over the path was approximated numerically by

$$\widehat{S}_{min} = \min_n \left(\widehat{S}_n - \beta^* b_n \sqrt{h} \right).$$

Here b_n is the volatility in the nth timestep, and $\beta^* \approx 0.5826$ is a constant which corrects the $O(h^{1/2})$ leading order error due to the discrete sampling of the path, and thereby restores $O(h)$ weak convergence [BGK97]. However, using this approximation, the difference between the computed minimum values and fine and coarse paths is $O(h_l^{1/2})$, and hence the variance V_l is $O(h_l)$, corresponding to $\beta = 1$. In the previous work, this was acceptable because $\beta = 1$ is the best that can be achieved in general with the Euler path discretisation which was used, but in this work we aim to achieve an improved convergence rate using the Milstein scheme.

To achieve this, we again approximate the behaviour within a timestep as simple Brownian motion, with constant drift and volatility, conditional on the computed values $\widehat{S}_n$. For the time interval $[t_n, t_{n+1}]$, standard Brownian Interpolation results (see section 6.4 in [Gla04]) give the minimum of Brownian motion, conditional on the end values, as

$$\widehat{S}_{n,min} = \tfrac{1}{2}\left(\widehat{S}_n + \widehat{S}_{n+1} - \sqrt{\left(\widehat{S}_{n+1} - \widehat{S}_n\right)^2 - 2\, b_n^2\, h \log U_n} \right), \qquad (6)$$

where b_n is the constant volatility and U_n is a uniform random variable on $[0,1]$.

The fine-path value $\widehat{P}_l^f$ is defined in this way using $b_n = b(\widehat{S}_n, t_n)$, and then taking the minimum over all timesteps to obtain the global minimum. However, for the coarse-path value $\widehat{P}_{l-1}^c$, we do something different. Again assuming simple Brownian motion conditional on the end-points, the value at the midpoint of the time interval $[t_n, t_{n+1}]$ is given by

$$\widehat{S}_{n+1/2} = \tfrac{1}{2}\left(\widehat{S}_n + \widehat{S}_{n+1} - b_n D_n\right), \tag{7}$$

where

$$D_n = W_{n+1} - 2W_{n+1/2} + W_n = (W_{n+1} - W_{n+1/2}) - (W_{n+1/2} - W_n),$$

is a $N(0,h)$ random variable which corresponds to a difference in the consecutive Brownian increments of a finer path with timestep $h/2$. Given this midpoint value, the minimum value over the full timestep is the smaller of the minima for each of the two half-timesteps,

$$\begin{aligned}\widehat{S}_{n,min} = \min\Bigg\{ &\tfrac{1}{2}\left(\widehat{S}_n + \widehat{S}_{n+1/2} - \sqrt{\left(\widehat{S}_{n+1/2} - \widehat{S}_n\right)^2 - b_n^2\, h \log U_{1,n}}\,\right), \\ &\tfrac{1}{2}\left(\widehat{S}_{n+1/2} + \widehat{S}_{n+1} - \sqrt{\left(\widehat{S}_{n+1} - \widehat{S}_{n+1/2}\right)^2 - b_n^2\, h \log U_{2,n}}\,\right)\Bigg\}.\end{aligned} \tag{8}$$

In computing $\widehat{P}_{l-1}^c$, we use the values for D_n, $U_{1,n}$ and $U_{2,n}$ that come from the fine-path simulation for $\widehat{P}_l^f$. D_n is the difference of the Brownian increments for the two fine-path timesteps, and $U_{1,n}$ and $U_{2,n}$ are the uniform random variables used to compute the minima for the two fine-path timesteps. Since these all have the correct probability distribution, it follows that the expected values of (6) and (8) are identical, and therefore equality (5) is satisfied.

Figure 2 shows the numerical results for parameters $S(0) = 1$, $T = 1$, $r = 0.05$, $\sigma = 0.2$. The top left plot shows that the variance is $O(h_l^2)$, corresponding to $\beta = 2$, while the top right plot shows that the mean correction is $O(h_l)$, corresponding to first order weak convergence, $\alpha = 1$. The bottom left plot shows that more levels are required to reduce the discretisation bias to the required level. Consequently, the savings relative to the standard Monte Carlo treatment are greater, up to a factor of approximately 200 for $\epsilon = 5\times 10^{-5}$. The computational cost of the multilevel method is almost perfectly proportional to ϵ^{-2}.

3.4 Barrier Option

The barrier option which is considered is a down-and-out call for which the discounted payoff is

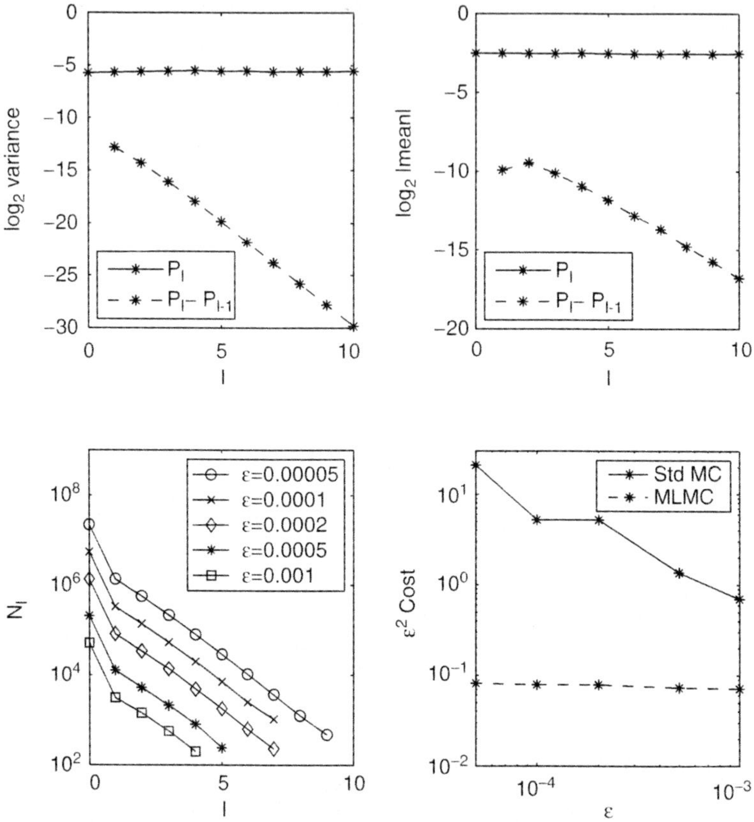

Fig. 2. Lookback option

$$P = \exp(-rT)\ (S(T)-K)^+\ \mathbf{1}\{\tau > T\},$$

where the notation $(S(T)-K)^+$ denotes $\max(0, S(T)-K)$, $\mathbf{1}(\tau > T)$ is an indicator function taking value 1 if the argument is true, and zero otherwise, and the crossing time τ is defined as

$$\tau = \inf_{t>0} \{S(t) < B\}.$$

Following a standard approach for continuously monitored barrier crossings (see section 6.4 in [Gla04]), for a particular Brownian path input sampled discretely at uniform intervals h, the conditional expectation of the payoff can be expressed as

$$\exp(-rT)\ (\widehat{S}_{n_T} - K)^+ \prod_{n=0}^{n_T-1} \widehat{p}_n,$$

where $n_T = T/h$ is again the number of timesteps, and $\widehat{p}_n$ represents the probability that the path did not cross the barrier during the nth timestep. If we again approximate the motion within each timestep as simple Brownian motion conditional on the endpoint values, then

$$\widehat{p}_n = 1 - \exp\left(\frac{-2\,(S_n - B)^+\,(S_{n+1} - B)^+}{b_n^2\,h}\right). \tag{9}$$

This is the expression used to define the payoff $\widehat{P}_l^f$ for the fine-path calculation, with b_n set equal to $b(\widehat{S}_n, t_n)$, as in the lookback calculation.

For the coarse path calculation, in which each timestep corresponds to two fine-path timesteps, we again use equation (7) to construct a midpoint value $\widehat{S}_{n+1/2}$. Given this value, the probability that the simple Brownian path does not cross the barrier is

$$\begin{aligned}\widehat{p}_n = & \left\{1 - \exp\left(\frac{-2\,(S_n - B)^+(S_{n+1/2} - B)^+}{b_n^2\,h}\right)\right\} \\ & \times \left\{1 - \exp\left(\frac{-2\,(S_{n+1/2} - B)^+(S_{n+1} - B)^+}{b_n^2\,h}\right)\right\}.\end{aligned} \tag{10}$$

The conditional expectation of (10) is equal to (9) and so equality (5) is satisfied.

Figure 3 shows the numerical results for parameters $S(0) = 1$, $K = 1$, $B = 0.85$, $T = 1$, $r = 0.05$, $\sigma = 0.2$. The top left plot shows that the variance is approximately $O(h_l^{\beta})$ for a value of β slightly less than 2. An explanation for this is that a small $O(h_l^{1/2})$ fraction of the paths have a minimum which lies within $O(h_l^{1/2})$ of the barrier, for which the product $\prod \widehat{p}_n$ is neither close to zero nor close to unity. The fine path and coarse path trajectories differ by $O(h_l)$, due to the first order strong convergence of the Milstein scheme. Since the $\widehat{p}_n$ have an $O(h_l^{-1/2})$ derivative, this results in the difference between $\prod \widehat{p}_n$ for this small subset of coarse and fine paths being $O(h_l^{1/2})$, giving a contribution to the variance which is $O(h_l^{3/2})$.

The top right plot shows that the mean correction is $O(h_l)$, corresponding to first order weak convergence, $\alpha = 1$. The bottom right plot shows that the computational cost of the multilevel method is again almost perfectly proportional to ϵ^{-2}, and for $\epsilon = 5 \times 10^{-5}$ it is over 100 times more efficient that the standard Monte Carlo method.

3.5 Digital Option

The digital option which is considered has the discounted payoff

$$P = \exp(-rT)\; \mathbf{1}\{S(T) > K\}.$$

The standard numerical discretisation would be to simulate the path of $S(t)$ right up to the final time T. This is the approach adopted previously

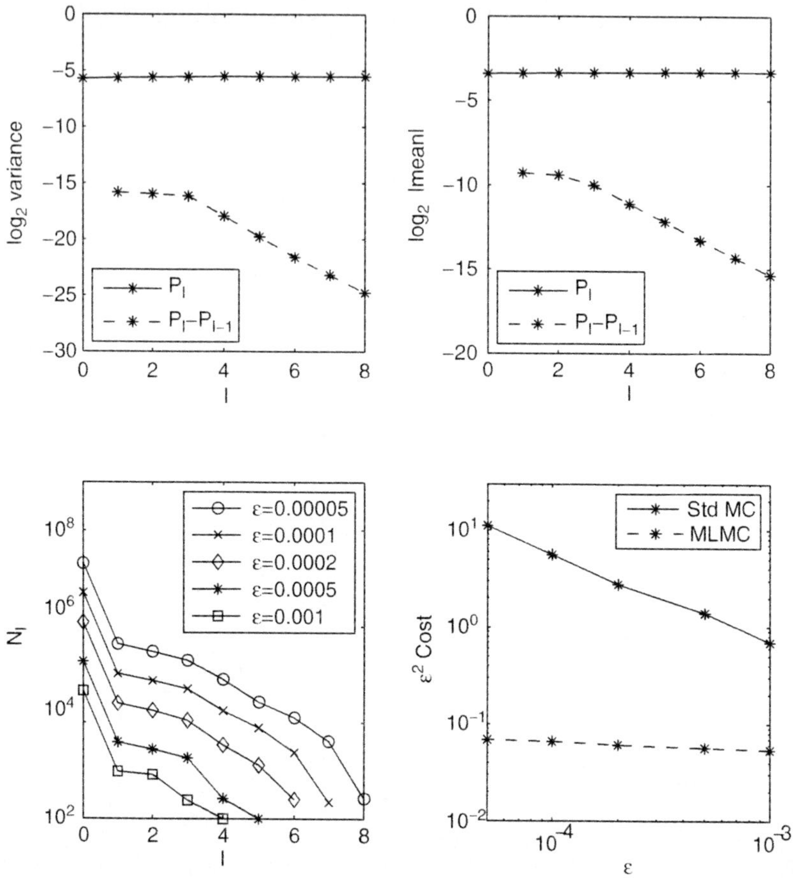

Fig. 3. Barrier option

for multilevel calculations using the Euler discretisation [Gil06]. In that case, the variance V_l was $O(h_l^{1/2})$, because $O(h_l^{1/2})$ of the paths terminate within $O(h_l^{1/2})$ of the strike K, and for these paths there is an $O(1)$ probability that the coarse and fine paths will terminate on opposite sides of the strike, giving an $O(1)$ value for $\widehat{P}_l - \widehat{P}_{l-1}$. Using the same approach with the Milstein method, there would be $O(h_l)$ of the paths terminating within $O(h_l)$ of the strike K, for which there would be an $O(1)$ probability that the coarse and fine paths would terminate on opposite sides of the strike. This would result in V_l being $O(h_l)$. This corresponds to $\beta = 1$ and would give a computational cost which is $O(\epsilon^{-2}(\log \epsilon)^2)$.

To achieve a better multilevel variance convergence rate, we instead smooth the payoff using the technique of conditional expectation (see section 7.2.3 in [Gla04]), terminating the path calculations one timestep before reaching the terminal time T. If $\widehat{S}_{n_T-1}$ denotes the value at this time, then if we approximate the motion thereafter as a simple Brownian motion with constant drift a_{n_T-1} and volatility b_{n_T-1}, the probability that $\widehat{S}_{n_T} > K$ after one further timestep is

$$\widehat{p} = \Phi\left(\frac{\widehat{S}_{n_T-1} + a_{n_T-1}h - K}{b_{n_T-1}\sqrt{h}}\right), \tag{11}$$

where Φ is the cumulative Normal distribution.

For the fine-path payoff $\widehat{P}_l^f$ we therefore use $\widehat{P}_l^f = \exp(-rT)\ \widehat{p}$, with $a_{n_T-1} = a(\widehat{S}_{n_T-1},\ T-h)$ and $b_{n_T-1} = b(\widehat{S}_{n_T-1}, T-h)$. For the coarse-path payoff, we note that given the Brownian increment ΔW for the first half of the Nth timestep, then the probability that $\widehat{S}_{n_T} > K$ is

$$\widehat{p} = \Phi\left(\frac{\widehat{S}_{n_T-1} + a_{n_T-1}h + b_{n_T-1}\Delta W - K}{b_{n_T-1}\sqrt{h/2}}\right). \tag{12}$$

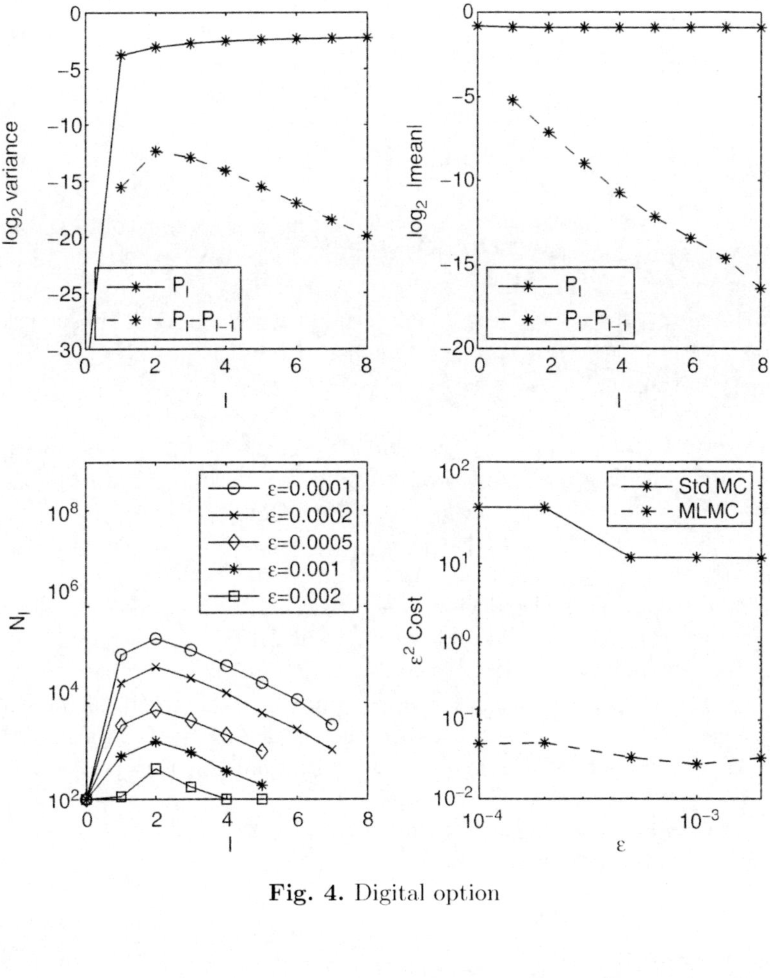

Fig. 4. Digital option

The value for ΔW is taken from the final timestep of the fine-path calculation, which corresponds to the first half of the Nth timestep in the coarse-path calculation. The conditional expectation of (12) is equal to (11), and so again equality (5) is satisfied.

Figure 4 shows the numerical results for parameters $S(0) = 1$, $K = 1$, $T = 1$, $r = 0.05$, $\sigma = 0.2$. The top left plot shows that the variance is approximately $O(h_l^{3/2})$, corresponding to $\beta = 1.5$. The reason for this is similar to the argument for the barrier option. $O(h_l^{1/2})$ of the paths have a minimum which lies within $O(h_l^{1/2})$ of the strike, for which the $\widehat{p}$ is neither close to zero nor close to unity. The fine path and coarse path trajectories differ by $O(h_l)$, due to the first order strong convergence of the Milstein scheme. Since $\widehat{p}$ has an $O(h_l^{-1/2})$ derivative, this results in the difference between $\widehat{p}$ for the coarse and fine paths being $O(h_l^{1/2})$, and that results in the variance being $O(h_l^{3/2})$.

One strikingly different feature is that the variance of the level 0 estimator, V_0, is zero. This is because at level $l = 0$ there would usually be only one timestep, and so here it is not simulated at all; one simply uses equation (11) to evaluate the payoff. This reduces the cost of the multilevel calculations even more than usual, leading to a factor 1000 computational savings for $\epsilon = 10^{-4}$.

4 Conclusions and Future Work

In this paper we have demonstrated numerically the ability of multilevel Monte Carlo path simulation using the Milstein discretisation to achieve an ϵ RMS error for a range of financial options at a computational cost which is $O(\epsilon^{-2})$. This requires the use of Brownian interpolation within each timestep for Asian, lookback and barrier options, and the use of conditional expectation to smooth the payoff of digital options.

There are three major directions for future research. The first is the theoretical analysis of the algorithms presented in this paper, to prove that they do indeed have variance convergence rates with $\beta > 1$. The analysis of earlier algorithms for lookback, barrier and digital options based on the Euler discretisation [Gil06] is currently being developed; it is hoped this can then be extended to the Milstein discretisation for scalar SDEs.

The second is the extension of the algorithms to multi-dimensional SDEs, for which the Milstein discretisation usually requires the simulation of Lévy areas [GL94, Gla04]. Current investigations indicate that this can be avoided for European options with a Lipschitz payoff through the use of antithetic variables. However, the extension to more difficult payoffs, such as the Asian, lookback, barrier and digital options considered in this paper, looks more challenging.

The third direction for future research is the use of quasi-Monte Carlo methods. The analysis in section 2 showed that the optimal number of samples on level l is proportional to $\sqrt{V_l h_l}$. If $V_l = O(h_l^\beta)$, then this number is proportional to $h_l^{(\beta+1)/2}$. Since the cost of an individual sample is proportional to the number of timesteps, and hence inversely proportional to h_l, the computational cost on level l is proportional to $h_l^{(\beta-1)/2}$. For $\beta > 1$, this shows that the computational effort decreases geometrically as one moves to finer levels of discretisation. Thus, when using the Milstein discretisation most of the computational effort is expended on the coarsest levels of the multilevel computation. For these low dimensional levels it is reasonable to expect that quasi-Monte Carlo methods [KS05, Ecu04, Nie92] will be very much more effective than the standard Monte Carlo methods used in this paper.

Acknowledgements

This research was funded in part by a research grant from Microsoft Corporation, and in part by a Springboard Fellowship from the UK Engineering and Physical Sciences Research Council.

References

[BGK97] M. Broadie, P. Glasserman, and S. Kou. A continuity correction for discrete barrier options. *Mathematical Finance*, 7(4):325–348, 1997.

[BT95] V. Bally and D. Talay. The law of the Euler scheme for stochastic differential equations, I: convergence rate of the distribution function. *Probability Theory and Related Fields*, 104(1):43–60, 1995.

[Ecu04] P. L'Ecuyer. Quasi-Monte Carlo methods in finance. In R.G. Ingalls, M.D. Rossetti, J.S. Smith, and B.A. Peters, editors, *Proceedings of the 2004 Winter Simulation Conference*, pages 1645–1655. IEEE Press, 2004.

[ES02] M. Emsermann and B. Simon. Improving simulation efficiency with quasi control variates. *Stochastic Models*, 18(3):425–448, 2002.

[Gil06] M.B. Giles. Multilevel Monte Carlo path simulation. Technical Report NA06/03, Oxford University Computing Laboratory, 2006 (to appear in *Operations Research*).

[GL94] J.G. Gaines and T.J. Lyons. Random generation of stochastic integrals. *SIAM J. Appl. Math.*, 54(4):1132–1146, 1994.

[Gla04] P. Glasserman. *Monte Carlo Methods in Financial Engineering*. Springer-Verlag, New York, 2004.

[Hei01] S. Heinrich. *Multilevel Monte Carlo Methods*, volume 2179 of *Lecture Notes in Computer Science*, pages 58–67. Springer-Verlag, 2001.

[Keb05] A. Kebaier. Statistical Romberg extrapolation: a new variance reduction method and applications to options pricing. *Annals of Applied Probability*, 14(4):2681–2705, 2005.

[KP92] P.E. Kloeden and E. Platen. *Numerical Solution of Stochastic Differential Equations*. Springer-Verlag, Berlin, 1992.

[KS05] F.Y. Kuo and I.H. Sloan. Lifting the curse of dimensionality. *Notices of the AMS*, 52(11):1320–1328, 2005.

[Nie92] H. Niederreiter. *Random Number Generation and Quasi-Monte Carlo Methods*. SIAM, 1992.

[Spe05] A. Speight. A multilevel approach to control variates. Working paper, Georgia State University, 2005.

[TT90] D. Talay and L. Tubaro. Expansion of the global error for numerical schemes solving stochastic differential equations. *Stochastic Analysis and Applications*, 8:483–509, 1990.

Generalized Tractability for Linear Functionals

Michael Gnewuch[1] and Henryk Woźniakowski[2]

[1] Institute of Computer Science, University of Kiel, Christian-Albrechts-Platz 4, 24098 Kiel, Germany
mig@informatik.uni-kiel.de

[2] Department of Computer Science, Columbia University, New York, NY 10027, USA, and
Institute of Applied Mathematics, University of Warsaw, ul. Banacha 2, 02-097 Warszawa, Poland
henryk@cs.columbia.edu

Summary. We study approximation of continuous linear functionals I_d defined over reproducing kernel weighted Hilbert spaces of d-variate functions. Let $n(\varepsilon, I_d)$ denote the minimal number of function values needed to solve the problem to within ε. There are many papers studying polynomial tractability for which $n(\varepsilon, I_d)$ is to be bounded by a polynomial in ε^{-1} and d. We study *generalized tractability* for which we want to guarantee that either $n(\varepsilon, I_d)$ is not exponentially dependent on ε^{-1} and d, which is called *weak tractability*, or is bounded by a power of $T(\varepsilon^{-1}, d)$ for $(\varepsilon^{-1}, d) \in \Omega \subseteq [1, \infty) \times \mathbb{N}$, which is called (T, Ω)-tractability. Here, the *tractability* function T is non-increasing in both arguments and does not depend exponentially on ε^{-1} and d.

We present necessary conditions on generalized tractability for arbitrary continuous linear functionals I_d defined on weighted Hilbert spaces whose kernel has a decomposable component, and sufficient conditions on generalized tractability for multivariate integration for general reproducing kernel Hilbert spaces. For some weighted Sobolev spaces these necessary and sufficient conditions coincide. They are expressed in terms of necessary and sufficient conditions on the weights of the underlying spaces.

1 Introduction

The study of approximation of continuous linear functionals I_d over spaces of d-variate functions has recently been a popular research subject especially for large d. The primary example of I_d is multivariate integration which occurs in many applications for huge d.

Let $n(\varepsilon, I_d)$ be the minimal number of function values needed to reduce the initial error by a factor $\varepsilon \in (0, 1)$ for functions from the unit ball of a given space. The initial error is the minimal error which can be achieved without sampling the functions and is equal to the norm of I_d. There are many papers, see [NW01a] for a survey, studying *polynomial tractability* for which $n(\varepsilon, I_d)$ is to be bounded by a polynomial in ε^{-1} and d for all $(\varepsilon^{-1}, d) \in [1, \infty) \times \mathbb{N}$. By now we know that polynomial tractability of multivariate integration holds for reproducing kernel weighted Hilbert spaces for sufficiently fast decaying weights. Let $\gamma_{d,j}$ be a weight which controls the influence of the jth variable for the d-dimensional case. A typical result is that multivariate integration is polynomially tractable iff

$$\limsup_{d\to\infty} \frac{\sum_{j=1}^{d} \gamma_{d,j}}{\ln d} < \infty.$$

There is also the notion of *strong polynomial tractability* for which $n(\varepsilon, I_d)$ is bounded by a polynomial only in ε^{-1} for all d. A typical result is that multivariate integration is strongly polynomially tractable iff

$$\limsup_{d\to\infty} \sum_{j=1}^{d} \gamma_{d,j} < \infty.$$

In this paper we study *generalized tractability*, see [GW06]. First of all, we possibly limit the set of all pairs (ε^{-1}, d) of our interest, by assuming that $(\varepsilon^{-1}, d) \in \Omega$ where $\Omega \subseteq [1, \infty) \times \mathbb{N}$. To have a meaningful problem, the set Ω is chosen such that at least one of the arguments ε^{-1} or d may go to infinity. Since our main emphasis is on large d, in many cases we assume that $[1, \varepsilon_0^{-1}) \times \mathbb{N} \subseteq \Omega$ for some $\varepsilon_0 \in (0, 1)$ which allows to take arbitrary large d.

We study *weak tractability* in Ω for which we want to check when $n(\varepsilon, I_d)$ is *not* exponentially dependent on ε^{-1} and d for $(\varepsilon^{-1}, d) \in \Omega$. We also study (T, Ω)-tractability. Here T is called a *tractability* function which means that T is non-increasing in both arguments and does not depend exponentially on ε^{-1} and d. In this case, we want to check when $n(\varepsilon, I_d)$ is bounded by a power of $T(\varepsilon^{-1}, d)$ for all $(\varepsilon^{-1}, d) \in \Omega$. *Strong* (T, Ω)-tractability means that $n(\varepsilon, I_d)$ is bounded by a power of $T(\varepsilon^{-1}, 1)$ for all $(\varepsilon^{-1}, d) \in \Omega$.

We present necessary and sufficient conditions on generalized tractability for $I = \{I_d\}$. Necessary conditions are obtained for arbitrary continuous linear functionals I_d defined over reproducing kernel weighted Hilbert spaces whose kernel has a decomposable component. We make heavy use of [NW01b] where this concept was introduced and polynomial tractability was studied for weights independent of d. We generalize the approach of [NW01b] by studying generalized tractability and weights which may depend on d. Sufficient conditions are obtained only for multivariate integration defined over general reproducing kernel weighted or unweighted Hilbert spaces. Sufficient conditions easily follow from upper bounds on $n(\varepsilon, I_d)$ which are obtained by using a known proof technique which can be found in, e.g., [SW98].

We prove that for some reproducing kernel weighted Hilbert spaces, such as some weighted Sobolev spaces, necessary and sufficient conditions for generalized tractability of multivariate integration coincide. These conditions are expressed in terms of the weights of the underlying space. A typical result is that weak tractability holds iff

$$\lim_{d\to\infty} \frac{\sum_{j=1}^{d} \gamma_{d,j}}{d} = 0.$$

If we compare this with polynomial tractability, we see that $\ln d$ is now replaced by d but the corresponding limit must be zero. Hence, the unweighted case, $\gamma_{d,j}$ = constant > 0, leads to the lack of weak tractability which is called *strong intractability*. To guarantee weak tractability we must take decaying weights. For example, for $\gamma_{d,j} = j^{-\beta}$ we have polynomial tractability iff $\beta \geq 1$ whereas we have weak tractability iff $\beta > 0$.

We obtain (T, Ω)-tractability iff

$$\limsup_{(\varepsilon^{-1},d)\,\in\,\Omega,\ \varepsilon^{-1}+d\to\infty} \frac{\sum_{j=1}^{d} \gamma_{d,j} \;+\; \ln \varepsilon^{-1}}{\ln\left(1+T(\varepsilon^{-1},d)\right)} \;<\; \infty.$$

and strong (T, Ω)-tractability iff

$$\limsup_{(\varepsilon^{-1},d)\,\in\,\Omega,\ \varepsilon^{-1}+d\to\infty} \frac{\sum_{j=1}^{d} \gamma_{d,j} \;+\; \ln \varepsilon^{-1}}{\ln\left(1+T(\varepsilon^{-1},1)\right)} \;<\; \infty.$$

We illustrate these conditions for $T(x,y) = x^{\beta_1} \exp(y^{\beta_2})$ with non-negative β_i and $\beta_2 < 1$ which is needed to guarantee that T is non-exponential. We consider two tractability domains $\Omega = \Omega_1 = [1,\infty) \times \mathbb{N}$ and $\Omega = \Omega_2 = [1,2] \times \mathbb{N}$. Then (T, Ω_1)-tractability holds iff

$$\beta_1 > 0 \;\text{ and }\; \limsup_{d} \frac{\sum_{j=1}^{d} \gamma_{d,j}}{d^{\beta_2}} < \infty,$$

and strong (T, Ω_1)-tractability holds iff

$$\beta_1 > 0 \;\text{ and }\; \limsup_{d} \sum_{j=1}^{d} \gamma_{d,j} < \infty.$$

For Ω_2, the dependence on ε is not important since $\varepsilon \geq \frac{1}{2}$, and (T, Ω_2)-tractability holds as before without assuming that $\beta_1 > 0$. That is, it holds even for $\beta_1 = 0$.

2 Approximation of Linear Functionals

For $d \in \mathbb{N}$, let F_d be a normed linear space of functions $f : D_d \subseteq \mathbb{R}^d \to \mathbb{R}$ for a Lebesgue measurable set D_d. Let

$$I_d \,:\, F_d \;\to\; \mathbb{R}$$

be a continuous linear functional. The primary example of I_d is *multivariate integration.* In this case, we assume that F_d is a space of Lebesgue measurable functions and

$$I_d f = \int_{D_d} \rho_d(x)\, f(x)\, \mathrm{d}x,$$

where ρ_d is a weight, i.e., $\rho_d \geq 0$ and $\int_{D_d} \rho_d(x)\, \mathrm{d}x = 1$. Obviously I_d is a linear functional. The norm of F_d is chosen such that I_d is also continuous.

Without loss of generality, see e.g., [TWW88], we approximate I_d by linear algorithms using function values, i.e., by algorithms of the form

$$Q_{n,d} f := \sum_{i=1}^{n} a_i f(z_i)$$

for some real coefficients a_i and deterministic sample points $z_i \in D_d$. Let

$$e(n, I_d) = \inf_{a_i, z_i\, ;\, i=1,2,\ldots,n} \ \sup_{f \in F_d,\ \|f\|_{F_d} \leq 1} \left| I_d f - \sum_{i=1}^{n} a_i f(z_i) \right|$$

be the nth minimal worst case error when we use n function values. In particular, for $n = 0$ we do not use function values and approximate I_d by zero. We then have the *initial error*,

$$e(0, I_d) = \|I_d\|,$$

where $\|\cdot\|$ is the operator norm induced by the norm of F_d. Let

$$n(\varepsilon, I_d) = \min\{\, n \mid e(n, I_d) \leq \varepsilon\, e(0, I_d)\,\}$$

denote the smallest number of sample points for which there exists an algorithm $Q_{n,d}$ such that $e(Q_{n,d}) := \|I_d - Q_{n,d}\|$ reduces the initial error by a factor at least ε.

3 Generalized Tractability

For polynomial tractability, we assume that $(\varepsilon^{-1}, d) \in [1, \infty) \times \mathbb{N}$. Sometimes it is natural, see [GW06], to assume that $(\varepsilon^{-1}, d) \in \Omega$, where Ω is a proper subset of $[1, \infty) \times \mathbb{N}$. As motivated in [GW06], it is natural to assume that Ω satisfies the following condition.

Let us define $[k] := \{1, 2, \ldots, k\}$ for arbitrary $k \in \mathbb{N}$ and $[0] := \emptyset$. A *tractability domain* Ω is a subset of $[1, \infty) \times \mathbb{N}$ satisfying

$$[1, \infty) \times [d^*] \cup [1, \varepsilon_0^{-1}) \times \mathbb{N} \subseteq \Omega \tag{1}$$

for some $d^* \in \mathbb{N} \cup \{0\}$ and some $\varepsilon_0 \in (0, 1]$ such that $d^* + (1 - \varepsilon_0) > 0$. This implies that at least one of the arguments ε^{-1} or d may go to infinity within Ω.

For polynomial tractability, $n(\varepsilon, I_d)$ is to be bounded by a polynomial in ε^{-1} and d. For generalized tractability, we replace this polynomial dependence by a tractability function T which does *not* depend exponentially on ε^{-1} and d. More precisely, as in [GW06], a function $T : [1,\infty) \times [1,\infty) \to [1,\infty)$ is called a *tractability function* if T is non-decreasing in x and y and

$$\lim_{(x,y)\in\Omega,\ x+y\to\infty} \frac{\ln T(x,y)}{x+y} = 0\,. \tag{2}$$

Let now Ω be a tractability domain and T a tractability function. The multivariate problem $I = \{I_d\}$ is (T,Ω)*-tractable* if there exist non-negative numbers C and t such that

$$n(\varepsilon, I_d) \le C\,T(\varepsilon^{-1}, d)^t \quad \text{for all } (\varepsilon^{-1}, d) \in \Omega. \tag{3}$$

The *exponent* $t^{\rm tra}$ *of* (T,Ω)*-tractability* is defined as the infimum of all non-negative t for which there exists a $C = C(t)$ such that (3) holds.

The multivariate problem I is *strongly* (T,Ω)*-tractable* if there exist non-negative numbers C and t such that

$$n(\varepsilon, I_d) \le C\,T(\varepsilon^{-1}, 1)^t \quad \text{for all } (\varepsilon^{-1}, d) \in \Omega. \tag{4}$$

The *exponent* $t^{\rm str}$ *of strong* (T,Ω)*-tractability* is the infimum of all non-negative t for which there exists a $C = C(t)$ such that (4) holds.

An extensive motivation of the notion of generalized tractability and many examples of tractability domains and functions can be found in [GW06].

Similarly as in [NW07], we say that $I = \{I_d\}$ is *weakly tractable* in Ω iff

$$\lim_{(\varepsilon^{-1},d)\in\Omega,\ \varepsilon^{-1}+d\to\infty} \frac{\ln n(\varepsilon, I_d)}{\varepsilon^{-1}+d} = 0.$$

If I is not weakly tractable in Ω then I is called *strongly intractable* in Ω.

The essence of weak tractability in Ω is to guarantee that $n(\varepsilon, I_d)$ is *not* exponential in ε^{-1} and d without specifying a bound on $n(\varepsilon, I_d)$. Note that if I is (T,Ω)-tractable then I is weakly tractable in Ω. Or equivalently, if I is strongly intractable in Ω then I is also not (T,Ω)-tractable for any tractability function T.

4 Hilbert Spaces with Reproducing Kernels

In this section we make specific assumptions on the spaces F_d. We slightly modify the approach proposed in [NW01b].

Let D_1 be a Lebesgue measurable subset of $\mathbb{R}$. Let $H(R_i)$, $i = 1, 2$, denote Hilbert spaces with reproducing kernels $R_i : D_1^2 \to \mathbb{R}$. We assume that

$$H(R_1) \cap H(R_2) = \{0\},$$

and define the reproducing kernel K_1 by $K_1 = R_1 + R_2$.

Let F_1 be the Hilbert space with reproducing kernel K_1. That is, for all $f \in F_1$ there exist uniquely determined $f_1 \in H(R_1)$ and $f_2 \in H(R_2)$ such that $f = f_1 + f_2$, and the inner product of F_1 is given by

$$\langle f, g \rangle_{F_1} = \langle f_1, g_1 \rangle_{H(R_1)} + \langle f_2, g_2 \rangle_{H(R_2)} \,.$$

Let I_1 be a continuous linear functional on F_1. Then there exists a $h_1 \in F_1$ such that

$$I_1 f \;=\; \langle f, h_1 \rangle_{F_1} \quad \text{for all} \;\; f \;\in\; F_1 .$$

The function h_1 has the unique decomposition

$$h_1 = h_{1,1} + h_{1,2} \quad \text{with } h_{1,i} \in H(R_i).$$

We assume that R_2 is *decomposable*, i.e., there exists an $a^* \in \mathbb{R}$ such that

$$R_2(x,t) \;=\; 0 \quad \text{for all} \;\; (x,t) \;\in\; D_{(0)} \times D_{(1)} \cup D_{(1)} \times D_{(0)},$$

where

$$D_{(0)} := \{x \in D_1 \,|\, x \le a^*\} \quad \text{and} \quad D_{(1)} := \{x \in D_1 \,|\, x \ge a^*\} \,.$$

Let $h_{1,2,(0)}$ and $h_{1,2,(1)}$ be functions defined on D_1 such that they are the restrictions of $h_{1,2}$ to the sets $D_{(0)}$ and $D_{(1)}$, respectively, and take zero values otherwise.

We also consider weighted tensor product problems. We define $F_{1,\gamma}$ as the Hilbert space determined by the weighted reproducing kernel $K_{1,\gamma}$,

$$K_{1,\gamma} = R_1 + \gamma R_2 \,, \tag{5}$$

where γ is positive. The case $\gamma = 0$ can be obtained by taking the limit of positive γ. Since the norms of $F_{1,\gamma} = F_1$ are equivalent, $I_{1,\gamma} = I_1$ is also a continuous linear functional on $F_{1,\gamma}$. It is easy to show that for

$$h_{1,\gamma} \;:=\; h_{1,1} + \gamma\, h_{1,2}$$

we have

$$I_{1,\gamma} f \;=\; I_1 f \;=\; \langle f, h_{1,\gamma} \rangle_{F_{1,\gamma}} \quad \text{for all } f \in F_{1,\gamma}$$

and

$$\|h_{1,\gamma}\|^2_{F_{1,\gamma}} = \|h_{1,1}\|^2_{H(R_1)} + \gamma \|h_{1,2}\|^2_{H(R_2)} \,. \tag{6}$$

For $d \ge 2$, let $F_{d,\gamma} = F_{1,\gamma_{d,1}} \otimes \cdots \otimes F_{1,\gamma_{d,d}}$ be the tensor product Hilbert space of the $F_{1,\gamma_{d,j}}$ for some positive weights $\gamma := \{\gamma_{d,j}\}$, $d \in \mathbb{N}$, $j \in [d]$. The case of a zero weight can be obtained, as before, by taking the limit of positive weights. Without loss of generality, we assume that

$$\gamma_{d,1} \ge \gamma_{d,2} \ge \cdots \ge \gamma_{d,d} \quad \text{for all} \;\; d \;\in\; \mathbb{N}.$$

To relate weights for different d, we assume that for fixed j the sequence $\{\gamma_{d,j}\}_{d=1}^{\infty}$ is non-increasing. Hence, for all $d \geq j$ we have

$$\gamma_{d,j} \leq \gamma_{j,j} \leq \gamma_{j,1} \leq \gamma_{1,1}.$$

Examples of weights include $\gamma_{d,j} = \gamma_j$ for $\gamma_j \geq \gamma_{j+1}$, as considered in [NW01b], or $\gamma_{d,j} = d^{-\beta}$ with $\beta \geq 0$.

Let us define the sequence of spaces $F_\gamma := \{F_{d,\gamma}\}$, where

$$K_{d,\gamma}(x,t) = \prod_{j=1}^{d} \left[R_1(x_j,t_j) + \gamma_{d,j} R_2(x_j,t_j)\right]$$

is the reproducing kernel of $F_{d,\gamma}$. We define the linear functional

$$I_{d,\gamma} \, : \, F_{d,\gamma} \; \to \mathbb{R}$$

as the d-fold tensor product of I_1. Then $I_{d,\gamma} f = \langle f, h_{d,\gamma}\rangle_{F_{d,\gamma}}$ with

$$h_{d,\gamma}(x) = \prod_{j=1}^{d} h_{1,\gamma_{d,j}}(x_j) = \prod_{j=1}^{d} \left[h_{1,1}(x_j) + \gamma_{d,j} h_{1,2}(x_j)\right] .$$

From (6) we have

$$e^2(0, I_{d,\gamma}) = \|h_{d,\gamma}\|_{F_{d,\gamma}}^2 = \prod_{j=1}^{d} \left[\|h_{1,1}\|_{H(R_1)}^2 + \gamma_{d,j} \|h_{1,2}\|_{H(R_2)}^2\right] .$$

Let $\alpha_1 := \|h_{1,1}\|_{H(R_1)}^2$, $\alpha_2 := \|h_{1,2}\|_{H(R_2)}^2$, $\alpha_3 := \|h_{1,2}\|_{H(R_2)}^2 \|h_{1,1}\|_{H(R_1)}^{-2}$, and

$$\alpha := \frac{\max\{\|h_{1,2,(0)}\|_{H(R_2)}^2, \|h_{1,2,(1)}\|_{H(R_2)}^2\}}{\|h_{1,2,(0)}\|_{H(R_2)}^2 + \|h_{1,2,(1)}\|_{H(R_2)}^2} .$$

Furthermore, let us define for $k = 1, 2, \ldots, d$,

$$C_{d,k} := \sum_{\mathfrak{u} \subseteq [d]; |\mathfrak{u}| = k} \prod_{j \in \mathfrak{u}} \gamma_{d,j} ,$$

and, by convention, $C_{d,0} := 1$. Then, Theorem 2 of [NW01b] states that

$$e^2(n, I_{d,\gamma}) \geq \sum_{k=0}^{d} C_{d,k} (1 - n\alpha^k)_+ \, \alpha_1^{d-k} \alpha_2^k , \tag{7}$$

where, by convention, $0^0 = 1$.

5 Necessary Conditions

We are ready to study generalized tractability of $I_\gamma = \{I_{d,\gamma}\}$. In this section, we consider lower bounds on the minimal errors $e(n, I_{d,\gamma})$ from which we obtain necessary conditions on generalized tractability.

The following theorem extends Theorem 3 from [NW01b].

Theorem 1. *Assume that $H(R_1) \cap H(R_2) = \{0\}$ and R_2 is decomposable. Assume that both $h_{1,2,(0)}$ and $h_{1,2,(1)}$ are non-zero. Let T be an arbitrary tractability function, and let Ω be a tractability domain with $[1, \varepsilon_0^{-1}) \times \mathbb{N} \subseteq \Omega$ for some $\varepsilon_0 \in (0,1)$.*

1. *Let $h_{1,1} = 0$. Then $I_\gamma = \{I_{d,\gamma}\}$ is strongly intractable in Ω and is not (T,Ω)-tractable.*
2. *Let $h_{1,1} \neq 0$ and*
$$\lim_{d\to\infty}{}^{*} \frac{\sum_{j=1}^{d} \gamma_{d,j}}{\ln(1+f(d))} = \infty$$
for some non-decreasing function $f\colon \mathbb{N} \to [1,\infty)$, where $\lim^* \in \{\lim, \limsup\}$. Then
$$\lim_{d\to\infty}{}^{*} \frac{e\left(\lfloor f(d)^q \rfloor, I_{d,\gamma}\right)}{e(0, I_{d,\gamma})} = 1 \quad \text{for all } q \in \mathbb{N}. \tag{8}$$
In particular,
$$\limsup_{d\to\infty} \sum_{j=1}^{d} \gamma_{d,j} = \infty$$
implies that I_γ is not strongly (T,Ω)-tractable, and
$$\limsup_{d\to\infty} \frac{\sum_{j=1}^{d} \gamma_{d,j}}{\ln(1+T(\varepsilon^{-1},d))} = \infty$$
for some $\varepsilon \in (\varepsilon_0, 1)$ implies that I_γ is not (T,Ω)-tractable.
3. *Let $h_{1,1} \neq 0$ and $\gamma_{d,d} \geq \gamma^* > 0$ for all $d \in \mathbb{N}$. Then*
$$\lim_{d\to\infty} \frac{e(\lfloor b^d \rfloor, I_{d,\gamma})}{e(0, I_{d,\gamma})} = 1 \quad \text{for all } b \in (1, \alpha^{-c}), \tag{9}$$
where $c \in (0,1)$ satisfies the following two inequalities
$$c \leq \alpha_3\gamma^* \quad \text{and} \quad (1 + \ln(\alpha_3\gamma^*) - \ln c)c < \ln(1 + \alpha_3\gamma^*).$$
Hence, I_γ is strongly intractable in Ω.
4. *Let $h_{1,1} \neq 0$ and*
$$\limsup_{d\to\infty} \frac{\sum_{j=1}^{d} \gamma_{d,j}}{d} > 0.$$
Then I_γ is strongly intractable in Ω.

Proof. We follow here the lines of the proof of [NW01b, Thm. 3]. Note that just now $\alpha \in [1/2, 1)$ since $h_{1,2,(0)} \neq 0 \neq h_{1,2,(1)}$.

Let us first prove statement 1. For $h_{1,1} = 0$, we have $\alpha_1 = 0$ and the only non-zero term in (7) is for $k = d$. Then

$$e(n, I_{d,\gamma}) \geq (1 - n\alpha^d)_+^{1/2}\, e(0, I_{d,\gamma}). \tag{10}$$

From (10) we conclude that

$$n(\varepsilon, I_{d,\gamma}) \geq (1 - \varepsilon^2)\alpha^{-d}.$$

Note that $[1, \varepsilon_0^{-1}) \times \mathbb{N} \subseteq \Omega$ for $\varepsilon_0 \in (0, 1)$ implies that, say, $((1+\varepsilon_0^{-1})/2, d) \in \Omega$ for all d. Therefore

$$\limsup_{(\varepsilon^{-1},d)\in\Omega,\, \varepsilon^{-1}+d\to\infty} \frac{\ln n(\varepsilon, I_{d,\gamma})}{\varepsilon^{-1} + d} \geq \ln \alpha^{-1} > 0\,,$$

which means that I_γ is strongly intractable in Ω and not (T, Ω)-tractable.

Let us now prove statement 2. For $h_{1,1} \neq 0$ we get from (7)

$$1 \geq \frac{e^2(n, I_{d,\gamma})}{e^2(0, I_{d,\gamma})} \geq \frac{\sum_{k=0}^{d} C_{d,k}\alpha_3^k(1 - n\alpha^k)_+}{\sum_{k=0}^{d} C_{d,k}\alpha_3^k}\,.$$

Define $\gamma'_{d,j} := \alpha_3\gamma_{d,j}$ and $C'_{d,k} := \alpha_3^k C_{d,k}$. Then we have

$$1 \geq \frac{e^2(n, I_{d,\gamma})}{e^2(0, I_{d,\gamma})} \geq \frac{\sum_{k=0}^{d} C'_{d,k}(1 - n\alpha^k)_+}{\sum_{k=0}^{d} C'_{d,k}}\,. \tag{11}$$

Now take $n = \lfloor f(d)^q \rfloor$ for an arbitrary $q \in \mathbb{N}$. For any $a \in (0, 1)$ there exist non-negative β_1 and β_2 such that

$$n\alpha^k \leq a \quad \text{for} \quad k \in [k(d, \beta), d], \quad \text{where} \quad k(d, \beta) = \lceil \beta_2 + \beta_1 \ln f(d) \rceil.$$

Let us denote

$$s_d := \sum_{j=1}^{d} \gamma'_{d,j} = C'_{d,1}.$$

Since $k(d, \beta) = O(\ln(1 + f(d)))$, the conditions

$$\lim_{d\to\infty}{}^{*} \frac{s_d}{\ln(1 + f(d))} = \infty \quad \text{and} \quad s_d = O(d)$$

imply that

$$s_d \geq k(d, \beta) \quad \text{and} \quad d \geq k(d, \beta)$$

for infinitely many d. We confine our analysis to those values of d. From (11) we conclude

$$\frac{e^2(n, I_{d,\gamma})}{e^2(0, I_{d,\gamma})} \geq (1-a)\,\frac{\sum_{k=k(d,\beta)+1}^{d} C'_{d,k}}{\sum_{k=0}^{d} C'_{d,k}} = (1-a)(1-\alpha_{d,\beta}),$$

where

$$\alpha_{d,\beta} = \frac{\sum_{k=0}^{k(d,\beta)} C'_{d,k}}{\sum_{k=0}^{d} C'_{d,k}}.$$

To prove (8) it is enough to show that $\alpha_{d,\beta}$ goes to zero as d tends to infinity since a can be arbitrarily small.

It is easy to see that $C'_{d,k} \leq s_d^k/k!$. Thus we have

$$\sum_{k=0}^{k(d,\beta)} C'_{d,k} \leq \sum_{k=0}^{k(d,\beta)} \frac{s_d^k}{k!}. \tag{12}$$

Observe that

$$\sum_{k=0}^{d} C'_{d,k} = \prod_{j=1}^{d} (1+\gamma'_{d,j}) = \exp\left(\sum_{j=1}^{d} \ln(1+\gamma'_{d,j})\right).$$

Assume for a moment that there exists a positive γ^* such that $\gamma_{d,d} \geq \gamma^* > 0$ for all $d \in \mathbb{N}$. Since $\ln(1+f(d)) = o(d)$ we have $\lfloor f(d)^q \rfloor \leq \lfloor b^d \rfloor$ for $b > 1$ and sufficiently large d, and therefore (8) follows from (9) which will be addressed in a moment.

Thus we may consider here only the case where $\lim_{d\to\infty} \gamma_{d,d} = 0$. Then for an arbitrary $\vartheta \in (0,1)$ there exists a positive constant c_ϑ with

$$\exp\left(\sum_{j=1}^{d} \ln(1+\gamma'_{d,j})\right) \geq c_\vartheta \exp\left(s_d(1-\vartheta)\right) \quad \text{for sufficiently large } d. \tag{13}$$

Indeed, let τ be such that $\vartheta = \tau/(1+\tau)$. It is easily seen that

$$x(1-\vartheta) \leq \ln(1+x) \quad \text{for all } x \in [0,\tau].$$

Since $\gamma'_{d,j} \leq \gamma'_{j,j}$ for $d > j$, there is an index j_τ such that $\gamma'_{d,j} \in [0,\tau]$ for all $d \geq j \geq j_\tau$. For $d \geq j_\tau$, we have

$$\begin{aligned}\sum_{j=1}^{d} \ln(1+\gamma'_{d,j}) &\geq \sum_{j=1}^{j_\tau-1} \ln(1+\gamma'_{d,j}) + \left(\sum_{j=j_\tau}^{d} \gamma'_{d,j}\right)(1-\vartheta)\\ &= \sum_{j=1}^{j_\tau-1} \ln(1+\gamma'_{d,j}) - (1-\vartheta)\sum_{j=1}^{j_\tau-1} \gamma'_{d,j} + s_d(1-\vartheta).\end{aligned}$$

Hence (13) holds for $d \geq j_\tau$ with

$$c_\vartheta = \exp\Big(-(1-\vartheta) \sum_{j=1}^{j_\tau - 1} \gamma'_{j_\tau,j} \Big).$$

Observe that $s_d^k/k!$ is an increasing function of k over the interval $[0, k^*]$ as long as $s_d \geq k^*$. Since $s_d \geq k(d,\beta)$ we get

$$\sum_{k=0}^{k(d,\beta)} \frac{s_d^k}{k!} \leq k(d,\beta) \frac{s_d^{k(d,\beta)}}{k(d,\beta)!} = \exp\left[k(d,\beta)\ln s_d - \ln((k(d,\beta)-1)!)\right].$$

Using the formula $k! \geq k^k e^{-k}$ and $k(d,\beta) = O(\ln(1+f(d)))$, we get

$$\begin{aligned}\sum_{k=0}^{k(d,\beta)} C'_{d,k} &\leq \exp\left[k(d,\beta)\ln s_d - (k(d,\beta)-1)(\ln(k(d,\beta)-1)-1)\right] \\ &\leq \exp\left[k(d,\beta)\ln\left(\frac{s_d}{k(d,\beta)-1}\right) + O(\ln(1+f(d)))\right] \\ &\leq \exp\left[O\left(\ln(1+f(d))\ln\left(\frac{s_d}{\ln(1+f(d))}\right)\right)\right].\end{aligned}$$

Let $\vartheta \in (0,1)$. Then for $d \geq j_\tau$,

$$\begin{aligned}\alpha_{d,\beta} &\geq c_\vartheta^{-1} \exp\left[-s_d(1-\vartheta) + O\left(\ln(1+f(d))\ln\left(\frac{s_d}{\ln(1+f(d))}\right)\right)\right] \\ &= c_\vartheta^{-1}\exp\left[-\ln(1+f(d))\left(\frac{s_d(1-\vartheta)}{\ln(1+f(d))} + O\left(\ln\left(\frac{s_d}{\ln(1+f(d))}\right)\right)\right)\right].\end{aligned}$$

Thus $\lim^*_{d\to\infty} \alpha_{d,\beta} = 0$, and the proof of (8) is completed.

Let $\lim^*_{d\to\infty} \sum_{j=1}^d \gamma_{d,j} = \infty$. Then $\lim^*_{d\to\infty} \sum_{j=1}^d \gamma_{d,j}/\ln(1+f(d)) = \infty$ for an arbitrary constant function f. According to (8), this results in

$$\lim_{d\to\infty}{}^* \frac{e(n, I_{d,\gamma})}{e(0, I_{d,\gamma})} = 1 \quad \text{for all } n \in \mathbb{N}.$$

Since $\{\varepsilon\} \times \mathbb{N} \subset \Omega$ for arbitrary $\varepsilon \in (\varepsilon_0, 1)$, we conclude that $n(\varepsilon, I_{d,\gamma})$ must go to infinity with d which means that I_d is not strongly (T,Ω)-tractable.

Finally assume that $\lim^*_{d\to\infty} \sum_{j=1}^d \gamma_{d,j}/\ln(1+T(\varepsilon^{-1},d)) = \infty$ for some $\varepsilon \in (\varepsilon_0, 1)$. This corresponds to $f(d) = T(\varepsilon^{-1}, d)$. If $T(\varepsilon^{-1}, d) = 1$ for all d, we are in the preceeding case and I_γ is not (T,Ω)-tractable. If $T(\varepsilon^{-1},d) > 1$ for some d, then (8) implies that for arbitrary positive constants C and t there exists a positive constant q and infinitely many d with

$$n(\varepsilon, I_{d,\gamma}) > T(\varepsilon^{-1}, d)^q \geq C T(\varepsilon^{-1}, d)^t.$$

This implies again that I_γ is not (T,Ω)-tractable.

We now address statement 3, which, apart from the slightly more general weights, is actually the fourth statement of [NW01b, Thm. 3]. If one checks the proof there and makes the obvious small modifications, one easily sees that (9) also holds. From it, we have $n(\varepsilon, I_{d,\gamma}) \geq \lfloor b^d \rfloor$ for a fixed $\varepsilon \in (\varepsilon_0, 1)$ and sufficiently large d. Hence, $n(\varepsilon, I_{d,\gamma})$ is exponential in d which implies strong intractability in Ω.

We turn to the last statement 4 and reduce it to statement 3. We now know that there exists a sequence $\{d_k\}$, with $\lim_k d_k = \infty$, and a positive c_1 such that

$$\sum_{j=1}^{d_k} \gamma_{d_k,j} \geq c_1 d_k \quad \text{for all} \quad k \in \mathbb{N}.$$

For $s \in [d_k]$ we have

$$\sum_{j=1}^{d_k} \gamma_{d_k,j} = \sum_{j=1}^{s-1} \gamma_{d_k,j} + \sum_{j=s}^{d_k} \gamma_{d_k,j} \leq (s-1)\gamma_{1,1} + (d_k - s + 1)\gamma_{s,s}.$$

For all $s \leq 1 + c_1 d_k/(2\gamma_{1,1})$ we obtain

$$\gamma_{s,s} \geq \gamma^* := c_1/2.$$

Since d_k goes to infinity this proves that $\gamma_{d,d} \geq \gamma^*$ for all d and the assumptions of statement 3 are satisfied. This completes the proof.

6 Sufficient Conditions for Integration

In this section we analyze sufficient conditions for generalized tractability of multivariate integration for a Hilbert space F_d with a general reproducing kernel $K_d : D_d \times D_d \to \mathbb{R}$. We assume that K_d is Lebesgue measurable and

$$\int_{D_d} \int_{D_d} \rho_d(x)\rho_d(y)\, K_d(x,y)\, \mathrm{d}x\, \mathrm{d}y \leq \int_{D_d} \rho_d(x)\, K_d(x,x)\, \mathrm{d}x < \infty,$$

where $\rho_d \geq 0$ and $\int_{D_d} \rho_d(x)\, \mathrm{d}x = 1$. Then multivariate integration

$$I_d f = \int_{D_d} \rho_d(x)\, f(x)\, \mathrm{d}x \quad \text{for all} \quad f \in F_d,$$

is a continuous linear functional, and $I_d f = \langle f, h_d \rangle_{F_d}$ with

$$h_d(x) = \int_{D_d} \rho_d(y)\, K_d(x,y)\, \mathrm{d}y.$$

Without loss of generality we assume that $h_d \neq 0$ since otherwise multivariate integration is trivial. The initial error is now of the form

$$
\begin{aligned}
e(0, I_d) &= \|I_d\| = \|h_d\|_{F_d} = \left(\int_{D_d} \rho_d(x)\, h_d(x)\, \mathrm{d}x\right)^{1/2} \\
&= \left(\int_{D_d}\int_{D_d} \rho_d(x)\rho_d(y)\, K_d(x,y)\, \mathrm{d}x\, \mathrm{d}y\right)^{1/2} > 0.
\end{aligned}
$$

For the algorithm $Q_{n,d}f = \sum_{i=1}^{n} a_i f(z_i)$ we have

$$
I_d f - Q_{n,d} f = \left\langle f, h_d - \sum_{i=1}^{n} a_i K_d(\cdot, z_i)\right\rangle_{F_d}.
$$

This yields a well know formula for the worst case error of $Q_{n,d}$,

$$
\begin{aligned}
e(Q_{n,d}) &= \sup_{f\in F_d,\, \|f\|_{F_d}\le 1} \left| I_d f - \sum_{i=1}^{n} a_i f(z_i)\right| = \left\| h_d - \sum_{i=1}^{n} a_i K_d(\cdot, z_i)\right\|_{F_d} \\
&= \left(\|h_d\|_{F_d}^2 - 2\sum_{i=1}^{n} a_i h_d(z_i) + \sum_{i,j=1}^{n} a_i a_j K_d(z_i, z_j)\right)^{1/2}.
\end{aligned}
$$

We now assume that $Q_{n,d}$ is a QMC algorithm, i.e., $a_i = n^{-1}$, and treat the sample points z_i as independent and identically distributed points over D_d with the density function ρ_d. We use the notation $e(Q_{n,d}) = e(Q_{n,d}, \{z_i\})$ to stress the dependence on the sample points z_i. Let

$$
E(n,d) = \int_{D_d^n} e^2(Q_{n,d}, \{z_i\})\rho_d(z_1)\, \rho_d(z_2) \cdots \rho_d(z_n)\, \mathrm{d}z_1\, \mathrm{d}z_2 \cdots \mathrm{d}z_n
$$

denote the average of the square of the worst case error of $Q_{n,d}$. It is easy to obtain an explicit formula for $E(n,d)$ which is also well known, see e.g., [SW98],

$$
\begin{aligned}
E(n,d) &= \|h_d\|^2 - 2\|h_d\|^2 + \frac{n^2-n}{n^2}\|h_d\|^2 + \frac{1}{n}\int_{D_d} \rho_d(x)\, K_d(x,x)\, \mathrm{d}x \\
&= \frac{\int_{D_d} \rho_d(x)\, K_d(x,x)\, \mathrm{d}x - \int_{D_d}\int_{D_d} \rho_d(x)\rho_d(y)\, K_d(x,y)\, \mathrm{d}x\, \mathrm{d}y}{n}.
\end{aligned}
$$

Here, $\|h_d\| = \|h_d\|_{F_d}$. By the mean value theorem we know that there are sample points z_i for which the square of the worst case error is at most $E(n,d)$. This proves that the square of the nth minimal error $e(n, I_d)$ is at most $E(n,d)$ and we have

$$
\frac{e(n, I_d)}{e(0, I_d)} \le \frac{1}{\sqrt{n}}\left(\frac{\int_{D_d} \rho_d(x)\, K_d(x,x)\, \mathrm{d}x}{\int_{D_d}\int_{D_d} \rho_d(x)\rho_d(y)\, K_d(x,y)\, \mathrm{d}x\, \mathrm{d}y} - 1\right)^{1/2}. \tag{14}
$$

From this estimate it is easy to conclude sufficient conditions on generalized tractability of multivariate integration.

Theorem 2. *Consider multivariate integration $I = \{I_d\}$ defined as in this section. Let T be an arbitrary tractability function, and let Ω be a tractability domain with $[1, \varepsilon_0^{-1}) \times \mathbb{N} \subseteq \Omega$ for some $\varepsilon_0 \in (0,1)$. Let*

$$\eta_d = \frac{\int_{D_d} \rho_d(x)\, K_d(x,x)\, \mathrm{d}x}{\int_{D_d}\int_{D_d} \rho_d(x)\rho_d(y)\, K_d(x,y)\, \mathrm{d}x\, \mathrm{d}y} - 1.$$

1. *We have*

$$n(\varepsilon, I_d) \le \left\lceil \frac{\eta_d}{\varepsilon^2} \right\rceil.$$

2. *If*

$$\lim_{d\to\infty} \frac{\ln \max(1,\eta_d)}{d} = 0$$

then I is weakly tractable in Ω.

3. *If*

$$t^* := \limsup_{(\varepsilon^{-1},d)\in\Omega,\ \varepsilon^{-1}+d\to\infty} \frac{\ln \max(1,\eta_d) + 2\ln \varepsilon^{-1}}{\ln\left(1 + T(\varepsilon^{-1},d)\right)} < \infty$$

then I is (T,Ω)-tractable with the exponent of (T,Ω)-tractability equal to at most t^*.

4. *If*

$$t^* := \limsup_{(\varepsilon^{-1},d)\in\Omega,\ \varepsilon^{-1}+d\to\infty} \frac{\ln \max(1,\eta_d) + 2\ln \varepsilon^{-1}}{\ln\left(1 + T(\varepsilon^{-1},1)\right)} < \infty$$

then I is strongly (T,Ω)-tractable with the exponent of strong (T,Ω)-tractability equal to at most t^*.

Proof. The proof is obvious. The bound on $n(\varepsilon,d)$ directly follows from (14). We have

$$n(\varepsilon, I_d) \le \left\lceil \max(1,\eta_d)\,\varepsilon^{-2} \right\rceil \le 2\max(1,\eta_d)\,\varepsilon^{-2}$$

since $\lceil x \rceil \le 2x$ for $x \ge 1$. Since pairs (ε,d) for $\varepsilon \in (\varepsilon_0,1)$ and $d \in \mathbb{N}$ belong to Ω, we have

$$\frac{\ln n(\varepsilon,I_d)}{\varepsilon^{-1}+d} \le \frac{\ln\max(1,\eta_d)}{d} + \frac{2\ln\varepsilon^{-1}}{\varepsilon^{-1}+d} + \frac{\ln 2}{\varepsilon^{-1}+d}$$

which goes to 0 if $\lim_{d\to\infty} \ln(\max(1,\eta_d))/d = 0$. This yields weak tractability in Ω. The rest follows from the fact that $n(\varepsilon,I_d) \le C\,T(\varepsilon^{-1},k_d)^t$, with $k_d = d$ when we consider (T,Ω)-tractability and $k_d = 1$ when we consider strong (T,Ω)-tractability, if

$$\frac{\ln\max(1,\eta_d) + 2\ln\varepsilon^{-1}}{\ln\left(1+T(\varepsilon^{-1},k_d)\right)} \le \frac{\ln(C/2)}{\ln\left(1+T(\varepsilon^{-1},k_d)\right)} + t.$$

For any $\delta \in (0,1)$ there exists C_δ such that for all $(\varepsilon^{-1},d) \in \Omega$ with $\varepsilon^{-1}+d \ge C_\delta$, the left hand side is at most $t^* + \delta$. Hence, we can take $t = t^* + \delta$ and C sufficiently large so that the last inequality holds for all $(\varepsilon^{-1},d) \in \Omega$. This proves (T,Ω)-tractability or strong (T,Ω)-tractability as well as the needed bounds on the exponents.

7 Examples

We illustrate Theorems 1 and 2 by a number of examples of spaces for multivariate integration.

Example 1: Sobolev Space for Bounded Domain

In this example we consider multivariate integration for the bounded domain, $D_d = [0,1]^d$ and for a specific Sobolev space. More precisely for $d = 1$, as in [NW01b], let $F_{1,\gamma}$ be the Sobolev space of absolutely continuous functions defined on $D_1 = [0,1]$ whose first derivatives are in $L_2([0,1])$ with the inner product

$$\langle f, g\rangle_{F_{1,\gamma}} = f(\tfrac{1}{2})g(\tfrac{1}{2}) + \gamma^{-1}\int_0^1 f'(x)g'(x)\,\mathrm{d}x.$$

The reproducing kernel is $K_{1,\gamma} = R_1 + \gamma\, R_2$ with

$$R_1 = 1 \quad\text{and}\quad R_2(x,y) = 1_M(x,y)\,\min\left(|x-\tfrac{1}{2}|, |y-\tfrac{1}{2}|\right)$$

with the characteristic function 1_M of $M = [0,\frac{1}{2}]\times[0,\frac{1}{2}] \,\cup\, [\frac{1}{2},1]\times[\frac{1}{2},1]$. Clearly, R_2 is decomposable with $a^* = \frac{1}{2}$. The kernel R_2 can also be written as

$$R_2(x,y) = \tfrac{1}{2}\left(|x-\tfrac{1}{2}| + |y-\tfrac{1}{2}| - |x-y|\right).$$

We have

$$\int_0^1\int_0^1 K_{1,\gamma}(x,y)\,\mathrm{d}x\,\mathrm{d}y = 1 + \tfrac{1}{12}\gamma \quad\text{and}\quad \int_0^1 K_{1,\gamma}(x,x)\,\mathrm{d}x = 1 + \tfrac{1}{4}\gamma.$$

For $d \geq 2$, we obtain the Sobolev space $F_{d,\gamma}$ with the inner product

$$\langle f, g\rangle_{F_{d,\gamma}} = \sum_{\mathfrak{u}\subseteq[d]} \gamma_{\mathfrak{u}}^{-1}\int_{[0,1]^{|\mathfrak{u}|}} \frac{\partial^{|\mathfrak{u}|}}{\partial x_{\mathfrak{u}}} f(x_{\mathfrak{u}},\tfrac{1}{2}) \frac{\partial^{|\mathfrak{u}|}}{\partial x_{\mathfrak{u}}} g(x_{\mathfrak{u}},\tfrac{1}{2})\,\mathrm{d}x_{\mathfrak{u}}.$$

Here $\gamma_\emptyset = 1$ and $\gamma_{\mathfrak{u}} = \prod_{j\in\mathfrak{u}}\gamma_{d,j}$ for non-empty $\mathfrak{u}$, and $x_{\mathfrak{u}}$ is the vector from $[0,1]^{|\mathfrak{u}|}$ whose components corresponding to indices in $\mathfrak{u}$ are the same as for the vector $x \in [0,1]^d$, and $(x_{\mathfrak{u}},\frac{1}{2})$ is the vector $x\in[0,1]^d$ with all components whose indices are not in $\mathfrak{u}$ replaced by $\frac{1}{2}$. Furthermore, we use the convention $\int_{[0,1]^{|\emptyset|}} dx_\emptyset = 1$.

Consider multivariate integration, $I_{d,\gamma}f = \int_{[0,1]^d} f(x)\,\mathrm{d}x = \langle f, h_{d,\gamma}\rangle_{F_{d,\gamma}}$ with

$$h_{d,\gamma}(x) = \prod_{j=1}^d \left[1 + \frac{1}{2}\gamma_{d,j}\left(|x_j - \tfrac{1}{2}| - \tfrac{1}{4} + x_j - x_j^2\right)\right].$$

We now have

$$\begin{aligned} h_{1,1} &= 1, \\ h_{1,2,(0)}(x) &= \tfrac{1}{2}(\tfrac{1}{2}-x)(\tfrac{1}{2}+x)\,1_{[0,\frac{1}{2}]}(x), \\ h_{1,2,(1)}(x) &= \tfrac{1}{2}(x-\tfrac{1}{2})(2-\tfrac{1}{2}-x)\,1_{[\frac{1}{2},1]}(x). \end{aligned}$$

Hence, $\|h_{1,2,(0)}\|_{H(R_2)} = \|h_{1,2,(1)}\|_{H(R_2)} = \frac{1}{24}$ and $\alpha = \frac{1}{2}$. Hence, the assumptions of Theorem 1 are satisfied. Furthermore, the initial error is

$$e(0, I_{d,\gamma}) = \prod_{j=1}^{d} \left(1 + \tfrac{1}{12}\,\gamma_{d,j}\right)^{1/2},$$

whereas

$$\int_{[0,1]^d} K_{d,\gamma}(x,x)\,\mathrm{d}x = \prod_{j=1}^{d} \left(1 + \tfrac{1}{4}\,\gamma_{d,j}\right).$$

Hence, the assumptions of Theorem 2 are also satisfied and

$$\eta_d = \prod_{j=1}^{d} \frac{1 + \frac{1}{4}\,\gamma_{d,j}}{1 + \frac{1}{12}\,\gamma_{d,j}} - 1 \le \prod_{j=1}^{d} \left(1 + \tfrac{1}{6}\,\gamma_{d,j}\right)$$

since $(1+b\,x)/(1+a\,x) \le 1 + (b-a)x$ for $b \ge a$ and $x \ge 0$.

Since $\ln(1+x) \le x$ for $x \ge 0$, we now have

$$\ln \max(1, \eta_d) \le \tfrac{1}{6} \sum_{j=1}^{d} \gamma_{d,j}. \tag{15}$$

Combining Theorems 1 and 2 we see that $\lim_d \sum_{j=1}^{d} \gamma_{d,j}/d = 0$ is necessary and sufficient for weak tractability.

For $d = 1$, it is known that the minimal error $e(n, I_{1,\gamma}) = \Theta(n^{-1})$, see e.g., [TWW88]. Hence, if $[1,\infty) \times [d^*] \subseteq \Omega$ with $d^* \ge 1$ then the limit superior of $(\ln \varepsilon^{-1})/\ln(1 + T(\varepsilon^{-1}, 1))$ must be finite if we want to have (T, Ω)-tractability.

We are ready to summarize results for generalized tractability of this multivariate integration problem.

Theorem 3. *Consider multivariate integration $I_\gamma = \{I_{d,\gamma}\}$ defined for the Sobolev space $F_{d,\gamma}$ as in this example. Let T be an arbitrary tractability function, and let Ω be a tractability domain with $[1, \varepsilon_0^{-1}) \times \mathbb{N} \subseteq \Omega$ for some $\varepsilon_0 \in (0,1)$. Then*

$$I_\gamma \text{ is weakly tractable in } \Omega \text{ iff } \lim_{d\to\infty} \frac{\sum_{j=1}^{d} \gamma_{d,j}}{d} = 0.$$

Assume additionally that $[1,\infty) \times [d^*] \subseteq \Omega$ for $d^* \ge 1$ and that

$$S(\varepsilon) := \sup_{d\in\mathbb{N}} \frac{\ln(1 + T(\varepsilon^{-1}, d))}{\ln(1 + T(1, d))} < \infty \quad \textit{for some } \varepsilon \in (\varepsilon_0, 1).$$

Then

1. *I_γ is (T,Ω)-tractable iff*

$$t_1^* := \limsup_{d\to\infty} \frac{\sum_{j=1}^d \gamma_{d,j}}{\ln(1+T(1,d))} < \infty,$$

$$t_2^* := \limsup_{\varepsilon^{-1}\to\infty} \frac{\ln \varepsilon^{-1}}{\ln(1+T(\varepsilon^{-1},1))} < \infty.$$

If this holds then for arbitrary $t > \frac{1}{6}\,t_1^* + 2t_2^*$ there exists a positive $C = C_t$ such that

$$n(\varepsilon, I_{d,\gamma}) \le C\,T(\varepsilon^{-1},d)^t \quad \textit{for all} \quad (\varepsilon^{-1},d) \in \Omega.$$

The exponent $t^{\rm tra}$ of (T,Ω)-tractability is in $[t_2^*, \frac{1}{6}\,t_1^* + 2t_2^*]$.

2. *I_γ is strongly (T,Ω)-tractable iff*

$$t_1^* := \limsup_{d\to\infty} \sum_{j=1}^d \gamma_{d,j} < \infty,$$

$$t_2^* := \limsup_{\varepsilon^{-1}\to\infty} \frac{\ln \varepsilon^{-1}}{\ln(1+T(\varepsilon^{-1},1))} < \infty.$$

If this holds then for arbitrary $t > 2t_2^*$ there exists a positive $C = C_t$ such that

$$n(\varepsilon, I_{d,\gamma}) \le C\,T(\varepsilon^{-1},1)^t \quad \textit{for all} \quad (\varepsilon^{-1},d) \in \Omega\,.$$

The exponent $t^{\rm str}$ of strong (T,Ω)-tractability is in $[t_2^*, 2t_2^*]$.

Proof. The statement on weak tractability follows from Theorems 1 and 2, and (15). Let us now assume that $[1,\infty)\times[d^*] \subseteq \Omega$ for some $d^* \ge 1$ and $S(\varepsilon) < \infty$ for some $\varepsilon \in (\varepsilon_0, 1)$.

We now address (T,Ω)-tractability. Let I_γ be (T,Ω)-tractable. Due to Theorem 1 we have

$$\limsup_{d\to\infty} \frac{\sum_{j=1}^d \gamma_{d,j}}{\ln(1+T(\varepsilon^{-1},d))} < \infty\,.$$

From $S(\varepsilon) < \infty$ it follows that $t_1^* < \infty$. From $e(n, I_{1,\gamma}) = \Theta(n^{-1})$, we get $n(\varepsilon, I_{1,\gamma}) = \Theta(\varepsilon^{-1})$. Thus we have $t_2^* < \infty$.

Let us now assume that t_1^* and t_2^* are finite. Let us first consider the case, where $\limsup_{d\to\infty} \sum_{j=1}^d \gamma_{d,j}$ is finite. Then, due to (15), η_d is uniformly bounded and Theorem 2 implies that for (T,Ω)-tractability it is sufficient to bound ε^{-2} by $CT(\varepsilon^{-1},1)^t$ which is possible if $t > 2t_2^*$. Let us now consider the case where $\limsup_{d\to\infty} \sum_{j=1}^d \gamma_{d,j}$ is infinite. Observe that this and the finiteness of t_1^* imply that $\lim_{d\to\infty} T(1,d) = \infty$. From Theorem 2 and (15) we know that for (T,Ω)-tractability it is sufficient to show

$$t' := \limsup_{(\varepsilon^{-1},d)\in\Omega,\, \varepsilon^{-1}+d\to\infty} \left\{ \frac{1}{6} \frac{\sum_{j=1}^d \gamma_{d,j}}{\ln(1+T(\varepsilon^{-1},d))} + 2\frac{\ln \varepsilon^{-1}}{\ln(1+T(\varepsilon^{-1},d))} \right\} < \infty .$$

Let us consider a sequence $\{(\varepsilon_k^{-1}, d_k)\}$ with $\varepsilon_k^{-1} \to \infty$ and $\{d_k\}$ bounded. For convenience we omit the indices k. Then

$$\limsup_{\{(\varepsilon^{-1},d)\}} \frac{\sum_{j=1}^d \gamma_{d,j}}{\ln(1+T(\varepsilon^{-1},d))} = \limsup_{\{(\varepsilon^{-1},d)\}} \frac{\sum_{j=1}^d \gamma_{d,j}}{\ln(1+T(1,d))} \frac{\ln(1+T(1,d))}{\ln(1+T(\varepsilon^{-1},d))} = 0 ,$$

since $\ln(1+T(\varepsilon^{-1},d)) \ge \ln(1+T(\varepsilon^{-1},1)) \to \infty$ as $\varepsilon^{-1} \to \infty$. Furthermore,

$$\limsup_{\{(\varepsilon^{-1},d)\}} \frac{\ln \varepsilon^{-1}}{\ln(1+T(\varepsilon^{-1},d))} \le t_2^* .$$

Let us now consider a sequence $\{(\varepsilon^{-1},d)\}$ with $d \to \infty$. Then

$$\limsup_{\{(\varepsilon^{-1},d)\}} \frac{\sum_{j=1}^d \gamma_{d,j}}{\ln(1+T(\varepsilon^{-1},d))} \le \limsup_{\{(\varepsilon^{-1},d)\}} \frac{\sum_{j=1}^d \gamma_{d,j}}{\ln(1+T(1,d))} \le t_1^* ,$$

and

$$\limsup_{\{(\varepsilon^{-1},d)\}} \frac{\ln \varepsilon^{-1}}{\ln(1+T(\varepsilon^{-1},d))} = \limsup_{\{(\varepsilon^{-1},d)\}} \frac{\ln \varepsilon^{-1}}{\ln(1+T(\varepsilon^{-1},1))} \frac{\ln(1+T(\varepsilon^{-1},1))}{\ln(1+T(\varepsilon^{-1},d))} .$$

Let us denote the last quantity by C_1. If $\varepsilon^{-1} \to \infty$ then obviously $C_1 \le t_2^*$. If $\{\varepsilon^{-1}\}$ is bounded from above by, say, C_2 then

$$\frac{\ln(1+T(\varepsilon^{-1},1))}{\ln(1+T(\varepsilon^{-1},d))} \le \frac{\ln(1+T(C_2,1))}{\ln(1+T(1,d))} ,$$

which converges to zero as d tends to infinity; hence $C_1 = 0$. This shows that $t' \le \frac{1}{6}t_1^* + 2t_2^*$. The statement concerning the exponents t and t^{tra} follows then from Theorem 2 and from the univariate case showing that $t^{\mathrm{tra}} \ge t_2^*$.

We now turn to strong (T,Ω)-tractability. If I_γ is strongly (T,Ω)-tractable then Theorem 1 implies that t_1^* is finite, whereas the univariate case $d=1$ implies that t_2^* is finite. Assume then that both t_1^* and t_2^* are finite. Then η_d is uniformly bounded and Theorem 2 implies that it is enough to bound ε^{-2} by $C\,T(\varepsilon^{-1},1)^t$ which is possible if $t > 2t_2^*$. The univariate case yields that the exponent t^{str} of strong (T,Ω)-tractability must be at least t_2^*. This completes the proof.

Remark 1. In Theorem 3 we made the additional assumptions $[1,\infty)\times[d^*] \subseteq \Omega$ for $d^* \ge 1$ and $S(\varepsilon) \le \infty$ to ensure that the conditions $t_1^* \le \infty$ and $t_2^* \le \infty$

in statement 1 and 2 are not only sufficient but also necessary for (T, Ω)-tractability and strong (T, Ω)-tractability, respectively. Observe that for a tractability function T of product form, i.e., $T(x, y) = f_1(x)f_2(y)$ with non-decreasing functions $f_i : [1, \infty) \to [1, \infty)$, and $\lim_{x\to\infty} \ln(f_i(x))/x = 0$ for $i = 1, 2$, we have $S(\varepsilon) < \infty$ for every $\varepsilon \in (\varepsilon_0, 1)$. On the other hand $S(\varepsilon) = \infty$ for all $\varepsilon \in (\varepsilon_0, 1)$ if, e.g., $T(x, y) = 1 + g_1(x)g_2(y)$, where the non-negative and non-decreasing functions g_i satisfy $g_1(1) = 0$, $g_1(x) > 0$ for $x > 0$, and $\lim_{y\to\infty} g_2(y) = \infty$. Observe also that one can always modify a tractability function T by putting $T(1, d) = 1$ for all d—it still remains a tractability function. In this case $S(\varepsilon) < \infty$ is equivalent to $\lim_{d\to\infty} T(\varepsilon^{-1}, d) < \infty$.

For this Sobolev space $F_{d,\gamma}$ and QMC algorithms $Q_{n,d}$ with $a_i = n^{-1}$, it is known that the worst case errors of $Q_{n,d}$ are equal to the centered discrepancy, see [Hic98, NW01b]. Since our upper bounds are based on QMC algorithms, the same estimates and conditions on generalized tractability presented in Theorem 3 are also valid for the centered discrepancy.

Example 2: Sobolev Space for Unbounded Domain

In this example we consider multivariate integration for the unbounded domain, $D_d = \mathbb{R}$, and for Sobolev spaces of smooth functions. More precisely let r be a positive integer. For $d = 1$, similarly as in [NW01b], let $F_{1,\gamma}$ be the Sobolev space of functions defined on $\mathbb{R}$ whose $(r-1)$st derivatives are absolutely continuous and whose rth derivatives belong to $L_2(\mathbb{R})$, and satisfy the conditions

$$f'(0) = f''(0) = \ldots = f^{(r-1)}(0) = 0.$$

The inner product of $F_{1,\gamma}$ is given by

$$\langle f, g\rangle_{F_{1,\gamma}} = f(0)g(0) + \gamma^{-1} \int_{\mathbb{R}} f^{(r)}(x)g^{(r)}(x)\,\mathrm{d}x.$$

The reproducing kernel of $F_{1,\gamma}$ is $K_{1,\gamma} = R_1 + \gamma R_2$ with $R_1 = 1$ and

$$R_2(x, y) = 1_M(x, y) \int_{\mathbb{R}_+} \frac{(|x| - u)_+^{r-1}}{(r-1)!} \frac{(|y| - u)_+^{r-1}}{(r-1)!}\,\mathrm{d}u,$$

where 1_M is the characteristic function of $M = \{(x, y) : xy \geq 0\}$. Clearly, R_2 is decomposable with $a^* = 0$. Consider univariate integration

$$I_{1,\gamma} f = \int_{\mathbb{R}} \rho(x)\, f(x)\,\mathrm{d}x,$$

where $\rho(x) = \rho(-x) \geq 0$, $\int_{\mathbb{R}} \rho(x)\,\mathrm{d}x = 1$, and $\int_{\mathbb{R}} \rho(x)\,|x|^{2r-1}\,\mathrm{d}x < \infty$. We now have $h_{1,1} = 1$ and $h_{1,2} = h_{1,2,(0)} + h_{1,2,(1)}$ with

$$h_{1,2,(0)}(x) = \int_{-\infty}^{0} \rho(y)\, R_2(x, y)\,\mathrm{d}y \quad\text{and}\quad h_{1,2,(1)}(x) = \int_{0}^{\infty} \rho(y)\, R_2(x, y)\,\mathrm{d}y.$$

Note that both $h_{1,2,(i)}$ are well defined since it can be checked that $R_2(x,x) = O(|x|^{2r-1})$ and therefore even the integral $\int_{\mathbb{R}} \rho(x)\, R_2(x,x)\, \mathrm{d}x < \infty$. Furthermore, the functions $h_{1,2,(i)}$ are not zero since ρ is symmetric and non zero. Hence, the assumptions of Theorem 1 are satisfied, and symmetry of ρ and R_2 yield that $\alpha = \frac{1}{2}$.

For $d \geq 2$, we obtain the Sobolev space $F_{d,\gamma}$ with the inner product

$$\langle f, g \rangle_{F_{d,\gamma}} = \sum_{\mathfrak{u} \subseteq [d]} \gamma_{\mathfrak{u}}^{-1} \int_{\mathbb{R}^{|\mathfrak{u}|}} \frac{\partial^{r|\mathfrak{u}|}}{\partial x_{\mathfrak{u}}^{r}} f(x_{\mathfrak{u}}, 0) \frac{\partial^{r|\mathfrak{u}|}}{\partial x_{\mathfrak{u}}^{r}} g(x_{\mathfrak{u}}, 0)\, \mathrm{d}x_{\mathfrak{u}}$$

with the same notation as in the previous example with the obvious exchange of $\frac{1}{2}$ to 0.

Consider multivariate integration,

$$I_{d,\gamma}(f) = \int_{\mathbb{R}^d} \rho_d(x)\, f(x)\, \mathrm{d}x = \langle f, h_d \rangle_{F_{d,\gamma}}$$

with $\rho_d(x) = \prod_{j=1}^{d} \rho(x_j)$ and

$$h_d(x) = \prod_{j=1}^{d} \left(1 + \gamma_{d,j}\, h_{1,2}(x)\right).$$

Furthermore, the initial error is

$$e(0, I_{d,\gamma}) = \prod_{j=1}^{d} \left(1 + \gamma_{d,j}\, A\right)^{1/2},$$

where

$$A := \int_{\mathbb{R}^2} \rho(x)\, \rho(y)\, R_2(x,y)\, \mathrm{d}x\, \mathrm{d}y.$$

We also have

$$\int_{\mathbb{R}^d} \rho_d(x)\, K_{d,\gamma}(x,x)\, \mathrm{d}x = \prod_{j=1}^{d} (1 + \gamma_{d,j}\, B) < \infty,$$

with

$$B := \int_{\mathbb{R}} \rho(x)\, R_2(x,x)\, \mathrm{d}x < \infty.$$

Hence, the assumptions of Theorem 2 are also satisfied and

$$\eta_d = \prod_{j=1}^{d} \frac{1 + \gamma_{d,j}\, B}{1 + \gamma_{d,j}\, A} - 1 \leq \prod_{j=1}^{d} \left(1 + \gamma_{d,j}\, (B - A)\right).$$

For $d = 1$ it is known that $e(n, I_{1,\gamma}) = \Theta(n^{-r})$, see again e.g., [TWW88]. Combining Theorems 1 and 2 and proceeding as for the previous example we obtain necessary and sufficient conditions on generalized tractability of this multivariate integration problem.

Theorem 4. *Consider multivariate integration $I_\gamma = \{I_{d,\gamma}\}$ defined for the Sobolev space $F_{d,\gamma}$ as in this example. Let T be an arbitrary tractability function, and let Ω be a tractability domain with $[1, \varepsilon_0^{-1}) \times \mathbb{N} \subseteq \Omega$ for some $\varepsilon_0 \in (0, 1)$. Then*

$$I_\gamma \text{ is weakly tractable in } \Omega \text{ iff } \lim_{d\to\infty} \frac{\sum_{j=1}^d \gamma_{d,j}}{d} = 0.$$

Assume additionally that $[1, \infty) \times [d^*] \subseteq \Omega$ for $d^* \geq 1$ and that

$$S(\varepsilon) := \sup_{d\in\mathbb{N}} \frac{\ln(1 + T(\varepsilon^{-1}, d))}{\ln(1 + T(1, d))} < \infty \quad \text{for some } \varepsilon \in (\varepsilon_0, 1).$$

Then

1. *I_γ is (T, Ω)-tractable iff*

$$t_1^* := \limsup_{d\to\infty} \frac{\sum_{j=1}^d \gamma_{d,j}}{\ln(1 + T(1, d))} < \infty,$$

$$t_2^* := \limsup_{\varepsilon^{-1}\to\infty} \frac{\ln \varepsilon^{-1}}{\ln(1 + T(\varepsilon^{-1}, 1))} < \infty.$$

If this holds then for arbitrary $t > (B - A)\, t_1^* + 2t_2^*$ there exists a positive $C = C_t$ such that

$$n(\varepsilon, I_{d,\gamma}) \leq C\, T(\varepsilon^{-1}, d)^t \quad \text{for all } (\varepsilon^{-1}, d) \in \Omega.$$

The exponent t^{tra} of (T, Ω)-tractability is in $[r^{-1} t_2^*, (B - A)\, t_1^* + 2t_2^*]$.

2. *I_γ is strongly (T, Ω)-tractable iff*

$$t_1^* := \limsup_{d\to\infty} \sum_{j=1}^d \gamma_{d,j} < \infty,$$

$$t_2^* := \limsup_{\varepsilon^{-1}\to\infty} \frac{\ln \varepsilon^{-1}}{\ln(1 + T(\varepsilon^{-1}, 1))} < \infty.$$

If this holds then for arbitrary $t > 2t_2^*$ there exists a positive $C = C_t$ such that

$$n(\varepsilon, I_{d,\gamma}) \leq C\, T(\varepsilon^{-1}, 1)^t \quad \text{for all } (\varepsilon^{-1}, d) \in \Omega.$$

The exponent t^{str} of strong (T, Ω)-tractability is in $[r^{-1} t_2^*, 2t_2^*]$.

Example 3: General Case with $R_1 = 1$

Based on the two previous examples, it is easy to see that we can obtain necessary and sufficient conditions for generalized tractability of multivariate integration for general spaces if we assume that the kernel $R_1 = 1$. Then

$H(R_1) = \mathrm{span}(1)$ and $H(R_1) \cap H(R_2) = \{0\}$ holds iff $1 \notin H(R_2)$. We now assume that R_2 is Lebesgue measurable and that

$$A := \int_{D_1^2} \rho(x)\, \rho(y)\, R_2(x,y)\, \mathrm{d}x\, \mathrm{d}y \le B := \int_{D_1} \rho(x)\, R_2(x,x)\, \mathrm{d}x < \infty.$$

As in Section 4 we assume that R_2 is decomposable and consider integration for the space $F_{1,\gamma}$,

$$I_{1,\gamma} f = \int_{D_1} \rho(x)\, f(x)\, \mathrm{d}x = \langle f, h_{1,\gamma} \rangle_{F_{1,\gamma}}$$

with

$$h_{1,\gamma}(x) = 1 + \gamma \left(h_{1,2,(0)}(x) + h_{1,2,(1)}(x) \right),$$

where

$$h_{1,2,(i)}(x) = 1_{D_{(i)}}(x) \int_{D_{(i)}} \rho(y)\, R_2(x,y)\, \mathrm{d}y.$$

We assume that both $h_{1,2,(i)}$ are non-zero. For $d = 1$ we assume that $e(n, I_{1,\gamma}) = \Theta(n^{-r})$ for some $r > 0$.

Then Theorems 1 and 2 and the analysis of the previous examples yield the following theorem.

Theorem 5. *Consider multivariate integration $I_\gamma = \{I_{d,\gamma}\}$ defined for the space $F_{d,\gamma}$ with $R_1 = 1$ as in this example. Let T be an arbitrary tractability function, and let Ω be a tractability domain with $[1, \varepsilon_0^{-1}) \times \mathbb{N} \subseteq \Omega$ for some $\varepsilon_0 \in (0,1)$. Then*

$$I_\gamma \text{ is weakly tractable in } \Omega \text{ iff } \lim_{d \to \infty} \frac{\sum_{j=1}^d \gamma_{d,j}}{d} = 0.$$

Assume additionally that $[1, \infty) \times [d^*] \subseteq \Omega$ for $d^* \ge 1$ and that

$$S(\varepsilon) := \sup_{d \in \mathbb{N}} \frac{\ln(1 + T(\varepsilon^{-1}, d))}{\ln(1 + T(1,d))} < \infty \quad \text{for some } \varepsilon \in (\varepsilon_0, 1).$$

Then

1. *I_γ is (T, Ω)-tractable iff*

$$t_1^* := \limsup_{d \to \infty} \frac{\sum_{j=1}^d \gamma_{d,j}}{\ln(1 + T(1,d))} < \infty,$$

$$t_2^* := \limsup_{\varepsilon^{-1} \to \infty} \frac{\ln \varepsilon^{-1}}{\ln(1 + T(\varepsilon^{-1}, 1))} < \infty.$$

If this holds then for arbitrary $t > (B - A)\, t_1^* + 2 t_2^*$ there exists a positive $C = C_t$ such that

$$n(\varepsilon, I_{d,\gamma}) \le C\, T(\varepsilon^{-1}, d)^t \quad \text{for all } (\varepsilon^{-1}, d) \in \Omega.$$

The exponent t^{tra} of (T, Ω)-tractability is in $[r^{-1} t_2^*, (B - A)\, t_1^* + 2 t_2^*]$.

2. I_γ is strongly (T,Ω)-tractable iff

$$t_1^* := \limsup_{d\to\infty} \sum_{j=1}^{d} \gamma_{d,j} < \infty,$$

$$t_2^* := \limsup_{\varepsilon^{-1}\to\infty} \frac{\ln \varepsilon^{-1}}{\ln(1+T(\varepsilon^{-1},1))} < \infty.$$

If this holds then for arbitrary $t > 2t_2^*$ there exists a positive $C = C_t$ such that

$$n(\varepsilon, I_{d,\gamma}) \le C\,T(\varepsilon^{-1},1)^t \quad \textit{for all} \quad (\varepsilon^{-1},d) \in \Omega.$$

The exponent t^{str} of strong (T,Ω)-tractability is in $[r^{-1}t_2^*, 2t_2^*]$.

Acknowledgement

We appreciate comments from Erich Novak.

Part of the work of the first author was done while he was at the Max Planck Institute for Mathematics in the Sciences in Leipzig.

The second author was partially supported by the National Sciences Foundation under Grant DMS-0608727, and by the Humboldt Foundation as a recipient of the Humboldt Research Award at the University of Jena.

References

[GW06] M. Gnewuch and H. Woźniakowski. Generalized Tractability for Multivariate Problems, Part I: Linear Tensor Product Problems and Linear Information. J. Complexity, **23**, 262–295 (2007)

[Hic98] F.J. Hickernell. A generalized discrepancy and quadrature error bound. Math. Comp., **67**, 299–322 (1998)

[NW01a] E. Novak and H. Woźniakowski. When are integration and discrepancy tractable? In: DeVore, R.A., Iserles, A., Süli, E. (eds) Foundation of Computational Mathematics, Oxford, 1999. Cambridge University, Cambridge, 211–266 (2001)

[NW01b] E. Novak and H. Woźniakowski. Intractability results for integration and discrepancy. J. Complexity, **17**, 388–441 (2001)

[NW07] E. Novak and H. Woźniakowski. Tractability of Multivariate Problems. In progress

[SW98] I.H. Sloan and H. Woźniakowski. When are quasi-Monte Carlo algorithms efficient for high dimensional integrals? J. Complexity, **14**, 1–33 (1998)

[TWW88] J.F. Traub, G.W. Wasilkowski, and H. Woźniakowski. Information-Based Complexity. Academic Press, New York (1988)

An Improved Implementation of Stochastic Particle Methods and Applications to Coagulation Equations

Flavius Guiaş

Department of Mathematics, University of Dortmund, Vogelpothsweg 87, 44221 Dortmund, Germany
flavius.guias@mathematik.uni-dortmund.de

Summary. We present a sampling method with applications to Monte Carlo simulations of particle systems which improves the efficiency in sampling from a table of probabilities with changing values and a variable number of elements. For this purpose an optimized partition of the set of events is constructed. The goal is to minimize the expected number of operations in choosing first an element of the partition and sampling afterwards the event from this set. The approach presented here computes an optimized grouping based only on the current structure of the table of events. It can be used as an universal tool, avoiding the experimental way for determining the best group structure, which depends on the problem parameters and on the number of particles. The method is tested numerically by simulations of coagulation processes.

1 Introduction

In particle-based Monte Carlo simulations the basic quantities which are involved are the empirical measures corresponding to the particle system. They are determined by the positions of the particles in the physical space or in a parameter space. Their support size is crucial for the efficiency of the stochastic algorithms. If we allow the positions of the particles to take only a discrete set of values, we can work directly with approximations of the macroscopic quantities (like densities, or mass corresponding to a given size or located at a given place) rather than treating each particle individually. We make therefore no distinction between particles located at the same point in the size space or in the physical space. The set of possible events can decrease significantly, while the state of the system continues to change according to the same particle dynamics. The main difference is that the set of events has now a variable number of elements, depending on the fluctuating support size, and not a fixed one (depending on the total number of particles).

In this paper we introduce introduce a method of sampling within the framework presented above, which optimizes the grouping principle used for computing transitions in Monte Carlo methods. The stochastic algorithms for simulating Markov processes have to face the following problem: to sample the next event E_k from a set $\mathcal{E} = \{E_1, \dots E_n\}$ of possible events with probabilities $p_i = P(E_i), i = 1, \dots n$. The state of the process is then changed according to the event E_k, which leads also to a change in the structure of the set $\mathcal{E}$ and of the probabilities of its elements. In many situations the new configuration can be computed by using only a few operations. This is the case in particle methods, where an "event" consists in modifying the position of only a small number of particles in a given space, while the total number of particles is usually very large.

In sampling from a set of events with changing probabilities we cannot use fast methods like the *alias-method* ([Fi96], p. 165). We have therefore to rely on the following two basic methods and on further improvements of them.

Consider a set of events $\mathcal{E}$ as above and denote $P = \sum_i p_i$, $p^{max} = \max_i p_i$.

- *the acceptance-rejection method*
  Let $\boxed{p_1}\boxed{p_2}\boxed{\dots}\boxed{p_n}$ be an array with a fixed number of elements which corresponds to the probability table and let $\bar{p}$ be an envelope for the table, for example $\bar{p} = p^{max}$.
  The sampling algorithm can be described as:
  1. choose uniformly an index i
  2. with probability $p_i/\bar{p}$ accept the event E_i
  3. if rejection, GOTO 1.

  The expected number of steps in the acceptance-rejection method is $\mathbf{E}_{ar} = n \cdot \frac{\bar{p}}{P}$. Note that if $\bar{p}/P = O(1/n)$ then we need $O(1)$ operations, while in the case $\bar{p}/P = O(1)$ we need $O(n)$ operations. Since the probabilities are subject to changes, we must always check if the modified value of one of the p_i's exceeds $\bar{p}$ and, if it is the case, we have to make the corresponding update. However, the last computed maximal value can also decrease during the simulations and it may be too costly to search every time for the new maximum. In this situation we continue the simulation with the current envelope $\bar{p}$ and recompute the maximum only periodically.

- *the inverse transform method*
  Let $p_1 \leftrightarrow p_2 \leftrightarrow \dots \leftrightarrow p_n$ be a doubly linked list with a variable number of elements.
  The sampling algorithm can be described as:
  1. simulate a uniformly distributed random variable U on $(0, 1)$
  2. compute for succesive k the sum $S = \sum_{i=1}^{k} p_i$
  3. if $S \geq U \cdot P$ choose the event E_k

The expected number of steps in the inverse transform method is $\mathbf{E}_{it} = \sum_{k=1}^{n} k \cdot \frac{p_k}{P}$. If $p^{max}/P = O(1/n)$ then this approach needs in the average $O(n)$ steps in sampling an event. However, if there are significant differences in the magnitude orders of the p_i's and they are sorted such that the large values are at the beginning of the list, in certain circumstances we may need only $O(1)$ steps.

In the situations when n is very large, for example in particle methods which use a large number of particles, one has to find a way to reduce the magnitude order of the required operations with the two methods presented above.

This issue is addressed in Section 2. The basic idea is to divide the events into a certain number of groups and computing the group probabilities. A group is chosen with respect to its probability and then the transition event is sampled only within that group. This principle is known and already used in Monte Carlo computations, for example in [EW00-1], [EW00-2] or [EW01]. There it was used in the context of acceptance-rejection techniques, while the groups were generated according to a power-law scale of the particle sizes. In this situation, the optimal number of groups which delivers the best performance of the implementation has to be determined experimentally from case to case, depending on the problem parameters and on the total number of particles. Our experiments show that this number influences strongly the overall performance of the program, so choosing a proper value for it is a crucial issue.

Having in mind more complex applications, like simulations of spatially inhomogeneous coagulation dynamics, the heuristic approach for determining the optimal grouping is not satisfactory. The structure of the empirical measures which describe the mass spectrum of the particles may be completely different in various regions, so we cannot choose overall the same value of the parameter which determines the group structure. Moreover, in the case of a large number of patches, one cannot afford to find the proper values by experimenting. The same problem arises in other applications, where the grouping according to a monotone (e.g. power-law) scale is not appropriate at all: how should the groups be chosen in such a case?

The aim of this paper is to present an approach for computing an optimized grouping, based only on the current structure of the table of events. The proposed approach can be therefore used as an universal tool, avoiding the experimental way for determining the best group structure.

The optimization of the group structure is done by a simple method based on binary partitions of the groups. Usually the events to be computed correspond to a list which is ordered in a natural way (e.g. the positions of the particles in the size space). Having a given group, we choose from all possible splittings in two consecutive parts the one which minimizes the number of expected operations needed to compute an event (to be precise: an approximation of it). We call this division a *binary partition* of the original group. The algorithm which computes the optimized structure starts with a single group which contains all possible events and follows then the next procedure: having the current group structure, do succesive binary partitions of the existing groups. If by replacing a group with the two parts the expected number of operations decreases, then keep this binary partition and update the group structure, otherwise reject it. We stop if the binary partition steps are rejected for all existing groups.

In the ideal situation, the "optimal" solution would be the absolute minimizer for the expected number of operations in sampling the next event. However, after performing one transition step, the set of events and the set of probabilities suffer changes. This requests eventually an update the group configuration, if we want to remain on the optimal track. Trying to do this in an "optimal" way at each step, will wipe away all advantages which come with this principle, due to the number of operations required for this optimization. The solution is to assign the new possible events to one of the existing groups in a straightforward and natural way, with a minimal number of operations. The recomputation of the group configuration takes place only from time to time, if the expected number of operations gets too far away from the optimum.

But, even if we compute the "optimal" structure only periodically, it will be immediately destroyed due to the changes in the event structure. Therefore it suffices a "nearly optimal" configuration which can be achieved as cheap as possible. A simpler and faster algorithm which leads to results which are close to the optimum should have priority upon the attempt to reach the exact minimum, but at a higher price. We must always have in mind the fact that our real "optimization problem" is the global Monte Carlo algorithm, which consists in computing the transition events, as well as the group structure of these events. The combination of these two aspects has to work as fast as possible. A complete solution to this problem is beyond our scope, but the binary partition algorithm presented in this paper is a satisfactory alternative, altough there is certainly room for improvement left.

Section 3 is dedicated to numerical experiments. The principles presented here are tested in the implementation of a stochastic algorithm for the numerical approximation of solutions of *coagulation equations.*

In coagulation phenomena, particles with size parameters x and y coalesce at rate $K(x, y)$ in order to form a particle of size $x + y$. By this conservation principle one arrives at the *coagulation equations* or *Smoluchowski equations.* If the size parameter is integer-valued, for example in the case of polymerization, where it equals the number of building blocks which form a polymer, these equations take the form

$$\frac{d}{dt}u^1(t) = -u^1 \sum_{i=1}^{\infty} K(1,i)u^i \tag{1}$$

$$\frac{d}{dt}u^k(t) = \frac{1}{2}\sum_{i=1}^{k-1} K(i,k-i)u^i u^{k-i} - u^k \sum_{i=1}^{\infty} K(k,i)u^i \quad k = 2, 3, \ldots$$

where u^k denotes the concentration of polymers of size k. The most usual initial condition is the *monodisperse* initial data: $u^1(0) = 1$, $u^k(0) = 0$, $k \geq 2$.

The *total mass* $M(t) = \sum_{k=1}^{\infty} k u^k(t)$ is formally conserved, since particles are neither created, nor destroyed. However, for large coagulation rates, e.g. $K(i,j) \geq (ij)^q$ with $q > \frac{1}{2}$ (see [Je98]), one can observe the so-called *gelation phenomenon*, which means the decay of the total mass: $M(t) < M(0)$ for $t > t_{gel}$ (the gelation time). This is related to the formation in finite time of infinitely large clusters, which are not described by the variables u^k.

For further details concerning the properties of coagulation equations we indicate the reference [BC90]. For the stochastic approach we recommend the papers [No99], [Al99], [Wa03] and the references within. The stochastic approach to the coagulation equations leads to existence results, as well as to derivation of qualitative properties and numerical schemes.

In this paper we are interested in comparing the runtimes of different implementation of the mass flow algorithm (see [EW01]) based on the same number of particles. The conclusion is that the the algorithm obtained by the optimized and periodically adapted grouping technique from Section 2 has a performance which is comparable to the best results provided by the implementations which use a power-law scale for grouping the particles. One should note that all implementations using the procedure introduced here have built in this part which requests additional computing time, which is not present in the cases where the group limits are defined apriori. However, under the ansatz of the power-law scale, one has to perform several experiments in order to determine the optimal value of the number of groups. These experiments show also that the efficiency of the implementation depends strongly on the choice of this parameter. The method of optimized grouping is however not so sensitive for parameter variations (the frequency of recomputing) within a meaningful range. This fact and its overall performance within the range of different implementations with "manual" choice of the group structure, speak for the various application possibilities of the method introduced here. The implementation of these principles to a more complex model, namely the coagulation-diffusion equations, is discussed in [Gu06]. It turns out that this approach leads to good numerical results at an affordable computational effort.

2 The Grouping Principle

In this section we address the question of dividing a set $\mathcal{E} = \{E_1, \ldots E_n\}$ of possible events with probabilities $p_i = P(E_i)$, $\sum_{i=1}^{n} p_i = 1$ into groups, in order to speed-up the computing of an event E_k according to the given distribution. The probabilities of the events, as well as their number, are subject to changes. The goal is a simple and fast algorithm which leads to a significant reduction of the expected number of operations, and *not* solving an optimization problem at a possibly higher price. An additional restriction is given by the fact that the events E_i are usually ordered in an array or in a list in a natural way, and we want to keep track of this order property.

We consider thus the following grouping of the events into m groups:

$$\{E_1, \dots E_{k_1} \mid E_{k_1+1}, \dots E_{k_2} \mid \dots \mid E_{k_{m-1}+1}, \dots E_{k_m}\}.$$

For $i = 1, \dots m$ denote the i-th group by $G_i = \{E_{k_{i-1}+1}, \dots E_{k_i}\}$, the number of its elements by n_i and its probability by $P_i = \sum_{j=k_{i-1}+1}^{k_i} p_j$.

The sampling of an event E_k according to the given distribution, based on a partition with a fixed number m of groups, is performed by the following algorithm :

1. choose the group G_l according to the probability table $\{P_i\}_{i=1}^m$ by the inverse transform method.
2. generate a uniformly distributed random variable U on $(0, 1)$.
3. for the group G_l compute succesively the sums $\sum_{j=k_{l-1}+1}^{k} p_j$, until the value for a given k exceeds $P_l U$.
4. the chosen event will be E_k.

Given the set of groups $\mathcal{G} = \{G_1, \dots G_m\}$, an upper bound for the expected number of additive operations needed to compute E_k is given by

$$M_{\mathcal{G}} = m + \sum_{i=1}^{m} P_i n_i.$$

Our goal is to get $M_{\mathcal{G}}$ as small as possible, by keeping at the same time things as simple as possible.

The easiest case is for $m = 2$. We have to divide the set of events $\mathcal{E}$ into two groups $\mathcal{G} = \{G_1, G_2\}$, such that $M_{\mathcal{G}}$ is minimal among all such possible divisions. This is done in a straightforward manner, by considering all possible partitions $\mathcal{E} = \{E_1, \dots E_k \mid E_{k+1}, \dots E_n\}$ and choosing the one which minimizes $M_{\mathcal{G}}$. We call this decomposition a *binary partition* of the original group, which in this case is the set $\mathcal{E}$ of all events.

The scheme of the algorithm is now the following:

1. Suppose we have a group structure $\mathcal{G} = \{G_1, \dots G_m\}$.
   For $i = 1$ to m do the following steps:
   a) perform a binary partition of the group G_i.
   b) if the replacing of G_i by the two new groups leads to a smaller value of $M_{\mathcal{G}}$, then keep this binary partition, otherwise reject it.

```
   c) if the binary partition of G_i was accepted, update the
      value of M_G.
2. If all binary partition steps were rejected, then STOP,
   otherwise perform another cycle of similar operations with
   the new group structure.
```

The simplicity of the formula for $M_{\mathcal{G}}$ is crucial for the efficient computation of the binary partitions of the groups, since we have to update its value while checking each possible splitting of the current group in two consecutive parts. The use of the majorants for the expected number of operations presented above (and not the exact formulas) leads to simple recursions in updating the value of $M_{\mathcal{G}}$.

Our next goal is to apply this principle at particle methods. An event consists in the modification of the position of a few number of particles in the state space. Consequently, the structure of events and their probabilities suffers changes, which in most situations are only of local nature. Due to efficiency reasons we do not perform a new division into groups after each transition step, but add the new event to one of the existing groups, as described below for the case of coagulation dynamics.

We compute the group structure anew only if for any of the groups, say for G_i, the quantity $P_i n_i$ becomes α times larger than its initial value, given at the moment of the previous computation of the group structure. $\alpha > 1$ is a parameter which one is free to choose, depending on the problem. If it is too close to 1, the algorithm for grouping the events will eventually be used too often, which slows down the global Monte Carlo algorithm. Choosing α too large may allow increasing groups sizes and group weights, without reorganizing them at the right moment. This leads again to a reduced efficiency of the global algorithm. In our applications a reasonable choice for α turned out to be in the range $(2, 6)$.

3 Numerical Examples: Applications to Coagulation Equations

In this section we will illustrate the application of the principle of group optimization at the mass flow algorithm for simulating coagulation dynamics. We consider N numerical particles located in the size space, which is the interval $(0, \infty)$. The presence of k_i particles at the location i means that the total mass of i-mers in the system is k_i/N. In the case of multiplicative coagulation kernels, i.e. for $K(i,j) = r(i)r(j)$, the mass flow algorithm needs to simulate independently the following events: "the first coagulating particle has size i" with probability proportional to $r(i)k_i/N$ and "the second coagulating particle has size j" with probability proportional to $r(j)k_j/(Nj)$. Then a particle at size i is removed and a new particle at size $i + j$ is added.

Note that this definition of events does not distinguish between the particles located at the same size, which brings another computational advantage over treating each particle individually. A smaller support leads to faster computations of the possible events and this is of advantage especially when one starts with monodisperse initial conditions, i.e. only with monomers. Figure 1 shows the time evolution of the support size of the empirical measure for a gelling and a non-gelling kernel. We note that in the case of the gelling kernel the maximal value is reached essentially at the gelation time (in this case $t = 1$) and formulate this as a conjecture which would be interesting to prove.

This approach leads straightforward to a data structure with a variable number of elements and to the use of the inverse transform method in sampling from such data structures. The number of events is continuously changing, but between two updates of the group structure the number of groups remains fixed. The data is organized as a doubly linked list. One element of the list contains the following fields: the size, the mass concentrated at the respective size, a pointer to the previous element, which corresponds to the next smallest particle size where we have mass, and a pointer to the next element, which corresponds to the next largest particle size where we have mass. Problems like: handling the head or the end of the list, removing a particle from a given size, removing an element form the list if the mass becomes 0, adding mass at an existing particle size or adding a new element to the list, are treated in a straightforward manner.

Since we have to compute two types of events, according to different distributions, we will need two group structures. This is implemented by considering two new lists or arrays, which contain only the first element from each group. We introduce thus two additional fields in our data type, which indicate the group to which the particles of that size belong, one field for each group structure. Since the original lists are ordered by the increasing particle size, the algorithm assigns a new generated element to one of the existing

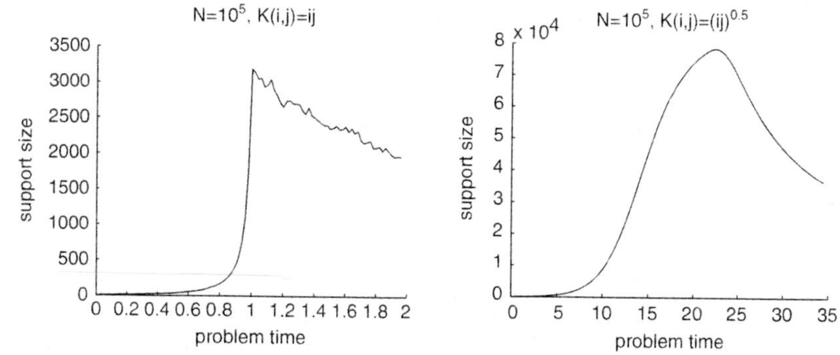

Fig. 1. Support size of the empirical measures for $N = 10^5$ particles and coagulation kernels $K(i, j) = ij$ (left) and $K(i, j) = (ij)^{0.5}$ (right)

groups in a natural way. A group consists of all elements with values of the size field larger or equal than that of the first element of the group (which remains fixed as long as we have particles located there), and less than that of the first element of the next group. Note that by the possibility of removing elements form the list (if the mass becomes 0), the first element of a group can be modified during the execution of the program, or groups may become empty. If in addition to coagulation we want to consider other possible events, for example fragmentation or dissapearance of particles, the number of group structures has to be increased accordingly.

3.1 Comparison of Computing Times

In this section we present a comparison of the CPU times for different implementations of the MFA which use various sampling techniques:

1. Method: inverse transform with optimized grouping
2. Method: inverse transform with grouping according to a power-law scale of the particle sizes
3. Method: acceptance-rejection with grouping according to a power-law scale of the particle sizes

We compare the method which uses the optimzed grouping described in the previous section with the inverse transform and the acceptance-rejection method which group together particles with sizes in the intervals $[b_N^{(i-1)/m}, b_N^{i/M})$, where b_N denotes the maximal particle size. In the mass flow algorithm with N numerical particles one usually takes $b_N = 100N$, while particles beyond this size are eliminated from the system and account for the gel phase. The parameter m gives therefore the maximal possible number of groups, the actual number being determined by the structure of the particle spectrum. The optimal value of m has to be determined experimentally for every coagulation kernel and for each number N of particles. It can be observed that the value of this parameter has a strong influence on the efficiency of the computation. In contrast to this, the method which uses optimized grouping computes the group structure based on the actual particle configuration. Moreover, it turns out to be more stable regarding the variations of the parameter α, which controls the frequency of the recomputation of the optimal group structure).

All computations were performed on a SUN workstation with UltraSPARC III processors at 900Mhz (using one processor). This performance corresponds to that of a mid-range PC.

Figure 2 presents the CPU times for different implementations of the MFA which use $N = 10^5$ particles for the mass-conserving kernel $K(i,j) = (ij)^{0.5}$.

Compare first the different variants of the inverse transform method: the one which uses optimized grouping for two different values of the parameter α, with the implementation which considers grouping based on the power-law scale for two different paramteters m. The choice $m = 1220$ is determined

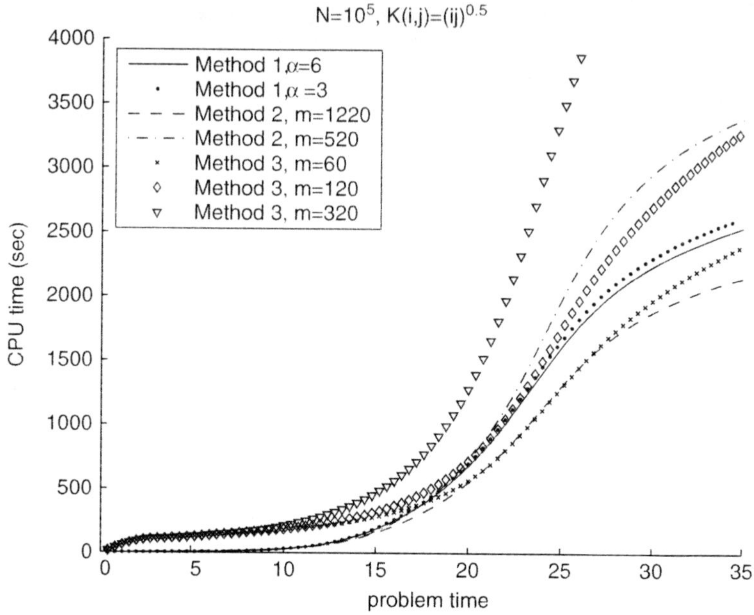

Fig. 2. CPU times for the mass-conserving kernel $K(i,j) = (ij)^{0.5}$

experimentally to be (close to) the optimal value. Taking this as a reference, we note that the method which computes adaptively the group structure reaches about 85% of the performance (speed) of the fastest implementation based on the inverse transform. A variation of the parameter α which determines the frequency of recomputing the group structure (within a meaningful range) has no significant influence on the performance. However, if we take the value $m = 520$, which lies again within a meaningful range for it, we remark a decrease in the efficiency of the implementation to about 60% of the best value.

Compare now the optimized grouping method with the acceptance-rejection method for different values of m. We note that on time intervals where the support size is not very large, the inverse transform method is significantly faster than all implementations of the acceptance-rejection technique. Nevertheless, on the time interval $t \in [15, 28]$, the increased support size of the mass spectrum (beween $0.5 \cdot N$ and $0.8 \cdot N$, see also Fig. 1) reduces the efficiency of the implementation based on the inverse-transform method. After the support sizes decreases below $0.5 \cdot N$, the inverse transform method becomes again more efficient. On the time interval $[0, 35]$ its overall performance is of about 95% of that of the fastest implementation of the acceptance-rejection method, which is delivered by the choice $m = 60$. Nevertheless, for $m = 120$ the performance decreases to 72% of it, while for $m = 320$ even to 38%.

Let us conclude these experiments in the case $K(i,j) = (ij)^{0.5}$ with some comments. It is known that for this kernel we have mass conservation. However, we observe numerical gelation at the moment $t = 21$. For an increasing number of particles in the simulation, this numerical gelation time will increase too and in the limit it will converge to infinity. From the point of view of studying the coagulation dynamics, the computations beyond this time are therefore not meaningful. Our purpose here is however to test the performance of different impelementations of stochastic particle methods. We consider this case as an example of an extreme situation, in which the support size of the empirical measure becomes very large and where the inverse transform method, which relies heavily on the reduced support size, reaches its limits. Even in this situation, its performance is close to that of the best implementation based on the acceptance-rejection technique. The main advantage of the method consists in the fact that we do not have to perform a series of numerical tests in order to determine the optimal parameter values. As the experiments indicate, the efficiency of the implementations (based either on inverse transform or acceptance-rejection) which use the power-law scale for grouping the particles depends strongly on the choice of the parameter which determines the number of groups.

Figure 3 presents the CPU times for different implementations of the MFA which use $N = 10^5$ particles for the gelling kernel $K(i,j) = (ij)^{0.7}$. The behaviour is similar as before: the optimized method has a performance of about 82% from that delivered by the best (experimental) choice of the

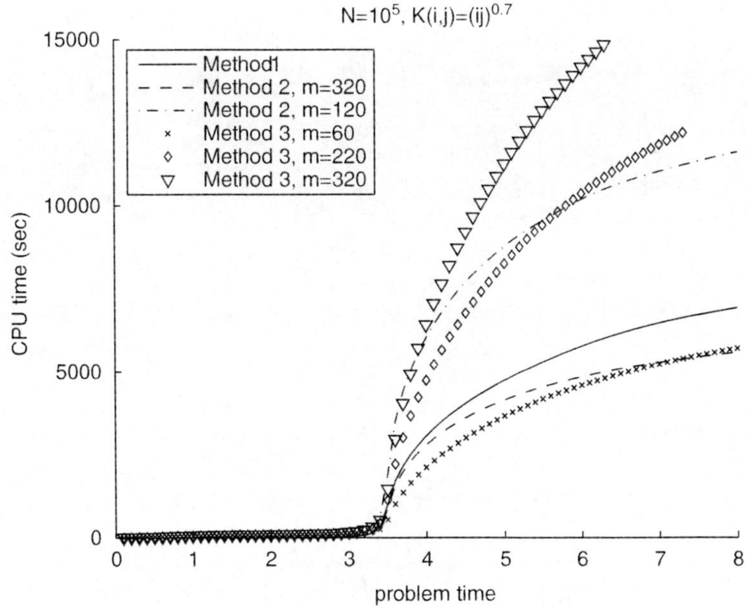

Fig. 3. CPU times for the gelling kernel $K(i,j) = (ij)^{0.7}$

parameter m of the power-law scale. Nevertheless, other values of m may lead to a decrease of the performance to about 45% or even below.

The behaviour of the CPU times for the kernel $K(i,j) = ij$ is illustrated in Figure 4. Taking a look at the support size in Figure 1, we note that in contrast to the case $K(i,j) = (ij)^{0.5}$ it is much smaller, altough the total number of particles is in both cases $N = 10^5$. As a consequence, the methods using the inverse transform (on the reduced support with a variable size) are generally much faster than the acceptance-rejection methods which have to deal with every individual particle. Comparing within the former class of methods, we note that the performance (speed) of the method with optimized grouping reaches in this case 92% of that of the reference implementation which groups particles according to a power-law scale for $m = 60$. Taking $m = 320$ we observe that the speed of the simulation decreases to about 45%, while the implementations which use the acceptance-rejection method (and individual particles) are even less efficient.

To conclude: these numerical simulations show that the algorithm based on the optimized group structure, which is periodically recomputed, delivers a performance which is comparable to that of the best choice (determined experimentally) of a grouping of the particles according to a power-law scale. It can be therefore employed to more complex simulations, where the heuristic choice of the optimal parameters cannot be afforded.

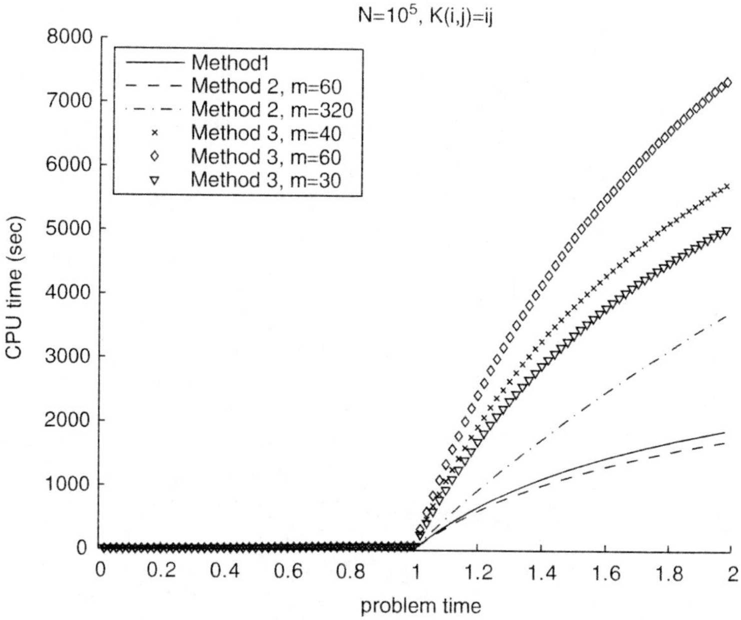

Fig. 4. CPU times for the gelling kernel $K(i,j) = ij$

References

[Al99] D. Aldous. Deterministic and stochastic models for coalescence (aggregation and coagulation): a review of the mean-field theory for probabilists. Bernoulli **5**(1), 3–48 (1999)

[BC90] J.M. Ball and J. Carr. The discrete coagulation-fragmentation equations: existence, uniqueness and density conservation. Commun. Math. Phys. **104**, 203–234, (1990)

[EW00-1] A. Eibeck and W. Wagner. An efficient stochastic algorithm for studying coagulation dynamics and gelation phenomena. SIAM J. Sci. Comput., **22(3)**, 802–821 (2000)

[EW00-2] A. Eibeck and W. Wagner. Approximative solution of the coagulation-fragmentation equation by stochastic particle systems. Stochastic Anal. Appl., **18(6)**, 921–948 (2000)

[EW01] A. Eibeck and W. Wagner. Stochastic particle approximations for Smoluchowski's coagulation equation. Ann. Appl. Prob., **11(4)**, 1137–1165 (2001)

[Fi96] G.S. Fishman. Monte Carlo Concepts, Algorithms and Applications. Springer, New York (1996)

[Gu06] F. Guiaş. Elements of a stochastic numerical method for transport and reaction problems. Habilitationsschrift, Universität Dortmund (2006)

[Je98] I. Jeon. Existence of gelling solutions for coagulation-fragmentation equations. Commun. Math. Phys. **194(3)**, 541–567 (1998)

[No99] J.R. Norris. Smoluchowski's coagulation equation: uniqueness, non-uniqueness and a hydrodynamic limit for the stochastic coalescent. Ann. Appl. Prob. **9(1)**, 78–109 (1999)

[Wa03] W. Wagner. Stochastic, analytic and numerical aspects of coagulation processes. Math. Comput. Sim., **62(3-6)**, 265–275 (2003)

(t, m, s)-Nets and Maximized Minimum Distance

Leonhard Grünschloß[1], Johannes Hanika[2], Ronnie Schwede[3], and Alexander Keller[4]

[1] Ulm University, Germany
leonhard.gruenschloss@googlemail.com
[2] Ulm University, Germany
johannes.hanika@uni-ulm.de
[3] Eawag, Swiss Federal Institute of Aquatic Science and Technology, Switzerland
ronnie.schwede@eawag.ch
[4] Ulm University, Germany
alexander.keller@uni-ulm.de

Summary. Many experiments in computer graphics imply that the average quality of quasi-Monte Carlo integro-approximation is improved as the minimal distance of the point set grows. While the definition of (t, m, s)-nets in base b guarantees extensive stratification properties, which are best for $t = 0$, sampling points can still lie arbitrarily close together. We remove this degree of freedom, report results of two computer searches for $(0, m, 2)$-nets in base 2 with maximized minimum distance, and present an inferred construction for general m. The findings are especially useful in computer graphics and, unexpectedly, some $(0, m, 2)$-nets with the best minimum distance properties cannot be generated in the classical way using generator matrices.

1 Introduction

Image synthesis can be considered as an integro-approximation problem [Kel06]

$$g(y) = \int_{[0,1)^s} f(x, y)dx \approx \frac{1}{n} \sum_{i=0}^{n-1} f(x_i, y), \tag{1}$$

where g is the image function with pixel coordinates y on the screen. One numerical method to compute approximations is the method of dependent tests, where one set $P_n := \{x_0, \ldots, x_{n-1}\}$ of points from the unit cube $[0, 1)^s$ is used to simultaneously estimate all averages $g(y)$. The accumulation buffer [HA90] for realistic image synthesis in computer graphics is a very popular realization of that scheme. In this application the image plane is tiled by one sampling point set P_n as illustrated in Figure 2.

2 (t, m, s)-Nets in Base 2

For the scope of this paper, only (t, m, s)-nets (for an extensive reference see [Nie92, Ch. 4]) in base 2 up to $s = 3$ are considered. While this may seem like a strong restriction, improving these patterns directly results in considerable performance gains in industrial rendering applications [HA90, CCC87, Kel03]. The considerations in base 2 go back to Sobol's LP_τ-nets and -sequences [Sob67]. This basic concept was generalized by Niederreiter [Nie92], which for base 2 is given by

Definition 1. *For two integers $0 \leq t \leq m$, a finite point set of 2^m s-dimensional points is a (t, m, s)-net in base 2, if every elementary interval of size $\lambda_s(E) = 2^{t-m}$ contains exactly 2^t points.*

The elementary intervals in base 2 are specified in

Definition 2. *For $l_j \in \mathbb{N}_0$ and integers $0 \leq a_j < 2^{l_j}$ the elementary interval is*

$$E := \prod_{j=1}^{s} \left[\frac{a_j}{2^{l_j}}, \frac{a_j + 1}{2^{l_j}} \right) \subseteq [0, 1)^s.$$

The parameter t controls the stratification properties of the net, which are best for $t = 0$, because then every elementary interval contains exactly one point (see Figure 1). Thus the pattern is both a Latin hypercube sample and stratified in the sense of [CPC84]. For base 2, $(0, m, s)$-nets can only exist up to dimension $s = 3$ [Nie92, Cor. 4.21].

2.1 Matrix-Generated (t, m, s)-Nets

The classical way to generate (t, m, s)-nets is the use of generator matrices. In the following we review the efficient generation of these nets and a method for checking whether $t = 0$.

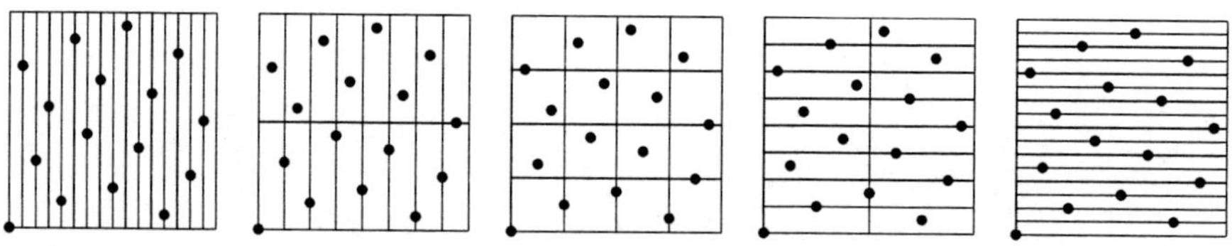

Fig. 1. The quality parameter $t = 0$ ensures extensive stratification: There is exactly one point inside each elementary interval of volume 2^{-4} of this $(0, 4, 2)$-net in base 2. There are exactly $4 + 1$ partitions into 2^4 elementary intervals. The leftmost and rightmost kind of elementary intervals constitute the Latin hypercube property.

Definition 3. *An s-dimensional point set $P_n := \{x_0, \ldots, x_{n-1}\}$ with $n = 2^m$ and $x_i := (x_i^{(1)}, \ldots, x_i^{(s)}) \in [0, 1)^s$ is called $C_1, \ldots, C_s$-generated point set in base 2 and dimension s, if $C_1, \ldots, C_s \in \mathbb{F}_2^{m \times m}$ with*

$$x_i^{(j)} = \left(\frac{1}{2} \cdots \frac{1}{2^m}\right) \left[C_j \begin{pmatrix} d_0(i) \\ \vdots \\ d_{m-1}(i) \end{pmatrix} \right], \text{ where } i = \sum_{k=0}^{m-1} d_k(i) 2^k$$

and the matrix-vector product in brackets is performed in $\mathbb{F}_2$. The matrices C_j for $j = 1, \ldots, s$ are called the generators of the point set P_n, more precisely, the matrix C_j is the generator of the j-th coordinate of the point set.

Computations in base 2 allow for exact representation of the coordinates in the IEEE floating point standard, as long as the mantissa holds enough bits (23 in the case of the 4-byte `float`). Interpreting unsigned integers as bit vectors, i.e. elements of $\mathbb{F}_2^{32}$ with $\mathbb{F}_2 = \mathbb{Z}/2\mathbb{Z}$, allows one to use standard bitwise operations for vector operations on $\mathbb{F}_2^{32}$. The matrix-vector multiplication $C_j\,(d_0(i), \ldots, d_{m-1}(i))$, which has to be performed in $\mathbb{F}_2$, then becomes very efficient. Exploiting this simple kind of parallelism results in the following algorithm in `C++`, which uses at most $\mathcal{O}(m)$ operations to compute $x_i^{(j)}$. It relies on the fact that addition corresponds to XOR in $\mathbb{F}_2$.

```
double x_j(unsigned int i)
{
  unsigned int result = 0;

  for (unsigned int k = 0; i; i >>= 1, k++)
    if(i & 1)
      // vector addition of (k+1)-th leftmost column of C_j
      result ^= C_j[k];

  return (double) result / (double) (1ULL << m);
}
```

Whether or not a given set of generator matrices produces a net with $t = 0$, can be checked using the following

Theorem 1 (see [Nie92, Thm. 4.28] and [LP01, Def. 1]). *Let P_n be a $C_1, \ldots, C_s$-generated $n = 2^m$-point set in base 2 and dimension s, then P_n is a $(0, m, s)$-net in base 2 if for all $\mathbf{d} = (d_1, \ldots, d_s)^T \in \mathbb{N}_0^s$ with $|\mathbf{d}| = m$ the following holds:*

$$\det\left(C^{(\sum_{j=1}^{s} d_j)}\right) \neq 0,$$

where

$$C^{(\sum_{j=1}^{s} d_j)} = \begin{pmatrix} c^1_{1,1} \cdots c^1_{1,m} \\ \vdots \\ c^1_{d_1,1} \cdots c^1_{d_1,m} \\ \vdots \\ c^s_{1,1} \cdots c^s_{1,m} \\ \vdots \\ c^s_{d_s,1} \cdots c^s_{d_s,m} \end{pmatrix} \in \mathbb{F}_2^{m \times m},$$

with $C_j = (c^j_{k,l})^m_{k,l=1}$ *for* $j = 1, \ldots, s$. *This means* $C^{(\sum_{j=1}^{s} d_j)}$ *is an* $m \times m$ *matrix consisting of the first* d_j *rows of the generator* C_j *for all* $j = 1, \ldots, s$.

As a consequence of Theorem 1, all generator matrices $C_1, \ldots, C_s$ must be regular. Remark (iv) in [LP01, p. 3] states that the $C_1, \ldots, C_s$-generated $(0, m, s)$-net in base 2 and the net generated by $C_1 D, \ldots, C_s D$, for any $D \in \mathrm{GL}_m(\mathbb{F}_2)$ contain exactly the same points, where $\mathrm{GL}_m(\mathbb{F}_2)$ denotes the general linear group of matrices over $\mathbb{F}_2$ of dimension $m \times m$. This means that the above mentioned nets are identical except for the numbering of their points. Using this remark we can fix one matrix of a matrix-generated $(0, m, s)$-net in base 2 by choosing

$$C_1 := \begin{pmatrix} 0 & & 1 \\ & \iddots & \\ 1 & & 0 \end{pmatrix} \tag{2}$$

which results in the first coordinate $x_i^{(1)} = \frac{i}{2^m}$.

For the special case, where $s = 2$ and C_1 as defined in (2), Theorem 1 then results in the following test for $t = 0$:

$$\det \underbrace{\begin{pmatrix} c^2_{1,1} \cdots c^2_{1,k} \\ \vdots \\ c^2_{k,1} \cdots c^2_{k,k} \end{pmatrix}}_{=:S_k} \neq 0 \ \forall k \in \{1, \ldots, m\} \Rightarrow C_1, C_2 \text{ generate a } (0, m, 2)\text{-net}, \tag{3}$$

because all possible vectors $\mathbf{d}$ as stated in the theorem are considered.

2.2 Permutation-Generated (t, m, s)-Nets

Definition 1 states that a (t, m, s)-net in base 2 is a Latin hypercube sample. This means that for all (t, m, s)-nets in base 2, the integer parts of the coordinates multiplied by 2^m must be a permutation of the numbers given by the integer parts of the first coordinates $x_i^{(1)}$ multiplied by 2^m.

The $(0, m, 2)$-nets in base 2 given by $x_i = \frac{1}{2^m}(i, \sigma(i))$, where σ is a permutation of the numbers $\{0, \ldots, 2^m - 1\}$, can be enumerated by using a modified version of the classic backtracking algorithm to solve the *n-Rooks* problem [Rol05] equipped with an additional test. This test checks whether the points fulfill the stratification conditions imposed by the structure of the elementary intervals (see Figure 1). While this seems to be a complicated test, it can in fact be realized with only a few lines of code in C, if the base is 2:

```
for (k = 1; k < m; k++)
{
  // combine k bits of i and m-k bits of j to form index
  idx = (i >> (m - k)) + (j & (0xFFFFFFFF << k));

  if(elementaryInterval[k][idx]++) // already one point there?
    break;  // t > 0 !
}
```

The code fragment tests whether a point given by the coordinates i and $j = \sigma(i)$ falls into an elementary interval that is already taken. In that case the points cannot be a $(0, m, 2)$-net in base 2. There are exactly $(m + 1)$ kinds of each 2^m elementary intervals (see Figure 1). The first and last kind of elementary intervals constitute the Latin hypercube property, which does not need to be checked, because the points are determined by a permutation. Therefore the variable `k` iterates from 1 to $m - 1$. The array `elementaryInterval` counts how many points are in the k-th kind of elementary interval addressed by the index `idx`, which is efficiently computed by using k bits of the first coordinate i and $m - k$ bits of the second coordinate $j = \sigma(i)$.

3 Low Discrepancy and Minimum Distance

Although $(0, m, s)$-nets have exhaustive stratification properties (see Defintion 1) that guarantee low discrepancy [Nie92], these properties do not avoid that points can lie arbitrarily close together across boundaries of elementary intervals.

Maximizing the minimum distance is a basic concept in nature [Yel83] and has been applied extensively in computer graphics [HA90, Gla95]. Minimum distance has been proposed as a measure for uniformity in [Pan04, Kel06], too. In addition it has been observed that scrambling [Owe95] can change minimum distance [Kel04] and that in fact points with maximized minimum distance perform better on the average. We therefore perform an exhaustive computer search for $(0, m, 2)$-nets in base 2 with maximized minimum distance.

The minimum distance

$$d_{\min}(P_n) := \min_{0 \le i < j < n} \|x_i - x_j\|$$

of a point set $P_n = \{x_0, \ldots, x_{n-1}\}$ is the smallest distance between any two distinct points of this set. However, the choice of the norm $\|\cdot\|$ is important.

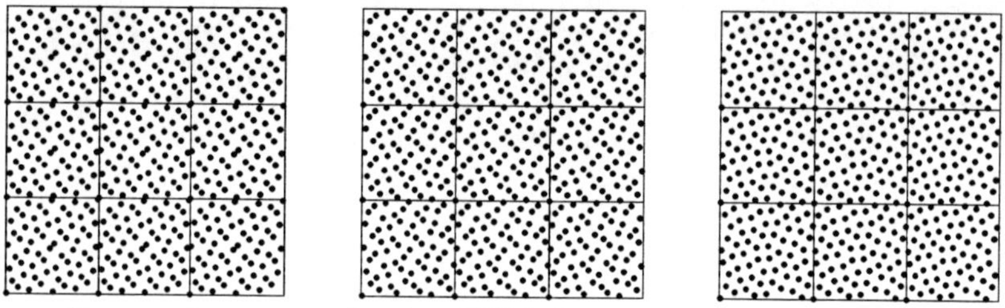

Hammersley Larcher-Pillichshammer new construction

Fig. 2. Three examples of 3×3 periodically tiled $(0, 6, 2)$-nets in base 2.

Choosing the Euclidean distance $\|x - y\| = \sqrt{\sum_{i=1}^{s} |x_i - y_i|^2}$ is not sufficient for our purposes, as for many graphics applications it is beneficial to be able to tile the same net periodically as can be seen in Figure 2 [KH01]. As we do not want the minimum distance to decrease for such a point set consisting of shifted copies of P_n we use the toroidal distance:

Definition 4. *For two points $x = (x_1, \ldots, x_s) \in [0, 1)^s$ and $y = (y_1, \ldots, y_s) \in [0, 1)^s$ their toroidal distance is defined as*

$$\|x - y\|_T := \sqrt{\sum_{i=1}^{s} (\min\{|x_i - y_i|, 1 - |x_i - y_i|\})^2}.$$

Randomization

Randomized quasi-Monte Carlo point sets can be used to reduce the variance of Monte Carlo estimators and allow for an unbiased error estimate on the class of square-integrable functions [Owe98]. While random scrambling [Owe95, FK02, KK02] does not alter the parameter t, it can decrease the minimum distance of a (t, m, s)-net dramatically [Kel04].

On the contrary a random shift on the unit torus, i.e. a Cranley-Patterson rotation [CP76], preserves minimum distance even for periodically tiled nets, which perfectly matches our optimization goal. While it is often argued that the parameter t can be affected by shifting, it also can be argued that just the integrand is shifted leaving the point set structure untouched.

3.1 Exhaustive Matrix Computer Search

Following Section 2.1, we consider matrix-generated $(0, m, 2)$-nets. As the search space of C_2 generator matrices is exponentially growing in m and the limiting factor for the performance is the minimum distance evaluation, we only want to do these calculations for matrices which fulfill the $t = 0$ property. To efficiently enumerate such matrices, we exploit the conclusions of

Table 1. Minimum distance $d_{\min}$ of $(0, m, 2)$-nets in base 2 for Hammersley, Larcher-Pillichshammer and the new construction.

m	Hammersley	Larcher-Pillichshammer	new construction
2	$\sqrt{2}/2^2 \approx 0.35355339$	$\sqrt{2}/2^2 \approx 0.35355339$	$\sqrt{2}/2^2 \approx 0.35355339$
3	$\sqrt{2}/2^3 \approx 0.17677670$	$\sqrt{5}/2^3 \approx 0.27950850$	$\sqrt{8}/2^3 \approx 0.35355339$
4	$\sqrt{2}/2^4 \approx 0.08838835$	$\sqrt{8}/2^4 \approx 0.17677670$	$\sqrt{13}/2^4 \approx 0.22534695$
5	$\sqrt{2}/2^5 \approx 0.04419417$	$\sqrt{18}/2^5 \approx 0.13258252$	$\sqrt{29}/2^5 \approx 0.16828640$
6	$\sqrt{2}/2^6 \approx 0.02209709$	$\sqrt{32}/2^6 \approx 0.08838835$	$\sqrt{52}/2^6 \approx 0.11267348$
7	$\sqrt{2}/2^7 \approx 0.01104854$	$\sqrt{72}/2^7 \approx 0.06629126$	$\sqrt{100}/2^7 \approx 0.07812500$
8	$\sqrt{2}/2^8 \approx 0.00552427$	$\sqrt{128}/2^8 \approx 0.04419417$	$\sqrt{208}/2^8 \approx 0.05633674$
9	$\sqrt{2}/2^9 \approx 0.00276214$	$\sqrt{265}/2^9 \approx 0.03179457$	$\sqrt{400}/2^9 \approx 0.03906250$
10	$\sqrt{2}/2^{10} \approx 0.00138107$	$\sqrt{512}/2^{10} \approx 0.02209709$	$\sqrt{832}/2^{10} \approx 0.02816837$
11	$\sqrt{2}/2^{11} \approx 0.00069053$	$\sqrt{1060}/2^{11} \approx 0.01589729$	$\sqrt{1600}/2^{11} \approx 0.01953125$
12	$\sqrt{2}/2^{12} \approx 0.00034527$	$\sqrt{2048}/2^{12} \approx 0.01104854$	$\sqrt{3328}/2^{12} \approx 0.01408418$
13	$\sqrt{2}/2^{13} \approx 0.00017263$	$\sqrt{4153}/2^{13} \approx 0.00786667$	$\sqrt{6385}/2^{13} \approx 0.00975417$
14	$\sqrt{2}/2^{14} \approx 0.00008632$	$\sqrt{8192}/2^{14} \approx 0.00552427$	$\sqrt{13312}/2^{14} \approx 0.00704209$
15	$\sqrt{2}/2^{15} \approx 0.00004316$	$\sqrt{16612}/2^{15} \approx 0.00393334$	$\sqrt{25313}/2^{15} \approx 0.00485536$
16	$\sqrt{2}/2^{16} \approx 0.00002158$	$\sqrt{32768}/2^{16} \approx 0.00276214$	$\sqrt{53248}/2^{16} \approx 0.00352105$

Table 2. Maximum obtainable minimum distance $d_{\min}$ for matrix-generated $(t, m, 2)$-nets with $t \geq 0$. Requiring $t = 0$ does not always allow for obtaining the maximum.

t	0	0	0, 1	0, 1	1, 2, 3
m	2	3	4	5	6
$d_{\min}$	$\sqrt{2}/2^2 \approx$ 0.35355339	$\sqrt{8}/2^3 \approx$ 0.35355339	$\sqrt{13}/2^4 \approx$ 0.22534695	$\sqrt{29}/2^5 \approx$ 0.16828640	$\sqrt{65}/2^6 \approx$ 0.12597278

Theorem 1 that are sufficient to ensure only one point per elementary interval. The matrices are enumerated using a backtracking algorithm that first checks, whether $\det(S_k) \neq 0$ (see Equation (3)) and in that case tries to extend the matrix S_k by a right column and a bottom row to form S_{k+1} (see Figure 3). If such an extension cannot be found with $\det(S_{k+1}) \neq 0$, the next S_k will be explored according to the backtracking search principle.

To compute the determinant, we apply the standard Gauss elimination scheme. The implementation exploits that permuting matrix rows does not change the determinant in $\mathbb{F}_2$. It exits on the first resulting zero on the diagonal

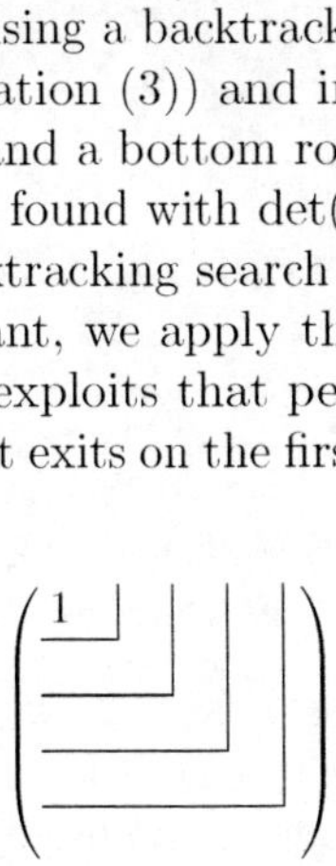

Fig. 3. Iterative construction of C_2 for a $(0, m, 2)$-net by sequences of S_k.

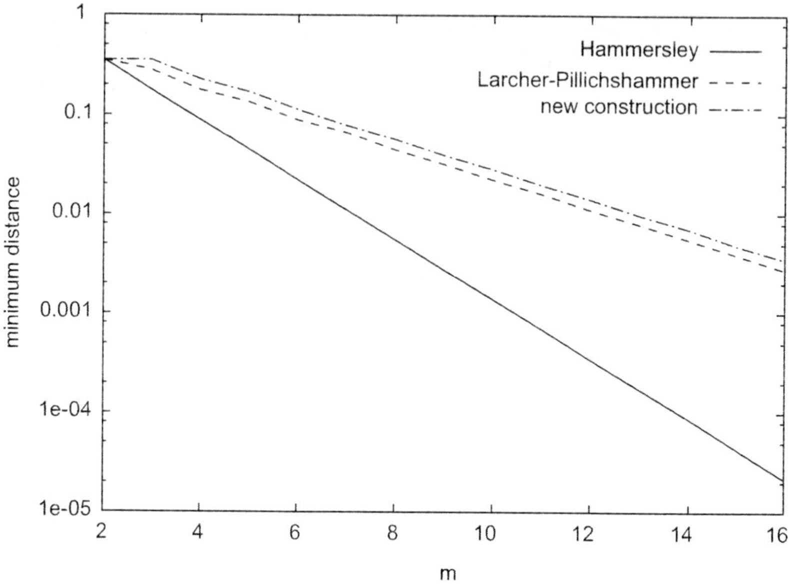

Fig. 4. Minimum distance $d_{\min}$ of a Hammersley-net, a Larcher-Pillichshammer-net and the new construction.

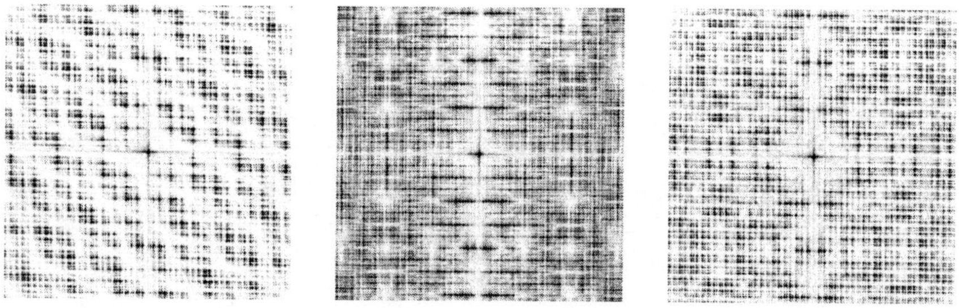

Fig. 5. Squared amplitudes of the Fourier transformation of the Hammersley-, Larcher-Pillichshammer- and the newly constructed $(0, 10, 2)$-net in base 2 (inverted, higher values are darker). A larger region of low frequencies gets attenuated as the minimum distance increases (similar to the blue-noise spectrum [Yel83]). This region is largest for the new construction and means that low frequencies are reproduced more precisely. Like the Larcher-Pillichshammer-net, the new construction is much more isotropic as compared to the Hammersley-net.

signaling a zero determinant. Using 32-bit integers together with the bitwise XOR operation to perform vector addition in $\mathbb{F}_2^{32}$, the algorithm runs in $\mathcal{O}(k^2)$. To further improve performance, we found it beneficial to use vector operations of modern computers (i.e. the SSE2 instruction set) for some values of m. A non-zero determinant implies $t = 0$.

Next, to calculate the minimum distance all points are enumerated according to the gray code numbering [PTVF92]. Computing square distances multiplied by 2^m allows for using integer arithmetic and guarantees to avoid any floating point arithmetic problem. The omission of the division by 2^m in the first code fragment (see Section 2.1) then results in integer coordinates.

For the efficient minimum distance computation the resulting points are sorted into a regular grid. This can be done very efficiently if m is even, because then one kind of elementary intervals must be squares forming a regular grid (see the middle plot in Figure 1). As mentioned before exactly one point will fall into each grid cell. Assigning a grid cell to a point is done by simply omitting the least $\lceil \frac{m}{2} \rceil$ significant bits of the point coordinates. To find the minimum distance of all points to one fixed point it is then sufficient to examine the eight points in the neighboring cells. If m is odd, we use a square regular grid with cells twice the area of the elementary intervals. So in this case there are two points per cell which gives the complexity of the algorithm a worse constant than in the even case. It still runs in $\mathcal{O}(n)$ where $n = 2^m$ is the number of points.

Search Results

For $m \leq 6$ Figures 6–10 show the $(0, m, 2)$-nets in base 2 along with the generator matrix C_2 for the second coordinate, where the minimum distance between the points on the torus is maximal. Note that C_1 always is the flipped unit matrix as defined in Equation (2). Figure 11 considers the case $m = 7$, where only a fraction of the search space could be explored and thus there could exist generator matrices resulting in a larger minimum distance.

By abandoning the $t = 0$ constraint, the stratification properties of $(0, m, s)$-nets are lost, but the minimum distance of resulting nets can be increased even further (see Table 2). Instead of testing for $t = 0$, now the quality parameter t is computed following the algorithm outlined in [PS01] after determining the minimum distance of the point set. The size $\mathcal{O}\left(2^{m^2}\right)$ of the search space is even larger, but a complete search is still feasible up to $m \leq 6$.

3.2 Restricted Computer Search

Our exhaustive matrix search is only computationally feasible up to $m \leq 6$. However, the search results allow one to infer a submatrix structure:

$$\text{for even } m: \begin{pmatrix} 1 & 1 \\ 0 & 1 \end{pmatrix} \rightsquigarrow \left(\begin{array}{c|cc|c} 1 & 0 & 1 & 1 \\ \cline{2-3} 1 & 1 & 1 & 1 \\ 0 & 0 & 1 & 1 \\ \cline{2-3} 0 & 0 & 0 & 1 \end{array}\right) \rightsquigarrow \left(\begin{array}{c|cccc|c} 1 & 0 & 0 & 0 & 0 & 0 \\ \cline{2-5} 0 & 1 & 0 & 1 & 1 & 0 \\ 0 & 1 & 1 & 1 & 1 & 0 \\ 0 & 0 & 0 & 1 & 1 & 0 \\ 0 & 0 & 0 & 0 & 1 & 1 \\ \cline{2-5} 0 & 0 & 0 & 0 & 0 & 1 \end{array}\right) \rightsquigarrow \ldots, \tag{4}$$

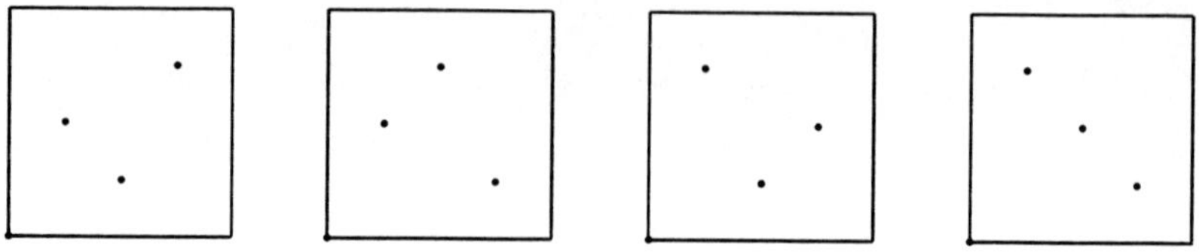

$$\begin{pmatrix}1&0\\0&1\end{pmatrix} \qquad \begin{pmatrix}1&1\\0&1\end{pmatrix} \qquad \begin{pmatrix}1&0\\1&1\end{pmatrix} \qquad \begin{pmatrix}1&1\\1&0\end{pmatrix}$$

Fig. 6. All matrix-generated $(0, 2, 2)$-nets in base 2 with maximal minimum distance $d_{\min} = \sqrt{2}/2^2 \approx 0.35355339$.

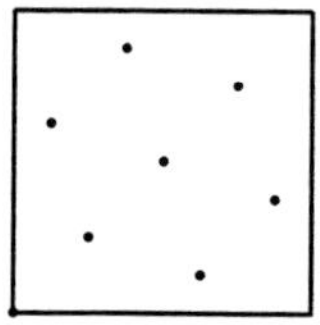

$$\begin{pmatrix}1&0&1\\0&1&0\\1&0&0\end{pmatrix}$$

Fig. 7. All matrix-generated $(0, 3, 2)$-nets in base 2 with maximal minimum distance $d_{\min} = \sqrt{8}/2^3 \approx 0.35355339$.

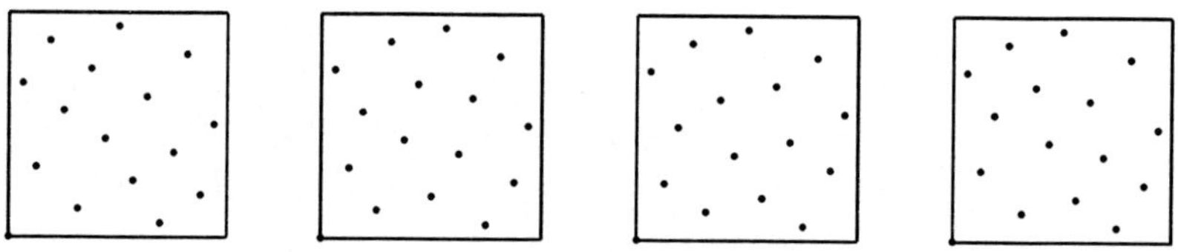

$$\begin{pmatrix}1&0&1&1\\0&1&0&1\\1&0&0&1\\1&1&1&1\end{pmatrix} \qquad \begin{pmatrix}1&0&0&0\\1&1&0&0\\0&0&1&1\\0&1&0&1\end{pmatrix} \qquad \begin{pmatrix}1&0&1&1\\1&1&1&1\\0&0&1&1\\0&0&0&1\end{pmatrix} \qquad \begin{pmatrix}1&0&1&1\\1&1&1&1\\0&0&1&1\\0&1&0&1\end{pmatrix}$$

Fig. 8. All matrix-generated $(0, 4, 2)$-nets in base 2 with maximal minimum distance $d_{\min} = \sqrt{13}/2^4 \approx 0.22534695$.

$$\text{for odd } m: \begin{pmatrix}1&0&1\\0&1&0\\1&0&0\end{pmatrix} \rightsquigarrow \left(\begin{array}{c|ccc|c}1&0&0&0&1\\ \hline 0&1&0&1&0\\0&0&1&0&0\\0&1&0&0&0\\ \hline 1&1&0&0&0\end{array}\right) \rightsquigarrow \dots . \tag{5}$$

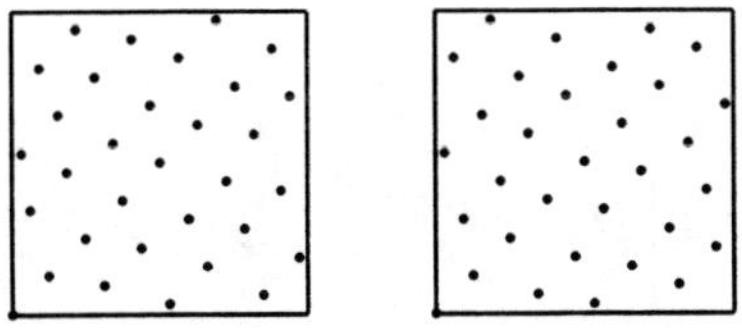

$$\begin{pmatrix} 1&0&0&0&1\\ 0&1&0&1&0\\ 0&0&1&0&0\\ 0&1&0&0&0\\ 1&1&0&0&0 \end{pmatrix} \qquad \begin{pmatrix} 1&1&0&0&1\\ 0&1&0&1&0\\ 0&0&1&0&0\\ 0&1&0&0&0\\ 1&1&0&0&0 \end{pmatrix}$$

Fig. 9. All matrix-generated $(0, 5, 2)$-nets in base 2 with maximal minimum distance $d_{\min} = \sqrt{29}/2^5 \approx 0.16828640$.

This observation motivated our approach to restrict the search to reduce the growth of the search space to order $\mathcal{O}(2^m)$. For this purpose, we define the following matrix structure.

We write a matrix $C \in \mathbb{F}_2^{m \times m}$ with $m \geq 4$ as matrix $\mathcal{I} = (i_{k,l})_{k,l=1}^{m-2}$ and vectors $\mathbf{u} = (u_1 \cdots u_m)$, $\mathbf{r} = (r_1, \ldots, r_{m-2}), \mathbf{b} = (b_1 \cdots b_m)$ and $\mathbf{l} = (l_1, \ldots, l_{m-2})$ in the following way:

$$C = \left(\begin{array}{c|ccc|c} u_1 & u_2 & \cdots & u_{m-1} & u_m \\ \hline l_1 & i_{1,1} & \cdots & i_{1,m-2} & r_1 \\ \vdots & \vdots & \ddots & \vdots & \vdots \\ l_{m-2} & i_{m-2,1} & \cdots & i_{m-2,m-2} & r_{m-2} \\ \hline b_1 & b_2 & \cdots & b_{m-1} & b_m \end{array} \right) = \begin{bmatrix} \mathbf{u} \\ \mathbf{l}\, \mathcal{I}\, \mathbf{r} \\ \mathbf{b} \end{bmatrix}.$$

This way we are able to continue the matrix expansion scheme (4) and (5) as follows:

$$C_2^{(m)} = \begin{bmatrix} \mathbf{u} \\ \mathbf{l}\, C_2^{(m-2)}\, \mathbf{r} \\ \mathbf{b} \end{bmatrix}.$$

Search Results

Our iterative search algorithm takes an $(m-2) \times (m-2)$-matrix and seeks for the best $\mathbf{u}, \mathbf{r}, \mathbf{b}, \mathbf{l}$ vectors by maximizing the minimum distance of the resulting $(0, m, 2)$-net. The resulting generator matrices for $m = 7, 8, 9$ are given in Figure 12.

However, this iterative approach does not yield generator matrices that obtain the maximal possible minimum distance, which already becomes apparent for $m = 7$. The largest minimum distance of the generator matrix that was found by the iterative search is $\sqrt{98}/2^7$ whereas the incomplete (see previous section) full matrix search already revealed a matrix with minimum distance $\sqrt{100}/2^7$ (see Figure 11).

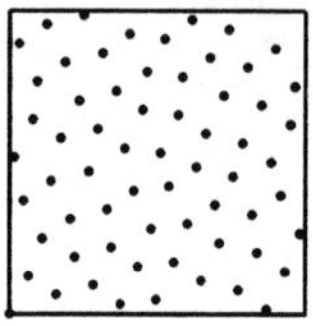
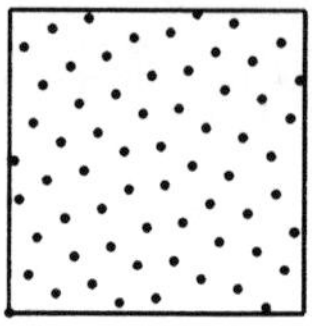
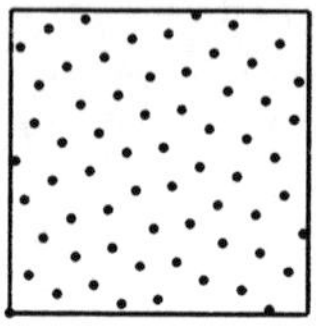
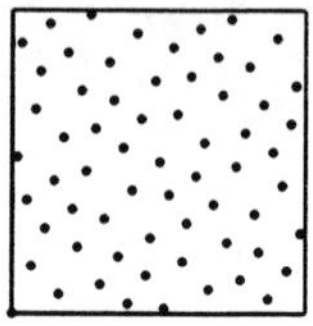

$$\begin{pmatrix} 1&1&0&1&1&0\\ 0&1&0&1&1&0\\ 0&1&1&1&1&0\\ 0&0&0&1&1&0\\ 0&0&0&0&1&1\\ 1&1&0&1&1&1 \end{pmatrix} \quad \begin{pmatrix} 1&0&0&0&0&0\\ 0&1&0&1&1&0\\ 0&1&1&1&1&0\\ 0&0&0&1&1&0\\ 0&0&0&0&1&1\\ 0&0&0&0&0&1 \end{pmatrix} \quad \begin{pmatrix} 1&1&0&1&1&0\\ 0&1&0&1&1&0\\ 0&1&1&1&1&0\\ 0&0&0&1&1&0\\ 0&0&0&0&1&1\\ 0&0&0&0&0&1 \end{pmatrix} \quad \begin{pmatrix} 1&1&0&1&0&1\\ 0&1&0&1&1&0\\ 0&1&1&1&1&0\\ 0&0&0&1&1&0\\ 0&0&1&0&1&0\\ 1&1&0&1&0&0 \end{pmatrix}$$

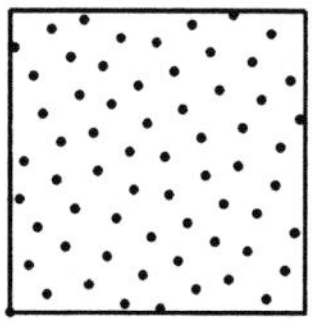
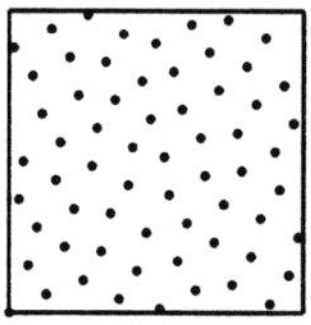
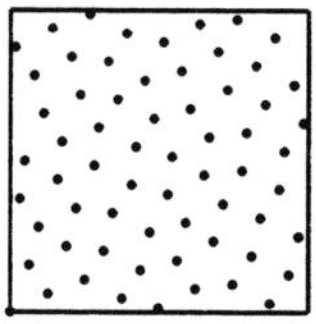
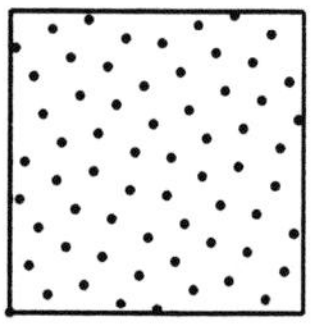

$$\begin{pmatrix} 1&0&0&0&1&1\\ 1&1&0&0&1&1\\ 1&1&1&0&1&1\\ 0&0&0&1&1&0\\ 0&0&1&0&1&0\\ 0&0&0&0&0&1 \end{pmatrix} \quad \begin{pmatrix} 1&0&0&0&0&1\\ 1&1&0&0&0&1\\ 1&1&1&0&0&1\\ 0&0&0&1&1&0\\ 0&0&1&0&1&0\\ 0&0&0&0&1&1 \end{pmatrix} \quad \begin{pmatrix} 1&0&0&0&0&0\\ 1&1&0&0&0&0\\ 1&1&1&0&0&0\\ 0&0&0&1&1&0\\ 0&0&1&0&1&0\\ 0&0&0&0&1&1 \end{pmatrix} \quad \underbrace{\begin{pmatrix} 1&0&0&0&0&0\\ 1&1&0&0&0&0\\ 1&1&1&0&0&0\\ 0&0&0&1&1&0\\ 0&0&1&0&1&0\\ 0&0&0&0&0&1 \end{pmatrix}}_{C_2^{(6)}}$$

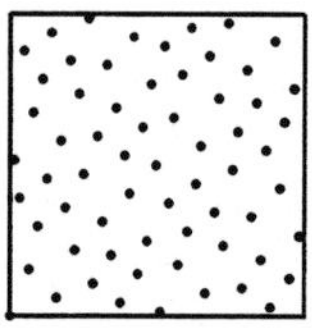
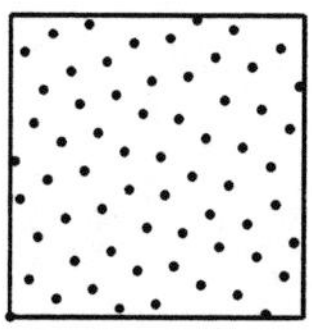

$$\begin{pmatrix} 1&0&0&0&0&1\\ 0&1&0&1&1&0\\ 0&1&1&1&1&0\\ 0&0&0&1&1&0\\ 0&0&1&0&1&0\\ 1&1&0&1&0&0 \end{pmatrix} \quad \begin{pmatrix} 1&0&0&0&0&0\\ 0&1&0&1&1&0\\ 0&1&1&1&1&0\\ 0&0&0&1&1&0\\ 0&0&0&0&1&1\\ 1&1&0&1&1&1 \end{pmatrix}$$

Fig. 10. All matrix-generated $(0, 6, 2)$-nets in base 2 with maximal minimum distance $d_{\min} = \sqrt{52}/2^6 \approx 0.11267348$.

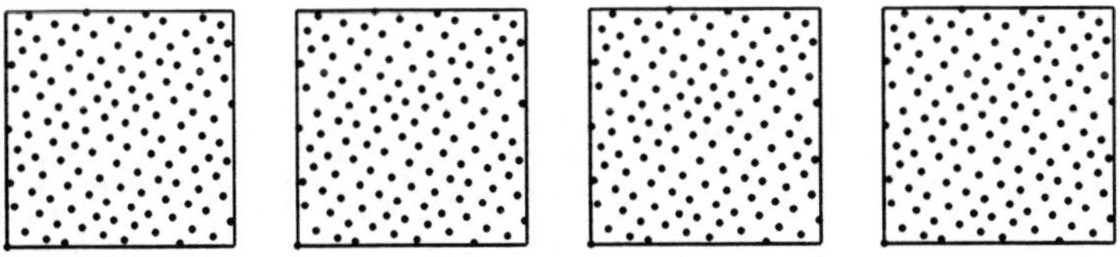

$$\begin{pmatrix} 1&0&0&0&0&0&0 \\ 0&1&0&1&1&0&1 \\ 0&0&1&0&1&0&0 \\ 0&0&0&1&0&0&0 \\ 0&0&1&0&0&0&0 \\ 0&1&1&1&0&1&0 \\ 0&1&0&1&1&0&0 \end{pmatrix} \begin{pmatrix} 1&1&0&1&1&0&0 \\ 0&1&0&1&1&0&1 \\ 0&0&1&0&1&0&0 \\ 0&0&0&1&0&0&0 \\ 0&0&1&0&0&0&0 \\ 0&1&1&1&0&1&0 \\ 0&1&0&1&1&0&0 \end{pmatrix} \begin{pmatrix} 1&0&0&0&0&0&1 \\ 0&1&0&1&1&0&1 \\ 0&0&1&0&1&0&0 \\ 0&0&0&1&0&0&0 \\ 0&0&1&0&0&0&0 \\ 0&1&1&1&0&1&0 \\ 0&1&0&1&1&0&0 \end{pmatrix} \underbrace{\begin{pmatrix} 1&1&0&1&1&0&1 \\ 0&1&0&1&1&0&1 \\ 0&0&1&0&1&0&0 \\ 0&0&0&1&0&0&0 \\ 0&0&1&0&0&0&0 \\ 0&1&1&1&0&1&0 \\ 0&1&0&1&1&0&0 \end{pmatrix}}_{C_2^{(7)}}$$

Fig. 11. $(0, 7, 2)$-nets in base 2 with minimum distance $d_{\min} = \sqrt{100}/2^7 = 0.078125$ (possibly not the maximum).

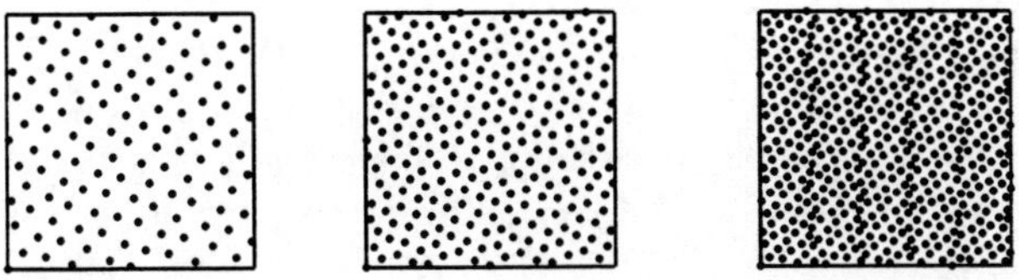

$m = 7$	$m = 8$	$m = 9$
$d_{\min} = \sqrt{98}/2^7$	$d_{\min} = \sqrt{208}/2^8$	$d_{\min} = \sqrt{392}/2^9$
≈ 0.07733980	≈ 0.05633674	≈ 0.03866990

$$\begin{pmatrix} 1&0&0&0&0&0&1 \\ 0&1&0&0&0&1&0 \\ 0&0&1&0&1&0&0 \\ 0&0&0&1&0&0&0 \\ 0&0&1&0&0&0&0 \\ 0&1&1&0&0&0&0 \\ 1&0&0&0&0&0&0 \end{pmatrix} \begin{pmatrix} 1&0&0&0&0&0&0&0 \\ 0&1&0&0&0&0&0&0 \\ 0&1&1&0&0&0&0&0 \\ 0&1&1&1&0&0&0&0 \\ 0&0&0&0&1&1&0&0 \\ 0&0&0&1&0&1&0&0 \\ 0&0&0&0&0&0&1&0 \\ 0&0&0&0&0&0&0&1 \end{pmatrix} \begin{pmatrix} 1&0&0&0&0&0&0&0&1 \\ 1&1&0&0&0&0&0&1&0 \\ 0&0&1&0&0&0&1&0&0 \\ 0&0&0&1&0&1&0&0&0 \\ 0&0&0&0&1&0&0&0&0 \\ 0&0&0&1&0&0&0&0&0 \\ 0&0&1&1&0&0&0&0&0 \\ 0&1&0&0&0&0&0&0&0 \\ 1&0&0&0&0&0&0&0&0 \end{pmatrix}$$

Fig. 12. Partial results of the restricted search for $(0, m, 2)$-nets in base 2.

3.3 Construction Inferred from Computer Search

For $m \geq 8$ we suggest the generator matrices

$$\underbrace{\begin{pmatrix} 1 & 0 & & \cdots & & & 0 \\ 0 & \ddots & & & & & \\ & & 1 & & & & \\ \vdots & & & \boxed{C_2^{(6)}} & & & \vdots \\ & & & & 1 & & \\ & & & & & \ddots & 0 \\ 0 & & & \cdots & & 0 & 1 \end{pmatrix}}_{\text{for even } m \geq 8} \quad \text{and} \quad \underbrace{\begin{pmatrix} 1 & 0 & & \cdots & & & 0 \\ 0 & \ddots & & & & & \\ & & 1 & & & & \\ \vdots & & & \boxed{C_2^{(7)}} & & & \vdots \\ & & & & 1 & & \\ & & & & & \ddots & 0 \\ 0 & & & \cdots & & 0 & 1 \end{pmatrix}}_{\text{for odd } m \geq 9,}$$

where $C_2^{(6)}$ and $C_2^{(7)}$ are the matrices for $m = 6, 7$ given in Figure 10 and Figure 11. The matrices extended by ones on the diagonals still fulfill the conditions of Theorem 1. Hence the resulting nets are $(0, m, 2)$-nets in base 2 for $m \geq 8$.

We would like to note that in general there are many matrices of size $(m-2) \times (m-2)$ with nets having the same minimum distance. However, when using these matrices for the inferred construction the resulting nets might have different minimum distances. The restricted search from the previous Section 3.2 was only computationally feasible up to $m = 9$, but did not find better results.

Resulting New Construction and Numerical Results

Combining the findings of the previous sections, we propose to use the results from

- the full matrix search for $m \leq 7$,
- and the inferred construction for $m \geq 8$.

The points generated by these matrices were compared to the Larcher-Pillichshammer- (for an implementation see [KK02]) and the Hammersley-net (see Table 1 and Figure 4). It turned out that the new construction performed better than the Larcher-Pillichshammer-net with respect to the toroidal minimum distance. In fact the Hammersley-net always generates the worst possible minimum distance of $\sqrt{2}/2^m$ for our constructions. This distance is found between the first and the last point of the net. The spectral properties of these nets can be visually compared in Figure 5.

3.4 Exhaustive Permutation Search

In order to verify the results, we searched the space of the permutation-generated $(0, m, 2)$-nets in base 2. For $m < 5$ the search produced nets with the same minimum distance as the new construction (see Table 1). However, for $m = 5$ points were found which cannot be generated using matrices. These also exhibited a better minimum distance ($\sqrt{32}/2^5 \approx 0.17677670$) compared to all matrix-generated nets (the maximum is $\sqrt{29}/2^5 \approx 0.16828640$ for the new construction). Unfortunately, for $m > 5$ this space is way too large for an exhaustive search.

4 Conclusion

By maximizing the minimum distance of a point set, we removed a degree of freedom from the definition of (t, m, s)-nets. Although an exhaustive computer search is infeasible, two interesting facts could be revealed by examples:

1. Nets with a quality parameter $t > 0$ can obtain a minimum distance larger than those with $t = 0$.
2. Permutation-generated nets can obtain a larger minimum distance than matrix-generated nets.

We will continue our research in this direction and search for $(0, m, 3)$-nets in base 2 with maximized minimum distance. We plan to investigate the practical benefits of the new concepts in the setting of computer graphics, namely the realistic simulation of light transport along the lines of [CCC87, KK02, WK07].

Acknowledgements

The first two authors have been funded by the project *information at your fingertips - Interaktive Visualisierung für Gigapixel Displays*, Forschungsverbund im Rahmen des Förderprogramms Informationstechnik in Baden-Württemberg (BW-FIT). Our special thanks go to Sun Microsystems GmbH who kindly supported our research with computing equipment.

References

[CCC87] R. Cook, L. Carpenter, and E. Catmull. The REYES Image Rendering Architecture. In *Computer Graphics (SIGGRAPH '87 Conference Proceedings)*, pages 95–102, July 1987.

[CP76] R. Cranley and T. Patterson. Randomization of number theoretic methods for multiple integration. *SIAM Journal on Numerical Analysis*, 13:904–914, 1976.

[CPC84] R. Cook, T. Porter, and L. Carpenter. Distributed Ray Tracing. In *Computer Graphics (SIGGRAPH '84 Conference Proceedings)*, pages 137–145, 1984.

[FK02] I. Friedel and A. Keller. Fast Generation of Randomized Low-Discrepancy Point Sets. In H. Niederreiter, K. Fang, and F. Hickernell, editors, *Monte Carlo and Quasi-Monte Carlo Methods 2000*, pages 257–273. Springer, 2002.

[Gla95] A. Glassner. *Principles of Digital Image Synthesis*. Morgan Kaufmann, 1995.

[HA90] P. Haeberli and K. Akeley. The Accumulation Buffer: Hardware Support for High-Quality Rendering. In *Computer Graphics (SIGGRAPH '90 Conference Proceedings)*, pages 309–318, 1990.

[Kel03] A. Keller. Strictly Deterministic Sampling Methods in Computer Graphics. *SIGGRAPH 2003 Course Notes, Course #44: Monte Carlo Ray Tracing*, 2003.

[Kel04] A. Keller. Trajectory Splitting by Restricted Replication. *Monte Carlo Methods and Applications*, 10(3-4):321–329, 2004.

[Kel06] A. Keller. Myths of Computer Graphics. In H. Niederreiter, editor, *Monte Carlo and Quasi-Monte Carlo Methods 2004*, pages 217–243. Springer, 2006.

[KH01] A. Keller and W. Heidrich. Interleaved Sampling. In K. Myszkowski and S. Gortler, editors, *Rendering Techniques 2001 (Proc. 12th Eurographics Workshop on Rendering)*, pages 269–276. Springer, 2001.

[KK02] T. Kollig and A. Keller. Efficient Multidimensional Sampling. *Computer Graphics Forum*, 21(3):557–563, September 2002.

[LP01] G. Larcher and F. Pillichshammer. Walsh Series Analysis of the L_2-Discrepancy of Symmetrisized Point Sets. *Monatsh. Math.*, 132:1–18, 2001.

[Nie92] H. Niederreiter. *Random Number Generation and Quasi-Monte Carlo Methods*. SIAM, Philadelphia, 1992.

[Owe95] A. Owen. Randomly Permuted (t, m, s)-Nets and (t, s)-Sequences. In H. Niederreiter and P. Shiue, editors, *Monte Carlo and Quasi-Monte Carlo Methods in Scientific Computing*, volume 106 of *Lecture Notes in Statistics*, pages 299–315. Springer, 1995.

[Owe98] A. Owen. Monte Carlo Extension of Quasi-Monte Carlo. In *Winter Simulation Conference*, pages 571–577. IEEE Press, 1998.

[Pan04] F. Panneton. *Construction d'ensembles de points basé sur une récurrence linéaire dans un corps fini de caractéristique 2 pour la simulation Monte Carlo et l'intégration quasi-Monte Carlo*. PhD thesis, Université de Montréal, 2004.

[PS01] G. Pirsic and W. Ch. Schmid. Calculation of the quality parameter of digital nets and application to their construction. *J. Complexity*, 17(4):827–839, 2001.

[PTVF92] H. Press, S. Teukolsky, T. Vetterling, and B. Flannery. *Numerical Recipes in C*. Cambridge University Press, 1992.

[Rol05] T. Rolfe. Optimal Queens - A classical problem solved by backtracking. *Dr. Dobb's Journal*, 30(372), May 2005.

[Sob67] I. Sobol'. On the Distribution of Points in a Cube and the approximate Evaluation of Integrals. *Zh. vychisl. Mat. mat. Fiz.*, 7(4):784–802, 1967.

[WK07] C. Wächter and A. Keller. Efficient Simultaneous Simulation of Markov Chains. In A. Keller, S. Heinrich, and H. Niederreiter, editors, *Monte Carlo and Quasi-Monte Carlo Methods 2006*, in this volume. Springer, 2007.

[Yel83] J. Yellot. Spectral Consequences of Photoreceptor Sampling in the Rhesus Retina. *Science*, 221:382–385, 1983.

Quasi-Monte Carlo Simulation of Discrete-Time Markov Chains on Multidimensional State Spaces

Rami El Haddad[1], Christian Lécot[2], and Pierre L'Ecuyer[3]

[1] Département de Mathématiques, Université Saint-Joseph, BP 11-514, Riad El Solh Beyrouth 1107 2050, Liban
Rami.El-Haddad@univ-savoie.fr

[2] LAMA, Université de Savoie, 73376 Le Bourget-du-Lac Cedex, France
Christian.Lecot@univ-savoie.fr

[3] DIRO, Université de Montréal, C.P. 6128, Succ. Centre-Ville, Montréal, H3C 3J7, Canada
lecuyer@iro.umontreal.ca

Summary. We propose and analyze a quasi-Monte Carlo (QMC) method for simulating a discrete-time Markov chain on a discrete state space of dimension $s \geq 1$. Several paths of the chain are simulated in parallel and reordered at each step, using a multidimensional matching between the QMC points and the copies of the chains. This method generalizes a technique proposed previously for the case where $s = 1$. We provide a convergence result when the number N of simulated paths increases toward infinity. Finally, we present the results of some numerical experiments showing that our QMC algorithm converges faster as a function of N, at least in some situations, than the corresponding Monte Carlo (MC) method.

1 Introduction

Markov chains are used in many fields such as physics, queueing theory, telecommunications, option pricing, etc. Frequently, we want to estimate the expectation of a cost function that depends on the sample path of the chain over several steps. In the simplest cases, if the chain has a small (finite) state space, we may use matrix equations to compute the state distribution at each step of the chain, and compute the expected cost from that. But very often, the state space is too large to allow such exact computations, and the only viable method to estimate the expected cost is MC simulation. Despite the versatility of MC methods, one drawback is their slow convergence: Based on the central limit theorem, their convergence rate is roughly of $\mathcal{O}(N^{-1/2})$ if N denotes the number of copies of the chain that are simulated.

One possible approach to accelerate the computation is to replace the random numbers by low discrepancy sequences, i.e., quasi-random numbers. This is the general idea of QMC methods, which outperform MC methods in some cases, but also have limitations. In general, the expected cost can be written as an integral over the s'-dimensional unit hypercube, where s' represents the total number of $\mathcal{U}(0,1)$ (uniform over (0,1)) random variates needed to realize a sample path. The classical QMC method would take an s'-dimensional low-discrepancy point set of cardinality N, use each point to simulate one copy of the chain, and estimate the expected cost by the average over these N copies. But this s' is usually very large, so we end up with an integration problem in a very large number of dimensions, in which case QMC is typically not very effective.

A QMC algorithm for the simulation of Markov chains with a one-dimensional state space was studied in [LT04]. Randomized variants of this method, and an extension to multidimensional state spaces, were proposed and examined in [LLT07]. The multidimensional extension of [LLT07] essentially maps the state space to a one-dimensional set by defining a sorting function that assigns a real number to each state. The choice of sorting function is crucial for the performance of the algorithm, and finding a good function can be difficult in general. Moreover, [LLT07] provide convergence proofs and variance bounds only for the case of a unidimensional state space.

In the present paper, we propose a different generalization of the QMC algorithm of [LT04], for Markov chains with multidimensional state spaces. This algorithm employs a low-discrepancy sequence of points with the property that each subsequence of length N starting at an index which is a multiple of N in the sequence is evenly distributed in a sense to be specified. At each step, it uses one such subsequence to advance the N chains by one step, after matching the chains with the points in a clever way. This matching is done by sorting both the chains and the points according to their successive coordinates. This multidimensional sort ensures theoretical convergence and often achieves better accuracy than one based on a real-valued sorting function, when N increases, because it preserves the relative proximities of the points in the space.

The remainder of the paper is organized as follows. In Section 2, we present the algorithm, which simulates the N sample paths of the chain in parallel using a low-discrepancy sequence. In Section 3 we adapt the basic concepts of QMC methods to the present study and we recall some definitions and properties related to the variation of multi-dimensional sequences. In Section 4, under a certain assumption on the transition matrix, we prove a convergence bound on the *worst-case error* for our method. This assumption could certainly be relaxed, at the expense of more complicated notation in the convergence proof. Finally, in Section 5, we present the results of numerical experiments that compare our method with standard MC. The convergence rate observed empirically for our method is much better than for MC and also much better than what is guaranteed by the worst-case bound.

2 The method

We consider a time-homogeneous discrete-time Markov chain $\{X_n, n \in \mathbb{N}\}$ with state space $\mathbf{E}$ of the form $\mathbf{E} := \prod_{r=1}^{s} E_r$, where $E_r \subseteq \mathbb{Z}$. The initial state X_0 has distribution λ_0, so $\mathbb{P}[X_0 = \mathbf{i}] = \lambda_0\{\mathbf{i}\}$ for each $\mathbf{i} = (i_1, \dots, i_s) \in \mathbf{E}$, and the transition matrix is $P = (p(\mathbf{i}, \mathbf{j}) : \mathbf{i}, \mathbf{j} \in \mathbf{E})$, where $p(\mathbf{i}, \mathbf{j}) = \mathbb{P}[X_n = \mathbf{j} \mid X_{n-1} = \mathbf{i}]$. The probability that the chain is in state $\mathbf{i}$ after n steps is

$$\lambda_n\{\mathbf{i}\} = \mathbb{P}[X_n = \mathbf{i}] = \lambda_0 P^n\{\mathbf{i}\}$$

for all $\mathbf{i} \in \mathbf{E}$. Our aim is to estimate the expected cost $\mathbb{E}[w(X_n)]$ at step n, for some bounded function $w : \mathbf{E} \to [0, \infty)$. This function is also called a *sequence with multivariate indices* (the indices are the elements of $\mathbf{E}$).

As an intermediate step for estimating this expectation, we will construct an approximation of λ_n for each n. Let $\delta_{\mathbf{i}}$ be the row vector of unit mass at $\mathbf{i} = (i_1, \dots, i_s)$ defined, for all $\mathbf{j} = (j_1, \dots, j_s) \in \mathbf{E}$, by

$$\delta_{\mathbf{i}}\{\mathbf{j}\} = \begin{cases} 1 \text{ if } i_1 = j_1, \dots, i_s = j_s, \\ 0 \text{ otherwise.} \end{cases}$$

We denote by $\delta_{\mathbf{i}} w$ the real number $w(\mathbf{i})$. Our approximation of λ_n will be an empirical distribution, of the form

$$\widehat{\lambda}_n := \frac{1}{N} \sum_{0 \le \ell < N} \delta_{\mathbf{i}_\ell^n}$$

for some integer N and for judiciously selected states $\mathbf{i}_0^n, \dots, \mathbf{i}_{N-1}^n \in \mathbf{E}$. Our aim is to have $\widehat{\lambda}_n \approx \lambda_n$ in the sense that the point set $\{\mathbf{i}_0^n, \dots, \mathbf{i}_{N-1}^n\}$ has a small star λ_n-discrepancy; this will be defined more precisely in Section 3.

A simple way of obtaining these N states is the *MC method*, which simulates N independent realizations (or copies) of the chain, and takes the corresponding states at step n. For each copy, if the chain is in state $X_{n-1} = \mathbf{i}$ before step n, we simply generate the next state X_n so that $\mathbf{P}[X_n = \mathbf{j}] = p(\mathbf{i}, \mathbf{j})$. The usual way of implementing this is as follows. For each $\mathbf{i} \in \mathbf{E}$, partition the interval $I = [0, 1)$ in subintervals $\mathcal{I}_{\mathbf{i},\mathbf{j}} := [m_{\mathbf{i},\mathbf{j}}, m'_{\mathbf{i},\mathbf{j}})$ where $m'_{\mathbf{i},\mathbf{j}} = m_{\mathbf{i},\mathbf{j}} + p(\mathbf{i}, \mathbf{j})$, for all $\mathbf{j} \in \mathbf{E}^{\mathbf{i}} := \{\mathbf{j} \in \mathbf{E} : p(\mathbf{i}, \mathbf{j}) > 0\}$. For any $y \in I$, let $\mathbf{j}(\mathbf{i}, y)$ denote the unique element $\mathbf{j} \in \mathbf{E}^{\mathbf{i}}$ such that $y \in \mathcal{I}_{\mathbf{i},\mathbf{j}}$. At step n, if $X_{n-1} = \mathbf{i}$, the MC method generates a $\mathcal{U}(0, 1)$ random variate U_n and puts $X_n = \mathbf{j}(\mathbf{i}, U_n)$.

The aim of our QMC approximation is essentially to obtain a more representative set of states $\mathbf{i}_0^n, \dots, \mathbf{i}_{N-1}^n$ at each step. Before defining this approximation, we recall the notion of (t, s)-sequence and (t, m, s)-net from [Nie92], page 48. Let $I^s := [0, 1)^s$ denotes the s-dimensional half open unit cube. For an integer $b \ge 2$, an *elementary interval in base* b is a subinterval of I^s of the form

$$\prod_{r=1}^{s} \left[\frac{a_r}{b^{d_r}}, \frac{a_r + 1}{b^{d_r}} \right),$$

for some integers $d_r \geq 0$ and $0 \leq a_r < b^{d_r}$ for all $1 \leq r \leq s$. If $0 \leq t \leq m$ are integers, a *(t, m, s)-net in base b* is a set Y of b^m points in I^s such that every elementary interval Q in base b with Lebesgue-measure (or volume) b^{t-m} contains exactly b^t points of Y. An infinite sequence of points $\mathbf{y}_0, \mathbf{y}_1, \ldots$ in I^s is a *(t, s)-sequence in base b* if for all integers $n \geq 0$ and $m > t$, the set $Y_n = \{\mathbf{y}_p : nb^m \leq p < (n+1)b^m\}$ is a (t, m, s)-net in base b.

Choose a base $b \geq 2$ and non-negative integers $d_1, \ldots, d_s$. Put $m := d_1 + \cdots + d_s$ and $N := b^m$. For the QMC approximation, we assume that $Y = \{\mathbf{y}_0, \mathbf{y}_1, \ldots\} \subset I^{s+1}$ is a $(t, s+1)$-sequence in base b for some integer $t \geq 0$, and such that if $\Pi : I^{s+1} \to I^s$ denotes the projection defined by

$$(x_1, \ldots, x_{s+1}) \stackrel{\Pi}{\longmapsto} (x_1, \ldots, x_s) =: \mathbf{x}',$$

then the point set $\Pi(Y_n)$ is a $(0, m, s)$-net in base b for each n. This implies that $b \geq s - 1$.

We first outline our algorithm; then we explain the different steps. Note that the N copies of the chain are simulated simultaneously.

1. Let $n \leftarrow 0$;
2. Use QMC sampling to obtain N initial states $\mathbf{i}_0^0, \ldots, \mathbf{i}_{N-1}^0 \in \mathbf{E}$;
3. Repeat until we have reached the desired number of steps:
   3.1 Let $n \leftarrow n + 1$;
   3.2 Relabel the states $\mathbf{i}_0^n, \ldots, \mathbf{i}_{N-1}^n \in \mathbf{E}$ according to their successive coordinates;
   3.3 Advance all the chains by one step using the point set Y_n, to obtain the states $\mathbf{i}_0^{n+1}, \cdots, \mathbf{i}_{N-1}^{n+1}$, and compute the average cost for this step;

Generating the Initial States (Step 2). This is done by mapping a $(0, m, s)$-net in base b on $\mathbf{E}$. The choice of the mapping depends on the initial distribution to be sampled.

Relabeling the States (Step 3.2). At step $n + 1$, given that we have a set Ξ_n of N states $\mathbf{i}_0^n, \ldots, \mathbf{i}_{N-1}^n$ such that $\widehat{\lambda}_n \approx \lambda_n$, we start computing $\widehat{\lambda}_{n+1}$ by sorting the set Ξ_n as we now explain. The states are labeled $\mathbf{i}_\mathbf{k}^n = (i_{\mathbf{k},1}^n, \ldots, i_{\mathbf{k},s}^n)$, using a multi-dimensional index $\mathbf{k} = (k_1, \ldots, k_s)$, with $0 \leq k_r < b^{d_r}$ for $1 \leq r \leq s$, such that:

$$\begin{aligned}
&\text{if } k_1 < l_1, \text{ then } i_{\mathbf{k},1}^n \leq i_{\mathbf{l},1}^n,\\
&\text{if } k_1 = l_1, k_2 < l_2, \text{ then } i_{\mathbf{k},2}^n \leq i_{\mathbf{l},2}^n,\\
&\qquad\vdots\\
&\text{if } k_1 = l_1, \ldots, k_{s-1} = l_{s-1}, k_s < l_s, \text{ then } i_{\mathbf{k},s}^n \leq i_{\mathbf{l},s}^n.
\end{aligned}$$

These conditions can be interpreted as follows. The N states are first sorted in b^{d_1} batches of size Nb^{-d_1} according to their first coordinates; then each batch is sorted in subgroups of b^{d_2} batches of size $Nb^{-d_1-d_2}$ by order of the second coordinates, and so on. At the last step of the sort, subgroups of size b^{d_s} are ordered according to the last coordinate of the state.

This type of hierarchical (or nested) sort was first introduced and motivated in [LC98]. It guarantees theoretical convergence. The idea is to match the s-dimensional empirical distribution of the states with the distribution of the s-dimensional points of low-discrepancy set $\Pi(Y_n)$.

A graphical illustration is given in Figure 1, with $b = 2$, $s = 2$, and $d_1 = d_2 = 2$. Here we have $N = b^{d_1+d_2} = 16$ points in $s = 2$ dimensions. We first sort these points in four groups of four points, according to their first coordinate (which corresponds to the horizontal axis in the figure), and then sort each group according to the second coordinate.

Advancing the Chains by one Step (Step 3.3). Let $\widetilde{\lambda}_{n+1} = \widehat{\lambda}_n P$. We have

$$\widetilde{\lambda}_{n+1} w = \frac{1}{N} \sum_{\mathbf{k} \in \mathcal{K}} (Pw)(\mathbf{i}_{\mathbf{k}}^n) = \frac{1}{N} \sum_{\mathbf{k} \in \mathcal{K}} \sum_{\mathbf{j} \in \mathbf{E}} p(\mathbf{i}_k^n, \mathbf{j}) w(\mathbf{j}) \tag{1}$$

where $\mathcal{K} = \{0, \ldots, b^{d_1} - 1\} \times \cdots \times \{0, \ldots, b^{d_s} - 1\}$. This expression could be seen as the expected average cost at the next step, given the current set of states Ξ_n. For $\mathbf{k} = (k_1, \ldots, k_s) \in \mathcal{K}$, denote by $\chi_{\mathbf{k}}$ the indicator function of the s-dimensional elementary interval

$$\mathcal{I}_{\mathbf{k}} = \prod_{r=1}^{s} \left[\frac{k_r}{b^{d_r}}, \frac{k_r + 1}{b^{d_r}} \right).$$

At any given step, if a chain in state $\mathbf{i}$ is matched with a point $\mathbf{y} = (\mathbf{y}', y_{s+1})$, then this chain will move to state $\mathbf{j}(\mathbf{i}, y_{s+1})$, where the latter is defined as in the MC method.

For any given point $\mathbf{y} = (\mathbf{y}', y_{s+1}) \in I^{s+1}$, let

$$\mathcal{G}^n w(\mathbf{y}) := \sum_{\mathbf{k} \in \mathcal{K}} \chi_{\mathbf{k}}(\mathbf{y}') w(\mathbf{j}(\mathbf{i}_{\mathbf{k}}^n, y_{s+1})), \tag{2}$$

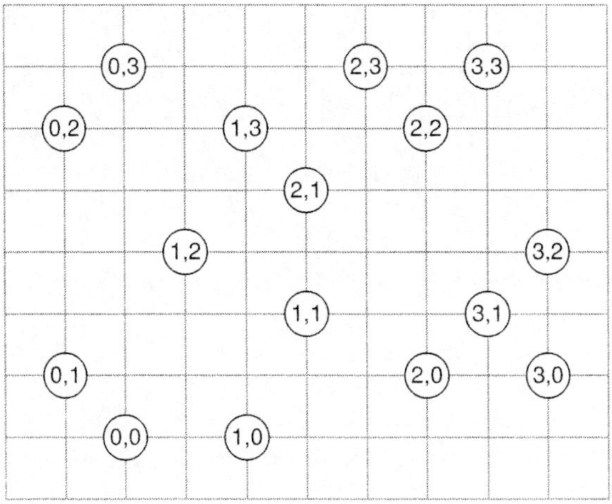

Fig. 1. Relabeling the states ($b = 2$, $s = 2$, $d_1 = d_2 = 2$)

which represents the cost at the next transition if we use $\mathbf{y}$ to move by one step the chain associated with the point of index $\mathbf{k}$. Integrating with respect to $\mathbf{y}$, we get

$$\widetilde{\lambda}_{n+1} w = \int_{I^{s+1}} \mathcal{G}^n w(\mathbf{y}) d\mathbf{y}, \tag{3}$$

whereas averaging over the point set used at step $n+1$, we obtain the QMC estimator $\widehat{\lambda}_{n+1}$ of λ_{n+1} defined by

$$\widehat{\lambda}_{n+1} w = \frac{1}{N} \sum_{nN \leq p < (n+1)N} \mathcal{G}^n w(\mathbf{y}_p).$$

For any $\mathbf{y}' = (y_1, \ldots, y_s) \in I^s$, we define the multidimensional index

$$\mathbf{k}(\mathbf{y}') := (\lfloor b^{d_1} y_1 \rfloor, \ldots, \lfloor b^{d_s} y_s \rfloor) \in \mathcal{K},$$

where $\lfloor x \rfloor$ denotes the largest integer $\leq x$. Because $\Pi(Y_n)$ is a $(0, m, s)$-net in base b, the function

$$p \in \{nN, nN+1, \ldots, (n+1)N - 1\} \longmapsto \mathbf{k}(\mathbf{y}'_p) \in \mathcal{K}$$

is one-to-one. Combining this with (2), we have

$$\frac{1}{N} \sum_{nN \leq p < (n+1)N} \mathcal{G}^n w(\mathbf{y}_p) = \frac{1}{N} \sum_{nN \leq p < (n+1)N} w(\mathbf{j}(\mathbf{i}^n_{\mathbf{k}(\mathbf{y}'_p)}, y_{p,s+1})). \tag{4}$$

Then, at step $n+1$, the point set $\Xi_{n+1} = \{\mathbf{i}_0^{n+1}, \ldots, \mathbf{i}_{N-1}^{n+1}\} \subset \mathbf{E}$ is computed according to

$$\mathbf{i}^{n+1}_{p-nN} = \mathbf{j}(\mathbf{i}^n_{\mathbf{k}(\mathbf{y}'_p)}, y_{p,s+1}), \qquad nN \leq p < (n+1)N.$$

This means that the projection $\mathbf{y}'_p$ of each point $\mathbf{y}_p$ of the low discrepancy sequence on the first s axes is used to select the state that is matched to this point at that step (i.e., which chain advances by one step), while the remaining component $y_{p,s+1}$ is used to determine the evolution (the next state).

3 Discrepancies and Variations of Sequences

The efficiency of a QMC method depends on the uniformity of the quasi-random points that are used. These points should form a low discrepancy point set. In this section, after recalling classical notions of discrepancy from [Nie92], we define and examine discrepancy measures adapted to the context of our method.

The Star Discrepancy. For an s-dimensional point set $Y = \{\mathbf{y}_0, \ldots, \mathbf{y}_{N-1}\} \subset I^s$ and for a Lebesgue-measurable subset Q of I^s we define the *local discrepancy* by

$$D(Q, Y) := \frac{1}{N} \sum_{0 \le p < N} \chi_Q(\mathbf{y}_p) - \int_{I^s} \chi_Q(\mathbf{x}) d\mathbf{x},$$

where χ_Q is the indicator function of Q. The *discrepancy* of the point set Y is defined by

$$D(Y) := \sup_Q |D(Q, Y)|,$$

the supremum being taken over all subintervals of I^s. The *star discrepancy* of Y is

$$D^*(Y) := \sup_{Q^*} |D(Q^*, Y)|,$$

where Q^* runs through all subintervals of I^s with one vertex at the origin.

The following result is shown in [Nie87].

Lemma 1. *Let Y be a (t, m, s)-net in base b. For any elementary interval $Q' \subset I^{s-1}$ in base b and for any $\xi \in \bar{I} := [0, 1]$, we have*

$$|D(Q' \times [0, \xi), Y)| \le b^{t-m}.$$

The Star λ-Discrepancy. Let λ be a distribution on $\mathbf{E}$ and consider a set of states $\Xi = \{\mathbf{i}_0, \ldots, \mathbf{i}_{N-1}\} \subset \mathbf{E}$. For an arbitrary $F \subset \mathbf{E}$, we define the *local λ-discrepancy* of Ξ for F by

$$D(F; \Xi, \lambda) := \frac{1}{N} \sum_{0 \le \ell < N} \chi_F(\mathbf{i}_\ell) - \sum_{\mathbf{i} \in F} \lambda\{\mathbf{i}\},$$

where χ_F denotes the function

$$\chi_F(\mathbf{i}) = \begin{cases} 1 \text{ if } \mathbf{i} \in F, \\ 0 \text{ otherwise.} \end{cases}$$

The *star λ-discrepancy* of the point set Ξ is defined by

$$D^*(\Xi, \lambda) := \sup_{\mathbf{h} \in \mathbf{E}} |D(F_\mathbf{h}; \Xi, \lambda)|,$$

where $F_\mathbf{h} = \prod_{r=1}^{s} ((-\infty, h_r) \cap E_r)$. If w is a nonnegative and bounded sequence, we set

$$D(w; \Xi, \lambda) := \frac{1}{N} \sum_{0 \le \ell < N} w(\mathbf{i}_\ell) - \sum_{\mathbf{i} \in \mathbf{E}} \lambda\{\mathbf{i}\} w(\mathbf{i}),$$

so that if $F \subset \mathbf{E}$, we have $D(F; \Xi, \lambda) = D(\chi_F; \Xi, \lambda)$. Similarly, if $Y' = \{\mathbf{y}'_0, \ldots, \mathbf{y}'_{N-1}\} \subset I^s$ and f is a nonnegative and bounded function defined on I^s, we put

$$D(f, Y') = \frac{1}{N} \sum_{0 \le \ell < N} f(\mathbf{y}'_\ell) - \int_{I^s} f(\mathbf{x}') d\mathbf{x}'.$$

If $Q \subset I^s$, then $D(\chi_Q, Y') = D(Q, Y')$.

Variations of Sequences. We must adapt to our case the definitions of variations: instead of considering functions defined on I^s, we have sequences defined on $\mathbf{E}$. We begin by giving the definition of the variation in the sense of Hardy-Krause, assuming that $\mathbf{0} = (0, \ldots, 0) \in \mathbf{E}$. For the case where $\mathbf{0} \notin \mathbf{E}$, it suffices to replace the origin $\mathbf{0}$ by some point $\mathbf{a} = (a_1, \ldots, a_s) \in \mathbf{E}$ in the definition. For $w : \mathbf{E} \to \mathbb{R}$ and $\mathbf{j}, \mathbf{j}' \in \mathbf{E}$, let $T^r_{\mathbf{j}} w$ and $\Delta^r_{\mathbf{j},\mathbf{j}'} w$ be the functions (or multivariate sequences) defined by

$$T^r_{\mathbf{j}} w(\mathbf{i}) := w(i_1, \ldots, i_{r-1}, j_r, i_{r+1}, \ldots, i_s) \quad \text{and} \quad \Delta^r_{\mathbf{j},\mathbf{j}'} w := T^r_{\mathbf{j}'} w - T^r_{\mathbf{j}} w.$$

If $R = \{r_1, \ldots, r_q\} \subset S = \{1, \ldots, s\}$, we denote

$$T^R_{\mathbf{j}} w := T^{r_1}_{\mathbf{j}} \ldots T^{r_q}_{\mathbf{j}} w \quad \text{and} \quad \Delta^R_{\mathbf{j},\mathbf{j}'} w := \Delta^{r_1}_{\mathbf{j},\mathbf{j}'} \ldots \Delta^{r_q}_{\mathbf{j},\mathbf{j}'} w.$$

For $\mathbf{j} \in \mathbf{E}$, let $\mathbf{j}+$ be the vector $(j_1 + 1, \ldots, j_s + 1)$; let $\mathbf{E}' = \{\mathbf{i} \in \mathbf{E} : \mathbf{i}+ \in \mathbf{E}\}$. The *variation in the sense of Vitali* of $w : \mathbf{E} \to \mathbb{R}$ is defined by

$$V^s(w) = \sum_{\mathbf{j} \in \mathbf{E}'} |\Delta^S_{\mathbf{j},\mathbf{j}+} w|,$$

and the *variation of w in the sense of Hardy and Krause* is the sum

$$V(w) = \sum_{r=1}^{s} \sum_{\substack{R \subset S \\ \#R = r}} V^r(T^{R^c}_{\mathbf{0}} w).$$

Let $\mathbf{M} = (M_1, \ldots, M_s)$ be the vector with coordinates $M_r = \sup\{i : i \in E_r\}$. If w can be extended to $\mathbf{M}$, we define the *upper variation* of w as

$$V^*(w) = \sum_{r=1}^{s} \sum_{\substack{R \subset S \\ \#R = r}} V^r(T^{R^c}_{\mathbf{M}} w).$$

One can prove that if w is of bounded variation in the sense of Hardy and Krause, then w may be extended to $\mathbf{M}$ and has a bounded upper variation. The next Lemma is a version of the classical Koksma-Hlawka inequality. The proof follows the general outline of the proof given by Zaremba [Zar68].

Lemma 2. *Let λ be a distribution on $\mathbf{E}$. If w is a sequence of bounded variation in the sense of Hardy and Krause and if $\Xi = \{\mathbf{i}_0, \ldots, \mathbf{i}_{N-1}\} \subset \mathbf{E}$, then*

$$|D(w; \Xi, \lambda)| \le V^*(w) D^*(\Xi, \lambda).$$

The following Lemma is an analogue of a result previously given in the continuous case in [Lec96]; it can be proved by similar arguments.

Lemma 3. *Let $w : \mathbf{E} \to \mathbb{R}$ be a sequence of bounded variation in the sense of Hardy and Krause and $p_1, \ldots, p_s$ be integers. We consider a nested partition of $\mathbf{E}$ of the form*

· $g_{0,1} \le g_{1,1} \le \dots \le g_{p_1,1} \in E_1$
· $g_{k_1,0,2} \le g_{k_1,1,2} \le \dots \le g_{k_1,p_2,2} \in E_2$ *for* $0 \le k_1 < p_1$.
· $\dots$
· $g_{\mathbf{k}',0,s} \le g_{\mathbf{k}',1,s} \le \dots \le g_{\mathbf{k}',p_s,s} \in E_s$ *for* $\mathbf{k}' = (k_1, \dots, k_{s-1})$ *with* $0 \le k_1 < p_1, \dots, 0 \le k_{s-1} < p_{s-1}$.

For each $\mathbf{k} = (k_1, \dots, k_s) \in \Lambda := \prod_{r=1}^{s} \{0, \dots, p_r - 1\}$, *let*

$$\mathbf{i_k}, \mathbf{j_k} \in \{g_{k_1,1}, \dots, g_{k_1+1,1}\} \times \{g_{k_1,k_2,2}, \dots, g_{k_1,k_2+1,2}\} \times \cdots \\ \cdots \times \{g_{\mathbf{k}',k_s,s}, \dots, g_{\mathbf{k}',k_s+1,s}\}.$$

Then we have the following inequality

$$\sum_{\mathbf{k} \in \Lambda} |w(\mathbf{j_k}) - w(\mathbf{i_k})| \le V^*(w) \prod_{r=1}^{s} p_r \sum_{r=1}^{s} \frac{1}{p_r}.$$

Equipped with these tools, we return to the convergence of the QMC algorithm.

4 Convergence analysis

We provide a convergence result under the following simplifying assumption, which means that at most one coordinate of the state can be changed at any step of the chain. We emphasize that this condition is not necessary for our method to apply. In our numerical experiments reported in the next section, we found that the method can be very effective even when this condition is not fulfilled. The reason why we outline the convergence proof only under this condition is to avoid excessively complicated notation.

Assumption 1 *We suppose that whenever there exists* $1 \le q \ne q' \le s$ *such that* $i_q \ne j_q$ *and* $i_{q'} \ne j_{q'}$, *we have* $p(\mathbf{i}, \mathbf{j}) = 0$.

For $\mathbf{i} \in \mathbf{E}$, denote $\mathbf{E}^{\mathbf{i}*} = \mathbf{E}^{\mathbf{i}} \setminus \{\mathbf{i}\}$. We suppose that $\mathcal{I}_{\mathbf{i},\mathbf{i}} = [0, p(\mathbf{i}, \mathbf{i}))$ if $\mathbf{i} \in \mathbf{E}^{\mathbf{i}}$, and $\mathcal{I}_{\mathbf{i},\mathbf{i}} = \emptyset$ otherwise. For $\mathbf{j} \in \mathbf{E}^{\mathbf{i}*}$, under Assumption 1, there exists a unique index $r \in \{1, \dots, s\}$ such that $i_r \ne j_r$, and we take

$$m_{\mathbf{i},\mathbf{j}} = p(\mathbf{i}, \mathbf{i}) + \sum_{\substack{\mathbf{g} \in \mathbf{E}^{\mathbf{i}*} \\ g_1 \ne i_1}} p(\mathbf{i}, \mathbf{g}) + \cdots + \sum_{\substack{\mathbf{g} \in \mathbf{E}^{\mathbf{i}*} \\ g_{r-1} \ne i_{r-1}}} p(\mathbf{i}, \mathbf{g}) + \sum_{\substack{\mathbf{g} \in \mathbf{E}^{\mathbf{i}*}, g_r < j_r \\ g_\ell = i_\ell, \ell \ne r}} p(\mathbf{i}, \mathbf{g})$$

in the definition of the intervals $\mathcal{I}_{\mathbf{i},\mathbf{j}}$. This means that, for $\mathbf{i} \in \mathbf{E}$, we add the probabilities $\{p(\mathbf{i}, \mathbf{g}) : \mathbf{g} \in \mathbf{E}^{\mathbf{i}*}\}$ according to the natural order of the axes. Let

$$q(\mathbf{i}) := p(\mathbf{i}, \mathbf{i}), \qquad p_r(\mathbf{i}) := \sum_{\substack{\mathbf{g} \in \mathbf{E}^{\mathbf{i}*} \\ g_r \ne i_r}} p(\mathbf{i}, \mathbf{g}),$$

$$p_{h_r,r}(\mathbf{i}) := \sum_{\substack{\mathbf{g}\in\mathbf{E}^{\mathbf{i}*}, g_r<h_r \\ g_\ell=i_\ell, \ell\neq r}} p(\mathbf{i},\mathbf{g}), \ \forall h_r \in E_r.$$

We assume that these sequences are of bounded variation in the sense of Hardy and Krause. We then have the following *worst-case* error bound.

Proposition 1. *If the transition matrix of the chain satisfies* $V^*(P\chi_{F_\mathbf{h}}) \leq 1$ *for all* $\mathbf{h} \in \mathbf{E}$ *and if there exist positive constants* c_0, $c_{1,r}$ *and* $c_{2,r}$, *for* $1 \leq r \leq s$, *such that*

$$V^*(q) \leq c_0, \ V^*(p_r) \leq c_{1,r}, \ \text{and}\ V^*(p_{h_r,r}) \leq c_{2,r},$$

then

$$D^*(\Xi_n, \lambda_n) \leq D^*(\Xi_0, \lambda_0) + \frac{(4s+1)n}{b^{\lfloor (d_s - t)/2 \rfloor}} + C(s)\, n \left(\frac{1}{b^{d_1}} + \cdots + \frac{1}{b^{d_{s-1}}} + \frac{1}{b^{\lfloor (d_s-t)/2 \rfloor}} \right),$$

where $C(s)$ *is a positive constant that depends only on* s.

Proof. The proof of this proposition is quite technical and will be given in [ElH]; we only sketch it in what follows. For any nonnegative and bounded sequence w, write

$$D(w; \Xi_{n+1}, \lambda_{n+1}) = D(Pw; \Xi_n, \lambda_n) + D(\mathcal{G}^n w, Y_n).$$

If we take $w = \chi_{F_\mathbf{h}}$, $\mathbf{h} \in \mathbf{E}$, we get

$$D(F_\mathbf{h}; \Xi_{n+1}, \lambda_{n+1}) = D(P\chi_{F_\mathbf{h}}; \Xi_n, \lambda_n) + D(\mathcal{G}^n \chi_{F_\mathbf{h}}, Y_n).$$

By Lemma 2 and using the inequality $V^*(P\chi_{F_\mathbf{h}}) \leq 1$, we have

$$|D(P\chi_{F_\mathbf{h}}; \Xi_n, \lambda_n)| \leq V^*(P\chi_{F_\mathbf{h}}) D^*(\Xi_n, \lambda_n) \leq D^*(\Xi_n, \lambda_n).$$

On the other hand, $\mathcal{G}^n \chi_{F_\mathbf{h}}$ is the indicator function of

$$Q^n_\mathbf{h} := \bigcup_{\mathbf{k}\in\mathcal{K}} (\mathcal{I}_\mathbf{k} \times \bigcup_{\substack{\mathbf{j}\in\mathbf{E}^{\mathbf{i}^n_\mathbf{k}} \\ \mathbf{j}<\mathbf{h}}} \mathcal{I}_{\mathbf{i}^n_\mathbf{k},\mathbf{j}}),$$

where $\mathbf{j} < \mathbf{h}$ means that $j_1 < h_1, \ldots, j_s < h_s$. Thus, $D(\mathcal{G}^n \chi_{F_\mathbf{h}}, Y_n) = D(Q^n_\mathbf{h}, Y_n)$. We decompose $Q^n_\mathbf{h}$ into the following $s+1$ disjoint subsets:

$$Q^n_\mathbf{h} = Q^n_{\mathbf{h},0} \cup Q^n_{\mathbf{h},1} \cup \cdots \cup Q^n_{\mathbf{h},s},$$

with

$$Q^n_{\mathbf{h},0} = \bigcup_{\substack{\mathbf{k}\in\mathcal{K} \\ \mathbf{i}^n_\mathbf{k}<\mathbf{h}}} \mathcal{I}_\mathbf{k} \times \mathcal{I}_{\mathbf{i}^n_\mathbf{k},\mathbf{i}^n_\mathbf{k}}$$

and

$$Q^n_{\mathbf{h},r} = \bigcup_{\mathbf{k} \in \mathcal{K}} \left(\mathcal{I}_{\mathbf{k}} \times \bigcup_{\substack{\mathbf{j} \in \mathbf{E}^{\mathbf{i}^n_{\mathbf{k}}*} \\ \mathbf{j} < \mathbf{h}, j_r \neq i^n_{\mathbf{k},r}}} \mathcal{I}_{\mathbf{i}^n_{\mathbf{k}},\mathbf{j}} \right)$$

for $1 \leq r \leq s$. Thus,

$$|D(Q^n_{\mathbf{h}}, U_n)| \leq \sum_{r=0}^{s} |D(Q^n_{\mathbf{h},r}, U_n)|.$$

By studying the local discrepancy for each one of the subsets apart using Lemmas 1 and 3, we get the following bound:

$$|D(Q^n_{\mathbf{h}}, Y_n)| \leq \frac{4s+1}{b^{\lfloor (d_s - t)/2 \rfloor}} + C(s) \left(\frac{1}{b^{d_1}} + \cdots + \frac{1}{b^{d_{s-1}}} + \frac{1}{b^{\lfloor (d_s - t)/2 \rfloor}} \right),$$

where $C(s)$ is a positive constant. The desired inequality is then obtained by induction on n. By taking

$$d_1 = \cdots = d_{s-1} = \left\lfloor \frac{m-t}{s+1} \right\rfloor \quad \text{and} \quad d_s = m - (s-1) \left\lfloor \frac{m-t}{s+1} \right\rfloor,$$

the proposition shows that the error converges as $\mathcal{O}(N^{-1/(s+1)})$ in the worst case.

5 Numerical examples

In this Section, we assess the accuracy of the QMC algorithm empirically, through three academic examples where exact solutions are known. We show the kind of improvement that our method can bring with respect to MC. For our experiment, we use Niederreiter's sequences in base 2. For MC, the pseudo-random numbers are produced by the generator MRG32k3a of [L'Ec99]. Convergence speed is assessed by looking at the absolute difference between the empirical and theoretical means, for several values of N.

5.1 A simple symmetric random walk on $\mathbb{Z}^2$

Our first example is a simple symmetric random walk on $\mathbb{Z}^2$; it is a Markov chain with transition probabilities

$$p(\mathbf{i}, \mathbf{j}) = \begin{cases} 1/4 \text{ if } |i_1 - j_1| + |i_2 - j_2| = 1, \\ 0 \quad \text{otherwise.} \end{cases}$$

At time $n = 0$, we start at a point $\mathbf{A} \in \mathbb{Z}^2$ and we want to estimate the probability $p_{\mathbf{A}}^n$ of returning to $\mathbf{A}$ at time n. One can prove that

$$p_{\mathbf{A}}^n = \begin{cases} \left(\binom{n}{n/2}(1/2)^n\right)^2 & \text{if n is even,} \\ 0 & \text{if n is odd.} \end{cases}$$

This example satisfies Assumption 1. We estimate the error for $n = 20$ as a function of N, say $\mathrm{Err}_{\mathrm{MC}}(N)$ for MC and say $\mathrm{Err}_{\mathrm{QMC}}(N)$ for our QMC method. The value of N varies from 2^8 to 2^{20}. Figure 2 shows the values of these errors, in log-log scale. Regression analysis gives the following convergence speed estimates:

$$\mathrm{Err}_{\mathrm{MC}} = \mathcal{O}(N^{-0.25}) \text{ and } \mathrm{Err}_{\mathrm{QMC}} = \mathcal{O}(N^{-0.89}),$$

showing a strong improvement when using QMC.

5.2 A bivariate Markovian asset valuation model

We consider a bivariate extension of Cox-Ross-Rubinstein's binomial single asset pricing model, as proposed in [HKY03]. At time n, the values of the two risky assets are denoted by S_n^1 and S_n^2, and $S_n = (S_n^1, S_n^2)^{\mathrm{t}}$ is the price vector (the "t" means "transposed"). The sequence $(S_n)_{n \geq 0}$ is defined in terms of two independent and identically distributed sequences $(V_n^1)_{n \geq 0}$ and $(V_n^2)_{n \geq 0}$: For all $n \in \mathbb{N}$, V_n^1 can take the two values a and b, where $-1 < a < b$, with probabilities p and $1 - p$, respectively, while V_n^2 can take the two values c and

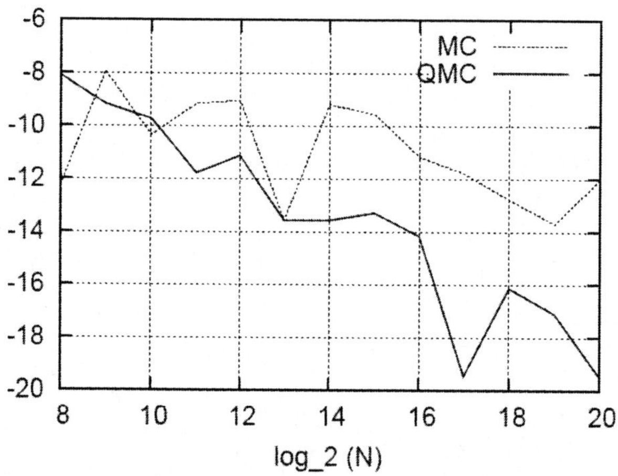

Fig. 2. Simple symmetric random walk on $\mathbb{Z}^2$. The error as a function of N on log-log scale (in base 2), with MC (thin line) and QMC (thick line).

d, where $-1 < c < d$, with probabilities q and $1-q$, respectively. We suppose that $0 < p, q < 1$. The model is a Markov chain whose state evolves as

$$\begin{pmatrix} S_n^1 \\ S_n^2 \end{pmatrix} = \begin{pmatrix} 1+V_n^1 & \varepsilon(V_n^2 - r) \\ \delta(V_n^1 - r) & 1+V_n^2 \end{pmatrix} \begin{pmatrix} S_{n-1}^1 \\ S_{n-1}^2 \end{pmatrix}, \tag{5}$$

where ε and δ are small numbers expressing the perturbation caused by S_n^2 on S_n^1 and by S_n^1 on S_n^2. We suppose $\varepsilon\delta \neq 1$ and $(\varepsilon, \delta) \neq (0,0)$. At each step, there are four possibilities for the vector (V_n^1, V_n^2).

For the QMC method, we partition the unit interval $[0,1)$ in four pieces; each one is assigned to one of the four possible outcomes (with length corresponding to the required probability), in this order: $[0, pq)$, $[pq, pq + p(1-q))$, $[pq + p(1-q), pq + p(1-q) + (1-p)q)$, and $[pq + p(1-q) + (1-p)q, 1)$. We want to estimate $\mathbb{E}(S_n^1)$ and $\mathbb{E}(S_n^2)$ for some fixed n. These (exact) values are

$$\mathbb{E}(S_n^1) = (1+r)^n\ \mathbb{E}(S_0^1) \text{ and } \mathbb{E}(S_n^2) = (1+r)^n\ \mathbb{E}(S_0^2) \tag{6}$$

but for the purpose of our experiment, we pretend that they are unknown and have to be estimated. Note that this bidimensional Markov chain does not satisfy Assumption 1, because *both* coordinates of the state are changed at each step.

We take the following parameters: $n = 20$, $a = 0.074$, $b = 0.141$, $c = 0.086$, $d = 0.182$, $r = 0.1$, $\varepsilon = 0.30$, and $\delta = 0.20$. The initial values are $S_0^1 = 120$ and $S_0^2 = 130$. The number of states N varies from 2^4 to 2^{20}, and we want to

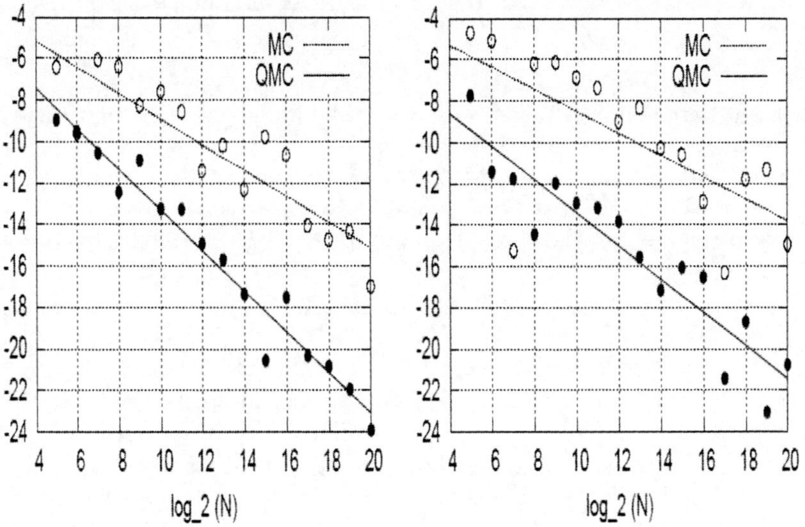

Fig. 3. Bivariate Markovian asset valuation model. Linear fits to the error as a function of N on log-log scale (in base 2), for S_1 (left) and S_2 (right), with MC (thin line) and QMC (thick line).

estimate the error as a function of N. For the errors, we use the same notations as in the previous example. Figure 3 shows the empirical values of these errors, in log-log scale, for both S_n^1 and S_n^2. A linear regression analysis with this data gives the following empirical convergence rates:

$$\text{For } S_n^1,\ \text{Err}_{\text{MC}} = \mathcal{O}(N^{-0.62}) \text{ and } \text{Err}_{\text{QMC}} = \mathcal{O}(N^{-0.98}).$$
$$\text{For } S_n^2,\ \text{Err}_{\text{MC}} = \mathcal{O}(N^{-0.53}) \text{ and } \text{Err}_{\text{QMC}} = \mathcal{O}(N^{-0.80}).$$

Clearly, the QMC method enjoys a much faster convergence than MC.

5.3 Pricing a European call on the maximum of two risky assets

For our second example, we consider the pricing of an European call option on the maximum of two risky assets, in a setting where the (continuous-time) evolution of the asset price vector is approximated by a (discrete-time) binomial lattice model. Again, the example is artificial and simplified, since the option price can be computed exactly in this case, but we want to use it as a benchmark to evaluate the viability of our method.

The original (continuous-time) model is a bivariate geometric Brownian motion (GBM) $\{S(t) = (S_1(t), S_2(t))^{\mathrm{t}},\ t \geq 0\}$ with *drift parameter* μ_i, *volatility parameter* σ_i, and *correlation* parameter ρ. Thus, for $i = 1, 2$,

$$S_i(t) = S_i(0) \exp\left[(\mu_i - \sigma_i^2/2)t + \sigma_i W_i(t)\right]$$

where W_i is a standard Brownian motion, and $\mathrm{Cov}[W_1(t+\delta) - W_1(t), W_2(t+\delta) - W_2(t)] = \rho\delta$ for all $\delta > 0$. The option has discounted payoff

$$e^{-rT} \max[\max(S_1(T), S_2(T)) - K, 0]$$

for some constants $K > 0$ (the strike price) and $T > 0$ (the maturity), where r is the riskless rate. The expected value C of this payoff, which is the exact value of the option, can be computed by formulas given in [Stu82, Joh87].

To estimate C, instead of simulating the GBM directly (which can be done by simulating the Brownian motion (W_1, W_2) at the desired observation times), here we simulate a numerical approximation based on the multivariate binomial lattice method developed in [BEG89]. This method is an extension of the Cox-Ross-Rubinstein approach [CRR79]. It proceeds as follows.

A discrete-time model with discrete probability distribution is constructed to approximate the bivariate lognormal distribution. In this model, at each time step, each asset price can only move up or down (only two possibilities), so there is four possible transitions for the process with probabilities p_1, p_2, p_3 and p_4. Let $h = T/m$ be the length of the time step, where m is the number of steps. The value of $S_i(t)$ is multiplied by $\exp(\sigma_i\sqrt{h})$ in an move up and divided by this same value in a down move. The formulas for the transition probabilities of up and down moves (and other details) can be found in [BEG89]. They are selected in a way that the characteristic function of the discrete distribution at any fixed time point converges to that of the lognormal at that point, when $h \to 0$. This model does not satisfy Assumption 1. As in the previous example, we partition the unit interval $[0, 1)$ in the following four pieces: $[0, p_1)$, $[p_1, p_1 + p_2)$, $[p_1 + p_2, p_1 + p_2 + p_3)$, and $[p_1 + p_2 + p_3, 1)$, so that each number in $[0, 1)$ corresponds to one of the four possibilities.

We use the following parameter values (time is measured in years): $S_1(0) = S_2(0) = 40$, $\sigma_1 = 0.2$, $\sigma_2 = 0.3$, $\rho = 0.5$, $r = 0.05$, $T = 7/12$, $K = 35$. Let $C_{N,m}$ be the QMC approximation of C with N paths and m time steps. We measure the error with the following discrete L^1 norm:

$$\mathrm{Err}_{N,m} = \frac{1}{20} \sum_{j=1}^{20} |C - C_{N,mj/20}|.$$

Figure 4 shows the value of $\mathrm{Err}_{N,m}$ for m varying from $2^2 \times 20$ to $2^6 \times 20$, and N varying from 2^8 to 2^{20}, for both the MC method (left panel), and QMC (right panel), in a log-log scale. Note that there are two sources of error here: (1) the discretization error and (2) the additional error due to using MC or QMC instead of solving the binomial lattice model exactly. The discretization error vanishes when $m \to \infty$, whereas the other source of error converges to zero when $N \to \infty$.

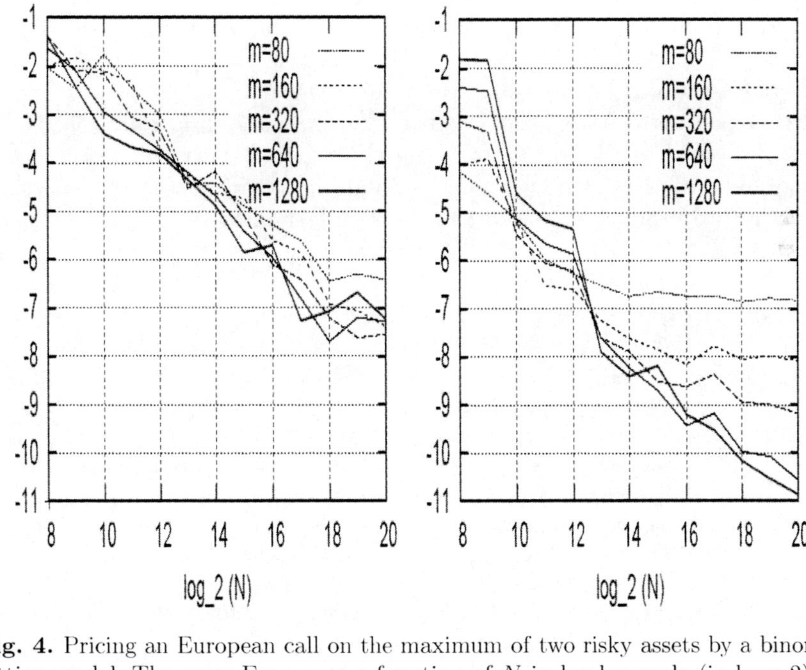

Fig. 4. Pricing an European call on the maximum of two risky assets by a binomial lattice model. The error $\mathrm{Err}_{N,m}$ as a function of N in log-log scale (in base 2) for different numbers of time steps m, for MC (left) and QMC (right).

For MC, the error does not seem to depend much on m, which means that the discretization error is small compared with the MC (statistical) error. For QMC, this is also true for the small values of N, but not for the large ones. For MC, we have a slope of about $-1/2$, so the error converges as $\mathcal{O}(N^{-1/2})$ as a function of N, as expected. For QMC, when m is large, the slope is steeper than $-1/2$, which indicates a faster convergence rate than for MC. For small values of m, the error eventually reaches a plateau and stops decreasing when we keep increasing N; this indicates that the QMC error eventually becomes negligible compared with the discretization error. This shows that a good strategy in this type of situation is to increase both m and N simultaneously. QMC clearly dominates MC in that case. For instance, for $m = 1\,280$ and $N = 2^{20}$, $\mathrm{Err}_{N,m}$ is about 2^{-7} for MC and 2^{-11} for QMC. As another example, for $m = 320$, the same error level is attained by QMC with $N = 8\,192$ and by MC with $N = 524\,288$.

6 Conclusion

We have proposed and analyzed a QMC method for the simulation of discrete-time Markov chains on a multi-dimensional state space. The method simulates several copies of the chain in parallel and reduces the error by a technique that sorts the chains in a special way, based on the several coordinates of their states, at each step. We have proved a convergence result for the worst-case error as the number of simulated paths increases, under a special condition. In our empirical experiments, the performance of the proposed method was clearly superior to MC. Directions for future research include the theoretical analysis of the method in more general settings, experiments with larger and more complicated models, and the analysis of a randomized version of the method to produce unbiased low-variance estimators.

References

[BEG89] P. P. Boyle, J. Evnine, and S. Gibbs. Numerical evaluation of multivariate contingent claims. The Review of Financial Studies, **2**, 241–250 (1989)

[CRR79] J.C. Cox, S.A. Ross, and M. Rubinstein. Option pricing : a simplified approach. Journal of Financial Economics, **7**, 229–263 (1979)

[ElH] R. El Haddad. Méthodes quasi-Monte Carlo pour la simulation des chaines de Markov. PhD thesis (in French), Université de Savoie, in preparation.

[HKY03] A. Hayfavi, H. Körezlioğlu, and K. Yildirak. Bivariate extension of Cox-Ross-Rubinstein model: model identification. 16th Annual Conference of Greek Statistical Institute. Kavala, Greece, 30 April – 3 May (2003)

[Joh87] H. Johnson. Options on the maximum or the minimum of several assets. Journal of Financial and Quantitative Analysis, **22**, 277–283 (1987)

[Lec96] C. Lécot. Error bounds for quasi-Monte Carlo integration with nets. Mathematics of Computation, **65**, 179–187 (1996)

[LC98] C. Lécot and I. Coulibaly. A quasi-Monte Carlo scheme using nets for a linear Boltzmann equation. SIAM Journal on Numerical Analysis, **35**, 51–70 (1998)

[LT04] C. Lécot and B. Tuffin. Quasi-Monte Carlo methods for estimating transient measures of discrete time Markov chains. In: Niederreiter, H. (ed.) Monte Carlo and Quasi Monte Carlo Methods 2002. Springer-Verlag, Berlin, 329–343 (2004)

[L'Ec99] P. L'Ecuyer. Good parameters and implementations for combined multiple recursive random number generators. Operations Research, **47**, 159–164 (1999)

[LLT07] P. L'Ecuyer, C. Lécot, and B. Tuffin. A randomized quasi-Monte Carlo simulation method for Markov chains. Operations Research, 2007, to appear.

[Nie87] H. Niederreiter. Point sets and sequences with small discrepancy. Monatshefte für Mathematik, **104**, 273–337 (1987)

[Nie92] H. Niederreiter. Random Number Generation and Quasi-Monte Carlo Methods. SIAM, Philadelphia, (1992)

[Stu82] R. M. Stulz. Options on the minimum or the maximum of two risky assets: analysis and applications. Journal of Financial Economics, **10**, 161–185, (1982)

[Zar68] S.K. Zaremba. Some applications of multidimensional integration by parts. Ann. Polon. Math, **21**, 85–96 (1968)

Computational Engine for a Virtual Tissue Simulator

Carole Hayakawa[1,2], Jerome Spanier[2], and Vasan Venugopalan[1,2,3]

[1] Dept. of Chemical Engineering and Materials Science, University of California, Irvine, CA
hayakawa@uci.edu
[2] Laser Microbeam and Medical Program, Beckman Laser Institute and Medical Clinic, University of California, Irvine, CA
[3] Dept. of Biomedical Engineering, University of California, Irvine, CA

We have developed a computational platform that simulates light transport in tissue in support of biomedical optics research. Although in its initial stage of development, this platform is being used to answer important questions regarding the detection of tissue changes, and the optimal design and positioning of optical probes to 'interrogate' the tissue best. We provide answers to such questions by applying perturbation and midway surface Monte Carlo techniques. Derivation of these methods makes rigorous use of the radiative transport equation which is essential if the methods are to provide accurate solutions for highly complex media such as biological tissue.

1 Introduction

Computational tools for modeling radiative transport in biological tissues have played a vital role in the development of optical techniques for the diagnosis and therapeutic treatment of tissues. These tools aid in the design of optical probes to detect noninvasively tissue transformations attributed to cancer and other abnormalities. To date, models of tissue have been confined to *very* simple geometries such as homogeneous and layered media. Recently, however, there is evidence that optical signals provided by multiply scattered light are sensitive to changes in tissue structure and composition on the mesoscopic (0.1–1 mm) spatial scales [FOV$^+$03, KWR$^+$03, MYLK05]. This realization has driven the need to (a) model tissue with greater spatial refinement, (b) understand the detectability of specific tissue changes, and (c) determine the tissue regions from which the detected light is remitted; i.e., the spatial and angular distribution of the light propagating from source to detector.

We are addressing these needs by developing a virtual tissue simulator (VTS). This computational platform allows the user to specify a probe configuration and define a voxelized tissue representation. A variety of probe configurations will eventually be incorporated. Here, we focus on one consisting of a fiber-optic source and detector at a fixed separation. The voxelized tissue representation can be provided from images provided by histology, CT or MRI. For a specified probe configuration and tissue definition, the user can run a conventional Monte Carlo simulation to model light transport through this system. The VTS also provides perturbation Monte Carlo capabilities that can be used to determine the change in the detected signal due to small changes in tissue structure and/or composition. Finally the VTS incorporates midway-surface Monte Carlo methods that couple forward and adjoint simulations at an intermediate surface to provide a spatial map of photon propagation from source to detector. Both perturbation and midway-surface Monte Carlo methods provide gains in efficiency and accuracy over conventional Monte Carlo methods.

The long-range goals for the VTS include the ability to solve inverse problems of two types. First, if a particular change in the detected optical signal is detected by a specific probe, we would like to determine the changes in the optical properties of the tissue that produced the optical signal change. We have already demonstrated the use of perturbation and differential Monte Carlo methods [Hay02, HS04, SYHV07] to solve this type of inverse problem and plan to extend this method to systems in which the tissue region is voxelized. Second, if we wish to target a specific tissue region, we would like to determine the probe design characteristics and probe placement that would optimally interrogate the targeted tissue region. Probe parameters such as the radius, numerical aperture or orientation of the source and detector fibers, source-detector (s-d) separation, and relationship of probe to target region can be varied to enhance detection of target tissue changes. Robust forward problem simulations are an essential part of accurate inverse problem solutions. With these ultimate goals in mind, we concentrate in this paper on the forward models for each of these inverse problems.

2 Monte Carlo Methods for Radiative Transport

The development of a Monte Carlo simulation of light transport in tissue follows from a probability model derived directly from the analytic radiative transport equation. This equation describes the physics of the problem and the probability model defines the probability space needed to solve the problem using Monte Carlo simulations. In the next sections, we present the linkages between the analytic and probabilistic formulations needed to establish the equivalence of these two formulations. The theoretical foundations presented in §2.1 support the Monte Carlo simulations that form the computational engine of the VTS.

2.1 Conventional Monte Carlo

The analytic model describing light transport through tissue is the radiative transport equation (RTE). In a closed, bounded subset $\mathbb{D}$ of $\mathbb{R}^3$, the integro-differential form of the RTE is

$$\nabla \cdot \mathbf{\Omega} \Phi(\mathbf{r}, \mathbf{\Omega}) + \mu_t(\mathbf{r})\Phi(\mathbf{r}, \mathbf{\Omega}) = \mu_s(\mathbf{r}) \int_{4\pi} p(\mathbf{\Omega}' \to \mathbf{\Omega})\Phi(\mathbf{r}, \mathbf{\Omega}')\, d\mathbf{\Omega}' + Q(\mathbf{r}, \mathbf{\Omega}) \tag{1}$$

where $\Phi(\mathbf{r}, \mathbf{\Omega})$ is the photon radiance, with $\mathbf{r}$ and $\mathbf{\Omega}$ representing position and unit direction vectors, respectively. $\mu_t(\mathbf{r}) = \mu_s(\mathbf{r}) + \mu_a(\mathbf{r})$ is the total attenuation coefficient, $\mu_s(\mathbf{r})$ is the scattering coefficient, $\mu_a(\mathbf{r})$ is the absorption coefficient, $p(\mathbf{\Omega}' \to \mathbf{\Omega})$ is the scattering phase function, and $Q(\mathbf{r}, \mathbf{\Omega})$ is the internal source function. A unique solution $\Phi(\mathbf{r}, \mathbf{\Omega})$ is assured for all $\mathbf{r} \in \mathbb{D}, \mathbf{\Omega} \in S^2$ by specifying appropriate conditions on the boundary $\partial\mathbb{D}$.

Detectors are often placed within the tissue system to measure a system response. For example, this system response could consist of reflected and/or transmitted light using of one or more detectors. The quantity describing the system response I in the context of the analytical model is

$$I = \int_{\Gamma} h(\mathbf{r}, \mathbf{\Omega})\Phi(\mathbf{r}, \mathbf{\Omega}) d\mathbf{r} d\mathbf{\Omega} \tag{2}$$

where h is a known 'detector' function, Φ is the solution to Eq. (1) and Γ is the phase space. Suppose, for example, that the amount of energy absorbed within a detector occupying a subregion V of phase space is of interest. In this case, the detector function is defined by $h(\mathbf{r}, \mathbf{\Omega}) = \frac{\mu_a(\mathbf{r})}{\mu_t(\mathbf{r})} \chi_V(\mathbf{r}, \mathbf{\Omega})$ where

$$\chi_V(\mathbf{r}, \mathbf{\Omega}) = \begin{cases} 1 \text{ for } (\mathbf{r}, \mathbf{\Omega}) \in V \\ 0 \text{ for } (\mathbf{r}, \mathbf{\Omega}) \notin V \end{cases} \tag{3}$$

and the integral in Eq. (2) measures the amount of photon absorption in the volume V. With this definition of h, I can be estimated as the ratio of the total number of photons absorbed in V and the total launched from the source. To represent a detector on the surface of the tissue measuring reflectance/transmittance, the same formalism contained in Eqs. (2) and (3) can be used. In this case, the volume V is treated as an infinite absorber outside of the tissue whose intersection with Γ is the surface of the detector.

The Monte Carlo solution of Eq. (1) is captured by a random variable ξ defined on the sample space of all random walks Ω whose expectation is

$$E[\xi] = \int_{\Omega} \xi \, d\mu = I \tag{4}$$

where μ is the analog measure that captures the physical model faithfully and I is given by Eq. (2). Details regarding such constructions may be found in [SG69].

2.2 Perturbation Monte Carlo

Once a tissue system of interest is defined within VTS and a Monte Carlo simulation is executed to estimate a desired system response I, the impact on I of changes in the tissue structure and/or composition can quickly be determined using perturbation Monte Carlo (pMC). We briefly review the pMC method next; details can be found in [Hay02, HS04, SYHV07, HSB$^+$01].

Perturbation Monte Carlo provides radiative transport solutions for multiple systems that can be expressed as a perturbation of a baseline system using a single set of random walks. A set of photon biographies is generated within a baseline tissue system of specified tissue properties. Using pMC, the change to the system response $\hat{I} = I + \Delta I$ due to perturbations in the optical properties of the baseline system can be determined. This is done by appropriately modifying the random variable used to estimate the reflected light in the baseline Monte Carlo simulation. The pMC method provides estimates of the responses in a 'perturbed' system with a computational cost that is orders of magnitude smaller than that required to run independent Monte Carlo simulations. In addition, the positive correlation between the baseline and perturbed system responses enables pMC to capture small changes ΔI in the system response with a much higher precision than would be obtained with independent simulations. The use of pMC can enable VTS users to determine rapidly the degree to which a diagnostic or therapeutic measurement will be sensitive to changes in tissue properties.

The pMC method can be described within a probability model in terms of the pair of random variables ξ, $\hat{\xi}$ where ξ is the system response of the baseline tissue system and $\hat{\xi}$ is the response in the perturbed tissue system. The derivations necessary to make this formulation correct are based on the identity

$$\int_P \hat{\xi}\, d\mu = \int_P \xi\, d\hat{\mu} \tag{5}$$

where

$$\hat{\xi} = \xi \frac{d\hat{\mu}}{d\mu} \tag{6}$$

and μ is the analog probability measure based on the baseline optical properties. The measure $\hat{\mu}$ incorporates the analog measure except in the perturbed region, where it uses the optical properties assumed for that region. The Radon-Nikodym derivative $(d\hat{\mu}/d\mu)$ expresses how the analog random variable ξ must be modified to produce an unbiased estimator $\hat{\xi}$ of the optical response in the perturbed system. Explicit formulas derived from Eq. (6) may be found in [Hay02, HSB$^+$01].

We apply the pMC method to study dysplasia within epithelial tissue. In the initial stages of epithelial tissue dysplasia, cells within the epithelium adjacent to the basal lamina exhibit a larger nucleus to cytoplasm ratio (see Fig. 1) in which the scattering of light is increased by a factor of three [CAM$^+$03, CFMRK05]. Our interest is in determining how the placement of the probe on the tissue surface relative to these dysplastic regions effects our ability to detect these changes. The baseline tissue is modelled as an epithelial region atop a stromal region with an undulating basal lamina interface. The entire tissue is divided into uniform voxels measuring $100\,\mu m \times 100\,\mu m$ in the x-z plane. The voxelized approximation of the undulating interface is shown with white line segments. The optical properties of the two tissue regions are typical of normal epithelial/stromal tissue [CAD$^+$04]: for the epithelial region, $\mu_a = 0.12/\text{mm}$, $\mu_s = 2.8/\text{mm}$ with g, the average cosine of the scattering phase function, set to 0.97 and n, the refractive index equal to 1.4. The stromal region optical properties are: $\mu_a = 0.09/\text{mm}$, $\mu_s = 17.5/\text{mm}$, $g = 0.8$ and $n = 1.4$. The probe configuration consists of a source and detector each with $200\mu m$ radius and 0.37 numerical aperture positioned 1 mm apart on the tissue surface. To model the initial stages of dysplastic transformation, one voxel within the epithelial region and positioned atop the two-region interface at $x = 0.45$ mm (outlined in black) is considered to be 'dysplastic' and assigned a scattering coefficient that exceeds that of the baseline tissue by a factor of 3. The probe is positioned so that the voxel with respect to the x-axis is approximately midway between source and detector: the source is centered at $x = 0$ mm and the detector at $x = 1$ mm. This placement is chosen because conventional wisdom suggests that the probe source and detector should straddle the target region.

Fig. 2 displays the detected reflected signal as a function of time, $R(t)$, for both the baseline tissue system and for this system with the dysplastic voxel at $x = 0.45$ mm atop the two region interface (as shown in Fig. 1). The two plots are visually indistinguishable. This leads us to the conclusion that this particular placement of the probe relative to the voxel exhibiting dysplasia is insensitive to this pre-cancerous tissue transformation.

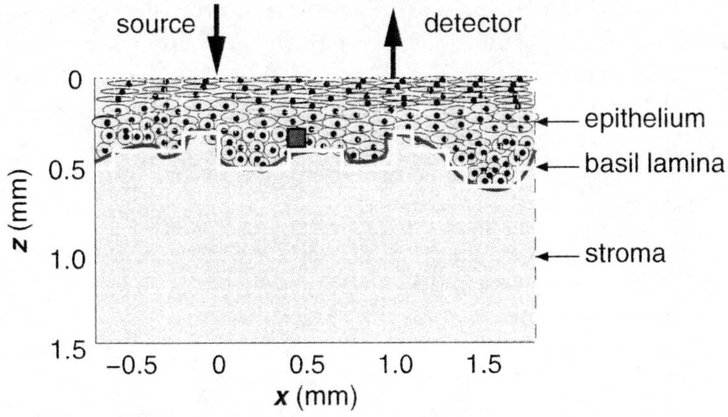

Fig. 1. Schematic of the tissue definition with upper epithelial and lower stromal regions separated by an undulating basal lamina interface. The white superimposed grid identifies the $100\,\mu m \times 100\,\mu m$ voxelization. The solid black box atop the interface designates a dysplastic voxel in which the scattering is increased three-fold.

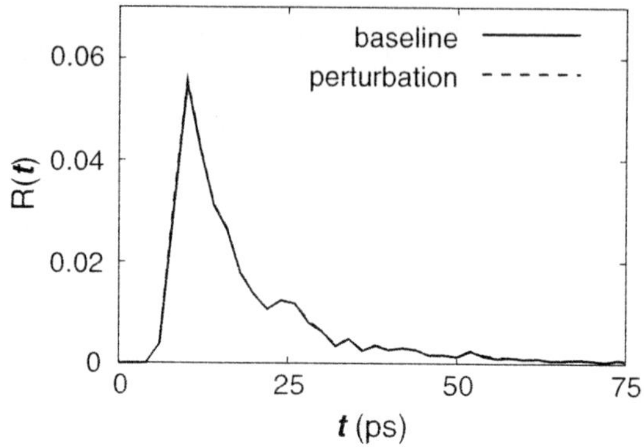

Fig. 2. Time-resolved reflectance for the baseline tissue system and for the perturbed system with a dysplastic voxel at lateral position $x = 0.45\,\text{mm}$ atop the two region interface with the probe source and detector centered at $x = 0\,\text{mm}$ and $x = 1\,\text{mm}$, respectively.

The absence of a significant change in the detected optical signal due to the appearance of the dysplastic voxel can be better understood if one has knowledge of how the light propagates from the source, through the tissue, to the detector. To date, available techniques to provide such information have been based mainly on the diffusion approximation to the radiative transport equation [BOCY97, FZC95, PAEWO95, SHL93]. However, the validity of diffusion-based models is compromised when: (a) s-d separations are small or (b) the tissue absorption is comparable to or greater than scattering. In addition, these models cannot, of course, provide transport-theoretic *quantitative* information.

We have recently developed a novel Monte Carlo method that couples forward and adjoint simulations to generate the spatial distribution of the migration of light from source to detector [HSVed]. This map provides valuable information regarding the relationship between a specific probe configuration and placement and the resulting tissue region that can be interrogated. These interrogation density maps are faithful to the radiative transport equation and therefore provide quantitative measures of the contribution of different tissue regions to the optical signal.

2.3 Midway Surface Monte Carlo

The magnitude and spatial extent to which an optical probe interrogates the tissue system under examination is of great interest when designing a diagnostic technique. For example, in the case of epithelial dysplasia considered above, we would like to determine the probe position on the tissue surface that would best interrogate the dysplastic voxel. A map that depicts the migration of the light from the source to the detector would aid in the probe placement.[1] To generate such a map, each voxel within the tissue representation is treated as a 'target' subvolume. Conventional Monte Carlo methods could be used to select those photon trajectories that have migrated from source to target region to detector either in a forward or adjoint simulation. However, when the source and detector are each small relative to the target subvolume, forward or adjoint simulations used alone engender low statistical signal-to-noise ratios. Such a situation is exceedingly common in biomedical optics.

Our approach to bypass this dilemma, is to break this problem into two components each of which can be determined rapidly through a Monte Carlo simulation. First we determine **(a)** the probability of photon visitation from the photon source to the target subvolume, $P(V)$ ('target visitation'), and then **(b)** the probability of photon detection conditioned by target visitation, $P(D|V)$, ('detection given target visitation'). These two probabilities can be combined using Bayes' Theorem to provide the joint transport probability of 'target visitation and detection':

$$P(V \cap D) = P(V) \cdot P(D|V). \tag{7}$$

We use a conventional Monte Carlo simulation to determine $P(V)$ for ***every*** voxel in the VTS representation of the tissue. This not only provides an estimate of the $P(V)$ term in Eq. (7), but also a spatially-resolved map of the absorbed or scattered light distribution within the tissue. To determine the second factor $P(D|V)$, we utilize an adjoint simulation to increase efficiency. This is done by modifying a generalized reciprocity principle that converts $P(D|V)$ to a coupled forward-adjoint computation at the 'midway' surface of the target subvolume.

'Midway' forward-adjoint coupling methods [Cra96, SJH98, UHK01, Wil91] have been successfully applied to increase efficiency in estimating detector responses. In essence, a midway surface between the source and detector is defined such that all detected radiation ***must*** pass through this surface. The coupling of a forward and adjoint simulation at this intermediate surface determines the detected response more efficiently, particularly in problems that involve deep penetration. The midway method is made rigorous by utilizing generalized reciprocity theory for transport equations.

We first present the analytical model describing generalized reciprocity and then show how it is modified to allow evaluation of a conditional probability to provide the desired interrogation maps. The intermediate derivations are presented to clarify the final application.

Generalized Reciprocity

Generalized reciprocity establishes the equivalence between the execution of a 'forward' Monte Carlo simulation from a source to detector and an adjoint

[1] Note that a solution of the radiative transport equation alone does not provide such a map because it omits any description of a detector.

simulation for 'backward-propagating' photons from the detector (adjoint source) to the source (adjoint detector). This result can be appreciated by first considering the equation adjoint to Eq. (1):

$$\begin{aligned} &- \nabla \cdot \mathbf{\Omega} \Phi^*(\mathbf{r}, \mathbf{\Omega}) + \mu_t(\mathbf{r}) \Phi^*(\mathbf{r}, \mathbf{\Omega}) \\ &= \mu_s(\mathbf{r}) \int_{4\pi} p(\mathbf{\Omega} \to \mathbf{\Omega}') \Phi^*(\mathbf{r}, \mathbf{\Omega}')\, d\mathbf{\Omega}' + Q^*(\mathbf{r}, \mathbf{\Omega}) \end{aligned} \tag{8}$$

where Φ^* is the adjoint photon radiance and Q^* is any adjoint source function. Let $\mathbb{V}_M$ be an arbitrary closed, bounded subset of $\mathbb{D}$ and $\partial \mathbb{V}_M$ its surface. If we multiply the radiative transport equation [Eq. (1)] by Φ^*, Eq. (8) by Φ, subtract the latter product from the former and integrate the difference over all locations and directions within V_M, we get

$$\int_{\mathbb{V}_M \times S^2} \nabla \cdot \mathbf{\Omega} \Phi \Phi^* = \int_{\mathbb{V}_M \times S^2} [Q \Phi^* - Q^* \Phi] \tag{9}$$

where the variables of integration are suppressed but are understood to be $(\mathbf{r}, \mathbf{\Omega})$ with the spatial vector $\mathbf{r}$ ranging over the volume $\mathbb{V}_M$. Using Green's theorem to replace the volume integral on the left side of Eq. (9) by a surface integral leads to:

$$\int_{\partial \mathbb{V}_M \times S^2} \mathbf{n}_M \cdot \mathbf{\Omega} \Phi \Phi^* = \int_{\mathbb{V}_M \times S^2} [Q \Phi^* - Q^* \Phi] \tag{10}$$

where $\mathbf{n}_M$ is the outward-pointing unit vector normal to $\partial \mathbb{V}_M$. Eq. (10) is often referred to as the global reciprocity theorem [WE77]. Note that if $\mathbb{V}_M = \mathbb{D}$ and the boundary conditions at the air-tissue interface cause the integral on the left hand side to vanish, we then arrive at the 'classical' statement of reciprocity:

$$\int_{\mathbb{V}_M \times S^2} [Q \Phi^* - Q^* \Phi] = 0. \tag{11}$$

Eq. (11) states that one can obtain the same transport estimates by performing either a forward simulation from the source Q to detector Q^* or by performing an adjoint simulation from adjoint source Q^* to adjoint detector Q.

While Eq. (10) is valid generally, it becomes particularly useful when $\mathbb{V}_M$ encloses either the source or the detector region. The surface of $\mathbb{V}_M$, $\partial \mathbb{V}_M$, can then be identified as a 'midway' surface between source and detector: every photon that is detected from the source ***must*** intersect the midway surface. The function $\Phi \Phi^*$ that occurs in Eq. (10) has been called a 'contributon' response function [Cra96, SJH98, UHK01, Wil91, WE77, SJH99, UH01] and can be used to define a function that characterizes transport from source to detector. If $\mathbb{V}_M$ encloses the source region, and $Q^* = 0$ in $\mathbb{V}_M$, the left hand side of Eq. (10) is positive, and equals $\int_{\mathbb{V}_M \times S^2} Q \Phi^*$, which is the adjoint representation of the detector response. If $\mathbb{V}_M$ encloses the detector region, and $Q = 0$ in $\mathbb{V}_M$, the left hand side of Eq. (10) is negative and equals $-\int_{\mathbb{V}_M \times S^2} Q^* \Phi$, which is the forward representation of the detector response.

$P(V \cap D)$ Maps

We extend this generalized reciprocity theory to determine the conditional probability $P(D|V)$ for an arbitrary target subvolume $\mathbb{V}$ enclosing neither the source nor the detector. Since we are interested not in the interrogation of surfaces but rather of subvolumes within a tissue domain, we slightly modify the midway method to facilitate the estimation of the conditional probability $P(D|V)$. We launch photons at a physical source Q that propagate until they exit the phase-space. Only photon trajectories that have intersected the target subvolume $\mathbb{V}$ contribute to the estimate of $P(V)$. These 'visiting' photons generate an induced source internal to $\mathbb{V}$ that produces a surface source $Q_{\partial\mathbb{V}}$ on $\partial\mathbb{V}$. This surface source is then paired with the adjoint radiance on $\partial\mathbb{V}$ in a bilinear integration that produces an estimate of $P(D|V)$. The product of the two probabilities $P(V)$ and $P(D|V)$ defines the probability that subvolumes within the phase-space are both visited and detected. We use this product to provide the key quantitative information used to assess and compare the characteristics of potential probe designs.

Let $Q_{\mathbb{V}}$ denote the source induced in $\mathbb{V}$ by photons launched according to the original optical source function $Q(\mathbf{r}, \mathbf{\Omega})$. This induced source internal to $\mathbb{V}$, $Q_{\mathbb{V}}$, generates a source on $\partial\mathbb{V}$. If we merely replace the source function Q by the source function $Q_{\mathbb{V}}(\mathbf{r}, \mathbf{\Omega})$ and repeat the derivation that led to Eq. (10), we obtain

$$\int_{\partial\mathbb{V}\times S^2} \mathbf{n}_{\mathbb{V}} \cdot \mathbf{\Omega}\tilde{\Phi}\Phi^* = \int_{\mathbb{V}\times S^2} \left[Q_{\mathbb{V}}\Phi^* - Q^*\tilde{\Phi}\right], \tag{12}$$

where the radiance $\tilde{\Phi}$ is the solution of the RTE [Eq. (1)] with the ***induced*** source function $Q_{\mathbb{V}}$. Recall, $Q^*(\mathbf{r}, \mathbf{\Omega})$ is a detector function; as such $Q^*(\mathbf{r}, \mathbf{\Omega}) = 0$ unless $\mathbf{r}$ is on the boundary of the tissue and $\mathbf{\Omega}$ points ***outward***. Replacing $Q^*(\mathbf{r}, \mathbf{\Omega})$ by $Q^*(\mathbf{r}, -\mathbf{\Omega})$, therefore, defines an adjoint source pointing ***into*** the tissue. This in turn generates an adjoint radiance, $\Phi^*(\mathbf{r}, -\mathbf{\Omega})$ ***inside*** the tissue. This reverses the direction in the arguments of Q^* and $\tilde{\Phi}^*$ in Eq. (12), which then reads

$$\begin{aligned}\int_{\partial\mathbb{V}\times S^2} \mathbf{n}_{\mathbb{V}} \cdot \mathbf{\Omega}\tilde{\Phi}(\mathbf{r}, \mathbf{\Omega})\Phi^*(\mathbf{r}, -\mathbf{\Omega}) &= \textstyle\int_{\mathbb{V}\times S^2} \big[Q_{\mathbb{V}}(\mathbf{r}, \mathbf{\Omega})\Phi^*(\mathbf{r}, -\mathbf{\Omega}) - \\ &\qquad Q^*(\mathbf{r}, -\mathbf{\Omega})\tilde{\Phi}(\mathbf{r}, \mathbf{\Omega})\big] \\ &= \textstyle\int_{\mathbb{V}\times S^2} Q_{\mathbb{V}}(\mathbf{r}, \mathbf{\Omega})\Phi^*(\mathbf{r}, -\mathbf{\Omega})\end{aligned} \tag{13}$$

since $Q^* = 0$ inside $\mathbb{V}$. The estimation of the right hand side of Eq. (13) is performed using an adjoint simulation and provides the detected response due to the induced source $Q_{\mathbb{V}}$, or $P(D|V)$.

The forward simulation of photons exiting an arbitrary target subvolume $\mathbb{V}$ is used to determine $P(V)$ and is matched at $\partial\mathbb{V}$ with the adjoint simulation estimate of $P(D|V)$. The joint probability of visitation and detection $P(V \cap D)$ [Eq. (7)] is formed by the product of these two factors.

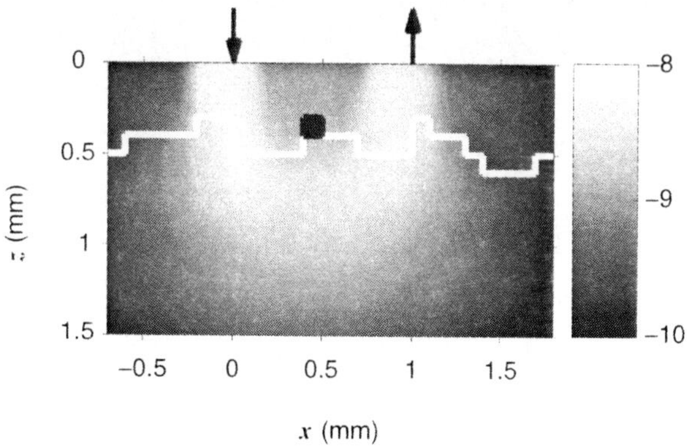

Fig. 3. Interrogation density $P(V \cap D)$ map of the baseline tissue system with a log-scale gray scale bar. The interface between the epithelial and stromal layers is shown using white line segments. A dysplastic voxel atop the interface at $x = 0.45$ mm with scattering increased threefold is shown in solid black.

Fig. 3 displays the resulting $P(V \cap D)$ map using our baseline epithelium/stroma tissue definition. Each voxel (shown by the superimposed grid in Fig 1) is treated as a target subvolume and the joint probability $P(V \cap D)$ is determined. Notice that the detected light field does not experience significant lateral dispersion within the epithelial region during its propagation from source to detector. This is due presumably to the high scattering asymmetry coefficient g and low scattering coefficient μ_s in that region whose combined effect is to produce minimal lateral dispersion. This behavior below the source and detector within the epithelial region suggests that perturbations within this region, in particular the voxels around positions $x = 0$ and $x = 1$, would have the greatest effect on the detected signal. On the other hand, the voxels midway between source and detector above the interface are fairly dark indicative of positions of small perturbative effect. This explains why the dysplastic voxel positioned at $x = 0.45$ mm did *not* noticably perturb the detected optical signal.

Based on this analysis, we reposition the probe to place the detector directly above the dysplastic voxel: the source is now centered at $x = -0.5$ mm and the detector at $x = 0.5$ mm, while the dysplastic voxel remains atop the two region interface at $x = 0.45$ mm. Fig. 4 displays $R(t)$ for the baseline tissue system with and without the dysplastic voxel. With the new probe position, the change in the detected signal due to the increased scattering in the dysplastic voxel can now be seen. Notice also the strong correlation between the baseline and perturbed plots. The small variations of the perturbed plot from the baseline capture the effect of the change with much more accuracy than if independent Monte Carlo simulations were used for the baseline and perturbed systems.

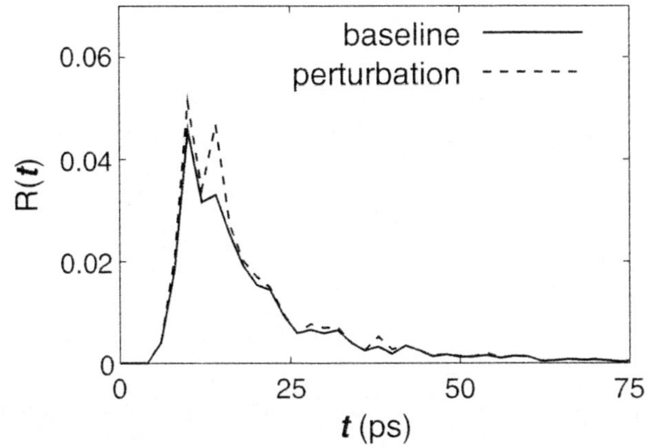

Fig. 4. Time-resolved reflectance for the baseline tissue system and for the perturbed system with a dysplastic voxel at lateral position $x = 0.45\,\mathrm{mm}$ atop the two region interface with the probe source and detector centered at $x = -0.5\,\mathrm{mm}$ and $x = 0.5\,\mathrm{mm}$, respectively.

3 Summary

We have completed the initial development of the computational engine of a virtual tissue simulator that incorporates efficient forward and adjoint Monte Carlo simulations in a voxelized representation of tissue. This new platform provides the biomedical optics researcher with a means to analyze better the size and location of detectable tissue anomalies and to design and position optical probes to capture these changes effectively. The engine is currently comprised of conventional, perturbation, and midway surface Monte Carlo techniques. In contrast to many other available methods, ours are radiative transport equation rigorous and therefore provide more accurate models that are essential for representing complex media such as biological tissue.

A robust forward model representation is a vital component of an inverse problem-solving capability. By adding differential Monte Carlo to the VTS, we plan next to extend the VTS platform to solve inverse problems. The first step in this direction will enable quantitative determination of optical properties in target regions based on measurements from a given probe design. For example, given baseline and perturbed measurements as shown in Fig. 4, we would like to determine what change in the absorption or scattering properties of the dysplastic voxel caused the perturbed measurement. Second, probe design parameters will be determined that best interrogate specific target regions within the tissue. Generating the change to the $P(V \cap D)$ maps as a function of probe design parameters will enable inverse solutions that identify the best probe configurations to target selected regions within the tissue.

We are grateful to the National Institutes of Health for support under Grants #P41-RR-00192 (Laser Microbeam and Medical Program) and #R01-EB-00345.

References

[BOCY97] D. A. Boas, M. A. O'Leary, B. Chance, and A. G. Yodh. Detection and characterization of optical inhomogeneities with diffuse photon density waves: a signal-to-noise analysis. *Appl. Opt.*, 36(1):75–92, 1997.

[CAD+04] S. K. Chang, D. Arifler, R. Drezek, M. Follen, and R. Richards-Kortum. Analytical model to describe fluorescence spectra or normal and preneoplastic epithelial tissue: comparison with Monte Carlo simulations and clinical measurements. *J. Biomed. Opt.*, 9(3):511–522, 2004.

[CAM+03] T. Collier, D. Arifler, A. Malpica, M. Follen, and R. Richards-Kortum. Determination of epithelial tissue scattering coefficient using confocal microscopy. *IEEE J. Sel. Top. Quantum Electron.*, 9(1):307–313, 2003.

[CFMRK05] T. Collier, M. Follen, A. Malpica, and R. Richards-Kortum. Sources of scattering in cervical tissue: determination of the scattering coefficient by confocal microscopy. *Appl. Opt.*, 44(11):2071–2081, 2005.

[Cra96] S. N. Cramer. Forward-adjoint Monte Carlo coupling with no statistical error propagation. *Nucl. Sci. Eng.*, 124:398–416, 1996.

[FOV+03] H. Fang, M. Ollero, E. Vitkin, L. M. Kimerer, P. B. Cipolloni, M. M. Zaman, S. D. Freedman, I. J. Bigio, I. Itzkan, E. B. Hanlon, and L. T. Perelman. Noninvasive sizing of subcellular organelles with light scattering spectroscopy. *IEEE J. Sel. Topics Quantum Elec.*, 9(2):267–276, 2003.

[FZC95] S. Feng, F. Zeng, and B. Chance. Photon migration in the presence of a single defect: a perturbation analysis. *Appl. Opt.*, 34(19):3826–3837, 1995.

[Hay02] C. K. Hayakawa. *Perturbation Monte Carlo Methods for the Solution of Inverse Problems*. PhD thesis, Claremont Graduate University, 2002.

[HS04] C. K. Hayakawa and J. Spanier. Perturbation Monte Carlo methods for the solution of inverse problems. In *Monte Carlo and Quasi-Monte Carlo Methods 2002*, pages 227–241. Springer, 2004.

[HSB+01] C. K. Hayakawa, J. Spanier, F. Bevilacqua, A. K. Dunn, J. S. You, B. J. Tromberg, and V. Venugopalan. Perturbation Monte Carlo methods to solve inverse photon migration problems in heterogeneous tissues. *Opt. Lett.*, 26(17):1335–1337, 2001.

[HSVed] C. K. Hayakawa, J. Spanier, and V. Venugopalan. Coupled forward-adjoint monte carlo simulations of radiative transport for the study of optical probe design in heterogeneous tissues. *SIAM J. on Appl. Math.*, accepted.

[KWR+03] Y. L. Kim, R. K. Wali, H. K. Roy, M. J. Goldberg, A. K. Kromin, K. Chen, and V. Backman. Simultaneous measurement of angular and spectral properties of light scattering for characterization of tissue microarchitecture and its alteration in early precancer. *IEEE J. Sel. Topics Quantum Elec.*, 9(2):243–256, 2003.

[MYLK05] C. J. Mann, L. Yu, C.-M. Lo, and M. K. Kim. High-resolution quantitative phase-constrast microscopy by digital holography. *Opt. Exp.*, 13(22):8693–8698, 2005.

[PAEWO95] M. S. Patterson, S. Andersson-Engels, B. C. Wilson, and E. K. Osei. Absorption spectroscopy in tissue-simulating materials: a theoretical and experimental study of photon paths. *Appl. Opt.*, 34(1):22–30, 1995.

[SG69] J. Spanier and E. Gelbard. *Monte Carlo Principles and Neutron Transport Problems*. Addison-Wesley, 1969.

[SHL93] J. C. Schotland, J. C. Haselgrove, and J. S. Leigh. Photon hitting density. *Appl. Opt.*, 32(4):448–453, 1993.

[SJH98] I. V. Serov, T. M. John, and J. E. Hoogenboom. A new effective Monte Carlo midway coupling method in MCNP applied to a well logging problem. *Appl. Radiat. Isot.*, 49(12):1737–1744, 1998.

[SJH99] I. V. Serov, T. M. John, and J. E. Hoogenboom. A midway forward-adjoint coupling method for neutron and photon Monte Carlo transport. *Nucl. Sci. Eng.*, 133:55–72, 1999.

[SYHV07] I. Seo, J. S. You, C. K. Hayakawa, and V. Venugopalan. Perturbation and differential Monte Carlo methods for measurement of optical properties in a layered epithelial tissue model. *J. Biomed. Opt.*, 12(1):014030, 2007.

[UH01] T. Ueki and J. E. Hoogenboom. Exact Monte Carlo perturbation analysis by forward-adjoint coupling in radiation transport calculations. *J. Comput. Phys.*, 171:509–533, 2001.

[UHK01] T. Ueki, J. E. Hoogenboom, and J. L. Kloosterman. Analysis of correlated coupling of Monte Carlo forward and adjoint histories. *Nucl. Sci. Eng.*, 137:117–145, 2001.

[WE77] M. L. Williams and W. W. Engle. The concept of spatial channel theory applied to reactor shielding analysis. *Nucl. Sci. Eng.*, 62:92–104, 1977.

[Wil91] M. L. Williams. Generalized contributon response theory. *Nucl. Sci. Eng.*, 108:355–383, 1991.

Randomized Approximation of Sobolev Embeddings

Stefan Heinrich

Department of Computer Science, University of Kaiserslautern, D-67653 Kaiserslautern, Germany
heinrich@informatik.uni-kl.de

Summary. We study approximation of functions belonging to Sobolev spaces $W_p^r(Q)$ by randomized algorithms based on function values. Here $1 \le p \le \infty$, $Q = [0,1]^d$, and $r, d \in \mathbf{N}$. The error is measured in $L_q(Q)$, with $1 \le q < \infty$, and we assume $r/d > 1/p - 1/q$, guaranteeing that $W_p^r(Q)$ is embedded into $L_q(Q)$. The optimal order of convergence for the case that $W_p^r(Q)$ is embedded even into $C(Q)$ is well-known. It is $n^{-r/d+\max(1/p-1/q,0)}$ (n the number of function evaluations). This rate is already reached by deterministic algorithms, and randomization gives no speedup.

In this paper we are concerned with the case that $W_p^r(Q)$ is not embedded into $C(Q)$ (but, of course, still into $L_q(Q)$). For this situation approximation based on function values was not studied before. We prove that for randomized algorithms the above rate also holds, while for deterministic algorithms no rate whatsoever is possible. Thus, in the case of low smoothness, Monte Carlo approximation algorithms reach a considerable speedup over deterministic ones (up to $n^{-1+\varepsilon}$ for any $\varepsilon > 0$).

We also give some applications to integration of functions and to approximation of solutions of elliptic PDE.

1 Introduction

Denote $\mathbf{N} = \{1, 2, \dots\}$, $\mathbf{N}_0 = \{0, 1, 2, \dots\}$, let $d \in \mathbf{N}$, and let $Q = [0,1]^d$ be the d-dimensional unit cube. For $1 \le p \le \infty$, let $L_p(Q)$ be the space of real-valued p-integrable functions, endowed with the norm

$$\|f\|_{L_p(Q)} = \left(\int_Q |f(x)|^p dx \right)^{1/p}$$

if $p < \infty$, and

$$\|f\|_{L_\infty(Q)} = \operatorname{ess\,sup}_{x \in Q} |f(x)|.$$

For $r \in \mathbf{N}$ the Sobolev space $W_p^r(Q)$ consists of all functions $f \in L_p(Q)$ such that for all $\alpha = (\alpha_1, \dots, \alpha_d) \in \mathbf{N}_0^d$ with $|\alpha| := \sum_{j=1}^d \alpha_j \le r$, the generalized partial derivative $D^\alpha f$ belongs to $L_p(Q)$. The norm on $W_p^r(Q)$ is defined as

$$\|f\|_{W_p^r(Q)} = \left(\sum_{|\alpha| \le r} \|D^\alpha f\|_{L_p(Q)}^p \right)^{1/p}$$

if $p < \infty$, and

$$\|f\|_{W_\infty^r(Q)} = \max_{|\alpha| \le r} \|D^\alpha f\|_{L_\infty(Q)}.$$

Let $C(Q)$ denote the space of continuous functions on Q, equipped with the supremum norm. Let $1 \le q < \infty$. By the Sobolev embedding theorem (see [Ada75], [Tri78]), for

$$r/d > 1/p - 1/q \tag{1}$$

$W_p^r(Q) \subset L_q(Q)$, and there is a constant $c > 0$ such that for each $f \in W_p^r(Q)$

$$\|f\|_{L_q(Q)} \le c\|f\|_{W_p^r(Q)}. \tag{2}$$

Consequently, the embedding operator $J_{pq} : W_p^r(Q) \to L_q(Q)$ defined by $J_{pq} f = f$ is bounded. We shall study optimal approximation of J_{pq} by randomized algorithms which use n function values.

For $n \in \mathbf{N}$ we consider the class $\mathcal{A}_n^{\text{ran}}$ of randomized algorithms which are of the form $A = (A_\omega)_{\omega \in \Omega}$, where

$$A_\omega(f) = \varphi_\omega(f(x_{1,\omega}), \dots, f(x_{n,\omega})), \tag{3}$$

$(\Omega, \Sigma, \mathbf{P})$ is a probability space, for each $\omega \in \Omega$, $x_{i,\omega}$ is an element of Q and φ_ω is a mapping from $\mathbf{R}^n$ to $L_q(Q)$, with the property that for each $f \in W_p^r(Q)$, the mapping

$$\omega \in \Omega \to A_\omega(f) = \varphi_\omega(f(x_{1,\omega}), \dots, f(x_{n,\omega}))$$

is a random variable with values in $L_q(Q)$ (that is, Σ-to-Borel measurable). Since elements of $W_p^r(Q)$ are equivalence classes of functions, relation (3) needs more explanation when $W_p^r(Q)$ is not embedded into $C(Q)$, and hence function values are, in general, not defined for such classes. We impose a further condition on the elements of $\mathcal{A}_n^{\text{ran}}$, let us call it consistency of the algorithm. We assume that whenever f_1 and f_2 are representatives of the same class $f \in W_p^r(Q)$, then

$$\varphi_\omega(f_1(x_{1,\omega}), \dots, f_1(x_{n,\omega})) = \varphi_\omega(f_2(x_{1,\omega}), \dots, f_2(x_{n,\omega})) \quad \mathbf{P}-\text{a.s.}$$

This means that $A_\omega(f_1)$ and $A_\omega(f_2)$ coincide almost surely, and in this sense we can take (3) as the definition of the random variable $A_\omega(f)$. A sufficient condition for consistency is obviously the following: For all i, the mapping

$\omega \to x_{i,\omega}$ is Lebesgue measurable and for each subset $Q_0 \subset Q$ of Lebesgue measure zero we have

$$\mathbf{P}\{\omega \in \Omega : x_{i,\omega} \in Q_0\} = 0,$$

or equivalently, the distribution of the $x_{i,\omega}$ is absolutely continuous with respect to the Lebesgue measure.

The class $\mathcal{A}_n^{\mathrm{ran}}$ contains the subclass of randomized linear algorithms – they are of the form above with linear φ_ω, thus

$$A_\omega(f) = \sum_{i=1}^{n} f(x_{i,\omega})\psi_{i,\omega}, \tag{4}$$

for certain $\psi_{i,\omega} \in L_q(Q)$.

Given any $1 \le s < \infty$, the error of an algorithm $A \in \mathcal{A}_n^{\mathrm{ran}}$ is defined as

$$e^{(s)}(J_{pq}, A, B_{W_p^r(Q)}) = \sup_{f \in B_{W_p^r(Q)}} \left(\mathbf{E}\, \|f - A_\omega(f)\|_{L_q(Q)}^s\right)^{1/s}, \tag{5}$$

where $\mathbf{E}$ is the expectation with respect to $\mathbf{P}$ and $B_{W_p^r(Q)}$ denotes the unit ball of $W_p^r(Q)$. The randomized n-th minimal error is defined as

$$e_n^{\mathrm{ran}}(J_{pq}, B_{W_p^r(Q)}) = \inf_{A \in \mathcal{A}_n} e^{(1)}(J_{pq}, A, B_{W_p^r(Q)}) \,.$$

Hence, no randomized algorithm that uses at most n function values can provide a smaller error than $e_n^{\mathrm{ran}}(J_{pq}, B_{W_p^r(Q)})$. We have chosen the $s = 1$ case for the minimal error, which is convenient for the sequel. Statements for other s can be read from the proofs below.

In terms of information-based complexity theory [TWW88], [Nov88], we consider randomized nonadaptive algorithms using standard information. In terms of [NT06], the e_n^{ran} can be viewed as randomized sampling numbers.

In the case that $W_p^r(Q)$ is embedded into $C(Q)$, the order of e_n^{ran} is well-known:

$$e_n^{\mathrm{ran}}(J_{pq}, B_{W_p^r(Q)}) \asymp n^{-r/d+\max(1/p-1/q,0)}, \tag{6}$$

where we used the following notation: for sequences (a_n) and (b_n) of nonnegative reals we write $a_n \asymp b_n$ if there are constants $c_1, c_2 > 0$ and an $n_0 \in \mathbf{N}$ such that $c_1 a_n \le b_n \le c_2 a_n$ for all $n \ge n_0$. Furthermore note that we often use the same symbols $c, c_1, \ldots$ for possibly different constants, also in sequences of relations.

The upper bound of (6) can be reached by deterministic methods, for example by piecewise polynomial approximation (see, e.g. [Cia78]). The lower bounds were shown in [Was89] ($p = q = \infty$), [Nov88] ($1 \le p \le \infty$, $q = \infty$), and [Mat91] ($1 \le p, q \le \infty$). Recall from [Ada75] that $W_p^r(Q)$ is embedded into $C(Q)$ iff

$$\left.\begin{array}{ll} p = 1 & \text{and} \quad r/d \ge 1 \\ \text{or} & \\ 1 < p < \infty & \text{and} \quad r/d > 1/p \\ \text{or} & \\ p = \infty. & \end{array}\right\} \qquad (7)$$

In this paper we are concerned with the case that (7) does *not* hold. Together with (1), this means we assume

$$\left.\begin{array}{lll} & 1 \le p, q < \infty & \\ \text{and} & & \\ & 1 - 1/q < r/d < 1 & \text{if} \quad p = 1, \\ & 1/p - 1/q < r/d \le 1/p & \text{if} \quad 1 < p < \infty. \end{array}\right\} \qquad (8)$$

Note that we demanded from the beginning that $W_p^r(Q)$ *is* embedded into $L_q(Q)$ (otherwise the operator J_{pq} would not even be defined). So here we study the case that $W_p^r(Q)$ is embedded into $L_q(Q)$ but *not* into $C(Q)$. In this case approximation based on function values was not studied before. In the deterministic case there is a certain reason for this: function values are not defined any longer! However, this can easily be overcome. Namely, let us consider approximation on $B_{W_p^r(Q)} \cap C(Q)$, a dense subset of $B_{W_p^r(Q)}$. Now function values *are* defined, and the question arises which rate can be reached on the basis of Sobolev smoothness. It turns out that in the deterministic setting no rate whatsoever is possible on $B_{W_p^r(Q)} \cap C(Q)$. We discuss this issue in section 4.

However, also the analysis of the randomized setting was restricted to the case (7) of embedding into $C(Q)$. In section 2 we present a randomized algorithm which reaches the rate (6) also in the case (8) of non-embedding into $C(Q)$. Hence, in contrast to the situation of embedding (7), randomized algorithms turn out to be superior to deterministic ones. We comment on this in more detail in section 4.

Some applications to randomized integration of functions from $W_p^r(Q)$ and to approximation of solution operators of elliptic partial differential equations are given in Section 3.

2 Randomized Approximation

The following is the main result of this paper. We state it for $1 \le p \le \infty$, $1 \le q < \infty$, since the proof works for all these cases, and thus shows that the method proposed also partly recovers the upper bound from (6) (having in mind though that the new part is the case in which (8) holds).

Theorem 1. *Let $r, d \in \mathbf{N}$, $1 \le p \le \infty$, $1 \le q < \infty$, and assume $r/d > 1/p - 1/q$. Then*

$$e_n^{\rm ran}(J_{pq}, B_{W_p^r(Q)}) \asymp n^{-r/d+\max(1/p-1/q,0)}. \tag{9}$$

We proceed as follows. Under the assumptions of Theorem 1, we develop a general scheme for randomized approximation of functions. Its convergence is then analysed in Proposition 1. Based on this, the proof of Theorem 1 is given at the end of this section.

Let $(\Omega, \Sigma, \mathbf{P})$ be a probability space, and for each $\omega \in \Omega$ let P_ω be any operator from $C(Q)$ to $L_q(Q)$ of the form

$$P_\omega f = \sum_{j=1}^{\kappa} f(x_{j,\omega})\psi_{j,\omega} \quad (f \in C(Q))$$

with $x_{j,\omega} \in Q$ and $\psi_{j,\omega} \in L_q(Q)$. We assume that the mappings

$$\omega \to x_{j,\omega}, \quad \omega \to \psi_{j,\omega}$$

are random variables, the distributions of the x_j being absolutely continuous with respect to the Lebesgue measure, that

$$(\mathbf{E}\,|f(x_{j,\omega})|^q)^{1/q} \le c\|f\|_{W_p^r(Q)} \quad (f \in W_p^r(Q)), \tag{10}$$

$$\operatorname{ess\,sup}_{\omega\in\Omega}\|\psi_{j,\omega}\|_{L_q(Q)} \le c \tag{11}$$

for $j = 1, \dots, \kappa$, and that for all $g \in \mathcal{P}_{r-1}(Q)$, the space of polynomials on Q of degree not exceeding $r-1$,

$$P_\omega g = g \qquad \mathbf{P} - \text{a.s.} \tag{12}$$

Families with such properties are easily constructed. For example, fix $0 < \delta < 1$ and let

$$P^{(1)} f = \sum_{j=1}^{\kappa} f(z_j)\psi_j \tag{13}$$

be for $d = 1$ the Lagrange interpolation operator of appropriate degree and for $d > 1$ its tensor product, with $(z_j)_{j=1}^{\kappa}$ the uniform grid on $[0, 1-\delta]^d$, and $(\psi_j)_{j=1}^{\kappa}$ the respective Lagrange polynomials, considered as functions on $\mathbf{R}^d$. Put $\Omega_1 = [0, \delta]^d$, Σ_1 the σ-algebra of Lebesgue measurable sets and $\mathbf{P}_1$ the normalized on $[0, \delta]^d$ Lebesgue measure. For $\omega_1 \in \Omega_1 = [0, \delta]^d$ and $f \in C(Q)$ put

$$x_{j,\omega_1} = z_j + \omega_1, \tag{14}$$

$$\psi_{j,\omega_1}(x) = \psi_j(x - \omega_1) \quad (x \in Q), \tag{15}$$

and

$$\left(P_{\omega_1}^{(1)} f\right)(x) = \sum_{j=1}^{\kappa} f(z_j + \omega_1)\psi_j(x - \omega_1). \tag{16}$$

Then we have for $f \in W_p^r(Q)$

$$(\mathbf{E}\,|f(x_{j,\omega_1})|^q)^{1/q} = \left(\delta^{-d}\int_{[0,\delta]^d} |f(z_j+y)|^q dy\right)^{1/q}$$
$$\leq \delta^{-d/q}\|f\|_{L_q(Q)} \leq c\delta^{-d/q}\|f\|_{W_p^r(Q)},$$

which shows that (10) is satisfied. It is readily checked that conditions (11) and (12) hold, as well.

Let $l \in \mathbf{N}_0$ and let

$$Q = \bigcup_{i=1}^{2^{dl}} Q_i$$

be the partition of Q into 2^{dl} cubes of sidelength 2^{-l} and of disjoint interior. Let x_i denote the point in Q_i with minimal coordinates. Define the operators E_i and R_i on $L_q(Q)$ by setting for $f \in L_q(Q)$ and $x \in Q$

$$(E_i f)(x) = f(x_i + 2^{-l}x)$$

and

$$(R_i f)(x) = \chi_{Q_i}(x) f(2^l(x-x_i)) = \begin{cases} f(2^l(x-x_i)) \text{ if } & x \in Q_i \\ 0 & \text{otherwise.}\end{cases}$$

For $\omega \in \Omega$ set

$$P_{l,\omega} f = \sum_{i=1}^{2^{dl}} R_i P_\omega E_i f = \sum_{i=1}^{2^{dl}} \sum_{j=1}^{\kappa} f(x_i + 2^{-l}x_{j,\omega}) R_i \psi_{j,\omega} \tag{17}$$

(observe that we use the same random variables $x_{j,\omega}$ for all i). It easily follows from the assumptions on $(P_\omega)_{\omega\in\Omega}$ that $(P_{l,\omega})_{\omega\in\Omega}$ is an algorithm from $\mathcal{A}_m^{\mathrm{ran}}$, where $m = \kappa 2^{dl}$. In fact it is a linear algorithm.

Proposition 1. *Let $r, d \in \mathbf{N}$, $1 \leq p \leq \infty$, $1 \leq q < \infty$, and assume $r/d > 1/p - 1/q$. Let $(P_\omega)_{\omega\in\Omega}$ be as above satisfying (10), (11), (12), and let $(P_{l,\omega})_{\omega\in\Omega}$ for $l \in \mathbf{N}_0$ be given by (17). Then there is a constant $c > 0$ such that for all $l \in \mathbf{N}_0$ and $f \in W_p^r(Q)$*

$$(\mathbf{E}\,\|f - P_{l,\omega} f\|_{L_q(Q)}^q)^{1/q} \leq c\, 2^{-rl + \max(1/p-1/q,0)dl}\|f\|_{W_p^r(Q)}. \tag{18}$$

Proof. It follows from (10) and (11) that for $f \in W_p^r(Q)$

$$(\mathbf{E}\,\|P_\omega f\|_{L_q(Q)}^q)^{1/q} \leq \left(\mathbf{E}\left(\sum_{j=1}^{\kappa} |f(x_{j,\omega})| \|\psi_{j,\omega}\|_{L_q(Q)}\right)^q\right)^{1/q}$$
$$\leq c \sum_{j=1}^{\kappa} (\mathbf{E}\,|f(x_{j,\omega})|^q)^{1/q} \leq c\|f\|_{W_p^r(Q)}. \tag{19}$$

We denote

$$|f|_{r,p,Q} = \left(\sum_{|\alpha|=r} \|D^\alpha f\|_{L_p(Q)}^p \right)^{1/p}$$

if $p < \infty$ and

$$|f|_{r,\infty,Q} = \max_{|\alpha|=r} \|D^\alpha f\|_{L_\infty(Q)}.$$

Next we apply Theorem 3.1.1 from [Cia78]: there is a constant $c > 0$ such that for all $f \in W_p^r(Q)$

$$\inf_{g \in \mathcal{P}_{r-1}(Q)} \|f - g\|_{W_p^r(Q)} \le c|f|_{r,p,Q}. \tag{20}$$

It follows from (2), (12), (19), and (20) that

$$\begin{aligned}(\mathbf{E}\,\|f - P_\omega f\|_{L_q(Q)}^q)^{1/q} &= \inf_{g \in \mathcal{P}_{r-1}(Q)} (\mathbf{E}\,\|(f-g) - P_\omega(f-g)\|_{L_q(Q)}^q)^{1/q} \\ &\le c \inf_{g \in \mathcal{P}_{r-1}(Q)} \|f - g\|_{W_p^r(Q)} \le c|f|_{r,p,Q}. \end{aligned} \tag{21}$$

Clearly,

$$\|R_i f\|_{L_q(Q)} = 2^{-dl/q} \|f\|_{L_q(Q)} \quad (f \in L_q(Q)). \tag{22}$$

From (21) and (22) we obtain for all $f \in W_p^r(Q)$,

$$\begin{aligned}(\mathbf{E}\,\|f - P_{l,\omega} f\|_{L_q(Q)}^q)^{1/q} &= \left(\mathbf{E} \left\| \sum_{i=1}^{2^{dl}} (R_i E_i f - R_i P_\omega E_i f) \right\|_{L_q(Q)}^q \right)^{1/q} \\ &= \left(\mathbf{E} \sum_{i=1}^{2^{dl}} \|R_i (E_i f - P_\omega E_i f)\|_{L_q(Q)}^q \right)^{1/q} \\ &= \left(2^{-dl} \sum_{i=1}^{2^{dl}} \mathbf{E}\,\|E_i f - P_\omega E_i f\|_{L_q(Q)}^q \right)^{1/q} \\ &\le c \left(2^{-dl} \sum_{i=1}^{2^{dl}} |E_i f|_{r,p,Q}^q \right)^{1/q} \\ &\le c\, 2^{\max(1/p-1/q,0)dl} \left(2^{-dl} \sum_{i=1}^{2^{dl}} |E_i f|_{r,p,Q}^p \right)^{1/p} \end{aligned}$$

and furthermore,

$$\begin{aligned}\left(2^{-dl}\sum_{i=1}^{2^{dl}}|E_i f|_{r,p,Q}^p\right)^{1/p} &= \left(2^{-dl}\sum_{i=1}^{2^{dl}}\sum_{|\alpha|=r}\int_Q |D^\alpha f(x_i+2^{-l}x)|^p\,dx\right)^{1/p}\\ &= 2^{-rl}\left(\sum_{i=1}^{2^{dl}}\sum_{|\alpha|=r}\int_{Q_i}|D^\alpha f(y)|^p\,dy\right)^{1/p}\\ &= 2^{-rl}|f|_{r,p,Q} \le 2^{-rl}\|f\|_{W_p^r(Q)}.\end{aligned}$$

(with the usual modifications for $p=\infty$). Combining the last two inequalities gives

$$(\mathbf{E}\,\|f - P_{l,\omega}f\|_{L_q(Q)}^q)^{1/q} \le c\,2^{-rl+\max(1/p-1/q,0)dl}\|f\|_{W_p^r(Q)},$$

which concludes the proof.

Proof of Theorem 1. Let $n\in\mathbf{N}$ and put

$$l = \left\lceil \frac{\log_2 n}{d}\right\rceil. \tag{23}$$

Then $(P_{l,\omega})_{\omega\in\Omega}$ belongs to $\mathcal{A}_m^{\mathrm{ran}}$ with $m=\kappa 2^{dl}\le cn$, so Proposition 1 together with (23) gives the upper bound in (9). The lower bound follows from standard techniques of information-based complexity (reduction to the average case on subsets formed by smooth bump functions) and is identical to that given in [Mat91] (see also [Nov88], [Hei93]). We omit it here.

3 Some Applications

First we consider integration. Let $Q=[0,1]^d$ and let $I: W_p^r(Q)\to\mathbf{R}$ be the integration operator

$$If = \int_Q f(x)dx.$$

Corollary 1. *Let $r,d\in\mathbf{N}$, $1\le p<\infty$, and put $\bar p=\min(p,2)$. Then*

$$e_n^{\mathrm{ran}}(I, B_{W_p^r(Q)}) \asymp n^{-r/d-1+1/\bar p}.$$

This result was shown by Novak for $p\ge 2$ and for $p<2$ and $r/d\ge 1/p-1/2$, that is, for the case that $W_p^r(Q)$ is embedded into $L_2(Q)$, see [Nov88], 2.2.9, and also the references therein for previous work. Our analysis supplies the remaining cases and a new technique: Novak used a result of [EZ60] on stochastic quadratures. For spaces embedded into $C(Q)$, another proof was given in [Hei93] using deterministic approximation as variance reduction (separation of the main part). What we present below might be viewed as a stochastic analogon of the latter.

Proof of Corollary 1. Fix $0 < \delta < 1$ and let $(\Omega_1, \Sigma_1, \mathbf{P}_1)$ be as defined above, following relation (13). Let $n \in \mathbf{N}$, let l be given by (23), and let $P_{l,\omega_1}^{(1)}$ for $\omega_1 \in \Omega_1$ be as defined in (13-17), that is, for $f \in W_p^r(Q)$ and $x \in Q$,

$$\left(P_{l,\omega_1}^{(1)} f\right)(x) = \sum_{i=1}^{2^{dl}} \sum_{j=1}^{\kappa} f(x_i + 2^{-l}(z_j + \omega_1))\chi_{Q_i}(x)\psi_j(2^l(x - x_i) - \omega_1). \quad (24)$$

Finally, let y_{k,ω_2} $(k = 1, \ldots, n)$ be independent, uniformly distributed on Q random variables over some probability space $(\Omega_2, \Sigma_2, \mathbf{P}_2)$. Define the algorithm $\left(A_{\omega_2}^{(2)}\right)_{\omega_2 \in \Omega_2}$ to be the usual Monte Carlo method

$$A_{\omega_2}^{(2)}(g) = \frac{1}{n} \sum_{k=1}^{n} g(y_{k,\omega_2}). \quad (25)$$

It is known that for $g \in L_p(Q)$

$$\left(\mathbf{E}_{\omega_2} |Ig - A_{\omega_2}^{(2)}(g)|^{\bar{p}}\right)^{1/\bar{p}} \leq 2^{2/\bar{p}-1} n^{-1+1/\bar{p}} \|g\|_{L_p(Q)} \quad (26)$$

(see [Hei93] for the case $1 \leq p < 2$). Now we put

$$(\Omega, \Sigma, \mathbf{P}) = (\Omega_1, \Sigma_1, \mathbf{P}_1) \times (\Omega_2, \Sigma_2, \mathbf{P}_2)$$

and define an algorithm $A = (A_\omega)_{\omega \in \Omega}$ by setting for $\omega = (\omega_1, \omega_2)$ and $f \in W_p^r(Q)$

$$A_\omega(f) = IP_{l,\omega_1}^{(1)} f + A_{\omega_2}^{(2)}(f - P_{l,\omega_1}^{(1)} f).$$

On the basis of (24) and (25) measurability and consistency readily follow, and we have $A \in \mathcal{A}_m^{\mathrm{ran}}$ for $m = \kappa 2^{dl} + n \leq cn$. Moreover,

$$If - A_\omega(f) = I(f - P_{l,\omega_1}^{(1)} f) - A_{\omega_2}^{(2)}(f - P_{l,\omega_1}^{(1)} f).$$

Using Fubini's theorem, (26), and Proposition 1 for $q = p$, we derive

$$\begin{aligned}
&\left(\mathbf{E}\,|If - A_\omega(f)|^{\bar{p}}\right)^{1/\bar{p}} \\
&= \left(\mathbf{E}_{\omega_1} \mathbf{E}_{\omega_2} \left|I\left(f - P_{l,\omega_1}^{(1)} f\right) - A_{\omega_2}^{(2)}\left(f - P_{l,\omega_1}^{(1)} f\right)\right|^{\bar{p}}\right)^{1/\bar{p}} \\
&\leq cn^{-1+1/\bar{p}} \left(\mathbf{E}_{\omega_1} \left\|f - P_{l,\omega_1}^{(1)} f\right\|_{L_p(Q)}^{\bar{p}}\right)^{1/\bar{p}} \\
&\leq cn^{-1+1/\bar{p}} \left(\mathbf{E}_{\omega_1} \left\|f - P_{l,\omega_1}^{(1)} f\right\|_{L_p(Q)}^{p}\right)^{1/p} \\
&\leq cn^{-1+1/\bar{p}-r/d} \|f\|_{W_p^r(Q)},
\end{aligned}$$

concluding the proof of the upper bound. The lower bound is already contained in [Nov88], 2.2.9, Proposition 1.

Now we turn to elliptic problems. First results on the randomized information complexity of elliptic partial differential equations were obtained in [Hei06b]. The results above have some direct consequences for certain instances of this problem. Let $d, m \in \mathbf{N}$, $d \geq 2$, let $Q_1 \subset \mathbf{R}^d$ be a bounded C^∞ domain (see, e.g., [Tri78] for the definition), and let $\mathcal{L}$ be an elliptic differential operator of order $2m$ on Q_1, that is

$$\mathcal{L}u = \sum_{|\alpha| \leq 2m} a_\alpha(x) D^\alpha u(x), \tag{27}$$

with boundary operators

$$\mathcal{B}_j u = \sum_{|\alpha| \leq m_j} b_{j\alpha}(x) D^\alpha u(x), \tag{28}$$

where $j = 1, \dots, m$, $m_j \leq 2m - 1$ and $a_\alpha \in C^\infty(Q_1)$ and $b_{j\alpha} \in C^\infty(\partial Q_1)$ are complex-valued infinitely differentiable functions. Consider the homogeneous boundary value problem

$$\mathcal{L}u(x) = f(x) \quad (x \in Q_1^0) \tag{29}$$

$$\mathcal{B}_j u(x) = 0 \quad (x \in \partial Q_1). \tag{30}$$

We asssume that $(\mathcal{L}, \{B_j\})$ is regularly elliptic (see [Tri78], 5.2.1/4, for the definition), and that 0 is not in the spectrum of $\mathcal{L}$, considered as an unbounded operator in $L_q(Q_1)$ with domain of definition $W^{2m}_{q,\{\mathcal{B}_j\}}(Q_1)$, where the latter denotes the subspace of $W^{2m}_q(Q_1)$ consisting of those f which satisfy (30). This implies that $\mathcal{L}$ is an isomorphism from $W^{2m}_{q,\{\mathcal{B}_j\}}(Q_1)$ to $L_q(Q_1)$ for $1 < q < \infty$, see [Tri78], Theorem 5.5.1(b). Now we put $S = \mathcal{L}^{-1} J_{pq}$, considered as an operator into $W^{2m}_q(Q_1)$, that is,

$$S : W^r_p(Q_1) \xrightarrow{J_{pq}} L_q(Q_1) \xrightarrow{\mathcal{L}^{-1}} W^{2m}_q(Q_1).$$

Hence S is the solution operator for the elliptic problem (29–30), where we consider the problem of approximating the full solution u, the right-hand side f is supposed to belong to $W^r_p(Q_1)$, and the error is measured in the norm of $W^{2m}_q(Q_1)$.

Corollary 2. *Let $r, d \in \mathbf{N}$, $1 \leq p \leq \infty$, $1 < q < \infty$, and $r/d > 1/p - 1/q$. Then*

$$e_n^{\mathrm{ran}}(S, B_{W^r_p(Q_1)}) \asymp n^{-r/d + \max(1/p - 1/q, 0)}. \tag{31}$$

Proof. Using local charts (like, e.g., in [Hei06b]) it is easy to extend Theorem 1 to smooth domains Q_1 in place of $Q = [0,1]^d$. It is also clear that the case of complex-valued functions is a direct consequence of the real case. From this and the above-mentioned fact that $\mathcal{L}^{-1}$ is an isomorphic embedding of $L_q(Q_1)$ into $W_q^{2m}(Q_1)$ the upper bound follows.

So does the lower bound if we verify that algorithms with values in $\mathcal{L}^{-1}(L_q(Q_1)) = W_{q,\{\mathcal{B}_j\}}^{2m}(Q_1)$, a subspace of $W_q^{2m}(Q_1)$ containing $S(W_p^r(Q_1))$, cannot be better (up to a constant) than algorithms with values in $W_q^{2m}(Q_1)$. This, however, follows, e.g., from the fact that $W_{q,\{\mathcal{B}_j\}}^{2m}(Q_1)$ is complemented in $W_q^{2m}(Q_1)$, see [Tri78], Theorem 5.5.2(b), completing the proof.

A similar approach (in the sense of using isomorphism properties to reduce approximation of solution operators to approximation of embeddings) was presented in [DNS06a, DNS06b] for the deterministic setting, with $q = 2$. There, however, more general classes of operators and, besides function values, also arbitrary linear functionals are considered.

4 Deterministic Approximation

We already mentioned that for those r, d and p for which $W_p^r(Q)$ is embedded into $C(Q)$, the order of the randomized n-th minimal error coincides with that of the deterministic one. If the embedding does not hold, that is, if (8) is satisfied, the situation is different. First of all, since function values are not well-defined, we replace $B_{W_p^r(Q)}$ by the (dense) subset $B_{W_p^r(Q)} \cap C(Q)$. In this section we show that, although now function values are defined, the Sobolev smoothness $W_p^r(Q)$ does not lead to any rate at all for the deterministic setting. That is, the deterministic n-th minimal error is bounded from below by a positive constant.

Put $F = B_{W_p^r(Q)} \cap C(Q)$, let $n \in \mathbf{N}$ and let $\mathcal{A}_n^{\det}$ be the class of all deterministic algorithms for the approximation of J_{pq} on F, which are of the form

$$A(f) = \varphi(f(x_1), \dots, f(x_n)), \tag{32}$$

where $x_i \in Q$ $(i = 1, \dots, n)$ and $\varphi : \mathbf{R}^n \to L_q(Q)$ is an arbitrary mapping. The error on F is defined as

$$e(J_{pq}, A, F) = \sup_{f \in F} \|S(f) - A(f)\|_{L_q(Q)}.$$

The deterministic n-th minimal error is defined as

$$e_n^{\det}(J_{pq}, F) = \inf_{A \in \mathcal{A}_n^{\det}} e(J_{pq}, A, F).$$

Proposition 2. *Let $1 \le p, q < \infty$ and $r/d > 1/p - 1/q$. Assume that either*

$$\frac{r}{d} < \frac{1}{p} \tag{33}$$

or

$$\frac{r}{d} = \frac{1}{p} \quad \text{and} \quad 1 < p < \infty. \tag{34}$$

Then

$$e_n^{\text{det}}(J_{pq}, B_{W_p^r(Q)} \cap C(Q)) \asymp 1. \tag{35}$$

For the proof we need the following lemma. Let $B(0, \varrho)$ denote the closed ball in $\mathbf{R}^d$ of radius ϱ around 0.

Lemma 1. *Assume that (33) or (34) holds. Then there is a sequence of functions*

$$(f_m)_{m=1}^{\infty} \subset W_p^r(\mathbf{R}^d) \cap C^\infty(\mathbf{R}^d)$$

such that for all m

$$f_m(0) = 1, \quad \operatorname{supp} f_m \subseteq B\left(0, \frac{1}{m}\right), \tag{36}$$

and

$$\lim_{m\to\infty} \|f_m\|_{W_p^r(\mathbf{R}^d)} = 0. \tag{37}$$

Proof. In case of (34) this is a combination of well-known facts from function space theory. Let $\psi \in C^\infty(\mathbf{R}^d)$ be such that $\psi \geq 0$,

$$\psi(x) = \begin{cases} 1 & \text{if} \quad |x| \leq \frac{1}{2}, \\ 0 & \text{if} \quad |x| \geq 1, \end{cases}$$

and

$$\int_{\mathbf{R}^d} \psi(x)dx = 1.$$

Put

$$g(x) = \begin{cases} \psi(x) \ln \ln \frac{3}{|x|} & \text{if} \quad 0 < |x| < 1, \\ 0 & \text{if} \quad |x| \geq 1. \end{cases}$$

Then

$$\|g\|_{W_p^r(\mathbf{R}^d)} < \infty, \quad \operatorname{supp} g \subseteq B(0,1), \quad \lim_{x\to 0} g(x) = +\infty, \tag{38}$$

see [Ada75], Example 5.26. Furthermore, setting

$$h_m = \psi_m * g \quad \text{with} \quad \psi_m(x) = m^d \psi(mx) \quad (x \in \mathbf{R}^d,\ m \in \mathbf{N}),$$

we get, using Lemma 3.15 of [Ada75] and (38),

$$h_m \in C^\infty(\mathbf{R}^d), \quad \operatorname{supp} h_m \subseteq B(0,2), \quad h_m(0) > 0 \quad (m \in \mathbf{N}),$$
$$\sup_{m\in\mathbf{N}} \|h_m\|_{W_p^r(\mathbf{R}^d)} < \infty, \quad \lim_{m\to\infty} h_m(0) = +\infty.$$

Finally we define f_m by

$$f_m(x) = h_m(0)^{-1} h_m(2mx) \quad (x \in \mathbf{R}^d,\ m \in \mathbf{N}).$$

Relation (34) implies

$$\|f_m\|_{W_p^r(\mathbf{R}^d)} \le h_m(0)^{-1} \|h_m(2m\,\cdot)\|_{W_p^r(\mathbf{R}^d)} \le c h_m(0)^{-1} \to 0,$$

while (36) is obviously fulfilled by definition. This completes the proof in the case of (34).

If (33) holds, we choose ψ as above and put

$$f_m(x) = \psi(mx) \quad (x \in \mathbf{R}^d,\ m \in \mathbf{N}).$$

Clearly, (36) is satisfied, and it follows from (33) that

$$\|f_m\|_{W_p^r(\mathbf{R}^d)} \le c m^{r-d/p} \to 0 \quad (m \to \infty).$$

Proof of Proposition 2. The upper bound follows from the fact that J_{pq} is bounded. To prove the lower bound, let $x_1, \dots, x_n$ be any fixed distinct points in Q. For $m \in \mathbf{N}$ consider the function $v_m \in C(Q)$ given by

$$v_m(x) = \left(1 + n\|f_m\|_{W_p^r(\mathbf{R}^d)}\right)^{-1} \left(1 - \sum_{i=1}^n f_m(x - x_i)\right) \quad (x \in Q).$$

We have

$$\|v_m\|_{W_p^r(Q)} \le 1 \tag{39}$$

and, using (37),

$$\begin{aligned} \|v_m\|_{L_q(Q)} &\ge \int_Q v_m(x)dx \\ &= \left(1 + n\|f_m\|_{W_p^r(\mathbf{R}^d)}\right)^{-1} \left(1 - \sum_{i=1}^n \int_Q f_m(x - x_i)dx\right) \\ &\ge \left(1 + n\|f_m\|_{W_p^r(\mathbf{R}^d)}\right)^{-1} \left(1 - n\|f_m\|_{W_p^r(\mathbf{R}^d)}\right) \to 1 \end{aligned} \tag{40}$$

as $m \to \infty$. Finally, by (36), for m large enough,

$$v_m(x_i) = 0 \quad (i = 1, \dots, n). \tag{41}$$

Now (39–41) combined with standard results from information-based complexity theory [TWW88], Ch. 3.1, prove Proposition 2.

Proposition 2 was independently obtained by Novak and Woźniakowski (unpublished notes).

5 Comments

Comparing the rates in Theorem 1 and in Proposition 2 for the case (8) of non-embedding into $C(Q)$, we see that on $B_{W_p^r(Q)} \cap C(Q)$ randomization gives a speedup over the deterministic setting of

$$\frac{e_n^{\mathrm{ran}}(J_{pq}, B_{W_p^r(Q)} \cap C(Q))}{e_n^{\mathrm{det}}(J_{pq}, B_{W_p^r(Q)} \cap C(Q))} \asymp n^{-r/d+\max(1/p-1/q,0)},$$

which is non-trivial in all cases, since $r/d > \max(1/p - 1/q, 0)$ by assumption. If $p = 1$, the maximal exponent of the speedup is r/d, reached for $q = 1$, and r/d can be arbitrarily close to 1. If $1 < p < \infty$, the maximal exponent of the speedup is again r/d and can now be as large as $1/p$, reached for $q \le p$.

Let us also mention that the lower bounds presented here for the randomized setting hold true for the more general case of adaptive (standard) information, as introduced e.g. in [Hei06a]. Similarly for the deterministic case. The latter follows from general results [TWW88].

As is readily seen from its proof, Proposition 2 remains true if J_{pq} is replaced by I of Corollary 1, the parameter q and the condition $r/d > 1/p - 1/q$ being omitted. Furthermore, the argument used in the proof of Corollary 2 shows that we can also replace J_{pq} in Proposition 2 by S.

References

[Ada75] R.A. Adams. Sobolev Spaces. Academic Press, New York (1975).

[Cia78] P.G. Ciarlet. The Finite Element Method for Elliptic Problems. North-Holland, Amsterdam (1978).

[DNS06a] S. Dahlke, E. Novak, and W. Sickel. Optimal approximation of elliptic problems by linear and nonlinear mappings I, J. Complexity **22**, 29–49 (2006).

[DNS06b] S. Dahlke, E. Novak, and W. Sickel. Optimal approximation of elliptic problems by linear and nonlinear mappings II, J. Complexity **22**, 549–603 (2006).

[EZ60] S.M. Ermakov and V.G. Zolotukhin. Polynomial approximation and the Monte Carlo method, Theor. Prob. Appl. **5**, 428–431 (1960).

[Hei93] S. Heinrich. Random approximation in numerical analysis. In: K.D. Bierstedt, A. Pietsch, W.M. Ruess, and D. Vogt (eds) Functional Analysis. Marcel Dekker, New York, 123–171 (1993).

[Hei06a] S. Heinrich. Monte Carlo approximation of weakly singular integral operators. J. Complexity **22**, 192–219 (2006).

[Hei06b] S. Heinrich. The randomized information complexity of elliptic PDE. J. Complexity **22**, 220–249 (2006).

[Mat91] P. Mathé. Random approximation of Sobolev embeddings. J. Complexity **7**, 261–281 (1991).

[Nov88] E. Novak. Deterministic and Stochastic Error Bounds in Numerical Analysis. Lecture Notes in Mathematics **1349**, Springer-Verlag, Berlin (1988).

[NT06] E. Novak and H. Triebel. Function spaces in Lipschitz domains and optimal rates of convergence for sampling. Constr. Approx. **23**, 325–350 (2006).

[TWW88] J.F. Traub, G.W. Wasilkowski, and H. Woźniakowski. Information-Based Complexity. Academic Press, New York (1988).

[Tri78] H. Triebel. Interpolation Theory, Function Spaces, Differential Operators. North-Holland, Amsterdam, New York, Oxford (1978).

[Was89] G.W. Wasilkowski. Randomization for continuous problems. J. Complexity **5**, 195–218 (1989).

Tractability of Linear Multivariate Problems in the Average Case Setting*

Fred Hickernell[1], Greg Wasilkowski[2], and Henryk Woźniakowski[3]

[1] Department of Applied Mathematics, Illinois Institute of Technology, Chicago, IL 60616, USA
hickernell@iit.edu

[2] Department of Computer Science, University of Kentucky, Lexington, KY, USA
greg@cs.uky.edu

[3] Department of Computer Science, Columbia University, New York, NY, USA, and Institute of Applied Mathematics, University of Warsaw, ul. Banacha 2, 02-097 Warszawa, Poland
henryk@cs.columbia.edu

Summary. We study the average case setting for linear multivariate problems defined over a separable Banach space of functions f of d variables. The Banach space is equipped with a Gaussian measure. We approximate linear multivariate problems by computing finitely many information evaluations. An information evaluation is defined as an evaluation of a continuous linear functional from a given class Λ. We consider two classes of information evaluations; the first class Λ^{all} consists of all continuous linear functionals, and the second class Λ^{std} consists of function evaluations.

We investigate the minimal number $n(\varepsilon, d, \Lambda)$ of information evaluations needed to reduce the initial average case error by a factor ε. The initial average case error is defined as the minimal error that can be achieved without any information evaluations.

We study tractability of linear multivariate problems in the average case setting. Tractability means that $n(\varepsilon, d, \Lambda)$ is bounded by a polynomial in both ε^{-1} and d, and strong tractability means that $n(\varepsilon, d, \Lambda)$ is bounded by a polynomial only in ε^{-1}.

For the class Λ^{all}, we provide necessary and sufficient conditions for tractability and strong tractability in terms of the eigenvalues of the covariance operator of a Gaussian measure on the space of solution elements. These conditions are simplified under additional assumptions on the measure. In particular, we consider measures with finite-order weights and product weights. For finite-order weights, we prove that linear multivariate problems are always tractable.

* The second and third authors were partially supported by the National Science Foundation under Grants DMS-0609703 and DMS-0308713, respectively.

For the class Λ^{std}, we consider weighted multivariate function approximation problems. We prove that for such problems the class Λ^{std} is as powerful as the class Λ^{all} in the following sense: if $n(\varepsilon, d, \Lambda^{\mathrm{all}}) = O(\varepsilon^{-p} d^{q})$ then

$$n(\varepsilon, d, \Lambda^{\mathrm{std}}) = O\left(\varepsilon^{-p} d^{q} \left[\ln \ln \left(\varepsilon^{-1} + d + 1\right)\right]^{1+p/2}\right).$$

Hence, modulo the double logarithm these classes are equally powerful. In particular, this means that strong tractability and tractability are equivalent in both classes.

1 Introduction

Tractability of multivariate problems has recently become the subject of intensive research. To give an idea of how this area has developed, note that [NW01] was able to survey tractability results before 2000, but that the number papers appearing after 200 is too many to cite. The initial results and early emphasis was on integration problems. There are also quite a few papers on the tractability of general linear multivariate problems, however the majority of them have studied only the worst case setting. Much less attention has been devoted to tractability of general linear multivariate problems in other than the worst case setting; for exceptions see, e.g., [HW00, KSW07, RW96, RW97, Was93, WW95, WW01]. This has led us to address tractability of linear multivariate problems in the average case setting.

More specifically, by a linear multivariate problem we mean the approximation of a linear operator S_d that maps a separable Banach space $\mathcal{F}_d$ of d-variate functions into a Hilbert space $\mathcal{G}_d$ for $d = 1, 2, \ldots$. An important example of such problems is provided by multivariate function approximation, $S_d f = f$, while other examples include linear partial differential equations and integral equations. In the average case setting, we assume that the space $\mathcal{F}_d$ is endowed with a zero-mean Gaussian probability measure μ_d. This allows us to measure the approximation errors in the corresponding mean-square sense. That is, if Af denotes the approximation to $S_d f$ given by an algorithm A, then the average case error of the algorithm A is given by

$$e^{\mathrm{avg}}(A; S_d) := \left(\int_{\mathcal{F}_d} \|S_d f - Af\|_{\mathcal{G}_d}^2 \, \mu_d(\mathrm{d}f)\right)^{1/2}.$$

The approximations Af use partial information about f that consists of a finite number n of information evaluations given by $L_1(f), L_2(f), \ldots, L_n(f)$. Here, the L_i are continuous linear functionals that belong to the class $\Lambda^{\mathrm{all}} = \mathcal{F}_d^*$ of all continuous linear functionals or to the class Λ^{std} of function evaluations. That is, $L_i \in \Lambda^{\mathrm{std}}$ iff there exists a point $\mathbf{x}_i$ such that $L_i(f) = f(\mathbf{x}_i)$ for all $f \in \mathcal{F}_d$.

It is important to know what algorithms and information evaluations are optimal. In particular, we would like to know the minimal average case error among all algorithms that use at most n information evaluations. For $n = 0$ this minimal error is called the initial error, which is the smallest average case error which can be achieved without any information evaluations. In our case, the initial error corresponds to the average case error of the zero algorithm $A \equiv 0$ and is equal to the square root of the trace of the covariance operator of the Gaussian measure $\nu_d = \mu_d S_d^{-1}$ of solution elements.

Knowledge of the minimal average case errors allows us to know the minimal number, $n = n(\varepsilon; S_d, \Lambda)$, of information evaluations for any information class $\Lambda \in \{\Lambda^{\text{all}}, \Lambda^{\text{std}}\}$ needed for the reduction of the initial error by a factor $\varepsilon \in (0, 1)$. Clearly, $n(\varepsilon; S_d, \Lambda)$ measures the intrinsic difficulty of the multivariate linear problem S_d.

The function $n(\varepsilon; S_d, \Lambda)$ has been extensively investigated in the literature, mainly for a fixed d and for ε tending to zero. For a number of important applications the number of variables d might be arbitrarily large and the error demand ε might be not too severe, see e.g., [TW98]. It is therefore also important to study the dependence of $n(\varepsilon; S_d, \Lambda)$ also on d. Controlling the dependence on d is the essence of tractability and strong tractability.

For the reader's convenience, we now recall the definition of tractability and strong tractability, see [Woź94]. We define $S := \{S_d\}$ and say that the problem S is *tractable* in the class Λ if there are non-negative constants C, p and q such that

$$n(\varepsilon; S_d, \Lambda) \leq C\, d^{\,q}\, \varepsilon^{-p} \quad \forall\, d \in \mathbb{N} \;\; \forall\, \varepsilon \in (0, 1).$$

It is *strongly tractable* if this bound above holds with $q = 0$. The exponent of strong tractability is defined as the infimum of p satisfying the bound above.

For the class Λ^{all}, the optimal algorithms, optimal linear functionals and $n(\varepsilon; S_d, \Lambda^{\text{all}})$ are known, see [PW90, Was86] for the original results which can be also found in [TWW88]. The number $n(\varepsilon; S_d, \Lambda^{\text{all}})$ is the square root of the truncated sum of eigenvalues of the covariance operator $\mathcal{W}_d : \mathcal{G}_d \to \mathcal{G}_d$ of the Gaussian measure ν_d of solution elements. From this, we obtain necessary and sufficient conditions for tractability and strong tractability in terms of eigenvalues of the operators $\mathcal{W}_d$; see Theorem 1 in Section 3.

In general, it is difficult to find the eigenvalues of the operator $\mathcal{W}_d$ for all d. Therefore, in the second part of Section 3, we assume that the eigenvalues of $\mathcal{W}_d$ can be expressed as weighted products of eigenvalues for the univariate $d = 1$ case, see (15). This corresponds to the assumption that the Hilbert space $\mathcal{G}_d$ is a tensor product of d copies of the Hilbert space $\mathcal{G}_1$ for the univariate case, and that the Fourier coefficients of the solutions with respect to the tensor product orthonormal system of $\mathcal{G}_d$ are independent Gaussian variables with zero means and whose variances are equal to weighted products of univariate variances. In Section 4 we present several linear multivariate problems for which the assumption (15) holds.

Roughly speaking, (15) corresponds to assuming that the Gaussian measure μ_d concentrates on functions of the form

$$f(\mathbf{x}) = \sum_{\mathfrak{u}\subseteq\{1,2,\dots,d\}} \gamma_{d,\mathfrak{u}}\, f_{\mathfrak{u}}(\mathbf{x}),$$

where $\boldsymbol{\gamma} = \{\gamma_{d,\mathfrak{u}}\}_{d,\mathfrak{u}}$ is a family of non-negative numbers, called weights, and the functions $f_{\mathfrak{u}}$ depend only on variables x_k with $k \in \mathfrak{u}$. We now briefly comment on the family of weights $\boldsymbol{\gamma}$; for more discussion see [SW98] where the concept of weights was introduced. Each weight $\gamma_{d,\mathfrak{u}}$ quantifies the importance of the group of variables x_j with $j \in \mathfrak{u}$. For instance, by taking small $\gamma_{d,\mathfrak{u}}$ for subsets of a large cardinality, $|\mathfrak{u}| > q^*$ say, we model problems with effective dimension at most q^*. By taking $\gamma_{d,\mathfrak{u}} = 0$ for $|\mathfrak{u}| > q^*$ we obtain the *finite-order* weights, see [DSWW06] where this concept was introduced. For finite-order weights, we force μ_d to be concentrated on the subspace of linear combinations of functions, each depending on at most q^* variables. Examples of such functions are provided by d-variate polynomials of degree at most q^*, or the Coulomb potential for which $q^* = 6$, see e.g., [WW05]. This idea is also behind the successful use of orthogonal arrays of fixed strength in experimental designs [DM99, HSS99]. Another important class of weights is provided by the *product* weights with $\gamma_{d,\mathfrak{u}} = \prod_{k\in\mathfrak{u}} \gamma_{d,k}$, first used in [WW01], or the *uniform product* weights with $\gamma_{d,k}$ independent of d, originally introduced in [SW98]. Such weights are used to model problems in which different variables have different significance; less important variables x_k correspond to smaller $\gamma_{d,k}$.

Using Theorem 1, we provide necessary and sufficient conditions for tractability and strong tractability for both finite-order and product weights for the class Λ^{all}. In particular, we show that problems with finite-order weights are always tractable; and that tractability is equivalent to strong tractability for uniform product weights.

For important applications, arbitrary linear functionals cannot be computed and, instead, only function evaluations $L_i(f) = f(\mathbf{x}_i)$ are available. This corresponds to the class Λ^{std}. For this class, optimal information evaluation and $n(\varepsilon; S_d, \Lambda^{\mathrm{std}})$ are much more difficult to characterize. In Section 5, we study the power of standard information for weighted multivariate function approximation problems in the average case setting. We assume that function evaluations are continuous functionals in $\mathcal{F}_d$, and that $\mathcal{F}_d$ is continuously embedded in $\mathcal{G}_d = L_{2,\rho}(D_d)$ with

$$\|g\|_{\mathcal{G}_d}^2 = \int_{D_d} |f(\mathbf{x})|^2\, \rho(\mathbf{x})\, \mathrm{d}\mathbf{x},$$

where ρ is a given probability density function on $D_d \subseteq \mathbb{R}^d$. The solution operator S_d is the embedding

$$S_d f = f \quad \forall f \in \mathcal{F}_d.$$

Assuming that $\Lambda^{\mathrm{std}} \subseteq \Lambda^{\mathrm{all}}$, we show that optimal standard information and optimal information from Λ^{all} have the same order of convergence if we have a polynomial rate of convergence for the class $\Lambda^{\rm all}$. That is, assume that there exists a sequence of algorithms A_n^*, each using n functionals from $\Lambda^{\rm all}$ and with the average case errors satisfying

$$e^{\rm avg}(A_n^*) \le \frac{C_0}{(n+1)^r} \qquad \forall n = 0, 1, \dots$$

for some positive C_0 and r. We prove that there exists a sequence of algorithms A_n, each using at most

$$n \left\lceil \frac{\ln(\ln(n))}{\ln(1+1/(2r))} \right\rceil \qquad \text{function evaluations,}$$

with the average case errors bounded by

$$e^{\rm avg}(A_n) \le \frac{e\, C_0}{(n+1)^r} \sqrt{2 + \frac{\ln(\ln(n))}{\ln(1+1/(2r))}}.$$

Furthermore, if multivariate approximation is (strongly) tractable in the class $\Lambda^{\rm all}$ then it is also (strongly) tractable in the class $\Lambda^{\rm std}$ with essentially the same exponents of d and ε^{-1}. That is, if

$$n(\varepsilon; S_d, \Lambda^{\rm all}) \le C\, d^q\, \varepsilon^{-p}$$

then

$$\begin{aligned} n(\varepsilon; S_d, \Lambda^{\rm std}) &\le \min_{k=1,2,\dots} k \left\lceil \left(2C(k+1)^{p/2}\, d^q\, \varepsilon^{-p}\right)^{1/(1-(1+p/2)^{-k})} \right\rceil \\ &= O\left(d^q\, \varepsilon^{-p}\, [\ln(\ln(\varepsilon^{-1}+d+1))]^{1+p/2}\right) \end{aligned}$$

with the factor in the big O notation independent of d and ε.

Unfortunately, the proofs in Section 5 are not fully constructive, i.e., they provide only semi-construction of the algorithms A_n. By semi-construction we mean that sample points used by the algorithm A_n are selected only probabilistically. If we want to guarantee that we succeed with probability $1-\delta$ then the cost of constructing good sample points is proportional to $\ln(\delta^{-1})$.

The lack of fully constructive algorithms achieving tractability error bounds in the average case setting is in contrast to the randomized setting, in which the construction of asymptotically optimal algorithms A_n using standard information for multivariate approximation is known, see [WW07].

Finally we would like to stress that the results of Section 5 do not necessarily mean that $\Lambda^{\rm std}$ is always as powerful as $\Lambda^{\rm all}$. Indeed, it is an open problem to characterize the convergence of nth minimal errors $n(\varepsilon; S, \Lambda^{\rm std})$ when $n(\varepsilon; S, \Lambda^{\rm all})$ converge to zero faster than polynomially; see Remark 1.

We summarize the content of the paper. Section 2 contains basic definitions and facts concerning average case setting and tractability. Section 3 deals with tractability of linear problems for the class $\Lambda^{\rm all}$. The assumptions and results of Section 3 are illustrated in Section 4 by four examples. Finally, Section 5 deals with approximation problems for the class $\Lambda^{\rm std}$ of function evaluations.

2 Problem Formulation

In this section we present general definitions and known results about the average case setting as well as assumptions specific to this paper. For more detailed discussion we refer the reader to [TWW88].

2.1 General Definitions

For $d = 1, 2, \ldots$, let $\mathcal{F}_d$ be a separable Banach space of functions $f : D_d \to \mathbb{R}$, where D_d is a Borel measurable subset of $\mathbb{R}^{md}$. Here and in the rest of this paper, m is a fixed positive integer. For many problems it is natural to assume that $m = 1$. There are, however, practically important problems for which $m > 1$, see e.g., [KS05, WW04]. We equip the space $\mathcal{F}_d$ with a zero mean Gaussian probability measure μ_d whose covariance operator is denoted by C_{μ_d}. For general properties of Gaussian measures on Banach spaces we refer to [Kuo75, Vak81], see also [TWW88, Appendix 2]. Here we only recall that

$$C_{\mu_d} : \mathcal{F}_d^* \to \mathcal{F}_d \quad \text{with} \quad L_1(C_{\mu_d} L_2) = \int_{\mathcal{F}_d} L_1(f) L_2(f)\, \mu_d(\mathrm{d}f) \quad \forall\, L_1, L_2 \in \mathcal{F}_d^*,$$

where $\mathcal{F}_d^*$ denotes the space of continuous linear functionals $L : \mathcal{F}_d \to \mathbb{R}$. Furthermore,

$$\int_{\mathcal{F}_d} \|f\|^2_{\mathcal{F}_d}\, \mu_d(\mathrm{d}f) < \infty. \tag{1}$$

Equivalently, $f \in \mathcal{F}_d$ can be viewed as a realization of a zero-mean Gaussian *stochastic process* with the covariance operator C_{μ_d}. When the function evaluations, $L_{\mathbf{x}}(f) := f(\mathbf{x})$, are continuous functionals for all $\mathbf{x}$, it is convenient to work with the covariance kernel, $K_d : D_d \times D_d \to \mathbb{R}$, which is defined in terms of the covariance operator applied to function evaluation functionals:

$$K_d(\mathbf{x}, \mathbf{y}) := L_{\mathbf{x}}(C_{\mu_d} L_{\mathbf{y}}) = \int_{\mathcal{F}_d} f(\mathbf{x}) f(\mathbf{y})\, \mu_d(\mathrm{d}f) \quad \forall\, \mathbf{x}, \mathbf{y} \in D_d. \tag{2}$$

We assume that

$$S_d : \mathcal{F}_d \to \mathcal{G}_d$$

is a continuous linear operator, and $\mathcal{G}_d$ is a separable Hilbert space. The operator S_d is called the *solution operator*. Let

$$\nu_d := \mu_d S_d^{-1}. \tag{3}$$

Then ν_d is a zero-mean Gaussian measure on $\mathcal{G}_d$ whose covariance operator $C_{\nu_d} : \mathcal{G}_d^* = \mathcal{G}_d \to \mathcal{G}_d$ is given by

$$C_{\nu_d} g = \int_{\mathcal{F}_d} S_d f \cdot \langle S_d f, g \rangle_{\mathcal{G}_d}\, \mu_d(\mathrm{d}f) = S_d\left(C_{\mu_d}\left(\langle S_d(\cdot), g\rangle_{\mathcal{G}_d}\right)\right) \qquad \forall\, g \in \mathcal{G}_d.$$

Equivalently,

$$C_{\nu_d} g = S_d \left(C_{\mu_d} (L_g S_d) \right) \qquad \forall g \in \mathcal{G}_d,$$

where $L_g(h) = \langle g, h \rangle_{\mathcal{G}_d}$, with $\langle \cdot, \cdot \rangle_{\mathcal{G}_d}$ standing for the inner product of $\mathcal{G}_d$. Then (1) and the continuity of S_d imply that the trace of C_{ν_d} is finite, i.e.,

$$\operatorname{trace}(C_{\nu_d}) = \int_{\mathcal{G}_d} \|g\|_{\mathcal{G}_d}^2 \, \nu_d(\mathrm{d}g) = \int_{\mathcal{F}_d} \|S_d f\|_{\mathcal{G}_d}^2 \, \mu_d(\mathrm{d}f) < \infty.$$

The computational problem addressed in this paper is to approximate $S_d f$ for $f \in \mathcal{F}_d$ by algorithms that use a finite number of evaluations of f. One evaluation is defined as computation of one continuous linear functional of f. For problems considered in this paper, it is known that adaptive choice of linear functionals as well as nonlinear algorithms do not essentially help, see [TWW88, Was86]. Hence, we can restrict our attention to *linear algorithms*, i.e., algorithms of the form

$$Af = \sum_{j=1}^{n} L_j(f)\, g_j,$$

where $g_j \in \mathcal{G}_d$ and L_j are linear functionals from a class Λ of *permissible functionals*. The number n is called the *cardinality* of A, and it is denoted by $\operatorname{card}(A) = n$. This number characterizes the cost of the algorithm A.

We will consider two classes of permissible information functionals. The first class is the class Λ^{all} of all continuous linear functionals, $\Lambda^{\mathrm{all}} = \mathcal{F}_d^*$. In particular, for any $g \in \mathcal{G}_d$, the functional $L_g(f) = \langle S_d f, g \rangle_{\mathcal{G}_d}$ is linear and continuous, and therefore $L_g \in \Lambda^{\mathrm{all}}$. Linear functionals of this kind are used to construct the optimal algorithm (6) below. The second class, which is probably the most important for practical applications, is the class Λ^{std} of function evaluations, which is called *standard information*. That is,

$$L \in \Lambda^{\mathrm{std}} \qquad \text{iff} \qquad \exists \mathbf{x} \in D_d \ \text{ such that } \ L(f) = f(\mathbf{x}) \ \ \forall f \in \mathcal{F}_d.$$

When we consider the class Λ^{std}, we assume that the norm in the space $\mathcal{F}_d$ is chosen so that the mapping $f \mapsto f(\mathbf{x})$ is a continuous linear functional for any $\mathbf{x} \in D_d$. Therefore, $\Lambda^{\mathrm{std}} \subset \Lambda^{\mathrm{all}}$.

In the average case setting the *error* of an algorithm A is defined as the root mean square:

$$e^{\mathrm{avg}}(A; S_d) := \left(\int_{\mathcal{F}_d} \|S_d f - Af\|_{\mathcal{G}_d}^2 \, \mu_d(\mathrm{d}f) \right)^{1/2}.$$

We will also use an abbreviated notation

$$\mathbb{E}_{\mu_d} \|S_d - A\|_{\mathcal{G}_d}^2 = \int_{\mathcal{F}_d} \|S_d f - Af\|_{\mathcal{G}_d}^2 \, \mu_d(\mathrm{d}f),$$

for the mean square error, where $\mathbb{E}_{\mu_d}$ denotes the corresponding expectation. This average case error is well defined and finite since $S_d - A$ is a continuous linear operator.

For $n = 0$, we formally set $A \equiv 0$; then $e^{\mathrm{avg}}(0; S_d)$ is the initial average case error which can be obtained without sampling the functions f. We have

$$[e^{\mathrm{avg}}(0; S_d)]^2 = \mathbb{E}_{\mu_d}\|S_d\|^2_{\mathcal{G}_d} = \int_{\mathcal{F}_d} \|S_d f\|^2_{\mathcal{G}_d}\, \mu_d(\mathrm{d}f) < \infty.$$

We want to reduce the initial average case error by a factor $\varepsilon \in (0,1)$ using an algorithm with the smallest cardinality. Let

$$\begin{aligned} n(\varepsilon; S_d, \Lambda) := \min\{\, & \mathrm{card}(A) \,:\, A \text{ uses } L_1, \dots, L_n \in \Lambda \text{ and} \\ & e^{\mathrm{avg}}(A; S_d) \le \varepsilon\, e^{\mathrm{avg}}(0; S_d)\,\} \end{aligned}$$

denote the minimal number of information evaluations needed for such a reduction.

Recall, that $S := \{S_d\}$. As in [Woź94], we say that the multivariate problem S is *tractable* in the class Λ if there exist non-negative numbers C, p and q such that

$$n(\varepsilon; S_d, \Lambda) \le C\, \varepsilon^{-p}\, d^q \qquad \forall\, \varepsilon \in (0,1),\ \forall\, d \ge 1. \tag{4}$$

Numbers $p = p^{\mathrm{tra}}(S, \Lambda)$ and $q = q^{\mathrm{tra}}(S, \Lambda)$ satisfying (4) are called *ε- and d-exponents of tractability*; we stress that they need not be uniquely defined.

Algorithms $A_{d,\varepsilon}$ are called *polynomial-time* algorithms if for every d and ε, the algorithm $A_{d,\varepsilon}$ uses at most $C\, \varepsilon^{-p}\, d^q$ functional evaluations from the class Λ and has the error bounded by $\varepsilon\, e^{\mathrm{avg}}(0; \mathcal{U}_d)$, i.e., if

$$\mathrm{card}(A_{d,\varepsilon}) \le C\, \varepsilon^{-p}\, d^q \qquad \text{and} \qquad e^{\mathrm{avg}}(A_{d,\varepsilon}; S_d) \le \varepsilon\, e^{\mathrm{avg}}(0; S_d). \tag{5}$$

If $q = 0$ in (4) then we say that the multivariate problem $S = \{S_d\}$ is *strongly tractable* in the class Λ. Moreover, algorithms $A_{d,\varepsilon}$ are called *strongly polynomial-time* algorithms if (5) holds with $q = 0$. The exponent $p^{\mathrm{str}}(S, \Lambda)$ of strong tractability is defined as the infimum of p satisfying (4) with $q = 0$.

3 Tractability for Linear Problems with Λ^{all}

For the class Λ^{all}, it is known that $n(\varepsilon; S_d, \Lambda^{\mathrm{all}})$ is fully characterized by the eigenvalues of the covariance operator C_{ν_d}. As we shall see, this characterization allows us to obtain necessary and sufficient conditions on tractability and strong tractability in terms of these eigenvalues.

We now briefly recall general results on optimal algorithms and the minimal cardinality number $n(\varepsilon; S_d, \Lambda^{\mathrm{all}})$, see [PW90, TWW88] for more details.

The covariance operator C_{ν_d} is self-adjoint and non-negative definite, and has a finite trace equal to $\mathbb{E}_{\mu_d}\|S_d\|^2_{\mathcal{G}_d}$. Consider the sequence, $\{(\lambda_{d,i}, \eta_{d,i})\}_i$, of eigenpairs, i.e.,

$$C_{\nu_d}\eta_{d,i} = \lambda_{d,i}\eta_{d,i},$$

where the eigenvalues are ordered and the eigenfunctions are orthonormal,

$$\lambda_{d,j} \geq \lambda_{d,j+1} \geq 0 \qquad \text{and} \qquad \langle \eta_{d,i}, \eta_{d,j} \rangle_{\mathcal{G}_d} = \delta_{i,j} \qquad \forall i,j \geq 1.$$

The eigenvalues $\lambda_{d,i}$ can be positive only for $\dim(S_d(\mathcal{F}_d))$ indices. In particular, if S_d is a continuous linear functional then $\lambda_{d,i} = 0$ for all $i \geq 2$. To make the problem non-trivial we assume that $\lambda_{d,1} > 0$. For finite dimensional spaces $\mathcal{G}_d$, we set $\lambda_{d,i} = 0$ for $i > \dim(\mathcal{G}_d)$ by convention. Recall that the square of the initial average case error is

$$[e^{\text{avg}}(0; S_d)]^2 = \text{trace}(C_{\nu_d}) = \sum_{\ell=1}^{\infty} \lambda_{d,\ell} < \infty.$$

Algorithm

$$A^*_{d,n} f = \sum_{\ell=1}^{n} \langle S_d f, \eta_{d,\ell} \rangle_{\mathcal{G}_d} \, \eta_{d,\ell} \qquad \forall f \in \mathcal{F}_d, \tag{6}$$

has the smallest average case error among all algorithms of cardinality at most n, and its square average case error is given by the truncated trace:

$$\mathbb{E}_{\mu_d} \|S_d - A^*_{d,n}\|^2_{\mathcal{G}_d} = \sum_{\ell=n+1}^{\infty} \lambda_{d,\ell}.$$

Hence, optimality of $A^*_{d,n}$ yields

$$n(\varepsilon; S_d, \Lambda^{\text{all}}) = \min \left\{ n : \sum_{\ell=n+1}^{\infty} \lambda_{d,\ell} \leq \varepsilon^2 \sum_{\ell=1}^{\infty} \lambda_{d,\ell} \right\}. \tag{7}$$

Furthermore, tractability and strong tractability of the problem S are equivalent to the polynomial-time and strong polynomial-time properties of $A^*_{d,n(\varepsilon;S_d,\Lambda^{\text{all}})}$, respectively.

Using (7), we now show conditions on tractability and strong tractability of the problem $S = \{S_d\}$. As in [HW00], for $r \geq 1$, define

$$M_{d,r} := \frac{\left[\sum_{\ell=1}^{\infty} \lambda_{d,\ell}^{1/r}\right]^r}{\sum_{\ell=1}^{\infty} \lambda_{d,\ell}} = \frac{\left[1 + \sum_{\ell=2}^{\infty} (\lambda_{d,\ell}/\lambda_{d,1})^{1/r}\right]^r}{1 + \sum_{\ell=2}^{\infty} \lambda_{d,\ell}/\lambda_{d,1}}. \tag{8}$$

Clearly $M_{d,1} = 1$. By Jensen's inequality, $M_{d,r} \geq 1$, and it may be infinite for some $r > 1$. Furthermore, for $r > 1$, $M_{d,r} = 1$ iff $\lambda_{d,\ell} = 0$ for all $\ell \geq 2$.

Lemma 1 *If there are $r > 1$ and $\alpha \geq 0$ such that*

$$M := \sup_d d^{-\alpha} M_{d,r} < \infty \tag{9}$$

then

$$n(\varepsilon; S_d, \Lambda^{\mathrm{all}}) \le \left\lceil \left(\frac{M}{r-1}\right)^{1/(r-1)} d^{\alpha/(r-1)} \left(\frac{1}{\varepsilon}\right)^{2/(r-1)} \right\rceil .$$

Hence, the problem $S = \{S_d\}$ is strongly tractable in the class Λ^{all} if $\alpha = 0$, and tractable in the class Λ^{all} if $\alpha > 0$.

Proof. We essentially repeat the proof of [HW00, Corollary 1], where only strong tractability is considered. Denote $M_{\mathrm{init}} := \sum_{\ell=1}^{\infty} \lambda_{d,\ell} = e^{\mathrm{avg}}(0; S_d)^2$. Since $\lambda_{d,j} \ge \lambda_{d,j+1}$ for all $j \ge 1$, we have $j\lambda_{d,j}^{1/r} \le \sum_{\ell=1}^{\infty} \lambda_{d,\ell}^{1/r} = (M_{d,r} M_{\mathrm{init}})^{1/r}$ $\le (d^{\alpha} M\, M_{\mathrm{init}})^{1/r}$, and therefore

$$\lambda_{d,j} \le \frac{M\, d^{\alpha}}{j^r} M_{\mathrm{init}} \qquad \forall\, j = 1, 2, \dots . \tag{10}$$

Hence

$$\sum_{\ell=n+1}^{\infty} \lambda_{d,\ell} \le M\, d^{\alpha}\, M_{\mathrm{init}} \sum_{\ell=n+1}^{\infty} j^{-r}.$$

Since $\sum_{\ell=n+1}^{\infty} j^{-r} \le \int_n^{\infty} x^{-r}\, \mathrm{d}x = (r-1)^{-1} n^{-(r-1)}$, we obtain

$$\sum_{\ell=n+1}^{\infty} \lambda_{d,\ell} \le \frac{M\, d^{\alpha}}{r-1} \left(\frac{1}{n}\right)^{r-1} M_{\mathrm{init}}.$$

From (7) we see that it is enough to choose n such that $M d^{\alpha} n^{-r+1}/(r-1) \le \varepsilon^2$ which yields the needed bound on $n(\varepsilon; S_d, \Lambda^{\mathrm{all}})$. □

Lemma 1 relates the numbers r and α to the exponents of tractability. We now show that an opposite estimate also holds. That is, the exponents of tractability yield numbers r and α for which (9) holds.

Lemma 2 *If for some non-negative C, p and q*

$$n(\varepsilon; S_d, \Lambda^{\mathrm{all}}) \le C\, \varepsilon^{-p}\, d^{q} \qquad \forall\, \varepsilon \in (0,1),\ \forall\, d = 1, 2, \dots$$

then

$$\sup_d d^{-q(r-1)} M_{d,r} < \infty \quad \forall\, r \in [1, 1+2/p).$$

Proof. Again as in the proof of [HW00, Corollary 1], let $m = m(\varepsilon, d) = \lceil C\varepsilon^{-p} d^{q} \rceil$. Since $\varepsilon \in (0,1)$ then $m \in [\lceil C d^{q} \rceil, \infty)$. We have $m \le C\varepsilon^{-p} d^{q} + 1$, and so

$$\varepsilon \le (C d^{q})^{1/p}\, (m-1)^{-1/p}$$

for $m \ge 2$. From (7) it follows that $m\lambda_{d,2m} \le \sum_{\ell=m+1}^{\infty} \lambda_{d,\ell} \le \varepsilon^2 M_{\mathrm{init}}$, with $M_{\mathrm{init}} = \sum_{\ell=1}^{\infty} \lambda_{d,\ell}$, which yields

$$\lambda_{d,2m} \le \left(\frac{1}{m-1}\right)^{1+2/p} (C\,d^q)^{2/p}\,M_{\rm init} \qquad \forall m \ge \lceil C\,d^q \rceil + 1.$$

We now estimate

$$\begin{aligned}\sum_{\ell=1}^{\infty} \lambda_{d,\ell}^{1/r} &= \lambda_{d,1}^{1/r} + \sum_{j=1}^{\infty}\left(\lambda_{d,2j}^{1/r} + \lambda_{d,2j+1}^{1/r}\right) \le \lambda_{d,1}^{1/r} + 2\sum_{j=1}^{\infty}\lambda_{d,2j}^{1/r}\\ &\le \lambda_{d,1}^{1/r} + 2\sum_{j=1}^{\lceil C\,d^q\rceil}\lambda_{d,2j}^{1/r} + 2\,(C\,d^q)^{2/(r\,p)}\\ &\quad\times\left(\sum_{j=\lceil C\,d^q\rceil+1}^{\infty}\left(\frac{1}{m-1}\right)^{(1+2/p)/r}\right) M_{\rm init}^{1/r}.\end{aligned}$$

Using $(a+b)^r \le 2^{r-1}(a^r+b^r)$ for non-negative a and b, we obtain

$$\begin{aligned}d^{-\alpha}\,M_{d,r} \le 2^{r-1}&\left(\frac{\left(\lambda_{d,1}^{1/r} + 2\sum_{j=1}^{\lceil C\,d^q\rceil}\lambda_{d,2j}^{1/r}\right)^r}{d^{\alpha}\sum_{j=1}^{\infty}\lambda_{d,j}}\right.\\ &\left.+\frac{(C\,d^q)^{2/p}}{d^{\alpha}\,\left[(1+2/p)/r-1\right]^r\,(C\,d^q)^{1+2/p-r}}\right).\end{aligned}$$

Since $\left(\sum_{j=1}^{\lceil C\,d^q\rceil}\lambda_{d,2j}^{1/r}\right)^r \le \left(\sum_{j=1}^{\lceil C\,d^q\rceil}\lambda_{d,2j}\right)(\lceil C\,d^q\rceil)^{r-1}$ by Hölder's inequality, we see that both terms in the displayed formula are uniformly bounded in d for $r\in[1,1+2/p)$ and $\alpha = q(r-1)$. This completes the proof. □

From Lemmas 1 and 2 imply necessary and sufficient conditions for the problem S to be strongly tractable or tractable. Define

$$r(S) = \sup\{\,r\ge 1:\ \sup_d M_{d,r} < \infty\,\}. \tag{11}$$

Note that $r(S)$ is well-defined since $M_{d,1}=1$. We have the following theorem.

Theorem 1

- *The problem S is strongly tractable in the class $\Lambda^{\rm all}$ iff $r(S)>1$. If this holds then the exponent of strong tractability is*

$$p^{\rm str}(S,\Lambda^{\rm all}) = \frac{2}{r(S)-1}.$$

- *The problem S is tractable in the class $\Lambda^{\rm all}$ iff there are numbers $r>1$ and $\alpha\ge 0$ such that*

$$M := \sup_d d^{-\alpha} M_{d,r} < \infty.$$

If this holds then the exponents of tractability are

$$p^{\rm tra}(S,\Lambda^{\rm all}) = \frac{2}{r-1} \quad \textit{and} \quad q^{\rm tra}(S,\Lambda^{\rm all}) = \frac{\alpha}{r-1},$$

and

$$n(\varepsilon; S_d, \Lambda^{\rm all}) \le \left\lceil \left(\frac{M}{r-1}\right)^{1/(r-1)} d^{\alpha/(r-1)} \left(\frac{1}{\varepsilon}\right)^{2/(r-1)} \right\rceil.$$

Proof. First, consider strong tractability. If S is strongly tractable then Lemma 2 holds with $\alpha = 0$ and r arbitrarily close to $1 + 2/p^{\rm str}(\Lambda^{\rm all})$. Then $r(S) \ge 1 + 2/p^{\rm str}(\Lambda^{\rm all}) > 1$, as claimed. The last inequality is equivalent to $p^{\rm str}(\Lambda^{\rm all}) \ge 2/(r(S)-1)$. Now assume that $r(S) > 1$. Then Lemma 1 holds with $\alpha = 0$ and r arbitrarily close to $r(S)$, and therefore it is larger than 1. Then $n(\varepsilon, S_d, \Lambda^{\rm all})$ is uniformly bounded in d by a polynomial in ε^{-1} of order arbitrarily close to $2/(r(S)-1)$. This yields strong tractability with $p^{\rm str}(\Lambda^{\rm all}) \le 2/(r(S)-1)$. This completes the proof of this part.

Consider now tractability. If S is tractable then Lemma 2 holds $r > 1$ and $\alpha \ge 0$ and yields that $d^{-\alpha} M_{d,r}$ is uniformly bounded in d, as claimed. On the other hand, if $d^{-\alpha} M_{d,r}$ is uniformly bounded in d for $r > 1$ and $\alpha \ge 0$ then Lemma 1 yields tractability of S and the bound on $n(\varepsilon; S_d, \Lambda^{\rm all})$. □

In general, it is difficult to find the eigenvalues $\lambda_{d,\ell}$ and to check, in particular, whether $r(S) > 1$. This problem is simplified if we assume the following additional properties of the multivariate problem.

Assume that $\mathcal{G}_d$ is a tensor product of d copies of an infinite dimensional separable Hilbert space $\mathcal{G}_1$, i.e., $\mathcal{G}_d = \bigotimes_{k=1}^d \mathcal{G}_1$. More specifically, $\mathcal{G}_d$ is a separable Hilbert space spanned by functions of the form $\bigotimes_{k=1}^d g_k$ with $g_k \in \mathcal{G}_1$, and the inner-product in $\mathcal{G}_d$ satisfies $\left\langle \bigotimes_{k=1}^d g_k, \bigotimes_{k=1}^d h_k \right\rangle_{\mathcal{G}_d} = \prod_{k=1}^d \langle g_k, h_k\rangle_{\mathcal{G}_1}$ for $f_k,\ g_k \in \mathcal{G}_1$. Let $\{\eta_i\}_i$ be an orthonormal system of $\mathcal{G}_1$. Then for $d \ge 1$ and $\mathbf{\imath} = [i_1, i_2, \dots, i_d]$ with $i_j \ge 1$, the system $\{\eta_{d,\mathbf{\imath}}\}$ with $\eta_{d,\mathbf{\imath}} = \bigotimes_{j=1}^d \eta_{i_j}$ is an orthonormal system of $\mathcal{G}_d$. Observe that

$$S_d f = \sum_{\mathbf{\imath}} \langle S_d f, \eta_{d,\mathbf{\imath}}\rangle_{\mathcal{G}_d}\, \eta_{d,\mathbf{\imath}} \qquad \text{and} \qquad \eta_{d,\mathbf{\imath}} = \bigotimes_{j=1}^d \eta_{i_j}. \tag{12}$$

It is enough to know the distribution of linear functionals $L_{\mathbf{\imath}}(f) = \langle S_d f, \eta_{d,\mathbf{\imath}}\rangle_{\mathcal{G}_d}$. Since μ_d has zero mean, we have

$$\mathbb{E}_{\mu_d} L_{\mathbf{\imath}}(f) = \int_{\mathcal{G}_d} \langle g, \eta_{d,\mathbf{\imath}}\rangle_{\mathcal{G}_d}\, \nu_d(\mathrm{d}g) = 0 \quad \forall\, \mathbf{\imath},$$

i.e., the expectations of these linear functionals vanish. We assume that for different values of $\mathbf{\imath}$, the functionals $L_{\mathbf{\imath}}$, viewed as random variables, are independent, i.e.,

$$\mathbb{E}_{\mu_d} L_{\imath}(f) L_{\jmath}(f) = \int_{\mathcal{G}_d} \langle g, \eta_{d,\imath} \rangle_{\mathcal{G}_d} \langle g, \eta_{d,\jmath} \rangle_{\mathcal{G}_d} \nu_d(\mathrm{d}g) = \lambda_{d,\imath}\, \delta_{\imath,\jmath}\,. \tag{13}$$

Hence,

$$C_{\nu_d} \eta_{d,\imath} = \lambda_{d,\imath}\, \eta_{d,\imath} \quad \forall\, \imath\,.$$

We consider a specific choice of eigenvalues $\lambda_{d,\imath}$, which correspond to the variances of $L_{\imath}(f)$. For $d = 1$, we assume that $\lambda_{1,i} = \lambda_i$ with $\lambda_1 \geq \lambda_2 \geq \cdots \geq 0$. Of course, we must have

$$\sum_{i=1}^{\infty} \lambda_i = \mathrm{trace}(C_{\nu_1}) < \infty. \tag{14}$$

For $d \geq 1$, let $\boldsymbol{\gamma} = \{\gamma_{d,\mathfrak{u}}\}$, where $\mathfrak{u}$ is an arbitrary subset of indices from $\{1, 2, \ldots, d\}$, with $\gamma_{1,\emptyset} = \gamma_{1,\{1\}} = 1$. For a given $\imath = [i_1, i_2, \ldots, i_d]$ with positive integer components, let $\mathfrak{u}_{\imath} = \{j : i_j > 1\}$ denote the set of indices larger than one. Of course, $\mathfrak{u}_{\imath} = \emptyset$ iff $\imath = \mathbf{1} = [1, \ldots, 1]$. We assume that the eigenvalues (variances), $\lambda_{d,\imath}$, have a product form like the eigenfunctions, $\eta_{d,\imath}$, do:

$$\lambda_{d,\imath} = \gamma_{d,\mathfrak{u}_{\imath}} \prod_{j=1}^{d} \lambda_{i_j} = \gamma_{d,\mathfrak{u}_{\imath}}\, \lambda_1^{d-|\mathfrak{u}_{\imath}|} \prod_{j\in\mathfrak{u}_{\imath}} \lambda_{i_j}. \tag{15}$$

We now explain the role of the weight family $\boldsymbol{\gamma} = \{\gamma_{d,\mathfrak{u}}\}$. Consider first $\gamma_{d,\mathfrak{u}} \equiv 1$. Then

$$\lambda_{d,\imath} = \prod_{j\notin\mathfrak{u}_{\imath}} \lambda_1 \prod_{j\in\mathfrak{u}_{\imath}} \lambda_{i_j} = \prod_{j=1}^{d} \lambda_{i_j}.$$

This corresponds to a tensor product structure of $\mathcal{F}_d$, μ_d and S_d, see the examples in the next section.

We now consider arbitrary $\gamma_{d,\mathfrak{u}}$. Suppose that $\mathfrak{u} = \emptyset$. Then

$$\lambda_{d,\mathbf{1}} = \gamma_{d,\emptyset}\, \lambda_1^d$$

which means that the variance of $L_{\mathbf{1}}(f)$ is weighted by $\gamma_{d,\emptyset}$. Now let $\mathfrak{u}$ be non-empty. Consider all indices $\imath$ for which $\mathfrak{u}_{\imath} = \mathfrak{u}$, i.e., $i_j > 1$ for all $j \in \mathfrak{u}$. Then

$$\lambda_{d,\imath} = \gamma_{d,\mathfrak{u}}\, \lambda_1^{d-|\mathfrak{u}|} \prod_{j\in\mathfrak{u}} \lambda_{i_j}$$

which means that the variances of $L_{\imath}(f)$ for all such $\imath$ are weighted by the same $\gamma_{d,\mathfrak{u}}$.

Suppose for a moment that $\gamma_{d,\mathfrak{u}} = 0$ for some $\mathfrak{u}$. Then the variances of $L_{\imath}(f)$ are zero for all indices $\imath$ for which $\mathfrak{u}_{\imath} = \mathfrak{u}$. Thus, with probability one, the elements $S_d f$ have zero coefficients $L_{\imath}(f) = \langle S_d f, \eta_{\imath} \rangle_{\mathcal{G}_d}$ in (12) for all $\imath$ with $\mathfrak{u}_{\imath} = \mathfrak{u}$. Similarly, for small $\gamma_{d,\mathfrak{u}}$ the coefficients $\langle S_d f, \eta_{\imath} \rangle_{\mathcal{G}_d}$ play a less significant role than coefficients corresponding to larger $\gamma_{d,\mathfrak{u}}$.

That is how the weight sequence $\boldsymbol{\gamma}$ allows us to model various dependence on groups of indices (variables) of the coefficients of $S_d f$ in (12). The weights allow us to make the class of problems small enough that there is a possibility of tractability or strong tractability, and that it can be verified.

As in the worst case setting, see [DSWW06, SWW04, WW04, WW05], we say that $\boldsymbol{\gamma} = \{\gamma_{d,\mathfrak{u}}\}$ is a family of *finite-order weights* if there exists an integer q such that

$$\gamma_{d,\mathfrak{u}} = 0 \qquad \text{for all } d \text{ and } \mathfrak{u} \text{ with } |\mathfrak{u}| > q. \tag{16}$$

The smallest integer q satisfying (16) is called the *order* of $\boldsymbol{\gamma}$, and is denoted by q^*. The essence of finite-order weights is that we know a priori that the coefficients $\langle S_d f, \eta_{\boldsymbol{\imath}} \rangle_{\mathcal{G}_d}$ are zero (with probability one) for all vectors $\boldsymbol{\imath}$ with more than q^* indices greater than one.

We say that $\boldsymbol{\gamma} = \{\gamma_{d,\mathfrak{u}}\}$ is a family of *product weights* if

$$\gamma_{d,\emptyset} = 1, \quad \gamma_{d,\mathfrak{u}} = \prod_{j \in \mathfrak{u}} \gamma_{d,j} \ \text{ for all non-empty } \mathfrak{u} \text{ and all } d,$$

for some $\gamma_{d,1} \ge \gamma_{d,2} \ge \cdots \ge \gamma_{d,d} \ge 0$. Finally, we say that $\boldsymbol{\gamma} = \{\gamma_{d,\mathfrak{u}}\}$ is a family of *uniform product weights* if $\gamma_{d,j}$ above do not depend on d, i.e., $\gamma_{d,j} = \gamma_j$ and

$$\gamma_{d,\emptyset} = 1, \quad \gamma_{d,\mathfrak{u}} = \prod_{j \in \mathfrak{u}} \gamma_j \quad \forall\, d, \forall\, \mathfrak{u} \neq \emptyset,$$

for some $\gamma_1 \ge \gamma_2 \ge \cdots \ge 0$. For product weights and uniform product weights we know a priori that the coefficients $\langle S_d f, \eta_{\boldsymbol{\imath}} \rangle_{\mathcal{G}_d}$ are weighted by $\gamma_{d,\mathfrak{u}_{\boldsymbol{\imath}}} = \prod_{j \in \mathfrak{u}_{\boldsymbol{\imath}}} \gamma_{d,j}$ or by $\gamma_{d,\mathfrak{u}_{\boldsymbol{\imath}}} = \prod_{j \in \mathfrak{u}_{\boldsymbol{\imath}}} \gamma_j$, respectively.

We are ready to study tractability for eigenvalues $\lambda_{d,\boldsymbol{\imath}}$ satisfying (15) with $\lambda_1 > 0$ and various weights $\gamma_{d,\mathfrak{u}}$. Observe that (15) implies that

$$\begin{aligned}\sum_{\boldsymbol{\imath} \in \mathbb{N}_+^d} \lambda_{d,\boldsymbol{\imath}} &= \gamma_{d,\emptyset} \lambda_1^d + \sum_{\emptyset \neq \mathfrak{u} \subseteq \{1,\ldots,d\}} \gamma_{d,\mathfrak{u}} \lambda_1^{d-|\mathfrak{u}|} \left(\sum_{j=2}^{\infty} \lambda_j \right)^{|\mathfrak{u}|} \\ &= \lambda_1^d \sum_{\mathfrak{u} \subseteq \{1,\ldots,d\}} \gamma_{d,\mathfrak{u}} \left(\sum_{j=2}^{\infty} (\lambda_j / \lambda_1) \right)^{|\mathfrak{u}|}\end{aligned}$$

with the convention that $0^0 = 1$. Therefore we have

$$M_{d,r} = \frac{\left(\sum_{\mathfrak{u} \subseteq \{1,\ldots,d\}} \gamma_{d,\mathfrak{u}}^{1/r} \left(\sum_{j=2}^{\infty} (\lambda_j/\lambda_1)^{1/r} \right)^{|\mathfrak{u}|} \right)^r}{\sum_{\mathfrak{u} \subseteq \{1,\ldots,d\}} \gamma_{d,\mathfrak{u}} \left(\sum_{j=2}^{\infty} (\lambda_j/\lambda_1) \right)^{|\mathfrak{u}|}}.$$

For $r \ge 1$, let

$$\beta_r := \sum_{j=2}^{\infty} \left(\frac{\lambda_j}{\lambda_1} \right)^{1/r}. \tag{17}$$

If $\lambda_i = 0$ for all $i \geq 2$, which happens when S_d is a continuous linear functional for all d, then $\beta_r = 0$ and $M_{r,d} = 1$ for all $r \geq 1$. Due to Theorem 1, this yields strong tractability with the exponent zero.

From now on we assume that $\lambda_1 \geq \lambda_2 > 0$, i.e., we have at least two positive eigenvalues of C_{ν_1}. Then $\beta_r > 0$. Note that $\beta_1 < \infty$ due to (14) and, depending on λ_j, we may have $\beta_r = \infty$ for some r. Indeed, if $\lambda_j = \Theta(j^{-\alpha^*})$ then to satisfy (14) we must assume that $\alpha^* > 1$. Then $\beta_r < \infty$ iff $r \in [1, \alpha^*)$. Note that β_r is increasing as a function of r, and by using Jensen's inequality we have $\beta_1^{1/r} \leq \beta_r$.

Assume that $\beta_r < \infty$ for some $r > 1$, i.e.,

$$r(\lambda) := \sup\{ r \geq 1 \, : \, \beta_r < \infty \} > 1. \tag{18}$$

For $r \in [1, r(\lambda))$, we have

$$M_{d,r} = \frac{\left(\sum_{\mathfrak{u} \subseteq \{1,\dots,d\}} \gamma_{d,\mathfrak{u}}^{1/r} \, \beta_r^{|\mathfrak{u}|} \right)^r}{\sum_{\mathfrak{u} \subseteq \{1,\dots,d\}} \gamma_{d,\mathfrak{u}} \, \beta_1^{|\mathfrak{u}|}}.$$

From Theorem 1 we easily conclude the following corollary, which expresses strong tractability and tractability in terms of $\gamma_{d,\mathfrak{u}}$ and β_r.

Corollary 1 *Consider the problem S with eigenvalues $\lambda_{d,\mathfrak{i}}$ given by (15) with $\lambda_1 \geq \lambda_2 > 0$ and $r(\lambda) > 1$.*

- *The problem S is strongly tractable in the class Λ^{all} iff*

$$r(S) = \sup \left\{ r \in [1, r(\lambda)) \, : \, \sup_d \frac{\left(\sum_{\mathfrak{u} \subseteq \{1,\dots,d\}} \gamma_{d,\mathfrak{u}}^{1/r} \, \beta_r^{|\mathfrak{u}|} \right)^r}{\sum_{\mathfrak{u} \subseteq \{1,\dots,d\}} \gamma_{d,\mathfrak{u}} \, \beta_1^{|\mathfrak{u}|}} < \infty \right\} > 1.$$

  If this holds then the exponent of strong tractability is $p^{\mathrm{str}}(S, \Lambda^{\mathrm{all}}) = 2/(r(S) - 1)$.
- *The problem S is tractable in the class Λ^{all} iff there are numbers $r \in (1, r(\lambda))$ and $\alpha \geq 0$ such that*

$$M := \sup_d d^{-\alpha} \frac{\left(\sum_{\mathfrak{u} \subseteq \{1,\dots,d\}} \gamma_{d,\mathfrak{u}}^{1/r} \, \beta_r^{|\mathfrak{u}|} \right)^r}{\sum_{\mathfrak{u} \subseteq \{1,\dots,d\}} \gamma_{d,\mathfrak{u}} \, \beta_1^{|\mathfrak{u}|}} < \infty.$$

  If this holds then the exponents of tractability are

$$p^{\mathrm{tra}}(S, \Lambda^{\mathrm{all}}) = \frac{2}{r-1} \quad \textit{and} \quad q^{\mathrm{tra}}(S, \Lambda^{\mathrm{all}}) = \frac{\alpha}{r-1},$$

  and we have

$$n(\varepsilon; S_d, \Lambda^{\mathrm{all}}) \leq \left\lceil \left(\frac{M}{r-1} \right)^{1/(r-1)} d^{\alpha/(r-1)} \left(\frac{1}{\varepsilon} \right)^{2/(r-1)} \right\rceil.$$

We now elaborate on finite-order and product weights. For finite-order weights of order q^*, strong tractability can be entirely expressed in terms of the family γ of weights, and the only dependence on the eigenvalues λ_j is through $r(\lambda)$.

Indeed, for $r \in (1, r(\lambda))$, denote

$$C^1_{r,q^*} = \frac{\min\{1, \beta_r^{q^* r}\}}{\max\{1, \beta_1^{q^*}\}}, \quad C^2_{r,q^*} = \frac{\max\{1, \beta_r^{q^* r}\}}{\min\{1, \beta_1^{q^*}\}}.$$

Then

$$C^1_{r,q^*} \frac{\left(\sum_{|\mathfrak{u}|\le q^*} \gamma_{d,\mathfrak{u}}^{1/r}\right)^r}{\sum_{|\mathfrak{u}|\le q^*} \gamma_{d,\mathfrak{u}}} \le M_{r,d} \le C^2_{r,q^*} \frac{\left(\sum_{|\mathfrak{u}|\le q^*} \gamma_{d,\mathfrak{u}}^{1/r}\right)^r}{\sum_{|\mathfrak{u}|\le q^*} \gamma_{d,\mathfrak{u}}}.$$

Hence, strong tractability holds iff $\sup_d \left(\sum_{|\mathfrak{u}|\le q^*} \gamma_{d,\mathfrak{u}}^{1/r}\right)^r / \sum_{|\mathfrak{u}|\le q^*} \gamma_{d,\mathfrak{u}} < \infty$ for some $r > 1$.

To obtain tractability observe that using Hölder's inequality we have

$$M_{d,r} = \frac{\left(\sum_{|\mathfrak{u}|\le q^*} \gamma_{d,\mathfrak{u}}^{1/r} \beta_1^{|\mathfrak{u}|/r} \beta_r^{|\mathfrak{u}|} \beta_1^{-|\mathfrak{u}|/r}\right)^r}{\sum_{|\mathfrak{u}|\le q^*} \gamma_{d,\mathfrak{u}} \beta_1^{|\mathfrak{u}|}} \le \left(\sum_{|\mathfrak{u}|\le q^*} \left(\frac{\beta_r}{\beta_1^{1/r}}\right)^{|\mathfrak{u}| r'}\right)^{r/r'},$$

where $1/r' + 1/r = 1$. Since $\beta_r/\beta_1^{1/r} \ge 1$ and the cardinality of the sum for $|\mathfrak{u}| \le q^*$ is

$$\binom{d}{0} + \binom{d}{1} + \cdots + \binom{d}{\min\{q^*, d\}} \le 2\, d^{q^*},$$

see [WW05], we have

$$M_{d,r} \le 2^{r-1} \left(\frac{\beta_r}{\beta_1^{1/r}}\right)^{q^* r} d^{q^*(r-1)} \quad \forall d.$$

Hence

$$\sup_d d^{-q^*(r-1)} M_{d,r} \le 2^{r-1} \left(\frac{\beta_r}{\beta_1^{1/r}}\right)^{q^* r}.$$

This and Theorem 1 imply the following corollary:

Corollary 2 *Consider the problem S with eigenvalues $\lambda_{d,\mathfrak{i}}$ given by (15) with $\lambda_1 \ge \lambda_2 > 0$ and $r(\lambda) > 1$. Let $\boldsymbol{\gamma} = \{\gamma_{d,\mathfrak{u}}\}$ be a sequence of finite-order weights of order q^*.*

- *The problem S is strongly tractable in the class Λ^{all} iff*

$$r(S) = \sup\left\{ r \in [1, r(\lambda)) : \sup_d \frac{\left(\sum_{|\mathfrak{u}|\le q^*} \gamma_{d,\mathfrak{u}}^{1/r}\right)^r}{\sum_{|\mathfrak{u}|\le q^*} \gamma_{d,\mathfrak{u}}} < \infty \right\} > 1. \quad (19)$$

If this holds then the exponent of strong tractability is $p^{\mathrm{str}}(S, \Lambda^{\mathrm{all}}) = 2/(r(S) - 1)$.

- *The problem S is tractable in the class $\Lambda^{\rm all}$. For $r \in (1, r(\lambda))$, the exponents of tractability are*

$$p^{\rm tra}(S, \Lambda^{\rm all}) = \frac{2}{r-1} \quad \textit{and} \quad q^{\rm tra}(S, \Lambda^{\rm all}) = q^*,$$

and

$$n(\varepsilon; S_d, \Lambda^{\rm all}) \le \left\lceil \left(\frac{2}{r-1}\left(\frac{\beta_r}{\beta_1^{1/r}}\right)^{q^* r}\right)^{1/(r-1)} d^{q^*} \left(\frac{1}{\varepsilon}\right)^{2/(r-1)} \right\rceil.$$

We stress that tractability holds for *arbitrary* finite-order weights with the d-exponent equal to the order q^* whereas the ε-exponent can be arbitrarily close to $2/(r(\lambda)-1)$.

We now turn to product and uniform product weights, where

$$M_{d,r} = \frac{\prod_{j=1}^d \left(1 + \gamma_{d,j}^{1/r}\beta_r\right)^r}{\prod_{j=1}^d (1 + \gamma_{d,j}\beta_1)}.$$

Then it is easy to check that for $r \in (1, r(\lambda))$ we have

$$\sup_d M_{d,r} < \infty \quad \text{iff} \quad \limsup_{d\to\infty} \sum_{j=1}^d \gamma_{d,j}^{1/r} < \infty$$

and

$$M_\alpha := \sup_d d^{-\alpha} M_{d,r} < \infty \text{ for some } \alpha \ge 0 \text{ iff } A_r := \limsup_{d\to\infty} \frac{\sum_{j=1}^d \gamma_{d,j}^{1/r}}{\ln(d+1)} < \infty.$$

In fact, $M_\alpha < \infty$ if $\alpha > r\beta_r A_r$.

We now show that the concepts of strong tractability and tractability are equivalent for uniform product weights, see [KSW07] where this point is also discussed. It is enough to show that tractability implies strong tractability. Assume then that $A_r < \infty$ for some $r \in (1, r(\lambda))$. We now have $\gamma_{d,j} = \gamma_j$ for non-increasing $\{\gamma_j\}$, and

$$\frac{k\,\gamma_k^{1/r}}{\ln(k+1)} \le \frac{\sum_{j=1}^k \gamma_j^{1/r}}{\ln(k+1)} \le A := \sup_d \frac{\sum_{j=1}^d \gamma_j^{1/r}}{\ln(d+1)} < \infty.$$

From this we have

$$\gamma_k \le \frac{(A\,\ln(k+1))^r}{k^r} \qquad \forall\, k = 1, 2, \dots, .$$

This implies that $\sum_{k=1}^\infty \gamma_k^{1/p} < \infty$ for all $p \in (1, r)$. Hence, $\sup_d M_{d,p} < \infty$, which implies strong tractability.

We stress that the concepts of strong tractability and tractability do not coincide for non-uniform product weights. Indeed, take $\gamma_{d,j} = 1$ for $j = 1, 2, \ldots, \lceil \ln(d+1) \rceil$, and $\gamma_{d,j} = d^{-1-\beta}$ for $j = \lceil \ln(d+1) \rceil + 1, \ldots, d$ and a positive β. Then $\limsup_d \sum_{j=1}^d \gamma_{d,j}^{1/r} = \infty$ for all $r > 1$, which implies the lack of strong tractability, whereas $A_r < \infty$ for all $r \in (1, 1+\beta]$, which implies tractability.

This discussion and Theorem 1 yield the following corollary:

Corollary 3 *Consider the problem S with the eigenvalues $\lambda_{d,\mathfrak{i}}$ given by (15) with $\lambda_1 \geq \lambda_2 > 0$ and $r(\lambda) > 1$. Let $\gamma = \{\gamma_{d,\mathfrak{u}}\}$ be a family of product or uniform product weights.*

- *The problem S is strongly tractable in the class Λ^{all} iff*

$$r(S) = \sup \left\{ r \in [1, r(\lambda)) : \limsup_{d\to\infty} \sum_{j=1}^{d} \gamma_{d,j}^{1/r} < \infty \right\} > 1.$$

  If this holds then the exponent of strong tractability is $p^{\text{str}}(S, \Lambda^{\text{all}}) = 2/(r(S)-1)$.
- *The problem S is tractable in the class Λ^{all} iff for some $r \in (1, r(\lambda))$, we have*

$$A_r := \limsup_{d\to\infty} \frac{\sum_{j=1}^{d} \gamma_{d,j}^{1/r}}{\ln(d+1)} < \infty.$$

  Then the exponents of tractability are

$$p^{\text{tra}}(S, \Lambda^{\text{all}}) = \frac{2}{r-1} \quad \text{and} \quad q^{\text{tra}}(S, \Lambda^{\text{all}}) = \alpha > \frac{r\beta_r}{r-1} A_r.$$

  For such r and α we have $M := \sup_d d^{-\alpha}(r-1)M_{d,r} < \infty$ and

$$n(\varepsilon; S_d, \Lambda^{\text{all}}) \leq \left\lceil \left(\frac{M}{r-1}\right)^{1/(r-1)} d^{\alpha} \left(\frac{1}{\varepsilon}\right)^{2/(r-1)} \right\rceil.$$

- *The concepts of strong tractability and tractability coincide for uniform product weights.*

4 Illustration

We illustrate the assumptions and results of the previous section by a several examples.

Example 1 (Continuous functions). We consider the class of continuous functions $\mathcal{F}_d = C(D_d)$ defined on the d-dimensional unit cube $D_d = [0,1]^d$, i.e., $m = 1$, with the norm $\|f\| = \max_{\mathbf{x}\in D_d} |f(\mathbf{x})|$. We analyze the approximation problem $S_d : \mathcal{F}_d \to \mathcal{G}_d = L_2([0,1]^d)$ given by $S_d f = f$. The space $\mathcal{F}_d$ is equipped with the measure μ_d which is zero-mean Gaussian with the covariance kernel

$$K_d(\mathbf{x},\mathbf{y}) = \gamma_{d,\emptyset} + \sum_{\mathfrak{u}\neq\emptyset} \gamma_{d,\mathfrak{u}}\, K_{d,\mathfrak{u}}(\mathbf{x},\mathbf{y}) \quad \forall \mathbf{x},\mathbf{y} \in D_d,$$

where

$$K_{d,\mathfrak{u}}(\mathbf{x},\mathbf{y}) = \prod_{k\in\mathfrak{u}} K(x_k,y_k) \quad \text{and} \quad K(x,y) = \min(x,y) - 3\left(x-\frac{x^2}{2}\right)\left(y-\frac{y^2}{2}\right).$$

The function K is the covariance kernel of the Gaussian measure μ on $C([0,1])$ obtained from the the classical Wiener measure[4] ω under the condition that $\int_0^1 f(x)\,\mathrm{d}x = 0$. That is, μ is the classical Wiener measure concentrated on the set of continuous functions with zero integral. Functions with nonzero integrals arise from the constant term in $K_d(\mathbf{x},\mathbf{y})$.

From the definition K_d, it follows that the functions from $\mathcal{F}_d$ can be viewed as a $\boldsymbol{\gamma}$-weighted sum

$$f = \sum_{\mathfrak{u}\subseteq\{1,\ldots,d\}} \gamma_{d,\mathfrak{u}}\, f_{\mathfrak{u}}$$

of independent Gaussian processes $f_{\mathfrak{u}}$, each with covariance kernel equal to $K_{d,\mathfrak{u}}$. Note also that each $f_{\mathfrak{u}}$ depends only on the variables x_k with $k \in \mathfrak{u}$. In particular, $f_\emptyset$ corresponds to a constant function whose value is a $\mathcal{N}(0,1)$ random variable.

Since S_d is the embedding operator, the measure ν_d has the same covariance kernel K_d. The covariance operator C_{ν_d} is of the form

$$(C_{\nu_d} g)(\mathbf{x}) = \int_{D_d} K_d(\mathbf{x},\mathbf{y}) g(\mathbf{y})\,\mathrm{d}\mathbf{y}.$$

Therefore the eigenpairs $(\lambda_{d,\mathfrak{i}}, \eta_{d,\mathfrak{i}})$ are the solutions of

$$\int_{D_d} K_d(\mathbf{x},\mathbf{y})\, \eta_{d,\mathfrak{i}}(\mathbf{y})\,\mathrm{d}\mathbf{y} = \lambda_{d,\mathfrak{i}}\, \eta_{d,\mathfrak{i}}(\mathbf{x}).$$

From the tensor product form of $K_{d,\mathfrak{u}}$ and the fact that the integral $\int_0^1 K(x,y)\,\mathrm{d}y = 0$ for any x, we see that the eigenvalues of C_{ν_d} are of the form (15), and the eigenpairs are

$$\eta_{d,\mathfrak{i}}(\mathbf{x}) = \prod_{k\in\mathfrak{u}_{\mathfrak{i}}} \eta_{i_k}(x_k) \quad \text{and} \quad \lambda_{d,\mathfrak{i}} = \gamma_{d,\mathfrak{u}_{\mathfrak{i}}}\, \lambda_1^{d-|\mathfrak{u}_{\mathfrak{i}}|} \prod_{k\in\mathfrak{u}_{\mathfrak{i}}} \lambda_{i_k},$$

[4] Property (15) does not hold for the classical Wiener measure ω; the average case for more general measures including ω will be dealt with in a future paper.

where $(\lambda_1, \eta_1) = (1, 1)$ and (λ_i, η_i), for $i \geq 2$, are the eigenpairs for the univariate case, i.e.,

$$\int_0^1 K(x, y)\, f(y)\, \mathrm{d}y = \lambda_i\, \eta_i(x).$$

This integral equation is equivalent to the following differential equation:

$$\lambda_i\, \eta_i''(x) = -\eta_i(x) + 3 \int_0^1 \left(y - \frac{1}{2}y^2\right) \eta_i(y) \mathrm{d}y \quad \text{with} \quad \eta_i(0) = 0 \text{ and } \eta_i'(1) = 0.$$

It can be verified that, for $i \geq 2$,

$$\eta_i = \frac{h_i}{\|h_i\|_{L_2([0,1])}} \quad \text{and} \quad \lambda_i = \alpha_i^{-2},$$

where

$$h_i(x) = \cos(\alpha_i(x-1)) - \cos(\alpha_i) \quad \text{and} \quad \|h_i\|_{L_2([0,1])} = \frac{\sqrt{2}\,|\sin(\alpha_i)|}{2}.$$

Here, α_i is the unique solution in $((i-1)\pi, (i-1/2)\pi)$ of the equation

$$\alpha_i = \tan(\alpha_i).$$

This means that

$$\lambda_i = \alpha_i^{-2} = \frac{1 + o(1)}{((i-1/2)\pi)^2} \quad \text{as} \quad i \to \infty.$$

For this example the key quantities in (17) and (18) are β_r and $r(\lambda)$, for which we have

$$\pi^{-2/r}(\zeta(2/r) - 1) \leq \beta_r \leq \pi^{-2/r}\zeta(2/r) \quad \text{and} \quad r(\lambda) = 2.$$

Here ζ denotes the Riemann zeta function.

We now apply Corollary 1 for

$$\gamma_{d,\mathfrak{u}} = d^{-s\,|\mathfrak{u}|} \qquad \forall\, \mathfrak{u} \in \{1, 2, \ldots, d\} \tag{20}$$

with a real number s. For $r \in [1, 2)$ we have

$$\begin{aligned} \sum_{\mathfrak{u} \subseteq \{1,\ldots,d\}} \gamma_{d,\mathfrak{u}}^{1/r} \beta_r^{|\mathfrak{u}|} &= \sum_{k=0}^{d} \binom{d}{k} \left(d^{-s/r}\beta_r\right)^k \\ &= \left(1 + d^{-s/r}\beta_r\right)^d = (1 + o(1)) \exp\left(\beta_r d^{1-s/r}\right), \end{aligned}$$

where the last expression is for large d. Hence, the sum above is uniformly bounded in d iff $s \geq r$, whereas for $s < r$, it goes to infinity faster than polynomially in d. From this we conclude that strong tractability and tractability are equivalent, holding iff $s > 1$. If $s > 1$ then $r(S) = \min(2, s)$, and the exponent of strong tractability is $2/(\min(2, s) - 1)$. For $s \geq 2$, the exponent of strong tractability is 2.

Corollaries 2 and 3 treat finite-order weights and product weights. For arbitrary finite-order weights, Corollary 2 states that we have tractability with an ε-exponent of tractability arbitrarily close to 2. We can have strong tractability iff $r(S) > 1$. To illustrate the application of these corollaries for this example, we consider specific choices of the weights $\gamma_{d,\mathfrak{u}}$.

For the case of finite-order weights let

$$\gamma_{d,\mathfrak{u}} = d^{-s\,|\mathfrak{u}|} \qquad \forall\, \mathfrak{u} \subseteq \{1, 2, \ldots, d\} \ \text{ with } \ |\mathfrak{u}| \leq q^*,$$

where $q^* \geq 1$. This is similar to the case (20) just considered but now the weights vanish for $|\mathfrak{u}| > q^*$. It is easy to check that in this case we also must assume that $s > 1$ to obtain strong tractability. For $s > 1$, the exponent of strong tractability is $2/(\min(2, s) - 1)$, as before. Thus, the restriction to finite-order does not enlarge the range of s for which the problem is strongly tractable.

For product weights, we apply Corollary 3 for the weights

$$\gamma_{d,j} = d^{-s_1}\, j^{-s_2} \qquad j = 1, 2, \ldots, d$$

for some non-negative numbers s_1 and s_2. The case $s_2 = 0$ corresponds to the case (20) considered above. We then have

$$a_r := \sum_{j=1}^{d} \gamma_{d,j}^{1/r} = d^{-s_1/r} \sum_{j=1}^{d} j^{-s_2/r} = \begin{cases} O(d^{1-(s_1+s_2)/r}), & r > s_2, \\ O(d^{-s_1/d} \ln(d)), & r = s_2, \\ O(d^{-s_1/r}), & r < s_2. \end{cases}$$

It easily follows that tractability and strong tractability are equivalent, holding iff $s_1 + s_2 > 1$. Indeed, if $s_1 + s_2 > 1$ then we can have two cases: $s_1 = 0$ or $s_1 > 0$. If $s_1 = 0$ then we take $r \in (1, s_2)$ and $a_r = O(d^{-s_1/r}) = O(1)$ is uniformly bounded, which yields strong tractability. If $s_1 > 0$, we take $r \in (s_2, s_1 + s_2)$ and $a_r = O(d^{1-(s_1+s+2)/r}) = O(1)$ is uniformly bounded and we again have strong tractability. On the other hand, if we have tractability, then $A_r < \infty$ for some $r > 1$. If $s_2 \leq 1$ then we have $r > s_2$ and $a_r / \ln(d+1) = O(d^{1-(s_1+s_2)/r} / \ln(d+1)$ can be bounded only if $s_1 + s_2 > 1$ which, in fact, also implies strong tractability. From this reasoning, it also follows that for $s_1 + s_2 > 1$ we have $r(S) = \min(2, s_1 + s_2)$, and the exponent of strong tractability is $2/(\min(2, s_1 + s_2) - 1)$.

It is well known that the measure μ of this example concentrates on functions that have no derivative at any point x. On the other hand, with probability one, functions f satisfy Hölder's condition with the exponent arbitrarily close to $1/2$. This explains why the ε-exponent of (strong) tractability is not greater than 2.

Example 2 (Weighted Korobov spaces of periodic functions). As in the previous section, we consider the approximation problem $S_d : \mathcal{F}_d \to \mathcal{G}_d$ with $S_d f = f$; however, we now take

$$\mathcal{F}_d = \mathcal{G}_d = L_2([0,1]^d).$$

Since function evaluation, $L_{\mathbf{x}}(f) = f(\mathbf{x})$, is not even well-defined, the probability measure cannot be introduced via its covariance kernel. Therefore we proceed as follows. Let

$$\eta_i(x) = \begin{cases} 1, & i = 1, \\ \sqrt{2}\sin(2\pi\lfloor i/2\rfloor x), & i \text{ even}, \\ \sqrt{2}\cos(2\pi\lfloor i/2\rfloor x), & i \text{ odd and } i \geq 3. \end{cases}$$

Of course, $\{\eta_i\}$ is an orthonormal system of $L_2([0,1])$, and the functions

$$\eta_{d,\imath}(\mathbf{x}) := \left(\bigotimes_{k=1}^{d} \eta_{i_k}\right)(\mathbf{x}) = \prod_{k=1}^{d} \eta_{i_k}(x_k) \quad \text{for} \quad \imath \in \mathbb{N}_+^d$$

form an orthonormal system of $L_2([0,1]^d)$.

We define the probability measure $\mu_d = \nu_d$ as the zero-mean Gaussian with the covariance operator

$$(C_{\nu_d} g)(\mathbf{x}) = \int_{[0,1]^d} K_{d,\alpha^*}(\mathbf{x},\mathbf{y})\, g(\mathbf{y})\, \mathrm{d}\mathbf{y} \qquad \forall g \in \mathcal{G}_d,$$

where

$$K_{d,\alpha^*}(\mathbf{x},\mathbf{y}) = \sum_{\mathfrak{u}\subseteq\{1,2,\ldots,d\}} \gamma_{d,\mathfrak{u}}\, 2^{|\mathfrak{u}|} \prod_{j\in\mathfrak{u}} \sum_{k=1}^{\infty} \frac{\cos(2\pi k(x_j - y_j))}{k^{\alpha^*}}. \tag{21}$$

Here the parameter $\alpha^* > 1$.

For even α^*, we may write K_{d,α^*} in terms of Bernoulli polynomials (see e.g., [AS64, Chapter 23]) as

$$K_{d,\alpha^*}(\mathbf{x},\mathbf{y}) = \sum_{\mathfrak{u}\subseteq\{1,\ldots,d\}} \gamma_{d,\mathfrak{u}} \left(\frac{-(-4\pi^2)^{\alpha^*/2}}{\alpha^*!}\right)^{|\mathfrak{u}|} \prod_{j\in\mathfrak{u}} B_{\alpha^*}(x_j - y_j). \tag{22}$$

A contour plot of $K_{1,2}(x,y)$ is given in Figure 1.

It is easy to find the eigenpairs of the covariance operator C_{ν_d} for arbitrary $\alpha^* > 1$. We have

$$C_{\nu_d}\eta_{d,\imath} = \lambda_{d,\imath}\, \eta_{d,\imath} \qquad \forall \imath$$

with the following eigenvalues. For $d = 1$, and $\gamma_{1,\emptyset} = \gamma_{1\{1\}} = 1$, we have $\lambda_{1,\imath} = \lambda_i$ with

$$\lambda_1 = 1 \qquad \text{and} \qquad \lambda_i = \lfloor i/2\rfloor^{-\alpha^*} \text{ for } i \geq 2.$$

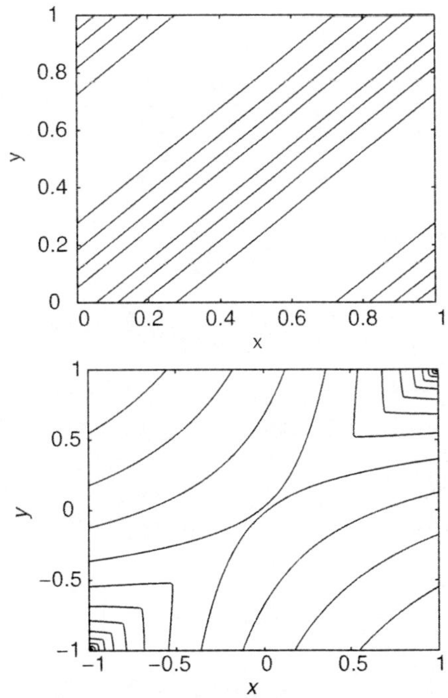

Fig. 1. Contour plots of $K_{1,2}(x,y)$ defined in (22) (left) and (23) (right) .

For $d \geq 2$, we have

$$\lambda_{d,\imath} \;=\; \gamma_{d,\mathfrak{u}_\imath} \prod_{j \in \mathfrak{u}_\imath} \lambda_{i_j},$$

The probability measure $\mu_d = \nu_d$ is such that (with probability one) the functions $f \in \mathcal{F}_d$ can be viewed as

$$f \;=\; \sum_{\imath \in \mathbb{N}_+^d} \gamma_{d,\mathfrak{u}_\imath}\, y_{d,\imath}\, \eta_{d,\imath}\,,$$

where the coefficients $y_{d,\imath}$'s are independent random variables, each with the normal $\mathcal{N}(0, \lambda_{d,\imath})$ distribution.

We now indicate, see also [KSW07], that the approximation problem of this example is essentially the same as the approximation problem for a weighted Korobov space. First, recall that for $\beta > 1$, the weighted Korobov space $H_{d,\beta}$, is a reproducing kernel Hilbert space of periodic functions f defined for $[0,1]^d$ for which $\|f\|_{H_{d,\beta}} = \langle f, f\rangle_{H_{d,\beta}}^{1/2} < \infty$ with the reproducing kernel $K_{d,\beta}$ of the form (21) with α^* replaced by β, see e.g., [DSWW06]. The inner product in the space $H_{d,\beta}$ is

$$\langle f, g\rangle_{H_{d,\beta}} \;=\; \sum_{\mathbf{h} \in \mathbb{Z}^d} r(\mathbf{h}, \gamma)\, \hat{f}(\mathbf{h})\, \overline{\hat{g}(\mathbf{h})},$$

where

$$\hat{f}(\mathbf{h}) = \int_{[0,1]^d} \exp\left(-2\pi\sqrt{-1}\,(h_1 x_1 + \cdots + h_d x_d)\right) f(\mathbf{x})\,\mathrm{d}\mathbf{x}$$

is a Fourier coefficient of f, and

$$r(\mathbf{h},\gamma) = \begin{cases} 1, & \mathbf{h} = \mathbf{0}, \\ \gamma_{d,\mathfrak{u}_\mathbf{h}}^{-1} \prod_{j\in\mathfrak{u}_\mathbf{h}} |h_j|^\beta, & \mathbf{h} \neq \mathbf{0}. \end{cases}$$

where $\mathfrak{u}_\mathbf{h} = \{j : h_j \neq 0\}$ We then have

$$f(\mathbf{x}) = \langle f, K_{d,\beta}\rangle_{H_{d,\beta}} \qquad \forall f \in H_{d,\beta}\ \forall \mathbf{x} \in [0,1]^d.$$

The parameter β of the weighted Korobov space determines the smoothness of the functions. Namely, for $f \in H_{d,\beta}$ the derivative $f^{(\alpha_1,\alpha_2,\ldots,\alpha_d)} \in L_2([0,1]^d)$ for non-negative integers α_i if $\max_i \alpha_i \leq \beta/2$.

Using the Kolmogorov zero-one principle it is easy to show, see also [KSW07], that

$$\mu_d(H_{d,\beta}) = \nu_d(H_{d,\beta}) = 1 \qquad \text{iff} \qquad \alpha^* > \beta + 1.$$

Hence if we assume that $\alpha^* > 2$ and $\beta \in (1, \alpha^* - 1)$, then the approximation problems for the spaces $L_2([0,1]^d)$ and $H_{d,\beta}$ are the same in the average case setting for the zero-mean Gaussian measure with the covariance operator given by (21). That is why we call this example the weighted Korobov space of periodic functions.

For this example the key quantities in (17) and (18) are $\beta_r = 2\zeta(\alpha^*/r)$ and $r(\lambda) = \alpha^*$. Corollaries 1, 2 and 3 provide necessary and sufficient conditions on the weights $\boldsymbol{\gamma} = \{\gamma_{d,\mathfrak{u}}\}$ for strong tractability and tractability. In particular, we know when the exponent of strong tractability and the ε-exponent of tractability are equal or arbitrarily close to $2/(\alpha^* - 1)$. For finite-order weights of order q^*, one obtains this exponent of strong tractability provided that condition (19) on the weights $\gamma_{d,\mathfrak{u}}$ is satisfied. This condition is not satisfied for $r > 1$ if, for example, $\gamma_{d,\mathfrak{u}} = 1$ for $|\mathfrak{u}| \leq q^*$. The ε-exponent of tractability can be arbitrarily close to $2/(\alpha^* - 1)$ with the d-exponent equal to the order q^* of the weights, and this holds for arbitrary finite-order weights $\gamma_{d,\mathfrak{u}}$. For uniform product weights with $\gamma_j = j^{-k^*}$ we obtain strong tractability (which is equivalent to tractability) iff $k^* > 1$ and then the exponent of strong tractability is $2/(\min(\alpha^*, k^*) - 1)$.

Example 3 (Spaces of non-periodic functions). The previous example of the approximation problem can be generalized by taking different selections of orthonormal systems of the space $\mathcal{F}_d = \mathcal{G}_d = L_2(D_d)$ with $D_d \subseteq \mathbb{R}^d$. For instance, we can take $D_d = [-1,1]^d$ and an orthonormal system

$$\eta_{d,\mathfrak{i}}(\mathbf{x}) = \prod_{k=1}^{d} \frac{P_{i_k-1}(x_k)}{\|P_{i_k-1}\|_{L_2([-1,1])}} = \prod_{k=1}^{d} \sqrt{i_k - 1/2}\, P_{i_k-1}(x_k),$$

where $\{P_k\}_{k=0}^{\infty}$ is the family of Legendre polynomials with $\|P_k\|_{L_2([-1,1])} = (k+1/2)^{-1/2}$, see e.g., [AS64, Chapter 8].

As before, we take zero-mean Gaussian $\mu_d = \nu_d$ with the covariance operator

$$(C_{\nu_d} g)(\mathbf{x}) = \int_{[-1,1]^d} K_{d,\alpha^*}(\mathbf{x},\mathbf{y})\,\mathrm{d}\mathbf{y},$$

now with

$$K_{d,\alpha^*}(\mathbf{x},\mathbf{y}) = \sum_{\mathfrak{u}\subseteq\{1,2,\dots,d\}} \gamma_{d,\mathfrak{u}} \prod_{j\in\mathfrak{u}} \sum_{k=1}^{\infty} \frac{(k+1/2)P_k(x)P_k(y)}{k^{\alpha^*}}. \tag{23}$$

To guarantee that $K_{d,\alpha^*}(\cdot,\mathbf{y})$ belongs to $L_2([-1,1]^d)$ we need to assume that $\alpha^* > 1$. A contour plot of $K_{1,2}$ is given in Figure 1.

The eigenvalues and eigenfunctions of the covariance operator C_{ν_d} are of the form (15) with

$$\lambda_i = \begin{cases} 1, & i = 1, \\ (i-1)^{-\alpha^*}, & i \geq 2, \end{cases} \qquad \eta_i(x) = \sqrt{i-1/2}\,P_{i-1}(x).$$

Similar to the previous example, the key quantities here are $\beta_r = \zeta(\alpha^*/r)$ and $r(\lambda) = \alpha^*$. The conditions for tractability and strong tractability are the same as in Example 2.

If we assume that $\alpha^* > 2$, then $K_{d,\alpha^*}(\mathbf{x},\mathbf{y})$ is pointwise well-defined since $\|P_k\|_{L_\infty([-1,1])} = 1$. Then, as in the previous example, $K_{d,\beta}$ with $\beta > 2$ could be viewed as a reproducing kernel generating a Hilbert space $H_{d,\beta}$ whose an orthonormal system is given by the functions $\sqrt{\lambda_{d,\mathfrak{i}}}\,\eta_{d,\mathfrak{i}}$ (of course with α^* replaced by β). The space $H_{d,\beta}$ is a subspace of $L_2([-1,1]^d)$. As before, using the Kolmogorov zero-one principle, we conclude that the probability measure $\mu_d = \nu_d$ is concentrated on $H_{d,\beta}$ iff $\alpha^* > \beta + 1$; otherwise $\mu_d(H_{d,\beta}) = 0$.

Example 4 (A More General Case). We end this section by a problem satisfying (15) that is an extension of the preceding three examples. As before, suppose that μ_d is such that the functions $f \in \mathcal{F}_d$ can be viewed as

$$f(\mathbf{x}) = \sum_{\mathfrak{i}\in\mathbb{N}_+^d} \gamma_{d,\mathfrak{u}_\mathfrak{i}}\, y_{d,\mathfrak{i}} \prod_{k=1}^{d} \xi_{i_k}(x_k),$$

where $y_{d\mathfrak{i}}$ are independent random variables each with a normal distribution $\mathcal{N}(0,\sigma_{d,\mathfrak{i}})$ with $\sigma_{d,\mathfrak{i}} = \prod_{k=1}^d \sigma_k$. The only assumption now about the functions $\xi_i \in \mathcal{F}_1$ is that they are linearly independent and $\xi_1 \equiv 1$.

Assume also that S_d is the tensor product of d copies of an operator $S_1 : \mathcal{F}_1 \to \mathcal{G}_1$, and that the $S\xi_i$ are orthogonal to $S\xi_1$:

$$\langle S\xi_1, S\xi_i\rangle_{\mathcal{G}_1} = 0 \qquad \forall i \geq 2. \tag{24}$$

Then, as can be verified, the condition (15) is satisfied; however, the eigenpairs of C_{ν_d} might be difficult to find.

The situation is further simplified if we additionally assume that the $S\xi_i$ are mutually orthogonal, i.e., that

$$\langle S\xi_j, S\xi_i\rangle_{\mathcal{G}_1} = \delta_{i,j} \,\|S\xi_i\|^2_{\mathcal{G}_1} \quad \forall i,j \geq 2 \tag{25}$$

When these hold, the eigenfunctions of C_{ν_d} take the explicit form

$$\eta_{d,\mathfrak{i}} = \frac{S_d\xi_{d,\mathfrak{i}}}{\|S_d\xi_{d,\mathfrak{i}}\|_{\mathcal{G}_d}} \quad \text{with} \quad S_d\xi_{d,\mathfrak{i}} = \bigotimes_{k=1}^{d} S\xi_{i_k}.$$

The corresponding eigenvalues are given by

$$\lambda_{d,\mathfrak{i}} = \gamma_{d,\mathfrak{u}_\mathfrak{i}} \left(\sigma_1 \|S\xi_1\|^2_{\mathcal{G}_1}\right)^{d-|\mathfrak{u}_\mathfrak{i}|} \prod_{k\in\mathfrak{u}_\mathfrak{i}} \sigma_{i_k} \|S\xi_{i_k}\|^2_{\mathcal{G}_1}.$$

In particular, $\lambda_{d,\mathbf{1}} = \gamma_{d,\emptyset} \left(\sigma_1 \|S\xi_1\|^2_{\mathcal{G}_1}\right)^d$. We are now in a position to apply Corollaries 1, 2 and 3 to deduce strong tractability and tractability of S in terms of the conditions on $\gamma_{d,\mathfrak{u}}$ and $\lambda_j = \sigma_j\|S\xi_j\|^2_{\mathcal{G}_1}$.

5 Tractability for Approximation with Λ^{std}

In this section we assume that $\mathcal{F}_d$ is continuously embedded in the space $\mathcal{G}_d = L_{2,\rho}(D_d)$, where the weight function $\rho : D_d \to \mathbb{R}^d$ is positive almost everywhere and $\int_{D_d} \rho(\mathbf{t})\,\mathrm{d}\mathbf{t} = 1$. For $g_1, g_2 \in \mathcal{G}_d$, we now have

$$\langle g_1, g_2\rangle_{\mathcal{G}_d} = \int_{D_d} g_1(\mathbf{t})g_2(\mathbf{t})\rho(\mathbf{t})\,\mathrm{d}\mathbf{t}.$$

We also assume that $L_\mathbf{x}(f) = f(\mathbf{x})$ is a continuous linear functional of $\mathcal{F}_d$ for all $\mathbf{x} \in D_d$. Then the class $\Lambda^{\mathrm{std}} = \{L_\mathbf{x} : \mathbf{x} \in D_d\}$ of standard information is well defined and is a subset of Λ^{all}. We now consider algorithms that use only function evaluations and are of the form

$$Af = \sum_{j=1} f(t_j)g_j$$

for some $t_j \in D_d$ and some $g_j \in \mathcal{G}_d$.

For the class Λ^{std}, we consider the multivariate approximation problem that is given by $S_d f = f$ for $f \in \mathcal{F}_d$. Combining the proof techniques of [HW00] and [WW07] we show that the class Λ^{std} is basically as powerful as the class Λ^{all} when we have a polynomial rate of convergence for the class Λ^{all}. That is, there are algorithms $A_{d,n}$ that use only function evaluations whose average case errors are essentially of the same order as the average case errors of the optimal algorithms $A^*_{d,n}$ given by (6). Recall that

$$A^*_{d,n}f = \sum_{\ell=1}^{n} \langle f, \eta_{d,\ell}\rangle_{\mathcal{G}_d}\, \eta_{d,\ell} \quad \text{and} \quad e^{\rm avg}(A^*_{n,d}; S_d) = \left(\sum_{\ell=n+1}^{\infty} \lambda_{d,\ell}\right)^{1/2}.$$

Without loss of generality, we may assume that all $\lambda_{d,i} > 0$ and then $\int_{D_d} \eta^2_{d,i}(\mathbf{t})\rho(\mathbf{t})\,\mathrm{d}\mathbf{t} = 1$ for all i. We assume that

$$\left(\sum_{\ell=n+1}^{\infty} \lambda_{d,\ell}\right)^{1/2} \le \frac{C_0}{(n+1)^p} \qquad \forall n = 0, 1, \dots \tag{26}$$

for some positive numbers p and C_0. Note that for $n = 0$ we know that

$$C_0 \ge \left(\sum_{\ell=n+1}^{\infty} \lambda_{d,\ell}\right)^{1/2} = e^{\rm avg}(0; S_d).$$

Obviously, we should choose the parameter p as large as possible to properly characterize the speed of convergence of the errors of the algorithms $A^*_{d,n}$.

Given d and n, we define a sequence of algorithms $\{A_k\}_k = \{A_{d,k,n}\}_k$ that use at most $k\,n$ function evaluations and whose average case errors are quickly approaching the average case error of $A^*_{d,n}$ as k goes to infinity. To define A_k we proceed as follows. Let

$$p_k := p\frac{2p_{k-1}+1}{2p+1} \quad \text{with} \quad p_0 = 0 \quad \text{and} \quad m_k := \lfloor n^{p_k/p} \rfloor. \tag{27}$$

It is easy to check that

$$p_k = p\left(1 - \left(\frac{2p}{2p+1}\right)^k\right).$$

The sequence $\{p_k\}$ is increasing and $\lim_k p_k = p$, whereas the sequence $\{m_k\}$ is non-decreasing and $m_k \le n$ for all k. We will use the functions

$$u_k(\mathbf{t}) := \frac{1}{m_k}\sum_{j=1}^{m_k} \eta^2_{d,j}(\mathbf{t}) \qquad \text{and} \qquad \omega_k(\mathbf{t}) := \rho(\mathbf{t})\, u_k(\mathbf{t}). \tag{28}$$

Observe that ω_k is non-negative and $\int_{D_d} \omega_k(\mathbf{t})\,\mathrm{d}\mathbf{t} = 1$. Hence, ω_k can be regarded as a probability density function.

We define the algorithms A_k inductively with respect to k. We set $A_0 = 0$, and then

$$A_k f = A_{k-1}f + \sum_{j=1}^{m_k}\left[\frac{1}{n}\sum_{\ell=1}^{n} (f - A_{k-1}(f))\,(\boldsymbol{\tau}_\ell)\,\frac{\eta_{d,j}(\boldsymbol{\tau}_\ell)}{u_k(\boldsymbol{\tau}_\ell)}\right] \eta_{d,j} \tag{29}$$

for some yet to be specified sample points $\boldsymbol{\tau}_\ell$ from D_d. We use the convention that $0/0 = 0$.

The algorithm A_k uses at most $(k-1)n$ function evaluations used by A_{k-1} and at most n function evaluations at the $\boldsymbol{\tau}_\ell$'s. Hence, the total number of function evaluations is at most $k\,n$. Note that $A_k f$ is orthogonal to $\eta_{d,j}$ for $j > m_k$. We are ready to estimate the average case error of the algorithm A_k.

Theorem 2 *Let (26) hold. Then for every n and k, we have*

$$e^{\text{avg}}(A_k; S_d) \le \frac{C_0}{n^{p_k}} \sqrt{k+1}.$$

Proof. The proof is by induction on k. For $k=0$ this trivially holds. For $k \ge 1$, we have

$$\begin{aligned} e^{\text{avg}}(A_k; S_d)^2 &= \int_{\mathcal{F}_d} \|f - A_k f\|_{\mathcal{G}_d}^2 \, \mu_d(\mathrm{d}f) = \int_{\mathcal{G}_d} \|f - A_k f\|_{\mathcal{G}_d}^2 \, \nu_d(\mathrm{d}f) \\ &= \sum_{i=1}^{\infty} \int_{\mathcal{G}_d} \langle f - A_k f, \eta_{d,i} \rangle_{\mathcal{G}_d}^2 \, \nu_d(\mathrm{d}f). \end{aligned}$$

Let $g_{k-1} = f - A_{k-1} f$. For $i \le m_k$, we have

$$\langle f - A_k f,\ \eta_{d,i} \rangle_{\mathcal{G}_d} = \int_{D_d} g_{k-1}(\mathbf{t}) \eta_{d,i}(\mathbf{t}) \rho(\mathbf{t}) \, \mathrm{d}\mathbf{t} \ - \ \frac{1}{n} \sum_{\ell=1}^{n} g_{k-1}(\boldsymbol{\tau}_\ell) \frac{\eta_{d,i}(\boldsymbol{\tau}_\ell)}{u_k(\boldsymbol{\tau}_\ell)},$$

whereas for $i > m_k$, we have

$$\langle f - A_k f,\ \eta_{d,i} \rangle_{\mathcal{G}_d} = \langle f, \eta_{d,i} \rangle_{\mathcal{G}_d} = \int_{D_d} f(\mathbf{t}) \eta_{d,i}(\mathbf{t}) \rho(\mathbf{t}) \, \mathrm{d}\mathbf{t}.$$

Hence, $e^{\text{avg}}(A_k; S_d)^2 = a_1 + a_2$, where

$$\begin{aligned} a_1 &= \sum_{i=1}^{m_k} \int_{\mathcal{G}_d} \left[\int_{D_d} g_{k-1}(\mathbf{t}) \eta_{d,i}(\mathbf{t}) \rho(\mathbf{t}) \, \mathrm{d}\mathbf{t} \ - \ \frac{1}{n} \sum_{\ell=1}^{n} g_{k-1}(\boldsymbol{\tau}_\ell) \frac{\eta_{d,i}(\boldsymbol{\tau}_\ell)}{u_k(\boldsymbol{\tau}_\ell)} \right]^2 \nu_d(\mathrm{d}f), \\ a_2 &= \sum_{i=m_k+1}^{\infty} \int_{\mathcal{G}_d} \langle f, \eta_{d,i} \rangle_{\mathcal{G}_d}^2 \, \nu_d(\mathrm{d}f) = \sum_{i=m_k+1}^{\infty} \lambda_{d,i}. \end{aligned}$$

From (26), we conclude that $a_2 \le C_0^2 \, (m_k+1)^{-2p}$. Since $m_k + 1 \ge n^{p_k/p}$ we obtain

$$a_2 \le \frac{C_0^2}{n^{2p_k}}.$$

To obtain a bound on a_1 we allow the sample points $\boldsymbol{\tau}_\ell$ to be independent random sample points distributed over D_d according to the measure with density function ω_k. We now take the expectation of $a_1 = a_1(\boldsymbol{\tau}_1, \ldots, \boldsymbol{\tau}_n)$ with respect to $\boldsymbol{\tau}_\ell$'s, and using the standard argument as for the classical Monte Carlo algorithm, we obtain

$$\int_{D_d}\cdots\int_{D_d}\Big[\int_{D_d} g_{k-1}(\mathbf{t})\eta_{d,i}(\mathbf{t})\rho(\mathbf{t})\,\mathrm{d}\mathbf{t} - \frac{1}{n}\sum_{\ell=1}^{n} g_{k-1}(\boldsymbol{\tau}_\ell)\frac{\eta_{d,i}(\boldsymbol{\tau}_\ell)}{u_k(\boldsymbol{\tau}_\ell)}\Big]^2\,\omega_k(\boldsymbol{\tau}_1)\cdots\omega_k(\boldsymbol{\tau}_n)\,\mathrm{d}\boldsymbol{\tau}_1\cdots\mathrm{d}\boldsymbol{\tau}_n$$

$$\le \frac{1}{n}\int_{D_d} g_{k-1}^2(\mathbf{t})\frac{\eta_{d,i}^2(\mathbf{t})}{u_k(\mathbf{t})}\,\rho(\mathbf{t})\,\mathrm{d}\mathbf{t}.$$

Therefore

$$\begin{aligned}
&\int_{D_d}\cdots\int_{D_d} a_1(\boldsymbol{\tau}_1,\dots,\boldsymbol{\tau}_n)\,\omega_k(\boldsymbol{\tau}_1)\cdots\omega_k(\boldsymbol{\tau}_n)\,\mathrm{d}\boldsymbol{\tau}_1\cdots\mathrm{d}\boldsymbol{\tau}_n\\
&\le \frac{m_k}{n}\int_{\mathcal{G}_d}\int_{D_d} g_{k-1}^2(\mathbf{t})\rho(\mathbf{t})\,\mathrm{d}\mathbf{t}\,\nu_d(\mathrm{d}f)\\
&= \frac{m_k}{n}\int_{\mathcal{G}_d}\|f-A_{k-1}f\|_{\mathcal{G}_d}^2\nu_d(\mathrm{d}f) = \frac{m_k}{n}\,e^{\mathrm{avg}}(A_{k-1};\,S_d)^2\\
&\le \frac{m_k\,C_0^2\,k}{n^{1+2p_{k-1}}}.
\end{aligned}$$

By the mean value theorem, we conclude that there are sample points $\boldsymbol{\tau}_1^*,\boldsymbol{\tau}_2^*,\dots,\boldsymbol{\tau}_n^*$ such that the square of the average case error is at most equal to the average of $a_1(\boldsymbol{\tau}_1,\dots,\boldsymbol{\tau}_n)+a_2$. Taking these $\boldsymbol{\tau}_\ell^*$'s in the definition of the algorithm A_k we obtain

$$e^{\mathrm{avg}}(A_k;S_d)^2 \le \frac{m_k\,C_0^2\,k}{n^{1+2p_{k-1}}} + \frac{C_0^2}{n^{2p_k}}.$$

Since $m_k \le n^{p_k/p}$ and $1+2p_{k-1}-p_k/p = 2p_k$, we finally get

$$e^{\mathrm{avg}}(A_k;S_d) \le \frac{C_0}{n^{p_k}}\,\sqrt{k+1},$$

as claimed. This completes the proof. □

We stress that the description of the algorithms $A_{d,k,n}$ is not constructive since we do not know how to choose the sample points $\boldsymbol{\tau}_\ell$ in (29). We only know that there exist sample points $\boldsymbol{\tau}_1,\dots,\boldsymbol{\tau}_n$ for which the average case error of $A_{d,k,n}$ enjoys the average case error bound in Theorem 2.

There is, however, a "semi"-construction of the algorithms $A_k = A_{n,k,n}$ based on the proof of Theorem 2. Indeed, assume inductively that A_{k-1} has been already constructed with the average case error bounded by $C\,C_0\sqrt{k}\,n^{-p_{k-1}}$ for some $C>1$. To construct A_k, we select sample points $\boldsymbol{\tau}_1,\dots,\boldsymbol{\tau}_n$ as independent random variables distributed according to the measure with density ω_k. Then we compute a_1 for these $\boldsymbol{\tau}_\ell$'s which is possible due to the explicit average case error formula. If $a_1 \le C^2\,m_k\,C_0^2 k\,n^{-(1+2p_{k-1})}$ then we are done since the average case error of A_k is at most $C\,C_0\sqrt{k+1}\,n^{-p_k}$.

If not we repeat the random selection of $\boldsymbol{\tau}_1, \dots, \boldsymbol{\tau}_n$. By Chebyshev's inequality, we know that the failure of finding proper $\boldsymbol{\tau}_\ell$'s with j selections is C^{-2j}. Hence, it is enough to repeat the selection of $\boldsymbol{\tau}_\ell$'s only a few times if C is large enough.

We return to the average case error of $A_{d,k,n}$ given by Theorem 2. For n such that $\ln(\ln(n)) > 1$, take

$$k = k^* = \left\lceil \frac{\ln(\ln(n))}{\ln(1+1/(2p))} \right\rceil .$$

It is easy to check that

$$e^{\mathrm{avg}}(A_{d,k^*,n}; S_d) \le e\, C_0\, n^{-p} \sqrt{2 + \frac{\ln(\ln(n))}{\ln(1+1/(2p))}} = O\left(n^{-p} \sqrt{\ln(\ln(n))}\right).$$

If $m = k^* n = \Theta(n \ln(\ln(n)))$ then the algorithm

$$\overline{A}_{d,m} := A_{d,k^*,n}$$

uses at most m function evaluations and

$$e^{\mathrm{avg}}(\overline{A}_{d,m}; S_d) = O\left(\frac{[\ln(\ln(m))]^{p+0.5}}{m^p}\right),$$

where the factor in the big O notation depends only on C_0 and p and is independent of m. Hence, modulo a power of $\ln(\ln(n))$, we obtain the same speed of convergence as for the algorithm $A^*_{n,d}$.

We now turn to tractability of multivariate approximation for the class Λ^{std}. Based on Theorem 2, it is obvious that tractability for the class Λ^{std} is equivalent to tractability for the class Λ^{all}. We provide a formal proof and estimates on $n(\varepsilon; S_d, \Lambda^{\mathrm{std}})$ in the following corollary:

Corollary 4 *Consider multivariate approximation in the classes Λ^{all} and Λ^{std} defined as in this section, so that $\Lambda^{\mathrm{std}} \subseteq \Lambda^{\mathrm{all}}$. Then strong tractability and tractability of multivariate approximation in the classes Λ^{all} and Λ^{std} are equivalent. Furthermore, the exponents of strong tractability and tractability are, modulo a power of the double logarithm of $\varepsilon^{-1} + d + 1$, the same in both classes. That is, if*

$$n(\varepsilon; S_d, \Lambda^{\mathrm{all}}) \le C\, \varepsilon^{-p_{\mathrm{err}}}\, d^{\,q_{\mathrm{dim}}} \quad \forall \varepsilon \in (0,1),\ d = 1, 2, \dots,$$

then for all $\varepsilon \in (0,1)$ and $d = 1, 2, \dots,$

$$n(\varepsilon; S_d, \Lambda^{\mathrm{std}}) \le \min_{k=1,2,\dots} k \left\lceil \left(2\, C\, (k+1)^{p_{\mathrm{err}}/2}\, \varepsilon^{-p_{\mathrm{err}}}\, d^{\,q_{\mathrm{dim}}}\right)^{\left[1 - \left(\frac{2}{2+p_{\mathrm{err}}}\right)^k\right]^{-1}} \right\rceil ,$$

and

$$n(\varepsilon; S_d, \Lambda^{\mathrm{std}}) = O\left(\left[\ln(\ln(\varepsilon^{-1} + d + 1))\right]^{1 + p_{\mathrm{err}}/2}\, \varepsilon^{-p_{\mathrm{err}}}\, d^{\,q_{\mathrm{dim}}} \right),$$

where the factor in the big O notation is independent of ε and d.

Proof. We know from (7) that

$$\sum_{\ell=n+1}^{\infty} \lambda_{d,\ell} \le \varepsilon^2 \sum_{\ell=1}^{\infty} \lambda_{d,\ell}$$

for $n = \lfloor C\,\varepsilon^{-p_{\rm err}}\, d^{\,q_{\rm dim}} \rfloor$. Varying $\varepsilon \in (0,1)$, we conclude that (26) holds with

$$p = p_{\rm err}^{-1} \quad \text{and} \quad C_0 = (2\,C\,d^{\,q_{\rm dim}})^p \left(\sum_{\ell=1}^{\infty} \lambda_{d,\ell}\right)^{1/2}.$$

From Theorem 2 we conclude that $e^{\rm avg}(A_{d,k,n}; S_d) \le \varepsilon \left(\sum_{\ell=1}^{\infty} \lambda_{d,\ell}\right)^{1/2}$ if we take

$$n = \left\lceil \varepsilon^{-1/p_k}\, (2\,C\,d^{\,q_{\rm dim}})^{p/p_k}\, (k+1)^{1/(2p_k)} \right\rceil.$$

Since

$$\frac{p}{p_k} = \left[1 - \left(\frac{2p}{2p+1}\right)^k\right]^{-1} = \left[1 - \left(\frac{2}{2+p_{\rm err}}\right)^k\right]^{-1},$$

we have

$$n = \left\lceil \left(2\,C\,(k+1)^{p_{\rm err}/2}\,\varepsilon^{-p_{\rm err}}\,d^{\,q_{\rm dim}}\right)^{\left[1-\left(\frac{2}{2+p_{\rm err}}\right)^k\right]^{-1}} \right\rceil.$$

This and the fact that $A_{d,k,n}$ uses at most $k\,n$ function evaluations completes the proof of the bound on $n(\varepsilon; S_d, \Lambda^{\rm std})$. We now take

$$k = \left\lceil \frac{\ln(\ln(\varepsilon^{-1}+d+1))}{\ln((2+p_{\rm err})/2)} \right\rceil.$$

Then $(2/(2+p_{\rm err}))^k \le 1/\ln(\varepsilon^{-1}+d+1)$ and

$$k\left(2\,C\,(k+1)^{p_{\rm err}/2}\,\varepsilon^{-p_{\rm err}}\,d^{\,q_{\rm dim}}\right)^{\left[1-\left(\frac{2}{2+p_{\rm err}}\right)^k\right]^{-1}}$$
$$= O\left(\left[\ln(\ln(\varepsilon^{-1}+d+1))\right]^{1+p_{\rm err}/2}\,\varepsilon^{-p_{\rm err}}\,d^{\,q_{\rm dim}}\right),$$

as claimed. This completes the proof. □

We end with the following remark.

Remark 1 Suppose that the nth minimal errors for the class $\Lambda^{\rm all}$ converge to zero exponentially fast, say,

$$e(n; S, \Lambda^{\rm all}) \le C_1 \exp\left(-\alpha\,(n+1)^{\beta}\right) \qquad \forall\, n = 0, 1, \dots,$$

for positive α and β. Of course, then (26) holds for any p with the constant $C_0 = C_0(p, \alpha, \beta)$ increasing super-exponentially with p, i.e.,

$$e(n; S, \Lambda^{\text{all}}) \leq \frac{C_0(p, \alpha, \beta)}{(n+1)^p} \quad \text{with} \quad C_0(p, \alpha, \beta) = C_1 \max\left(e^{-\alpha}, \left(\frac{p}{e\alpha\beta}\right)^{p/\beta}\right).$$

Hence, for every p, at the expense of a huge multiplicative constant for large p, one could construct algorithms using standard information with the rate of convergence proportional to n^{-p}. However, it is not clear if $e(n; S, \Lambda^{\text{std}})$ are proportional to $e(n; S, \Lambda^{\text{all}})$, or even if $e(n; S, \Lambda^{\text{std}})$ converges to zero faster than polynomially.

Acknowledgments

We are grateful for many helpful comments on the draft manuscript by Prof. Art Werschulz and fruitful discussions with several colleagues. The second and third authors were partially supported by the National Science Foundation under Grants DMS-0609703 and DMS-0308713, respectively.

References

[AS64] M. Abramowitz and I. A. Stegun, editors. *Handbook of Mathematical Functions with Formulas, Graphs and Mathematical Tables.* U.S. Government Printing Office, Washington, DC, 1964.

[DM99] A. Dey and R. Mukerjee. *Fractional Factorial Plans.* John Wiley & Sons, New York, 1999.

[DSWW06] J. Dick, I. H. Sloan, X. Wang, and H. Woźniakowski. Good lattice rules in weighted Korobov spaces with general weights. *Numer. Math.*, 103(1):63–97, 2006.

[HSS99] A. S. Hedayat, N. J. A. Sloane, and J. Stufken. *Orthogonal Arrays: Theory and Applications.* Springer Series in Statistics. Springer-Verlag, New York, 1999.

[HW00] F. J. Hickernell and H. Woźniakowski. Integration and approximation in arbitrary dimensions. *Adv. Comput. Math.*, 12:25–58, 2000.

[KS05] F. Y. Kuo and I. H. Sloan. Quasi-Monte Carlo methods can be efficient for integration over product spheres. *J. Complexity*, 21:196–210, 2005.

[KSW07] F. Y. Kuo, I. H. Sloan, and H. Woźniakowski. Lattice rule algorithms for multivariate approximation in the average case setting. *J. Complexity*, 2007. to appear.

[Kuo75] H. H. Kuo. *Gaussian Measures in Banach Spaces.* Number 463 in Lecture Notes in Mathematics. Springer-Verlag, Berlin, 1975.

[NW01] E. Novak and H. Woźniakowski. When are integration and discrepancy tractable? In *Foundations of Computational Mathematics*, volume 284 of *London Math. Soc. Lecture Note Ser.*, pages 211–266. Cambridge University Press, Cambridge, 2001.

[PW90] A. Papageorgiou and G. W. Wasilkowski. On the average complexity of multivariate problems. *J. Complexity*, 6:1–23, 1990.

[RW96] K. Ritter and G. W. Wasilkowski. On the average case complexity of solving Poisson equations. In J. Renegar, M. Shub, and S. Smale, editors, *The mathematics of numerical analysis*, volume 32 of *Lectures in Appl. Math.*, pages 677–687. American Mathematical Society, Providence, Rhode Island, 1996.

[RW97] K. Ritter and G. W. Wasilkowski. Integration and L_2 approximation: Average case setting with isotropic Wiener measure for smooth functions. *Rocky Mountain J. Math.*, 26:1541–1557, 1997.

[SW98] I. H. Sloan and H. Woźniakowski. When are quasi-Monte Carlo algorithms efficient for high dimensional integrals. *J. Complexity*, 14:1–33, 1998.

[SWW04] I. H. Sloan, X. Wang, and H. Woźniakowski. Finite-order weights imply tractability of multivariate integration. *J. Complexity*, 20:46–74, 2004.

[TW98] J. F. Traub and A. G. Werschulz. *Complexity and Information*. Cambridge University Press, Cambridge, 1998.

[TWW88] J. F. Traub, G. W. Wasilkowski, and H. Woźniakowski. *Information-Based Complexity*. Academic Press, Boston, 1988.

[Vak81] N. N. Vakhania. *Probability Distributions on Linear Spaces*. North-Holland, New York, 1981.

[Was86] G. W. Wasilkowski. Information of varying cardinality. *J. Complexity*, 2:204–228, 1986.

[Was93] G. W. Wasilkowski. Integration and approximation of multivariate functions: Average case complexity with isotropic Wiener measure. *Bull. Amer. Math. Soc.*, 28:308–314, 1993.

[Woź94] H. Woźniakowski. Tractability and strong tractability of linear multivariate problems. *J. Complexity*, 10:96–128, 1994.

[WW95] G. W. Wasilkowski and H. Woźniakowski. Explicit cost bounds for multivariate tensor product problems. *J. Complexity*, 11:1–56, 1995.

[WW01] G. W. Wasilkowski and H. Woźniakowski. On the power of standard information for weighted approximation. *Found. Comput. Math.*, 1:417–434, 2001.

[WW04] G. W. Wasilkowski and H. Woźniakowski. Finite-order weights imply tractability of linear multivariate problems. *J. Approx. Theory*, 130:57–77, 2004.

[WW05] G. W. Wasilkowski and H. Woźniakowski. Polynomial-time algorithms for multivariate problems with finite-order weights; worst case setting. *Found. Comput. Math.*, 5:451–491, 2005.

[WW07] G. W. Wasilkowski and H. Woźniakowski. The power of standard information for multivariate approximation in the randomized setting. *Math. Comp.*, 76:965–988, 2007.

Zinterhof Sequences in GRID-Based Numerical Integration

Heinz Hofbauer[1], Andreas Uhl[2], and Peter Zinterhof[3]

[1] Salzburg University, Department of Computer Sciences
hhofbaue@cosy.sbg.ac.at
[2] Salzburg University, Department of Computer Sciences
uhl@cosy.sbg.ac.at
[3] Salzburg University, Department of Computer Sciences
peter.zinterhof@sbg.ac.at

Summary. The appropriateness of Zinterhof sequences to be used in GRID-based QMC integration is discussed. Theoretical considerations as well as experimental investigations are conducted comparing and assessing different strategies for an efficient and reliable usage. The high robustness and ease of construction exhibited by those sequences qualifies them as excellent QMC point set candidates for heterogeneous environments like the GRID.

1 Introduction

High dimensional numerical integration problems may require a significant amount of computational effort. Therefore, substantial effort has been invested in finding techniques for performing these calculations on all kinds of high performance computing platforms (see e.g. [KÜ94, SU03]). GRID environments are highly beneficial but exhibit specifically challenging properties for numerical integration techniques. This class of computing facilities show extreme heterogeneity in terms of computing speed (caused by different memory capacity, cache sizes, and processor speed of the involved compute nodes) and network connections, moreover the available computing resources may change over time even during ongoing computations. These hardware properties require the employed integration routines to exhibit certain features:

- Variety in computing speed requires dynamic load balancing capability.
- Variety in network bandwidth and latency requires load balancing strategies without central organization and a minimal number of control messages exchanged among the computing nodes.
- Failure in hardware resources requires tolerance to lost partial results.
- Additional resources becoming available require a possibility to assign workload to these resources (i.e. by redistributing or redefining workload).

Additionally, error bounds and numerical results should preferable carry over from sequential execution, also reproducibility is considered an important issue.

Quasi-Monte Carlo (QMC) algorithms are among the most efficient techniques for evaluating high-dimensional integrals. Consequently, recent work has been devoted to apply this numerical integration approach in GRID environments [LKd05, LM05, HUZ06], however, many QMC techniques investigated for heterogeneous distributed systems may be used in the GRID context as well (e.g. [ÖS02, SU01, dZCG00]).

In this work we investigate a special type of QMC sequences, so-called Zinterhof sequences, for their applicability in GRID environments. In Section 2, we discuss the use of Zinterhof sequences in the general (sequential) QMC setting. Section 3 reviews strategies for using QMC techniques on parallel or distributed architectures. The main contribution of this work is presented in Section 4 where we give theoretical as well as experimental results on the use of Zinterhof sequences in GRID-type environments. Section 5 concludes the paper.

2 QMC Integration using Zinterhof Sequences

The basic concept of any QMC method for numerical integration is to approximate the integral by a finite sum, such that

$$I(f) := \int_{I^s} f(x)dx \approx \frac{1}{N}\sum_{n=1}^{N} f(x_n) =: I'_N(f)$$

where x_n are suitably chosen and I^s is the unit interval. To identify suitable, i.e. uniformly distributed, points x_n with low star discrepancy are selected in order to exploit the Koksma-Hlawka inequality [Nie92]:

$$E_N(f) \leq V(f)D_N^*(f),$$

where $E_N(f) := |I(f) - I'_N(f)|$ is the integration error.

2.1 Zinterhof Sequences

Zinterhof sequences [Zin69] are a special case of Weyl sequences. Weyl sequences are defined by

$$x_n = n\boldsymbol{\theta} = (\{n\theta_1\}, \{n\theta_2\}, \ldots, \{n\theta_s\}) \;\; n = 1, 2, 3, \ldots$$

where s is the dimension and $\{x\}$ is the fractional part of x. It is well known that a Weyl sequence is uniformly distributed if and only if θ_i are independent irrational numbers. An important issue with respect to their quality in terms of uniformity of distribution is the amount or degree of irrationality of the employed starting vector $\Theta = (\theta_1, \ldots, \theta_s)$. See [KY81][Theorem 4.15] for an estimation of discrepancy for this type of sequences. For the Zinterhof sequence we set $\theta_i = e^{1/i}$ and consequently:

$$x_n = (\{ne^{1/1}\}, \ldots, \{ne^{1/s}\}) \ n = 1, 2, 3, \ldots . \tag{1}$$

Note that due to their simplicity these sequences are extremely easy to generate and may be used by non-experts in a straightforward way without specific knowledge (which is not the case for all types of QMC sequences).

2.2 Numerical Integration with Zinterhof Sequences

Consider for dimension s the Fourier series expansion of the function $f(\mathbf{x})$ to be numerically integrated

$$f(\mathbf{x}) = \sum_{m_1, \ldots, m_s = -\infty}^{\infty} C(\mathbf{m}) e^{2\pi i (m_1 x_1 + \cdots + m_s x_s)} \tag{2}$$

with the integration error

$$E_N(\boldsymbol{\theta}) = \frac{1}{N} \sum_{n=1}^{N} f(n\boldsymbol{\theta}) - \int_0^1 \cdots \int_0^1 f(\mathbf{x}) dx_1 \ldots dx_s \tag{3}$$

where $\mathbf{x} = (x_1, \ldots, x_s)$, $\mathbf{m} = (m_1, \ldots, m_s)$ and $\boldsymbol{\theta} = (\theta_1, \ldots, \theta_s)$.

For absolute convergent Fourier series the error is

$$E_N(\boldsymbol{\theta}) = \sum_{\substack{m_1, \ldots, m_s = -\infty \\ \mathbf{m} \neq \mathbf{0}}}^{\infty} C(\mathbf{m}) \frac{1}{N} \sum_{n=1}^{N} e^{2\pi i (\theta_1 m_1 + \cdots + \theta_s m_s) n} = \sum_{\mathbf{m} \neq \mathbf{0}} C(\mathbf{m}) S_N(\boldsymbol{\theta}).$$

Thus to determine the quality of the integration method we have to estimate $S_N(\boldsymbol{\theta})$. Clearly $\theta_1, \ldots, \theta_s$ must be rational independent unless $\theta_1 m_1 + \cdots + \theta_s m_s \in \mathbb{Z}$ and thus $S_N(\boldsymbol{\theta}) = 1$. Furthermore, by using Weyl's criterion, we know that for independent irrational numbers $\boldsymbol{\theta}$ it holds that

$$\lim_{N \to \infty} S_N(\boldsymbol{\theta}) \to 0 \quad \forall \mathbf{m} \in \mathbb{Z}^s \backslash \{0\}.$$

Since $S_N(\boldsymbol{\theta})$ is a geometric series we can write

$$S_N(\boldsymbol{\theta}) = \frac{1}{N} e^{2\pi i (m_1 \theta_1 + \cdots + m_s)} \frac{1 - e^{2\pi i (m_1 \theta_1 + \cdots + m_s) N}}{1 - e^{2\pi i (m_1 \theta_1 + \cdots + m_s)}}.$$

For the rational independent $\theta_1, \ldots, \theta_s$ with the equality $e^{ix} = \cos(x) + i \sin(x)$ and the basic approximation $|\sin(\pi x)| \geq 2 \ll x \gg$, where $\ll x \gg$ is the distance of x to the nearest integer, we can approximate

$$|S_N(\boldsymbol{\theta})| \leq 1/N \frac{1}{2 \ll m_1 \theta_1 + \cdots + m_s \theta_s \gg}.$$

Consider for $\alpha > 1$ the class $E^s_\alpha(C) = \{f(\mathbf{x}) : |C(\mathbf{m})| \leq \frac{C}{||\mathbf{m}||^\alpha}\}$ then

$$|E_N(\boldsymbol{\theta})| \leq 1/N \sum_{\mathbf{m} \neq \mathbf{0}} \frac{C}{||\mathbf{m}||^\alpha} \frac{1}{2 \ll m_1\theta_1 + \cdots + m_s\theta_s \gg}$$

where $||\mathbf{m}|| = \prod_{i=1}^s \max(1, |m_i|)$.

Now for $\theta_1 = e^{r_1}, \ldots, \theta_s = e^{r_s}$, $r_i \neq r_j$ for $i \neq j$, $r_i \in \mathbb{Q}$ the subsequent result follows from an approximation by A. Baker (c.f. [KY81]):

$$\ll m_1\theta_1 + \ldots + m_s\theta_s \gg \geq \frac{C(\boldsymbol{\theta})}{||\mathbf{m}||\psi(\mathbf{m})},$$

where $\psi(\mathbf{m})$ weakly converges towards ∞ for $||\mathbf{m}|| \to \infty$.

Since there is no irrational vector $\boldsymbol{\theta}$ such that for all $\mathbf{m} \ll m_1\theta_1 + \cdots + m_s\theta_s \gg \geq \frac{C(\boldsymbol{\theta})}{||\mathbf{m}||}$ holds, we obtain the final error approximation for $\alpha > 2$ (Zinterhof provides the same error magnitude even for $\alpha > 1$ [Zin69])

$$|E_N(\boldsymbol{\theta})| \leq 1/N \sum_{\mathbf{m} \neq \mathbf{0}} \frac{C||\mathbf{m}||}{||\mathbf{m}||^\alpha} \frac{\psi(\mathbf{m})}{2C(\boldsymbol{\theta})}.$$

To give an illustration of the excellent actual integration performance, Fig. 1 shows a comparison of numerical integration accuracy among several QMC sequences for two of the test functions used in Section 4 (we plot the integration error versus sample size).

It can be clearly seen that for each of the two test scenarios there is a single QMC sequence which shows very poor integration results, the Halton sequence in the first case and the Faure sequence in the second case. While being the top performing sequence considered for some test functions (compare also [HUZ06] and Fig. 5), Zinterhof sequences are at least always competitive to the best sequences available and lead to consistently low integration errors.

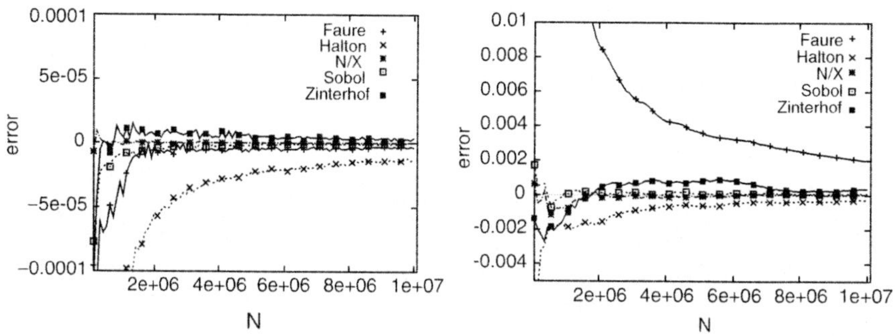

$g(\mathbf{x})$, $s = 15$ $h(\mathbf{x})$, $s = 20$

Fig. 1. Comparison of Zinterhof, Halton, Sobol, Niederreiter/Xing (N/X) and Faure sequences.

This fact taken together with the available error estimates and the simplicity of their construction and generation makes these sequences attractive candidates for practical QMC integration applications.

3 QMC Techniques in GRID Environments

The generation of the points of a QMC sequence on a single machine and the subsequent distribution of the generated points generates a significant bottleneck for the integration application. When considering GRID properties, the constraint of the unknown network capacity can become a problem, as such a fast processing element (PE) behind a slow link would be wasted. Likewise, if the point generating PE is behind a slow network link all other PEs are penalized when they have to wait for new points. Thus, rather than distributing the points, the generation of the points itself is distributed in such a fashion that each PE can generate the points nearly independently of other PEs.

So far, two entirely different strategies have been discussed in literature to employ QMC sequences in parallel and distributed environments (see [HUZ06] for an exhaustive literature review and a detailed assessment of the effectiveness of the different strategies in GRID environments).

1. Splitting a given QMC sequence into separately initialized and disjoint parts which are then used independently on the PEs. This strategy comes in two flavors (assuming availability of p PEs):

   Blocking: p disjoint contiguous blocks of maximal length l of the original sequence are used on the PEs. This is achieved by simply using a different starting point on each PE (e.g., PE$_i$, $i = 0, \ldots, p-1$, generates the vectors $\mathbf{x}_{il}, \mathbf{x}_{il+1}, \mathbf{x}_{il+2}, \ldots, \mathbf{x}_{il+l-1}$) ("big blocks" scenario). In case a large number of smaller blocks is used index j is assigned dynamically to PE$_i$ which generates the vectors $\mathbf{x}_j, \mathbf{x}_{j+1}, \ldots, \mathbf{x}_{j+l-1}$ (where j is incremented in steps of size l to avoid overlap – "small blocks" scenario). See [LM05, SU01] for investigations and applications with respect to the blocking approach.

   Leaping: interleaved streams of the original sequence are used on the PEs. Each PE skips those points consumed by other PEs (*leap-frogging*) (e. g. employing p PEs, PE$_i$, $i = 0, \ldots, p-1$, generates the vectors $\mathbf{x}_i, \mathbf{x}_{i+p}, \mathbf{x}_{i+2p}, \ldots$). Usually a QMC point set is partitioned into p interleaved substreams if p PEs are available. However, if more PEs become available during the computation, there is no additional substream available in this scenario. A way to handle this situation is to partition a given QMC point set into $I > p$ substreams in case of p PEs are available. The $I - p$ substreams are not used by default but kept as additional work share in case additional PEs become available. See [Bro96, SU01, ESSU03] for investigations and applications with respect to the leaping approach.
2. Using inherently independent sequences on the different PEs (denoted as "parameterization" which can be realized for example by randomizations

of a given QMC sequence). The most important difference (and also disadvantage) of parameterization as compared to blocking and leaping is that the QMC point set used in parallel or distributed computation does not correspond to a single (sequentially used) point set. Therefore, the investigation of the results' quality when using this technique is of great importance since it is not clear a priori how results from different point sets will interact in the final result. See [Cd02, ÖS02] for investigations and applications with respect to the parameterization approach.

4 Zinterhof Sequences in GRID Environments

In this section we investigate whether Zinterhof sequences are sensible candidates for use in GRID environments. We present theoretical as well as experimental results with respect to the three approaches for distributed generation of QMC point sets as discussed in the previous section.

4.1 Theoretical Results

Leaping

For estimating the integration error resulting from using leaped Zinterhof sequences, we replace $\boldsymbol{\theta}$ in equation (3) by $L\theta_1, \ldots, L\theta_s$ for leap size $L \in \mathbb{N}$. Then instead of $S_N(\boldsymbol{\theta})$ we have

$$S_N(L\boldsymbol{\theta}) = \frac{1}{N} \sum_{n=1}^{N} e^{2\pi i (Lm_1\theta_1 + \cdots + Lm_s\theta_s)n}.$$

By analogy to the general case we can approximate the integration error, however this approximation is worse since instead of $\mathbf{m}$ we now have $L\mathbf{m}$ in all formulas. Thus with $||L\mathbf{m}|| = \prod_{i=1}^{s} \max(1, |Lm_i|) \leq L^s ||\mathbf{m}||$ and $\psi(L\mathbf{m})$ instead of $\psi(\mathbf{m})$ we get

$$|E_N(L\boldsymbol{\theta})| \leq 1/N \sum_{\mathbf{m} \neq \mathbf{0}} \frac{L^s C ||\mathbf{m}||}{||\mathbf{m}||^\alpha} \frac{\psi(L\mathbf{m})}{2C(\boldsymbol{\theta})}.$$

Considering that $\psi(\mathbf{m})$ grows only logarithmically for $\boldsymbol{\theta} = (\theta_1^{r_1}, \ldots, \theta_s^{r_s})$ and likewise for $\psi(L\mathbf{m})$ the difference of $\psi(\mathbf{m})$ to $\psi(L\mathbf{m})$ plays hardly any role. Thus the error approximation for leaping with leap size L is worse by the factor L^s than the error approximation for the unleaped sequence. This indicates a potentially significant deterioration of the results independent of the specific leap value (note that contrasting to this result we have derived poor discrepancy estimates only for 2^n type leaped (t,s)-sequence substreams in earlier work [SU01]).

Blocking

Again, consider the Fourier series given in Equation (2) and the error given in Equation (3) with the same parameters.

Then we have for $f \in E_\alpha^s(C)$ with $\alpha > 3/2$ and $x_1, \ldots, x_N \in I^s := [0,1]^s$ the approximation ([DT97, Theorem 1.35])

$$\left| \frac{1}{N} \sum_{n=1}^{N} f(x_n) - \int_{I^s} f(\mathbf{x}) dx_1, \ldots, dx_s \right| \leq C \left(\frac{4\alpha - 4}{2\alpha - 3} \right)^{s/2} F_N(x_n) \quad (4)$$

where $F_N(x_n)$ is the diaphony of $x_1, \ldots, x_N$.

It is known [Zin76] that for the Zinterhof sequence the estimation of the diaphony

$$F_N(n\boldsymbol{\theta}) = O(1/N^{1-\epsilon}), \quad (5)$$

for $\epsilon > 0$ holds, since $\boldsymbol{\theta}$ is of the form $\theta_1 = e^{r_1}, \ldots, \theta_s = e^{r_s}$ where the $r_i \in \mathbb{Q}\ \forall i = 1, \ldots, s$ are rationally independent.

The definition of the diaphony F_N for a general s-dimensional sequence $\mathbf{x_1}, \ldots, \mathbf{x_N}$ is

$$F_N^2(\mathbf{x_n}) = \frac{1}{N} \sum_{i,j=1}^{N} H_2(\mathbf{x_i} - \mathbf{x_j}), \quad (6)$$

with

$$H_2(\mathbf{x}) = \prod_{i=1}^{s} h_2(x_i) - 1,$$

and h_2 being the normed Bernoulli polynomial of degree 2,

$$h_2 = 1 + 2\pi^2 \left(\{x\}^2 - \{x\} + \frac{1}{6} \right),$$

where $\{x\}$ is the fractional part of x.

The diaphony F_N of the sequence $x_1, \ldots, x_N$ is translation invariant, which follows directly from Equation (6) where we get $H_2((a + \mathbf{x_i}) - (a + \mathbf{x_j})) = H_2(\mathbf{x_i} - \mathbf{x_j})$, thus for any $a = (a_1, \ldots, a_s)$

$$F_N(x_n) = F_N(a + x_n)$$

holds.

For the Zinterhof sequence we can choose $a = x_B = (B\theta_1, \ldots, B\theta_s)$ such that we obtain

$$F_N(x_n) = F_N(x_B + x_n) = F_N(x_{B+n})$$

where $n = 1, \ldots, N$.

Thus when using the error approximation (4) we see that we can use an arbitrary block of length N instead of the first N points without deterioration of the integration error. Note that this corresponds well to an earlier result on (t,s)-sequences where we showed that discrepancy estimates of arbitrary blocks do not degrade as compared to estimates of entire (t,s)-sequences [SU01].

Now, similar to [ÖS02], let us consider the general case of blocking with block size b where new blocks are handed out as requested ("small blocks"). The classical blocking scheme, which we call "big blocks", is essentially a subset of this general case. When using p PEs we have always p continuous subsets, each subset of points ends where a block is still unfinished. So we have p sequences each generating an approximation of the integral I

$$I'^i = \frac{1}{c_i} \sum_{\lambda(c_i)} f(x_i)$$

where $i = 1, \ldots, p$, c_i is the number of vectors in sequence i and $\lambda(c_i)$ is the set of indices of vectors of the original Zinterhof sequence which generates sequence i and the numbering be such that $c_p \leq c_{p-1} \leq \cdots \leq c_1$ holds. Figure 2 illustrates this for three PEs, when an PE finishes with block 3 the former c_1 and c_2 collapse to form the new c_1. Also when one block is finished another block is assigned and an PE starts to work on it, this forms a new sequence c_3.

Now with N the total number of points, we get

$$I'_N = \frac{1}{N} \sum_{\lambda(N)} f(x_i) = \sum_{i=1}^{p} \frac{c_i}{N} \frac{1}{c_i} \sum_{\lambda(c_i)} f(x_i) = \sum_{i=1}^{p} \frac{c_i}{N} I'^i,$$

which gives us the overall estimate from the estimates of the individual sequences.

We can consider the error

$$E_N(f) = |I'_N - I(f)| \leq \sum_{i=1}^{p} \frac{c_i}{N} |I'^i - I(f)| \leq \sum_{i=1}^{p} \frac{c_i}{N} D^*_{c_i} V(f).$$

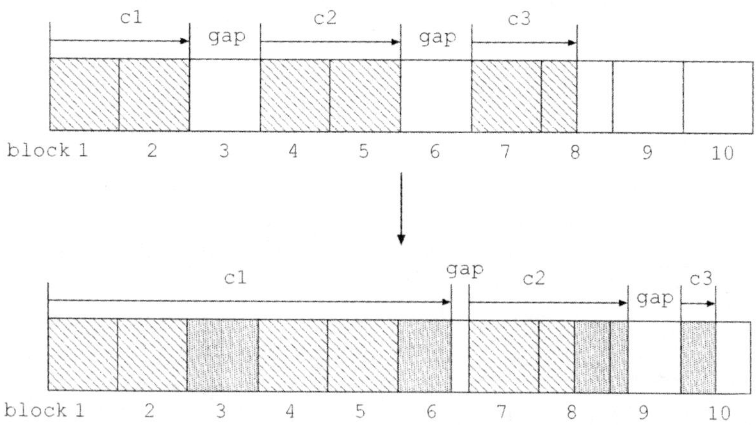

Fig. 2. Growth of subsequences with small blocks.

When looking at blocking with a small block size, i.e. not the big block scenario, it is clear that the first sequence grows continuously as more intermediate blocks are finished and likewise new sequences are introduced at the end with a very small c_p. From the above error estimate we see that the weighted average of the discrepancies is used, but since c_1 continually grows for $N \to \infty$ we get $c_i/c_1 \to 0$ for $1 < i \leq p$. Since $c_p \leq c_i \leq c_2$ for $i = 3, \ldots, p-1$, we get

$$E_N(f) \leq V(f)(\frac{c_1}{N} D^*_{c_1} + \frac{(p-1)c_2}{N} D^*_{c_p}).$$

For very big N the error estimation thus becomes approximately

$$E_N(f) \leq V(f) D^*_{c_1}$$

where $c_1 \approx N$.

Clearly the smaller the blocks are the faster they become insignificant and the faster the first sequence grows. For big blocks we have the same problem as with parameterization since unlike normal blocking no sequence becomes insignificant and for the error we can only get the general error estimate. Given a homogenous environment where $c_1 = \cdots = c_p$ we get only

$$E_N(f) \leq V(f) D^*_{N/p}$$

which shows no advantage over using a single machine.

Parameterization

A result with respect to a possible parameterization of Zinterhof sequences may be found in [HUZ07], which provides a set of almost uncorrelated sequences.

For almost all collections of P specimen of s-dimensional Weyl sequences

$$\begin{aligned} f_1^{(k)} &= (\{k\theta_1\}, \ldots, \{k\theta_s\}), \ldots, \\ f_i^{(k)} &= (\{k\theta_{(i-1)s+1}\}, \ldots, \{k\theta_{is}\}), \ldots, \\ f_P^{(k)} &= (\{k\theta_{(P-1)s+1}\}, \ldots, \{k\theta_{Ps}\}),\ k = 1, 2, \ldots, N, \ldots \end{aligned}$$

the estimations for covariance and correlation

$$\mathrm{cov}_N(f_1, \ldots, f_P) = \frac{s}{12} I^P + \mathrm{O}(N^{\epsilon - 1})$$

and

$$\mathrm{cor}_N(f_1, \ldots, f_P) = I^P + \mathrm{O}(N^{\epsilon - 1})$$

hold, where I^P is the $P \times P$ unit matrix with entries $e_{ii} = 1$ and $e_{jk} = 0$ for $j \neq k$ and $i, j, k = 1, \ldots, P$. The estimations hold especially for the collections of P sequences of the Weyl type having generators $\theta_1, \ldots, \theta_s, \theta_{s+1}, \ldots, \theta_{Ps}$ of the form $\theta_u = e^{r_u}, r_u \in \mathbb{Q}, r_u \neq r_v \neq 0, 1 \leq u, v \leq Ps$.

Since the Zinterhof sequences are of the form given above, we have well distributed s-dimensional point generating sequences which are nearly uncorrelated. Essentially this allows us to use Zinterhof sequences for a parameterization approach where PE_n uses $x_n = (\{ke^{1/(n-1)s+1}\}, \ldots, \{ke^{1/ns}\})$, $k = 1, 2, \ldots$.

4.2 Numerical Experiments

Settings

For the Zinterhof sequences, we use our own custom implementation. In order to be able to assess the accuracy of our results, we also employ different QMC sequences. Overall we used Zinterhof, Sobol', Halton, Faure and Niederreiter/Xing sequences, due to page limitation only the more interesting results are shown, however, we will always comment on the remaining sequences. For generating the Sobol', Halton, Faure and Niederreiter-Xing sequences we use the implementation of the "High-dimensional Integration Library" HIntLib[4]. The HIntLib uses an implementation of construction 6, 8 and 18 from [MMN95] for the Sobol', Faure and Niederreiter/Xing sequences respectively. For Halton it uses the construction which was introduced in [Hal60]. For more information on Sobol', Faure, and Niederreiter-Xing sequences see [SS], and for Halton sequences see [Hal60].

The numerical experiments have been conducted by integrating the following test functions:

$$f(\mathbf{x}) = \prod_{i=1}^{s} \frac{1}{x_i^{0.5}}, \tag{7}$$

$$g(\mathbf{x}) = \sqrt{\frac{45}{4s}} \left(\sum_{i=1}^{s} x_i^2 - \frac{s}{3} \right), \tag{8}$$

$$h(\mathbf{x}) = \prod_{i=i}^{s} \left(x_i^3 - \frac{1}{4} + 1 \right). \tag{9}$$

All three functions have been employed extensively in experimental evaluations, e.g. in own earlier work function (7) in [HUZ06] and functions (8) and (9) in [SU01]. The function $f(\mathbf{x})$ is unbounded due to the singularity in 0, the value of the integral is 2^s (we use the identical integration routine as outlined in [HUZ06]). The integral for $g(\mathbf{x})$ and $h(\mathbf{x})$ is 0 in both cases, therefore we display an absolute integration error on the ordinate of the plots instead of a relative error as for $f(\mathbf{x})$ (the abscissa shows the number of points used in numerical integration).

For all experiments we used a randomly chosen mixture of machines using AMD CPUs with 1200, 1600, and 2000 MHz interconnected by 100Mbit ethernet [HUZ06]. The actual number of machines used is given for each experiment.

Results

For leaping, the experimental results (fortunately) do not confirm the poor error estimate. Figure 3 shows integration results when single leaps (with

[4] Available at: http://www.cosy.sbg.ac.at/~rschuer/hintlib/

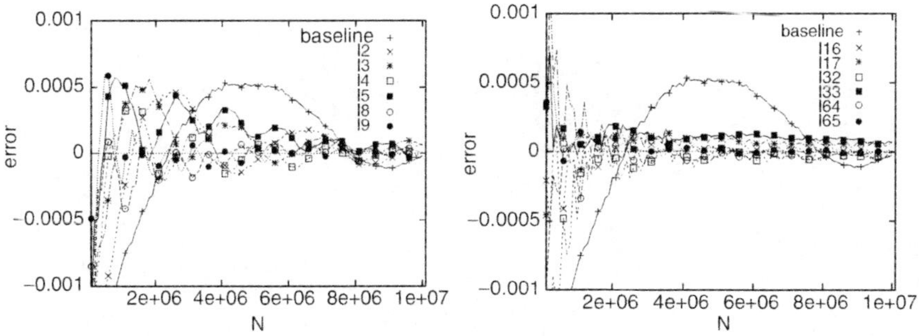

Fig. 3. Comparison of different leap sizes for the Zinterhof sequence, function $h(\mathbf{x})$, $s = 10$.

different leap sizes) are used in sequential execution instead of the unleaped sequence. Surprisingly it turns out that the leaped substreams actually improve the integration result and lead to significantly faster convergence as compared to the baseline case. A similar behavior (except for leap 65) has been observed for $f(\mathbf{x})$ (see also [HUZ06]), $g(\mathbf{x})$ is very easy to handle and therefore almost no differences show up between the original and leaped versions of the Zinterhof sequence.

Of the remaining sequences only the Niederreiter-Xing sequences show comparable (and even better) stability with respect to splitting. All other sequences show significant result degradation for one or more leaps.

In order to relate these results to an execution in a GRID like environment, we simulate failure of PEs in the following manner: "client-server" is numerically identical to the sequential result (but executed in the distributed environment), "leap" is the standard case where one stream is assigned each PE, and "one-fast" and "one slow" are scenarios where one PE is speed up or slowed down by a factor of 10^3: consequently, one-slow simulates the case of one failing PE, whereas one-fast simulates the rather unlikely case that all but one PE fail after an initial start.

Figure 4 gives the results for leap 11 (on 11 PEs) and compares them to the result employing the Sobol' sequence with leap $2^3 + 1 = 9$ which is of the form $2^n + 1$ (which is known to have a good star discrepancy estimate [SU01]).

It is clearly visible, that the Zinterhof sequence is very robust against PE failure and that the integration errors are significantly smaller as compared to the Sobol' sequence. However, the Sobol' case shows severely degraded results, especially in the highly probable one-slow case. In this scenario we have a systematic missing of equally spaced points which can not be grasped by a discrepancy estimate of a single leaped substream and obviously leads to significant problems in this type of point sets. Comparable results are found for $h(\mathbf{x})$.

For the other sequences, except for the Niederreiter/Xing sequence, we find problems comparable to those seen with the Sobol' sequence. Overall, a high robustness in the case of employing leaped substreams can be stated for Zinterhof sequences.

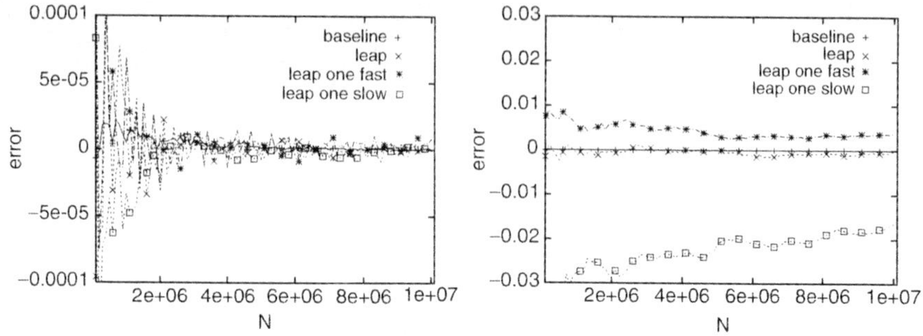

Zinterhof (leap 11)

Sobol' (leap 9)

Fig. 4. Comparison of Zinterhof and Sobol' sequences, $g(x)$, $s = 10$.

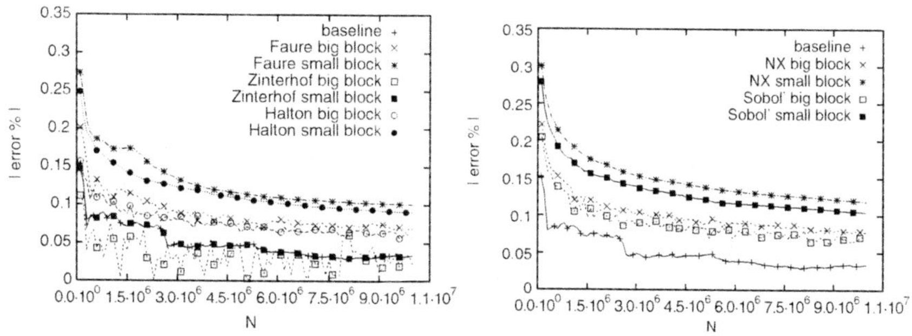

Faure, Zinterhof, Halton

Niederreiter/Xing, Sobol'

Fig. 5. Comparison of blocking (big and small blocks) using Zinterhof, Halton, Faure, Niederreiter/Xing and Sobol' sequences, $f(\mathbf{x})$, $s = 10$.

Contrasting to the leaping case, robust behavior is to be expected due to the theoretical result when using contiguous blocks of Zinterhof sequences in distributed integration.

Figure 5 compares the results of different QMC sequences for blocking using 11 machines. For small blocks we use block size 500 and for big blocks the relatively bad case of block size N is chosen, which results in gaps roughly nine times the size of the actually used blocks. As a baseline for comparison we show integration with the Zinterhof sequence using all N points in consecutive manner.

The first things to note is that the three results corresponding to the Zinterhof sequence are the best ones in terms of error magnitude. The small blocks' result is in fact identical to the baseline version (which holds true for all sequences and was to be expected since the error estimate for small blocks is practically independent of block discrepancy for high N and thus valid for all sequences) whereas the big blocks' result shows lower error but a higher degree of result fluctuations in the Zinterhof sequence case.

For all cases (except the Zinterhof sequence between 8×10^6 and 9×10^6 integration points) the big blocks case is better than the small blocks case, and thus the baseline. While this isn't generally the case (see Fig. 6) it seems

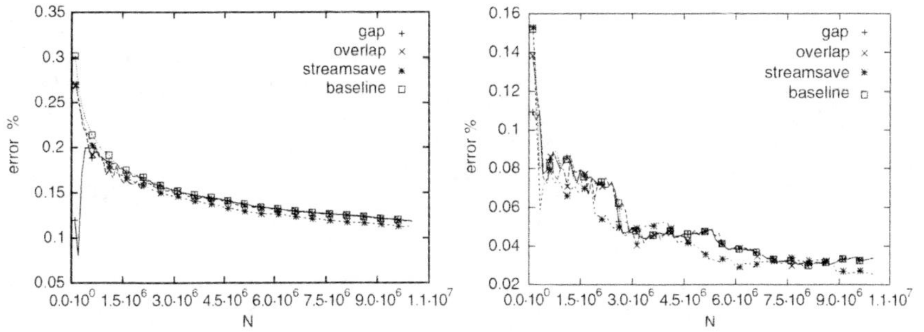

Niederreiter/Xing Zinterhof

Fig. 6. Comparison Niederreiter/Xing and Zinterhof sequences regarding overlap, gaps and unused streams for $f(\mathbf{x})$, $s = 10$.

that the integration error estimation gained from our approach for big blocks is not the best possible.

Figure 6 shows results corresponding to the more realistic case that we employ big blocks with a gap between blocks that cover 20% of the size of a block and big blocks resulting in an overlap where about 30% of one block overlaps the following block. Additionally, we investigate the "Streamsave" scenario, where we use small blocks with a block size of 75 and between the blocks there is a 25 point gap. This simulates the synchronized use of substreams with leap 100 where the last 25 out of 100 streams are reserved for PEs which become available during the computation (but are not used in the experiment). 71 PEs are used for these computations.

The result shows that both considered sequences can cope well with gaps, overlap, and the "streamsave" scenario, no degradation of the result is observed.

To relate parameterization behavior to blocking and leaping effects, we compare all three approaches in the following. Results with respect to functions $f(\mathbf{x})$ and $h(\mathbf{x})$ for dimension $s = 10$ (not shown) raise doubts about the reliability of parameterization since the error seems to decrease slower for increasing N as compared to other techniques. In order to facilitate a fair comparison we increase the dimension and employ 10 PEs and less favorable conditions for leaping and blocking: For blocking the block size is set to N and for leaping we use leap 100 resulting in 90% gaps for blocking and likewise to 90% unused streams for leaping (note that the gaps are distributed differently in both variants).

As observed in Fig. 7, for higher dimension the approximation becomes worse, which is to be expected since we use the same number of points independent of dimension (compare e.g. Figs. 3 and 4). The results of blocking and leaping are almost identical to the baseline version, although the conditions are much more difficult in the current setting. On the other hand, parameterization shows a larger error and a slower convergence towards the correct solution.

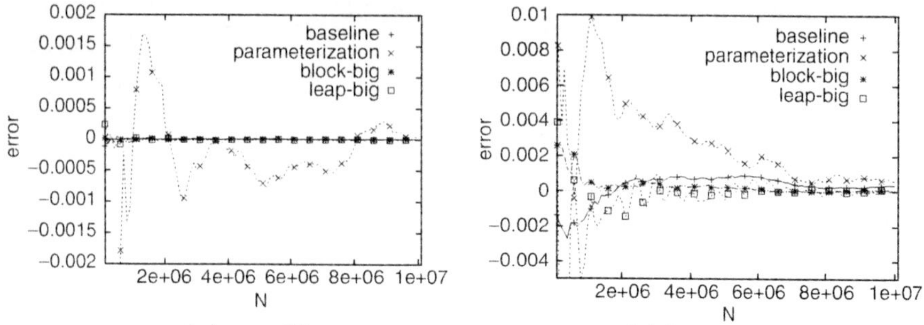

$g(\mathbf{x})$, $s = 15$ $h(\mathbf{x})$, $s = 20$

Fig. 7. Comparison of leaping (with large leapsize), blocking (large gaps) and parameterization for Zinterhof sequences.

5 Conclusion

Overall, we have shown that Zinterhof sequences are well suited for numerical integration in GRID environments. Whereas the error estimation for leaped substreams suggests worse integration errors as compared to sequential usage, we have found no experimental evidence corresponding to this result. In contrary, leaping turns out to behave very reliable and robust even to hardware failures and may be used in a flexible way.

For the case of using contiguous blocks for integration the theoretical prediction suggesting behavior equal to the sequential case is supported by experimental results. Similar to leaping, high robustness against gaps between blocks and against overlap has been observed.

While the suggested parameterization scheme works in principle, the results show clearly slower convergence as compared to the leaping or blocking strategies, respectively. Parameterization (at least in the proposed manner) should only be used if this effect is acceptable.

Concluding we may state that Zinterhof sequences have been shown to exhibit excellent behavior when using separately initialized and disjoint substreams for distributed numerical integration and they excel by their ease of construction and implementation even for non-specialists.

References

[Bro96] B.C. Bromley. Quasirandom number generators for parallel Monte Carlo algorithms. *Journal of Parallel and Distributed Computing*, 38:101–104, 1996.

[Cd02] L. Cucos and E. deDoncker. Distributed QMC algorithms: new strategies for and performance evaluation. In *Proceedings of the High Performance Computing Symposium 2002 (HPC'02)/Advanced Simulation Techniques Conference*, pages 155–159, 2002.

[DT97] M. Drmota and R.F. Tichy. *Sequences, discrepancies and applications*, volume 1651 of *Lect. Notes in Math.* Springer-Verlag, 1997.

[dZCG00] E. deDoncker, R. Zanny, M. Ciobanu, and Y. Guan. Asynchronous quasi-Monte Carlo methods. In *Proceedings of the High Performance Computing Symposium 2000 (HPC'00)*, pages 130–135, 2000.

[ESSU03] K. Entacher, T. Schell, W. Ch. Schmid, and A. Uhl. Defects in parallel Monte Carlo and quasi-Monte Carlo integration using the leap-frog technique. *Parallel Algorithms and Applications*, 18(1–2):27–47, 2003.

[Hal60] J. H. Halton. On the efficiency of certain quasi-random sequences of points in evaluating multi-dimension integrals. *Numer. Math.*, 2:84–90, 1960. Berichtigung, ibid., (1960), p. 196.

[HUZ06] H. Hofbauer, A. Uhl, and P. Zinterhof. Quasi Monte Carlo Integration in GRID Environments: Further Leaping Effects. *Parallel Processing Letters*, 16(3):285–311, 2006.

[HUZ07] H. Hofbauer, A. Uhl, and P. Zinterhof. Parameterization of Zinterhof Sequences for GRID-based QMC Integration. In J. Volkert, T. Fahringer, D. Kranzlmüller, and W. Schreiner, editors, *Proceedings of the 2nd Austrian Grid Symposium*, volume 221 of *books@ocg.at*, pages 91–105, Innsbruck, Austria, 2007. Austrian Computer Society.

[KÜ94] A.R. Krommer and C.W. Überhuber. *Numerical Integration on Advanced Computer Systems*, volume 848 of *Lecture Notes in Computer Science*. Springer, Berlin, 1994.

[KY81] H. L. Keng and W. Yuan. *Applications of Number Theory to Numerical Analysis*. Springer Verlag, Science Press, 1981.

[LKd05] S. Li, K. Kaugars, and E. deDoncker. Grid-based numerical integration and visualization. In *Sixth International Conference on Computational Intelligence and Multimedia Applications (ICCIMA'05)*, pages 260–265. IEEE Computer Society Press, 2005.

[LM05] Y. Li and M. Mascagni. Grid-based Quasi-Monte Carlo applications. *Monte Carlo Methods and Appl.*, 11(1):39–55, 2005.

[MMN95] G. L. Mullen, A. Mahalanabis, and H. Niederreiter. Tables of (t, m, s)-net and (t, s)-sequence parameters. In Harald Niederreiter and P. J.-S. Shiue, editors, *Monte Carlo and Quasi-Monte Carlo Methods in Scientific Computing*, volume Lecture Notes in Statistics of *106*, pages 58–86. Springer-Verlag, 1995.

[Nie92] H. Niederreiter. *Random Number Generation and Quasi-Monte Carlo Methods*, volume 62 of *CBMS-NSF Regional Conference Series in Applied Mathematics*. Society for Industrial and Applied Mathematics (SIAM), 1992.

[ÖS02] G. Ökten and A. Srinivasan. Parallel quasi-Monte Carlo methods on a heterogeneous cluster. In K. T. Fang, F. J. Hickernell, and H. Niederreiter, editors, *Monte Carlo and Quasi-Monte Carlo Methods 2000*, pages 406–421. pub-springer, 2002.

[SS] R. Schürer and W. Ch. Schmid. Mint - the database of optimal (t, m, s)-net parameters. online: `http://mint.sbg.ac.at/`.

[SU01] W. Ch. Schmid and A. Uhl. Techniques for parallel quasi-Monte Carlo integration with digital sequences and associated problems. *Mathematics and Computers in Simulation*, 55:249–257, 2001.

[SU03] R. Schürer and A. Uhl. An evaluation of adaptive numerical integration algorithms on parallel systems. *Parallel Algorithms and Applications*, 18(1–2):13–26, 2003.

[Zin69] P. Zinterhof. Einige zahlentheoretische Methoden zur numerischen Quadratur und Interpolation. *Sitzungsberichte der Österreichischen Akademie der Wissenschaften, math.-nat.wiss. Klasse Abt. II*, 177:51–77, 1969.

[Zin76] P. Zinterhof. Über einige Abschätzungen bei der Approximation von Funktionen mit Gleichverteilungsmethoden. *Sitzungsber. Österr. Akad. Wiss. Math.-Natur. Kl. II*, 185:121–132, 1976.

A Pragmatic View on Numerical Integration of Unbounded Functions

Heinz Hofbauer[1], Andreas Uhl[2], and Peter Zinterhof[3]

[1] Salzburg University, Department of Computer Sciences
hhofbaue@cosy.sbg.ac.at
[2] Salzburg University, Department of Computer Sciences
uhl@cosy.sbg.ac.at
[3] Salzburg University, Department of Computer Sciences
peter.zinterhof@sbg.ac.at

Summary. We take a pragmatic approach to numerical integration of unbounded functions. In this context we discuss and evaluate the practical application of a method suited also for non-specialists and application developers. We will show that this method can be applied to a rich body of functions, and evaluate it's merits in comparison to other methods for integration of unbounded integrals. Furthermore, we will give experimental results to illustrate certain issues in the actual application and to confirm theoretic results.

1 Introduction

The basic concept of any QMC method for numerical integration is to approximate the integral by a finite sum, such that

$$I(f) := \int_{U^s} f(x)dx \approx \frac{1}{N}\sum_{n=1}^{N} f(x_n) =: I'_N(f)$$

where x_n are suitably chosen and U^s is the unit cube. To identify suitable, i.e. uniformly distributed, points x_n the star discrepancy is defined as

$$D_N^* := D_N^*(x_1, \ldots, x_n) = \sup_{J\in\mathcal{F}} \left\| \frac{\#\{x|x \in J\}}{N} - m(J) \right\|$$

where $\mathcal{F}$ is the family of all subintervals of the form $J = \prod_{i=1}^{d}[0, t_i) \in U^s$ with volume $m(J)$. The approximation error

$$E_N(f) := |I'_N(f) - I(f)|$$

depends on D_N^* and the variation $V(f)$ in the sense of Hardy and Krause, see [Nie92] or [Owe05] for details, of the function f. The dependency is stated in the Koksma-Hlawka inequality

$$E_N(f) \leq V(f) D_N^*(f)$$

which is the fundamental error bound for quasi-Monte Carlo methods.

A big problem with numerical integration is the fact that the variation is rather restrictive. Even simple functions like $m(x) = max(x_1 + x_2 + x_3 - 1, 0)$ are of unbounded variation $V(m) = \infty$, see [Owe05]. Also if a function is unbounded the variation is unbounded resulting in an error estimate $E_N(f) \leq \infty$.

There have already been a number of methods proposed in literature which aim at tackling the problem of numerically integrating unbounded functions, these will be described in Section 2. The method proposed in this work is discussed in detail in Section 3. In Section 4 we give experimental results which indicate that the method may even be applied to functions not contained in the restrictive class of functions it is proved for, associated problems are also discussed.

2 Methods for the Numerical Integration of Unbounded Functions

In the case of singularities the Koksma-Hlawka inequality becomes meaningless since functions containing singularities are unbound and thus of infinite variation. When examining methods of numerical integration for integrands with singularities usually the distinction is made whether the singularities are in the interior of the unit cube or on the boundary.

2.1 Singularities on the Boundary

Sobol' [Sob73] investigated a number of functions which have singularities in the origin. By restricting the growth of the integral by

$$D_N \int_{a_N}^{1} |f(x)| dx = o(N)$$

for $N \to \infty$, where $a_N = \min_{1 \leq i \leq N} x_i$, he shows that

$$\lim_{N \to \infty} \frac{1}{N} \sum_{\mu=1}^{N} f(P_\mu) = \int_{U^s} f(P) dP$$

holds, but unfortunately fails to give an error bound. He allows one dimensional functions $f(x)$ to have a rational singularity $0 < \xi < 1$ but reduces them to functions with singularities in the origin. In the multi dimensional case

Sobol's test function is $f(x) = x_1^{-\beta_1} \cdots x_s^{-\beta_s}$ where the growth condition holds if $\forall i\, \beta_i < 1$.

An important class of methods deals with point generation sequences which avoid the corners. Owen [Owe06] deals with singularities by replacing the part of the function which is not touched by the numerical integration, i.e. the part lying in the hyperbolic or L-shaped region avoided by the Halton sequence, by a bounded extension of the function. This way he attains a finite variation for the function and can prove error bounds for the numerical integration.

Definition 1. *Let $1 > \epsilon > 0$ then the following regions are subsets of the s-dimensional unit cube $U^s = [0,1]^s$*

$$H_o(\epsilon) = \{x \in U^s | \prod_{i=1}^{s} x_i \geq \epsilon\} \qquad H(\epsilon) = \{x \in U^s | \prod_{i=1}^{s} \min(x_i, 1 - x_i) \geq \epsilon\}$$

$$L_o(\epsilon) = \{x \in U^s | \min_{1 \leq i \leq s}^{s} x_i \geq \epsilon\} \quad L(\epsilon) = \{x \in U^s | \min_{1 \leq i \leq s} \min(x_i, 1 - x_i) \geq \epsilon\}.$$

The region H_o excludes a hyperbolic region *near the origin and H excludes hyperbolic regions near all corners of the unit cube. Likewise, L_o excludes a* L-shaped region *near the origin and L near all corners.*

Definition 2. *Let $1 > \epsilon > 0$, then if a sequence of N points P_N which fulfills*

$$\forall x \in P_N \Rightarrow x \in H_o(\epsilon)$$

we say the sequence avoids the origin in a hyperbolic fashion. *The same holds for H (*avoids all corners*), L_o (*in an L-shaped fashion*) and L (*avoids all corners*).*

Remark 1. It is not unusual to differentiate between corners since a given sequence usually doesn't avoid all corners to the same degree, i.e. with the same ϵ.

Owen shows that the Halton sequences avoid all corners in a hyperbolical sense. He also shows that for a finite $C > 0$ the Halton points $x_1, \ldots, x_n$ avoid the hyperbolic region $\{x | \prod_j x^j \leq Cn^{-1}\}$, while independent uniform points x_n enter that region infinitely often, with probability one. It is also shown that while points from the Halton sequence avoid the 0 and 1 corner stronger than independent uniform points do this doesn't hold for all other corners. To show an error bound for numerical integration with point sequences which avoid the origin in a hyperbolic $\{x \in [0,1]^d | \prod_{1 \leq i \leq d} x^i \geq \epsilon\}$ or L-shaped $\{x \in [0,1]^d | \min_{1 \leq i \leq d} x^i \geq \epsilon\}$ way Owen also imposes growth conditions on the functions.

Hartinger et al. show in [HKZ05] that generalized Niederreiter sequences possess corner avoidance properties similar to Halton sequences around the origin. They also show the corner avoidance rates for Halton and Faure sequences for corners different than the origin. To get efficient QMC rules for the integrands one has to find point sets satisfying the condition $\prod_{i=1}^{s} x_n^{(i)} \geq cN^{-r}$ with small r as stated in [Owe06] or for an all corner case when the avoidance condition is written as $\min_{1 \leq n \leq N} \prod_{i=1}^{s} \min(1 - x_n^{(i)}, x_n^{(i)}) \geq cN^{-r}$.

They show that for each point $\mathbf{x}_n$, $0 \leq n < b^l$, of a generalized Niederreiter (t,s)-sequence in base b the bound $\prod_{i=1}^{s} x_n^{(i)} \geq b^{-l-t-s}$ holds.

Kainhofer, Hartinger, and Tichy [HKT04] also dealt with QMC methods for multidimensional integrals with respect to a measure other than the uniform distribution. They allow the integrand to be unbounded on the lower boundary of the interval and justify the "strategy of ignoring the singularity" by using weighted integration with a non-uniform distribution. This means integration problems of the form $I_{[\mathbf{a},\mathbf{b}]} := \int_{[\mathbf{a},\mathbf{b}]} f(\mathbf{x}) dH(\mathbf{x})$ where H denotes a s-dimensional distribution with support $K = [\mathbf{a}, \mathbf{b}] \subset \mathbb{R}^s$ and f is a function with singularities on the left boundary of K. To use a generalized version of the Koksma-Hlawka inequality they have to define a H-discrepancy of $\omega = (y_1, \ldots)$ which measures the distribution properties of the sequence. It is defined as $D_{N,H}(\omega) = \sup_{J \subset K} |N^{-1} A_N(J, \omega) - H(J)|$ where A_N counts the number of elements in $(y_1, \ldots, y_N)$ falling into the interval J, e.g. $A_N(J, \omega) = \sum_{n=1}^{s} \chi_J(y_n)$, and $H(J)$ denotes the probability of $J \subset K$ under H. With this $D_{N,H}$ they can define the Koksma-Hlawka inequality for this case as $|I_K - N^{-1} \sum_{n=1}^{N} f(y_n)| \leq V(f) D_{N,H}(\omega)$.

While the authors state that there is a certain lack of sequences with low H-discrepancy they also propose a technique for constructing such sequences by using the Hlawka and Mück method [HM72]. However, for such a sequence $\tilde{\omega}$ there might be some elements $\tilde{y}_k$ which attain 0. Since the singularities of $f(x)$ are on the lower boundary these sequences are not directly suited to be used with the numerical integration, however a simple change is proposed, generating a new sequence $\bar{\omega}$.

For the multi-dimensional case the idea is basically the same, Kainhofer et al. state a convergence criterion as follows. Similar to the one-dimensional case the construction of H-distributed sequences leads to problems when using Hlawkas method. However an adjustment to the generated sequence is given by the authors.

In [HK05] Hartinger and Kainhofer deal with the problem of generating low discrepancy sequences with an arbitrary distribution H. While they did so before ([HKT04]) they identify some disadvantages which carry over to the transformed sequence they proposed. They specifically deal with the property of the Hlawka-Mück method that for some applications the points of the generated sequence of a set with cardinality N lie on a lattice with spacing $1/N$. Their solution is to use a smoothed approximation where the values between the jumps are interpolated in the empirical distribution function.

In order to integrate functions with singularities at the boundary it will be convenient to shift the interpolated sequences in an appropriate way to avoid regions that lie too close to that singularity. The authors define how to generate a new sequence $\hat{\omega}$ from the constructed sequence $\bar{\omega}$ which has the same distance to the boundaries as the original sequence ω.

They show, by utilizing the same techniques as Owen in [Owe06], that the sequence can be used to integrate improper integrals which have a singularity in the corner. They also show an error estimate for the L-shaped and hyperbolic corner avoidance cases.

DeDoncker [dDG03] reduces the error rate of the methods of Klinger and Sobol' by constructing extensions which reduce the approximation error. She looks at the leading asymptotic order of the error and generates extrapolations for such functions in such a way that error terms vanish. She has shown that for one dimensional functions with algebraic end point singularities her method works very well. Furthermore, it gains significant convergence acceleration when applied to some logarithmic and interior algebraic singularities. Additionally an asymptotic error expansion was derived for integrands with algebraic singularities at the boundaries of the d-dimensional unit cube.

The improvements were found to occur in stages, as each error term vanishes. She also states that further research is needed to determine conditions for which an exact order of leading error terms can be established, and thus a proper extrapolation can be made.

2.2 Singularities in the Interior

In [Owe04] Owen applied and extended his results from [Owe06] to singularities $z \in [1, 0]^d$ inside the unit cube. However since the sequences do not avoid the region of the singularity, which can be in the interior of the unit cube, he proposes using the extended function $\tilde{f}$ instead of the original function f for numerical integration. Like in [Owe06] he requires the function to obey a growth condition. He then defines an extendible region K around singularity z for which $||x - z||_p \geq \nu$ holds for some $\nu > 0$, additionally he defines an anchor $c \in K$ for which $\text{rect}[c, y] \subset K \; \forall y \in K$ holds, where $\text{rect}[x, y] = \prod_{i=1}^{s} [\min(x_i, y_i), \max(x_i, y_i)]$ is the rectangular hull of x and y, thus he can use Sobol's extension $\tilde{f}(x)$. With the help of the extendible region, $\tilde{f}$ and the growth condition he gives an error estimate for any Lebesgue measurable function f for the integration.

However, Owen states in the conclusion that it is not clear if such a good extension to f can be found for arbitrary level sets.

Klinger [Kli97] shows that the numerical integration of a function is still possible when it has a singularity in the origin, or can be transformed such that the singularity is in the origin, by removing the point closest to the origin from the integration. This basically excludes an elemental interval containing the origin from the estimation, for Halton sequences he defines similar intervals. Only Halton and $(0, s)$-sequences are used and Klinger uses the properties of elemental intervals to find points which are near the singularity and thus not included in the numerical integration. While the $(0, s)$-sequence is a Niederreiter sequence, and thus a notion of elemental intervals already exits, Klinger needs to define a similar notion for Halton sequences:

Let a_k be positive rational numbers which satisfy $\sum_{k=1}^{s} 1/a_k \geq 1$ and define positive numbers, coprime integers p_k, q_k by $p_k/q_k := a_k$. Now let $R = \prod_{k=1}^{s} [\delta_k(\delta_k - CN^{-1/a_k}), (1-\delta_k)CN^{-1/a_k} + \delta_k)$ where $\delta_k \in \{0,1\}$ and $C = \min_{1\leq k\leq s} b_k^{-q_k}$. Then at most one of the first N points $x_0, \ldots, x_N$ of Halton's sequence falls into R.

Consequently Klinger states that since $x_0 = \mathbf{0}$, this argument also shows that none of the first N points of a Halton sequence x_n fall into the above interval when $\delta_k = 0$ for all $1 \leq k \leq s$. He gives the error bounds for Halton and $(0, s)$-sequences.

The computational experiments included in the paper compare the error bounds of the Halton, Sobol' and Niederreiter sequences. The error bounds are shown to be reasonable and also that Halton sequences have, for low dimension, basically the same characteristics as Sobol' or Niederreiter sequences but are less computationally expensive. For high dimension, shown for dimension 10 in the experiments, Halton sequences are worse for at least moderate N.

The method proposed in the next section is rather simple in application, and can deal with arbitrary patterns of singularities. However, this entails a rather problematic (at least theoretical) restriction to the class of functions to which it can be applied.

3 A Pragmatic Approach

A number of people, starting with Sobol' in [Sob73], have conducted research for error bounds for improper integrals. One of the more recent results is by Zinterhof [Zin02].

Definition 3. *For a function $f(x)$ and a $B > 0$ the functions $f_B(x)$ and $\hat{f}_B(x)$ are defined as*

$$f_B(x) = \begin{cases} f(x) & |f(x)| \leq B \\ 0 & |f(x)| > B \end{cases}$$

$$\hat{f}_B(x) = \begin{cases} 0 & |f(x)| \leq B \\ f(x) & |f(x)| > B. \end{cases}$$

Definition 4. *Consider the class of s-variate functions, $f(x_1, \ldots, x_s)$ $0 \leq x_i \leq 1$ $i = 1, \ldots, s$, consisting of all functions which fulfill*

(a) $I(|\hat{f}_B|) = O(B^{-\beta})$ *for some* $\beta > 0$
(b) $V(f_B) = O(B^{\gamma})$ *for some* $\gamma \geq 1$.

This class will be called $C(\beta, \gamma)$.

Theorem 1. *Let $f \in C(\beta, \gamma)$, D_N^* be the discrepancy of the set of nodes $\mathbf{x}_1, \ldots, \mathbf{x}_n$ and $B = {D_N^*}^{-1/(\beta+\gamma)}$, then the estimate*

$$I(f) = \frac{1}{N} \sum_{n=1}^{N} f_B(\mathbf{x}_n) + O({D_N^*}^{\beta/(\beta+\gamma)})$$

holds, where $I(f) = I(f_B) + I(\hat{f}_B)$.

Proof. From

$$I(f) = I(f_B) + I(\hat{f}_B)$$

using the Hlawka-Koksma inequality we get

$$\left| I(f_B) - \frac{1}{N} \sum_{n=1}^{N} f_B(x_n) \right| \leq V(f_B) D_N^* \leq C_1(f) B^\gamma D_N^*$$

and from condition (a) we get

$$|I(\hat{f}_B)| \leq C_2(f) B^{-\beta}.$$

Consequently

$$E_N = \left| I(f) - \frac{1}{N} \sum_{n=1}^{N} f_B(x_n) \right| \leq C_1(f) B^\gamma D_N^* + C_2(f) B^{-\beta},$$

which takes it's minimum of order when using

$$B = {D_N^*}^{\frac{-1}{\beta+\gamma}}.$$

Thus for an error estimate we get

$$E_N \leq C(f) {D_N^*}^{\beta/(\beta+\gamma)}$$

where $C(f)$ is a constant depending on f.

Remark 2. Zinterhof [Zin02] also shows that the error bound is optimal.

Remark 3. Since the optimal B is depending on D_N^*, β and γ it is in any case depending on N. Also, if either of D_N^*, β or γ is depending on s then B is also depending on s.

3.1 The Class $D(\beta, \gamma)$

An important issue remains: the richness of the class $C(\beta, \gamma)$. Generally if the jump line, i.e. $f(x) = B$, $x \in U^s$, is not axis parallel the variation is unbounded and consequently $f \notin C(\beta,\gamma)$ since condition (b) (in Definition 4) is violated[4]. It can clearly be seen that the class $C(\beta,\gamma)$ is very restrictive.

Consider the class T of bounded step functions $t \in T$, $t(x_1,\dots,x_s) : U^s \to \mathbb{C}$. These are functions which are piecewise constant on U^s where U^s is partitioned into a finite number of, pairwise disjoint, intervals of the form $\prod_{i=1}^{s}[a_i,b_i)$. All functions $t \in T$ are of bounded variation.

Definition 5. *Let the class $D(\beta,\gamma)$ be defined as $D(\beta,\gamma) := \{f | f \in C(\beta,\gamma)$ and $f_B \in T\}$. If it is clear from context we will abbreviate and write D.*

Remark 4. Using the definition of D we can easily state $T \subset D \subset C := C(\beta,\gamma)$.

It is well known that if $g \in L_1$, which implies $\int_{U^s} |g(\mathbf{x})|d\mathbf{x} < \infty$, then there exists for every $\epsilon > 0$ a $t_\epsilon(x) \in T$ such that

$$\int_{U^s} |g(\mathbf{x}) - t_\epsilon(\mathbf{x})|d\mathbf{x} < \epsilon.$$

Also since $T \subset D$ we can easily write

$$\int_{U^s} |g(\mathbf{x}) - d_\epsilon(\mathbf{x})|d\mathbf{x} < \epsilon,$$

with $d_\epsilon(x) \in D$. Generally, the functions $\in D$ will approximate a given function $g \in C \subset L_1$ better than functions $\in T$.

In any case, C is rich since T is rich and $T \subset C$. The restrictiveness of $C(\beta,\gamma)$ is a direct result of the restrictiveness of the variation in the sense of Hardy and Krause.

3.2 The Function $f(\mathbf{x}) = \max(x_1,\dots,x_s)^{-\beta}$

Now let us consider the function $f = \frac{1}{\max(x_1,\dots,x_s)^\beta}$ with $0 < \beta < 1$. Certainly $f \notin T$ and $\lim_{\mathbf{x}\to 0} f(\mathbf{x}) = \infty$, thus if we can show that $f \in C$ we have $T \subsetneq C$. To do this we need to estimate the integral value and variation of the function f to see if conditions (a) and (b) in Definition 4 are met.

Integral Value

Theorem 2. *Let $f = \frac{1}{\max(x_1,\dots,x_s)^\beta}$ with $0 < \beta < 1$, then $\int_{I^s} f(\mathbf{x})d\mathbf{x} = \frac{s}{s-\beta}$ $0 < \beta < 1$.*

Proof. With induction. For $s = 1$ the claim holds since $\int_0^1 x_1^{-\beta}dx_1 = \frac{1}{-\beta+1}x_1^{-\beta+1}|_0^1 = \frac{1}{1-\beta}$.

[4] Thanks to Reinhold Kainhofer for pointing this out.

Now

$$\int_0^1 \cdots \int_0^1 \max(x_1, \ldots, x_s)^{-\beta} dx_1 \ldots dx_s =$$

$$= \int_0^1 \left(\int_0^1 \cdots \int_0^1 \max(x_1, \ldots, x_s)^{-\beta} dx_1 \ldots dx_{s-1} \right) dx_s =$$

$$= \int_0^1 \cdots \int_0^1 dx_1 \ldots dx_{s-1} \int_0^1 \max(x_1, \ldots, x_s)^{-\beta} dx_s.$$

Let us consider $\int_0^1 \max(x_1, \ldots, x_{s-1}, x_s)^{-\beta} dx_s$, and let $\hat{x}_s := \max(x_1, \ldots, x_{s-1})$, thus $\int_0^1 \max(x_1, \ldots, x_{s-1}, x_s)^{-\beta} dx_s = \int_0^1 \max(\hat{x}_s, x_s)^{-\beta} dx_s$ where

$$\max(\hat{x}_s, x_s) = \begin{cases} x_s & x_s \geq \hat{x}_s \\ \hat{x}_s & x_s < \hat{x}_s. \end{cases}$$

Thus

$$\int_0^1 \max(\hat{x}_s, x_s)^{-\beta} dx_s = \int_0^{\hat{x}_s} \hat{x}_s^{-\beta} dx_s + \int_{\hat{x}_s}^1 x_s^{-\beta} dx_s = \frac{-\beta \hat{x}_s^{-\beta+1} + 1}{1 - \beta}.$$

Now we have

$$\int_0^1 \cdots \int_0^1 dx_1 \ldots dx_{s-1} \int_0^1 \max(x_1, \ldots, x_s)^{-\beta} dx_s =$$

$$= \int_0^1 \cdots \int_0^1 dx_1 \ldots dx_{s-1} \frac{1 - \beta \hat{x}_s^{-\beta+1}}{1 - \beta} =$$

$$= \frac{1}{1 - \beta} \left[1 - \beta \int_0^1 \cdots \int_0^1 \max(x_1, \ldots, x_{s-1}^{-\beta+1} dx_1 \ldots dx_{s-1} \right].$$

Now using the induction hypothesis we get

$$\int_0^1 \cdots \int_0^1 dx_1 \ldots dx_{s-1} \int_0^1 \max(x_1, \ldots, x_s)^{-\beta} dx_s =$$

$$= \frac{1}{1-\beta} \left[1 - \beta \frac{s-1}{s-1-\beta+1} \right] = \frac{s - \beta - s\beta + \beta}{(1-\beta)(s-\beta)} = \frac{s}{s-\beta}.$$

Remark 5. In a similar fashion we obtain $\int_{I^s} \hat{f}_B(\mathbf{x}) d\mathbf{x} = \frac{s}{s-\beta} \left(\frac{1}{B}\right)^{\frac{s-\beta}{\beta}}$.

Variation

Definition 6. *Let* $\mathbf{P}$ *be the set of all partitions of the s-dimensional unit cube* I^s *then the variation of a function* f *in the sense of Vitali is defined as*

$$V_V(f) := \sup_{P \in \mathbf{P}} \sum_{p \in P} |\Delta(f; p)|,$$

where $\Delta(f;p)$ *is the s-fold alternate sum, i.e. adjacent corners have opposite sign, of the function values on the corners of the interval* p.

Definition 7. *Let* $n \in \mathbf{N}$ *and* $0 = t_0 < \ldots < t_{n-1} < t_n = 1$, $t_i \in I^s\, 0 \leq i \leq n$. *Now let* $Z_s(t_0,\ldots,t_n) = \{\{t_1,\ldots,t_n\}^s\}$ *be the set of s-tuples formed by* $t_0,\ldots,t_n$. *A partition* P *of s-dimensional unit cube* $I^s = [0,1)^s$ *is called* valid partition *if there is exists a* $Z_s(t_0,\ldots,t_n)$ *such that* $P = \{\{[t_0,t_1),[t_1,t_2),\ldots,[t_{n-1},t_n)\}^s\}$ *where* $[t_{k-1},t_k)[t_{l-1},t_l) = [t_{k-1},t_k) \times [t_{l-1},t_l)$. *We say the partition* P *belongs to* $Z_s(t_0,\ldots,t_n)$ *and write* $P(Z_s(t_0,\ldots,t_n))$.

Remark 6. From the construction of $Z_s(t_0,\ldots,t_n)$ it follows that for a valid partition only the intervals of the form $[t_{k-1},t_k) \times \cdots \times [t_{k-1},t_k)$ $1 \leq k \leq n$ cross the principal diagonal, i.e. the restriction of the principal diagonal of the unit cube to such an interval is the principal diagonal of the interval.

Remark 7. Every partition of I^s can be refined to a valid partition.

Lemma 1. $V_V(\max(x_1,\ldots,x_s)) = 1$ *for* $(x_1,\ldots,x_s) \in I^s$.

Proof. The function $f(x_1,\ldots,x_s) = \max(x_1,\ldots,x_s)$ fulfills $\max(x_1,\ldots,x_{k-1},x_k,x_{k+1},\ldots,x_s) = x_k$ for $\max(x_1,\ldots,x_{k-1},x_{k+1},\ldots,x_s) \leq x_k$. Let $I_{n_1,\ldots,n_s} = [t_{n_1-1},t_{n_1}) \times \cdots \times [t_{n_s-1},t_{n_s})$ be an interval of the valid partition $P(Z_s(t_0,\ldots,t_n))$, which doesn't cross the principal diagonal $(x_1,\ldots,x_s) = t(1,\ldots,1)\, 0 \leq t \leq 1$. Now we can write $V_V(f;I_{n_1,\ldots,n_s}) = |\sum_{\tau_1=0}^1 \cdots \sum_{\tau_s=0}^1 (-1)^{\tau_1+\cdots+\tau_s} \max(t_{n_1-1}+\tau_1(t_{n_1}-t_{n_1-1}),\ldots,t_{n_s-1}+\tau_s(t_{n_s}-t_{n_s-1})|$. Since $I_{n_1,\ldots,n_s}$ is not on the principal diagonal of I^s there is a k_0, $1 \leq k_0 \leq s$ such that

$$\max(t_{n_1-1}+\tau_1(t_{n_1}-t_{n_1-1}),\ldots,t_{n_{k_0}-1}+ \tau_{k_0}(t_{n_{k_0}}-t_{n_{k_0}-1}),t_{n_{k_0}},t_{n_{k_0}+1}+\tau_{k_0+1}(t_{n_{k_0}}-t_{n_{k_0}-1}),\ldots) = t_{n_{k_0}}$$

and

$$\max\left(t_{n_1-1}+\tau_1(t_{n_1}-t_{n_1-1}),\ldots,t_{n_s-1}+\tau_s(t_{n_s}-t_{n_s-1})\right) = = t_{n_{k_0}-1}+\tau_{k_0}(t_{n_{k_0}}-t_{n_{k_0}-1})$$

for all $\tau_1,\ldots,\tau_{k_0-1},\tau_{k_0+1},\ldots,\tau_s$, $\forall i: \tau_i \in \{0,1\}$.

It follows that

$$V_V(f;I_{n_1,\ldots,n_s}) = \left|\sum_{\tau_1=0}^{1}\cdots\sum_{\tau_{k_0-1}=0}^{1}\sum_{\tau_{k_0+1}=0}^{1}\cdots\sum_{\tau_s=0}^{1} (-1)^{\tau_1+\cdots+\tau_{k_0-1}+\tau_{k_0+1}+\cdots+\tau_s}(t_{n_{k_0}}-t_{n_{k_0}-1})\right| = = (t_{n_{k_0}}-t_{n_{k_0}-1})\left|\sum_{\tau_1=0}^{1}\cdots\sum_{\tau_{k_0-1}=0}^{1}\sum_{\tau_{k_0+1}=0}^{1}\cdots\sum_{\tau_s=0}^{1} (-1)^{\tau_1+\cdots+\tau_{k_0-1}+\tau_{k_0+1}+\cdots+\tau_s}\right| = 0.$$

If on the other hand $I_{n_1,\dots,n_s}$ lies on the principal diagonal of I^s, then $n_1 = \dots = n_s = n_0$ and $I_{n_1,\dots,n_s} = I_{n_0,\dots,n_0} = [t_{n_0-1}, t_{n_0}) \times [t_{n_0-1}, t_{n_0}) \times \dots \times [t_{n_0-1}, t_{n_0})$, then

$$
\begin{aligned}
V_V(f; I_{n_0,\dots,n_0}) &= \left| \sum_{\tau_1=0}^{1} \dots \sum_{\tau_s=0}^{1} (-1)^{\tau_1+\dots+\tau_s} \max\left(t_{n_0-1} + \tau_1(t_{n_0} - t_{n_0-1}), \dots, \right.\right. \\
&\qquad \left.\left. t_{n_0-1} + \tau_s(t_{n_0} - t_{n_0-1})\right) \right| = \\
&= \left| \sum_{\tau_1,\dots,\tau_s=0}^{1} (-1)^{\tau_1+\dots+\tau_s} t_{n_0-1} \max(\tau_1, \dots, \tau_s)(t_{n_0} - t_{n_0-1}) \right| = \\
&= \left| t_{n_0-1} \sum_{\tau_1,\dots,\tau_s=0}^{1} (-1)^{\tau_1+\dots+\tau_s} + (t_{n_0} - t_{n_0-1}) \right. \\
&\qquad \left. \times \left(\sum_{\tau_1,\dots,\tau_s=0}^{1} (-1)^{\tau_1+\dots+\tau_s} - \sum_{\tau_1,\dots,\tau_s=0}^{1} (-1)^{\tau_1+\dots+\tau_s} \right) \right| = \\
&= |t_{n_0-1} 0 + (t_{n_0} - t_{n_0-1})(0 - 1)| = t_{n_0} - t_{n_0-1}.
\end{aligned}
$$

Then holds $V_V(f; I^s) = \sum_{n_0=1}^{n} (t_{n_0} - t_{n_0-1}) = 1$, where $V_V(f; I^s)$ is attained already at the principal diagonal of I^s.

Remark 8. If $g(x)$ in $[0, 1]$ is monotone or of finite variation $\mathrm{Var}(g)$ then $V_V(g(\max(x_1, \dots, x_s); I^s)) = |g(1) - g(0)|$ or $V_V(g(\max(x_1, \dots, x_s); I^s)) = \mathrm{Var}(g)$.

Remark 9. Let $0 \le a_k < b_k$, $1 \le k \le s$ then for $I_{a,b} = \prod_{k=1}^{s} [a_k, b_k)$ we analogously get $V_V(g(\max(x_1, \dots, x_s)); I_{a,b}) = |g(b) - g(a)|$ for $a_1 = \dots = a_s$ and $b_1 = \dots = b_s$, otherwise $V_V(g(\max(x_1, \dots, x_s)); I_{a,b}) = 0$. The variation in the sense of Vitali of functions $g(\max(x_1, \dots, x_s))$ is concentrated on the principal diagonal of the unit cube.

Remark 10. For functions $f = g(\max(x_1, \dots, x_s))$

$$V_{HK}(f; I^s) = V_V(f; I^s)$$

holds. The variation in the sense of Hardy and Krause is defined as

$$V_{HK}(f; I_{a,b}) = \sum_{J \subset I^s} V_V(f; J)$$

where J are all k-dimensional faces $\{(u_1, \dots, u_s) \in I^s | u_j = 1, j \ne i_1, \dots, i_k\}$ with $1 \le k \le s$ and $1 \le i_1 < \dots < i_k \le s$. Since for $k < s$ there are some $u_j = 1$, the function $f = g(\max(x_1, \dots, x_{t-1}, 1, x_{t+1}, \dots, x_s)) = g(1)$, $t \ne i_1, \dots, i_k$, is constant and consequently $V_V(f; J) = 0$, $\forall J \subsetneq I^s$.

Let now $g(x) = 1/x^\beta$, $0 < x \leq 1$, $0 < \beta < 1$ and $f_\beta(x_1, \ldots, x_s) = 1/\max(x_1, \ldots, x_s)^\beta$, $(x_1, \ldots, x_s) \in I^s$. Now let

$$\hat{f}_{\beta,B} = \begin{cases} 0 & f_\beta(x_1,\ldots,x_s) > B,\ \max(x_1,\ldots,x_s) < 1/B^\beta = B' \\ f_\beta(x_1,\ldots,x_s) & f_\beta(x_1,\ldots,x_s) \leq B,\ \max(x_1,\ldots,x_s) \geq 1/B^\beta = B', \end{cases}$$

and

$$\tilde{f}_{\beta,B} = \begin{cases} f_\beta(B',\ldots,B') = B & f_\beta(x_1,\ldots,x_s) > B \\ f_\beta(x_1,\ldots,x_s) & f_\beta(x_1,\ldots,x_s) \leq B, \end{cases}$$

and

$$\chi_{\beta,B} = \begin{cases} B & f_\beta(x_1,\ldots,x_s) > B \\ 0 & f_\beta(x_1,\ldots,x_s) \leq B, \end{cases}$$

and clearly $\tilde{f}_{\beta,B} = \hat{f}_{\beta,B} + \chi_{\beta,B}$. It can be easily seen that $V_V(\chi_{\beta,B}; I^s) = B$ and from the remarks before we know that $V_V(\tilde{f}_{\beta,B}) = |g(0) - g(1)| = B - 1$. Consequently, $V_V(\hat{f}_{\beta,B}; I^s) = V_V(\tilde{f}_{\beta,B} - \chi_{\beta,B}; I^s) \leq V_V(\tilde{f}_{\beta,B}; I^s) + V_V(\chi_{\beta,B}; I^s) = 2B - 1$.

Thus we have finally shown that $f(\mathbf{x})(= \max(\mathbf{x})^{-\beta})$, $0 < \beta < 1$ is in $C(\frac{s-\beta}{\beta}, 1)$.

Remark 11. Since $\max(\mathbf{x})^{-\beta} \in C(\frac{s-\beta}{\beta}, 1)$, $0 < \beta < 1$, we also know that the error takes it's minimum when $B = D_N^{*\,\frac{-\beta}{s}}$ (c.f.: proof of Theorem1).

4 Experimental Results

First, we want to investigate the behavior of function $f(\mathbf{x}) = \max(x_1, \ldots, x_s)^{-\beta}$ as discussed in the last section in numerical experiments. As point sequence we used the Zinterhof sequence [Zin69], which is a special case of Weyl sequences defined as follows

$$x_n = (\{ne^{1/1}\}, \ldots, \{ne^{1/s}\}),\ \ n = 1, 2, 3, \ldots,$$

for points $n = 1, 2, \ldots$ and dimension s. Note that the Zinterhof sequence has certain corner avoidance properties as well, which is due to the high degree of irrationality of the generated points. Caused by corresponding diophantine properties this is true not only for the origin but for all rational points as well. For the calculation of the bound B we use the bound of the discrepancy given by LeVeques inequality [KY81] and the diaphony of the Zinterhof sequence [Zin69].

Figure 1 displays the results for the original and integral preserving transformed function (transformed in such a way as to get singularities in the interior of the unit interval as well as on the border)

$$\max = \max(x_1, \ldots, x_s)^{-0.5}, \qquad \max{}' = \max(\{5x_1\}, \ldots, \{5x_s\})^{-0.5}$$

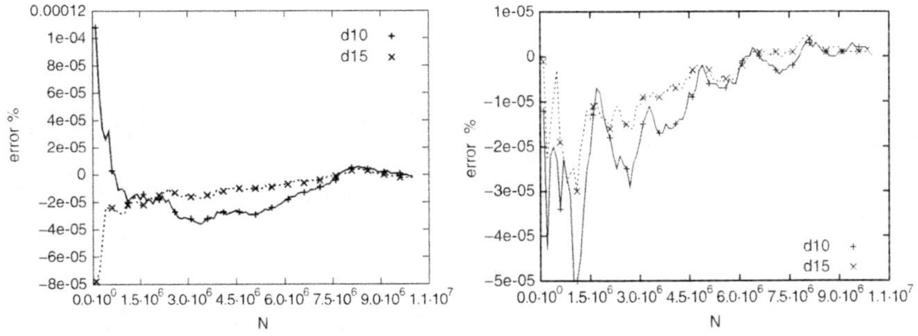

max max′

Fig. 1. Functions max and max′ for dimension 10 and 15 with $B = 2$, relative error over N.

respectively, where $\{x\}$ is the remainder of x. We let N run and hold $B = 2$ (according to Remark 11) fixed for dimensions 10 and 15 (labeled `d10` and `d15` respectively).

Remark 12. We hold B fixed at a value which is calculated for $N = 10^7$ so that we will get the best result towards the end of calculation. If we wanted the lowest error for each N we would have to let B vary accordingly.

As expected, the error rates are very good, especially towards higher N.

Theoretically, we are restricted to functions of class C, practically however the method can be applied to a wider range of functions. Consider the functions

$$f^1 = \prod_{i=1}^{s} \frac{1}{x_i^{0.5}}, \qquad f^2 = \prod_{i=1}^{s} \frac{1}{\ln(\frac{1}{x_i})^{0.5}}$$

where f_B^1 and f_B^2 both have non axis parallel jump curves and consequently infinity variation.

Remark 13. The integral values over the s-dimensional unit cube for $f_\alpha^1(\mathbf{x}) = \prod_{i=1}^{s} x_i^{-\alpha}$ and $f_\alpha^2(\mathbf{x}) = \prod_{i=1}^{s} \ln(1/x_i)^{\alpha-1}$, $0 < \alpha < 1$, are $\int_{U^s} f_\alpha^1(\mathbf{x})d\mathbf{x} = (1/(1-\alpha))^s$ and $\int_{U^s} f_\alpha^2(\mathbf{x})d\mathbf{x} = \Gamma(\alpha)^s$ respectively.

Experimentally, functions f_1 and f_2 can be integrated using our technique, even though their bound representations for this method have infinite variation, c.f. Definition 4 condition (b). For a test we used a fixed $B = 10^9$, which also hints at a serious problem with this method if $f \notin C$. Since the variations of f_B^1 and f_B^2 are infinite we can not obtain a β and thus no optimal bound B using dimension 10 and 15.

Figure 2 left hand side shows the results for function f^1 over the number of points N, and the right side shows the results for the integral preserving transformation

$$f'^1 = \prod_{i=1}^{s} \frac{1}{\{5x_i\}^{0.5}}$$

where $\{x\}$ is again the remainder of x.

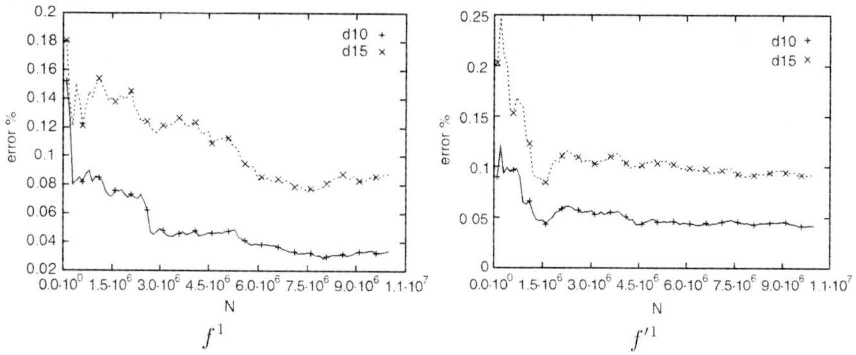

Fig. 2. Functions f^1 and f'^1 for dimension 10 and 15 with $B = 10^9$, relative error over N.

The figures show that the estimation converges toward a fixed error, this is to be expected since we will by construction always miss $I(\hat{f}_B)$ (see Theorem 1) since we kept B fixed while it is actually a function of the discrepancy and thus of N. The difference in error between dimension 10 and 15 is a well known phenomenon (curse of dimensionality). However, given that we can somehow obtain the proper bound B for the number of points N used for the integration the error converges even though f^1 (and f_B^1) is of unbounded variation.

Keeping the bound B fixed and again using dimensions 10 and 15 we will likely experience problems in the integration when we turn to another function. To illustrate this we used function f^2, and an integral preserving transformation as follows

$$f'^2 = \prod_{i=1}^{s} \frac{1}{\ln(\frac{1}{\{5x_i\}})^{0.5}},$$

shown in Fig. 3 left and right hand side respectively. When the bound is chosen too low the results usually becomes stable quickly with a high error, stemming again from $\hat{f}_B^2$. In this case the bound was chosen too high, i.e. we would need to use more points N to get to the region where B is optimal. This can be seen from the overshoots, usually high error rates at the beginning, due to points falling near the jump curve, thus early introducing high values, i.e. close to B, to the estimation. These will usually vanish when the number of points is high enough to get a fine grained sampling of the unit cube but will stay visible a long time. So while the method works, experimentally, even for functions not in C this poses the problem of estimating a proper B to be used in the integration.

To illustrate the effect B has on the integration we use a fixed number of points $N = 10^6$ and let B vary. The result of this test, again for functions f^1 and f^2 in dimensions 10 and 15, is given in Fig. 4. What can be seen is that the bound depends not only on the number of points but also on the

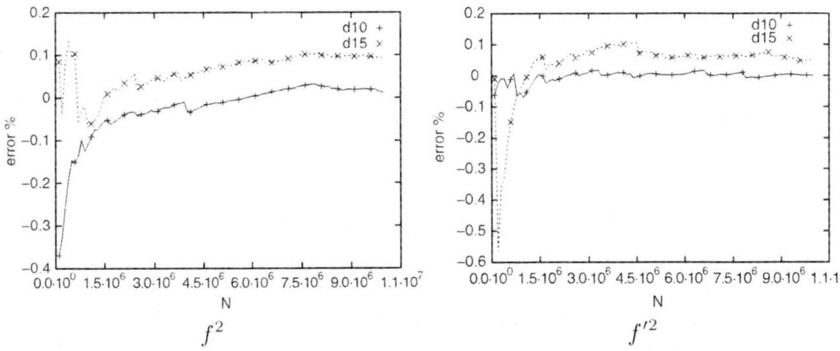

Fig. 3. Functions f^2 and f'^2 for dimension 10 and 15 with $B = 10^9$, relative error over N.

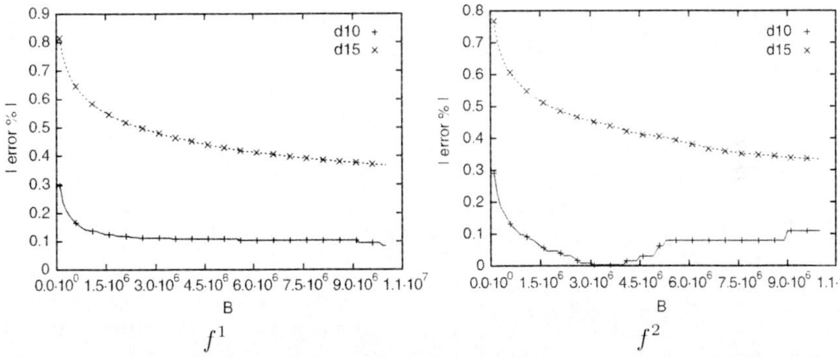

Fig. 4. Functions f^1 and f^2 for dimension 10 and 15 with $N = 10^6$, relative error over B.

dimension, which is not surprising since it depends on the discrepancy. Also, for f^2, there is an interval of B where the integration holds, while on the other hand we get an increase in error as we move away from that interval. Also, since the bound is depending on the discrepancy, which in turn depends on the number of points and the dimension, the bound is a function of the dimension leading to quite some error in dimension 15 where the approximation was very close for dimension 10.

Finally, we consider a function which can not be reduced to singularities along the border or in the corner. Consider the function (again with integral preserving transformation)

$$m = \sum_{i=1}^{s-1} \frac{1}{|x_i - x_{i+1}|^{0.5}}, \qquad m' = \sum_{i=1}^{s-1} \frac{1}{|\{5x_i\} - \{5x_{i+1}\}|^{0.5}}.$$

Remark 14. The integral value of $m_\alpha(\mathbf{x}) = \sum_{i=1}^{s-1} |x_i - x_{i+1}|^{-\alpha}$, $0 < \alpha < 1$ over the s-dimensional unit cube U^s is $\int_{U^s} m_\alpha(\mathbf{x}) d\mathbf{x} = 2(s-1)/(1-\alpha)(2-\alpha)$.

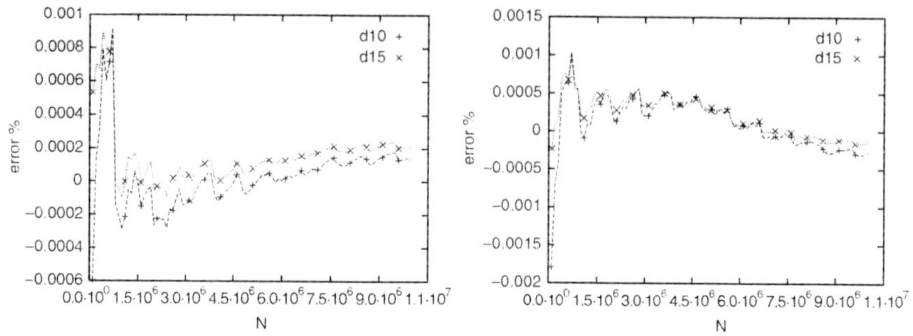

m m'

Fig. 5. Functions m and m' for dimension 10 and 15 with $B = 10^4$, relative error over N.

Remark 15. The function m_α has a singularity along the $s-1$ dimensional manifold $R = \cup_{i=1,\dots,s-1} R_i$, where $R_i = \{\mathbf{x}|\mathbf{x} \in U^s, x_i = x_{i+1}\}$, $i = 1,\dots,s-1$.

Remark 16. If we use a step function m_M to approximate m with M intervals we can give a bound for the variation as $V_{HK}(m_M) \leq M2^s s\binom{s}{\lfloor s/2 \rfloor} B = \mathrm{O}(B)$. Furthermore the integral of $\hat{m}_B$ can easily calculated as $\int_{U^s} \hat{m}_{\alpha B}(\mathbf{x})d\mathbf{x} = (2(s-1)/(1-\alpha)(2-\alpha))B^{-2\alpha+\alpha^2} = \mathrm{O}(B^{-(2\alpha-\alpha^2)})$ thus $m \in D(\alpha^2 - 2\alpha, 1)$.

For m and m' we used $B = 10^4$ (according to Remark 16), again for both dimension 10 and 15, and the results are given in Fig. 5. As expected (since we use a single fixed B) the error at the beginning is quite high but swiftly falls to "normal" levels. Somewhat more important is the rather good convergence when considering that this functions singularity is more severe than the singularity of the two previous functions.

Since we only used an estimation for B we can not expect the numerical integration to be optimal, so now we will assess how good the approximation actually is. For this purpose, we integrated m in dimension 10 for different values of B. The results are given in Fig. 6, the left side gives the relative error over N for different values of B and the right side gives the number of points for which the function value exceeded B for a fixed $N = 10^7$. On the left side we see that for a lower B the error is increased and if we set B to high the error increases again. Overall our estimated B seems to be a bit too low and some value between 10^4 and $2\ 10^4$ would have been the best fit. This is affirmed by the right hand side of the figure, where we see that the difference between 10^4 and $2\ 10^4$ is more than one point (otherwise it would not be possible to get results between these two values on the left hand side). Overall we see that with a function approximation using class $D(\beta, \gamma)$ it is hard to get the best approximation but it is certainly possible to get a good approximation. Also, when we look at the left hand side we see that for $B = 10^4$ and $B = 2\ 10^4$ the error increases again after about $N = 6\ 10^6$. This is a further evidence that the choice of B is vital for the integration.

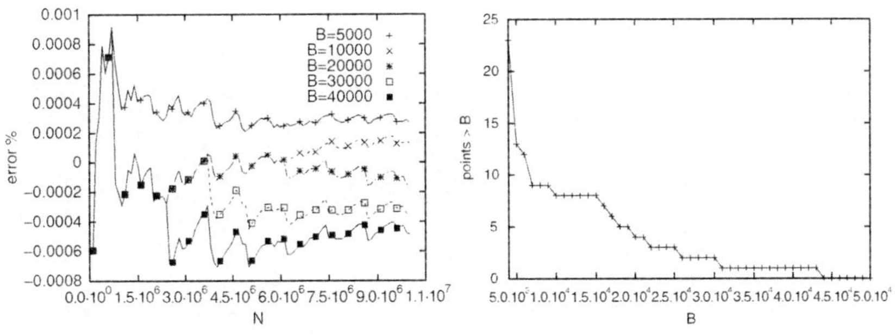

Fig. 6. The effects of different values of B for the function m.

5 Conclusion

We have shown that the proposed method can be used to numerically integrate over a rich class of functions $C(\beta, \gamma)$. The method also works experimentally on an even bigger class of functions with the problem that some parameters, i.e. the bound B, cannot be chosen specifically for the function. Furthermore, the bound can be applied during runtime, and thus the method can be applied to the function directly. Also, the method is not restricted to singularities on the boundary or in the corner. Thus this method is extremely easy to implement and apply, even for non specialists.

However, even if a function is of class C we face the problem that we have to know the number of points beforehand to choose an optimal B. Also, since we need β and γ to choose optimal parameters for the numerical integration the function must be well known. This is theoretically of no importance but practically can prevent (optimal) integration.

References

[dDG03] E. deDoncker and Y. Guan. Error bounds for integration of singular functions using equidistributed sequences. *Journal of Complexity*, 19(3):259–271, 2003.

[HK05] J. Hartinger and R. F. Kainhofer. Non-uniform low-discrepancy sequence generation and integration of singular integrands. In H. Niederreiter and D. Talay, editors, *Proceedings of MC2QMC2004, Juan-Les-Pins France, June 2004*. Springer Verlag, June 2005.

[HKT04] J. Hartinger, R. F. Kainhofer, and R. F. Tichy. Quasi-monte carlo algorithms for unbound, weighted integration problems. *Journal of Complexity*, 20(5):654–668, 2004.

[HKZ05] Jürgen Hartinger, Reinhold Kainhofer, and Volker Ziegler. On the corner avoidance properties of various low- discrepancy sequences. *INTEGERS: Electronic Journal of Combinatorial Number Theory*, 5(3), 2005. paper A10.

[HM72] E. Hlawka and R. Mück. Über eine Transformation von gleichverteilten Folgen. *Computing*, 9:127–138, 1972.

[Kli97] B. Klinger. Numerical Integration of Singular Integrands Using Low-Discrepancy Sequences. *Computing*, 59:223–236, March 1997.

[KY81] H. L. Keng and W. Yuan. *Applications of Number Theory to Numerical Analysis*. Springer Verlag, Science Press, 1981.

[Nie92] H. Niederreiter. *Random Number Generation and Quasi-Monte Carlo Methods*, volume 62 of *CBMS-NSF Regional Conference Series in Applied Mathematics*. Society for Industrial and Applied Mathematics (SIAM), 1992.

[Owe04] Art B. Owen. Quasi-Monte Carlo for integrands with point singularities at unknown locations. Technical Report 26, Stanford University, 2004.

[Owe05] A. B. Owen. Multidimensional variation for quasi-Monte Carlo. In Jianqing Fan and Gang Li, editors, *International Conference on Statistics in honour of Professor Kai-Tai Fang's 65th birthday*, pages 49–74, 2005.

[Owe06] Art B. Owen. Halton sequences avoid the origin. *SIAM Rev.*, 48(3):487–503, 2006.

[Sob73] I.M. Sobol'. Calculation of improper integrals using uniformly distributed sequences. *Soviet Math Dokl.*, 14(3):734–738, July 1973.

[Zin69] P. Zinterhof. Einige zahlentheoretische Methoden zur numerischen Quadratur und Interpolation. *Sitzungsberichte der Österreichischen Akademie der Wissenschaften, math.-nat.wiss. Klasse Abt. II*, 177:51–77, 1969.

[Zin02] P. Zinterhof. High dimensional improper integration procedures. In Civil Engineering Faculty Technical University of Košice, editor, *Proceedings of Contributions of the 7th International Scientific Conference*, pages 109–115, Hroncova 5, 04001 Košice, SLOVAKIA, May 2002. TULIP.

Assessment of Genetic Association using Haplotypes Inferred with Uncertainty via Markov Chain Monte Carlo

Raquel Iniesta and Victor Moreno

Cancer Epidemiology Service, Catalan Institute of Oncology, Barcelona, Spain
riniesta@iconcologia.net

Summary. In the last years, haplotypic information has become an important subject in the context of molecular genetic studies. Assuming that some genetic mutations take part in the etiology of some diseases, it could be of great interest to compare sets of genetic variations among different unrelated individuals, inherited in block from their parents, in order to conclude if there is some association between variations and a disease. The main problem is that, in the absence of family data, obtaining haplotypic information is not straightforward: individuals having more than one polymorphic heterozygous locus have uncertain haplotypes.
We have developed a Markov Chain Monte Carlo method to estimate simultaneously the sample frequency of each possible haplotype and the association between haplotypes and a disease.

1 Introduction

Nowadays, haplotypic information has become vitally important in the context of association studies. Association studies deal with the relationship between genetic information and the etiology of some particular disease. Comparing DNA of healthy and diseased individuals, we can find changes in the sequence that could modify the risk of suffering from the disease [Bal06].

DNA variations we are going to deal with are the changes in only one nucleotide, called SNP (Single Nucleotide Polymorphism).

The knowledge of haplotypes corresponding to a sample of genotypes observed for some SNPs of a set of unrelated individuals is very helpful to better describe this association. Unfortunately, in the absence of family data, obtaining haplotypic information is not straightforward. Since every cell of the human organism contains 22 pairs of homologous chromosomes, plus the sexual chromosomes, for each chromosomical location at the autosomal chromosomes there are two bases, one for each homologous chromosome at the same position of the DNA sequence. Given that current lab techniques usually only report genotypic data and do not provide the chromosome for each base, individuals with two or more heterozygous sites have uncertain haplotypes because there is more than one possible haplotype pair compatible with their genotype.

Methods of Haplotypic Reconstruction

In the last years several methods of haplotypic reconstruction have been developed in order to overcome this lack of information. Since Clark, in 1990 [Cla90], developed a parsimony algorithm to estimate haplotype frequencies from a sample of genotypes, quite a large number of methods have been developed. Most of them rely on the use of different techniques to calculate the Maximum Likelihood Estimator (MLE).

In 1995, Excoffier and Slatkin [ES95] adapted the Expectation–Maximization algorithm, an iterative algorithm of maximization developed by Dempster in 1977 [DLR77] to maximize the likelihood function of the haplotypes given the genotypes at specific loci. This method has some limitations and convergence to a local maximum may occur in some situations (Celeux and Diebolt, [CD85]).

Some authors have attempted to minimize these limitations in their works, like Qin *et al.* [QNL02] using *Divide and conquer* strategies, or David Clayton, implementing an EM-algorithm (*snphap* software) which adds SNPs one by one and estimates haplotype frequencies, discarding haplotypes with low frequency as it progresses. Besides, other techniques have been considered, too. In the context of Bayesian statistics, Stephens *et al.* in 2001 proposed an algorithm based on coalescent theory [SSD01] with a especial prior based on the general mutational model. Niu *et al.* [NQXL02] implemented another Bayesian approach using a Markov Chain Monte Carlo method. In general, algorithms dealing with Bayesian models are suitable to infer haplotypes from genotypes having a large number of polymorphisms.

The most recent methods work with clusters of haplotypes in order to avoid the major limitations of many current haplotype-based approaches [WWB06].

Once the frequencies have been estimated by any of the methods mentioned above, the next goal is to test the association between haplotypes and a disease. The most accurate strategy in order to take into account the uncertainty of the sample is to estimate simultaneously haplotype frequencies and haplotype effects. There are some works in this direction (Tanck *et al.* [TKJ+03], Tregouet *et al.* [TET+04]).

2 Methodologies

The algorithm we have developed makes the simultaneous estimation of haplotype frequencies and haplotype effects within the frame of Bayesian models. We aim to compute the Maximum Likelihood Estimator of the parameters using Markov Chain Monte Carlo techniques. To do so, it is first required to define the models for both cases in order to deduce the two associated likelihood functions.

2.1 Notation

Consider a sample of individuals of size N, and let be G_i the genotype for the i-th individual, $i = 0, \ldots, N$. Each individual has a finite number of haplotypes compatible with his genotype G_i. If this genotype has at most 1 heterozygous locus, there is only one possible pair of haplotypes compatible with it and there is no uncertainty. Let be m the number of heterozygous loci. If $m \geq 2$, the genotype has 2^m different haplotypes compatibles with it. Let be H_i, $i = 1, \ldots, 2^m$ the set of compatible haplotypes with the genotype of each individual. Assuming that in the whole sample there are M possible haplotypes, h_j denotes the j-th haplotype, with $j = 0, \ldots, M$. The sample frequency for each haplotype is denoted by f_{h_j}.

2.2 Likelihood for Genotypes Sample

Now, assuming *Hardy-Weinberg equilibrium*, the sample frequency for each G_i can be expressed by the product of the frequencies of every haplotype in H_i. For example, if an individual is certain, H_i only has two elements h_r and h_s, $r, s \in \mathbf{(1, \ldots, 2^m)}$, then $F_{G_i} = f_{h_r} \times f_{h_s}$. But for individuals with uncertain haplotypes, we have to consider the sum over all the possible pairs:

$$F_{G_i} = \sum_{h_r, h_s \in H_i} c_{rs} f_{h_r} f_{h_s} \tag{1}$$

where c_{rs} is a constant value, equal to 1 if $h_r = h_s$ and 2 if $h_r \neq h_s$. Now, taking the product of (1) over all the individuals, the likelihood function $\ell(F)$ of the sample of genotypes can be written as Excoffier and Slatkin stated in [ES95]:

$$\ell(F) = \prod_{i=1}^{N} F_{G_i} = \prod_{i=1}^{N} \sum_{h_r, h_s \in H_i} c_{rs} f_{h_r} f_{h_s} \tag{2}$$

where $F = \{F_{G_i}, i = 0, \ldots, N\}$.

2.3 Estimation of Haplotype Effects. Logistic Regression Model

The estimation of haplotype effects can be done with a case–control design if a binary status variable is known. A suitable model is the logistic regression model, which has related to its coefficients the definition of a useful measure of association, the odds ratio. Let be Y_i the binary variable and y_i the values of Y_i over each individual in the sample. Y_i is equal to 0 for a healthy individuals and 1 for diseased ones:

$$\begin{cases} y_i = 1 \; with \; p_i \\ y_i = 0 \; with \; 1 - p_i \end{cases}$$

Now, consider H_i as the covariate for the model. The conditional probability for Y can be expressed like:

$$p_i = P(Y_i = 1 \mid H_i) = \frac{exp(\alpha + \beta_1 h_{1,i} + \cdots + \beta_{M-1} h_{M-1,i})}{1 + exp(\alpha + \beta_1 h_{1,i} + \cdots + \beta_{M-1} h_{M-1,i})} \quad (3)$$

where $\beta = (\alpha, \beta_1, \cdots, \beta_{M-1}) \in \mathrm{R}^M$ is the parameter vector of the model. β has at most 2^m non zero entries.
Taking the product over all the individuals in the sample, the likelihood for the logistic regression model with haplotypic covariate is:

$$\prod_{i=1}^{N} p_i^{y_i} (1 - p_i)^{1-y_i} \quad (4)$$

Then, $\mathcal{E}^\beta$ are odds ratios (OR), measuring the possible association between Y_i and H_i. Taking as reference a base haplotype (usually the most frequent in the sample), the odds ratio quantifies the effect of a given haplotype by comparison with the effect of the reference haplotype.

2.4 Estimating Parameters

To estimate the parameters of both likelihood functions, independence among the parameters for the two models is assumed. Then, two Markov Chains are created, one for each likelihood function, with stationary distribution the distribution of the unknown parameters. The way the chain is created depends on the model:

- For the estimation of the haplotype frequencies in (2), a particular case of the Metropolis-Hastings algorithm, the *Random walk*, is a simple and efficient method.
- To estimate the parameters of the logistic regression model (3), the sampling will be generated using another particular case of the Metropolis-Hastings algorithm, the *Gibbs Sampler.*

The Algorithm

Rebuilding the Haplotypes Sample
It starts with a sample of genotypes of N individuals, both cases or controls for a particular disease (variable Y_i). The algorithm begins taking an initial seed for the haplotype frequencies and for the regression coefficients. Then, the i-th step of the algorithm is described as follows:

Let be $f^{(i-1)} = (f_{h_1}^{(i-1)}, f_{h_2}^{(i-1)}, \ldots, f_{h_M}^{(i-1)})$ the previous state of the chain. Then, a new state $f^{(i)}$ is generated using Random Walk sampling, with invariant distribution proportional to (2):

1. $f^{(i)} = f^{(i-1)} + u$ where $u = (u_1, \ldots, u_M)$ such as $u_i \sim Unif(0, s)$ or $u_i \sim N(0, s)$ $i = 1, \ldots, M$ where s is chosen empirically.

2. Then, a value v is generated from a $Unif(0, 1)$ distribution.
3. if $v < \ell(f^{(i)})/\ell(f^{(i-1)})$ where ℓ is defined as in (2), the new state is accepted. If it is not, $f^{(i)} = f^{(i-1)}$.

After that, haplotypes for the uncertain individuals are rebuilt, drawing a value from a categorical distribution taking the frequencies of the previous state. For example, if an individual has a genotype compatible with the haplotypic pair $H_1 = (h_1, h_2)$ and also with $H_2 = (h_3, h_4)$, then $p_1 = P(H_1)$ and $p_2 = P(H_2)$. Now, a value from a $cat(p_1, p_2)$ is drawn, where $p_1 = f_{h_1} f_{h_2}/(f_{h_1} f_{h_2} + f_{h_3} f_{h_4})$ and $p_2 = f_{h_3} f_{h_4}/(f_{h_1} f_{h_2} + f_{h_3} f_{h_4})$.

Estimation of Haplotype Effects

After having the rebuilt haplotypes for the whole sample, they are passed as a covariate inside the logistic regression model and a new state of the chain for its coefficients is generated. This new state $\beta^{(i)}$ is sampled with a Gibbs sampler simulation:

1. The Gibbs sampler is a sampling method which draws values from the full conditional distribution of the model. Let be $\pi(\cdot \mid \beta)$ the full conditional function for the logistic regression model (3). Then, the Gibbs Sampler makes $2^m + 1$ samples to generate the new state β^i of the chain, i.e.:

$$\beta_{k_j}^{(i)} \sim \pi(\beta_{k_j} | \alpha^{(i)}, \ldots, \beta_{k_{j-1}}^{(i)}, \beta_{k_{j+1}}^{(i-1)}, \ldots, \beta_{k_{2^m}}^{(i-1)})$$

   Notice that drawing the value $\beta_{k_j}^{(i)}$ is not straightforward. Since $\pi(\cdot \mid \beta)$ is log-concave we use DFARS (*Derivative Free Adaptive Rejection Sampling*) [Gil92] a rejection method for sampling from log-concave distributions, using an envelope function.
2. Hence, $\beta^{(i)}$ is a new state of the chain.

This is a complete stage of the algorithm. Now, return to the first step and generate a new value for the chain of the haplotype frequencies.

2.5 Limiting Distribution

The constructed Markov Chains are both irreducible and ergodic (i.e. aperiodic and positive recurrent), and so the limiting distribution is unique. This limiting distribution is the stationary distribution of the chain, and so it is the distribution of our parameters. Since the chain values are a sample of the parameters distribution, the posterior mean for f and β can be estimated by the arithmetic average of sample values and it can be taken as the MLE for the parameters. Furthermore, sample values allow us to calculate different estimators such as the median, the symmetry, etc. The variances for these estimators can also be calculated from the chain.

3 Results

Performed simulations show that with a burn-in period of about 500 iterations and a sample of 1000, the convergence of the chains is remarkably good, even in presence of a large quantity of polymorphisms.

Another good feature to point out is that for haplotypes with low frequency ($<1/100$), the MCMC algorithm seems to be able to make a good estimation of the effect, while other commonly used algorithms of numerical optimization may have more difficulties to solve it. Results have also shown that the simultaneous algorithm diminishes the possibility of converging to a local minimum.

3.1 Variance of the Estimators

The considered simultaneous method of sampling gives a good estimation for the variance of β parameter, which is capturing the uncertainty of the haplotype sample. The alternative generation of two chains could make every rebuilding of the haplotype sample different at each step of the algorithm. Thus, individuals with more than two elements in H_i may be rebuilt in a different way depending on the f generated and the covariate value inside the logistic model will then change. Therefore, for samples with a great number of ambiguous individuals, the variance of the β distribution generated with the MCMC algorithm is larger than with non-simultaneous methods. Hence, the latter ones may resolve an odds ratio as significant, while the former may not do it.

4 Conclusions

Markov Chain Monte Carlo techniques can be successfully applied in the context of haplotype effects estimation. These techniques allow us to generate the distribution for each parameter, i.e. to have all the information about each one. This is an improvement over other commonly used methods like the EM algorithm, which only reports point estimators. Furthermore, for small sample sizes, estimations made with MCMC capture the possible asymmetry of the sample distribution, while methods based on asymptotic estimators do not. MCMC also seems to perform quite well for haplotypes having low frequency in the sample. Finally, the simultaneous estimation we have considered diminishes the possibility of convergence to a local minimum, so it makes the algorithm suitable to be applied over samples with a large number of polymorphisms.

References

[Bal06] D. Balding. A tutorial on statistical methods for population association studies. *Nature reviews (Genetics)*, 7:781–791, 2006.

[CD85] G. Celeux and J. Diebolt. The sem algorithm: a probabilistic teacher derived from the em algorithm for the mixture problem. *Computer Statistics Quarterly*, 2:73–82, 1985.

[Cla90] A.G. Clark. Inference of haplotypes from pcr-amplified samples of diploid populations. *Molecular Biology Evolution*, 7:111–122, 1990.

[DLR77] A.P. Dempster, N.M. Laird, and D.B. Rubin. Maximum likelihood from incomplete data via the em-algorithm. *Journal of the Royal Satistical Society*, 39:1–38, 1977.

[ES95] L. Excoffier and M. Slatkin. Maximum-likelihood estimation of molecular haplotypes frequencies in a diploid population. *Molecular biologic evolution*, 5:921–927, 1995.

[Gil92] W.R. Gilks. Derivative-free adaptive rejection sampling for gibbs sampling. In J. Bernardo, J. Berger, A.P. Dawid, and A.F.M. Smith, editors, *Bayesian Statistics 4*. Oxford University Press, 1992.

[NQXL02] T. Niu, Z.S. Qin, X. Xu, and J.S. Liu. Bayesian haplotype inference for multiple linked single-nucleotide polymorphisms. *American Journal of human genetics*, 70:157–169, 2002.

[QNL02] Z.S. Qin, T. Niu, and J.S. Liu. Partition-ligation-expectation-maximization algorithm for haplotype inference with single-nucleotide polymorphisms. *American Journal of human genetics*, 71:1242–1247, 2002.

[SSD01] M. Stephens, N.J. Smith, and P. Donnelly. A new statistical method for haplotype reconstruction from population data. *American Journal of human genetics*, 68:978–989, 2001.

[TET+04] D.A. Tregouet, S. Escolano, L. Tiret, A. Mallet, and J.L. Golmard. A new algorithm for haplotype-based association analysis: the stochastic-em algorithm. *Annals of Human genetics*, 68:165–177, 2004.

[TKJ+03] M.W. Tanck, A.H. Klerkx, J.W. Jukema, P. De Knijff, J.J. Kastelein, and A.H. Zwinderman. Estimation of multilocus haplotype effects using weighted penalised log-likelihood: analysis of five sequence variations at the cholesteryl ester transfer protein gene locus. *Annals of human genetics*, 67:175–184, 2003.

[WWB06] E.R. Waldron, J.C. Whittaker, and D.J. Balding. Fine mapping of disease genes via haplotype clustering. *Genetic Epidemiology*, 30:170–179, 2006.

The Generalized Gibbs Sampler and the Neighborhood Sampler

Jonathan Keith[1], George Sofronov[2], and Dirk Kroese[3]

[1] School of Mathematical Sciences, Queensland University of Technology, GPO Box 2434, Brisbane, Qld. 4001, Australia, and
Department of Mathematics, University of Queensland, St. Lucia, Qld. 4072, Australia
j.keith@qut.edu.au

[2] Department of Mathematics, University of Queensland, St. Lucia, Qld. 4072, Australia
georges@maths.uq.edu.au

[3] Department of Mathematics, University of Queensland, St. Lucia, Qld. 4072, Australia
kroese@maths.uq.edu.au

The Generalized Gibbs Sampler (GGS) is a recently proposed Markov chain Monte Carlo (MCMC) technique that is particularly useful for sampling from distributions defined on spaces in which the dimension varies from point to point or in which points are not easily defined in terms of co-ordinates. Such spaces arise in problems involving model selection and model averaging and in a number of interesting problems in computational biology. Such problems have hitherto been the domain of the Reversible-jump Sampler, but the method described here, which generalizes the well-known conventional Gibbs Sampler, provides an alternative that is easy to implement and often highly efficient.

The GGS provides a very general framework for MCMC simulation. Not only the conventional Gibbs Sampler, but also a variety of other well known samplers emerge as special cases. These include the Metropolis-Hastings sampler, the Reversible-jump Sampler and the Slice Sampler. We also present a new special case of the GGS, called the Neighborhood Sampler (NS), which does not conform to any of the other existing MCMC frameworks. We illustrate use of the GGS and the NS with a number of examples. In particular, we use the NS to sample from a discrete state space represented as a graph in which nodes have varying degree. Finally, we introduce a technique for improving convergence and mixing between sub-spaces of different dimension.

1 Markov Samplers

MCMC is a technique for approximately sampling from a target distribution with pdf f. Almost any distribution can be sampled via MCMC, and often it is the only technique capable of generating a sample from a given distribution. There are numerous types of MCMC sampler, but the two most frequently encountered in practice are the Metropolis-Hastings algorithm [MRRT53, Has70] and the Gibbs sampler [GG84, GS90]. Another sampler, which is distinct from these and gaining in importance, is the slice sampler [Nea03]. The slice sampler is often the most efficient MCMC method for sampling from one-dimensional distributions, and it is used as the standard technique for sampling from a general distribution in a one-dimensional sub-space in the popular MCMC software package BUGS (available at `http://www.mrc-bsu.cam.ac.uk/bugs/`).

1.1 The Discrete Case

All of the above-mentioned MCMC algorithms can be described within a framework that we call the Generalized Gibbs Sampler (GGS), although we have also called it the Generalized Markov Sampler [KKB04]. The authors have used the GGS to develop highly efficient new samplers for a range of problems arising in computational biology [KAB⁺02, KAB⁺03, KKB04, KAB⁺04, KARB05, Kei06].

Consider a Markov chain $\{(\mathbf{X}_n, \mathbf{Y}_n), n = 0, 1, 2, \ldots\}$ on the set $\mathscr{X} \times \mathscr{Y}$, where $\mathscr{X}$ is the target set and $\mathscr{Y}$ is an auxiliary set. For the sake of simplicity, we initially assume that both $\mathscr{X}$ and $\mathscr{Y}$ are finite. We extend the GGS to the general case in Section 1.2. Let $f(\mathbf{x})$ be the target pdf, defined on $\mathscr{X}$. Each transition of the Markov chain consists of two parts. The first is $(\mathbf{x}, \tilde{\mathbf{y}}) \to (\mathbf{x}, \mathbf{y})$, according to a transition matrix $\mathbf{Q}$; the second is $(\mathbf{x}, \mathbf{y}) \to (\mathbf{x}', \mathbf{y}')$, according to a transition matrix $\mathbf{R}$. In other words, the transition matrix $\mathbf{P}$ of the Markov chain is given by the product $\mathbf{Q}\,\mathbf{R}$. Both steps are illustrated in Figure 1, and further explained below.

The first step, the *Q-step*, only changes the $\mathbf{y}$-coordinate, but leaves the $\mathbf{x}$-coordinate as it is. In particular, $\mathbf{Q}$ is of the form

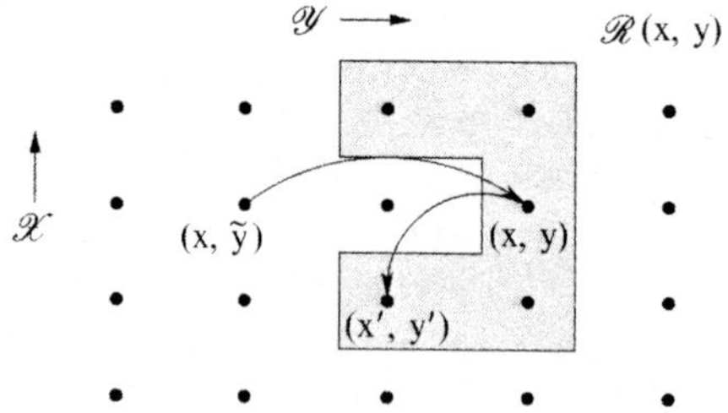

Fig. 1. Each transition of the Markov chain consists of two steps: the Q-step, followed by the R-step.

$\mathbf{Q}((\mathbf{x}, \tilde{\mathbf{y}}), (\mathbf{x}, \mathbf{y})) = \mathbf{Q}_{\mathbf{x}}(\tilde{\mathbf{y}}, \mathbf{y})$, where $\mathbf{Q}_{\mathbf{x}}$ is a transition matrix on $\mathscr{Y}$. Let $q_{\mathbf{x}}$ be a stationary distribution for $\mathbf{Q}_{\mathbf{x}}$, assuming that this exists.

The second step, the *R-step*, is determined by (a) the stationary distribution $q_{\mathbf{x}}$ and (b) a partition of the set $\mathscr{X} \times \mathscr{Y}$. Specifically, we define for each point $(\mathbf{x}, \mathbf{y})$ a set $\mathscr{R}(\mathbf{x}, \mathbf{y})$ containing $(\mathbf{x}, \mathbf{y})$ such that *if* $(\mathbf{x}', \mathbf{y}') \in \mathscr{R}(\mathbf{x}, \mathbf{y})$ *then* $\mathscr{R}(\mathbf{x}', \mathbf{y}') = \mathscr{R}(\mathbf{x}, \mathbf{y})$; see Figure 1, where the shaded area indicates the neighborhood set of $(\mathbf{x}, \mathbf{y})$. The crucial step is now to define the transition matrix $\mathbf{R}$ as

$$\mathbf{R}[(\mathbf{x}, \mathbf{y}), (\mathbf{x}', \mathbf{y}')] = c(\mathbf{x}, \mathbf{y})\, f(\mathbf{x}')\, q_{\mathbf{x}'}(\mathbf{y}'), \quad \text{for all} \quad (\mathbf{x}', \mathbf{y}') \in \mathscr{R}(\mathbf{x}, \mathbf{y}),$$

where $c(\mathbf{x}, \mathbf{y})$ is a normalisation constant. Note that $c(\mathbf{x}', \mathbf{y}') = c(\mathbf{x}, \mathbf{y})$ if $(\mathbf{x}', \mathbf{y}') \in \mathscr{R}(\mathbf{x}, \mathbf{y})$. The distribution

$$\mu(\mathbf{x}, \mathbf{y}) = f(\mathbf{x})\, q_{\mathbf{x}}(\mathbf{y}), \tag{1}$$

is trivially stationary with respect to $\mathbf{Q}$, and satisfies detailed balance with respect to $\mathbf{R}$, and hence is a stationary distribution with respect to the Markov chain. It will also be the limiting distribution, provided that the chain is ergodic. In particular, by ignoring the $\mathbf{y}$-coordinate, we see that the limiting pdf of $\mathbf{X}_n$ is the required target $f(\mathbf{x})$. This leads to the following algorithm:

Algorithm 1.1 (Generalized Gibbs Sampler) Starting with an arbitrary $(\mathbf{X}_0, \mathbf{Y}_0)$, perform the following steps iteratively:

[*Q*-**step:**] Given $(\mathbf{X}_n, \mathbf{Y}_n)$, generate $\mathbf{Y}$ from $\mathbf{Q}_{\mathbf{x}}(\mathbf{Y}_n, \mathbf{y})$.
[*R*-**step:**] Given $\mathbf{Y}$ generate $(\mathbf{X}_{n+1}, \mathbf{Y}_{n+1})$ from $\mathbf{R}[(\mathbf{X}_n, \mathbf{Y}), (\mathbf{x}, \mathbf{y})]$.

Denoting $\mathscr{R}^-(\mathbf{x}, \mathbf{y}) = \mathscr{R}(\mathbf{x}, \mathbf{y}) \setminus \{(\mathbf{x}, \mathbf{y})\}$, the sampler can be generalized further (without disturbing detailed balance) by redefining $\mathbf{R}$ as:

$$\mathbf{R}[(\mathbf{x}, \mathbf{y}), (\mathbf{x}', \mathbf{y}')] = \begin{cases} s((\mathbf{x}, \mathbf{y}), (\mathbf{x}', \mathbf{y}'))\, f(\mathbf{x}')\, q_{\mathbf{x}'}(\mathbf{y}') & \text{if } (\mathbf{x}', \mathbf{y}') \in \mathscr{R}^-(\mathbf{x}, \mathbf{y}) \\ 1 - \displaystyle\sum_{(\mathbf{z}, \mathbf{w}) \in \mathscr{R}^-(\mathbf{x}, \mathbf{y})} \mathbf{R}[(\mathbf{x}, \mathbf{y}), (\mathbf{z}, \mathbf{w})] & \text{if } (\mathbf{x}', \mathbf{y}') = (\mathbf{x}, \mathbf{y}) \end{cases} \tag{2}$$

where s is any symmetric function such that the quantities above are indeed probabilities.

1.2 GGS in a General Space

In order to relax the requirement that $\mathscr{X}$ and $\mathscr{Y}$ be finite, one must first specify the reference measure ϕ with respect to which the target density f is defined, and the reference measures $\psi_{\mathbf{x}}$ with respect to which the densities $\mathbf{Q}_{\mathbf{x}}$ and $q_{\mathbf{x}}$ are defined. Moreover, one must specify reference measures η_r

with respect to which the density $\mathbf{R}[(\mathbf{x},\mathbf{y}),(\mathbf{x}',\mathbf{y}')]$ will be defined for each set $r = \mathscr{R}(\mathbf{x},\mathbf{y})$. It will be necessary to make the following assumption:

Assumption: There exists a measure ζ for the set $\mathscr{R} = \{\mathscr{R}(\mathbf{x},\mathbf{y}) : (\mathbf{x},\mathbf{y}) \in \mathscr{X} \times \mathscr{Y}\}$ such that the measure $\phi(d\mathbf{x})\psi_{\mathbf{x}}(d\mathbf{y})$ has a finite density $g(\mathbf{x},\mathbf{y})$ with respect to the measure $\zeta(dr)\eta_r(d\mathbf{x},d\mathbf{y})$.

We can now define

$$\mathbf{R}[(\mathbf{x},\mathbf{y}),(\mathbf{x}',\mathbf{y}')] = c(\mathbf{x},\mathbf{y})\, f(\mathbf{x}')\, q_{\mathbf{x}'}(\mathbf{y}')\, g(\mathbf{x}',\mathbf{y}'), \quad \text{for all} \quad (\mathbf{x}',\mathbf{y}') \in \mathscr{R}(\mathbf{x},\mathbf{y}),$$

where again $c(\mathbf{x},\mathbf{y})$ is a normalisation constant. More generally, we can define:

$$\mathbf{R}[(\mathbf{x},\mathbf{y}),\ (\mathbf{x}',\mathbf{y}')] = \begin{cases} s((\mathbf{x},\mathbf{y}),(\mathbf{x}',\mathbf{y}'))\, f(\mathbf{x}')\, q_{\mathbf{x}'}(\mathbf{y}')\, g(\mathbf{x},\mathbf{y}) & \text{if } (\mathbf{x}',\mathbf{y}') \in \mathscr{R}^-(\mathbf{x},\mathbf{y}) \\ 1 - \displaystyle\sum_{(\mathbf{z},\mathbf{w}) \in \mathscr{R}^-(\mathbf{x},\mathbf{y})\}} \mathbf{R}[(\mathbf{x},\mathbf{y}),(\mathbf{z},\mathbf{w})] & \text{if } (\mathbf{x}',\mathbf{y}') = (\mathbf{x},\mathbf{y}) \end{cases} \tag{3}$$

where s is any symmetric function such that $\mathbf{R}$ is indeed a pdf.

Note that if $\mathscr{X}$ and $\mathscr{Y}$ are finite, then all of the above measures may be assumed to be counting measures. Moreover, in that case ζ always exists (it, too, is a counting measure) and $g(\mathbf{x},\mathbf{y}) = 1$.

2 Special cases

The GGS framework makes it possible to obtain many different samplers in a simple and unified manner; we give the slice sampler as an example. Other instances of the GCS include (see [KKB04]) the Metropolis-Hastings sampler, the Gibbs sampler and the Reversible-jump sampler [Gre95]. Here, we also introduce a new sampler — the Neighborhood sampler — and use it to sample from a discrete space represented as a graph.

2.1 Slice Sampler

The slice sampler [Nea03] has numerous variants, all of which can be conveniently described within the GGS framework. Here we discuss a fairly general form of the slice sampler. Suppose we wish to generate samples from the pdf

$$f(\mathbf{x}) = b \prod_{k=1}^{m} f_k(\mathbf{x}), \tag{4}$$

where b is a known or unknown constant, and the $\{f_k\}$ are known positive functions — not necessarily densities. We employ Algorithm 1.1, where at the Q-step we generate, for a given $\mathbf{X} = \mathbf{x}$, a vector $\mathbf{Y} = (Y_1,\ldots,Y_m)$ by independently drawing each component Y_k from the uniform distribution on $[0, f_k(\mathbf{x})]$. Thus, $q_{\mathbf{x}}(\mathbf{y}) = 1/\prod_{k=1}^{m} f_k(\mathbf{x}) = b/f(\mathbf{x})$. Secondly, we let $\mathscr{R}(\mathbf{x},\mathbf{y}) = \{(\mathbf{x}',\mathbf{y}) : f_k(\mathbf{x}') \geqslant y_k,\ k = 1,\ldots,m\}$. Then, (note that $f(\mathbf{x}')\, q_{\mathbf{x}'}(\mathbf{y}) = b$)

$$\mathbf{R}[(\mathbf{x},\mathbf{y}),(\mathbf{x}',\mathbf{y})] = \frac{1}{|\mathscr{R}(\mathbf{x},\mathbf{y})|}\ .$$

where $|\mathscr{A}|$ means the measure of set $\mathscr{A}$: the cardinality in the discrete case, or the area/volume in the continuous case. In other words, in the R-step, given $\mathbf{x}$ and $\mathbf{y}$, we draw $\mathbf{X}'$ uniformly from the set $\{\mathbf{x}' : f_k(\mathbf{x}') \geqslant y_k, k = 1,\ldots,m\}$. This gives the following *slice sampler*, in which N is a predetermined number of iterations:

Algorithm 2.1 (Slice Sampler) *Let $f(\mathbf{x})$ be of the form (4).*

1. *Initialize $\mathbf{X}_0$. Set $t = 1$.*
2. *For $k = 1$ to m draw $U_k \sim \mathsf{U}(0,1)$ and let $Y_k = U_k\, f_k(\mathbf{x}_{t-1})$.*
3. *Draw $\mathbf{X}_t$ uniformly from the set $\{\mathbf{x} : f_k(\mathbf{x}) \geqslant Y_k, k = 1,\ldots,m\}$.*
4. *If $t = N$ stop. Otherwise set $t = t+1$ and repeat from step 2.*

Suppose we want to generate a sample from the target pdf

$$f(x) = c\,\frac{x\,\mathrm{e}^{-x}}{1+x}, \qquad x \geqslant 0\,,$$

using the slice sampler with $f_1(x) = x/(1+x)$ and $f_2(x) = \mathrm{e}^{-x}$. Suppose that at iteration t, $X_{t-1} = z$, and u_1 and u_2 are generated in step 2. In step 3, X_t is drawn uniformly from the set $\{x : f_1(x)/f_1(z) \geqslant u_1,\ f_2(x)/f_2(z) \geqslant u_2\}$, which implies the bounds $x \geqslant \frac{u_1 z}{1+z-u_1 z}$, and $x \leqslant z - \ln u_2$. Since for $z > 0$ and $0 \leqslant u_1, u_2 \leqslant 1$, the latter bound is larger than the former, the interval to be drawn from in step 3 is $(\frac{u_1 z}{1+z-u_1 z},\ z - \ln u_2)$. Figure 2 depicts a histogram of $N = 10^5$ samples generated via the slice sampler, along with the true pdf $f(x)$. We see that the two are in close agreement.

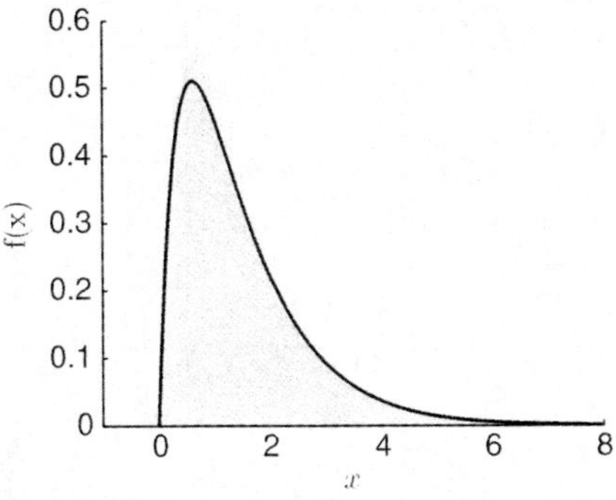

Fig. 2. True density and histogram of samples produced by the slice sampler.

2.2 The Neighborhood Sampler

The Neighborhood Sampler (NS) is a new instance of the GGS that resembles the slice sampler and in certain cases corresponds to it. The NS can be used to sample from a target distribution f on some measure space $(\mathscr{X}, \Sigma, \mu)$ consisting of a target space $\mathscr{X}$ with σ-algebra Σ and measure μ. The aim of the NS to reduce sampling from a complicated distribution function f to sampling from uniform distributions over local neighborhoods. To construct a NS, we must first assign a unique neighborhood $\mathcal{N}_{\mathbf{x}}$ to each element $\mathbf{x} \in \mathscr{X}$. These neighborhoods must satisfy the following three conditions:

1. $\mathbf{x} \in \mathcal{N}_{\mathbf{x}}$ for all $\mathbf{x} \in \mathscr{X}$,
2. $0 < \mu(\mathcal{N}_{\mathbf{x}}) < \infty$ for all $\mathbf{x} \in \mathscr{X}$, and
3. $\mathbf{y} \in \mathcal{N}_{\mathbf{x}}$ iff $\mathbf{x} \in \mathcal{N}_{\mathbf{y}}$ for all $\mathbf{x}, \mathbf{y} \in \mathscr{X}$.

In what follows we use the notation $\mathcal{N}(\mathbf{x})$ synonymously with $\mathcal{N}_{\mathbf{x}}$ to avoid placing subscripts on subscripts.

To sample from an arbitrary distribution having density f with respect to μ, the NS consists of the following steps performed iteratively, starting with an arbitrary element $\mathbf{x}_0$ and with $t = 0$:

Algorithm 2.2 (Neighborhood Sampler)
Given the current state $\mathbf{X}_t = \mathbf{x}$:

1. *Generate* $\mathbf{Y} \sim \mathsf{U}(\mathcal{N}_{\mathbf{x}})$ *where* $\mathsf{U}(\mathcal{N}_{\mathbf{x}})$ *is the uniform distribution (with respect to* μ*) on* $\mathcal{N}_{\mathbf{x}}$. *Set* $H = \mathcal{N}_{\mathbf{Y}}$.
2. *Generate* $\mathbf{U} \sim \mathsf{U}(0, f(\mathbf{x})/\mu[\mathcal{N}_{\mathbf{x}}])$.
3. *Generate* $\mathbf{Z}_1 \sim \mathsf{U}(H)$.
4. *Set* $k = 1$ *and iterate the following steps until* $f(\mathbf{Z}_k)/\mu[\mathcal{N}(\mathbf{Z}_k)] \geqslant \mathbf{U}$:
   a) *Optionally reduce* H *so that it excludes* $\mathbf{Z}_k$ *while still containing* $\mathbf{x}$.
   b) *Generate* $\mathbf{Z}_{k+1} \sim \mathsf{U}(H)$ *and set* $k := k + 1$.
5. *Set* $\mathbf{X}_{t+1} = \mathbf{Z}_k$.

The reduction in Step 4a) must be done so that $H(\mathbf{x}', \mathbf{y}, \mathbf{z}_1, \ldots, \mathbf{z}_k) = H(\mathbf{x}, \mathbf{y}, \mathbf{z}_1, \ldots, \mathbf{z}_k)$ for all $\mathbf{x}' \in H(\mathbf{x}, \mathbf{y}, \mathbf{z}_1, \ldots, \mathbf{z}_k)$, where the notation $H(\mathbf{x}, \mathbf{y}, \mathbf{z}_1, \ldots, \mathbf{z}_k)$ indicates the neighborhood obtained by reducing $\mathcal{N}_{\mathbf{y}}$ in such a way as to include $\mathbf{x}$ and exclude $\mathbf{z}_1, \ldots, \mathbf{z}_k$. The details of this reduction depend on the specific application. We give some examples below.

If we do not implement the reduction of H in Step 4a), then the Q-step in this algorithm consists of selecting the pair $(\mathbf{Y}, \mathbf{U})$ in Steps 1 and 2. The R-step consists of uniform sampling of the subset of H for which $f(\mathbf{z})/\mu[\mathcal{N}(\mathbf{z})] \geqslant \mathbf{U}$ in Steps 3 to 5. If we do implement the reduction of H, then the Q-step consists of selecting $(\mathbf{Y}, \mathbf{U}, \mathbf{Z}_1, \ldots, \mathbf{Z}_k)$ in Steps 1 to 4 and the R-step consists of accepting $\mathbf{Z}_k$ with probability 1 in Step 5.

It is not difficult to show that if $\mu(\mathcal{N}_{\mathbf{x}})$ is constant for all $\mathbf{x}$, then the denominators in Steps 2 and 4 above can be replaced by 1. With $\mathcal{N}_{\mathbf{x}} = \mathscr{X}$ for all $\mathbf{x}$, the NS reduces to the variant of the slice sampler described in Section 2.1

with $m = 1$. If $\mathscr{X}$ is $\mathbb{R}^n$ with Lebesbue measure and $\mathcal{N}_{\mathbf{x}}$ is a hypercube of side s for all $\mathbf{x} \in \mathscr{X}$, then the NS reduces to a variant of the slice sampler described in [Nea03]. In this case, H is reduced to a smaller hyper-rectangle with $\mathbf{Z}_k$ as a corner in Step 4a). Another interesting case is obtained if we suppose that $\mathscr{X}$ is a discrete space represented by a connected graph in which nodes represent states and edges represent allowed transitions. Let μ be counting measure and let $\mathcal{N}_{\mathbf{x}}$ consist of $\mathbf{x}$ and all of its neighbors, that is, all nodes adjacent to $\mathbf{x}$. Then $\mathbf{U}$ is chosen between 0 and $f(\mathbf{x})/|\mathcal{N}_{\mathbf{x}}|$ at step 2. The optional reduction of H in step 4a can be achieved by simply excluding $\mathbf{Z}_k$. The fact that previously rejected elements are excluded from being chosen a second time at Step 4a) should in principle make the neighborhood sampler faster than a random walk sampler with a uniform proposal function on the same neighborhoods, since the amount of computation is otherwise comparable.

2.3 Resequencing

Resequencing is the practice of determining the sequence of a biological molecule — usually DNA — by assembling short sub-sequences using related sequences that are already known to aid the assembly. For example, sequencing of some part of the genome of an individual human can be achieved by assembling short sub-sequences using the corresponding part of the already sequenced reference genome to guide the assembly. Similarly, parts of the genomes of other species can be assembled with reference to known genomes of related species. The problem of resequencing is important because modern high-throughput sequencing technologies determine only very short sub-sequences, or reads. For example, the technology known as *Sequencing By Hybridization* (SBH) determines the subsequence content of an unknown DNA by identifying all probes of a given length (often around 10 nucleotides) that bind to it [DDS+93]. More recently, a number of groups have developed fast, high-throughput technologies that use short reads [MEA+05, SPR+05]. The use of sequences known to be similar can greatly facilitate the assembly process.

We propose the following idealized model of resequencing. Suppose that we have a known sequence S of length L as shown in Figure 3. Suppose further that we model the generation of the unknown sequence $\mathbf{x}$ as the result of independent transitions at each base, with a known transition matrix M. Finally, suppose that we know all contiguous sub-sequences of length k contained in the unknown sequence. Let the set of such sub-sequences be denoted D. The target space $\mathscr{X}$ here thus consists of all sequences of length L with precisely the same set of length k sub-sequences. The posterior distribution on this space, given the sub-sequences, is the restriction to $\mathscr{X}$ of the distribution over all sequences of length L defined by the transition matrix. That is:

$$p(\mathbf{x}|D, S, M) = \prod_{i=1}^{L} M(S_i, \mathbf{x}_i)$$

restricted to $\mathscr{X}$.

← L →

```
ACTTAGCACCGGTGACAACCCTACT
            ↓
ACCTAACACAGGTGATAACCCCGCT
            ↓
ACCTAACACA AACACAGGTG CAGGTGATAA TGATAACCCC
CCTAACACAG ACACAGGTGA AGGTGATAAC GATAACCCCG
CTAACACAGG CACAGGTGAT GGTGATAACC ATAACCCCGC
TAACACAGGT ACAGGTGATA GTGATAACCC TAACCCCGCT
```

Fig. 3. An idealized picture of resequencing. The sequence in the centre is unknown. A similar sequence, shown at the top, is known, as is the complete set of subsequences of length 10.

To avoid having to estimate the transition matrix M, we can specify a prior probability distribution for each row of the transition matrix. Here we use a Dirichlet distribution:

$$g(M_i) \propto M_{i1}^{\alpha-1} M_{i2}^{\alpha-1} M_{i3}^{\alpha-1} M_{i4}^{\alpha-1}$$

for each row i, with $\alpha = 0.001$. Integrating over M_{ij} for $i, j = 1, \ldots, 4$ results in the target distribution:

$$f(\mathbf{x}|D, S) \propto \prod_{i=1}^{4} \frac{\Gamma(C_{i1}+\alpha)\Gamma(C_{i2}+\alpha)\Gamma(C_{i3}+\alpha)\Gamma(C_{i4}+\alpha)}{\Gamma(C_{i1}+C_{i2}+C_{i3}+C_{i4}+4\alpha)}$$

where $C_{ij} = C_{ij}(\mathbf{x})$ is the number of positions at which sequence S has character i and sequence $\mathbf{x}$ has character j.

We sample from the distribution using the NS for a discrete space described above. Let the nodes of the graph be all sequences that have the same set of length k sub-sequences. The edges of the graph connect sequences that are related by transformations of the following form:

1. **Transpositions:** Sequences of the form $y_1 z_1 y_2 z_2 y_3 z_1 y_4 z_2 y_5$ where z_1 and z_2 are length $k-1$ subsequences and y_1, y_2, y_3, y_4 and y_5 are sequences of any length can be *transposed* by swapping the sequences y_2 and y_4. Sequences of the form $y_1 z_1 y_2 z_1 y_4 z_1 y_5$ can also be transposed by swapping the sequences y_2 and y_4.
2. **Rotations:** Sequences of the form $z_1 y_1 z_2 y_2 z_1$ where z_1 and z_2 are length $k-1$ subsequences and y_1 and y_2 are subsequences of any length can be *rotated* by forming the sequence $z_2 y_2 z_1 y_1 z_2$.

Some subtlety is required here: the sub-sequences labelled y may be null sequences, and in fact the sequences labelled z may even overlap. It has been shown [Pev95] that sequences related by these transformations have the same set of length k subsequences and that graphs formed as described above are connected. Note that rotations are only possible if the sequence begins and

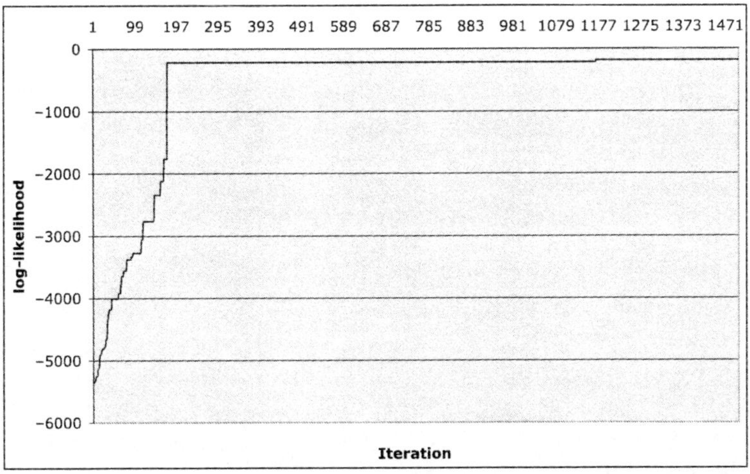

Fig. 4. log-likelihood for resequencing application of the Neighborhood sampler.

ends with the same $k-1$ characters. Here we consider only sequences that do not begin and end with the same sub-sequence, so that we need only consider transpositions. Hence the neighborhood $N_{\mathbf{x}}$ of a sequence $\mathbf{x}$ consists of $\mathbf{x}$ plus all sequences that can be obtained from $\mathbf{x}$ by transpositions.

We implemented the NS for discrete spaces for this problem and tested it on a known human DNA sequence of length 4009 nucleotides containing an exon of the breast cancer associated BRCA1 gene (specifically locus 38,496,578-38,500,586 of the March 2006 assembly of human chromosome 17). We also obtained a known sequence of the chimpanzee genome of the same length, aligned to this section of BRCA1 without any insertions or deletions. The human sequence was used as the reference and the chimpanzee sequence was treated as unknown. Figure 4 shows the log-likelihood values for 400 iterations of the algorithm, showing rapid convergence to a single sequence which on inspection turns out to be identical to the known chimpanzee sequence.

Note, however, that the sampler spends many iterations at a nearly optimal sequence, only making the final transposition at iteration 1170. It is not clear whether it is possible for this sampler to become stuck in a local mode for infeasible lengths of time. However, one possible way to improve mixing would be to expand each neighborhood $N_{\mathbf{x}}$ to include sequences that can be obtained from $\mathbf{x}$ by two successive transpositions.

3 Discussion

The GGS provides a general framework within which all of the commonly used MCMC samplers can be described. This generality raises the interesting possibility of theoretically determining the sampler within this framework with optimal convergence and/or mixing rate for a specific distribution or family of distributions. The GGS also supplies a framework within which new samplers can be generated by exploring various possibilities for the sets $\mathscr{R}(\mathbf{x},\mathbf{y})$ and other parameters. The Neighborhood Sampler is an example of a new sampler generated in this manner.

To conclude this paper, we describe a technique that we recently developed to improve the convergence and mixing rate of the genome segmentation sampler that we described in [Kei06]. The technique is based on the fact that typical trans-dimensional sampling (that is, sampling of spaces in which the dimension varies from point to point) involves conditional sampling of sets $\mathscr{R}(\mathbf{x},\mathbf{y})$ in which the dimension of the x component varies by at most one. In other words, each set $\mathscr{R}(\mathbf{x},\mathbf{y})$ can be subdivided naturally into a set $\mathscr{R}_1(\mathbf{x},\mathbf{y})$ in which all x components have dimension k, say, and a set $\mathscr{R}_2(\mathbf{x},\mathbf{y})$ in which all x components have dimension $k+1$.

We can now define a symmetric function s as follows. Let $s[(\mathbf{x},\mathbf{y}),(\mathbf{x}',\mathbf{y}')]=0$ if $(\mathbf{x},\mathbf{y})$ and $(\mathbf{x}',\mathbf{y}')$ are both in $\mathscr{R}_1(\mathbf{x},\mathbf{y})$ or both in $\mathscr{R}_2(\mathbf{x},\mathbf{y})$. Otherwise, let

$$s[(\mathbf{x},\mathbf{y}),(\mathbf{x}',\mathbf{y}')]=\frac{1}{\max\{\sum_{(\mathbf{z},\mathbf{w})\in\mathscr{R}_1(\mathbf{x},\mathbf{y})} f(\mathbf{z})q_{\mathbf{z}}(\mathbf{w}),\ \sum_{(\mathbf{z},\mathbf{w})\in\mathscr{R}_2(\mathbf{x},\mathbf{y})} f(\mathbf{z})q_{\mathbf{z}}(\mathbf{w})\}}$$

if $(\mathbf{x}',\mathbf{y}')\in\mathscr{R}^-(\mathbf{x},\mathbf{y})$. Consequently, if $(\mathbf{x},\mathbf{y})\in\mathscr{R}_1(\mathbf{x},\mathbf{y})$ and

$$\sum_{(\mathbf{z},\mathbf{w})\in\mathscr{R}_1(\mathbf{x},\mathbf{y})} f(\mathbf{z})q_{\mathbf{z}}(\mathbf{w})\leqslant\sum_{(\mathbf{z},\mathbf{w})\in\mathscr{R}_2(\mathbf{x},\mathbf{y})} f(\mathbf{z})q_{\mathbf{z}}(\mathbf{w})$$

then the probability of transition to $\mathscr{R}_2(\mathbf{x},\mathbf{y})$, obtained by summing (2) over $\mathscr{R}_2(\mathbf{x},\mathbf{y})$, is one. Similarly, if $(\mathbf{x},\mathbf{y})\in\mathscr{R}_1(\mathbf{x},\mathbf{y})$ and

$$\sum_{(\mathbf{z},\mathbf{w})\in\mathscr{R}_1(\mathbf{x},\mathbf{y})} f(\mathbf{z})q_{\mathbf{z}}(\mathbf{w})\geqslant\sum_{(\mathbf{z},\mathbf{w})\in\mathscr{R}_2(\mathbf{x},\mathbf{y})} f(\mathbf{z})q_{\mathbf{z}}(\mathbf{w})$$

then the probability of transition to $\mathscr{R}_1(\mathbf{x},\mathbf{y})$ is one. The probability of a change in dimension is therefore high.

Acknowledgements

The authors acknowledge the support of an Australian Research Council discovery grant (DP0556631). The first author would like to acknowledge the support of an Australian Research Council discovery grant (DP0452412) and a National Medical and Health Research Council grant "Statistical methods and algorithms for analysis of high-throughput genetics and genomics platforms" (389892).

References

[DDS+93] R. Drmanac, S. Drmanac, A. Strezoska, T. Paunesku, I. Labat, M. Zeremski, J. Snoddy, W. K. Funkhouser, B. Koop, L. Hood, and R. Crkvenjakov. DNA sequence determination by hybridization: a strategy for efficient large-scale sequencing. *Science*, 260:1649–1952, 1993.

[GG84] S. Geman and D. Geman. Stochastic relaxation, Gibbs distributions, and the Bayesian restoration of images. *IEEE T. Pattern Anal.*, 6:721–741, 1984.

[Gre95] P. J. Green. Reversible jump Markov chain Monte Carlo computation and Bayesian model determination. *Biometrika*, 82:711–732, 1995.

[GS90] A. F. Gelfand and A. F. M. Smith. Sampling-based approaches to calculating marginal densities. *J. Am. Stat. Assoc.*, 85:398–409, 1990.

[Has70] W. K. Hastings. Monte Carlo sampling methods using Markov chains and their applications. *Biometrika*, 57:97–109, 1970.

[KAB+02] J. M. Keith, P. Adams, D. Bryant, K. R. Mitchelson, D. A. E. Cochran, and G. H. Lala. A simulated annealing algorithm for finding consensus sequences. *Bioinformatics*, 18:1494–1499, 2002.

[KAB+03] J. M. Keith, P. Adams, D. Bryant, K. R. Mitchelson, D. A. E. Cochran, and G. H. Lala. Inferring an original sequence from erroneous copies: a Bayesian approach. In *Proceedings of the First Asia-Pacific Bioinformatics Conference*, volume 19 of *Conferences in Research and Practice in Information Technology*, pages 23–28, 2003.

[KAB+04] J. M. Keith, P. Adams, D. Bryant, D. A. E. Cochran, G. H. Lala, and K. R. Mitchelson. Algorithms for sequence analysis via mutagenesis. *Bioinformatics*, 20:2401–2410, 2004.

[KARB05] J. M. Keith, P. Adams, M. A. Ragan, and D. Bryant. Sampling phylogenetic tree space with the generalized Gibbs sampler. *Molecular Phylogenetics and Evolution*, 34:459–468, 2005.

[Kei06] J. M. Keith. Segmenting eukaryotic genomes with the generalized Gibbs sampler. *Journal of Computational Biology*, 2006. In press.

[KKB04] J. M. Keith, D. P. Kroese, and D. Bryant. A Generalized Markov sampler. *Methodology and Computing in Applied Probability*, 6:29–53, 2004.

[MEA+05] M. Margulies, M. Egholm, W. E. Altman, S. Attiya, et al. Genome sequencing in microfabricated high-density picolitre reactors. *Nature*, 437:376–380, 2005.

[MRRT53] N. Metropolis, A. W. Rosenbluth, M. N. Rosenbluth, and A. H. Teller. Equations of state calculations by fast computing machines. *J. Chem. Phys.*, 21:1087–1092, 1953.

[Nea03] R. M. Neal. Slice sampling. *Ann. Stat.*, 31(3):705–767, 2003.

[Pev95] P. A. Pevzner. DNA Physical Mapping and Alternating Eulerian Cycles in Colored Graphs. *Algorithmica*, 13:77–105, 1995.

[SPR+05] J. Shendure, G. J. Porreca, N. B. Reppas, X. Lin, J. P. McCutcheon, A. M. Rosenbaum, M. D. Wang, K. Zhang, R. D. Mitra, and G. M. Church. Accurate Multiplex Polony Sequencing of an Evolved Bacterial Genome. *Science*, 309:1728–1732, 2005.

The Weighted Dyadic Diaphony of Digital Sequences

Peter Kritzer[1] and Friedrich Pillichshammer[2]*

[1] Fachbereich Mathematik, Universität Salzburg, Hellbrunnerstraße 34, A-5020 Salzburg, Austria
peter.kritzer@sbg.ac.at

[2] Institut für Finanzmathematik, Universität Linz, Altenbergerstraße 69, A-4040 Linz, Austria
friedrich.pillichshammer@jku.at

Summary. The (weighted) dyadic diaphony is a measure for the irregularity of distribution modulo one of a sequence. Recently it has been shown that the (weighted) dyadic diaphony can be interpreted as the worst-case error for QMC integration in a certain Hilbert space of functions. In this paper we give upper bounds on the weighted dyadic diaphony of digital (t, s)-sequences over $\mathbb{Z}_2$.

1 Introduction

Motivated by Weyl's criterion for the uniform distribution modulo one of a sequence (see [DT97, KN74, SP05]), Zinterhof [Zin76] introduced the diaphony as a quantitative measure for the irregularity of distribution of a sequence (see also [DT97, KN74]). In [HL97] Hellekalek and Leeb introduced the notion of dyadic diaphony which is similar to the classical diaphony but with the trigonometric functions replaced by Walsh functions. As for the classical diaphony it can be shown that a sequence ω is uniformly distributed modulo one if and only if the the dyadic diaphony of the first N elements of the sequence ω tends to zero as N goes to infinity (see [HL97, Theorem 3.1]). Recently it was shown in [DP05a] that the dyadic diaphony is—up to a factor depending only on the dimension s—the worst-case error for quasi-Monte Carlo integration of functions from a certain Hilbert space. This motivates the introduction of the more general notion of weighted dyadic diaphony as the worst-case error for quasi-Monte Carlo integration of functions from a certain weighted Hilbert space (see [DP05a, DP05b, Gro06]). This function space has been introduced (in a slightly

* The authors are supported by the Austrian Science Foundation (FWF), Project S9609, that is part of the Austrian National Research Network "Analytic Combinatorics and Probabilistic Number Theory".

more general setting) in [DP05b]. We give the definition in Section 3. In Section 4 we analyze the weighted dyadic diaphony of digital (t,s)-sequences over $\mathbb{Z}_2$.

2 Digital Sequences

In this paper we consider the weighted dyadic diaphony of digital (t,s)-sequences as introduced by Niederreiter in [Nie87]. We only deal with digital (t,s)-sequences over the finite field $\mathbb{Z}_2$. For the definition of the general case see, for example, [Lar98, LNS96, Nie87, Nie92].

Before we give the definition of digital (t,s)-sequences we introduce some notation: for a vector $\mathbf{c} = (c_1, c_2, \ldots) \in \mathbb{Z}_2^\infty$ and for $m \in \mathbb{N}$ we denote the vector in $\mathbb{Z}_2^m$ consisting of the first m components of $\mathbf{c}$ by $\mathbf{c}(m)$, i.e., $\mathbf{c}(m) = (c_1, \ldots, c_m)$. Further for an $\mathbb{N} \times \mathbb{N}$ matrix C over $\mathbb{Z}_2$ and for $m \in \mathbb{N}$ we denote by $C(m)$ the left upper $m \times m$ submatrix of C.

Definition 1. For $s \in \mathbb{N}$ and $t \in \mathbb{N}_0$, choose s $\mathbb{N} \times \mathbb{N}$ matrices $C_1, \ldots, C_s$ over $\mathbb{Z}_2$ with the following property: for every $m \in \mathbb{N}$, $m \geq t$ and every $d_1, \ldots, d_s \in \mathbb{N}_0$ with $d_1 + \cdots + d_s = m - t$ the vectors

$$\mathbf{c}_1^{(1)}(m), \ldots, \mathbf{c}_{d_1}^{(1)}(m), \ldots, \mathbf{c}_1^{(s)}(m), \ldots, \mathbf{c}_{d_s}^{(s)}(m)$$

are linearly independent in $\mathbb{Z}_2^m$. Here $\mathbf{c}_i^{(j)}$ is the i-th row vector of the matrix C_j.

For $n \geq 0$ let $n = n_0 + n_1 2 + n_2 2^2 + \cdots$ be the base 2 representation of n. For $j \in \{1, \ldots, s\}$ multiply the vector $\mathbf{n} = (n_0, n_1, \ldots)^\top$ by the matrix C_j,

$$C_j \cdot \mathbf{n} =: (x_n^j(1), x_n^j(2), \ldots)^\top \in \mathbb{Z}_2^\infty,$$

and set

$$x_n^{(j)} := \frac{x_n^j(1)}{2} + \frac{x_n^j(2)}{2^2} + \cdots .$$

Finally set $\boldsymbol{x}_n := (x_n^{(1)}, \ldots, x_n^{(s)})$.

Every sequence $(\boldsymbol{x}_n)_{n \geq 0}$ constructed in this way is called *digital* (t,s)-*sequence over* $\mathbb{Z}_2$. The matrices $C_1, \ldots, C_s$ are called the *generator matrices* of the sequence.

To guarantee that the points $\boldsymbol{x}_n$ belong to $[0,1)^s$ (and not just to $[0,1]^s$) and also for the analysis of the sequence we need the condition that for each $n \geq 0$ and $1 \leq j \leq s$, we have $x_n^j(i) = 0$ for infinitely many i. This condition is always satisfied if we assume that for each $1 \leq j \leq s$ and $w \geq 0$ we have $c_{v,w}^{(j)} = 0$ for all sufficiently large v, where $c_{v,w}^{(j)}$ are the entries of the matrix C_j. Throughout this paper we assume that the generator matrices fulfill this condition (see [Nie92, p. 72] where this condition is called (S6)).

It is well known that any digital (t,s)-sequence over $\mathbb{Z}_2$ is uniformly distrubuted modulo one (in fact, it is even well-distributed modulo one). Bounds on the star discrepancy (which is another very popular quantitative measure for the irregularity of distribution of a sequence) of (not necessarily digital) (t,s)-sequences can be found in [Nie92, Chapter 4] or [Nie87], see also [Kri06].

3 Walsh Functions and the Hilbert Space $H_{\mathrm{wal},s,\alpha,\gamma}$

Throughout this paper let $\mathbb{N}_0$ denote the set of non-negative integers. For $k \in \mathbb{N}_0$ with base 2 representation $k = \kappa_{a-1}2^{a-1} + \cdots + \kappa_1 2 + \kappa_0$, with $\kappa_i \in \{0,1\}$, we define the *(dyadic) Walsh function* $\mathrm{wal}_k : \mathbb{R} \to \{-1,1\}$, periodic with period one, by

$$\mathrm{wal}_k(x) := (-1)^{\xi_1\kappa_0 + \cdots + \xi_a\kappa_{a-1}},$$

for $x \in [0,1)$ with base 2 representation $x = \xi_1/2 + \xi_2/2^2 + \cdots$ (unique in the sense that infinitely many of the ξ_i must be zero). For dimension $s \geq 2$, $x_1, \ldots, x_s \in [0,1)$ and $k_1, \ldots, k_s \in \mathbb{N}_0$ we define $\mathrm{wal}_{k_1,\ldots,k_s} : \mathbb{R}^s \to \{-1,1\}$ by

$$\mathrm{wal}_{k_1,\ldots,k_s}(x_1,\ldots,x_s) := \prod_{j=1}^{s} \mathrm{wal}_{k_j}(x_j).$$

For vectors $\boldsymbol{k} = (k_1,\ldots,k_s) \in \mathbb{N}_0^s$ and $\boldsymbol{x} = (x_1,\ldots,x_s) \in \mathbb{R}^s$ we write

$$\mathrm{wal}_{\boldsymbol{k}}(\boldsymbol{x}) := \mathrm{wal}_{k_1,\ldots,k_s}(x_1,\ldots,x_s).$$

It is clear from the definitions that Walsh functions are piecewise constant. It can be shown that for any integer $s \geq 1$ the system $\{\mathrm{wal}_{k_1,\ldots,k_s} : k_1,\ldots,k_s \geq 0\}$ is a complete orthonormal system in $L_2([0,1)^s)$, see for example [Chr55, Nie82] or [Pir95, Satz 1]. More information on Walsh functions can be found in [Chr55, Pir95, Wal23].

For a natural number k with $2^a \leq k < 2^{a+1}$, $a \in \mathbb{N}_0$, let $\psi(k) = a$. For $\alpha > 1$ and $\gamma > 0$ we define

$$r(\alpha,\gamma,k) = \begin{cases} 1 & \text{if } k = 0, \\ \gamma 2^{-\alpha\psi(k)} & \text{if } k \neq 0. \end{cases}$$

For a vector $\boldsymbol{k} = (k_1,\ldots,k_s)$ and a sequence $\boldsymbol{\gamma} = (\gamma_i)_{i\geq 1}$ of positive reals we write $r(\alpha,\boldsymbol{\gamma},\boldsymbol{k}) = \prod_{j=1}^{s} r(\alpha,\gamma_j,k_j)$.

The function space $H_{\mathrm{wal},s,\alpha,\boldsymbol{\gamma}}$ is defined as a reproducing kernel Hilbert space with reproducing kernel given by

$$K(\boldsymbol{x},\boldsymbol{y}) = \sum_{\boldsymbol{k}\in\mathbb{N}_0^s} r(\alpha,\boldsymbol{\gamma},\boldsymbol{k})\mathrm{wal}_{\boldsymbol{k}}(\boldsymbol{x})\mathrm{wal}_{\boldsymbol{k}}(\boldsymbol{y}).$$

The inner product in this space is defined by

$$\langle f,g\rangle_{\mathrm{wal},s,\boldsymbol{\gamma}} = \sum_{\boldsymbol{k}\in\mathbb{N}_0^s} r(\alpha,\boldsymbol{\gamma},\boldsymbol{k})^{-1}\widehat{f}_{\mathrm{wal}}(\boldsymbol{k})\widehat{g}_{\mathrm{wal}}(\boldsymbol{k}),$$

with $\boldsymbol{k} = (k_1,\ldots,k_s)$ and

$$\widehat{f}_{\mathrm{wal}}(\boldsymbol{k}) := \int_{[0,1]^s} f(\boldsymbol{x})\mathrm{wal}_{\boldsymbol{k}}(\boldsymbol{x})\,\mathrm{d}x.$$

The corresponding norm is given by $\|f\|_{\mathrm{wal},s,\boldsymbol{\gamma}} = \langle f,f\rangle^{1/2}_{\mathrm{wal},s,\boldsymbol{\gamma}}$. Note that $H_{\mathrm{wal},s,\alpha,\boldsymbol{\gamma}}$ is the space of all absolutely convergent Walsh series with finite norm $\|\cdot\|_{\mathrm{wal},s,\boldsymbol{\gamma}}$. For more information on this space and generalizations thereof we refer to [DP05b].

4 Worst-Case Error for QMC Integration in $H_{\mathrm{wal},s,\alpha,\boldsymbol{\gamma}}$

We want to approximate the integral $I_s(f) = \int_{[0,1]^s} f(\boldsymbol{x})\,\mathrm{d}\boldsymbol{x}$ for $f \in H_{\mathrm{wal},s,\alpha,\boldsymbol{\gamma}}$ by a *quasi-Monte Carlo (QMC) rule* of the form $Q(f;\mathcal{P}) = \frac{1}{N}\sum_{n=0}^{N-1} f(\boldsymbol{x}_n)$, where $\mathcal{P} = \{\boldsymbol{x}_0,\ldots,\boldsymbol{x}_{N-1}\} \subset [0,1)^s$ is a well chosen point set (here we use the first N elements of a digital (t,s)-sequence over $\mathbb{Z}_2$). We define the *worst-case error* for QMC integration in the space $H_{\mathrm{wal},s,\alpha,\boldsymbol{\gamma}}$ by

$$e_{N,s,\alpha,\boldsymbol{\gamma}} := e(\mathcal{P}; H_{\mathrm{wal},s,\alpha,\boldsymbol{\gamma}}) = \sup_{\substack{f\in H_{\mathrm{wal},s,\alpha,\boldsymbol{\gamma}} \\ \|f\|_{\mathrm{wal},s,\boldsymbol{\gamma}}\le 1}} |I_s(f) - Q(f;\mathcal{P})|.$$

It has been shown in [DP05b] that for any point set $\mathcal{P} = \{\boldsymbol{x}_0,\ldots,\boldsymbol{x}_{N-1}\} \subset [0,1)^s$ we have

$$e^2_{N,s,\alpha,\boldsymbol{\gamma}} = \sum_{\boldsymbol{k}\in\mathbb{N}_0^s\setminus\{\boldsymbol{0}\}} r(\alpha,\boldsymbol{\gamma},\boldsymbol{k}) \left|\frac{1}{N}\sum_{n=0}^{N-1} \mathrm{wal}_{\boldsymbol{k}}(\boldsymbol{x}_n)\right|^2.$$

For $\alpha = 2$ and $\boldsymbol{\gamma} = (1)_{j\ge 1}$ this is, up to the factor $1/(3^s-1)$, the squared *dyadic diaphony* $F_{2,N}$ of $\mathcal{P}$ as introduced by Hellekalek and Leeb [HL97]. I.e.,

$$F_{2,N}(\mathcal{P}) = \frac{e_{N,s,2,(1)_{j\ge 1}}}{\sqrt{3^s-1}}.$$

For this reason we refer to $e_{N,s,\alpha,\boldsymbol{\gamma}}$ as *weighted dyadic diaphony* (see also [Gro06]). It has been shown by Grozdanov [Gro06, Theorem 1] that a sequence ω is uniformly distributed modulo one if and only if $\lim_{N\to\infty} e_{N,s,\alpha,\boldsymbol{\gamma}}(\omega_N) = 0$ holds for an arbitrary real $\alpha > 1$ and an arbitrary sequence $\boldsymbol{\gamma}$ of positive weights, where ω_N is the point set consisting of the first N points of the sequence ω. See also [HL97] for a proof of this fact in the special case $\alpha = 2$ and $\boldsymbol{\gamma} = (1)_{j\ge 1}$.

In the following we study the worst-case error $e_{N,s,\alpha,\boldsymbol{\gamma}}$ if the integration nodes stem from a digital (t,s)-sequence over $\mathbb{Z}_2$.

Before we state our results we introduce some notation: for a subset $\mathfrak{u} \subseteq \{1,\ldots,s\} =: D_s$ we define $\gamma_{\mathfrak{u}} := \prod_{j\in\mathfrak{u}} \gamma_j$. Further we denote by $\|\cdot\|$ the distance to the nearest integer function, i.e., $\|x\| = \min(x-\lfloor x\rfloor, 1-(x-\lfloor x\rfloor))$.

Theorem 1. *Let $\alpha > 1$, let $\boldsymbol{\gamma}$ be an arbitrary sequence of positive weights and let $N \in \mathbb{N}$. For the first N elements of a digital (t,s)-sequence over $\mathbb{Z}_2$ it is true that*

$$e_{N,s,\alpha,\gamma}^2 \leq \frac{2^{2t}}{N^2} \sum_{\emptyset \neq \mathfrak{u} \subseteq D_s} \gamma_{\mathfrak{u}} 2^{2|\mathfrak{u}|} \left(\frac{2^{\alpha-1}}{2^{\alpha-1}-1} \right)^{|\mathfrak{u}|}$$
$$\times \left(1 + 2^{(t+|\mathfrak{u}|)(\alpha-2)} \sum_{v=t+|\mathfrak{u}|+1}^{\infty} \left\| \frac{N}{2^v} \right\|^2 \frac{(v-t-1)^{|\mathfrak{u}|-1}}{2^{v(\alpha-2)}} \right).$$

Remark 1. Observe that the upper bound in Theorem 1 converges to infinity if α approaches one. Furthermore, the bound is independent of the specific choice of the generating matrices of the digital (t,s)-sequence.

From Theorem 1 we obtain

Corollary 1. *Let ω be a digital (t,s)-sequence over $\mathbb{Z}_2$ and let $2^{m-1} < N \leq 2^m$, with $m > t+s$. Then we have the following results for the first N elements of the sequence.*

(a) If $\alpha = 2$, we have

$$e_{N,s,2,\gamma}^2 \leq \frac{2^{2t}}{N^2} \sum_{\emptyset \neq \mathfrak{u} \subseteq D_s} \gamma_{\mathfrak{u}} 2^{3|\mathfrak{u}|} \left(\frac{(m-t-1)^{|\mathfrak{u}|}}{4} + (m-t-1)^{|\mathfrak{u}|-1} c_{|\mathfrak{u}|} + 1 \right),$$

where $c_{|\mathfrak{u}|} = \sum_{v=1}^{\infty} \frac{(v-1)^{|\mathfrak{u}|-1}}{2^{2v}}$.
(b) If $\alpha > 2$, we have

$$e_{N,s,\alpha,\gamma}^2 \leq \frac{2^{2t}}{N^2} \sum_{\emptyset \neq \mathfrak{u} \subseteq D_s} \gamma_{\mathfrak{u}} 2^{3|\mathfrak{u}|}$$
$$\times \left((m-t-1)^{|\mathfrak{u}|-1} \left(\frac{1}{2^\alpha - 4} \left(1 - \frac{1}{2^{(\alpha-2)(m-t-|\mathfrak{u}|)}} \right) \right.\right.$$
$$\left.\left. + \frac{1}{2^{(m-t-|\mathfrak{u}|)(\alpha-2)}} \widetilde{c}_{\alpha,|\mathfrak{u}|} \right) + 1 \right),$$

where $\widetilde{c}_{\alpha,|\mathfrak{u}|} = \sum_{v=1}^{\infty} \frac{(v+1)^{|\mathfrak{u}|-1}}{2^{\alpha v}}$.

From Corollary 1 (a), we see that the dyadic diaphony of the first N elements of a digital (t,s)-sequence over $\mathbb{Z}_2$ is of order $O(2^t (\log N)^{s/2}/N)$. More exactly we have

Corollary 2. *Let ω be a digital (t,s)-sequence over $\mathbb{Z}_2$ and let $F_{2,N}(\omega)$ denote the dyadic diaphony of the first N elements of ω. Then we have*

$$\limsup_{N\to\infty} \frac{N F_{2,N}(\omega)}{(\log N)^{s/2}} \leq 2^t \left(\frac{8}{\log 2} \right)^{s/2} (3^s - 1)^{-1/2}.$$

See [Pil07] for very precise results on the dyadic diaphony of digital $(0,s)$-sequences over $\mathbb{Z}_2$ for $s = 1, 2$.

For the proof of Theorem 1 we need the following lemma.

Lemma 1. *Let the non-negative integer N have binary expansion $N = N_0 + N_1 2 + \cdots + N_{m-1}2^{m-1}$. For any non-negative integer $n \le N-1$ let $n = n_0 + n_1 2 + \cdots + n_{m-1}2^{m-1}$ be the binary representation of n. Let $b_0, b_1, \ldots, b_{m-1}$ be arbitrary elements of $\mathbb{Z}_2$, not all zero. Then*

$$\sum_{n=0}^{N-1} (-1)^{b_0 n_0 + \cdots + b_{m-1} n_{m-1}} = (-1)^{b_{w+1}N_{w+1} + \cdots + b_{m-1}N_{m-1}} 2^{w+1} \left\| \frac{N}{2^{w+1}} \right\|,$$

where w is minimal such that $b_w = 1$.

Proof. See [Pil07, Lemma 4.1].

We give the proof of Theorem 1.

Proof. For a point $\boldsymbol{x}_n$ of ω and for $\emptyset \neq \mathfrak{u} \subseteq D_s$, we define $\boldsymbol{x}_n^{(\mathfrak{u})}$ as the projection of $\boldsymbol{x}_n$ onto the coordinates in $\mathfrak{u}$. Further, we define $\boldsymbol{\gamma}^{(\mathfrak{u})}$ as the collection of those γ_i with $i \in \mathfrak{u}$.

We then have, for the first N points of ω,

$$\begin{aligned}
(Ne_{N,s,\alpha,\boldsymbol{\gamma}})^2 &= \sum_{\boldsymbol{k} \in \mathbb{N}_0^s \setminus \{\boldsymbol{0}\}} r(\alpha, \boldsymbol{\gamma}, \boldsymbol{k}) \left| \sum_{n=0}^{N-1} \mathrm{wal}_{\boldsymbol{k}}(\boldsymbol{x}_n) \right|^2 \\
&= \sum_{\emptyset \neq \mathfrak{u} \subseteq D_s} \sum_{\boldsymbol{k} \in \mathbb{N}^{|\mathfrak{u}|}} r(\alpha, \boldsymbol{\gamma}^{(\mathfrak{u})}, \boldsymbol{k}) \left| \sum_{n=0}^{N-1} \mathrm{wal}_{\boldsymbol{k}}(\boldsymbol{x}_n^{(\mathfrak{u})}) \right|^2 \\
&= \sum_{\substack{\emptyset \neq \mathfrak{u} \subseteq D_s \\ \mathfrak{u} = \{w_1, \ldots, w_{|\mathfrak{u}|}\}}} \sum_{k_{w_1}=1}^{\infty} \cdots \sum_{k_{w_{|\mathfrak{u}|}}=1}^{\infty} \left(\prod_{i \in \mathfrak{u}} r(\alpha, \gamma_i, k_i) \right) \left| \sum_{n=0}^{N-1} \mathrm{wal}_{(k_{w_1}, \ldots, k_{w_{|\mathfrak{u}|}})}(\boldsymbol{x}_n^{(\mathfrak{u})}) \right|^2 \\
&= \sum_{\substack{\emptyset \neq \mathfrak{u} \subseteq D_s \\ \mathfrak{u} = \{w_1, \ldots, w_{|\mathfrak{u}|}\}}} \sum_{k_{w_1}=1}^{\infty} \cdots \sum_{k_{w_{|\mathfrak{u}|}}=1}^{\infty} \left(\prod_{i \in \mathfrak{u}} \frac{\gamma_i}{2^{\alpha\psi(k_i)}} \right) \left| \sum_{n=0}^{N-1} \mathrm{wal}_{(k_{w_1}, \ldots, k_{w_{|\mathfrak{u}|}})}(\boldsymbol{x}_n^{(\mathfrak{u})}) \right|^2 .
\end{aligned}$$

Let now $\emptyset \neq \mathfrak{u} = \{w_1, \ldots, w_{|\mathfrak{u}|}\} \subseteq D_s$ be fixed. We have to study

$$\Sigma(\mathfrak{u}) = \sum_{k_{w_1}=1}^{\infty} \cdots \sum_{k_{w_{|\mathfrak{u}|}}=1}^{\infty} \left(\prod_{i \in \mathfrak{u}} \frac{\gamma_i}{2^{\alpha\psi(k_i)}} \right) \left| \sum_{n=0}^{N-1} \mathrm{wal}_{(k_{w_1}, \ldots, k_{w_{|\mathfrak{u}|}})}(\boldsymbol{x}_n^{(\mathfrak{u})}) \right|^2 .$$

For the sake of simplicity we assume in the following $\mathfrak{u} = \{1, \ldots, \sigma\}$, $1 \le \sigma \le s$. The other cases are dealt with in a similar fashion. We have

$$\Sigma(\{1,\ldots,\sigma\}) := \sum_{k_1=1}^{\infty} \cdots \sum_{k_\sigma=1}^{\infty} \left(\prod_{j=1}^{\sigma} \frac{\gamma_j}{2^{\alpha\psi(k_j)}} \right) \left| \sum_{n=0}^{N-1} \mathrm{wal}_{(k_1,\ldots,k_\sigma)}(\boldsymbol{x}_n^{(\{1,\ldots,\sigma\})}) \right|^2 .$$

For $1 \le j \le \sigma$, let $2^{a_j} \le k_j < 2^{a_j+1}$, then $k_j = \kappa_0^{(j)} + \kappa_1^{(j)} 2 + \cdots + \kappa_{a_j}^{(j)} 2^{a_j}$ with $\kappa_v^{(j)} \in \{0,1\}$, $0 \le v < a_j$, and $\kappa_{a_j}^{(j)} = 1$.

Let $\mathbf{c}_i^{(j)}$ be the i-th row vector of the generator matrix C_j, $1 \le j \le \sigma$. Since the i-th digit $x_n^{(j)}(i)$ of $x_n^{(j)}$ is given by $\left\langle \mathbf{c}_i^{(j)}, \mathbf{n} \right\rangle$, we have

$$\begin{aligned} \sum_{n=0}^{N-1} \mathrm{wal}_{(k_1,\ldots,k_\sigma)}(\boldsymbol{x}_n^{(\{1,\ldots,\sigma\})}) &= \sum_{n=0}^{N-1} (-1)^{\sum_{j=1}^{\sigma}\left(\kappa_0^{(j)}\left\langle \mathbf{c}_1^{(j)},\mathbf{n}\right\rangle + \cdots + \kappa_{a_j}^{(j)}\left\langle \mathbf{c}_{a_j+1}^{(j)},\mathbf{n}\right\rangle\right)} \\ &= \sum_{n=0}^{N-1} (-1)^{\left\langle \sum_{j=1}^{\sigma}(\kappa_0^{(j)}\mathbf{c}_1^{(j)} + \cdots + \kappa_{a_j}^{(j)}\mathbf{c}_{a_j+1}^{(j)}),\mathbf{n}\right\rangle} . \end{aligned}$$

Let $C_j = (c_{v,w}^{(j)})_{v,w\ge 1}$. Define

$$u(k_1,\ldots,k_\sigma) := \min\left\{ p \ge 1 : \sum_{j=1}^{\sigma} (\kappa_0^{(j)} c_{1,p}^{(j)} + \cdots + \kappa_{a_j}^{(j)} c_{a_j+1,p}^{(j)}) = 1 \right\} .$$

Since $C_1, \ldots, C_s$ generate a digital (t,s)-sequence over $\mathbb{Z}_2$, it is easy to verify that $u(k_1,\ldots,k_\sigma) \le \sum_{j=1}^{\sigma} a_j + \sigma + t =: R_\sigma + \sigma + t$. Indeed, let

$$A := \begin{pmatrix} c_{1,1}^{(1)} & \cdots & c_{a_1+1,1}^{(1)} & \cdots\cdots & c_{1,1}^{(\sigma)} & \cdots & c_{a_\sigma+1,1}^{(\sigma)} \\ c_{1,2}^{(1)} & \cdots & c_{a_1+1,2}^{(1)} & \cdots\cdots & c_{1,2}^{(\sigma)} & \cdots & c_{a_\sigma+1,2}^{(\sigma)} \\ \vdots & & \vdots & & \vdots & & \vdots \\ c_{1,R_\sigma+\sigma+t}^{(1)} & \cdots & c_{a_1+1,R_\sigma+\sigma+t}^{(1)} & \cdots\cdots & c_{1,R_\sigma+\sigma+t}^{(\sigma)} & \cdots & c_{a_\sigma+1,R_\sigma+\sigma+t}^{(\sigma)} \end{pmatrix} .$$

Note that $A = A(a_1,\ldots,a_\sigma)$ is an $(R_\sigma + \sigma + t) \times (R_\sigma + \sigma)$ matrix. Since $C_1, \ldots, C_s$ generate a digital (t,s)-sequence over $\mathbb{Z}_2$, it follows that $A(a_1,\ldots,a_\sigma)$ has rank $R_\sigma + \sigma$. If, however, $u(k_1,\ldots,k_\sigma) > R_\sigma + \sigma + t$, we would have

$$\begin{aligned} A &\cdot (\kappa_0^{(1)},\ldots,\kappa_{a_1-1}^{(1)},1,\kappa_0^{(2)},\ldots,\kappa_{a_2-1}^{(2)},1,\ldots,\ldots,\kappa_0^{(\sigma)},\ldots,\kappa_{a_\sigma-1}^{(\sigma)},1)^\top \\ &= (0,\ldots,0)^\top , \end{aligned} \tag{1}$$

which would lead to a contradiction since the matrix $A(a_1,\ldots,a_\sigma)$ has full rank and is multiplied by a non-zero vector in (1).

Therefore we obtain, by applying Lemma 1,

$$\begin{aligned}
&\Sigma(\{1,\ldots,\sigma\})\\
&= \sum_{k_1=1}^{\infty}\cdots\sum_{k_\sigma=1}^{\infty}\left(\prod_{j=1}^{\sigma}\frac{\gamma_j}{2^{\alpha\psi(k_j)}}\right)2^{2u(k_1,\ldots,k_s)}\left\|\frac{N}{2^{u(k_1,\ldots,k_\sigma)}}\right\|^2\\
&= \sum_{a_1=0}^{\infty}\cdots\sum_{a_\sigma=0}^{\infty}\left(\prod_{j=1}^{\sigma}\frac{\gamma_j}{2^{\alpha a_j}}\right)\sum_{k_1=2^{a_1}}^{2^{a_1+1}-1}\cdots\sum_{k_\sigma=2^{a_\sigma}}^{2^{a_\sigma+1}-1}2^{2u(k_1,\ldots,k_\sigma)}\left\|\frac{N}{2^{u(k_1,\ldots,k_\sigma)}}\right\|^2\\
&= \sum_{a_1=0}^{\infty}\cdots\sum_{a_\sigma=0}^{\infty}\frac{\gamma_{\{1,\ldots,\sigma\}}}{2^{\alpha R_\sigma}}\sum_{u=1}^{R_\sigma+\sigma+t}2^{2u}\left\|\frac{N}{2^u}\right\|^2\underbrace{\sum_{k_1=2^{a_1}}^{2^{a_1+1}-1}\cdots\sum_{k_\sigma=2^{a_\sigma}}^{2^{a_\sigma+1}-1}}_{u(k_1,\ldots,k_\sigma)=u}1.
\end{aligned}$$

We need to estimate the sum

$$\underbrace{\sum_{k_1=2^{a_1}}^{2^{a_1+1}-1}\cdots\sum_{k_\sigma=2^{a_\sigma}}^{2^{a_\sigma+1}-1}}_{u(k_1,\ldots,k_\sigma)=u}1$$

for $1\leq u\leq R_\sigma+\sigma+t$. This is the number of $\kappa_0^{(1)},\ldots,\kappa_{a_1-1}^{(1)},\kappa_0^{(2)},\ldots,\kappa_{a_2-1}^{(2)},\ldots,\kappa_0^{(\sigma)},\ldots,\kappa_{a_\sigma-1}^{(\sigma)}\in\mathbb{Z}_2$ such that

$$\begin{aligned}
A&\cdot(\kappa_0^{(1)},\ldots,\kappa_{a_1-1}^{(1)},1,\kappa_0^{(2)},\ldots,\kappa_{a_2-1}^{(2)},1,\ldots,\ldots,\kappa_0^{(\sigma)},\ldots,\kappa_{a_\sigma-1}^{(\sigma)},1)^\top\\
&=(0,\ldots,0,1,x_{u+1},\ldots,x_{R_\sigma+\sigma+t})^\top,
\end{aligned}\tag{2}$$

where A is defined as above, for arbitrary $x_{u+1},\ldots,x_{R_\sigma+\sigma+t}\in\mathbb{Z}_2$. Let us rewrite system (2) as

$$B\cdot\begin{pmatrix}\kappa_0^{(1)}\\ \vdots\\ \kappa_{a_1-1}^{(1)}\\ \kappa_0^{(2)}\\ \vdots\\ \kappa_{a_2-1}^{(2)}\\ \vdots\\ \vdots\\ \kappa_0^{(\sigma)}\\ \vdots\\ \kappa_{a_\sigma-1}^{(\sigma)}\end{pmatrix}=\begin{pmatrix}0\\ \vdots\\ 0\\ 1\\ x_{u+1}\\ \vdots\\ x_{R_\sigma+\sigma+t}\end{pmatrix}+\begin{pmatrix}c_{a_1+1,1}^{(1)}\\ \vdots\\ \vdots\\ \vdots\\ \vdots\\ c_{a_1+1,R_\sigma+\sigma+t}^{(1)}\end{pmatrix}+\cdots+\begin{pmatrix}c_{a_\sigma+1,1}^{(\sigma)}\\ \vdots\\ \vdots\\ \vdots\\ \vdots\\ c_{a_\sigma+1,R_\sigma+\sigma+t}^{(\sigma)}\end{pmatrix},\tag{3}$$

where

$$B := \begin{pmatrix} c_{1,1}^{(1)} & \cdots & c_{a_1,1}^{(1)} & \cdots\cdots & c_{1,1}^{(\sigma)} & \cdots & c_{a_\sigma,1}^{(\sigma)} \\ c_{1,2}^{(1)} & \cdots & c_{a_1,2}^{(1)} & \cdots\cdots & c_{1,2}^{(\sigma)} & \cdots & c_{a_\sigma,2}^{(\sigma)} \\ \vdots & & \vdots & & \vdots & & \vdots \\ c_{1,R_\sigma+\sigma+t}^{(1)} & \cdots & c_{a_1,R_\sigma+\sigma+t}^{(1)} & \cdots\cdots & c_{1,R_\sigma+\sigma+t}^{(\sigma)} & \cdots & c_{a_\sigma,R_\sigma+\sigma+t}^{(\sigma)} \end{pmatrix}.$$

Obviously, the matrix B has rank R_σ. Let now $1 \le u \le R_\sigma + \sigma + t$ be fixed. For a fixed choice of $x_{u+1}, \ldots, x_{R_\sigma+\sigma+t}$, it is clear that we have at most one solution of system (3). Therefore we have

$$\underbrace{\sum_{k_1=2^{a_1}}^{2^{a_1+1}-1} \cdots \sum_{k_\sigma=2^{a_\sigma}}^{2^{a_\sigma+1}-1}}_{u(k_1,\ldots,k_\sigma)=u} 1 \le 2^{R_\sigma+\sigma+t-u}.$$

Now we have

$$\begin{aligned} \Sigma(\{1,\ldots,\sigma\}) &\le \gamma_{\{1,\ldots,\sigma\}} \sum_{a_1=0}^{\infty} \cdots \sum_{a_\sigma=0}^{\infty} \frac{1}{2^{\alpha R_\sigma}} \sum_{u=1}^{R_\sigma+\sigma+t} 2^{2u} \left\| \frac{N}{2^u} \right\|^2 2^{R_\sigma+\sigma+t-u} \\ &= \gamma_{\{1,\ldots,\sigma\}} \sum_{a_1=0}^{\infty} \cdots \sum_{a_\sigma=0}^{\infty} \frac{1}{2^{(\alpha-1)R_\sigma}} 2^{\sigma+t} \sum_{u=1}^{R_\sigma+\sigma+t} 2^u \left\| \frac{N}{2^u} \right\|^2 \\ &= \gamma_{\{1,\ldots,\sigma\}} 2^{\sigma+t} \sum_{u=1}^{\infty} 2^u \left\| \frac{N}{2^u} \right\|^2 \sum_{\substack{a_1,\ldots,a_\sigma=0 \\ R_\sigma \ge \max\{u-t-\sigma,0\}}}^{\infty} \frac{1}{2^{(\alpha-1)R_\sigma}} \\ &= \gamma_{\{1,\ldots,\sigma\}} 2^{\sigma+t} \left(\sum_{u=1}^{t+\sigma} 2^u \left\| \frac{N}{2^u} \right\|^2 \sum_{a_1,\ldots,a_\sigma=0}^{\infty} \frac{1}{2^{(\alpha-1)R_\sigma}} \right. \\ &\qquad \left. + \sum_{u=t+\sigma+1}^{\infty} 2^u \left\| \frac{N}{2^u} \right\|^2 \sum_{\substack{a_1,\ldots,a_\sigma=0 \\ R_\sigma \ge u-t-\sigma}}^{\infty} \frac{1}{2^{(\alpha-1)R_\sigma}} \right) \\ &=: \gamma_{\{1,\ldots,\sigma\}} 2^{\sigma+t} (\Sigma_1 + \Sigma_2). \end{aligned}$$

For Σ_1 we have

$$\begin{aligned} \Sigma_1 &= \sum_{u=1}^{t+\sigma} 2^u \left\| \frac{N}{2^u} \right\|^2 \sum_{a_1,\ldots,a_\sigma=0}^{\infty} \frac{1}{2^{(a_1+\cdots+a_\sigma)(\alpha-1)}} \\ &\le \frac{1}{4} \left(\sum_{a=0}^{\infty} \frac{1}{2^{a(\alpha-1)}} \right)^{\sigma} \sum_{u=1}^{t+\sigma} 2^u \le 2^{t+\sigma-1} \left(\frac{2^{\alpha-1}}{2^{\alpha-1}-1} \right)^{\sigma}. \end{aligned}$$

For Σ_2 we have

$$\Sigma_2 = \sum_{u=t+\sigma+1}^{\infty} 2^u \left\| \frac{N}{2^u} \right\|^2 \sum_{\substack{a_1,\dots,a_\sigma=0 \\ R_\sigma \geq u-t-\sigma}}^{\infty} \frac{1}{2^{R_\sigma(\alpha-1)}}$$
$$= \sum_{u=t+\sigma+1}^{\infty} 2^u \left\| \frac{N}{2^u} \right\|^2 \sum_{w=u-t-\sigma}^{\infty} \frac{1}{2^{w(\alpha-1)}} \binom{w+\sigma-1}{\sigma-1}.$$

We now use [DP05c, Lemma 6] to obtain

$$\Sigma_2 \leq \sum_{u=t+\sigma+1}^{\infty} 2^u \left\| \frac{N}{2^u} \right\|^2 \frac{1}{2^{(u-t-\sigma)(\alpha-1)}} \binom{u-t-1}{\sigma-1} \left(\frac{2^{\alpha-1}}{2^{\alpha-1}-1} \right)^{\sigma}$$
$$= 2^{(t+\sigma)(\alpha-1)} \left(\frac{2^{\alpha-1}}{2^{\alpha-1}-1} \right)^{\sigma} \sum_{u=t+\sigma+1}^{\infty} \left\| \frac{N}{2^u} \right\|^2 \binom{u-t-1}{\sigma-1} \frac{2^u}{2^{u(\alpha-1)}}$$
$$\leq 2^{(t+\sigma)(\alpha-1)} \left(\frac{2^{\alpha-1}}{2^{\alpha-1}-1} \right)^{\sigma} \sum_{u=t+\sigma+1}^{\infty} \left\| \frac{N}{2^u} \right\|^2 \frac{(u-t-1)^{\sigma-1}}{2^{u(\alpha-2)}}.$$

This yields

$$\Sigma(\{1,\dots,\sigma\}) \leq \gamma_{\{1,\dots,\sigma\}} 2^{2t+2\sigma}$$
$$\times \left(\frac{2^{\alpha-1}}{2^{\alpha-1}-1} \right)^{\sigma} \left(1 + 2^{(t+\sigma)(\alpha-2)} \sum_{u=t+\sigma+1}^{\infty} \left\| \frac{N}{2^u} \right\|^2 \frac{(u-t-1)^{\sigma-1}}{2^{u(\alpha-2)}} \right).$$

The result follows. □

We give the proof of Corollary 1.

Proof. Suppose first that $\alpha > 2$ and $2^{m-1} < N \leq 2^m$ with $m > t+s$, then

$$2^{(t+|\mathfrak{u}|)(\alpha-2)} \sum_{v=t+|\mathfrak{u}|+1}^{\infty} \left\| \frac{N}{2^v} \right\|^2 \frac{(v-t-1)^{|\mathfrak{u}|-1}}{2^{v(\alpha-2)}}$$
$$= \sum_{v=t+|\mathfrak{u}|+1}^{m} \left\| \frac{N}{2^v} \right\|^2 \frac{(v-t-1)^{|\mathfrak{u}|-1}}{2^{(\alpha-2)(v-t-|\mathfrak{u}|)}} + \sum_{v=m+1}^{\infty} \left\| \frac{N}{2^v} \right\|^2 \frac{(v-t-1)^{|\mathfrak{u}|-1}}{2^{(\alpha-2)(v-t-|\mathfrak{u}|)}}$$
$$=: \Sigma_3 + \Sigma_4.$$

Now,

$$\Sigma_3 \leq \frac{1}{4} \sum_{v=t+|\mathfrak{u}|+1}^{m} \frac{(v-t-1)^{|\mathfrak{u}|-1}}{2^{(\alpha-2)(v-t-|\mathfrak{u}|)}} \leq \frac{(m-t-1)^{|\mathfrak{u}|-1}}{4} \sum_{v=1}^{m-t-|\mathfrak{u}|} \frac{1}{2^{(\alpha-2)v}}$$
$$= (m-t-1)^{|\mathfrak{u}|-1} \frac{1}{2^{\alpha}-4} \left(1 - \frac{1}{2^{(\alpha-2)(m-t-|\mathfrak{u}|)}} \right)$$

and

$$\begin{aligned}
\Sigma_4 &= 2^{(t+|u|)(\alpha-2)} \sum_{v=m+1}^{\infty} \left\| \frac{N}{2^v} \right\|^2 \frac{(v-t-1)^{|\mathfrak{u}|-1}}{2^{v(\alpha-2)}} \\
&= 2^{(t+|u|)(\alpha-2)} \sum_{v=m+1}^{\infty} \left(\frac{N}{2^v} \right)^2 \frac{(v-t-1)^{|\mathfrak{u}|-1}}{2^{v(\alpha-2)}} \\
&= 2^{(t+|u|)(\alpha-2)} \left(\frac{N}{2^m} \right)^2 \sum_{v=1}^{\infty} \frac{1}{2^{2v}} \frac{(v+m-t-1)^{|\mathfrak{u}|-1}}{2^{(v+m)(\alpha-2)}} \\
&\leq \frac{1}{2^{(m-t-|\mathfrak{u}|)(\alpha-2)}} \sum_{v=1}^{\infty} \frac{(v+m-t-1)^{|\mathfrak{u}|-1}}{2^{\alpha v}} \\
&= \frac{1}{2^{(m-t-|\mathfrak{u}|)(\alpha-2)}} \sum_{v=1}^{\infty} \frac{1}{2^{\alpha v}} \sum_{k=0}^{|\mathfrak{u}|-1} \binom{|\mathfrak{u}|-1}{k} (m-t-1)^k v^{|\mathfrak{u}|-1-k} \\
&\leq \frac{1}{2^{(m-t-|\mathfrak{u}|)(\alpha-2)}} (m-t-1)^{|\mathfrak{u}|-1} \sum_{v=1}^{\infty} \frac{1}{2^{\alpha v}} \sum_{k=0}^{|\mathfrak{u}|-1} \binom{|\mathfrak{u}|-1}{k} v^{|\mathfrak{u}|-1-k} \\
&= \frac{1}{2^{(m-t-|\mathfrak{u}|)(\alpha-2)}} (m-t-1)^{|\mathfrak{u}|-1} \sum_{v=1}^{\infty} \frac{(v+1)^{|\mathfrak{u}|-1}}{2^{\alpha v}}.
\end{aligned}$$

The result for $\alpha > 2$ follows from Theorem 1.

If we choose $\alpha = 2$ in Theorem 1, we obtain

$$e_{N,s,2,\gamma}^2 \leq \frac{2^{2t}}{N^2} \sum_{\emptyset \neq \mathfrak{u} \subseteq D_s} \gamma_{\mathfrak{u}} 2^{3|\mathfrak{u}|} \left(1 + \sum_{v=t+|\mathfrak{u}|+1}^{\infty} \left\| \frac{N}{2^v} \right\|^2 (v-t-1)^{|\mathfrak{u}|-1} \right).$$

If we again assume $2^{m-1} < N \leq 2^m$ with $m > t+s$, then it can be shown, in a similar way as in the case $\alpha > 2$, that

$$\sum_{v=t+|\mathfrak{u}|+1}^{\infty} \left\| \frac{N}{2^v} \right\|^2 (v-t-1)^{|\mathfrak{u}|-1} \leq \frac{(m-t-1)^{|\mathfrak{u}|}}{4} + (m-t-1)^{|\mathfrak{u}|-1} \sum_{v=1}^{\infty} \frac{(v+1)^{|\mathfrak{u}|-1}}{2^{2v}}.$$

The result follows. □

References

[Chr55] H.E. Chrestenson. A class of generalized Walsh functions. *Pacific J. Math.*, 5:17–31, 1955.

[DP05a] J. Dick and F. Pillichshammer. Diaphony, discrepancy, spectral test and worst-case error. *Math. Comput. Simulation*, 70:159–171, 2005.

[DP05b] J. Dick and F. Pillichshammer. Multivariate integration in weighted Hilbert spaces based on Walsh functions and weighted Sobolev spaces. *J. Complexity*, 21(2):149–195, 2005.

[DP05c] J. Dick and F. Pillichshammer. On the mean square weighted $\mathcal{L}_2$ discrepancy of randomized digital (t, m, s)-nets over $\mathbb{Z}_2$. *Acta Arith.*, 117:371–403, 2005.

[DT97] M. Drmota and R. F. Tichy. *Sequences, discrepancies and applications.* Lecture Notes in Mathematics. 1651. Berlin: Springer., 1997.

[Gro06] V. Grozdanov. The weighted b-adic diaphony. *J. Complexity*, 22:490–513, 2006.

[HL97] P. Hellekalek and H. Leeb. Dyadic diaphony. *Acta Arith.*, 80:187–196, 1997.

[KN74] L. Kuipers and H. Niederreiter. *Uniform distribution of sequences.* Pure and Applied Mathematics. New York etc.: John Wiley & Sons., 1974.

[Kri06] P. Kritzer. Improved upper bounds on the star discrepancy of (t, m, s)-nets and (t, s)-sequences. *J. Complexity*, 22:336–347, 2006.

[Lar98] G. Larcher. Digital point sets: Analysis and application. Hellekalek, Peter (ed.) et al., Random and quasi-random point sets. New York, NY: Springer. Lect. Notes Stat., Springer-Verlag. 138, 167–222, 1998.

[LNS96] G. Larcher, H. Niederreiter, and W. Ch. Schmid. Digital nets and sequences constructed over finite rings and their application to quasi-Monte Carlo integration. *Monatsh. Math.*, 121:231–253, 1996.

[Nie82] K. Niederdrenk. *Die endliche Fourier- und Walsh-Transformation mit einer Einführung in die Bildverarbeitung.* Braunschweig - Wiesbaden: Vieweg, 1982.

[Nie87] H. Niederreiter. Point sets and sequences with small discrepancy. *Monatsh. Math.*, 104:273–337, 1987.

[Nie92] H. Niederreiter. *Random number generation and quasi-Monte Carlo methods.* CBMS-NSF Regional Conference Series in Applied Mathematics. 63. Philadelphia, SIAM, Society for Industrial and Applied Mathematics., 1992.

[Pil07] F. Pillichshammer. Dyadic diaphony of digital sequences. J. Théor. Nombres Bordx. (to appear), 2007.

[Pir95] G. Pirsic. Schnell konvergierende Walshreihen über Gruppen. Diplomarbeit, University of Salzburg, 1995.

[SP05] O. Strauch and Š. Porubský. *Distribution of sequences. A sampler.* Schriftenreihe der Slowakischen Akademie der Wissenschaften 1. Bern: Peter Lang, 2005.

[Wal23] J. L. Walsh. A closed set of normal orthogonal functions. *American J. Math.*, 45:5–24, 1923.

[Zin76] P. Zinterhof. Über einige Abschätzungen bei der Approximation von Funktionen mit Gleichverteilungsmethoden. *Sitzungsber. Österr. Akad. Wiss. Math.-Natur. Kl. II*, 185:121–132, 1976.

A New Criterion for Finiteness of Weight Estimator Variance in Statistical Simulation

Ilya Medvedev[1] and Gennadii Mikhailov[2]

[1] Institute of Computational Mathematics and Mathematical Geophysics, pr. Akademika Lavrentjeva, 6, Novosibirsk, 630090, Russia
medvedev@gorodok.net and medvedev79@ngs.ru

[2] Institute of Computational Mathematics and Mathematical Geophysics, pr. Akademika Lavrentjeva, 6, Novosibirsk, 630090, Russia
gam@sscc.ru

Summary. It has been found recently that an increase in phase space dimension by including simulated auxiliary random variables in the number of phase coordinates can be effective for the construction of weight modifications. In this paper the effectiveness of "value" and partial "value" modelling is considered. These types of modelling are related to the construction of simulated distribution for some auxiliary random variable by multiplying the initial density by the "value" function which is usually corresponds to the solution of adjoint integral equation of the second kind. It is proved that the weight estimator variance in case of the partial value modelling is finite. On the basis of this fact a new criterion based on the use of majorant adjoint equation was proposed for finiteness of the weight estimator variance. Using this criterion the classical "exponential transformation" method is studied for the free path simulation in one and three dimensional modifications.

1 Introduction

Let us consider a terminating homogeneous Markov chain $x_0, x_1, \ldots, x_N$, defined by the distribution density $f(x)$ of the initial state x_0 and the substochastic generalized transition density $k(x_0, x)$ which satisfies $\int k(x', x)dx = q(x') \leq 1 - \delta < 1$ where δ is a real number and $0 < \delta < 1$. Here and further $x \in X$, X is m-dimensional Euclidean space and N is the number of the state at which the trajectory terminates (i.e., the termination moment). The total distribution density of the phase states of the chain $\varphi(x) = \sum\limits_{n=0}^{\infty} \varphi_n(x)$ represents the Neumann series for the following integral equation of the second kind $\varphi(x) = \int\limits_X k(x', x)\varphi(x')\, dx' + f(x)$, where functions $f(x), k(x', x)$ and $\varphi(x')$ belong to space $N_1(X)$ of generalized densities of bounded variation measures [EM82]. For a given function $h \in C_b(X)$, where $C_b(X)$ is a set of all nonnegative continuous functions, we also consider the conjugate equation

$$\varphi^*(x) = \int_X k(x', x)\varphi(x')\, dx' + h, \quad \text{or} \quad \varphi^* = K^*\varphi^* + h. \tag{1}$$

We suppose that $K^* \in [C_b(X) \to C_b(X)]$.

Under general unbiasedness conditions [EM82] let us introduce transition density $p(x', x)$, initial density $\pi(x)$ and the auxiliary weights by the formulas $Q_0(x_0) = f(x_0)/\pi(x_0)$, $Q_n = Q_{n-1}k(x_{n-1}, x_n)/p(x_{n-1}, x_n)$. It is well-known, that for the weighted "collision estimator"

$$\xi = \frac{f(x_0)}{\pi(x_0)}\xi_{x_0}, \qquad \xi_x = h(x) + \sum_{n=1}^{N} Q_n h(x_n), \tag{2}$$

we obtain $I_h = (\varphi, h) = (f, \varphi^*) = \mathrm{E}\xi$ [EM82]. In the theory of weighted Monte Carlo methods, the function $\varphi^*(\cdot)$ is conventionally called the "value function" in connection with its probability representation: $\varphi^*(x) = \mathrm{E}\xi_x$, moreover, for the estimator ξ_x the variable $\mathrm{E}\xi_x^2$ is defined by the Neumann series for the following integral equation [Mik92]

$$g = K_p^* g + h(2\varphi^* - h), \tag{3}$$

where symbol K_p^* denotes the operator with the kernel $k^2(x, x')/p(x, x')$. It is well-known [EM82] that if the spectral radius $\rho(K_p) < 1$ then $\mathrm{D}\xi_x < +\infty$. Moreover if $h(x) \geq 0$ and

$$p(x', x) = \frac{k(x', x)\varphi^*(x)}{[K^*\varphi^*](x')}, \quad \pi(x) = \frac{f(x)\varphi^*(x)}{(f, \varphi^*)}, \tag{4}$$

then $\mathrm{D}\xi = 0$ [EM82],[Mik92].

As usual, the transition $x' \to x$ is realized by choosing a set of the values of the auxiliary random variables, for example a collusion number type, scattering angles and a free path in transport simulation process.

Let us represent $\mathbf{t} = (t_1, t_2) \in T = T_1 \times T_2$ as a set of two auxiliary random variables (vector variables, in general) which are simulated for the transition $x \to x'$ in Markov chain. In new phase space $T \times X = \{(\mathbf{t}, x)\}$ the substochastic kernel has the form [Mik03]

$$\mathbf{k}((\mathbf{t}, x), (\mathbf{t}', x')) = \delta(x' - x'(x, \mathbf{t}'))k_1(x, t_1')k_2((x, t_1'), t_2'),$$

where $x'(x, \mathbf{t}')$ is the function which defines new euclidean coordinates via x and values of auxiliary variables $\mathbf{t}'$.

Due to the definition of collision estimator we have

$$\mathrm{E}\xi^2 = \int_X \frac{f^2(x)}{\pi(x)}\mathrm{E}\xi_x^2 dx. \tag{5}$$

Therefore it is worth to consider the problem of uniform minimization of $\mathrm{E}\xi_x^2$ for $\forall x \in \mathrm{X}$. In [Mik03] it was shown that "value" modelling of all the elementary auxiliary transitions in accordance with (4) gives an estimator with zero variance. In practice such global optimization of modelling is rather difficult; then it is important to consider the possibility of variance decrease by means of optimal choice of the distribution density of a part of the auxiliary random variables, for example, t_1'.

Let us consider a Markov chain with the substochastic transition density

$$\mathbf{p}((\mathbf{t},x),(\mathbf{t}',x')) = \delta(x'-x'(x,\mathbf{t}'))p_1(x,t_1')p_2((x,t_1'),t_2').$$

In addition we assume that

$$\int\limits_{T_1} k_1(x,t_1')dt_1' \equiv \int\limits_{T_1} p_1(x,t_1')dt_1' \equiv 1, \quad \int\limits_{T_2} \frac{k_2^2((x,t_1'),t_2')}{p_2((x,t_1'),t_2')}\, dt_2' \le q < 1. \tag{6}$$

Theorem 1. [MM04] *The equation*

$$g(x) = h(x)[2\varphi^*(x) - h(x)] + \\ + \left\{ \int\limits_{T_1} k_1(x,t_1') \left[\int\limits_{T_2} \frac{k_2^2((x,t_1'),t_2')}{p_2((x,t_1'),t_2')} g(x')dt_2' \right]^{1/2} dt_1' \right\}^2. \tag{7}$$

has unique solution and the density

$$p_1(x,t_1') = \frac{k_1(x,t_1') \left[\int\limits_{T_2} \frac{k_2^2((x,t_1'),t_2')}{p_2((x,t_1'),t_2')} g(x')\, dt_2' \right]^{1/2}}{\int\limits_{T_1} k_1(x,t_1') \left[\int\limits_{T_2} \frac{k_2^2((x,t_1'),t_2')}{p_2((x,t_1'),t_2')} g(x')\, dt_2' \right]^{1/2} dt_1'} \tag{8}$$

gives the minimum value of $\mathrm{E}\xi_x^2 \in C_b(X)$.

In [MM04] for one-velocity problem in transport theory it was shown how Theorem 1 can be used for approximated estimation of optimal parameter in "exponential transformation" method. However Theorem 1 can be practically useless for solving more complex problems. Therefore it is worth investigating the possibility of using more simple modifications of simulation t_1' which are connected with the use of approximate solutions of conjugate integral equation. Such modifications will be called "partial value" modelling. It turned out that in some cases the use of partial value modelling can increase the estimator variance as compared with direct one. Moreover it was discovered that under value modelling of free path in the important applied problem of estimating the probability of particles take-off from the half-space $-\infty < z \le H$ standard criterion $\rho(K_p^*) < 1$ of variance finiteness is not valid any more [MM03].

The idea to define conditional transition distribution densities proportional to the integral transformation kernel is not new. For example in [Dim91] general conditions for $k(x', x)$ and $f(x)$ are derived under which the choice $p(x', x) = C \times k(x', x)$ can be considered as optimal (with a minimal variance). Here we consider different problem of defining conditions for $p(x', x)$ by formulas (4) that guarantee variance finiteness of weight estimator.

In this paper it is proved that the weight estimator variance under partial value modelling of the part of auxiliary variables is finite. On the basis of this fact we propose a new criterion of estimator variance finiteness without investigation of spectral radius.

2 New Criterion of Estimator Variance Finiteness

Let the random variable t'_2 be simulated according to given distribution and random variable t'_1 be simulated with the use of auxiliary value function [Mik03], i.e., corresponding conditional transition distribution densities are

$$p_2((x, t'_1), t'_2) \equiv k_2((x, t'_1), t'_2), \tag{9}$$

$$p_1(x, t'_1) = \frac{k_1(x, t'_1)\varphi_1^*(x, t'_1)}{[K^*\varphi^*](x)} = \frac{k_1(x, t'_1)\varphi_1^*(x, t'_1)}{\varphi^*(x) - h(x)}, \tag{10}$$

where auxiliary value function has the form:

$$\begin{aligned} \varphi_1^*(x, t'_1) &= \int\limits_{T_2}\int\limits_{X} \delta(x' - x'(x, \mathbf{t}'))k_2((x, t'_1), t'_2)\varphi^*(x')dx'dt'_2 \\ &= \int\limits_{T_2} k_2((x, t'_1), t'_2)\varphi^*(x'(x, \mathbf{t}'))dt'_2. \end{aligned} \tag{11}$$

In addition we suppose that $\forall x \in X$

$$\int\limits_{T_1} k_1(x, t'_1)dt'_1 = 1 - \alpha(x) \leq 1.$$

Theorem 2. *The variance of the collision estimator ξ_x under exact partial value modelling of t'_1 (i.e. according to (9), (10)) is finite.*

Proof. Since $h, \varphi^* \in C_b(X)$ the following statement is true [Mik92]:

$$\mathrm{E}\xi_x^2 = \sum_{n=0}^{\infty} K_p^{*n}\Big(h[2\varphi^* - h]\Big)(x). \tag{12}$$

Let us show that the series (12) is converged for all $h \in C_b(X)$.

Taking into account (9)–(11) and condition on $k_1(x, t'_1)$ it is easy to verify the following equality by forward substitution:

$$[K_p^*\varphi^*](x) = \int\limits_{T_1} \frac{k_1^2(x,t_1')}{p_1(x,t_1'))}\left[\int\limits_{T_2}\int\limits_{X} k_2((x,t_1'),t_2')\delta(x'-x'(x,\mathbf{t}'))\varphi^*(x')dx'dt_2'\right]dt_1'$$

$$= \int\limits_{T_1} \frac{k_1(x,t_1')[K^*\varphi^*](x)}{\varphi_1^*(x,t_1')}\left[\int\limits_{T_2} k_2((x,t_1'),t_2')\varphi^*(x'(x,\mathbf{t}')dt_2'\right]dt_1'$$

$$= [K^*\varphi^*](x)\int\limits_{T_1} k_1(x,t_1')dt_1' = [K^*\varphi^*](x)(1-\alpha(x)) = (\varphi^*(x)-h(x))(1-\alpha(x)).$$

Note that the last equality can be rewritten as $\varphi^* = K_p^*\varphi^* + \alpha(\varphi^* - h) + h$. Substituting the function φ^* for its equivalent function $K_p^*\varphi^* + \alpha(\varphi^* - h) + h$ under the operator K_p^* in the last equality we obtain

$$\varphi^* = K_p^{*2}\varphi^* + K_p^*\Big(\alpha(\varphi^* - h) + h\Big) + \alpha(\varphi^* - h) + h.$$

Let us make the same substitution in the last equality under the operator K_p^{*2} and so on. Since all the functions are nonnegative as a result we have

$$\varphi^* = \lim_{n\to\infty}\left[K_p^{*n}\varphi^* + \sum_{k=0}^{n-1} K_p^{*k}\Big(\alpha(\varphi^* - h) + h\Big)\right].$$

Therefore $\sum\limits_k K_p^{*k}\Big(\alpha(\varphi^* - h) + h\Big)$ is convergent series since all the functions are nonnegative and thus the series $\sum\limits_k K_p^{*k}h$ is also convergent. Since $h(2\varphi^* - h) \le h2\varphi^* \le 2hC$ then the series (12) is converged which required to be proved.

Let us assume that conditional transition distribution density for t_1 has the form

$$p_1^*(x,t_1) = \frac{k_1(x,t_1')\varphi_1^*(x,t_1')}{\varphi^*(x)}. \tag{13}$$

In this case the following statement is true.

Theorem 3. *If*

$$h(x)/\alpha(x) \le C < \infty, \forall x \in \text{supp } h \subseteq \text{supp } \alpha, \tag{14}$$

then the variance of the collision estimator ξ_x under partial value modelling of t_1' (i.e. according to (9), (13)*) is finite.*

Proof. Taking into account (9), (11), (13) and condition on $k_1(x,t_1')$, it is easy to verify the following equality by forward substitution:

$$[K_p^*\varphi^*](x) = \varphi^*(x)(1-\alpha(x)) \qquad \varphi^*(x) = [K_p^*\varphi^*](x) + \varphi^*(x)\alpha(x).$$

As above in the proof of Theorem 2 it easy to show that the Neumann series $\sum_k K_p^{*k}\varphi^*\alpha$ for the last equation is convergent since all the functions are nonnegative. Due to the condition (14) the series $\sum_k K_p^{*k}\varphi^* h$ is also convergent. As $h(2\varphi^* - h) \leq h 2\varphi^*$ then the series (12) is converged.

From the proof of Theorem 3 it follows replaced by more simple one.

Theorem 4. *If $\forall x \in X$ we have $1 - \alpha(x) \leq 1 - \epsilon < 1$, then the variance of the collision estimator ξ_x under partial value modelling of t_1' (i.e. according to* (9), (13)*) is finite.*

Let us assume that

$$\int_{T_2} k_2((x, t_1'), t_2') dt_2' = 1 - \beta(x, t_1') \leq 1.$$

In the case of exact partial value modelling of t_2' transition densities have the following form

$$p_1(x, t_1') \equiv k_1(x, t_1'), \tag{15}$$

$$p_2((x, t_1'), t_2') = \frac{k_2((x, t_1'), t_2')\varphi_2^*(x, t_1', t_2')}{\varphi_1^*(x, t_1')}, \tag{16}$$

where

$$\varphi_2^*(x, t_1', t_2') = \int_X \delta(x' - x'(x, \mathbf{t}'))\varphi^*(x')dx' = \varphi^*(x').$$

$$\begin{aligned} \varphi_1^*(x, t_1') &= \int_{T_2}\int_X \delta(x' - x'(x, \mathbf{t}'))k_2((x, t_1'), t_2')\varphi^*(x')dx'dt_2' \\ &= \int_{T_2} k_2((x, t_1'), t_2')\varphi_2^*(x, t_1', t_2')dt_2'. \end{aligned} \tag{17}$$

By substituting transition densities (15), (16) in the expression for $K_p^*\varphi^*$ it is easy to verify the validity of the following analog of Theorem 2.

Theorem 5. *The variance of the collision estimator ξ_x under exact partial value modelling of t_2' (i.e. according to* (15), (16)*) is finite.*

If transition density has the form

$$p_2((x, t_1'), t_2') = \frac{k_2((x, t_1'), t_2')\varphi_2^*(x, t_1', t_2')}{\hat{\varphi_1^*}(x, t_1')}, \tag{18}$$

where $\hat{\varphi_1^*} \geq \varphi_1^*$ then from the Theorem 5 follows the analog of Theorems 3, 4.

Theorem 6. *If for the function $\hat{\varphi_1^*}$ holds*

$$\varphi^*(x) \geq \int_{T_1} k_1(x, t_1')\hat{\varphi_1^*}(x, t_1')(1 - \beta(x, t_1'))dt_1', \tag{19}$$

then the variance of the collision estimator ξ_x under partial value modelling of t_2' (i.e. according to (15), (18)) is finite.

Let us note that Theorems 3, 4, 5, 6 are valid even if we use auxiliary value function $u(x) + C_1$, $C_1 > 0$ where $u \in C_b(X)$ and satisfies to the *majorant conjugate equation*

$$u = K^*u + \hat{h},$$

with obligatory conditions:

$$\text{supp } h \subseteq \text{supp } \hat{h}, \quad \frac{h}{\hat{h}} \leq C_2 < \infty \quad \forall x \in \text{supp } \hat{h}. \tag{20}$$

As a generalization of the last note and Theorem 3, 4, 5, 6 we propose new criterion: *under conditions of the Theorem 3, 4, 5, 6 variance for the weight collision estimator is finite if modified transition density for auxiliary variable equals to the product of initial density and some function $u(x) + C_1, C_1 > 0$ where u satisfies to majorant conjugate equation.*

3 Investigation of Weight Estimator Variance in "Exponential Transformation" Method

Let us consider *one-dimensional* problem of estimating the escape probability from the half-space $-\infty < z \leq H$. We assume that an absolute absorbent fills the outside of this half-space and the mean free path $\sigma^{-1} = 1$ in the entire space. The scattering of a particle at a collision point is described by the symmetric normalized density $\omega(\mu, \mu')$, where μ is the cosine of the angle between the particle path and the axis z. The probability of survival in the collision at the point $x = (z, \mu)$, $z < H$ is $q < 1$. Let us suppose that distribution density of the mean free path l is e^l. Then we have $\mathbf{t} = (l, \mu)$, $z' = z + \mu l$ and the equation (1) has the form

$$\varphi^*(z, \mu) = q \int_0^A e^{-l} \int_{-1}^1 w(\mu, \mu')\varphi^*(z', \mu')\, d\mu' dl + h(z, \mu), \quad z < H, \tag{21}$$

where $A = +\infty$ for $\mu < 0$ and $A = (H - z)/\mu$ for $\mu > 0$. The free term in equation (21) is

$$h(z, \mu) = \begin{cases} \exp\{-(H - z)/\mu\}, & \text{for } z < H \text{ and } \mu > 0, \\ 0 & \text{otherwise.} \end{cases} \tag{22}$$

The quantity $E\xi_x = \varphi^*(x)$ is equal to the desired probability of escape of the particle that started at the point $x = (z, \mu)$.

Additionally consider (21) with the free term $h_a(z, \mu) = a(\mu)h(z, \mu)$, where $a(\mu)$ and the parameter $c = 1/L$ satisfy the characteristic Milne equation [Dav60].

$$(1 - c\mu)a(\mu) = q \int_{-1}^{1} w(\mu, \mu')a(\mu')d\mu'. \tag{23}$$

For the real scattering indicatrix, we have $a(\mu) \geq \epsilon > 0$. Using a substitution, it is easily verified (see, e.g., [Mik92], Section 2.4) that

$$\varphi_a^*(z, \mu) = E\xi_x(a) = a(\mu)\exp\{-(H - z)/L\}. \tag{24}$$

It is well known that simulation of the mean free path with the density $e^{-l}e^{c\mu l}$ leads to the exponential transformation for which $\sigma' = 1 - c\mu$ [Mik92]. Moreover, if the termination of the trajectory is modelled physically and $c = 1/L$ then the exponential transformation is equivalent to the partial value modeling of the first (see (13)) auxiliary variable l (the mean free path with the use of (24)). Note that if $\sigma \neq 1$ then the value weighted modelling is constructed for $\sigma = \sigma(1-\mu/L)$. It turned out that, if the exponential transformation is used, there are some doubts whether the variance is bounded or not. For example for $1-p = q$ and $c = 1/L$ we have $\rho(K_p^*) = 1$ in the case of an isotropic scattering therefore we cannot be sure that the variance is bounded in this case.

However, the criterion for the investigation of the variance boundedness proposed in Section 2 makes it possible to establish the practically important fact that the variance in the exponential transformation method is bounded for $0 < c \leq 1/L$. It is known (see [Dav60]) that for the real indicatrix $w(\mu, \mu')$ the equation (23) establishes the correspondence $c \leftrightarrow \{q(c), a_c(\mu)\}$ and $q(c) \geq q$ if $c \in (0, 1/L]$. Consider the value function

$$\varphi_c^*(z, \mu) = a_c(\mu)\exp\{-(H - z)c\}, \tag{25}$$

for which

$$\varphi_c^* = K^*\varphi_c^* + \varphi_c^*(1 - \frac{q}{q(c)}(1 - e^{-A(1-c\mu)})) = K^*\varphi_c^* + \hat{h},$$

and

$$\frac{h}{\hat{h}} \leq \frac{e^{-(H-z)(\frac{1}{\mu}+c)}}{a_c(\mu)(1 - \frac{q}{q(c)}(1 - e^{-A(1-c\mu)}))} < \infty, \quad \forall x \in \text{supp}\,\hat{h}.$$

Note that, if the probability of survival at the collision point is $q(c)$, then the exponential transformation is equivalent to the partial value modelling with value function (25). When the initial probability of survival is q, the exponential transformation with $c \in (0, 1/L]$ is equivalent to the use of the partial value density multiplied by $q(c)/q$. Note that the proposed method makes it possible to use such a density because the expression for $K_p^* \varphi_c^*$ includes the factor $q(c)/q < 1$ in this case (see the proof of Theorem 3). If $q/q(c) = 1$, i.e., for $c = 1/L$, it is easy to verify that the conditions of Theorem 3 for the functions h, α are true.

We numerically estimated the probability that a particle goes beyond the boundary $H = 0$ from the point $(z_0, \mu_0) = (-20, 1)$ using (22) with $q = 0.7$ for the isotropic scattering. In this case, the exponential transformation with $c = 0$ is equivalent to the direct simulation of the mean free path; for $c = \sqrt{1-q} \approx 0,548$, it is asymptotically (as $H \to \infty$) optimal (see [MM04]); and, for $c = 1/L \approx 0.829$, it is equivalent to the approximate value modeling of the mean free path with the value function defined by (24) and, correspondingly,

$$Q_n = \frac{e^{-lc\mu}}{(1-c\mu)} Q_{n-1}. \tag{26}$$

Table 1 presents the results of the computations (obtained by modelling 10^7 trajectories of particles) based on the estimate with respect to scattering (2).

The rows in this table correspond to the direct, asymptotically value, and asymptotically optimal variants of modelling (see (26)) of the mean free path. To make the comparison easier, the same set of pseudorandom numbers was used so that the modelling of the trajectories with the same index in different variants started from the same pseudorandom numbers produced by the multiplicative congruent generator with the parameters $M = 5^{17}, m = 2^{40}$ [EM82]. The following notation is used: c is the parameter of the exponential transformation, $\tilde{\varphi^*}$ is the statistical estimate of φ^* for $x_0 = (-20, 1)$, and $\tilde{\sigma}$ is the corresponding estimate of the root-mean-square probability error.

Now, we consider the *three-dimensional problems* concerning the transport of particles inside a sphere of radius R. We assume that the space outside the sphere is filled with an absolute absorber and $\sigma = const$ in the entire space. As a result of a collision in the interior point $\mathbf{r} = (x, y, z)$ of the sphere the particle is scattered with the probability q along the random unit vector ω' indicating the new direction of the particle motion. The spherical variant of the exponential transformation is constructed using the substitution

$$\Phi(\mathbf{r}, \omega) = e^{cr} \Phi_1(\mathbf{r}, \omega), \ |c| < \sigma, \ r = |\mathbf{r}|$$

in the integro-differential transport equation (see [EM82])

$$(\omega, \mathrm{grad}\Phi) + \sigma\Phi(\mathbf{r}, \omega) = \int \Phi(\mathbf{r}, \omega')\sigma \mathrm{w}(\omega, \omega', \mathbf{r}) d\omega' + \Phi_0(\mathbf{r}, \omega). \tag{27}$$

Table 1. Estimator with respect to scattering

c	$\tilde{\varphi^*}$	$\tilde{\sigma}$
0	$9.191 \cdot 10^{-8}$	$3.74 \cdot 10^{-8}$
0.829	$1.120 \cdot 10^{-7}$	$6.35 \cdot 10^{-9}$
0.548	$1.032 \cdot 10^{-7}$	$3.90 \cdot 10^{-9}$

This variant is reduced to the use of the modified cross section

$$\sigma'(\mathbf{r},\omega) = \sigma - c\mu(\mathbf{r},\omega), \quad \text{where} \quad \mu(\mathbf{r},\omega) = (\mathbf{r},\omega)/r. \tag{28}$$

Note that the quantity $\mu(\mathbf{r}(l),\omega) = \mu(l)$ along the direction of the particle motion $\mathbf{r}(l) = \mathbf{r} + l\omega$ is nothing else than $\cos v(l)$, where $v(l)$ is the angle between $\mathbf{r}(l)$ and ω. It is easy to verify that

$$\mu(l) = \frac{\partial r(l)}{\partial l}, \quad r(l) = \sqrt{r^2 + l^2 + 2lr\mu(0)}.$$

In what follows we assume for the simplicity of presentation that $\sigma = 1$. Then, cross section (28) corresponds to the modified density of distribution of the mean free path

$$f_l(\mathbf{r},w,l) = \sigma'(r(l),\omega)\exp\left\{\int_0^l \sigma'(\mathbf{r}(t),\omega)dt\right\}$$

$$= e^{-l}(1 - c\mu(l))e^{cr(l)-cr}.$$

If we use the auxiliary value function

$$g(\mathbf{r},\omega) = [1 - c\mu(\mathbf{r},\omega)]e^{cr}, r > 0,$$

in the collision scheme (see [EM82]), then the density f_l is given by (18); and it is the partial value density of the distribution of the second auxiliary variable l if the function g satisfies condition (19) in Theorem 6 and, as a consequence, the majorizing adjoint equation (20). Let us verify this condition. We denote by the symbol $l^* = l^*(\mathbf{r},\omega')$ the distance from the point $\mathbf{r}$ to the sphere of radius R along the given direction ω'. Then, we have

$$g(\mathbf{r},\omega) - q\int_{-1}^{1} w(\omega,\omega')e^{cr}(1 - e^{cR-l^*})d\omega' \geq (1 - c\mu)e^{cr} - qe^{cr}\int_{-1}^{1} w(\omega,\omega')d\omega'$$

$$\geq (1 - |c| - q)e^{cr}.$$

Therefore, the condition of Theorem 6 is satisfied and according to the proposed method (see Section 2) the variance of the weighted estimate is bounded only for $|c| \in (0, 1 - q)$.

In this problem, the admissible value of c can be increased by evaluating the spectral radius of the operator K_p^*. We have the following inequality:

$$\rho(K_p^*) \leq \| K_p^* \| = \sup_r q\int_{-1}^{1} w(\omega,\omega')\int_0^{\infty} \frac{e^{-2l}}{e^{-l}(1 - c\mu(l))e^{cr(l)-cr}}dld\omega'$$

$$= \sup_r q \int_{-1}^{1} w(\omega, \omega') \int_0^\infty \frac{e^{-l-cr(l)+cr}(1+c\mu(l))}{(1-c^2\mu^2(l))} dl d\omega' \leq \frac{q}{1-c^2}.$$

Therefore, when the modified cross section (28) is used, the standard criterion guarantees that the variance is bounded for $|c| \in (0, \sqrt{1-q})$.

Note that the admissible range of values of c for cross section (28) can be significantly increased by choosing another auxiliary value function. For example, if we use the function $g(\mathbf{r}, \omega) = g(\mathbf{r}) = e^{-cr}$ with $c > 0$ and simulate the mean free path according to density (16), then some empirical considerations suggest that the function $[g - K^*g](\mathbf{r})$ monotonically decreases with increasing r. Consider the case of an isotropic scattering; i.e., let $w(\omega, \omega') \equiv 1/4\pi$. It is easy to verify that the following inequality holds at $\mathbf{r} = 0$:

$$g(0) - [K^*g](0) = 1 - q \int_{-1}^{1} \frac{1}{4\pi} \int_0^{l^*} e^{-l} e^{-cl} dl d\omega'$$

$$= 1 - \frac{q}{1+c}\left(1 - e^{-R(1+c)}\right) \geq 1 - \frac{q}{1+c} > 0.$$

On the sphere of radius R at the point $\mathbf{r}$ we have the inequality

$$g(\mathbf{r}) - [K^*g](\mathbf{r}) \equiv e^{-cR} - q \int_{-1}^{0} \frac{1}{2} \int_0^{l^*} e^{-l} e^{-c\sqrt{R^2+l^2+2Rl\mu'}} dl d\omega'$$

$$\geq e^{-cR} - q \int_{-1}^{0} \frac{1}{2} \int_0^{2R} e^{-l} e^{-c|R-l|} dl d\omega' \geq \left(1 - \frac{q}{2(1-c)}\right)$$

where the last expression is positive for $c \in (0, 1 - q/2)$. It is very difficult to analytically prove that $[g - K^*g](\mathbf{r})$ is monotonically decreasing. For this reason we used the trapezoid method with a small step to verify that the function $g(\mathbf{r}) - [K^*g](\mathbf{r})$ is monotonically decreasing with increasing r for $c = 0, 0.1, 0.2, \ldots, 1 - q/2$. For example, for the sphere of radius $R = 10$, the results of the computations on the grid with the size 0.001 with respect to l and μ for $q = 0.7$ and $c = 0.6$ are presented in the Table 2.

Thus for an isotropic scattering with the use of the auxiliary value function e^{-cr} the variance of the weighted estimator is bounded if

Table 2. Decrease of the function $[g - K^*u](\mathbf{r})$ r

r	1	2	5	7	8	9	10
$[g - K^*g](\mathbf{r})$	0.9999	0.5488	0.0907	0.0149	0.0082	0.0045	0.0024

$c \in (0, 1-q/2)$. The construction of an efficient direct simulation method for the density

$$p_2((\mathbf{r}, \omega'), l) = \frac{e^{-l} e^{-cr(l)}}{\int\limits_0^{l^*} e^{-l} e^{-cr(l)} dl} \tag{29}$$

is a difficult problem. For this reason, we used the rejection method (see [EM82]) to simulate the mean free path. Since

$$\mu'(l) = \frac{\partial^2 r}{\partial^2 l} = \frac{r^2(1-\mu^2(0))}{r^3(l)} \geq 0,$$

the function $\mu(l)$ is monotonically nondecreasing with increasing l for any fixed $\mathbf{r}$ and ω; therefore, we have the inequality

$$e^{-cr(l)} \leq \frac{(1+c\mu(l))}{1+c\mu(0)} e^{-cr(l)}.$$

With regard for these remarks, the mean free path l is simulated as follows:

- a random value of $\tilde{l}$ in the interval $[0, l^*]$ is chosen according to the distribution density
$$\frac{(1+c\mu(l))e^{-l}e^{-cr(l)}}{(e^{-cr} - e^{-cR-l^*})}$$
- the random number γ uniformly distributed in the interval $[0,1]$ is realized.
- if $\gamma \frac{1+c\mu(\tilde{l})}{1+c\mu(0)} e^{-cr(l)} < e^{-cr(l)}$, or, which is equivalent, $\gamma \frac{1+c\mu(\tilde{l})}{1+c\mu(0)} < 1$, then set $l = \tilde{l}$, otherwise, the process is repeated.

Under this modelling algorithm the weight is transformed by the formula

$$Q_n = Q_{n-1} e^{cr(l)} \int\limits_0^{l^*} e^{-l} e^{-cr(l)} dl.$$

Note that the computation of the normalizing constant$C(r, \mu(0)) = \int\limits_0^{l^*} e^{-l} e^{-cr(l)} dl$ at each collision point can significantly increase the computation time. To reduce it, it is reasonable to precalculate the quantity $C(r, \mu(0))$ at the points (r_k, μ_i), $(0 \leq k, i \leq N)$ of the grid built in the domain $[0, R] \times [-1, 1]$. Given an arbitrary point $(r, \mu(0))$, we can find the appropriate grid points

$$r_i \leq r \leq r_{i+1} \quad \mu_i \leq \mu(0) \leq \mu_{i+1},$$

and determine $C(r, \mu(0))$ using the linear interpolation by the formulas

$$\tilde{C}(r, \mu_i) \approx C(r_{k+1}, \mu_i) + \frac{r_{k+1} - r}{r_{k+1} - r_k} [C(r_k, \mu_i) - C(r_{k+1}, \mu_i)],$$

$$\tilde{C}(r, \mu_{i+1}) \approx C(r_{k+1}, \mu_{i+1}) + \frac{r_{k+1} - r}{r_{k+1} - r_k} [C(r_k, \mu_{i+1}) - C(r_{k+1}, \mu_{i+1})],$$

$$C(r, \mu(0)) \approx \tilde{C}(r, \mu_{i+1}) + \frac{\mu_{i+1} - \mu}{\mu_{i+1} - \mu_i} [\tilde{C}(r, \mu_i) - \tilde{C}(r, \mu_{i+1})].$$

Let us consider some problems concerning the passage of particles through an optically thick spherical layer.

Assume that a black ball of radius $R_1 < R$ is placed inside the given sphere. When a particle hits the surface of this ball, it is absorbed with the unit probability. We assume that the scattering of particles inside the layer $R_1 \leq r \leq R$ is isotropic and the probability of survival after a collision is $q = 0.7$.

Let a point source that emits particles according to the Lambert law (see [EM82]) is placed on the interior surface of the sphere of radius and

$$w(\omega_0, \omega') = 2(\omega_0, \omega'), \quad (\omega_0, \omega') \geq 0,$$

where ω_0 is the inner normal to the sphere of radius R. We want to estimate the *probability* $P = P(R_1)$ *that a particle is absorbed on the surface of the black ball of radius* R_1.

We must calculate the passage of the particles through the optically thick $R_1 \leq r \leq R$ to the internal sphere of radius R_1. According to the reasoning in Section 3, it is expedient to simulate the mean free path using the modified cross section (28) with $c < 0$.

We numerically solved the problem of estimating the functional $P(R_1)$ for $R = 10$ and $R_1 = 2$. To estimate the probability $P(R_1)$ it is sufficient to evaluate the integral [EM82]:

$$h(\mathbf{r}, \omega) = q \int_D \frac{1}{4\pi} \int_0^{-l^*_{R_1}(\mathbf{r},\omega')} e^{-l} dl d\omega',$$

at each collision point $(\mathbf{r}, \omega)$ where $l^*_{R_1}(\mathbf{r}, \omega')$ is the distance from $\mathbf{r}$ to the sphere of radius R_1 along the given direction ω' and

$$D = \left\{ \omega' : \frac{(\omega', \mathbf{r})}{r} \leq -\frac{\sqrt{r^2 - R_1^2}}{r} \right\}$$

is the cone of directions with the vertex at $\mathbf{r}$ "subtended" by the internal sphere of radius R_1. The evaluation of integral $h(\mathbf{r}, \omega)$ at each collision point is a computationally costly procedure. For such problem it is convenient to use the randomized collision estimator (see [EM82]). More precisely after simulating the new direction ω' we recommend to calculate the quantity

$$h^*(\mathbf{r},\omega') = \begin{cases} qe^{l^*_{R_1}(\mathbf{r},\omega')} & \text{for } \omega' \in D, \\ 0 & \text{otherwise.} \end{cases} \tag{30}$$

Quantity (30) is also calculated and added to the estimator when the particle is absorbed at the point $\mathbf{r}$.

The results of the computations based on the randomized collision estimator in the exponential transformation without escape are presented at the Table 3. In this modification, the mean free path l inside the layer $R_1 \leq r \leq R$ is simulated according to the density $C(1+\mu(l))e^{-l-cr(l)}$, therefore the trajectory can terminate only as a result of absorption with the probability $1-q$. After the scattering is modelled at the point $(\mathbf{r}_n, \omega')$, the weight is determined by

$$Q_n = Q_{n-1}\frac{(e^{-cr_{n-1}} - e^{-cR(l^*_m)-l^*_m})e^{cr(l)}}{1+c\mu(l)},$$

where l^*_m is either l^* or $l^*_{R_1}$ depending on the direction of motion ω. Let us note that in this case the time needed to modell one trajectory depends only on the given probability of absorption $1-q$, therefore the average time needed to modell one trajectory is the same for any value of c.

It is seen that the partial value modelling based on the majorizing adjoint equation ($c = 0.3$) significantly reduces the variance compared to the direct simulation, while it is slightly inferior to the exponential transformation with the maximally admissible $c = 0.548$. When the boundedness of the variance is not guaranteed theoretically ($c = 0.65, 0.829$) the modified modelling is significantly more efficient than the direct simulation. For this reason we believe that the variance of the weighted estimator is still bounded.

For comparison, at Table 4 we present the results of the computations of $P(R_1)$ based on the randomized collision estimator and the partial value modelling with the auxiliary value function e^{-cr}. In this modification, the mean free path inside the layer $R_1 \leq r \leq R$ is simulated according to density (29) by the rejection method and the weight is transformed by the formula

$$Q_n = Q_{n-1}C(r_{n-1}, \mu_{n-1}(0))\, e^{cr_n}.$$

Table 3. Randomized collisions estimator of $P(R_1)$ with the use of exponential transformation without escape

c	0	0.3	0.548	0.65	0.829
$\tilde{P}(R_1)$	$1.456 \cdot 10^{-4}$	$1.463 \cdot 10^{-4}$	$1.469 \cdot 10^{-4}$	$1.471 \cdot 10^{-4}$	$1.461 \cdot 10^{-4}$
$\tilde{\sigma}^*$	$1.85 \cdot 10^{-6}$	$9.42 \cdot 10^{-7}$	$7.03 \cdot 10^{-7}$	$7.11 \cdot 10^{-7}$	$8.01 \cdot 10^{-7}$

Table 4. Randomized collisions estimator of $P(R_1)$ under partial value modelling with the auxiliary value function e^{-cr}

c	0	0.3	0.548	0.65	0.829
$\tilde{P}(R_1)$	$1.473 \cdot 10^{-4}$	$1.466 \cdot 10^{-4}$	$1.465 \cdot 10^{-4}$	$1.474 \cdot 10^{-4}$	$1.482 \cdot 10^{-4}$
$\tilde{\sigma}^*$	$1.88 \cdot 10^{-6}$	$9.79 \cdot 10^{-7}$	$7.04 \cdot 10^{-7}$	$6.65 \cdot 10^{-7}$	$7.41 \cdot 10^{-7}$

The normalizing constant $C(r_{n-1}, \mu_{n-1}(0))$ was calculated according to rejection method in Section 3 with the use the linear interpolation by the precalculated values of $C(r, \mu)$ at the grid points:

$$2 = r_0 < r_1 < \ldots < r_{80} = 10, \quad -1 = \mu_0 < \mu_1 < \ldots < \mu_{100} = 1,$$

$$r_k - r_{k-1} = 0.1, \quad \mu_i - \mu_{i-1} = 0.02 \ .$$

At each grid point $C(r_k, \mu_i)$ was calculated using the trapezoid method with the step 0.005.

It is seen that the partial value modelling based on the majorizing adjoint equation ($c = 0.3, 0.548$) significantly reduces the variance compared to the direct simulation. The partial value modelling with the maximally admissible $c = 1 - q/2 = 0.65$ also significantly reduces the variance. When the boundedness of the variance is not guaranteed theoretically ($c = 0.829$) the modified modelling is significantly more efficient than the direct simulation. For this reason we believe that the variance of the weighted estimator is still bounded in this case.

We see that the variances of the weighted estimators obtained with use of the exponential transformation without the escape (algorithm A) and using the partial value modeling with the auxiliary value function e^{-cr} (algorithm B) are very close to each other. On the other hand, modelling of the mean free path by the rejection method increases the computation time. For example, for $c = 0.548$, the average time needed to modell one trajectory in algorithm B is $2.5 \cdot 10^{-5}$ s; and in algorithm A, it is $2 \cdot 10^{-5}$ s.

Thus, when $P(R_1)$ is estimated in the case of the isotropic scattering with $q = 0.7$, the simple standard algorithm A is as good as the complex value algorithm B. However, as $q \to 1$, the situation changes because algorithm B admits the values $c < 1 - q/2 \approx 0.5$, while algorithm A admits only $c < \sqrt{1-q} \to 0$. Let us note that the value algorithms can be useful for solving more complicated problems with an anisotropic scattering and variable density, as well as for multivelocity problems.

References

[Dav60] B. Davidson. Neutron Transport Theory. *Atomizdat*, Moscow, 1960 (in Russian).

[Dim91] I. Dimov. Minimization of the probable error for some Monte Carlo methods. In *Proc. of the Summer School on Mathematical Modelling and Scientific Computations*, 23–28.09.1990, Albena, Bulgaria, Sofia, Publishing House of the Bulgarian Academy of Sciences, pp. 159–170, 1991.

[EM82] S. Ermakov and G. Mikhailov. *Statistical Simulation.* Nauka, Moscow, 1982 (in Russian).

[Med05] I. Medvedev. The variance of weight estimate in the case of partial "value" modelling. In *Proceedings of the 5th St. Petersburg Workshop on Simulations.* NII Chemistry St.Petersburg University Publishers, pp. 465–470, 2005.

[Mik92] G. Mikhailov. Optimization of Monte-Carlo Weighted Methods. Springer-Verlag, 1992.

[Mik03] G. Mikhailov. Construction of Weighted Monte Carlo Methods via an Increase in the Phase Space Dimension. *Dokl. Ross. Akad. Nauk.*, Vol. 389, No. 4, pp. 461–464, 2003.

[MM04] G. Mikhailov and I. Medvedev. Optimization of weighted Monte-Carlo methods with respect to auxiliary variables. *Siberian Mathematical Journal*, Vol. 45, No. 2, pp. 331–340, 2004.

[MM03] I. Medvedev and G. Mikhailov. Optimization of weighted methods for part of variables. *Russ. J. Numer. Anal. Math. Modelling.* Vol. 18, No. 5, pp. 397–412, 2003.

Optimal Pointwise Approximation of a Linear Stochastic Heat Equation with Additive Space-Time White Noise

Thomas Müller-Gronbach[1], Klaus Ritter[2], and Tim Wagner[3]

[1] Fakultät für Mathematik und Informatik, FernUniversität Hagen, Lützowstraße 125, 58084 Hagen, Germany
Thomas.Mueller-Gronbach@FernUni-Hagen.de

[2] Fachbereich Mathematik, Technische Universität Darmstadt, Schloßgartenstraße 7, 64289 Darmstadt, Germany
ritter@mathematik.tu-darmstadt.de

[3] Fachbereich Mathematik, Technische Universität Darmstadt, Schloßgartenstraße 7, 64289 Darmstadt, Germany
twagner@mathematik.tu-darmstadt.de

We consider a linear stochastic heat equation on the spatial domain $]0,1[$ with additive space-time white noise, and we study approximation of the mild solution at a fixed time instance. We show that a drift-implicit Euler scheme with a non-equidistant time discretization achieves the order of convergence $N^{-1/2}$, where N is the total number of evaluations of one-dimensional components of the driving Wiener process. This order is best possible and cannot be achieved with an equidistant time discretization.

1 Introduction

We consider a linear stochastic heat equation

$$\begin{aligned} dX(t) &= \Delta X(t)\, dt + dW(t), \\ X(0) &= \xi \end{aligned} \tag{1}$$

on the Hilbert space $H = L_2(]0,1[)$ with a deterministic initial value $\xi \in H$. Here Δ denotes the Laplace operator with Dirichlet boundary conditions. Moreover, $W = (\langle W(t), h\rangle)_{t\geq 0, h\in H}$ is a cylindrical Brownian motion on H with the identity operator as its covariance. See [DPZ92].

Note that $\Delta h_i = -\mu_i \cdot h_i$ with

$$h_i(u) = 2^{1/2} \cdot \sin(i\pi u)$$

and

$$\mu_i = \pi^2 \cdot i^2$$

for $i \in \mathbb{N}$, and put

$$\beta_i(t) = \langle W(t), h_i \rangle$$

for $t \geq 0$. Then $(\beta_i)_{i\in\mathbb{N}}$ is an independent family of standard one-dimensional Brownian motions. The mild solution X of equation (1) is given by

$$X(t) = \sum_{i\in\mathbb{N}} Y_i(t) \cdot h_i,$$

where the real-valued processes Y_i are independent Ornstein-Uhlenbeck processes satisfying

$$\begin{aligned} dY_i(t) &= -\mu_i Y_i(t)\,dt + d\beta_i(t), \\ Y_i(0) &= \langle \xi, h_i \rangle \end{aligned} \tag{2}$$

for every $i \in \mathbb{N}$.

Let $T > 0$. We study approximation of $X(T)$ on the basis of evaluations of finitely many scalar Brownian motions β_i at a finite number of points in $[0, T]$. The selection and evaluation of the scalar Brownian motions β_i, i.e., the discretization of the cylindrical Brownian motion W, is specified by a non-empty finite set

$$\mathcal{I} \subseteq \mathbb{N},$$

a collection

$$\nu = (\nu_i)_{i\in\mathcal{I}} \in \mathbb{N}^{\#\mathcal{I}}$$

of integers, and nodes

$$0 < t_{1,i} < \cdots < t_{\nu_i,i} \leq T$$

for every $i \in \mathcal{I}$. Every Brownian motion β_i with $i \in \mathcal{I}$ is evaluated at the corresponding nodes $t_{\ell,i}$, and the total number of evaluations is given by

$$|\nu|_1 = \sum_{i\in\mathcal{I}} \nu_i.$$

An approximation $\widehat{X}(T)$ to $X(T)$ is given by

$$\widehat{X}(T) = \phi(\beta_{i_1}(t_{1,i_1}), \ldots, \beta_{i_1}(t_{\nu_{i_1},i_1}), \ldots, \beta_{i_k}(t_{1,i_k}), \ldots, \beta_{i_k}(t_{\nu_{i_k},i_k})), \tag{3}$$

where

$$\phi : \mathbb{R}^{|\nu|_1} \to H$$

is any measurable mapping and $\mathcal{I} = \{i_1, \ldots, i_k\}$. The error of $\widehat{X}(T)$ is defined by

$$e(\widehat{X}(T)) = \left(E\|X(T) - \widehat{X}(T)\|^2\right)^{1/2},$$

where $\|\cdot\| = \|\cdot\|_H$.

Let $\mathfrak{X}_N$ denote the class of all algorithms (3) that use at most a total of N evaluations of the scalar Brownian motions β_i, i.e., $|\nu|_1 \leq N$. We wish to minimize the error in this class, and hence we study the Nth minimal error

$$e_N = \inf_{\widehat{X}(T) \in \mathfrak{X}_N} e(\widehat{X}(T)).$$

As a subclass $\mathfrak{X}_N^{\text{equi}} \subset \mathfrak{X}_N$ we consider all methods $\widehat{X}(T) \in \mathfrak{X}_N$ that use equidistant nodes for evaluation of the scalar Brownian motions β_i, i.e., $|\nu|_1 \leq N$ and $t_{\ell,i} = \ell/\nu_i \cdot T$ for every $i \in \mathcal{I}$. Furthermore, we consider the subclass $\mathfrak{X}_N^{\text{uni}} \subset \mathfrak{X}_N^{\text{equi}}$ of methods $\widehat{X}(T) \in \mathfrak{X}_N^{\text{equi}}$ that use the same number of equidistant nodes for every scalar Brownian motion β_i, i.e., $\nu_i = n$ and $t_{\ell,i} = \ell/n \cdot T$ for all $i \in \mathcal{I}$ and some $n \in \mathbb{N}$ with $n \cdot \#\mathcal{I} \leq N$. The definition of the corresponding minimal errors e_N^{equi} and e_N^{uni} is canonical. Clearly,

$$e_N \leq e_N^{\text{equi}} \leq e_N^{\text{uni}}.$$

Construction and analysis of algorithm for stochastic heat equations or, more generally, stochastic evolution equations, started with the work by [GK96] and [GN97]. A partial list of further references includes [ANZ98], [DZ02], [H03], [S99], and [Y05]. Lower bounds and optimality of algorithms has first been studied by [DG01], see [MGR07a] and [MGR07b] for further results.

2 Results and Remarks

For two sequences $(a_k)_{k\in\mathbb{N}}$ and $(b_k)_{k\in\mathbb{N}}$ of positive real numbers we write $a_k \preceq b_k$ if $\sup_{k\in\mathbb{N}} a_k/b_k < \infty$. Additionally, $a_k \asymp b_k$ means $a_k \preceq b_k$ and $b_k \preceq a_k$.

We determine the asymptotic behaviour of the minimal errors for the classes $\mathfrak{X}_N$, $\mathfrak{X}_N^{\text{equi}}$ and $\mathfrak{X}_N^{\text{uni}}$. See Sections 3.2 and 3.3 for the proof.

Theorem 1. *The Nth minimal errors satisfy*

$$e_N \asymp N^{-1/2}$$

and

$$e_N^{\text{equi}} \asymp e_N^{\text{uni}} \asymp N^{-1/6}$$

for every $\xi \in H$.

Theorem 1 states that $N^{-1/2}$ is the best possible order of convergence for any sequence of methods $\widehat{X}_N(T) \in \mathfrak{X}_N$. Moreover, this rate cannot be achieved by using equidistant nodes for evaluation of the scalar Brownian motions β_i.

Now we construct an implicit Euler scheme, which performs asymptotically optimal in the class $\mathfrak{X}_N$, up to a constant. Fix $N \in \mathbb{N}$. For every $i = 1, \ldots, N$

we apply a drift-implicit Euler scheme to the corresponding equation (2) for Y_i to obtain an approximation $\widehat{Y}_{i,N}(T)$ to $Y_i(T)$. To this end we put

$$\nu_i = \lceil \mu_i^{-1/3} \cdot N^{2/3} \rceil, \tag{4}$$

and we define nodes $t_{\ell,i}$ by

$$\int_0^{t_{\ell,i}} \exp(-\mu_i/3 \cdot (T-t))\, dt = \frac{\ell}{\nu_i} \cdot \int_0^T \exp(-\mu_i/3 \cdot (T-t))\, dt \tag{5}$$

for $\ell = 0, \ldots, \nu_i$. The corresponding drift-implicit Euler scheme is given by

$$\widehat{Y}_{i,N}(0) = \langle \xi, h_i \rangle$$

and

$$\widehat{Y}_{i,N}(t_{\ell,i}) = \widehat{Y}_{i,N}(t_{\ell-1,i}) - \mu_i \cdot \widehat{Y}_{i,N}(t_{\ell,i}) \cdot (t_{\ell,i} - t_{\ell-1,i}) + \beta_i(t_{\ell,i}) - \beta_i(t_{\ell-1,i}) \tag{6}$$

for $\ell = 1, \ldots, \nu_i$. Finally, we use

$$\widehat{X}_N(T) = \sum_{i=1}^N \widehat{Y}_{i,N}(T) \cdot h_i$$

as an approximation to $X(T)$.

For the proof of the following error bound we refer to Section 3.3.

Theorem 2. *Suppose that $\xi \in C^1([0,1])$. Then the error of the algorithm $\widehat{X}_N(T)$ satisfies*

$$e(\widehat{X}_N(T)) \preceq N^{-1/2}.$$

The algorithm $\widehat{X}_N(T)$ is of the form (3) with $\mathcal{I} = \{1, \ldots, N\}$. Moreover, its total number of evaluations of scalar Brownian motions satisfies

$$\begin{aligned} |\nu|_1 = \sum_{i=1}^N \lceil \mu_i^{-1/3} \cdot N^{2/3} \rceil &\le N + (N/\pi)^{2/3} \sum_{i=1}^N i^{-2/3} \\ &\le N + (N/\pi)^{2/3} \int_0^N x^{-2/3}\, dx \le 2N. \end{aligned}$$

Consequently, $\widehat{X}_N(T)$ belongs to the class $\mathfrak{X}_{2N}$, and combining Theorem 1 with Theorem 2 we obtain the following optimality result.

Corollary 1. *Suppose that $\xi \in C^1([0,1])$. Then the sequence of algorithms $\widehat{X}_N(T)$ is asymptotically optimal, i.e.,*

$$e(\widehat{X}_N(T)) \asymp e_{2N}.$$

Remark 1. Due to Theorem 1, non-equidistant time discretizations of the cylindrical Brownian motion W are superior to equidistant ones for equation (1). Whether this superiority carries over to more general equations with space-time white noise is an open question. The result on the minimal errors e_N in Theorem 1 does not carry over in general. For example,

$$e_N \succeq N^{-1/6}$$

holds for the equation

$$dX(t) = \Delta X(t)\,dt + X(t)\,dW(t)$$

on H, see [DG01].

Remark 2. Minimal errors are studied, too, for approximation of stochastic heat equations

$$\begin{aligned} dX(t) &= \Delta X(t)\,dt + B(t, X(t))\,dW(t), \\ X(0) &= \xi \end{aligned} \tag{7}$$

on spaces $H = L_2(]0,1[^d)$ w.r.t. to the error criterion

$$e(\widehat{X}) = \left(E \int_0^T \|X(t) - \widehat{X}(t)\|^2\,dt \right)^{1/2}. \tag{8}$$

The latter takes into account the quality of an approximation $\widehat{X}$ on the whole time interval $[0,T]$. We add that (1) corresponds to (7) with $B(t,x) = \mathrm{id}$ and $d = 1$.

We briefly survey results that hold under suitable assumptions on the noise, the initial value ξ, and the operator-valued mapping B, see [MGR07a] and [MGR07b]. These findings significantly differ from the results on approximation of X at the single point T.

For equations with space-time white noise as well as nuclear noise approximations based on equidistant discretizations turn out to be asymptotically optimal, i.e., $e_N \asymp e_N^{\mathrm{equi}}$ for the respective minimal errors based on the error criterion (8). Furthermore, for $d = 1$ and space-time white noise, $e_N \asymp e_N^{\mathrm{uni}} \asymp e_N^{\mathrm{equi}} \asymp N^{-1/6}$. On the other hand, for equations with nuclear noise uniform discretizations are suboptimal, asymptotically, at least for the specific equation (7) with $B(t,x) = \mathrm{id}$ and $d \in \mathbb{N}$.

Remark 3. For a fixed index $i \in \mathbb{N}$ and every choice of ν_i the nodes $t_{\ell,i}$ given by (5) are $1/\nu_i$-quantiles w.r.t. a fixed probability density. Sequences of discretizations of this kind are called regular. For approximation of stochastic differential equations regular sequences of discretizations have first been used by [HC96]. See, e.g., [R00] for further results and references.

3 Proofs

3.1 One-Point Approximation of Ornstein-Uhlenbeck Processes

We start with lower and upper error bounds for the approximation of Ornstein-Uhlenbeck processes at the single point $T > 0$. In the sequel we use c to denote unspecified positive constants that may only depend on T.

Fix $y_0 \in \mathbb{R}$, $\mu \geq 1$, as well as a standard one-dimensional Brownian motion β, and consider the Ornstein-Uhlenbeck process $Y = (Y(t))_{t \geq 0}$ given by

$$\begin{aligned} Y(t) &= y_0 \cdot \exp(-\mu \cdot t) + \int_0^t \exp(-\mu \cdot (t-s))\, d\beta(s) \\ &= y_0 \cdot \exp(-\mu \cdot t) + \beta(t) - \mu \cdot \int_0^t \exp(-\mu \cdot (t-s)) \cdot \beta(s)\, ds. \end{aligned}$$

Thus approximation of $Y(T)$ based on a finite number of values of β is an integration problem for a Brownian motion with weight function $t \mapsto \mu \cdot \exp(-\mu \cdot (T-t))$. For this problem asymptotic results, where the number ν of evaluations of β tends to infinity but μ remains fixed, are well known, see, e.g., [R00] for results and references. In the present context, however, we need error bounds that uniformly hold for μ and ν.

Lemma 1. *Let $\nu \in \mathbb{N}$. For all $0 < t_1 < \ldots < t_\nu \leq T$,*

$$E\big(Y(T) - E(Y(T)|\beta(t_1), \ldots, \beta(t_\nu))\big)^2 \geq c \cdot 1/(\mu\nu^2).$$

Moreover, for equidistant nodes $t_\ell = \ell/\nu \cdot T$,

$$E\big(Y(T) - E(Y(T)|\beta(t_1), \ldots, \beta(t_\nu))\big)^2 \geq c \cdot \min(1/\mu, \mu/\nu^2).$$

Proof. Define $m \in \{\nu, \nu+1, \nu+2\}$ and nodes $0 \leq s_1 < \ldots < s_m = T$ by

$$\{s_1, \ldots, s_m\} = \{t_1, \ldots, t_\nu\} \cup \{T - T/\mu, T\}.$$

Clearly,

$$\begin{aligned} &E\big(Y(T) - E(Y(T)|\beta(t_1), \ldots, \beta(t_\nu))\big)^2 \\ &\qquad \geq E\big(Y(T) - E(Y(T)|\beta(s_1), \ldots, \beta(s_m))\big)^2. \end{aligned}$$

Put

$$Z(t) = \beta(t) - E(\beta(t)|\beta(s_1), \ldots, \beta(s_m))$$

for $t \geq 0$. Then

$$Y(T) - E(Y(T)|\beta(s_1), \ldots, \beta(s_m)) = -\mu \cdot \int_0^T \exp(-\mu(T-t)) \cdot Z(t)\, dt.$$

Put $s_0 = 0$ and note that

$$E(Z(s) \cdot Z(t)) = \sum_{k=1}^{m} \frac{(s_k - \max(s,t)) \cdot (\min(s,t) - s_{k-1})}{s_k - s_{k-1}} \cdot 1_{[s_{k-1},s_k]^2}(s,t).$$

Hence

$$\begin{aligned}
&E\big(Y(T) - E(Y(T)|\beta(s_1),\ldots,\beta(s_m))\big)^2 \\
&= \mu^2 \cdot \int_0^T \int_0^T \exp(-\mu(2T - s - t)) \cdot E(Z(s)Z(t))\, ds\, dt \\
&\geq \mu^2 \cdot \exp(-2T) \cdot \int_{T-T/\mu}^T \int_{T-T/\mu}^T E(Z(s)Z(t))\, ds\, dt \\
&= \mu^2 \cdot \exp(-2T) \cdot \sum_{s_k > T-T/\mu} \frac{(s_k - s_{k-1})^3}{12}.
\end{aligned}$$

Let

$$K = \#\{k \in \{1,\ldots,m\} : \ s_k > T - T/\mu\}.$$

By the Hölder inequality,

$$\sum_{s_k > T-T/\mu} (s_k - s_{k-1})^3 \geq T^3/(\mu^3 \cdot K^2),$$

and summarizing we obtain

$$E\big(Y(T) - E(Y(T)|\beta(t_1),\ldots,\beta(t_\nu))\big)^2 \geq 1/12 \cdot \exp(-2T) \cdot T^3 \cdot 1/(\mu \cdot K^2).$$

Now the first statement follows from $K \leq m \leq 3\nu$. In the case of equidistant nodes t_ℓ we have $K \leq \nu/\mu + 1$, which yields the second statement.

For $\nu \in \mathbb{N}$ and $0 = t_0 < \ldots < t_\nu = T$ let

$$\widetilde{Y}(T) = y_0 \cdot \exp(-\mu \cdot T) + \sum_{\ell=1}^{\nu} \exp(-\mu \cdot (T - t_{\ell-1})) \cdot (\beta(t_\ell) - \beta(t_{\ell-1})). \quad (9)$$

We establish upper error bounds for the approximation $\widetilde{Y}(T)$ to $Y(T)$ in the case of equidistant nodes $t_\ell = \ell/\nu \cdot T$ as well as in the case of nodes t_ℓ defined by

$$\int_0^{t_\ell} \exp(-\mu/3 \cdot (T - t))\, dt = \frac{\ell}{\nu} \cdot \int_0^T \exp(-\mu/3 \cdot (T - t))\, dt, \quad (10)$$

cf. (5).

Lemma 2. *For the nodes given by (10),*

$$E(Y(T) - \widetilde{Y}(T))^2 \leq c \cdot 1/(\mu\nu^2).$$

For equidistant nodes,

$$E(Y(T) - \widetilde{Y}(T))^2 \leq c \cdot \min(1/\mu, \mu/\nu^2).$$

Proof. We have

$$
\begin{aligned}
&E(Y(T)-\widetilde{Y}(T))^2\\
&\quad=\sum_{\ell=1}^{\nu}\int_{t_{\ell-1}}^{t_\ell}\left(\exp(-\mu\cdot(T-t))-\exp(-\mu\cdot(T-t_{\ell-1}))\right)^2 dt\\
&\quad=\mu^2\cdot\sum_{\ell=1}^{\nu}\int_{t_{\ell-1}}^{t_\ell}\left(\int_{t_{\ell-1}}^{t}\exp(-\mu\cdot(T-s))\,ds\right)^2 dt\\
&\quad\le\mu^2\cdot\sum_{\ell=1}^{\nu}\int_{t_{\ell-1}}^{t_\ell}\left(\exp(-2\mu/3\cdot(T-t))\cdot\int_{t_{\ell-1}}^{t_\ell}\exp(-\mu/3\cdot(T-s))\,ds\right)^2 dt\\
&\quad\le\mu^2\cdot\max_{\ell=1,\dots,\nu}\left(\int_{t_{\ell-1}}^{t_\ell}\exp(-\mu/3\cdot(T-s))\,ds\right)^2\cdot\int_0^T\exp(-4\mu/3\cdot(T-t))\,dt\\
&\quad\le\mu\cdot\max_{\ell=1,\dots,\nu}\left(\int_{t_{\ell-1}}^{t_\ell}\exp(-\mu/3\cdot(T-s))\,ds\right)^2.
\end{aligned}
$$

It remains to observe that

$$\int_{t_{\ell-1}}^{t_\ell}\exp(-\mu/3\cdot(T-s))\,ds=\frac{1}{\nu}\cdot\int_0^T\exp(-\mu/3\cdot(T-s))\,ds\le 3/(\mu\nu)$$

in case of (10), while

$$\int_{t_{\ell-1}}^{t_\ell}\exp(-\mu/3\cdot(T-s))\,ds\le\min(T/\nu,3/\mu)$$

for equidistant nodes.

Let $\nu\in\mathbb{N}$ and $0=t_0<\ldots<t_\nu=T$. The corresponding drift-implicit Euler scheme for the process Y is given by $\widehat{Y}(0)=y_0$ and

$$\widehat{Y}(t_\ell)=\widehat{Y}(t_{\ell-1})-\mu\cdot\widehat{Y}(t_\ell)\cdot(t_\ell-t_{\ell-1})+\beta(t_\ell)-\beta(t_{\ell-1})$$

for $\ell=1,\ldots,\nu$, cf. (6).

Lemma 3. *For the nodes given by (10),*

$$E(Y(T)-\widehat{Y}(T))^2\le c\cdot 1/\nu^2\cdot(y_0^2+1/\mu).$$

Proof. Due to Lemma 2 it suffices to prove

$$E(\widetilde{Y}(T)-\widehat{Y}(T))^2\le c\cdot 1/\nu^2\cdot(y_0^2+1/\mu).$$

Put $\Delta_\ell=t_\ell-t_{\ell-1}$,

$$\gamma_\ell=\prod_{k=\ell}^{\nu}\frac{1}{1+\mu\cdot\Delta_k}$$

and

$$\delta_\ell = \gamma_\ell - \exp(-\mu \cdot (T - t_{\ell-1}))$$

for $\ell = 1, \dots, \nu$. Then

$$\widehat{Y}(T) = y_0 \cdot \gamma_1 + \sum_{\ell=1}^{\nu} \gamma_\ell \cdot (\beta(t_\ell) - \beta(t_{\ell-1})),$$

and consequently

$$E(\widetilde{Y}(T) - \widehat{Y}(T))^2 = y_0^2 \cdot \delta_1^2 + \sum_{\ell=1}^{\nu} \delta_\ell^2 \cdot \Delta_\ell. \tag{11}$$

Note that

$$\begin{aligned} \delta_\ell &= \frac{\delta_{\ell+1}}{1 + \mu \cdot \Delta_\ell} + \exp(-\mu \cdot (T - t_\ell)) \cdot \Big(\frac{1}{1 + \mu \cdot \Delta_\ell} - \exp(-\mu \cdot \Delta_\ell) \Big) \\ &= \frac{1}{1 + \mu \cdot \Delta_\ell} \cdot \Big(\delta_{\ell+1} + \exp(-\mu \cdot (T - t_\ell)) \cdot \int_0^{\mu \cdot \Delta_\ell} t \cdot \exp(-t)\, dt \Big) \end{aligned} \tag{12}$$

for $\ell = 1, \dots, \nu$, where $\delta_{\nu+1} = 0$. To estimate the quantities δ_ℓ we use the recursion (12) and the fact that the nodes (10) satisfy

$$\mu \cdot \Delta_\ell \leq 3/\nu \cdot \exp(\mu/3 \cdot (T - t_{\ell-1})), \qquad \ell = 1, \dots, \nu, \tag{13}$$

as well as

$$\mu \cdot \Delta_\ell \leq 3 \ln 2, \qquad \ell = 2, \dots, \nu. \tag{14}$$

Assume $\ell \geq 2$. By (13) and (14),

$$\begin{aligned} \exp(-\mu \cdot (T - t_\ell)) \cdot \int_0^{\mu \cdot \Delta_\ell} t \cdot \exp(-t)\, dt &\leq 8 \exp(-\mu \cdot (T - t_{\ell-1})) \cdot (\mu \cdot \Delta_\ell)^2 \\ &\leq 72/\nu^2 \cdot \exp(-\mu/3 \cdot (T - t_{\ell-1})). \end{aligned}$$

Moreover, due to (14),

$$\frac{1}{1 + \mu \cdot \Delta_\ell} \leq \frac{1}{1 + 1/(3 \ln 2) \cdot \mu \cdot \Delta_\ell} \leq \exp(-\kappa \cdot \mu \cdot \Delta_\ell)$$

with $\kappa = 1/(6 \ln 2)$. Employing (12) we obtain

$$\delta_\ell \leq \exp(-\kappa \cdot \mu \cdot \Delta_\ell) \cdot \delta_{\ell+1} + 72/\nu^2 \cdot \exp(-\kappa \cdot \mu \cdot (T - t_{\ell-1})),$$

and by induction we conclude that

$$\delta_\ell \leq 72(\nu - \ell + 1)/\nu^2 \cdot \exp(-\kappa \cdot \mu \cdot (T - t_{\ell-1})) \leq c/\nu \cdot \exp(-\kappa \cdot \mu \cdot (T - t_{\ell-1})). \tag{15}$$

Consequently,

$$\sum_{\ell=2}^{\nu} \delta_\ell^2 \cdot \Delta_\ell \le c/\nu^2 \cdot \int_0^1 \exp(-2\kappa \cdot \mu \cdot (T-t))\, dt \le c \cdot 1/(\mu\nu^2).$$

In view of (11), for finishing the proof it suffices to show that

$$\delta_1 \le c \cdot \frac{1}{\nu \cdot (1 + \mu \cdot \Delta_1)}. \tag{16}$$

Note that

$$\sup_{t \ge 0} t \cdot \exp(-2/3 \cdot t) \le 1.$$

Hence, by definition of the node t_1,

$$\begin{aligned}
&\exp(-\mu \cdot (T - t_1)) \cdot \int_0^{\mu \cdot \Delta_1} t \cdot \exp(-t)\, dt \\
&\le \exp(-\mu/3 \cdot (T - t_1)) \cdot \int_0^{\mu \cdot \Delta_1} \exp(-t/3)\, dt \\
&= \mu \cdot \int_0^{t_1} \exp(-\mu/3 \cdot (T - t))\, dt \\
&\le 3/\nu.
\end{aligned}$$

Due to (15) we have $\delta_2 \le c/\nu$. Now, apply (12) to obtain (16).

3.2 Proof of the Lower Bounds in Theorem 1

Let $N \in \mathbb{N}$ and consider an arbitrary method $\widehat{X}(T) \in \mathfrak{X}_N$ of the form (3). Clearly,

$$\begin{aligned}
E\|X(T) - \widehat{X}(T)\|^2 &= \sum_{i \in \mathbb{N}} E(Y_i(T) - \langle \widehat{X}(T), h_i \rangle)^2 \\
&\ge \sum_{i \in \mathbb{N}} E\big(Y(T) - E(Y(T) | \beta_{i_1}(t_{1,i_1}), \ldots, \beta_{i_k}(t_{\nu_{i_k}, i_k}))\big)^2 \\
&= \sum_{i \in \mathcal{I}} E\big(Y(T) - E(Y(T) | \beta_i(t_{1,i}), \ldots, \beta_i(t_{\nu_i, i}))\big)^2 \\
&\quad + \sum_{i \notin \mathcal{I}} E\big(Y_i^2(T)\big).
\end{aligned}$$

Due to the first statement in Lemma 1

$$\sum_{i \notin \mathcal{I}} E\big(Y_i^2(T)\big) \ge c \cdot \sum_{i \notin \mathcal{I}} 1/\mu_i \ge c \cdot \sum_{i > \#\mathcal{I}} i^{-2} \ge c/(\#\mathcal{I} + 1) \ge c/N,$$

which yields the lower bound for e_N in Theorem 1.

Next, assume that $\widehat{X}(T)$ uses equidistant nodes $t_{\ell,i} = \ell/\nu_i$ for evaluation of β_i with $i \in \mathcal{I}$. Put

$$\mathcal{J} = \{i \in \mathcal{I} : \nu_i > \mu_i\}$$

and use the second statement in Lemma 2 to obtain

$$\begin{aligned} E\|X(T) - \widehat{X}(T)\|^2 &\geq c \cdot \sum_{i \in \mathcal{I}} \min(\mu_i/\nu_i^2, 1/\mu_i) + c \cdot \sum_{i \notin \mathcal{I}} 1/\mu_i \\ &= c \cdot \sum_{i \in \mathcal{J}} \mu_i/\nu_i^2 + c \cdot \sum_{i \notin \mathcal{J}} 1/\mu_i. \end{aligned}$$

By the Hölder inequality

$$\begin{aligned} N^2 \cdot \sum_{i \in \mathcal{J}} \mu_i/\nu_i^2 &\geq \Big(\sum_{i \in \mathcal{J}} \nu_i\Big)^2 \cdot \sum_{i \in \mathcal{J}} i^2/\nu_i^2 \geq \Big(\sum_{i \in \mathcal{J}} i^{2/3}\Big)^{1/3} \\ &\geq \Big(\sum_{i \leq \#\mathcal{J}} i^{2/3}\Big)^{1/3} \geq c \cdot (\#\mathcal{J})^5. \end{aligned}$$

Thus, if $\#\mathcal{J} > N^{1/3}$ then

$$\sum_{i \in \mathcal{J}} \mu_i/\nu_i^2 \geq c \cdot N^{-1/3}.$$

If $\#\mathcal{J} \leq N^{1/3}$ then

$$\sum_{i \notin \mathcal{J}} 1/\mu_i \geq c \cdot 1/(\#\mathcal{J} + 1) \geq c \cdot N^{-1/3},$$

which finishes the proof of the lower bound for e_N^{equi} and e_N^{uni} in Theorem 1.

3.3 Proof of the Upper Bounds in Theorem 1 and Theorem 2

Let $N \in \mathbb{N}$ and consider the the drift-implicit Euler scheme $\widehat{X}_N(T)$ from Section (2). By definition,

$$E\|X(T) - \widehat{X}_N(T)\|^2 = \sum_{i \leq N} E\big(Y_i(T) - \widehat{Y}_{i,N}(T)\big)^2 + \sum_{i > N} E(Y_i(T))^2.$$

Assume $\xi \in C^1([0,1])$ and put $c(\xi) = \max(1, \|\xi\|_\infty + \|\xi'\|_\infty)$. Then

$$\langle \xi, h_i \rangle \leq c \cdot c(\xi) \cdot 1/\mu_i^{1/2}.$$

Hence

$$E(Y_i(T))^2 \leq \langle \xi, h_i \rangle^2 + 1/\mu_i \leq c \cdot (c(\xi))^2 \cdot 1/\mu_i \tag{17}$$

for $i \in \mathbb{N}$, and by Lemma 3

$$E\big(Y_i(T) - \widehat{Y}_{i,N}(T)\big)^2 \leq c \cdot 1/\nu_i^2 \cdot (\langle \xi, h_i\rangle^2 + 1/\mu_i) \leq c \cdot (c(\xi))^2 \cdot 1/(\mu_i \cdot \nu_i^2)$$

for $i \leq N$. Recall that $\nu_i \geq \mu_i^{-1/3} \cdot N^{2/3}$ by definition (4). It follows that

$$\begin{aligned} E\|X(T) - \widehat{X}_N(T)\|^2 &\leq c \cdot (c(\xi))^2 \cdot \Big(N^{-4/3} \cdot \sum_{i \leq N} i^{-2/3} + \sum_{i>N} i^{-2}\Big) \\ &\leq c \cdot (c(\xi))^2 \cdot 1/N, \end{aligned}$$

which finishes the proof of Theorem 2.

For the proof of the upper bounds in Theorem 1 we may assume $\xi = 0$. Hence $e_N \leq c \cdot N^{-1/2}$ follows from Theorem 2. It remains to show that

$$e_N^{\mathrm{uni}} \leq c \cdot N^{-1/6}. \tag{18}$$

Consider the approximation

$$\widetilde{X}_N(T) = \sum_{i=1}^{\lfloor N^{1/3} \rfloor} \widetilde{Y}_{i,N}(T) \cdot h_i,$$

where $\widetilde{Y}_{i,N}(T)$ is defined by (9) with $y_0 = 0$, $\nu = \lfloor N^{2/3} \rfloor$ and $t_\ell = \ell/\nu$. Obviously, $\widetilde{X}_N(T) \in \mathfrak{X}_N^{\mathrm{uni}}$. Use Lemma 2 and observe (17) to obtain

$$\begin{aligned} E\|X(T) - \widetilde{X}_N(T)\|^2 &= \sum_{i \leq \lfloor N^{1/3} \rfloor} E\big(Y_i(T) - \widetilde{Y}_{i,N}(T)\big)^2 + \sum_{i > \lfloor N^{1/3} \rfloor} E(Y_i(T))^2 \\ &\leq c \cdot \sum_{i \leq \lfloor N^{1/3} \rfloor} \min(\mu_i/\nu^2, 1/\mu_i) + c \cdot \sum_{i > \lfloor N^{1/3} \rfloor} 1/\mu_i \\ &\leq c \cdot N^{-4/3} \sum_{i \leq \lfloor N^{1/3} \rfloor} i^2 + c \cdot \sum_{i > \lfloor N^{1/3} \rfloor} i^{-2} \\ &\leq c \cdot N^{-1/3}, \end{aligned}$$

which implies (18).

Acknowledgements

This work is supported by the Deutsche Forschungsgemeinschaft (DFG).

References

[ANZ98] E. J. Allen, S. J. Novosel, and Z. Zhang. Finite element and difference approximation of some linear stochastic partial differential equations, Stochastics Stochastics Rep. **64**, 117–142 (1998)

[DPZ92] G. Da Prato and J. Zabczyk. Stochastic Equations in Infinite Dimensions, Cambridge Univ. Press, Cambridge (1992)

[DG01] A. M. Davie and J. Gaines. Convergence of numerical schemes for the solution of parabolic partial differential equations, Math. Comp. **70**, 121–134 (2001)

[DZ02] Q. Du and T. Q. Zhang. Numerical approximation of some linear stochastic partial differential equations driven by special additive noises, SIAM J. Numer. Anal. **40**, 1421–1445 (2002)

[GK96] W. Grecksch and P. E. Kloeden. Time-discretised Galerkin approximations of parabolic stochastic PDEs, Bull. Austr. Math. Soc. **54**, 79–85 (1996)

[GN97] I. Gyöngy and D. Nualart. Implicit scheme for stochastic parabolic partial differential equations driven by space-time white noise, Potential Analysis **7**, 725–757 (1997)

[H03] E. Hausenblas. Approximation for semilinear stochastic evolution equations, Potential Analysis **18**, 141–186 (2003)

[HC96] Y. Hu and S. Cambanis. Exact convergence rate of the Euler-Maruyama scheme, with application to sampling design, Stochastics Stochastics Rep. **59**, 211–240 (1996)

[MGR07a] T. Müller-Gronbach and K. Ritter. Lower bounds and nonuniform time discretization for approximation of stochastic heat equations. Found. Comput. Math. **7**, 135–181 (2007)

[MGR07b] T. Müller-Gronbach and K. Ritter. An implicit Euler scheme with non-uniform time discretization for heat equations with multiplicative noise. BIT **47**, 393–418 (2007)

[R00] K. Ritter. Average-Case Analysis of Numerical Problems, Lect. Notes in Math. **1733**, Springer-Verlag, Berlin (2000)

[S99] T. Shardlow. Numerical methods for stochastic parabolic PDEs, Numer. Funct. Anal. Optim. **20**, 121–145 (1999)

[Y05] Y. Yan. Galerkin finite element methods for stochastic parabolic partial differential equations, SIAM J. Numer. Anal. **43**, 1363–1384 (2005)

Unbiased Global Illumination with Participating Media

Matthias Raab[1], Daniel Seibert[2], and Alexander Keller[3]

[1] Ulm University, Germany
matthias.raab@uni-ulm.de
[2] mental images GmbH, Fasanenstr. 81, D-10623 Berlin, Germany
daniel@mental.com
[3] Ulm University, Germany
alexander.keller@uni-ulm.de

Summary. We present techniques that enhance global illumination algorithms by incorporating the effects of participating media. Instead of ray marching we use a sophisticated Monte Carlo method for the realization of propagation events and transmittance estimations. The presented techniques lead to unbiased estimators of the light transport equation with participating media.

1 Introduction

A complete computation of all illumination effects is critical for the synthesis of photorealistic images. This means that the *global illumination* problem has to be solved. Global illumination is an important problem in graphics and of high interest for applications like architecture, industrial design, and even production. Accordingly, there are a great number of approaches that attempt to solve the task of simulating light transport through virtual three-dimensional scenes in an unbiased and physically correct way. Nevertheless, most algorithms lack the ability to correctly estimate the effects caused by interactions with media like smoke, fog, or dust.

The most sophisticated unbiased approaches are Bidirectional Path Tracing [LW93, VG94] and the Metropolis Light Transport algorithm [VG97, KSKAC02]. All of these algorithms are robust and can capture a wide variety of illumination effects, albeit with varying efficiency. Additionally, extensions to scenes with participating media were presented in [LW96, PKK00]. Note, however, that these techniques are not unbiased as they rely on ray marching techniques [PH89] to sample distances and to approximate the transmittance of participating media.

2 Light Transport with Participating Media

In the theory of radiative transfer there is generally a distinction between solid objects, i.e. those that do not allow light to pass through them, and diaphanous media like gases and liquids. Scenes are therefore often modeled as a volume $\mathcal{V}$ and its boundary, i.e. the surface of solid objects $\partial\mathcal{V}$. Note that we assume $\mathcal{V}$ to be an open set, so that $\mathcal{V} \cap \partial\mathcal{V} = \emptyset$. On the surface $\partial\mathcal{V}$ the *local scattering equation* governs light transport

$$L(x, \omega) = L_{e,\partial\mathcal{V}}(x, \omega) + \int_{\mathcal{S}^2} f_s(\omega, x, \omega') L(x, \omega') |\cos \theta_x| d\sigma(\omega'). \tag{1}$$

Here, $\mathcal{S}^2$ is the set of all directions, f_s is the *bidirectional scattering distribution function* (BSDF), which describes the scattering behavior at x, and $\cos \theta_x$ is the cosine of the angle between direction ω' and the surface normal in x. This formulation is sufficient for scenes in a vacuum, where all interaction events occur on the surface.

In order to account for effects caused by participating media inside the volume, we have to consider the *equation of transfer*

$$\begin{aligned} \frac{\partial}{\partial \omega} L(x, \omega) &= L_{e,\mathcal{V}}(x, \omega) - \sigma_t(x) L(x, \omega) \\ &\quad + \sigma_s(x) \int_{\mathcal{S}^2} f_p(\omega, x, \omega') L(x, \omega') d\sigma(\omega'), \end{aligned} \tag{2}$$

which describes the radiance change at position x in direction ω due to volume emission and interaction events. The medium's scattering and absorption characteristics are given by the phase function f_p, the scattering coefficient σ_s, and the absorption coefficient σ_a. The latter two form the extinction coefficient $\sigma_t := \sigma_s + \sigma_a$. Usually, equation (2) is integrated along straight light rays to the next surface interaction point $x_{\mathcal{S}} = h(x, -\omega)$, which is found by the *ray casting function* h. This approach yields a Fredholm integral equation of the second kind, which can be handled by Monte Carlo techniques.

2.1 Path Integral Formulation

A very general formulation of light transport is given in [Vea97] and has been extended to handle participating media in [PKK00] and [Kol04]. The local description by equations (1) and (2) is recursively expanded to obtain an integral over an abstract space of complete transport paths.

The path space $\mathcal{P}$ can be modeled as the union of spaces containing paths of a specific finite length, i.e.

$$\mathcal{P} := \bigcup_{k \in \mathbb{N}} \mathcal{P}_k \text{ where } \mathcal{P}_k := \left\{ \bar{x} = x_0 \ldots x_k : x_i \in \mathbb{R}^3 \right\}.$$

For each of these spaces we use a product measure μ_k, defined for a set $\mathcal{M}_k \subseteq \mathcal{P}_k$ by

$$\mu_k(\mathcal{M}_k) := \int_{\mathcal{M}_k} d\lambda(x_0)d\lambda(x_1)\cdots d\lambda(x_k),$$

of the corresponding Lebesgue measure on the volume and on the surface, i.e.

$$d\lambda(x) := \begin{cases} dA(x) & \text{if } x \in \partial\mathcal{V} \\ dV(x) & \text{if } x \in \mathcal{V}. \end{cases}$$

The measure μ of a set $\mathcal{M} \subseteq \mathcal{P}$ then is the natural expansion of those disjoint spaces' measures

$$\mu(\mathcal{M}) := \sum_{k \in \mathbb{N}} \mu_k\left(\mathcal{M} \cap \mathcal{P}_k\right).$$

In this context, the sensor response I_j of pixel j can be expressed as an integral over $\mathcal{P}$,

$$I_j = \int_{\mathcal{P}} f_j(\bar{x})d\mu(\bar{x}), \tag{3}$$

where f_j is called the *measurement contribution function*. In order to find this function, we describe light transport in a slightly different manner. Let $L(y \to z)$ denote the radiance scattered and emitted from point y in direction $\overrightarrow{yz} := \frac{z-y}{\|z-y\|}$. Inside the volume this quantity is given by

$$L(y \to z) = L_{e,\mathcal{V}}(y, \overrightarrow{yz}) + \int_{\mathcal{S}^2} \sigma_s(y) f_p(\overrightarrow{yz}, y, \omega) L(y, \omega) d\sigma(\omega). \tag{4}$$

On the surface we obtain $L(y \to z)$ directly by equation (1), i.e. $L(y \to z) = L(y, \overrightarrow{yz})\ \forall y \in \partial\mathcal{V}$.

Using these notions in the integration of (2) with boundary condition (1) and changing the integration domain to $\mathbb{R}^3$ yields the *three point form*:

$$\begin{aligned} L(y \to z) &= L_e(y \to z) \\ &\quad + \lim \int_{\mathbb{R}^3} L(x \to y) f(x \to y \to z) G(x \leftrightarrow y) V(x \leftrightarrow y) d\lambda(x), \end{aligned} \tag{5}$$

where the following abbreviations are used (see [Kol04] for a full derivation):

- Three point scattering function

$$f(x \to y \to z) := \begin{cases} f_r(\overrightarrow{yz}, y, \overrightarrow{xy}) & \text{if } y \in \partial\mathcal{V} \\ \sigma_s(y) f_p(\overrightarrow{yz}, y, \overrightarrow{xy}) & \text{if } y \in \mathcal{V} \end{cases} \tag{6}$$

- Source radiance distribution

$$L_e(x \to y) := \begin{cases} L_{e,\partial\mathcal{V}}(x, \overrightarrow{xy}) & \text{if } y \in \partial\mathcal{V} \\ L_{e,\mathcal{V}}(x, \overrightarrow{xy}) & \text{if } y \in \mathcal{V} \end{cases} \tag{7}$$

- Geometric term

$$G(x \leftrightarrow y) := \begin{cases} \frac{|\cos\theta_x||\cos\theta_y|}{\|y-x\|^2} & \text{if } x, y \in \partial\mathcal{V} \\ \frac{|\cos\theta_x|}{\|y-x\|^2} & \text{if } x \in \partial\mathcal{V}, y \in \mathcal{V} \\ \frac{|\cos\theta_y|}{\|y-x\|^2} & \text{if } y \in \partial\mathcal{V}, x \in \mathcal{V} \\ \frac{1}{\|y-x\|^2} & \text{if } x, y \in \mathcal{V} \end{cases} \tag{8}$$

with θ_x and θ_y the angles between $\overrightarrow{xy}$ and the surface normals at the respective points
- Attenuated visibility function

$$\begin{aligned} V(x \leftrightarrow y) &:= V'(x \leftrightarrow y)\tau(x \leftrightarrow y) \\ &= V'(x \leftrightarrow y)e^{-\int_0^{\|y-x\|} \sigma_t(x+t\overrightarrow{xy})dt} \end{aligned} \tag{9}$$

with the standard binary visibility function

$$V'(x \leftrightarrow y) = \begin{cases} 1 & \text{if } \|y-x\| \leq \|h(x, \overrightarrow{xy}) - x\| \\ 0 & \text{otherwise} \end{cases}$$

Finally, we need a function $\psi_j(x_{k-1} \to x_k)$ to describe the sensor response of a pixel j to radiance arriving from x_{k-1} at the point x_k on the image plane. Along with the recursive expansion of the three point form (5) this yields the measurement contribution function for a path $\bar{x} = x_0 \ldots x_k$ beginning on a light source and ending on the image plane:

$$\begin{aligned} f_j(\bar{x}) := {} & L_e(x_0 \to x_1)G(x_0 \leftrightarrow x_1)V(x_0 \leftrightarrow x_1) \\ & \cdot \prod_{l=1}^{k-1} \left(f(x_{l-1} \to x_l \to x_{l+1})G(x_l \leftrightarrow x_{l+1})V(x_l \leftrightarrow x_{l+1})\right) \\ & \cdot \psi_j(x_{k-1} \to x_k) \end{aligned} \tag{10}$$

3 Unbiased Techniques for Transport Path Sampling

In a vacuum the next interaction point is fixed as the closest surface point along the ray. With participating media, however, the position of this point is no longer given deterministically but described by a stochastic process. From the structure of equation (10) it is obvious that we have two separate operators that describe the transport: one governing the distance to the next interaction point (the τ term) and the other governing the scattering behavior (the phase function or the BSDF). We treat these factors independently as in [PKK00] by first sampling a distance and then sampling a direction, as described in the next section. Since the processes are clearly independent, the resulting density is simply the product of the two densities involved.

Homogeneous Media

In the case of a homogeneous medium with $\sigma_t(x) \equiv \sigma_t$, the transmittance is proportional to the exponential distribution's density $p(t) = \sigma_t e^{-\sigma_t t}$. Applying the inversion method we realize the desired distance as

$$t = \frac{-\ln(1-\xi)}{\sigma_t} \tag{11}$$

for a uniformly distributed random number $\xi \in [0,1)$.

Heterogeneous Media

In the general case, the density proportional to $\tau = e^{-K(t)}$ with $K(t) := \int_0^t \sigma_t(x_0 + t'\omega)dt'$ is given by $p(t) = \left(\frac{d}{dt}K(t)\right)e^{-K(t)} = \sigma_t(x_0 + t\omega)e^{-K(t)}$. Things are more complicated here since the inversion method only yields the implicit equation

$$K(t) = \int_0^t \sigma_t(x_0 + t'\omega)dt' = -\ln(1-\xi) \tag{12}$$

for a uniformly distributed random number $\xi \in [0,1)$. As the inversion method cannot be applied directly and there is no straightforward Monte Carlo estimator available (as we cannot evaluate τ), this distance is usually sampled using the classic ray marching algorithm [PH89]. Note that this method, which is frequently used in computer graphics, is biased.

A more sophisticated and unbiased approach to sampling this distance can be found in [Col68]. Let σ_t be a constant with $\sigma_t \geq \sigma_t(x)$ for all $x \in \mathcal{V}$ and let $t_1, t_2, \ldots$ be independent random distances sampled according to equation (11) with parameter σ_t. Let, furthermore, $\xi_1, \xi_2, \ldots \in [0,1]$ be independent uniformly distributed random numbers. Then, the first distance $T_{i_0} = \sum_{i=1}^{i_0} t_i$ satisfying $\xi_{i_0} \leq \frac{\sigma_t(x_0 + T_{i_0}\omega)}{\sigma_t}$ is distributed as desired. Algorithm 1 implements this procedure and thus provides an unbiased distance sampling routine for arbitrary media.

In fact, the condition $\sigma_t \geq \sigma_t(x)$ only has to be satisfied for points along the ray. Choosing a larger σ_t simply yields more iterations in Algorithm 1. However, determining the maximum $\sigma_t = \sup_{x\in\mathcal{V}} \sigma_t(x)$ in the whole volume data set during a preprocessing step is usually easy while determining the maximum along a single ray may be quite complicated.

3.1 Line Integral along a Ray

An explicit estimation of the transmittance τ is quite important in many algorithms. Furthermore, splitting along the primary ray is often beneficial when rendering scenes that contain participating media. We therefore

```
float sampleDistance(Point x0, Direction ω)
{
    //sample with the maximum extinction σt
    float t = -log(rand()) / σt;

    while (σt(x0+tω)/σt < rand())
        t -= log(rand()) / σt;

    return t;
}
```

Algorithm 1. Unbiased distance sampling for arbitrary media.

Fig. 1. Samples along a ray $x_0 + t\omega$ for $t \in [0, t_{\partial\mathcal{V}})$: transformed random points (top) and transformed equidistant points (bottom). The latter yield a better distribution for the initial points in Algorihm 1.

generalize the one-dimensional integration along a ray in the context of Algorithm 1 in order to obtain a generic and unbiased solution.

Assume that we have to estimate a transmittance-weighted integral C along a light ray $(x_{\partial\mathcal{V}}, \omega)$ starting at the surface intersection point $x_{\partial\mathcal{V}}$ as present in the transport equations,

$$C = c_{\partial\mathcal{V}}(x_{\partial\mathcal{V}})\,\tau(x \leftrightarrow x_{\partial\mathcal{V}}) + \int_0^{\|x_{\partial\mathcal{V}} - x\|} c_{\mathcal{V}}(x - t\omega)\tau(x \leftrightarrow x - t\omega)dt, \qquad (13)$$

where $c_{\mathcal{V}}$ and $c_{\partial\mathcal{V}}$ are volume and surface contributions, respectively. For the simple case $c_{\partial\mathcal{V}} \equiv 1$ and $c_{\mathcal{V}} \equiv 0$ we obtain an estimate of the transmittance itself. We can estimate C by applying Algorithm 1 repeatedly and averaging the corresponding contributions. Instead of using random numbers for the starting points of Algorithm 1 the convergence can be improved by transforming equidistant samples that are shifted by one random offset (see Figure 1). Algorithm 2 is a formulation of this approach where we additionally have seperated what we know from the maximum extinction.

A comparison for splitting along the primary ray in the scenario of shading from a large set of point light sources representing global illumination is given in Figure 2.

1. Determine the surface intersection point $x_{\partial V} = h(x, -\omega)$, the distance $t_{\partial V} = \|x - x_{\partial V}\|$, and the maximum volume extinction σ_t
2. Compute the probabilities $p_V = 1 - e^{-\sigma_t t_{\partial V}}$ and $p_{\partial V} = 1 - p_V$
3. Estimate the surface contribution $c_{\partial V} = L(x_{\partial V}, \omega)$ and set $C = p_{\partial V} \cdot c_{\partial V}$
4. Generate n randomly shifted equidistant sample points $\Delta \subset [0, p_V)$

$$\Delta := \left\{ \frac{(k + \xi)p_V}{n} : k = 0, \ldots, n-1 \right\}$$

for one uniformly distributed random number $\xi \in [0, 1)$
5. Use Δ as initial random numbers for n random walks according to Algorithm 1 resulting in distances $t_1, \ldots, t_n$
6. Add contributions

$$C = C + \begin{cases} \frac{1}{n} \cdot p_V \cdot \frac{c(x - t_k\omega, \omega)}{\sigma_t(x - t_k\omega)}, & \text{if } t_k < t_{\partial V} \\ \frac{1}{n} \cdot p_V \cdot c_{\partial V}, & \text{else} \end{cases}$$

for $k = 1, \ldots, n$

Algorithm 2. Unbiased line integration along a ray.

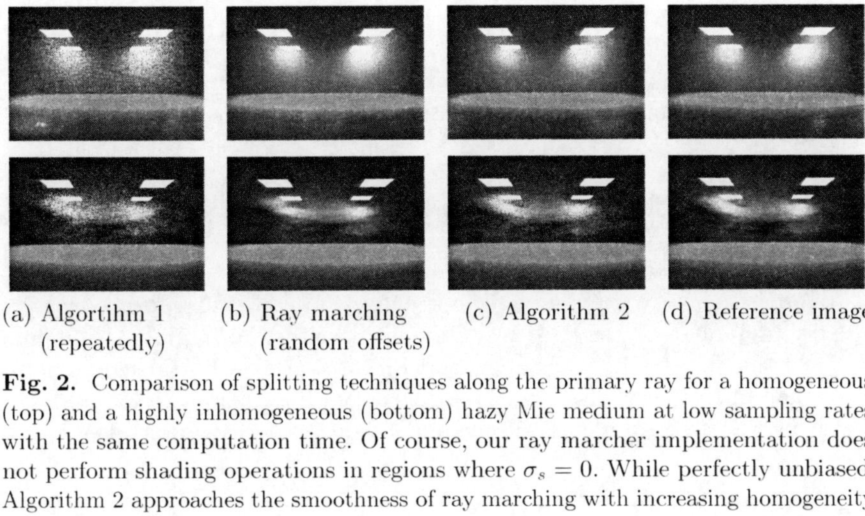

(a) Algortihm 1 (repeatedly) (b) Ray marching (random offsets) (c) Algorithm 2 (d) Reference image

Fig. 2. Comparison of splitting techniques along the primary ray for a homogeneous (top) and a highly inhomogeneous (bottom) hazy Mie medium at low sampling rates with the same computation time. Of course, our ray marcher implementation does not perform shading operations in regions where $\sigma_s = 0$. While perfectly unbiased, Algorithm 2 approaches the smoothness of ray marching with increasing homogeneity along the ray. The reference image has been computed using Algorithm 1 with a vast amount of samples.

3.2 Handling Multiple Wavelengths

It is a common notion in computer graphics that solving the transport equations separately for each wavelength is less efficient than simulating various wavelengths at once. While this may be true for moderately saturated colors, the general setting requires some additional considerations for sampling the BSDF. Color-dependent implementations of Russian roulette [SSKK03] can help but still cannot avoid infinite variance in general and are impractical for bidirectional path tracing, as the probability density evalutions for the heuristics become extremely complicated.

Using Algorithm 1 in a context where σ_t depends on the wavelength is also problematic when computing a single solution for several wavelengths. Consider sampling equation (13) with $c_V \equiv 1$ in an infinite homogeneous medium for two wavelengths λ_1 and λ_2 with $\sigma_{t,\lambda_2} > \sigma_{t,\lambda_1}$ simultaneously, using the density $p_{\lambda_2}(t) = \sigma_{t,\lambda_2} e^{-\sigma_{t,\lambda_2} t}$. The variance on wavelength λ_2 is now zero, whereas the variance on λ_1 can be arbitrarily high:

$$V\left(\frac{\tau_{\lambda_1}(x \leftrightarrow y)}{p_{\lambda_2}}\right) = \begin{cases} \infty & \text{if } \sigma_{t,\lambda_2} \geq 2\sigma_{t,\lambda_1} \\ \left(2\sigma_{t,\lambda_1}\sigma_{t,\lambda_2} - \sigma_{t,\lambda_2}^2\right)^{-1} - (\sigma_{t,\lambda_1})^{-2} & \text{otherwise.} \end{cases}$$

The problem can be avoided by limiting the extinction coefficient to a scalar value, i.e. by forcing $\sigma_{a,\lambda_1}(x) + \sigma_{s,\lambda_1}(x) = \sigma_{a,\lambda_2}(x) + \sigma_{s,\lambda_2}(x) = \sigma_t(x)$ for every pair of wavelengths λ_1, λ_2. This corresponds to situations where the extinction coefficent is a quanitity depending on volume particle density and size only. Then, the color of a medium is due to the reflection, refraction, and absorption probabilities of the particles themselves (in analogy to the BSDF) and may vary within this constraint. However, in scenarios where this restriction cannot be applied (e.g. atmospheric scattering) solving the transport equations for each wavelength separately should be preferred.

4 Applications

In order to obtain truly unbiased estimators for light transport, we incorporate the presented techniques into several Monte Carlo global illumination algorithms. These include simple and Bidirectional Path Tracing, an unbiased variant of Instant Radiosity for Participating Media, and a very robust version of the Metropolis Light Transport algorithm.

4.1 Path Tracing

Path Tracing [Kaj86] is one of the most basic global illumination algorithms. Due to its simplicity, it is still used frequently to compute reference solutions or handle complex scenes. A recently presented variant called Adjoint Photon Tracing [MBE+06] is capable of rendering difficult settings with participating media accurately. The simple form of *pure* Path Tracing without next event estimation, which does not estimate direct illumination at each path vertex explicitly, can be extended to handle participating media without much effort. The only modification is an additional distance sampling call for each ray that is cast.

The spatial sampling found in Path Tracing with next event estimation requires an estimate of the transmittance. This quantity is easily obtained with Algorithm 2. Note that the stability of Path Tracing may be compromised in this case: a path vertex can be sampled arbitrarily close to a light source, in which case the geometric term of the connection yields a weakly singular integrand with infinite variance. This is generally a problem for algorithms with next event estimation, though it is often avoided in a vacuum by only modeling light sources with a certain minimum distance to other surface points. With participating media the same problems arise from non-vacuum volume points near light sources. Such cases are usually much harder to avoid. However, the techniques presented below can be applied to handle the weak singularity in an unbiased manner.

4.2 Instant Radiosity

Instant Radiosity [Kel97] is a popular global illumination algorithm for scenes with predominantly diffuse surfaces. A set of transport paths is started from the light source in a preprocessing pass and point light sources are stored at each path vertex. The point lights, which represent path space samples, are then used to shade each camera ray's first interaction point.

Adapting this process for participating media using Algorithm 1 is straightforward. Note that the volume point light sources have to be equipped with a phase function instead of a BSDF. Shading the primary ray can be done by sampling a first interaction or, preferably, applying some splitting in the sense of Algorithm 2.

Solutions computed with Instant Radiosity can converge quite quickly, given that the weak singularity found in the shading path is avoided by bounding the geometric term [Kol04]. However, this approach introduces bias.

Bias Compensation

In order to handle the singularity without introducing bias we extend the method from [KK04] to participating media. We set the geometric term G as defined in equation (8) to G' by letting

$$G'(x \leftrightarrow y) := \begin{cases} G(x \leftrightarrow y) & \text{if } G(x \leftrightarrow y) < b \\ b & \text{otherwise} \end{cases} \tag{14}$$

for an arbitrary positive bound $b \in \mathbb{R}^+$. The bias introduced into the evaluation of the three point form $L(y \to z)$ as defined in equation (5) by replacing G with G' is

$$
\begin{aligned}
&L(y \to z) - L'(y \to z) \\
&= \int_{\mathbb{R}^3} L(x \to y) f(x \to y \to z) \cdot \max\{G(x \leftrightarrow y) - b, 0\} V(x \leftrightarrow y) d\lambda(x) \\
&= \int_{\mathbb{R}^3} L(x \to y) f(x \to y \to z) \frac{\max\{G(x \leftrightarrow y) - b, 0\}}{G(x \leftrightarrow y)} \\
&\quad \cdot G(x \leftrightarrow y) V(x \leftrightarrow y) d\lambda(x) \\
&= \int_{S^2} \int_0^{\|h(y, -\omega) - y\|} \tau(y - t\omega \leftrightarrow y) L(y - t\omega \to y) f(y - t\omega \to y \to z) \\
&\quad \cdot \frac{\max\{G(y - t\omega \leftrightarrow y) - b, 0\}}{G(y - t\omega \leftrightarrow y)} dt d\sigma^*(\omega). \qquad (15)
\end{aligned}
$$

In the last step we have changed the integration domain from $\mathbb{R}^3$ back to spherical coordinates, and depending on whether y is a surface or a volume point we have

$$
d\sigma^*(\omega) = \begin{cases} |\cos \theta_y| d\sigma(\omega) & \text{if } y \in \partial\mathcal{V} \\ d\sigma(\omega) & \text{if } y \in \mathcal{V}. \end{cases}
$$

This reformulation directly leads to a recursive algorithm for computing the bias which does not suffer from any singularities. We simply sample a direction ω according to the projected BSDF or the phase function, and a distance t according to τ. At the resulting next vertex $x = y - t\omega$, we recursively estimate $L(x \to y)$ and weight the result by $\frac{\max\{G(x \leftrightarrow y) - b, 0\}}{G(x \leftrightarrow y)}$. This weight is 0 for $G(x \leftrightarrow y) \leq b$. If the next vertex is not close enough, we can thus terminate the path from the camera. In fact, the bounding and bias compensation step can be interpreted as s special case of Bidirectional Path Tracing where the weighting functions are constructed with respect to the value of the geometric term.

The efficiency of the approach is, of course, highly dependent on the choice of the bound b. Generally, one wants avoid bright spots in weakly illuminated areas caused by close by point lights. Options to achieve this include bounding point light contributions and bias compensation to the same maximum value [KK04] or using radiance estimates (e.g. based on direct illumination) and bound contributions of the point light sources to values below. Note that the contribution of the bias compensation step is always bounded by the brightness of the scene's light source.

4.3 Bidirectional Path Tracing

Combining Path Tracing and its adjoint approach, Light Tracing, leads to Bidirectional Path Tracing (BDPT) [LW93, VG94]. The algorithm uses a whole family of sampling techniques (Path Tracing and Instant Radiosity are two of them), which are combined using *multiple importance sampling* [VG95]. The *multi-sample estimator*

$$F = \sum_{i=1}^{n} \frac{1}{n_i} \sum_{j=1}^{n_i} w_i(x_{i,j}) \frac{f(x)}{p_i(x_{i,j})} \tag{16}$$

joins the samples $x_{i,j}$ that were created according to the density p_i. The estimator is unbiased as long as the weights $w_i(x)$ sum up to 1 for $f(x) \neq 0$ and are 0 for $p_i(x) = 0$. In fact, this is also true if the weights are only normalized in expectation, i.e.

$$E_y\Big(\sum_{i=1}^{n} w_i(x,y)\Big) = \int_{\Omega} \Big(\sum_{i=1}^{n} w_i(x,y)\Big) d\mu(y) = 1 \qquad \forall x. \tag{17}$$

A good choice for a weighting function that satisfies these conditions is the *power heuristic*

$$w_s(x) = \frac{p_s^{\beta}(x)}{\sum_i p_i^{\beta}(x)} \qquad \text{for } \beta \in \mathbb{R}^+. \tag{18}$$

For $\beta = 1$ we obtain the *balance heuristic* and $\beta \to \infty$ results in the *maximum heuristic*. As shown in [Vea97], these heuristics guarantee a fairly low variance.

The application of BDPT to participating media is given in [LW96] and can easily be modified to handle propagation events and transmittance estimations in an unbiased way by utilizing Algorithm 1 and Algorithm 2. However, the weighting heuristics need some additional consideration. For heterogeneous media, the densities $p_i \propto \tau$ due to propagation are not analytically computable and must therefore be approximated. Using Algorithm 2 for this purpose is generally not an option, as equation (17) fails for all but the maximum heuristic. However, note that no bias is introduced if we approximate τ by a deterministic quadrature rule, since the weights produced by equation (18) will definitely sum up to one then. Presuming a decent approximation, the good properties of the heuristics are preserved.

4.4 Metropolis Light Transport

Metropolis Light Transport (MLT) is a powerful alternative to the previous rendering approaches. The algorithm, which was first presented in [VG97] and strongly modified in [KSKAC02], uses Metropolis sampling [MRR⁺53] to sample the path space. Whereas ordinary importance sampling only considers factors of the measurement contribution in the previous algorithms, the flexibility of Metropolis sampling allows us to generate paths according to

$$p(\bar{x}) = \frac{f(\bar{x})}{b} \quad \text{with } b = \int_{\mathcal{P}} f(\bar{y})\, d\mu(\bar{y}).$$

This yields the importance sampling estimator

$$F_{j,N} = \frac{1}{N} \sum_{i=1}^{N} h_j(\bar{X}_i)\, b,$$

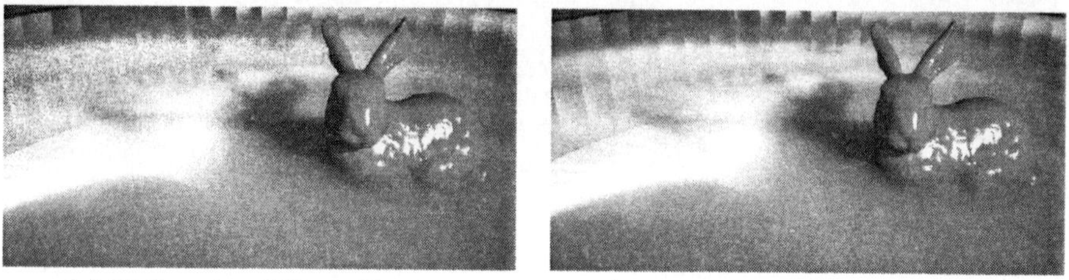

(a) Bidirectional Path Tracing (b) Metropolis Light Transport

Fig. 3. A scene with a heterogeneous medium featuring caustics seen indirectly through a reflection. Both images were rendered with the same number of samples in approximately the same time. Splitting along the primary ray was employed to speed up the estimation of direct light for MLT.

where the measurement contribution function $f_j(\bar{x})$ has been split into a pixel filter function $h_j(\bar{x})$ and a remainder $f(\bar{x})$, which is the same for every pixel.

The value of b can be estimated using any suitable rendering algorithm and a fairly low number of samples.Usually, 10^4–10^5 samples yield a sufficiently accurate estimate, which makes the cost of the initialization phase negligible. One of the approximation samples is selected as the initial state for the Metropolis phase of the algorithm with probability proportional to f/p. The selected sample is expected to be distributed according to the stationary density and thus avoids start-up bias without the need to discard any samples.

Once the first state has been selected, each new state is found by generating a tentative sample $\bar{y}$ from the current sample $\bar{x}$ according to the tentative transition function $T(\bar{x} \to \bar{y})$ and either accepting or rejecting the proposal according to the acceptance function

$$a(\bar{x} \to \bar{y}) := \min\left\{1, \frac{f(\bar{y})T(\bar{y} \to \bar{x})}{f(\bar{x})T(\bar{x} \to \bar{y})}\right\}.$$

A good proposal strategy is of paramount importance to the success of the algorithm. Key features of a sound strategy are the ability to exploit the coherence in the scene and a low correlation between subsequent samples.

A number of important optimizations of the basic algorithm may be found in [Vea97]. While they will not be discussed here, they are of great importance to a successful implementation of MLT.

Adaptive Mutation

Kelemen et al. present a novel implementation of the MLT algorithm in [KSKAC02], which we adapt to handle participating media. The new mutation is simpler to implement than that proposed by Veach, reduces the correlation between samples, and is considerably more robust. Furthermore, the inclusion of participating media and other phenomena does not require extensive modifications to the mutation.

For any path tracing algorithm—e.g. classic Path Tracing or BDPT—, a path is uniquely defined by the set of random numbers used to create it. These numbers can be interpreted as a point in the infinitely-dimensional unit cube $[0,1)^\infty$, called the *primary sample space* $\mathcal{U}$. The transformation between the spaces, $s\colon \mathcal{U} \to \mathcal{P}$, is determined by the path tracing algorithm. The path integral presented in equation (3) can then be transformed to an integral over the primary sample space:

$$I_j = \int_{\mathcal{P}} f_j(\bar{x})\, d\mu(\bar{x}) = \int_{\mathcal{U}} \frac{f_j(s(\bar{u}))}{p(s(\bar{u}))} d\mu^*(\bar{u}) \,,$$

where f and p are defined as before.

If the chosen path tracing algorithm properly employs importance sampling, the transformed integrand, $f^*(\bar{u}) := f(s(\bar{u}))\,/p(s(\bar{u}))$, will vary only moderately. Any mutation performed in the primary sample space will thus automatically adapt to the modalities of the integrand.

The proposed mutation generates a new path by perturbing the primary sample point that corresponds to the current path by a small exponentially distributed amount. This perturbation is symmetric, so that the acceptance probability simplifies to $a(\bar{u} \to \bar{v}) = \min\{1, f^*(\bar{v})\,/f^*(\bar{u})\}$.

Like Veach's original perturbations, the new mutation cannot ensure ergodicity by itself. To guarantee that the algorithm cannot get stuck in isolated areas of the scene, independent paths are generated at random intervals by the underlying path tracer. This is done by simply feeding a fresh set of uniform random numbers into the algorithm. The mutation type is chosen at random before each mutation step.

Transformation

The presented Metropolis algorithm only works efficiently if the transformation $s\colon \mathcal{U} \to \mathcal{P}$ ensures that a small change in $\mathcal{U}$ corresponds to a small change in the path space. Because of the robustness and efficiency of the approach, we use BDPT as the basic sample generation technique. Due to the separate generation of subpaths by BDPT and the inclusion of participating media, it is not possible to use values from successive dimensions of $\mathcal{U}$ as random numbers for the generation of samples and still satisfy the requirement of corresponding small changes. Rather, the values driving distinct parts of the path generation must be separated from each other.

Random numbers used for the generation of eye and light subpaths can simply be separated e.g. by assigning positive indices to one type of subpath and negative indices to the other type. While the number of random values needed to sample distances in homogeneous media is fixed, the amount of random input needed to drive the presented distance sampling routine cannot be determined in advance in scenes that contain heterogeneous media. In such cases, each deterministic connection between two subpaths also needs an

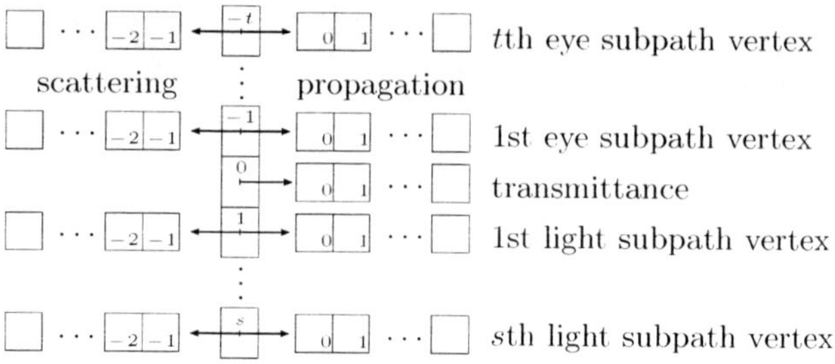

Fig. 4. A primary sample stored as an array of arrays. Random numbers are separated between light and eye subpaths, as well as between scattering and propagation.

additional unknown amount of random values to estimate the transmittance as described in section 3.1. Finally, some scattering models may need more input than others when sampling a direction.

We therefore propose storing the current primary sample $\bar{u}$ in an array of arrays as outlined in Figure 4. The ith row vector provides input to the sampling that is done at the $|i|$th vertex of the respective subpath. The elements of each vector are accessed in sequential order. This separation ensures that the random numbers that generated each vertex remain associated to that vertex. The estimation of the various path transmittance values may be driven by the row at index 0. A further separation is not necessary because the transmittance is fairly smooth in realistic settings. Large variations in the random values running the estimation thus have very little effect on the result.

5 Conclusion

We have presented unbiased global illumination algorithms for scenes with participating media. We thus close a gap in computer graphics where many algorithms are labeled as unbiased despite the fact that this claim was previously not true for heterogeneous media. The new approaches presented here allow for the physically accurate and efficient visualization of a wide range of scenes.

Acknowledgements

The first author has been funded by the project *information at your fingertips - Interaktive Visualisierung für Gigapixel Displays*, Forschungsverbund im Rahmen des Förderprogramms Informationstechnik in Baden-Württemberg (BW-FIT).

References

[Col68] W. Coleman. Mathematical Verification of a certain Monte Carlo Sampling Technique and Applications of the Technique to Radiation Transport Problems. *Nuclear Science and Engineering*, 32:76–81, 1968.

[Kaj86] J. Kajiya. The Rendering Equation. In *SIGGRAPH 86 Conference Proceedings*, volume 20 of *Computer Graphics*, pages 143–150, 1986.

[Kel97] A. Keller. Instant Radiosity. In *SIGGRAPH 97 Conference Proceedings*, Annual Conference Series, pages 49–56, 1997.

[KK04] T. Kollig and A. Keller. Illumination in the Presence of Weak Singularities. In D. Talay and H. Niederreiter, editors, *Monte Carlo and Quasi-Monte Carlo Methods*, pages 243–256. Springer, 2004.

[Kol04] T. Kollig. *Efficient Sampling and Robust Algorithms for Photorealistic Image Synthesis*. PhD thesis, University of Kaiserslautern, Germany, 2004.

[KSKAC02] C. Kelemen, L. Szirmay-Kalos, G. Antal, and F. Csonka. A Simple and Robust Mutation Strategy for the Metropolis Light Transport Algorithm. *Computer Graphics Forum*, 21(3):531–540, September 2002.

[LW93] E. Lafortune and Y. Willems. Bidirectional Path Tracing. In *Proc. 3rd International Conference on Computational Graphics and Visualization Techniques (Compugraphics)*, pages 145–153, 1993.

[LW96] E. Lafortune and Y. Willems. Rendering Participating Media with Bidirectional Path Tracing. *Rendering Techniques '96 (Proc. 7th Eurographics Workshop on Rendering)*, pages 91–100, 1996.

[MBE+06] R. Morley, S. Boulos, D. Edwards, J. Johnson, P. Shirley, M. Ashikhmin, and S. Premože. Image Synthesis using Adjoint Photons. In *Graphics Interface '06*, pages 179–186, June 2006.

[MRR+53] N. Metropolis, A. Rosenbluth, M. Rosenbluth, A. Teller, and E. Teller. Equations of State Calculations by Fast Computing Machine. *Journal of Chemical Physics*, 21:1087–1091, 1953.

[PH89] K. Perlin and E. Hoffert. Hypertexture. In *SIGGRAPH '89: Proceedings of the 16th annual conference on Computer graphics and interactive techniques*, pages 253–262, 1989.

[PKK00] M. Pauly, T. Kollig, and A. Keller. Metropolis Light Transport for Participating Media. In B. Péroche and H. Rushmeier, editors, *Rendering Techniques 2000 (Proc. 11th Eurographics Workshop on Rendering)*, pages 11–22. Springer, 2000.

[SSKK03] L. Szécsi, L. Szirmay-Kalos, and C. Kelemen. Variance Reduction for Russian Roulette. *Journal of WSCG*, 2003.

[Vea97] E. Veach. *Robust Monte Carlo Methods for Light Transport Simulation*. PhD thesis, Stanford University, 1997.

[VG94] E. Veach and L. Guibas. Bidirectional Estimators for Light Transport. In *Proc. 5th Eurographics Worshop on Rendering*, pages 147–161, Darmstadt, Germany, June 1994.

[VG95] E. Veach and L. Guibas. Optimally Combining Sampling Rechniques for Monte Carlo Rendering. In *SIGGRAPH '95: Proceedings of the 22nd annual conference on Computer graphics and interactive techniques*, pages 419–428, New York, NY, USA, 1995. ACM Press.

[VG97] E. Veach and L. Guibas. Metropolis Light Transport. *Computer Graphics*, 31:65–76, 1997.

SIMD-Oriented Fast Mersenne Twister: a 128-bit Pseudorandom Number Generator

Mutsuo Saito[1] and Makoto Matsumoto[2]

[1] Dept. of Math., Hiroshima University, Japan
saito@math.sci.hiroshima-u.ac.jp
[2] Dept. of Math., Hiroshima University, Japan
m-mat@math.sci.hiroshima-u.ac.jp

Summary. Mersenne Twister (MT) is a widely-used fast pseudorandom number generator (PRNG) with a long period of $2^{19937}-1$, designed 10 years ago based on 32-bit operations. In this decade, CPUs for personal computers have acquired new features, such as Single Instruction Multiple Data (SIMD) operations (i.e., 128-bit operations) and multi-stage pipelines. Here we propose a 128-bit based PRNG, named SIMD-oriented Fast Mersenne Twister (SFMT), which is analogous to MT but making full use of these features. Its recursion fits pipeline processing better than MT, and it is roughly twice as fast as optimised MT using SIMD operations. Moreover, the dimension of equidistribution of SFMT is better than MT.

We also introduce a block-generation function, which fills an array of 32-bit integers in one call. It speeds up the generation by a factor of two. A speed comparison with other modern generators, such as multiplicative recursive generators, shows an advantage of SFMT. The implemented C-codes are downloadable from `http://www.math.sci.hiroshima-u.ac.jp/~m-mat/MT/SFMT/index.html`.

1 Introduction

Recently, the scale of simulations is getting larger, and faster pseudorandom number generators (PRNGs) are required. The power of CPUs for usual personal computers are now sufficiently strong for such purposes, and the necessity of efficient PRNGs for CPUs on PCs is increasing. One such generator is Mersenne Twister (MT) [MN98], which is based on a linear recursion modulo 2 over 32-bit words. An implementation MT19937 has the period of $2^{19937}-1$. MT was designed 10 years ago, and the architectures of CPUs, such as Pentium and PowerPC, have changed. They have Single Instruction Multiple Data (SIMD) operations, which may be regarded as operations on 128-bit registers. Also, they have more registers and automatic parallelisms by multi-stage pipelining. These are not reflected in the design of MT.

In this article, we propose an MT-like pseudorandom number generator that makes full use of these new features: SFMT, a SIMD-oriented Fast Mersenne

Twister. We implemented an SFMT with the period a multiple of $2^{19937}-1$, named SFMT19937, which has a better equidistribution property than MT. SFMT is much faster than MT, even without using SIMD instructions.

There is an argument that the CPU time consumed for function calls to PRNG routines occupies a large part of the random number generation. This is not always the case: one can avoid the function calls by (1) inline-expansion and/or (2) generation of pseudorandom numbers in an array in one call. Actually some demanding users re-coded MT to avoid the function call; see the homepage of [MN98]. In this article, we introduce a block-generation scheme which is much faster than using function calls.

2 SIMD-Oriented Fast Mersenne Twister

We propose a SIMD-oriented Fast Mersenne Twister (SFMT) pseudorandom number generator. It is a Linear Feedbacked Shift Register (LFSR) generator based on a recursion over $\mathbb{F}_2^{128}$. We identify the set of bits $\{0,1\}$ with the two element field $\mathbb{F}_2$. This means that every arithmetic operation is done modulo 2. A w-bit integer is identified with a horizontal vector in $\mathbb{F}_2^w$, and $+$ denotes the sum as vectors (i.e., bit-wise exor), not as integers. We consider three cases: w is 32, 64 or 128.

2.1 LFSR Generators

A LFSR method is to generate a sequence $\mathbf{x}_0, \mathbf{x}_1, \mathbf{x}_2, \ldots$ of elements $\mathbb{F}_2^w$ by a recursion

$$\mathbf{x}_{i+N} := g(\mathbf{x}_i, \mathbf{x}_{i+1}, \ldots, \mathbf{x}_{i+N-1}), \tag{1}$$

where $\mathbf{x}_i \in \mathbb{F}_2^w$ and $g : (\mathbb{F}_2^w)^N \to \mathbb{F}_2^w$ is an $\mathbb{F}_2$-linear function (i.e., the multiplication of a $(wN \times w)$-matrix from the right to a wN-dimensional vector) and use it as a pseudorandom w-bit integer sequence. In the implementation, this recursion is computed by using an array `W[0..N-1]` of N integers of w-bit size, by the simultaneous substitutions

$$\begin{aligned}&\mathtt{W[0]} \leftarrow \mathtt{W[1]},\ \mathtt{W[1]} \leftarrow \mathtt{W[2]},\ \ldots, \mathtt{W[N-2]} \leftarrow \mathtt{W[N-1]},\\ &\mathtt{W[N-1]} \leftarrow g(\mathtt{W[0]}, \ldots, \mathtt{W[N-1]}).\end{aligned}$$

The first $N-1$ substitutions shift the content of the array, hence the name of LFSR. Note that in the implementation we may use an indexing technique to avoid computing these substitutions, see [Knu97, P.28 Algorithm A]. The array `W[0..N-1]` is called the state array. Before starting the generation, we need to set some values to the state array, which is called the initialization.

Mersenne Twister (MT) [MN98] is an example with

$$g(\mathbf{w}_0, \ldots, \mathbf{w}_{N-1}) = (\mathbf{w}_0|\mathbf{w}_1)A + \mathbf{w}_M,$$

where $(\mathbf{w}_0|\mathbf{w}_1)$ denotes the concatenation of the $32 - r$ most significant bits (MSBs) of $\mathbf{w}_0$ and the r least significant bits (LSBs) of $\mathbf{w}_1$, A is a (32×32)-matrix for which the multiplication $\mathbf{w}A$ is computable by a few bit-operations, and M is an integer $(1 < M < N)$. Its period is $2^{32N-r} - 1$, chosen to be a Mersenne prime. To obtain a better equidistribution property, MT transforms the sequence by a suitably chosen (32×32) matrix T, namely, MT outputs $\mathbf{x}_0T, \mathbf{x}_1T, \mathbf{x}_2T, \ldots$ (called tempering).

2.2 New Features of Modern CPUs for Personal Computers

Modern CPUs for personal computers (e.g. Pentium and PowerPC) have new features such as (1) fast integer multiplication instructions (2) fast floating point operations (3) SIMD operations (4) multi-stage pipelining. These were not common to standard PC CPUs, when MT was designed.

An advantage of $\mathbb{F}_2$-linear generators over integer multiplication generators (such as Linear Congruential Generators [Knu97] or Multiple Recursive Generators [L'E93]) was high-speed generation by avoiding multiplications. This advantage is now smaller, since 32-bit integer multiplication is now quite fast.

Among the new features, (3) and (4) fit $\mathbb{F}_2$-linear generators. Our idea is simple: to design a 128-bit integer PRNG, considering the benefit of such parallelism in the recursion.

2.3 The Recursion of SFMT

We choose g in the recursion (1) as

$$g(\mathbf{w}_0, \ldots, \mathbf{w}_{N-1}) = \mathbf{w}_0A + \mathbf{w}_MB + \mathbf{w}_{N-2}C + \mathbf{w}_{N-1}D, \tag{2}$$

where $\mathbf{w}_0, \mathbf{w}_M, \ldots$ are $w(= 128)$-bit integers (= horizontal vectors in $\mathbb{F}_2^{128}$), and A, B, C, D are sparse 128×128 matrices over $\mathbb{F}_2$ for which $\mathbf{w}A, \mathbf{w}B, \mathbf{w}C, \mathbf{w}D$ can be computed by a few SIMD bit-operations. The choice of the suffixes $N - 1$, $N - 2$ is for speed: in the implementation of g, `W[0]` and `W[M]` are read from the array `W`, while the copies of `W[N-2]` and `W[N-1]` are kept in two 128-bit registers in the CPU, say `r1` and `r2`. Concretely speaking, we assign `r2` ← `r1` and `r1` ← "the result of (2)" at every generation, then `r2` (`r1`) keeps a copy of `W[N-2]` (`W[N-1]`, respectively). The merit of doing this is to use the pipeline effectively. To fetch `W[0]` and `W[M]` from memory takes some time. In the meantime, the CPU can compute $\mathbf{w}_{N-2}C$ and $\mathbf{w}_{N-1}D$, because copies of $\mathbf{w}_{N-2}$ and $\mathbf{w}_{N-1}$ are kept in the registers. This selection was made through experiments on the speed of generation.

By trial and error, we searched for a set of parameters of SFMT, with the period being a multiple of $2^{19937} - 1$ and having good equidistribution properties. The degree of recursion N is $\lceil 19937/128 \rceil = 156$, and the linear transformations A, B, C, D are as follows.

- $\mathbf{w}A := (\mathbf{w} \overset{128}{<<} 8) + \mathbf{w}$.
  This notation means that $\mathbf{w}$ is regarded as a single 128-bit integer, and $\mathbf{w}A$ is the result of the left-shift of $\mathbf{w}$ by 8 bits exor-ed with $\mathbf{w}$. There are such SIMD operations in both Pentium SSE2 and PowerPC AltiVec SIMD instruction sets (SSE2 permits only a multiple of 8 as the amount of shifting). Note that the notation + means the exclusive-or in this article.
- $\mathbf{w}B := (\mathbf{w} \overset{32}{>>} 11)\&$(`BFFFFFF6 BFFAFFFF DDFECB7F DFFFFFEF`).
  This notation means that $\mathbf{w}$ is considered to be a quadruple of 32-bit integers, and each 32-bit integer is shifted to the right by 11 bits, (thus the eleven most significant bits are filled with 0s, for each 32-bit integer). The C-like notation & means the bitwise AND with a constant 128-bit integer, denoted in the hexadecimal form.
  In the search, this constant is generated as follows. Each bit in the 128-bit integer is independently randomly chosen, with the probability to choose 1 being 7/8. This is because we prefer to have more 1's for a denser feedback.
- $\mathbf{w}C := (\mathbf{w} \overset{128}{>>} 8)$.
  This is the right shift of an 128-bit integer by 8 bits, similar to the first.
- $\mathbf{w}D := (\mathbf{w} \overset{32}{<<} 18)$.
  Similar to the second, $\mathbf{w}$ is cut into four pieces of 32-bit integers, and each of these is shifted by 18 bits to the left.

All these instructions are available in both Intel Pentium's SSE2 and PowerPC's AltiVec SIMD instruction sets. Figure 1 shows a concrete description of SFMT19937 generator with period a multiple of $2^{19937} - 1$.

2.4 Endianness

Let $\mathbf{x}[0..3]$ be an array of 32-bit integers of size four. There are two natural ways to convert the array to a 128-bit integer. One is to concatenate in the order of $\mathbf{x}[3]\mathbf{x}[2]\mathbf{x}[1]\mathbf{x}[0]$, from MSBs to LSBs, which is called the little-endian system, adopted in Pentium. The converse is the big-endian system adopted in PowerPC, see [Wik].

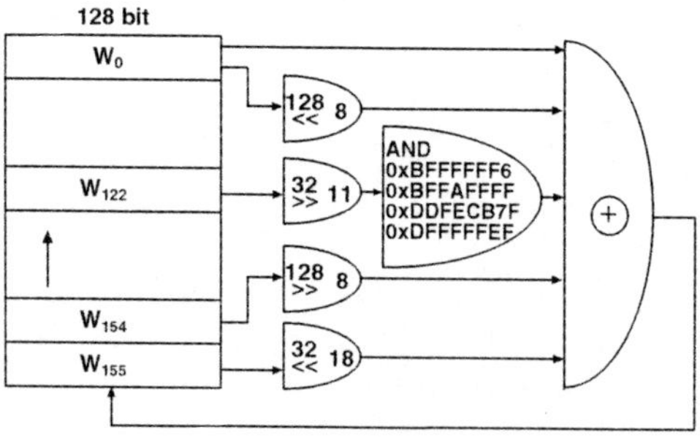

Fig. 1. A circuit-like description of SFMT19937.

The descriptions in this article is based on the former. To assure the portability for both endian systems, we implemented two codes: one is for little-endian system (SSE2 of Pentium) and the other is for big-endian system (AltiVec of PowerPC), to assure the exactly same outputs as 32-bit integer generators. In the latter code, the recursion (2) is considered as a recursion on quadruples of 32-bit integers, rather than 128-bit integers, so that the content of the state array coincides both for little and big endian systems, as an array of 32-bit integers (not as 128-bit integers). Thus, shift-operations on 128-bit integers in the little-endian system is different from that in the big-endian system. PowerPC supports arbitrary permutations of 16 blocks of 8-bit integers in a 128-bit register, which can emulate the shift in (2).

2.5 Block-Generation

In the block-generation scheme, the user of the PRNG specifies an array of w-bit integers of the length L, where $w = 32$, 64 or 128 and L is specified by the user. In the case of SFMT19937, wL should be a multiple of 128 and no less than $N \times 128$, since the array needs to accommodate the state space (note that $N = 156$). By calling the block generation function with the pointer to this array, w, and L, the routine fills up the array with pseudorandom integers, as follows. SFMT19937 keeps the state space S in an internal array of 128-bit integers of length 156. We concatenate this state array with the user-specified array, using the indexing technique. Then, the routine generates 128-bit integers in the user-specified array by recursion (2), as described in Figure 2, until it fills up the array. The last 156 128-bit integers are copied back to the internal

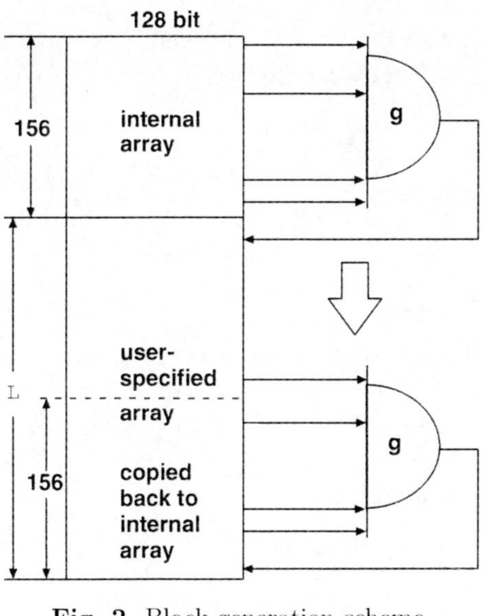

Fig. 2. Block-generation scheme.

CPU/compiler	return	MT	MT(SIMD)	SFMT	SFMT(SIMD)
Pentium-M	block	1.122	0.627	0.689	0.298
1.4GHz	(ratio)	3.77	2.10	2.31	1.00
Intel C/C++	seq	1.511	1.221	1.017	0.597
ver. 9.0	(ratio)	5.07	4.10	3.41	2.00
Pentium IV	block	0.633	0.391	0.412	0.217
3GHz	(ratio)	2.92	1.80	1.90	1.00
Intel C/C++	seq	1.014	0.757	0.736	0.412
ver. 9.0	(ratio)	4.67	3.49	3.39	1.90
Athlon 64 3800+	block	0.686	0.376	0.318	0.156
2.4GHz	(ratio)	4.40	2.41	2.04	1.00
gcc	seq	0.756	0.607	0.552	0.428
ver. 4.0.2	(ratio)	4.85	3.89	3.54	2.74
PowerPC G4	block	1.089	0.490	0.914	0.235
1.33GHz	(ratio)	4.63	2.09	3.89	1.00
gcc	seq	1.794	1.358	1.645	0.701
ver. 4.0.0	(ratio)	7.63	5.78	7.00	2.98

Table 1. The CPU time (sec.) for 10^8 generations of 32-bit integers, for four different CPUs and two different return-value methods. The ratio to the SFMT coded in SIMD is listed, too.

array of SFMT19937. This makes the generation much faster than sequential generation (i.e., one generation per one call) as shown in Table 1.

3 How to Select the Recursion and Parameters.

We wrote a code to compute the period and the dimensions of equidistribution (DE, see §3.2). Then, we searched for a recursion with good DE admitting a fast implementation.

3.1 Computation of the Period

An LFSR that obeys the recursion (1) may be considered as an automaton, with the state space $S = (\mathbb{F}_2^w)^N$ and the state transition function $f : S \to S$ given by $(\mathbf{w}_0, \ldots, \mathbf{w}_{N-1}) \mapsto (\mathbf{w}_1, \ldots, \mathbf{w}_{N-1}, g(\mathbf{w}_0, \ldots, \mathbf{w}_{N-1}))$. As a w-bit integer generator, the output function is $o : S \to \mathbb{F}_2^w$, $(\mathbf{w}_0, \ldots, \mathbf{w}_{N-1}) \mapsto \mathbf{w}_0$.

Let χ_f be the characteristic polynomial of $f : S \to S$. If χ_f is primitive, then the period of the state transition takes the maximal value $2^{\dim(S)} - 1$ [Knu97, §3.2.2]. However, to check the primitivity, we need the integer factorization of this number, which is often hard for $\dim(S) = nw > 10000$. On the other hand, the primarity test is much easier than the factorization, so many huge primes of the form $2^p - 1$ have been found. Such a prime is called a Mersenne prime, and p is called the Mersenne exponent, which itself is a prime.

MT and WELL[PLM06] discard r specific bits from the array S, so that $nw - r$ is a Mersenne exponent. Then, the primitivity of χ_f is easily checked by the algorithm in [Knu97, §3.2.2], avoiding the integer factorization.

SFMT adopted another method to avoid the integer factorization, the reducible transition method (RTM), which uses a reducible characteristic polynomial with a large primitive factor. This idea appeared in [Fus90] [BZ03][BZ04], and applications in the present context are discussed in detail in another article [SHP+06], therefore we only briefly recall it.

Let p be the Mersenne exponent, and $N := \lceil p/w \rceil$. Then, we randomly choose parameters for the recursion of LFSR (1). By applying the Berlekamp-Massey Algorithm to the output sequence, we obtain $\chi_f(t)$. (Note that a direct computation of $\det(tI - f)$ is time-consuming because $\dim(S) = 19968$.)

By using a sieve, we remove all factors of small degree from χ_f, until we know that it has no irreducible factor of degree p, or that it has a (possibly reducible) factor of degree p. In the latter case, the factor is passed to the primitivity test described in [Knu97, §3.2.2].

Suppose that we found a recursion with an irreducible factor of desired degree p in $\chi_f(t)$. Then, we have a factorization

$$\chi_f = \phi_p \phi_r,$$

where ϕ_p is a primitive polynomial of degree p and ϕ_r is a polynomial of degree $r = wN - p$. These are coprime, since we assume $p > r$. Let $\mathrm{Ker}(g)$ denote the kernel of a linear transformation g. By putting $V_p := \mathrm{Ker}\,(\phi_p(f))$ and $V_r := \mathrm{Ker}\,(\phi_r(f))$, we have a decomposition into f-invariant subspaces

$$S = V_p \oplus V_r \quad (\dim V_p = p,\ \dim V_r = r).$$

Note that the characteristic polynomial of the restriction f_p of f to V_p is $\phi_p(t)$, and that of the restriction f_r to V_r is $\phi_r(t)$. For any state $s \in S$, we denote $s = s_p + s_r$ for the corresponding decomposition with $s_p \in V_p$ and $s_r \in V_r$. Then, the k-th state $f^k(s)$ is equal to $f_p^k(s_p) + f_r^k(s_r)$. This implies that the automaton is equivalent to the sum of two automata $f_p : V_p \to V_p$ and $f_r : V_r \to V_r$. To combine two linear automata by sum is well-studied as combined Tausworthe generators or combined LFSRs, see [CLT93] [L'E96] [L'E99]. Their purpose is to obtain a good PRNG from several simple generators, which is different from ours.

The period length of the state transition is the least common multiple of that started from s_p and that started from s_r. Hence, if $s_p \neq 0$, then the period is a nonzero multiple of $2^p - 1$. We checked the following.

Proposition 1. *The period of SFMT19937 as a 128-bit integer generator is a nonzero multiple of* $2^{19937} - 1$, *if the 32 MSBs of* $\mathbf{w}_0$ *are set to the value* `6d736d6d` *in hexadecimal form.*

This value of $\mathbf{w}_0$ assures that $s_p \neq 0$, see [SHP+06] for a way to find such a value.

Remark 1. The number of non-zero terms in $\chi_f(t)$ is an index measuring the amount of bit-mixing. In the case of SFMT19937, the number of nonzero terms is 6711, which is much larger than 135 of MT, but smaller than 8585 of WELL19937c [PLM06].

3.2 Computation of the Dimension of Equidistribution

We briefly recall the definition of dimension of equidistribution (cf. [CLT93][L'E96]).

Definition 1. *A periodic sequence with period P*

$$\chi := \mathbf{x}_0, \mathbf{x}_1, \ldots, \mathbf{x}_{P-1}, \mathbf{x}_P = \mathbf{x}_0, \ldots$$

*of v-bit integers is said to be k-*dimensionally equidistributed *if any kv-bit pattern occurs equally often as a k-tuple*

$$(\mathbf{x}_i, \mathbf{x}_{i+1}, \ldots, \mathbf{x}_{i+k-1})$$

for a period $i = 0, \ldots, P-1$. We allow an exception for the all-zero pattern, which may occur once less often. (This last loosening of the condition is technically necessary, because the zero state does not occur in an $\mathbb{F}_2$-linear generator). The largest value of such k is called the dimension of equidistribution (DE).

We want to generalize this definition slightly. We define the k-window set of the periodic sequence χ as

$$W_k(\chi) := \{(\mathbf{x}_i, \mathbf{x}_{i+1}, \ldots, \mathbf{x}_{i+k-1}) | i = 0, 1, \ldots, P-1\},$$

which is considered as a *multi-set*, namely, the multiplicity of each element is considered.

For a positive integer m and a multi-set T, let us denote by $m \cdot T$ the multi-set where the multiplicity of each element in T is multiplied by m. Then, the above definition of equidistribution is equivalent to

$$W_k(\chi) = (m \cdot \mathbb{F}_2^{vk}) \setminus \{\mathbf{0}\},$$

where m is the multiplicity of the occurrences, and the operator $\setminus$ means that the multiplicity of $\mathbf{0}$ is subtracted by one.

Definition 2. *In the above setting, if there exist a positive integer m and a multi-subset $D \subset (m \cdot \mathbb{F}_2^{vk})$ such that*

$$W_k(\chi) = (m \cdot \mathbb{F}_2^{vk}) \setminus D,$$

we say that χ is k-dimensionally equidistributed with defect ratio $\#(D)/\#(m \cdot \mathbb{F}_2^{vk})$, where the cardinality is counted with multiplicity.

Thus, in Definition 1, the defect ratio up to $1/(P+1)$ is allowed to claim the dimension of equidistribution. If $P = 2^{19937} - 1$, then $1/(P+1) = 2^{-19937}$. In the following, the dimension of equidistribution allows the defect ratio up to 2^{-19937}.

For a w-bit integer sequence, its *dimension of equidistribution at v-bit accuracy* $k(v)$ is defined as the DE of the v-bit sequence, obtained by extracting the v MSBs from each of the w-bit integers. If the defect ratio is $1/(P+1)$, then there is an upper bound

$$k(v) \leq \lfloor \log_2(P+1)/v \rfloor.$$

The gap between the realized $k(v)$ and the upper bound is called the dimension defect at v of the sequence, and denoted by

$$d(v) := \lfloor \log_2(P+1)/v \rfloor - k(v).$$

The summation of all the dimension defects at $1 \leq v \leq 32$ is called the total dimension defect, denoted by Δ.

There is a difficulty in computing $k(v)$ when a 128-bit integer generator is used as a 32-bit (or 64-bit) integer generator. SFMT generates a sequence $\mathbf{x}_0, \mathbf{x}_1, \mathbf{x}_2, \ldots$ of 128-bit integers. Then, they are converted to a sequence of 32-bit integers $\mathbf{x}_0[0], \mathbf{x}_0[1], \mathbf{x}_0[2], \mathbf{x}_0[3], \mathbf{x}_1[0], \mathbf{x}_1[1], \ldots$, where $\mathbf{x}[0]$ is the 32 LSBs of $\mathbf{x}$, $\mathbf{x}[1]$ is the 33rd–64th bits, $\mathbf{x}[2]$ is the 65rd–96th bits, and $\mathbf{x}[3]$ is the 32 MSBs.

Then, we need to modify the model automaton as follows. The state space is $S' := S \times \{0, 1, 2, 3\}$, the state transition function $f' : S' \to S'$ is

$$f'(s, i) := \begin{cases} (s, i+1) \ (\text{ if } i < 3), \\ (f(s), 0) \ (\text{ if } i = 3) \end{cases}$$

and the output function is

$$o' : S' \to \mathbb{F}_2^{32}, \ ((\mathbf{w}_0, \ldots, \mathbf{w}_{N-1}), i) \mapsto \mathbf{w}_0[i].$$

We fix $1 \leq v \leq w$, and let $o_k(s, i)$ be the k-tuple of the v MSBs of the consecutive k-outputs from the state (s, i).

Proposition 2. *Assume that f is bijective. Let $k' = k'(v)$ denote the maximum k such that*

$$o_k(-, i) : V_p \to \mathbb{F}_2^{kv}, \quad s \mapsto o_k(s, i) \tag{3}$$

are surjective for all $i = 0, 1, 2, 3$. Take an initial state s satisfying $s_p \neq 0$. Then, the 32-bit output sequence is at least $k'(v)$-dimensionally equidistributed with v-bit accuracy with defect ratio 2^{-p}.

Moreover, if $4 < k'(v)+1$*, then for any initial state with* $s = s_p \neq 0$ *(hence* $s_r = 0$*), the dimension of equidistribution with defect ratio* 2^{-p} *is exactly* $k'(v)$*.*

Proof. Take $s \in S$ with $s_p \neq 0$. Then, the orbit of s by f has the form of $(V_p - \{0\}) \times U \subset V_p \times V_r$, since $p > r$ and $2^p - 1$ is a prime. The surjectivity of the linear mapping $o_{k'}(-,i)$ implies that the image of

$$o_{k'}(-,i) : V_p \times U \to \mathbb{F}_2^{kv}$$

is $m \cdot \mathbb{F}_2^{kv}$ as a multi-set for some m. The defect comes from $0 \in V_p$, whose ratio in V_p is 2^{-p}. Then the first statement follows, since $W_{k'}(\chi)$ is the union of the images $o_{k'}(-,i)((V_p - \{0\}) \times U)$ for $i = 0, 1, 2, 3$.

For the latter half, we define L_i as the multiset of the image of $o_{k'+1}(-,i) : V_p \to \mathbb{F}_2^{(k'+1)v}$. Because of $s_r = 0$, we have $U = \{0\}$, and the union of $(L_i - \{0\})$ $(i = 0, 1, 2, 3)$ as a multi-set is $W_{k'+1}(\chi)$. If the sequence is $(k'+1)$-dimensionally equidistributed, then the multiplicity of each element in $W_{k'+1}(\chi)$ is at most $2^p \times 4/2^{(k'+1)v}$.

On the other hand, the multiplicity of an element in L_i is equal to the cardinality of the kernel of $o_{k'+1}(-,i)$. Let d_i be its dimension. Then by the dimension theorem, we have $d_i \geq p - (k'+1)v$, and the equality holds if and only if $o_{k'+1}(-,i)$ is surjective. Thus, if there is a nonzero element $x \in \cap_{i=0}^{3} L_i$, then its multiplicity in $W_{k'+1}(\chi)$ is no less than $4 \times 2^{p-(k'+1)v}$, and since one of $o_{k'+1}(-,i)$ is not surjective by the definition of k', its multiplicity actually exceeds $4 \times 2^{p-(k'+1)v}$, which implies that the sequence is not $(k'+1)$-dimensionally equidistributed, and the proposition follows. Since the codimension of L_i is at most v, that of $\cap_{i=0}^{3} L_i$ is at most $4v$. The assumed inequality on k' implies the existence of nonzero element in the intersection.

The dimension of equidistribution $k(v)$ depends on the choice of the initial state s. The above proposition implies that $k'(v)$ coincides with $k(v)$ for the worst choice of s under the condition $s_p \neq 0$. Thus, we adopt the following definition (analogously to t_l in [L'E96]).

Definition 3. *Let* k *be the maximum such that (3) is satisfied. We call this the dimension of equidistribution of* v*-bit accuracy, and denote it simply by* $k(v)$*. We have an upper bound* $k(v) \leq \lfloor p/v \rfloor$*.*

We define the dimension defect at v *by*

$$d(v) := \lfloor p/v \rfloor - k(v) \text{ and } \Delta := \sum_{v=1}^{w} d(v).$$

We may compute $k(v)$ by standard linear algebra. We used a more efficient algorithm based on a weighted norm, generalizing [CLT93]. This will be written somewhere else, because of lack of space.

4 Comparison of Speed

We compared two algorithms: MT19937 and SFMT19937, with implementations using and without using SIMD instructions.

We measured the speeds for four different CPUs: Pentium M 1.4GHz, Pentium IV 3GHz, AMD Athlon 64 3800+, and PowerPC G4 1.33GHz. In returning the random values, we used two different methods. One is sequential generation, where one 32-bit random integer is returned for one call. The other is block generation, where an array of random integers is generated for one call (cf. [Knu97]). For detail, see §2.5 above.

We measured the consumed CPU time in second, for 10^8 generations of 32-bit integers. More precisely, in case of the block generation, we generate 10^5 of 32-bit random integers by one call, and this is iterated for 10^3 times. For sequential generation, the same 10^8 32-bit integers are generated, one per call. We used the inline declaration `inline` to avoid the function call, and unsigned 32-bit, 64-bit integer types `uint32_t`, `uint64_t` defined in INTERNATIONAL STANDARD ISO/IEC 9899: 1999(E) Programming Language-C, Second Edition (which we shall refer to as C99 in the rest of this article). Implementations without SIMD are written in C99, whereas those with SIMD use some standard SIMD extension of C99 supported by the compilers icl (Intel C compiler) and gcc.

Table 1 summarises the speed comparisons. The first four lines list the CPU time (in seconds) needed to generate 10^8 32-bit integers, for a Pentium-M CPU with the Intel C/C++ compiler. The first line lists the seconds for the block-generation scheme. The second line shows the ratio of CPU time to that of SFMT(SIMD). Thus, SFMT coded in SIMD is 2.10 times faster than MT coded in SIMD, and 3.77 times faster than MT without SIMD. The third line lists the seconds for the sequential generation scheme. The fourth line lists the ratio, with the basis taken at SFMT(SIMD) block-generation (not sequential). Thus, the block-generation of SFMT(SIMD) is 2.00 times faster than the sequential-generation of SFMT(SIMD).

Roughly speaking, in the block generation, SFMT(SIMD) is twice as fast as MT(SIMD), and four times faster than MT without using SIMD. Even in the sequential generation case, SFMT(SIMD) is still considerably faster than MT(SIMD).

Table 2 lists the CPU time for generating 10^8 32-bit integers, for four PRNGs from the GNU Scientific Library and two recent generators. They are re-coded with inline specification. Generators examined were: a multiple recursive generator `mrg` [L'E93], linear congruential generators `rand48` and `rand`, a lagged fibonacci generator `random256g2`, a WELL generator `well` (WELL19937c in [PLM06]), and a XORSHIFT generator `xor3` [PL05] [Mar03]. The table shows that SFMT(SIMD) is faster than these PRNGs, except for the outdated linear congruential generator `rand`, the lagged-fibonacci generator `random256g2` (which is known to have poor randomness, cf. [MN03]), and `xor3` with a Pentium-M.

CPU	return	mrg	rand48	rand	random256g2	well	xor3
Pentium M	block	3.277	1.417	0.453	0.230	1.970	0.296
	seq	3.255	1.417	0.527	0.610	2.266	1.018
Pentium IV	block	2.295	1.285	0.416	0.121	0.919	0.328
	seq	2.395	1.304	0.413	0.392	1.033	0.702
Athlon	block	1.781	0.770	0.249	0.208	0.753	0.294
	seq	1.798	0.591	0.250	0.277	0.874	0.496
PowerPC	block	2.558	1.141	0.411	0.653	1.792	0.618
	seq	2.508	1.132	0.378	1.072	1.762	1.153

Table 2. The CPU time (sec.) for 10^8 generations of 32-bit integers, by six other PRNGs.

v	MT	SFMT	v	MT	SFMT	v	MT	SFMT	v	MT	SFMT
$d(1)$	0	0	$d(9)$	346	1	$d(17)$	549	543	$d(25)$	174	173
$d(2)$	0	*2	$d(10)$	124	0	$d(18)$	484	478	$d(26)$	143	142
$d(3)$	405	1	$d(11)$	564	0	$d(19)$	426	425	$d(27)$	115	114
$d(4)$	0	*2	$d(12)$	415	117	$d(20)$	373	372	$d(28)$	89	88
$d(5)$	249	2	$d(13)$	287	285	$d(21)$	326	325	$d(29)$	64	63
$d(6)$	207	0	$d(14)$	178	176	$d(22)$	283	282	$d(30)$	41	40
$d(7)$	355	1	$d(15)$	83	*85	$d(23)$	243	242	$d(31)$	20	19
$d(8)$	0	*1	$d(16)$	0	*2	$d(24)$	207	206	$d(32)$	0	*1

Table 3. Dimension defects $d(v)$ of MT19937 and SFMT19937 as a 32-bit integer generator. The mark * means that MT has a smaller defect than SFMT at that accuracy.

5 Dimension of Equidistribution

Table 3 lists the dimension defects $d(v)$ of SFMT19937 (as a 32-bit integer generator) and of MT19937, for $v = 1, 2, \ldots, 32$. SFMT has smaller values of the defect $d(v)$ at 26 values of v. The converse holds for 6 values of v, but the difference is small. The total dimension defect Δ of SFMT19937 as a 32-bit integer generator is 4188, which is smaller than the total dimension defect 6750 of MT19937.

We also computed the dimension defects of SFMT19937 as a 64-bit (128-bit) integer generator, and the total dimension defect Δ is 14089 (28676, respectively). In some applications, the distribution of LSBs is important. To check them, we inverted the order of the bits (i.e. the i-th bit is exchanged with the $(w-i)$-th bit) in each integer, and computed the total dimension defect. It is 10328 (21337, 34577, respectively) as a 32-bit (64-bit, 128-bit, respectively) integer generator. Throughout the experiments, $d'(v)$ is very small for $v \leq 10$. We consider that these values are satisfactorily small, since they are comparable with MT for which no statistical deviation related to the dimension defect has been reported, as far as we know.

6 Recovery from 0-Excess States

For an LFSR with a sparse feedback function g, we observe the following phenomenon: if the bits in the state space contain too many 0's and few 1's (called a 0-excess state), then this tendency continues for many steps, since only a small part is changed in the state array at one step, and the change is not well-reflected to the next setp because of the sparseness.

We measure the recovery time from 0-excess states, by the method introduced in [PLM06], as follows.

1. Choose an initial state with only one bit being 1.
2. Generate k pseudorandom numbers, and discard them.
3. Compute the ratio of 1's among the next 32000 bits of outputs (i.e., in the next 1000 pseudorandom 32-bit integers).
4. Let γ_k be the average of the ratio over all such initial states.

We draw graphs of these ratio γ_k ($1 \le k \le 20000$) in Figure 3 for the following generators: (1) WELL19937c, (2) PMT19937 [SHP+06], (3) SFMT19937, and (4) MT19937. Because of its dense feedback, WELL19937c shows the fastest recovery among the compared generators. SFMT is better than MT, since its recursion refers to two most recently computed words (`W[N-1]` and `W[N-2]`) that acquire new 1s, while MT refers only to the words generated long before (`W[M]` and `W[0]`). PMT19937 shows faster recovery than SFMT19937, since PMT19937 has two feedback loops. The speed of recovery from 0-excess states is a trade-off with the speed of generation. Such 0-excess states will not happen practically, since the probability that 19937 random bits have less than 19937×0.4 of 1's is about 5.7×10^{-177}. The only plausible case would

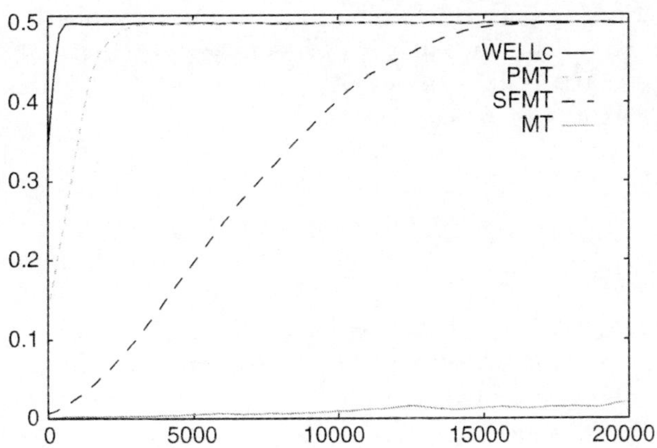

Fig. 3. γ_k ($k = 0, \ldots, 20000$): Starting from extreme 0-excess states, discard the first k outputs and then measure the ratio γ_k of 1's in the next 1000 outputs. In the order of the recovery speed: (1) WELL19937c, (2) PMT19937, (3) SFMT19937, and (4) MT19937.

be that a poor initialization scheme gives a 0-excess initial state (or gives two initial states whose Hamming distance is too small). In a typical simulation, the number of initializations is far smaller than the number of generations, therefore we may spend more CPU time in the initialization than the generation. Under the assumption that a good initialization scheme is provided, the slower recovery of SFMT compared to WELL would perhaps not be a great issue.

7 Concluding Remarks

We proposed the SFMT pseudorandom number generator, which is a very fast generator with satisfactorily high-dimensional equidistribution property.

It is difficult to measure the generation speed of a PRNG in a fair way, since it depends heavily on the circumstances. The WELL [PLM06] generators have the best possible dimensions of equidistribution (i.e. $\Delta = 0$) for various periods ($2^{1024} - 1$ to $2^{19937} - 1$). If we use the function call to the PRNG for each generation, then a large part of the CPU time is consumed for handling the function call, and in the experiments in [PLM06] or [PL05], WELL is not much slower than MT. On the other hand, if we avoid the function call, WELL is slower than MT for some CPUs, as seen in Table 1.

Since $\Delta = 0$, WELL has a better quality than MT or SFMT in a theoretical sense. However, one may argue whether this difference is observable or not. In the case of an $\mathbb{F}_2$-linear generator, the dimension of equidistribution $k(v)$ of v-bit accuracy means that there is no constant linear relation among the kv bits, but there exists a linear relation among the $(k+1)v$ bits, where kv bits ($(k+1)v$ bits) are taken from all the consecutive k integers ($k+1$ integers, respectively) by extracting the v MSBs from each. However, the existence of a linear relation does not necessarily mean the existence of some observable bias. According to [MN02], it requires 10^{28} samples to detect an $\mathbb{F}_2$-linear relation with 15 (or more) terms among 521 bits, by weight distribution test. If the number of bits is increased, the necessary sample size is increased rapidly. Thus, it seems that $k(v)$ of SFMT19937 is sufficiently large, far beyond the level of the observable bias. On the other hand, the speed of the generator is observable. Thus, SFMT focuses more on the speed, for applications that require fast generations. (Note: the referee pointed out that statistical tests based on the rank of $\mathbb{F}_2$-matrix is sensitive to the linear relations [LS06], so the above observation is not necessarily true.)

There is a trade-off between the speed and portability. We prepared (1) a standard C code of SFMT, which uses functions specified in C99 only, (2) an optimized C code for Intel Pentium SSE2, and (3) an optimized C code for PowerPC AltiVec. The optimized codes require the icl (Intel C Compiler) or gcc compiler with suitable options. We had put and will keep the newest version of the codes in the homepage [SM].

Acknowledgments

This study is partially supported by JSPS/MEXT Grant-in-Aid for Scientific Research No. 18654021, No. 16204002, and JSPS Core-to-Core Program No. 18005.

References

[BZ03] R.P. Brent and P. Zimmermann. Random number generators with period divisible by a Mersenne prime. In *Computational Science and its Applications - ICCSA 2003*, volume 2667, pages 1–10, 2003.

[BZ04] R.P. Brent and P. Zimmermann. Algorithms for finding almost irreducible and almost primitive trinomials. *Fields Inst. Commun.*, 41:91–102, 2004.

[CLT93] R. Couture, P. L'Ecuyer, and S. Tezuka. On the distribution of k-dimensional vectors for simple and combined Tausworthe sequences. *Math. Comp.*, 60(202):749–761, 1993.

[Fus90] M. Fushimi. Random number generation with the recursion $x_t = x_{t-3p} \oplus x_{t-3q}$. *Journal of Computational and Applied Mathematics*, 31:105–118, 1990.

[Knu97] D.E. Knuth. *The Art of Computer Programming. Vol. 2. Seminumerical Algorithms.* Addison-Wesley, Reading, Mass., 3rd edition, 1997.

[L'E93] P. L'Ecuyer. A search for good multiple recursive random number genarators. *ACM Transactions on Modeling and Computer Simulation*, 3(2):87–98, April 1993.

[L'E96] P. L'Ecuyer. Maximally equidistributed combined tausworthe generators. *Math. Comp.*, 65(213):203–213, 1996.

[L'E99] P. L'Ecuyer. Tables of maximally equidistributed combined LFSR generators. *Math. Comp.*, 68(225):261–269, 1999.

[LS06] P. L'Ecuyer and R. Simard. TestU01: A C library for empirical testing of random number generators. *ACM Transactions on Mathematical Software*, 2006. to appear.

[Mar03] G. Marsaglia. Xorshift RNGs. *Journal of Statistical Software*, 8(14):1–6, 2003.

[MN98] M. Matsumoto and T. Nishimura. Mersenne twister: A 623-dimensionally equidistributed uniform pseudorandom number generator. *ACM Trans. on Modeling and Computer Simulation*, 8(1):3–30, January 1998. `http://www.math.sci.hiroshima-u.ac.jp/~m-mat/MT/emt.html`.

[MN02] M. Matsumoto and T. Nishimura. A nonempirical test on the weight of pseudorandom number generators. In *Monte Carlo and Quasi-Monte Carlo methods 2000*, pages 381–395. Springer-Verlag, 2002.

[MN03] M. Matsumoto and T. Nishimura. Sum-discrepancy test on pseudorandom number generators. *Mathematics and Computers in Simulation*, 62(3–6):431–442, 2003.

[PL05] F. Panneton and P. L'Ecuyer. On the Xorshift random number generators. *ACM Transactions on Modeling and Computer Simulation*, 15(4):346–361, 2005.

[PLM06] F. Panneton, P. L'Ecuyer, and M. Matsumoto. Improved long-period generators based on linear reccurences modulo 2. *ACM Transactions on Mathematical Software*, 32(1):1–16, 2006.

[SHP+06] M. Saito, H. Haramoto, F. Panneton, T. Nishimura, and M. Matsumoto. Pulmonary LFSR: pseudorandom number generators with multiple feedbacks and reducible transitions. 2006. submitted.

[SM] M. Saito and M. Matsumoto.
SFMT Homepage.
http://www.math.sci.hiroshima-u.ac.jp/~m-mat/MT/SFMT/index.html.

[Wik] Endianness from Wikipedia, the free encyclopedia.
http://en.wikipedia.org/wiki/Endianness.

A New Lower Bound on the t-Parameter of (t, s)-Sequences

Rudolf Schürer

Department of Mathematics, University of Salzburg, Austria
rudolf.schuerer@sbg.ac.at

Summary. Let $t_b(s)$ denote the least t such that a (t, s)-sequence in base b exists. We present a new lower bound on $t_b(s)$, namely

$$t_b(s) > \frac{1}{b-1}s - \frac{1}{b-1} - \frac{1}{\log b} - \log_b\left(1 + s\left(1 - 1/b + \log_b s\right)\log b\right),$$

which leads to the new asymptotic result

$$L_b^* := \liminf_{s\to\infty} \frac{t_b(s)}{s} \geq \frac{1}{b-1}.$$

The best previously known result has been $L_b^* \geq 1/b$ for arbitrary $b \geq 2$ and $L_2^* \geq \log_2 3 - 1$.

1 Introduction

(t, m, s)-nets and (t, s)-sequences [Nie87, Nie92] are among the best methods for constructing low-discrepancy point sets in the s-dimensional unit cube. We present a new lower bound on the t parameter of (t, s)-sequences in arbitrary bases b. This new bound (Theorem 1) is an explicit formula depending on b and s, stronger than all previously known results both for typical values of s and b as well as in an asymptotic sense.

For every base $b \geq 2$ and every dimension $s \geq 1$ let $t_b(s)$ denote the least t such that a (t, s)-sequence in base b exists. It is well known that t_b grows linearly for all bases b. The fact that $t_b(s) = \mathcal{O}(s)$ is established in [XN95]. If b is a prime power this follows from applying Niederreiter–Xing sequence construction III to a tower of algebraic function fields over $\mathbb{F}_b$ related to the García–Stichtenoth tower [GS95] over $\mathbb{F}_{b^2}$. Using a direct product of such digital sequences over appropriate finite fields [NX96a] yields $t_b(s) = \mathcal{O}(s)$ for all bases $b \geq 2$.

The first lower bound on t that grows linearly in s is given by Larcher and Schmid in [LS95]: All digital (t, s)-sequences over $\mathbb{Z}_b$ with $t \geq 1$ satisfy

$$t > \frac{p-1}{p^3}(s+1), \tag{1}$$

where p is the smallest prime factor of b. A stronger lower bound applicable also to the non-digital case is established by Niederreiter and Xing in [NX96b]. There it is shown that

$$t_b(s) \geq \frac{1}{b}\, s - \log_b \frac{(b-1)s+b+1}{2} = \frac{1}{b}\, s - \mathcal{O}(\log s) \tag{2}$$

based on a result for (t, m, s)-nets by Lawrence [Law95].

Therefore $t_b(s)$ grows linearly in s for all $b \geq 2$ and one is interested in upper and lower bounds on its slope. Since it is unknown whether $\lim_{s\to\infty} t_b(s)/s$ exists, one examines the upper and lower limit of this quotient. In the following we discuss only the lower limit, which is denoted by

$$L_b^* := \liminf_{s\to\infty} \frac{t_b(s)}{s}$$

for every integer $b \geq 2$.

It follows from (2) that

$$L_b^* \geq 1/b\,, \tag{3}$$

which for $b \geq 3$ has been the best result known for the last decade. Only for $b = 2$ the following improvement has been known: In [Sch98] Schmid shows that

$$t > (\log_2 3 - 1)\, s - \mathcal{O}(\log s)$$

for all digital (t, s)-sequences over $\mathbb{Z}_2$ based on a bound for digital (t, m, s)-nets in [SW97]. In [MS99b] a formally equivalent bound for arbitrary (t, m, s)-nets is given, which (when substituted in the proof in [Sch98]) yields

$$L_2^* \geq \log_2 3 - 1 \approx 0.58496.$$

This is slightly better than $L_2^* \geq 0.5$ obtained from (3).

Before we discuss the new lower bound on L_b^* we recapitulate some results on upper bounds, i.e., on asymptotic existence results for (t, s)-sequences. For prime power bases good sequences can be constructed based on algebraic function fields with many rational places using methods due to Niederreiter and Xing. It is shown in [NX96a] that

$$L_q^* \leq 1 \Big/ \limsup_{g\to\infty} \frac{N_q(g)}{g} \tag{4}$$

for all prime powers q, with $N_q(g)$ denoting the maximal number of rational places of an algebraic function field with genus g and full constant field $\mathbb{F}_q$. Using appropriate towers of function fields it can be shown (see, e.g. the survey [NX98]) that

$$L_q^* \leq \frac{c}{\log q}$$

for all prime powers q, with an effective absolute constant $c > 0$. Furthermore,

$$L_q^* \leq \frac{1}{\sqrt{q} - 1}$$

if q is a square, and

$$L_q^* \leq \frac{q^{1/3} + 2}{2\left(q^{2/3} - 1\right)}$$

if q is the cube of a prime. Bounds for $q = p^e$ with p prime and $e \geq 3$ odd can also be found in [NX98].

An upper bound for L_q^* not derived from (4) is

$$L_q^* \leq \frac{q+1}{q-1}$$

from [XN95], which is especially useful for small nonsquares q.

2 The New Bound

An improved lower bound on L_b^* is implied by the following theorem:

Theorem 1. *Let $b \geq 2$ and $s \geq 2$. A $(t, s-1)$-sequence in base b can only exist if*

$$t > \frac{1}{b-1}s - \frac{1}{b-1} - \frac{1}{\log b} - \log_b\bigl(1 + s\,(1 - 1/b + \log_b s)\log b\bigr).$$

We prove this theorem in Section 3. Once it is established, an improved lower bound on L_b^* follows easily.

Corollary 1. *For all integers $b \geq 2$,*

$$t_b(s) > \frac{1}{b-1}\,s - \mathcal{O}(\log s),$$

where the constants in the $\mathcal{O}$-term depend only on b.

Proof. For every (t, s)-sequence in base b it follows from Theorem 1 that

$$t > \frac{s}{b-1} - \frac{1}{\log b} - \log_b\bigl(1 + (s+1)\,(1 - 1/b + \log_b(s+1))\log b\bigr).$$

The first term is already the correct leading term, the second term is constant. Using the inequality $\log_b x < x/\log b$, one shows that the logarithmic term is indeed in $\mathcal{O}(\log s)$. □

Corollary 2. *For all integers $b \geq 2$,*

$$L_b^* \geq \frac{1}{b-1}\,.$$

Proof. Follows trivially from Corollary 1. □

3 The Proof of the Theorem

The remainder of this article discusses the proof of Theorem 1. It is based on the theory of ordered orthogonal arrays (OOAs) introduced independently in [Law96] and [MS96].

3.1 The Plotkin Bound for Ordered Orthogonal Arrays

Definition 1. *Let $M \geq 1$, $s \geq 1$, $b \geq 2$, $T \geq 1$, and $0 \leq k \leq sT$ denote integers. Let S_b denote an arbitrary set of cardinality b.*

Furthermore, let $\Omega = \Omega^{(s,T)} := \{1, \dots, s\} \times \{1, \dots, T\}$ and let $\Xi_k^{(s,T)} \subseteq \mathcal{P}(\Omega)$ denote the set of those subsets ξ of Ω with

$$\sum_{i=1}^{s} \max_{(i,j)\in\xi} j \leq k ,$$

where an empty maximum is treated as 0.

Let $\mathcal{N}$ denote an array with M rows and sT columns indexed by the elements of Ω and (for some $\xi \in \mathcal{P}(\Omega)$) let $\xi(\mathcal{N})$ denote the projection of $\mathcal{N}$ on the coordinates in ξ.

Then $\mathcal{N}$ is called an ordered orthogonal array *(OOA) with strength k (denoted as $\mathrm{OOA}(M, s, S_b, T, k)$) if for every $\xi \in \Xi_k^{(s,T)}$ each of the $b^{|\xi|}$ possible vectors of $S_b^{|\xi|}$ appears the same number of times as a row in $\xi(\mathcal{N})$.*

Remark 1. Note that OOAs are a generalization of *orthogonal arrays*. An $\mathrm{OA}(M, s, S_b, k)$ is an $\mathrm{OOA}(M, s, S_b, 1, k)$ and vice versa.

There is a close relation between (t, m, s)-nets and the OOAs obtained from the digit expansion of the coordinates of the points of these nets. In particular it is shown in [Law96, MS96] that a (t, m, s)-net in base b exists if and only if an $\mathrm{OOA}(b^m, s, S_b, m - t, m - t)$ exists.

The basic ingredient for the proof of Theorem 1 is the following result by Martin and Visentin:

Theorem 2 (Theorem 5 in [MV07]). *An $\mathrm{OOA}(M, s, S_b, T, k)$ can only exist if*

$$M \geq b^{Ts} \left(1 - s\frac{\varrho_T}{k+1}\right), \tag{5}$$

where ϱ_T is defined as

$$\varrho_T := T - \sum_{i=1}^{T} \frac{1}{b^i}.$$

This result generalizes the dual Plotkin bound for OAs to OOAs and is derived from the linear programming bound for OOAs [MS99a]. An alternative proof is given by Bierbrauer in [Bie07].

Remark 2. A formally equivalent result for the special case of linear OOAs can also be obtained based on the Plotkin bound for generalized codes in [RT97] and the duality of these codes to linear OOAs established in [NP01]. Even though this result has been available for a number of years, it has not been applied to digital (t, s)-sequences so far.

The value ϱ_T plays an important role in the following discussion. Therefore it is convenient to establish some of its basic properties. We define

$$\varrho_\tau := \tau - \frac{1}{b-1} + \frac{1}{(b-1)b^\tau} \tag{6}$$

for all positive real numbers τ and note that this definition coincides with the definition given in Theorem 2 for all positive integers T. Furthermore, note that ϱ_τ is strictly increasing in τ for a fixed base b.

Since ϱ_T is always positive the right hand side of (5) becomes negative if s is large compared to k. In this case (5) is trivially satisfied and Theorem 2 does not give any information. Thus Theorem 2 can be expected to give good bounds only for OOAs with large strength k and a small number of factors s. Since the OOAs derived from (t, s)-sequences can have arbitrarily large strength k for fixed s, it is no surprise that Theorem 2 leads to good bounds for (t, s)-sequences.

Lemma 1. *Let $b \geq 2$ and $s \geq 2$ be integers. A $(t, s-1)$-sequence in base b can only exist if*

$$t \geq Ts - k + \log_b \left(1 - s\frac{\varrho_T}{k+1}\right) \tag{7}$$

for all integers T and k with $T \geq 1$ and $k > s\varrho_T - 1$.

Proof. Assume that a $(t, s-1)$-sequence in base b exists. From this sequence we can construct $(t, t+k, s)$-nets in base b for all integers $k \geq 1$ according to [Nie87, Lemma 5.15]. As discussed above each of these nets is equivalent to an OOA(b^{t+k}, s, S_b, k, k). From this OOA an OOA(b^{t+k}, s, S_b, T, k) can be obtained for all integers $T \geq k/s$ by either discarding columns (if $T < k$) or by appending columns with arbitrary content (if $T > k$).

For each of these OOAs Theorem 2 must hold, i.e., we have

$$b^{t+k} \geq \left(1 - s\frac{\varrho_T}{k+1}\right) b^{sT}$$

for all integer $T \geq k/s$ and $k \geq 1$. Solving for t yields (7), assuming that the argument of the logarithm is positive. This is the case if and only if $k > s\varrho_T - 1$. Therefore (7) holds for all integers k, T with $T \geq 1$ and $s\varrho_T - 1 < k \leq sT$.

If $k > sT$, the right hand side of (7) is negative and yields only a trivial bound on t. Therefore the condition $k \leq sT$ can be dropped. □

In order to establish a lower bound on the t-parameter of (t, s)-sequences the integers k and T must be chosen depending on b and s. It turns out that choosing these numbers properly is crucial, otherwise the resulting bound does not yield an asymptotic rate of $L_b^* \geq 1/(b-1)$.

Since k and T must be integers, we study the discretization error resulting from replacing arbitrary real values by integers:

Lemma 2. *Let $b \geq 2$ and $s \geq 2$ be integers. A $(t, s-1)$-sequence in base b can only exist if*

$$t > \frac{1}{b-1}s - \frac{s}{(b-1)b^\tau} - \varepsilon - \log_b\left(1 + s\varrho_{\tau+1}/\varepsilon\right) \tag{8}$$

for all positive real numbers τ and ε.

Proof. Lemma 1 states that

$$t \geq Ts - k + \log_b\left(1 - s\frac{\varrho_T}{k+1}\right)$$

for all integers T and k with $T \geq 1$ and $k > s\varrho_T - 1$. The integers $T = \lceil \tau \rceil$ and $k = \lceil s\varrho_T - 1 + \varepsilon \rceil$ satisfy these conditions.

Obviously $s\varrho_T - 1 + \varepsilon \leq k < s\varrho_T + \varepsilon$. Therefore

$$\begin{aligned} t &\geq Ts - \lceil s\varrho_T - 1 + \varepsilon \rceil + \log_b\left(1 - s\frac{\varrho_T}{\lceil s\varrho_T - 1 + \varepsilon \rceil + 1}\right) \\ &> Ts - s\varrho_T - \varepsilon + \log_b\left(1 - \frac{s\varrho_T}{s\varrho_T - 1 + \varepsilon + 1}\right) \\ &= s\left(T - \varrho_T\right) - \varepsilon + \log_b\left(\frac{\varepsilon}{s\varrho_T + \varepsilon}\right) \\ &= s\left(T - \varrho_T\right) - \varepsilon - \log_b\left(1 + s\varrho_T/\varepsilon\right). \end{aligned}$$

Using

$$T - \varrho_T = \frac{1}{b-1} - \frac{1}{(b-1)\,b^T}$$

from (6) and substituting $T = \lceil \tau \rceil$ yields

$$t > \frac{1}{b-1}s - \frac{s}{(b-1)b^{\lceil \tau \rceil}} - \varepsilon - \log_b\left(1 + s\varrho_{\lceil \tau \rceil}/\varepsilon\right),$$

which, together with $\tau \leq T < \tau + 1$ and the fact that ϱ_τ is increasing in τ, establishes (8). □

Lemma 2 yields a function f of two variables ε and τ which is equal to the right hand side of (8), bounded by $s/(b-1)$. We are looking for the maximum of f since $t > f(\varepsilon, \tau)$ for all (ε, τ). According to classical calculus, a necessary condition is $\frac{\partial f}{\partial \varepsilon}(\varepsilon, \tau) = 0$ and $\frac{\partial f}{\partial \tau}(\varepsilon, \tau) = 0$. Solving analytically this system of two equations in the unknowns ε and τ is impossible because of the second equation (see Section 3.3). Therefore, one has to find good approximations for such ε and τ. This approximation has to be made carefully, because for many choices the resulting bound is weak or even trivial.

3.2 Choosing ε

The optimal value for ε can be determined analytically. The derivative of the right hand side of (8) with respect to ε is

$$\frac{\partial}{\partial\varepsilon}\left(s\frac{1}{b-1} - \frac{s}{(b-1)b^\tau} - \varepsilon - \log_b\left(1 + s\varrho_{\tau+1}/\varepsilon\right)\right)$$
$$= -1 + \frac{s\varrho_{\tau+1}/\varepsilon^2}{(1 + s\varrho_{\tau+1}/\varepsilon)\log b}$$
$$= -1 + \frac{s\varrho_{\tau+1}}{(\varepsilon^2 + \varepsilon s\varrho_{\tau+1})\log b}\,.$$

Thus the ε-coordinates of the extremal points of (8) are given by the quadratic equation

$$\varepsilon^2 + s\varrho_{\tau+1}\,\varepsilon - \frac{s\varrho_{\tau+1}}{\log b} = 0$$

with solutions

$$\varepsilon_\pm = -\frac{s\varrho_{\tau+1}}{2} \pm \sqrt{\left(\frac{s\varrho_{\tau+1}}{2}\right)^2 + \frac{s\varrho_{\tau+1}}{\log b}}\,. \tag{9}$$

Since $\varepsilon_- < 0$ and $\varepsilon_-\varepsilon_+ < 0$, we get $\varepsilon_+ > 0$, which together with

$$\varepsilon_+ < -\frac{s\varrho_{\tau+1}}{2} + \sqrt{\left(\frac{s\varrho_{\tau+1}}{2}\right)^2 + \frac{s\varrho_{\tau+1}}{\log b} + \left(\frac{1}{\log b}\right)^2} = \frac{1}{\log b} \tag{10}$$

gives $\varepsilon_+ \in (0, 1/\log b)$. Moreover, the second-order Taylor expansion of $\sqrt{1+x}$ at $x = 0$ is $\sqrt{1+x} = 1 + x/2 - x^2/8 + \mathcal{O}(x^3)$, which yields

$$\varepsilon_+ = -\frac{s\varrho_{\tau+1}}{2} + \frac{s\varrho_{\tau+1}}{2}\sqrt{1 + \frac{4}{s\varrho_{\tau+1}\log b}}$$
$$= \frac{1}{\log b} - \frac{1}{s\varrho_{\tau+1}\log^2 b} + \mathcal{O}\left(\frac{1}{s^2\tau^2\log^3 b}\right).$$

Therefore a good choice for ε is $\varepsilon_1 := 1/\log b$ (moreover independent of τ and s). Including also the quadratic term yields

$$\varepsilon_2 := \frac{1}{\log b} - \frac{1}{s\varrho_T\log^2 b}.$$

For $b = 2$ the choice $\varepsilon_{\text{one}} := 1 \in (0, 1/\log 2)$ gives a suboptimal, but particularly simple result.

Since the resulting value for k is given by $k = \lceil s\varrho_T - 1 + \varepsilon\rceil$, we have

$$k \in (s\varrho_T - 1, s\varrho_T + \log_b).$$

In other words, the optimal k is always a small, bounded amount larger than $s\varrho_T - 1$ and asymptotically equal to $s\varrho_T$ when s turns towards infinity.

3.3 Choosing τ

The optimal $\tau = \tau_b(s)$ cannot be determined analytically, therefore an appropriate approximation has to be found. If we use $\varepsilon = \varepsilon_1$ or $\varepsilon = \varepsilon_{\text{one}}$ in (8), then ε is independent of s and τ. The first term $\frac{1}{b-1}s$ of (8) is already the sought-after leading term, but s appears also in the second term

$$A_b(s) := \frac{s}{(b-1)\, b^{\tau_b(s)}}$$

and in the fourth term, which can be bounded by

$$B_b(s) := \log_b\left(1 + s\,(\tau_b(s)+1)\,/\varepsilon\right) > \log_b\left(1 + s\varrho_{\tau_b(s)+1}/\varepsilon\right).$$

An obvious choice for τ is

$$\tau_b(s) = \log_b s, \tag{11}$$

because then $A_b(s) = 1/(b-1) \leq 1$ is bounded. We will use this choice for establishing Theorem 1.

Note that $\tau_b(s)$ must be unbounded. In particular, for $\tau_b(s) = 1$ we obtain

$$t \geq \frac{s}{b-1} - \frac{s}{(b-1)\,b} - \varepsilon - \log_b\left(1 + 2s/\varepsilon\right) = \frac{1}{b}\, s - \varepsilon - \log_b\left(1 + 2s/\varepsilon\right),$$

and therefore only $L_b^* \geq 1/b$, which is a rediscovery of Niederreiter and Xing's result (2). In general, if $\tau_b(s)$ is bounded, $A_b(s)$ grows linearly in s and therefore an asymptotic result of the form $L_b^* \geq 1/(b-1)$ cannot be obtained.

A quick calculation shows that the exact growth rate of $\tau_b(s)$ is not important for deriving Corollaries 1 and 2. It is sufficient to have

$$c + \log_b s \leq \tau_b(s) \leq C s^n$$

for some $c, C \in \mathbb{R}$, $n \in \mathbb{N}$, and s sufficiently large. However, in order to establish a strong bound, τ must be chosen carefully in addition to one of the ε's discussed in the previous section. Numerical experiments show that

$$\tau_b(s) = -\frac{1}{b} + \log_b s + \log_b\left(1 + \log_b s\right) \tag{12}$$

is a perfect fit for the optimal τ, resulting in a bound that is slightly stronger than the one obtained using (11).

3.4 The Proof of Theorem 1 and Further Remarks

Proof of Theorem 1. We use Lemma 2 with

$$\varepsilon = \varepsilon_1 = \frac{1}{\log b} \qquad \text{and} \qquad \tau = \log_b s.$$

Note that $\tau > 0$ for all $s \geq 2$. By substituting ε in (8), bounding $\varrho_{\tau+1}$ in the log-term by $\varrho_{\tau+1} \leq \tau + 1 - 1/b$, and substituting τ we get

$$\begin{aligned} t &> \frac{1}{b-1}s - \frac{s}{(b-1)\,b^{\tau}} - \frac{1}{\log b} - \log_b\bigl(1 + s\,(\tau + 1 - 1/b)\log b\bigr) \\ &= \frac{1}{b-1}s - \frac{1}{b-1} - \frac{1}{\log b} - \log_b\bigl(1 + s\,(1 - 1/b + \log_b s)\log b\bigr), \end{aligned}$$

which completes the proof. □

Remark 3. Using $\varepsilon = \varepsilon_{\text{one}} = 1$, $\tau_b(s) = \log_b s$, and bounding $\varrho_{\tau+1} < \tau + 1 = 1 + \log_b s$ yields the particularly simple (and asymptotically equivalent) bound

$$t > \frac{1}{b-1}s - \frac{b}{b-1} - \log_b\left(1 + s + s\log_b s\right)$$

for all $(t, s-1)$-sequences in base b.

Remark 4. Using

$$\varepsilon = \varepsilon_1 = \frac{1}{\log b}, \qquad \tau = -\frac{1}{b} + \log_b s + \log_b\left(1 + \log_b s\right)$$

according to (12), and the exact value for $\varrho_{\tau+1}$ in the log-term yields

$$\begin{aligned} t > s\frac{1}{b-1} - \frac{1}{\log b} - \frac{\sqrt[b]{b}}{(b-1)\,(1+\log_b s)} \\ - \log_b\Biggl(s\biggl(\log\bigl(s\,(1+\log_b s)\bigr) + \frac{(b^2-3b+1)\log b}{b\,(b-1)}\biggr) \\ + 1 + \frac{\sqrt[b]{b}\log b}{b\,(b-1)\,(1+\log_b s)}\Biggr) \end{aligned}$$

for all $(t, s-1)$-sequences in base b.

Acknowledgment

The author is supported by the Austrian Science Foundation (FWF), project no. P18455-N18. He would also like to thank the anonymous referee for several suggestions that improved the conciseness of the paper considerably.

References

[Bie07] J. Bierbrauer. A direct approach to linear programming bounds for codes and tms-nets. *Des. Codes Cryptogr.*, 42:127–143, 2007.

[GS95] A. García and H. Stichtenoth. A tower of Artin–Schreier extensions of function fields attaining the Drinfeld–Vladut bound. *Invent. Math.*, 121:211–222, 1995.

[Law95] K. M. Lawrence. *Combinatorial Bounds and Constructions in the Theory of Uniform Point Distributions in Unit Cubes, Connections with Orthogonal Arrays and a Poset Generalization of a Related Problem in Coding Theory.* PhD thesis, University of Wisconsin, Madison, 1995.

[Law96] K. M. Lawrence. A combinatorial characterization of (t, m, s)-nets in base b. *J. Combin. Des.*, 4:275–293, 1996.

[LS95] G. Larcher and W. Ch. Schmid. Multivariate Walsh series, digital nets and quasi-Monte Carlo integration. In H. Niederreiter and P. J.-S. Shiue, editors, *Monte Carlo and Quasi-Monte Carlo Methods in Scientific Computing*, volume 106 of *Lecture Notes in Statistics*, pages 252–262. Springer-Verlag, Berlin, 1995.

[MS96] G. L. Mullen and W. Ch. Schmid. An equivalence between (t, m, s)-nets and strongly orthogonal hypercubes. *J. Combin. Theory Ser. A*, 76:164–174, 1996.

[MS99a] W. J. Martin and D. R. Stinson. Association schemes for ordered orthogonal arrays and (t, m, s)-nets. *Canad. J. Math.*, 51:326–346, 1999.

[MS99b] W. J. Martin and D. R. Stinson. A generalized Rao bound for ordered orthogonal arrays and (t, m, s)-nets. *Canad. Math. Bull.*, 42:359–370, 1999.

[MV07] W. J. Martin and T. I. Visentin. A dual Plotkin bound for (T, M, S)-nets. *IEEE Trans. Inform. Theory*, 53:411–415, 2007.

[Nie87] H. Niederreiter. Point sets and sequences with small discrepancy. *Monatsh. Math.*, 104:273–337, 1987.

[Nie92] H. Niederreiter. *Random Number Generation and Quasi-Monte Carlo Methods.* SIAM, Philadelphia, 1992.

[NP01] H. Niederreiter and G. Pirsic. Duality for digital nets and its applications. *Acta Arith.*, 97:173–182, 2001.

[NX96a] H. Niederreiter and C. P. Xing. Low-discrepancy sequences and global function fields with many rational places. *Finite Fields Appl.*, 2:241–273, 1996.

[NX96b] H. Niederreiter and C. P. Xing. Quasirandom points and global function fields. In S. Cohen and H. Niederreiter, editors, *Finite Fields and Applications*, volume 233 of *London Math. Soc. Lecture Note Ser.*, pages 269–296. Cambridge University Press, Cambridge, 1996.

[NX98] H. Niederreiter and C. P. Xing. Nets, (t, s)-sequences, and algebraic geometry. In P. Hellekalek and G. Larcher, editors, *Random and Quasi-Random Point Sets*, volume 138 of *Lecture Notes in Statistics*, pages 267–302. Springer-Verlag, Berlin, 1998.

[RT97] M. Y. Rosenbloom and M. A. Tsfasman. Codes for the m-metric. *Probl. Inf. Transm.*, 33:55–63, 1997.

[Sch98] W. Ch. Schmid. Shift-nets: a new class of binary digital (t, m, s)-nets. In H. Niederreiter et al., editors, *Monte Carlo and Quasi-Monte Carlo Methods 1996*, volume 127 of *Lecture Notes in Statistics*, pages 369–381. Springer-Verlag, Berlin, 1998.

[SW97] W. Ch. Schmid and R. Wolf. Bounds for digital nets and sequences. *Acta Arith.*, 78:377–399, 1997.

[XN95] C. P. Xing and H. Niederreiter. A construction of low-discrepancy sequences using global function fields. *Acta Arith.*, 73:87–102, 1995.

Walk-on-Spheres Algorithm for Solving Boundary-Value Problems with Continuity Flux Conditions

Nikolai Simonov[1,2]

[1] Institute of Computational Mathematics and Mathematical Geophysics SB RAS, Novosibirsk, Russia
nas@osmf.sscc.ru

[2] Institute of Molecular Biophysics, Florida State University, Tallahassee, FL, USA
simonov@scs.fsu.edu

Summary. We consider boundary-value problem for elliptic equations with constant coefficients related through the continuity conditions on the boundary between the domains. To take into account conditions involving the solution's normal derivative, we apply a new mean-value relation written down at a boundary point. This integral relation is exact and provides a possibility to get rid of the bias caused by usually used finite-difference approximation. Randomization of this mean-value relation makes it possible to continue simulating walk-on-spheres trajectory after it hits the boundary. We prove the convergence of the algorithm and determine its rate. In conclusion, we present the results of some model computations.

1 Statement of the Problem

We consider the boundary-value problem for a function, $u(x)$, that satisfies different elliptic equations with constant coefficients inside a bounded simple-connected domain, $G_i \subset \mathbb{R}^3$, and in its exterior, $G_e = \mathbb{R}^3 \setminus \overline{G_i}$.

Denote, for convenience, by $u_i(x)$ and $u_e(x)$ the restrictions of function $u(x)$ to G_i and G_e, respectively. Let the first function satisfy the Poisson equation

$$\epsilon_i \Delta u_i = -\rho, \tag{1}$$

and the second one satisfy the linearized Poisson-Boltzmann equation

$$\epsilon_e \Delta u_e - \epsilon_e \kappa^2 u_e = 0. \tag{2}$$

The continuity conditions on the piecewise smooth boundary, Γ, relate limiting values of solutions, and their fluxes as well:

$$u_i(y) = u_e(y)\ ,\ \epsilon_i \frac{\partial u_i}{\partial n}(y) = \epsilon_e \frac{\partial u_e}{\partial n}(y)\ ,\ y \in \Gamma. \tag{3}$$

Here, the normal vector, n, is pointed out into G_e; and $u_e(x) \to 0$ as $|x|$ goes to infinity. We assume the parameters of the problem to guarantee that the unique solution exists [Mir70]. Problems of this kind arise in molecular biophysics applications [DM90]. In this case, G_i can be thought of as a molecule in aqueous solution. In the framework of the implicit solvent model, only the geometric structure of this molecule is described explicitly, whereas the surrounding water with ions dissolved is considered a continuous medium.

2 Monte Carlo Methods

To solve numerically the problem (1), (2), (3), we propose to use a Monte Carlo method. There are several reasons for such a choice. Finite-difference and other deterministic computational methods that are commonly used for solving elliptic boundary-value problems encounter with apparent complications when applied to calculating electrostatic properties of molecules in solvent. Most of these difficulties are caused by the complexity of molecular surface. On the other hand, the most efficient and commonly used Monte Carlo methods such as the walk-on-spheres (WOS) algorithm [M̈56, EM82, EKMS80] and the random walk on the boundary algorithm [SS94] can analytically take the geometry into account. The latter can be applied to solving not only Dirichlet, Neumann and third boundary-value problems, but also for the problems with continuity boundary conditions in the case when $\kappa = 0$ [KMS04b, KMS04a]. This method works well for small molecules, but becomes computationally expensive for larger structures, which means that it needs substantial optimization and further algorithmic development. It is well known that the WOS algorithm is designed to work with the Dirichlet boundary conditions. With this method, the common way of treating flux conditions is to simulate reflection from the boundary in accordance with the finite-difference approximation to the normal derivative [HSS66, Kro84, MM97, MS04]. Such an approach has a drawback. It introduces a bias into the estimate and substantially elongates simulation of Markov chain trajectories.

Recently [Sim06], we described a new approach to constructing Monte Carlo algorithms for solving elliptic boundary-value problems with flux conditions. This approach is based on the mean-value relation written for the value of the solution at a point, which lies exactly on the boundary. It provides a possibility to get rid of the bias when using Green's function based random walk methods and treating algorithmically boundary conditions that involve normal derivatives.

Consider the problem of computing the solution to (1), (2), (3) at a fixed point, x_0. Suppose, for definiteness that $x_0 \in G_i$. To estimate $u(x_0)$, we represent it as a sum of the regular part and the volume potential: $u(x_0) = u^0(x_0) + g(x_0)$. Here, $g(x_0) = \int_{G_i} \frac{1}{4\pi\epsilon_i} \frac{1}{|x_0 - y|} \rho(y)dy$. In molecular electrostatic problems, charges are considered to be concentrated at a finite number of fixed points. Hence, $g(x_0) = \sum_{m=1}^{M} \frac{1}{4\pi\epsilon_i} \frac{1}{|x_0 - x_{c,m}|} \rho_m$. Usually, in these problems, x_0 coincides with one of $x_{c,m}$.

The regular part of the solution satisfies the Laplace equation in G_i. Therefore, we have a possibility to use the WOS algorithm to find it. Let x_0 be the starting point and $d(x_i)$ be the distance from the given point, x_i, to the boundary, Γ. Generally, WOS Markov chain is defined by the recursive relation: $x_{i+1} = x_i + d(x_i)\omega_i$, where $\{\omega_0, \omega_1, \ldots\}$ is a sequence of independent isotropic unit vectors. With probability one, this chain converges to the boundary [EKMS80]. Let x_k be the first point of the Markov chain that hit Γ_ε, the ε-strip near the boundary. Denote by $x_k^* \in \Gamma$ the point nearest to x_k. Clearly, the sequence $\{u^0(x_i), i = 0, 1, \ldots\}$ forms a martingale. Therefore, $u^0(x_0) = \mathbb{E}u^0(x_k) = \mathbb{E}(u(x_k) - g(x_k)) = \mathbb{E}(u(x_k^*) - g(x_k^*) + \phi)$, where $\phi = O(\varepsilon)$, for elliptic boundary points, x_k^*.

Mean number of steps in the WOS Markov chain before it for the first time hits ε-strip near the boundary is $O(\log \varepsilon)$ [EKMS80]. In the molecular electrostatics problems, however, it is natural to use the more efficient way of simulating exit points of Brownian motion on the boundary, Γ. The algorithm is based on the walk-in-subdomains approach [Sim83]. Such construction utilizes the commonly used representation of a molecular structure in the form of a union of intersecting spheres. This version of the WOS algorithm converges geometrically, and there is no bias in the estimate, since the last point of Markov chain lies exactly on Γ.

Note, that to compute the estimate for $u^0(x_0)$, we use the unknown boundary values of the solution. With the Monte Carlo calculations, we can use estimates instead of these values. In the next section we derive the mean-value relation, which is then used in Section 4 to construct such an estimate.

3 Integral Representation at a Boundary Point

To elucidate the approach we use, we consider first the exterior Neumann problem for the Poisson-Boltzmann equation (2) in G_e:

$$\frac{\partial u_e}{\partial n}(y) = f(y) \, , \; y \in \Gamma. \tag{4}$$

Let $x \in \Gamma$ be an elliptic point on the boundary. To construct an integral representation for the solution value at this point, consider the ball, $B(x, a)$, of radius a, and x being its center.

Let $\Phi_{\kappa,a}(x, y) = -\frac{1}{4\pi} \frac{\sinh(\kappa(a - |x - y|))}{|x - y| \sinh(\kappa a)}$ be the Green's function of the Dirichlet problem for the Poisson-Boltzmann equation (2) considered in the ball, $B(x, a)$, and taken at the central point of this ball. Denote by $B_e(x, a) = B(x, a) \bigcap G_e$ the external part of the ball, and let $S_e(x, a)$ be the part of the spherical surface that lies in G_e. Next, we exclude from this ball a small vicinity of the point, x. From here it follows that, for arbitrary $\epsilon < a$, both functions, u_e and $\Phi_{\kappa,a}$, satisfy the Poisson-Boltzmann equation in $B_e(x,a)\backslash B(x,\epsilon)$. Therefore, it is possible to use the Green's formula for this pair of functions in this domain. Taking the limit of this formula as $\epsilon \to 0$, we have

$$u_e(x) = \int_{S_e} 2\frac{\partial \Phi_{\kappa,a}}{\partial n} u_e \, ds - \int_{\Gamma \bigcap B(x,a)\backslash\{x\}} 2\frac{\partial \Phi_{\kappa,a}}{\partial n} u_e \, ds + \int_{\Gamma \bigcap B(x,a)\backslash\{x\}} 2\Phi_{\kappa,a}\frac{\partial u_e}{\partial n} \, ds. \quad (5)$$

Here, in the second and third integrals, we took into account that, on Γ, the normal vector external with respect to $B_e(x,a)\setminus B(x,\epsilon)$ has the direction opposite to the normal vector we use in the boundary conditions (4). The normal derivative of the Green's function can be written down explicitly. We have $\frac{\partial \Phi_{\kappa,a}}{\partial n}(y) = Q_{\kappa,a}(r)\frac{\partial \Phi_{0,a}}{\partial n}(y)$. Here, $Q_{\kappa,a}(r) = \frac{\sinh(\kappa(a-r)) + \kappa r \cosh(\kappa(a-r)}{\sinh(\kappa a)} < 1$, $r = |x-y|$, and $2\frac{\partial \Phi_{0,a}}{\partial n}(y) = \frac{1}{2\pi}\frac{\cos\phi_{yx}}{|x-y|^2}$, where ϕ_{yx} is the angle between $n(y)$ and $y-x$. $\Phi_{0,a}(x,y) = -\frac{1}{4\pi}\left(\frac{1}{|x-y|} - \frac{1}{a}\right)$ is the Green's function for the Laplace equation.

Application of the Green's formula to the pair of functions, u_i and $\Phi_{\kappa,a}$, in $B_i(x,a)\setminus B(x,\epsilon)$ provides the analogous result. To have a possibility to do this, we have to suppose that there are no charges in this part of the interior domain. From here it follows that the total potential, u_i, satisfies the Laplace equation. Thus using the Green's formula and taking the limit as $\epsilon \to 0$, we obtain

$$u_i(x) = \int_{S_i} 2\frac{\partial \Phi_{\kappa,a}}{\partial n} u_i \, ds + \int_{\Gamma \bigcap B(x,a)\backslash\{x\}} 2\frac{\partial \Phi_{\kappa,a}}{\partial n} u_i \, ds + \int_{B_i(x,a)} [-2\kappa^2 \Phi_{\kappa,a}] u_i \, dy - \int_{\Gamma \bigcap B(x,a)\backslash\{x\}} 2\Phi_{\kappa,a}\frac{\partial u_i}{\partial n} \, ds. \quad (6)$$

Note the additional volume integral in this representation.

To make use of the continuity boundary conditions (3), we multiply (5) by ϵ_e and (6) by ϵ_i, respectively, and sum up the results. This gives us

$$
\begin{aligned}
u(x) = {} & \frac{\epsilon_e}{\epsilon_e+\epsilon_i}\int_{S_e(x,a)} \frac{1}{2\pi a^2}\frac{\kappa a}{\sinh(\kappa a)} u_e \, ds \\
& + \frac{\epsilon_i}{\epsilon_e+\epsilon_i}\int_{S_i(x,a)} \frac{1}{2\pi a^2}\frac{\kappa a}{\sinh(\kappa a)} u_i \, ds \\
& - \frac{(\epsilon_e-\epsilon_i)}{\epsilon_e+\epsilon_i}\int_{\Gamma\bigcap B(x,a)\setminus\{x\}} \frac{1}{2\pi}\frac{\cos\phi_{yx}}{|x-y|^2} Q_{\kappa,a} u \, ds \\
& + \frac{\epsilon_i}{\epsilon_e+\epsilon_i}\int_{B_i(x,a)} [-2\kappa^2 \Phi_{\kappa,a}] u_i \, dy. \qquad (7)
\end{aligned}
$$

In the limiting case, $\kappa = 0$, this relation simplifies:

$$
\begin{aligned}
u(x) = {} & \frac{\epsilon_e}{\epsilon_e+\epsilon_i}\int_{S_e(x,a)} \frac{1}{2\pi a^2} u_e \, ds \\
& + \frac{\epsilon_i}{\epsilon_e+\epsilon_i}\int_{S_i(x,a)} \frac{1}{2\pi a^2} u_i \, ds \\
& - \frac{\epsilon_e-\epsilon_i}{\epsilon_e+\epsilon_i}\int_{\Gamma\bigcap B(x,a)\setminus\{x\}} \frac{1}{2\pi}\frac{\cos\phi_{yx}}{|x-y|^2} u \, dy. \qquad (8)
\end{aligned}
$$

Note that for a plane boundary, $\cos\phi_{yx} = 0$, and the integral over Γ vanishes.

4 Estimate for the Boundary Value

To construct an estimate for the solution of the boundary-value problem, we need an estimate for the unknown boundary value. To do this, we use the mean-value relation constructed in the previous section. It is not so simple as it may seem at first sight, since we have to iterate integral operators standing in the right-hand sides of these representations. The kernels of these operators can be alternating, and the convergence of the Neumann series after replacing the kernel by its modulus cannot be ensured. As a consequence, the direct randomization of integral relations (5), (7), (8) cannot be used to constructing a Monte Carlo estimate [EM82, ENS89]. However, as we see, the only part of the integral operator, which can be negative, is the integral over the boundary. Thus if Γ consists of planes, then the direct randomization is really possible.

In the general case, we can use a simple probabilistic approximation. Let $x = x_k^*$ be the point on the boundary nearest to the last point of the already constructed random walk.

Consider first the exterior Neumann problem. If Γ is concave everywhere in $B(x,a)$ then the kernel of the integral operator in (5) is positive, and the next point of the Markov chain, x_{k+1}, can be sampled isotropically in solid angle with x being its vertex. However, it would be more natural to suppose that the boundary was convex. In this case, we draw the tangent plane to Γ at this point, and sample x_{k+1} isotropically on the external hemisphere, $S^+(x_k^*, a)$. Then, for sufficiently smooth function, u, we have:

Lemma 1.

$$u(x_k^*) = \mathbb{E}\{u(x_{k+1}) \mid x_k^*\} + \phi_\Gamma,$$

where $\phi_\Gamma = O\left(\frac{a}{2R}\right)^3$ *as* $a/2R \to 0$. *Here,* R *is the minimal radius of curvature at the point,* x_k^*.

This statement is easily verified by expansion of u into series and direct integration.

The same approach works in the case of continuity boundary conditions (3). For $\kappa = 0$, with probability $p_e = \frac{\epsilon_e}{\epsilon_i + \epsilon_e}$ we sample the next point of the Markov chain, x_{k+1}, isotropically on $S^+(x_k^*, a)$. With the complementary probability, p_i, the next point is chosen on the other hemisphere, $S^-(x_k^*, a)$.

If $\kappa > 0$, with probability p_e we sample the next point on $S^+(x_k^*, a)$ and treat the coefficient $q(\kappa, a) = \frac{\kappa a}{\sinh(\kappa a)}$ as the survival probability. With probability p_i we sample the direction of the vector, $x_{k+1} - x$, isotropically pointing to $S^-(x_k^*, a)$. Next, with probability $q(\kappa, a)$ ($Q_{\kappa,a}$, if this vector intersects Γ), the next point is taken on the surface of the sphere, and with the complementary probability it is sampled inside the ball, $B^-(x_k^*, a)$. The simulation density of $r = x_{k+1} - x$ is taken to be consistent with $\sinh(\kappa(a-r))$.

It is easy to prove that Lemma 1 is also valid for the proposed randomized approach to treating continuity boundary conditions.

It is essential to note that randomization of the finite-difference approximation for the normal derivative with step, h, provides an $O(h^2)$ bias, both for the Neumann and continuity boundary conditions.

5 Construction of the Algorithm and its Convergence

To complete the construction, we need a Monte Carlo estimate for the solution value at an arbitrary point in the exterior domain, G_e. It is natural to use the estimate based on simulation of the walk-on-spheres Markov chain. Every step of the algorithm is the direct randomization of the mean-value formula for the Poisson-Boltzmann equation: $u(x) = \int_{S(x,d)} \frac{q(\kappa, d)}{4\pi d^2} u \, ds$, where $S(x,d)$ is the surface of a ball totally contained in G_e, and d is usually taken to be the maximum possible, i.e. equal to the distance from x to Γ. For $\kappa = 0$, we use the modification of the walk-on-spheres algorithm with the direct simulation of jump to the absorbing state of the Markov chain at infinity [ENS89]. The conditional mean number of steps for the chain to hit the ε-strip near the boundary is $O(\log \varepsilon)$. For positive κ, we consider $q(\kappa, d)$ the survival probability. In this case the walk-on-spheres either comes to Γ or terminates at some finite absorbing state in G_e.

To prove the convergence of the algorithm, we reformulate the problem in terms of integral equations with generalized kernels [EM82, EKMS80].

Inside the domains, G_i and G_e, we consider the mean-value formulas as such equations. In Γ_ε we use the approximation $u(x) = u(x^*) + \phi(x, x^*)$ and substitute the integral representation for $u(x^*)$ at a boundary point, x^*. Hence, the described random-walk construction corresponds to so-called direct simulation of (approximated) integral equation [EM82]. This means that the resulting estimate for the solution's value at a point, $x = x_0 \in \mathbb{R}^3$, is

$$\xi[u](x) = \sum_{i=0}^{N} \xi[F](x_i). \tag{9}$$

Here, N is the random length of Markov chain, and $\xi[F]$ are estimates for the right-hand side of the integral equation. For the external Neumann problem, this function is

$$\begin{aligned} F(x) &= 0 \text{ when } x \in G_e \setminus \overline{\Gamma}_\varepsilon \,; \\ &= -\int_{\Gamma \bigcap B(x^*,a)\setminus\{x^*\}} \frac{1}{2\pi r}\left(1 - \frac{r}{a}\right) Q^1_{\kappa,a}(r)\, f(y)\, ds(y) \\ &\quad + \phi_\Gamma + \phi(x, x^*)\ , \text{ when } x \in \overline{\Gamma}_\varepsilon, \end{aligned} \tag{10}$$

where $Q^1_{\kappa,a}(r) = \dfrac{\sinh(\kappa(a-r))}{a-r}\dfrac{a}{\sinh(\kappa a)}$, $r = |y - x^*|$.

For Markov chains based on the direct simulation, finiteness of the mean number of steps, $\mathbb{E}N < \infty$, is equivalent to convergence of the Neumann series for the corresponding integral operator, which kernel coincides with the transition density of this Markov chain. Besides that, the kernel of the integral operator that defines the second moment of the estimate is also equal to this density [EM82]. It means that for exactly known free term, F, the estimate (9) is unbiased and has finite variance. The same is true if estimates, $\xi[F](x_i)$, are unbiased and have uniformly in x_i bounded second moments. It is clear that we can easily choose such density that estimate for the integral in (10) will have the requested properties.

To prove that the mean number of steps is finite, we consider the auxiliary boundary-value problem:

$$\Delta p_0(x) - \kappa^2 p_0(x) = 0, x \in G_e, \quad \frac{\partial p_0}{\partial n}|_\Gamma = -1. \tag{11}$$

For this problem, the integral in (10) equals $(\cosh(\kappa a) - 1)/(\kappa \sinh(\kappa a))$ when Γ is a plane or a sphere. This is equal to $\dfrac{a}{2}\left(1 - \dfrac{(\kappa a)^2}{24} + O(\kappa a)^4\right)$, as $\kappa a \to 0$. Setting $\varepsilon = O(a/2R)^3$ we have

Lemma 2. *The mean number of boundary hits in the walk-on-spheres solving the exterior Neumann problem is* $\mathbb{E}N^* = \dfrac{2p_0}{a}\left(1 + O(a^2)\right)$.

In the full analogy we obtain the following.

Lemma 3. *The mean number of boundary hits of the walk-on-spheres algorithm solving continuity boundary-value problem is* $\mathbb{E}N^* = \frac{2p_1}{a}(1+O(a^2))$. *Here* p_1 *is a bounded solution to the problem (1), (2) with no charges in* G_i *and with boundary condition* $\epsilon_i \frac{\partial p_{1,i}}{\partial n}(y) = \epsilon_e \frac{\partial p_{1,e}}{\partial n}(y) + 1$.

Denote by $\{x^*_{k,j} \in \Gamma, j = 1, \ldots, N^*_i\}$ the sequence of exit points from G_i for the walk-on-spheres Markov chain used to calculate the solution of the continuity boundary-value problem. Let $\{x_{k+1,j} \in G_i, j = 1, \ldots, N^*_i - 1\}$ be the sequence of return points. Clearly, $\mathbb{E}N^*_i = p_i \mathbb{E}N^*$. Then we have

Theorem 1. *The quantity*

$$\xi[u](x_0) = g(x_0) - g(x^*_{k,1}) + \sum_{j=1}^{N^*_i - 1} \left[g(x_{k+1,j}) - g(x^*_{k,j+1})\right] \tag{12}$$

is the estimate for solution of the boundary-value problem (1), (2), (3). For $\varepsilon = (a/2R)^3$, *the bias of this estimate is* $O(a/2R)^2$ *as* $a \to 0$. *Variance of this estimate is finite, and computational cost is* $O(\log(\delta)\ \delta^{-5/2})$, *for a given accuracy,* δ.

The variance is finite because the algorithm is based on the direct simulation of the transformed integral equation [EM82]. Logarithmic factor in the estimate comes from the mean number of steps in the WOS Markov chain until it hits for the first time the ε-strip near the boundary [EKMS80, ENS89].

It is essential to note that with the finite-difference approximation of the normal derivative using step h, the mean number of boundary hits is $O(h^{-1}$. Therefore, bias of the resulting estimate is $O(h)$, and computational cost $O(\log(\delta)\ \delta^{-3})$ [Kro84, MM97]. Thus even the simplest approximation to the (exact!) integral relation we constructed substantially improves the efficiency of the WOS algorithm.

6 Results of Computations and Discussion

To show the efficiency and dependence of the proposed algorithm on parameters, we consider simple model problem with known analytical solution. For the sphere, $G_i = \{x : r \equiv |x| < R\}$, with one point charge, Q, at its center, the solution is $u_i(x) = \frac{Q}{4\pi}\left[\frac{1}{\epsilon_i}\left(\frac{1}{r} - \frac{1}{R}\right) + \frac{1}{\epsilon_e R(1+\kappa R)}\right]$, and $u_e(x) = \frac{Q}{4\pi\epsilon_e}\frac{\exp(-\kappa(r-R)}{r(1+\kappa R)}$. To elucidate the behavior of the algorithm in the exterior, we also considered the model Neumann problem (11). Its analytical solution is $p_0 = \exp(-\kappa(r-R))\dfrac{R^2}{r(1+\kappa R)}$. The statistical error of the computed result is given as σ, which is calculated as the square root of the estimate's variance divided by $(N_{samples}-1)^{-1/2}$.

The results in Table 1 clearly show the conformity with the theoretically predicted behavior. In particular, we see that the mean number of random walks returning to the boundary linearly depends on $(a/2R)^{-1}$.

The results for the regular part of solution to the continuity boundary-value problem given in Table 2 also show the theoretically predicted behavior of number of boundary hits. Note, however, that for this problem, the bias is substantially smaller then expected. This can be explained by the particularly smooth behavior of the solution.

The results given in Table 3 show that the behavior of both numbers can be approximated with good accuracy by const $* \, p_0(r)$, as it is predicted by Lemma 2.

Mean number of points in the WOS trajectory, as is clearly seen from Table 4, even for non-zero κ, to the high accuracy can be approximated by logarithmic law. The bias, as we know, should linearly depend on ε. However,

Table 1. Solution of Neumann problem ($\kappa = 0.1$, $\varepsilon = 10^{-6}$) at $|x| = R$. Dependence of bias on radius a of auxiliary sphere

a	bias (sigma)	theoretical bias	mean number of boundary hits
0.05	0.000957 (0.000897)	0.000545	36.40
0.1	0.002098 (0.000886)	0.002273	18.22
0.2	0.009586 (0.000867)	0.009187	9.18

Table 2. Regular solution of continuity problem at the center of sphere ($\kappa = 0.1$, $\varepsilon = 10^{-6}$). Dependence of bias (multiplied by 4π) on radius a of auxiliary sphere

a	bias (sigma)	theoretical bias	mean number of boundary hits
0.05	0.000011 (0.000015)	0.000598	38.24
0.1	0.000030 (0.000018)	0.002389	19.16
0.2	0.000106 (0.000024)	0.009646	9.64

Table 3. Solution of Neumann problem ($\kappa = 0.1$, $a = 0.1$, $\varepsilon = 10^{-6}$) at $|x| = r$. Dependence of number of points in WOS on r

r	1.0	1.1	1.2	1.5	2.0	3.0	4.0	6.0
N points	697.6	666.3	608.6	476.9	345.1	210.7	144.7	80.8
N boundary hits	18.22	16.39	14.89	11.55	8.27	4.97	3.37	1.84

Table 4. Regular solution of continuity problem ($\kappa = 0.1$, $\varepsilon = 10^{-6}$) at $|x| = R$. Dependence of number of points in WOS and bias ($\sigma = 0.00050$) on the width of strip near the boundary

ε	10^{-6}	10^{-5}	10^{-4}	10^{-3}	10^{-2}
N points	356.8	289.5	222.6	155.2	93.1
bias	0.00035	0.00098	0.00285	0.00523	0.05209

we see that for small values of width of the boundary strip, its behavior is screened by the statistical error.

The results of the model computations we presented confirm theoretically predicted behavior of the algorithm described. Currently we are in the process of adjusting this Monte Carlo method to computing electrostatic properties of molecules in solvent. We already found out that complicated geometry of these molecules requires special treatment and further development of the algorithm. The work on these issues and on extending the Monte Carlo approach to solving non-linear Poisson-Boltzmann equation is still in progress.

Acknowledgments

I would like to acknowledge the kind hospitality of the Institute of Molecular Biophysics at the Florida State University.

References

[DM90] M.E. Davis and J.A. McCammon. Electrostatics in biomolecular structure and dynamics. *Chem. Rev.*, 90:509–521, 1990.

[EKMS80] B.S. Elepov, A.A. Kronberg, G.A. Mikhailov, and K.K. Sabelfeld. *Reshenie kraevyh zadach metodom Monte-Carlo [Solution of boundary value problems by the Monte Carlo method].* Nauka, Novosibirsk, 1980. (Russian).

[EM82] S.M. Ermakov and G.A. Mikhailov. *Statisticheskoe modelirovanie [Statistical simulation].* Nauka, Moscow, 1982. (Russian).

[ENS89] S.M. Ermakov, V.V. Nekrutkin, and A.S. Sipin. *Random processes for classical equations of mathematical physics.* Kluwer Academic Publishers, Dodrecht, The Netherlands, 1989.

[HSS66] A. Haji-Sheikh and E.M. Sparrow. The floating random walk and its application to Monte Carlo solutions of heat equations. *SIAM J. Appl. Math.*, 14(2):570–589, 1966.

[KMS04a] A. Karaivanova, M. Mascagni, and N.A. Simonov. Parallel quasi-random walks on the boundary. *Monte Carlo Methods and Applications*, 10(3-4):311–320, 2004.

[KMS04b] A. Karaivanova, M. Mascagni, and N.A. Simonov. Solving BVPs using quasirandom walks on the boundary. *Lecture Notes in Computer Science*, 2907:162–169, 2004.

[Kro84] A.A. Kronberg. On algorithms for statistical simulation of the solution of boundary value problems of elliptic type. *Zh. Vychisl. Mat. i Mat. Phyz.*, 84(10):1531–1537, 1984. (Russian).

[M5̈6] M.E. Müller. Some continuous Monte Carlo methods for the Dirichlet problem. *Ann. Math. Statistics*, 27(3):569–589, 1956.

[Mir70] C. Miranda. *Partial differential equations of elliptic type.* Springer-Verlag, New York, 1970.

[MM97] G.A. Mikhailov and R.N. Makarov. Solution of boundary value problems by the "random walk on spheres" method with reflection from the boundary. *Dokl. Akad. Nauk*, 353(6):720–722, 1997. (Russian).

[MS04] M. Mascagni and N.A. Simonov. Monte Carlo methods for calculating some physical properties of large molecules. *SIAM J. Sci. Comp.*, 26(1):339–357, 2004.

[Sim83] N.A. Simonov. A random walk algorithm for solving boundary value problems with partition into subdomains. In *Metody i algoritmy statisticheskogo modelirovanija [Methods and algorithms for statistical modelling]*, pages 48–58. Akad. Nauk SSSR Sibirsk. Otdel., Vychisl. Tsentr, Novosibirsk, 1983. (Russian).

[Sim06] N.A. Simonov. Monte Carlo methods for solving elliptic equations with boundary conditions containing the normal derivative. *Doklady Mathematics*, 74:656–659, 2006.

[SS94] K.K. Sabelfeld and N.A. Simonov. *Random Walks on Boundary for solving PDEs.* VSP, Utrecht, The Netherlands, 1994.

Good Lattice Rules with a Composite Number of Points Based on the Product Weighted Star Discrepancy

Vasile Sinescu[1] and Stephen Joe[2]

[1] Department of Mathematics, University of Waikato, Private Bag 3105, Hamilton, New Zealand
vs27@math.waikato.ac.nz
[2] Department of Mathematics, University of Waikato, Private Bag 3105, Hamilton, New Zealand
stephenj@math.waikato.ac.nz

Summary. Rank-1 lattice rules based on a weighted star discrepancy with weights of a product form have been previously constructed under the assumption that the number of points is prime. Here, we extend these results to the non-prime case. We show that if the weights are summable, there exist lattice rules whose weighted star discrepancy is $O(n^{-1+\delta})$, for any $\delta > 0$, with the implied constant independent of the dimension and the number of lattice points, but dependent on δ and the weights. Then we show that the generating vector of such a rule can be constructed using a component-by-component (CBC) technique. The cost of the CBC construction is analysed in the final part of the paper.

Key words: Rank-1 lattice rules, weighted star discrepancy, component-by-component construction.

1 Introduction

We consider rank-1 lattice rules for the approximation of integrals over the d-dimensional unit cube given by

$$I_d(f) = \int_{[0,1]^d} f(\boldsymbol{x})\,\mathrm{d}\boldsymbol{x}.$$

These rank-1 lattice rules are quadrature rules of the form

$$Q_{n,d}(f) = \frac{1}{n}\sum_{k=0}^{n-1} f\left(\left\{\frac{k\boldsymbol{z}}{n}\right\}\right),$$

where $z \in \mathbb{Z}^d$ is the generating vector having all the components conveniently assumed to be relatively prime with n, while the braces around a vector indicate that we take the fractional part of each component of the vector.

In this paper we are looking to extend the recent results in [Joe06] by constructing rank-1 lattice rules with a composite number of points. Hence, the same assumptions as in [Joe06] will be used here with the main difference that n is assumed to be just a positive integer. The vast majority of earlier research papers have assumed that n was a prime number; an assumption which simplifies the analysis of the problem.

However there are some known results in the non-prime case. For instance, it has been proven in [Dis90], [Nie78], or [Nie92, Chapter 5] that good lattice rules with a non-prime number of points do exist. Several measures of goodness were used in those works, but under the assumptions that variables are equally important. Here, we assume that variables are arranged in the decreasing order of their importance and we employ a weighted star discrepancy as a criterion of goodness. An unweighted star discrepancy (corresponding to an L_∞ maximum error) has been previously used in [Joe04] and in more general works such as [Nie92] or [SJ94], while the weighted star discrepancy has been used in [HN03], [Joe06], and [SJ07].

A constructive approach in the non-prime case has been proposed in [KJ02], where the integrands were assumed to belong to certain reproducing kernel Hilbert spaces such as weighted Korobov spaces of periodic functions or weighted Sobolev spaces with square-integrable mixed first derivatives. Here we require the integrands to have the weaker requirement of integrable mixed first derivatives. Let us remark that in [Kuo03] it was proven that in the reproducing kernel Hilbert spaces of [KJ02], the component-by-component construction (used also here) achieves the optimal rate of convergence. In [Dic04], the results in [Kuo03] were further improved and then extended to the non-prime case.

Let us also mention that lattice rules with a composite number of points have become more interesting since the introduction of extensible lattice rules in [HH97]. Later, in [HN03], it was shown that extensible lattice rules in number of points with a low weighted star discrepancy do exist, but the proof was non-constructive. More recently, in [DPW07], a possible way of constructing extensible lattice rules was proposed. Therein, it was assumed that n is of the form p^m with $p \geq 2$ an arbitrary prime. For such a case, it has been shown that lattice rules extensible in number of points based on the weighted star discrepancy can be constructed, but the results were not generalised to arbitrary integers as we propose here.

2 Weighted Star Discrepancy

As mentioned in the previous section, throughout this paper we make similar assumptions as in [Joe06] and we start by recalling some of those results and assumptions.

In order to introduce the general weighted star discrepancy, let us consider first the point set $P_n(\boldsymbol{z}) := \{\{k\boldsymbol{z}/n\},\ 0 \le k \le n-1\}$. Then the local discrepancy of the point set $P_n(\boldsymbol{z})$ at $\boldsymbol{x} \in [0,1]^d$ is defined by

$$\operatorname{discr}(\boldsymbol{x}, P_n) := \frac{A([\mathbf{0}, \boldsymbol{x}), P_n)}{n} - \prod_{j=1}^{d} x_j .$$

Here $A([\mathbf{0}, \boldsymbol{x}), P_n)$ represents the counting function, namely the number of points in $P_n(\boldsymbol{z})$ which lie in $[\mathbf{0}, \boldsymbol{x})$ with $\boldsymbol{x} = (x_1, x_2, \ldots, x_d)$.

Let now $\mathfrak{u}$ be an arbitrary non-empty subset of $\mathcal{D} := \{1, 2, \ldots, d-1, d\}$ and denote its cardinality by $|\mathfrak{u}|$. For the vector $\boldsymbol{x} \in [0,1]^d$, let $\boldsymbol{x}_\mathfrak{u}$ denote the vector from $[0,1]^{|\mathfrak{u}|}$ containing the components of $\boldsymbol{x}$ whose indices belong to $\mathfrak{u}$. By $(\boldsymbol{x}_\mathfrak{u}, \mathbf{1})$ we mean the vector from $[0,1]^d$ whose j-th component is x_j if $j \in \mathfrak{u}$ and 1 if $j \notin \mathfrak{u}$. From Zaremba's identity (see for instance [SW98] or [Zar68]), we obtain

$$Q_{n,d}(f) - I_d(f) = \sum_{\mathfrak{u} \subseteq \mathcal{D}} (-1)^{|\mathfrak{u}|} \int_{[0,1]^{|\mathfrak{u}|}} \operatorname{discr}((\boldsymbol{x}_\mathfrak{u}, \mathbf{1}), P_n) \frac{\partial^{|\mathfrak{u}|} f((\boldsymbol{x}_\mathfrak{u}, \mathbf{1}))}{\partial \boldsymbol{x}_\mathfrak{u}} \, \mathrm{d}\boldsymbol{x}_\mathfrak{u} . \tag{1}$$

Now let us introduce a sequence of positive weights $\{\gamma_j\}_{j=1}^{\infty}$, which describe the decreasing importance of the successive coordinates x_j and consider $\boldsymbol{\gamma}_\mathfrak{u}$ as the weight associated with the set $\mathfrak{u}$. In this paper, we assume that the weights $\{\boldsymbol{\gamma}_\mathfrak{u}\}$ are "product", that is

$$\boldsymbol{\gamma}_\mathfrak{u} = \prod_{j \in \mathfrak{u}} \gamma_j ,$$

for any subset $\mathfrak{u} \subseteq \mathcal{D}$. Such assumptions on the weights have been made in [HN03], [Joe06], [SW98] and in other research papers. Using (1) we see that we can write

$$\begin{aligned} & Q_{n,d}(f) - I_d(f) \\ & = \sum_{\mathfrak{u} \subseteq \mathcal{D}} (-1)^{|\mathfrak{u}|} \boldsymbol{\gamma}_\mathfrak{u} \int_{[0,1]^{|\mathfrak{u}|}} \operatorname{discr}((\boldsymbol{x}_\mathfrak{u}, \mathbf{1}), P_n) \boldsymbol{\gamma}_\mathfrak{u}^{-1} \frac{\partial^{|\mathfrak{u}|} f((\boldsymbol{x}_\mathfrak{u}, \mathbf{1}))}{\partial \boldsymbol{x}_\mathfrak{u}} \, \mathrm{d}\boldsymbol{x}_\mathfrak{u} . \end{aligned}$$

Applying Hölder's inequality for integrals and sums, we obtain

$$\begin{aligned} |Q_{n,d}(f) - I_d(f)| \le & \left(\sum_{\mathfrak{u} \subseteq \mathcal{D}} \sup_{\boldsymbol{x}_\mathfrak{u} \in [0,1]^{|\mathfrak{u}|}} \boldsymbol{\gamma}_\mathfrak{u} \left| \operatorname{discr}((\boldsymbol{x}_\mathfrak{u}, \mathbf{1}), P_n) \right| \right) \\ & \times \left(\max_{\mathfrak{u} \subseteq \mathcal{D}} \boldsymbol{\gamma}_\mathfrak{u}^{-1} \int_{[0,1]^{|\mathfrak{u}|}} \left| \frac{\partial^{|\mathfrak{u}|}}{\partial \boldsymbol{x}_\mathfrak{u}} f((\boldsymbol{x}_\mathfrak{u}, \mathbf{1})) \right| \mathrm{d}\boldsymbol{x}_\mathfrak{u} \right) . \end{aligned}$$

Thus we can define a weighted star discrepancy $D^*_{n,\boldsymbol{\gamma}}(\boldsymbol{z})$ by

$$D^*_{n,\boldsymbol{\gamma}}(\boldsymbol{z}) := \sum_{\mathfrak{u} \subseteq \mathcal{D}} \boldsymbol{\gamma}_\mathfrak{u} \sup_{\boldsymbol{x}_\mathfrak{u} \in [0,1]^{|\mathfrak{u}|}} \left| \operatorname{discr}((\boldsymbol{x}_\mathfrak{u}, \mathbf{1}), P_n) \right| . \tag{2}$$

From [Nie92, Theorem 3.10 and Theorem 5.6], we obtain the following inequality:

$$\sup_{\boldsymbol{x}_{\mathfrak{u}} \in [0,1]^{|\mathfrak{u}|}} |\mathrm{discr}((\boldsymbol{x}_{\mathfrak{u}}, \mathbf{1}), P_n)| \leq 1 - (1 - 1/n)^{|\mathfrak{u}|} + \frac{R_n(\boldsymbol{z}, \mathfrak{u})}{2},$$

where

$$R_n(\boldsymbol{z}, \mathfrak{u}) = \sum_{\substack{\boldsymbol{h} \cdot \boldsymbol{z}_{\mathfrak{u}} \equiv 0 \ (\bmod\ n) \\ \boldsymbol{h} \in E^*_{n,|\mathfrak{u}|}}} \prod_{j \in \mathfrak{u}} \frac{1}{\max(1, |h_j|)}.$$

Here $\boldsymbol{z}_{\mathfrak{u}}$ denotes the vector consisting of the components of $\boldsymbol{z}$ whose indices belong to $\mathfrak{u}$, while

$$E^*_{n,m} = \{\boldsymbol{h} \in \mathbb{Z}^m,\ \boldsymbol{h} \neq \mathbf{0} : -n/2 < h_j \leq n/2,\ 1 \leq j \leq m\}.$$

This result, together with (2) shows that the general weighted star discrepancy satisfies the inequality

$$D^*_{n,\boldsymbol{\gamma}}(\boldsymbol{z}) \leq \sum_{\mathfrak{u} \subseteq \mathcal{D}} \gamma_{\mathfrak{u}} \left(1 - (1 - 1/n)^{|\mathfrak{u}|} + \frac{R_n(\boldsymbol{z}, \mathfrak{u})}{2}\right). \quad (3)$$

For calculation purposes, the theory of lattice rules, (for example, see [Nie92] or [SJ94]) shows that we may write $R_n(\boldsymbol{z}, \mathfrak{u})$ as

$$R_n(\boldsymbol{z}, \mathfrak{u}) = \frac{1}{n} \sum_{k=0}^{n-1} \prod_{j \in \mathfrak{u}} \left(1 + \sum_{-n/2 < h \leq n/2}{}' \frac{e^{2\pi i h k z_j / n}}{|h|}\right) - 1,$$

where the $'$ in the sum indicates we omit the $h = 0$ term. It is easy to see that $R_n(\boldsymbol{z}, \mathfrak{u})$ represents the quadrature error produced by applying the lattice rule to the integrand

$$\prod_{j \in \mathfrak{u}} \left(1 + \sum_{-n/2 < h \leq n/2}{}' \frac{e^{2\pi i h x_j}}{|h|}\right).$$

It is also easy to check that if $|\mathfrak{u}| = 1$, then the corresponding error $R_n(\boldsymbol{z}, \mathfrak{u}) = 0$ for each subset of $\mathcal{D}$ with only one element.

3 Bounds on the Weighted Star Discrepancy

To obtain bounds on $D^*_{n,\boldsymbol{\gamma}}(\boldsymbol{z})$, we see from (3) that we need to bound the quantity

$$\sum_{\mathfrak{u} \subseteq \mathcal{D}} \gamma_{\mathfrak{u}} \left(1 - (1 - 1/n)^{|\mathfrak{u}|}\right)$$

and the quantity

$$e_{n,d}^2(\boldsymbol{z}) := \sum_{\mathfrak{u}\subseteq\mathcal{D}} \gamma_{\mathfrak{u}} R_n(\boldsymbol{z},\mathfrak{u}). \tag{4}$$

Under the assumption that the weights are summable, that is $\sum_{j=1}^{\infty}\gamma_j < \infty$, it follows from [Joe06, Lemma 1] that

$$\sum_{\mathfrak{u}\subseteq\mathcal{D}} \gamma_{\mathfrak{u}} \left(1-(1-1/n)^{|\mathfrak{u}|}\right) = O(n^{-1}), \tag{5}$$

with the implied constant depending on the weights, but independent of d and n.

We now consider $e_{n,d}^2(\boldsymbol{z})$ in more detail and by expanding the quadrature error defined by (4) as in [Joe06], we obtain

$$e_{n,d}^2(\boldsymbol{z}) = \frac{1}{n}\sum_{k=0}^{n-1}\prod_{j=1}^{d}(1+\gamma_j+\gamma_j C_k(z_j)) - \prod_{j=1}^{d}(1+\gamma_j),$$

where

$$C_k(z) = \sum_{-n/2<h\le n/2}{}' \frac{e^{2\pi i h k z/n}}{|h|}.$$

By setting $\beta_j = 1+\gamma_j$, we obtain

$$e_{n,d}^2(\boldsymbol{z}) = \frac{1}{n}\sum_{k=0}^{n-1}\prod_{j=1}^{d}(\beta_j+\gamma_j C_k(z_j)) - \prod_{j=1}^{d}\beta_j. \tag{6}$$

We can obtain a bound on $e_{n,d}^2(\boldsymbol{z})$ by obtaining a bound on a certain mean value of $e_{n,d}^2(\boldsymbol{z})$. The mean $M_{n,d,\gamma}$ is defined by

$$M_{n,d,\gamma} := \frac{1}{\varphi(n)^d}\sum_{\boldsymbol{z}\in\mathcal{Z}_n^d} e_{n,d}^2(\boldsymbol{z}),$$

where φ is Euler's totient function and

$$\mathcal{Z}_n = \{z : z\in\{1,2,\ldots,n-1\},\ (z,n)=1\}$$

has cardinality $\varphi(n)$. Here $(z,n)=\gcd(z,n)$. A bound on the mean $M_{n,d,\gamma}$ is given next.

Theorem 1. *Let $n\ge 2$ be an integer and let*

$$S_n = \sum_{-n/2<h\le n/2}{}' \frac{1}{|h|}.$$

If the weights $\{\gamma_j\}_{j=1}^{\infty}$ are summable, then

$$M_{n,d,\gamma} \le \frac{1}{n}\prod_{j=1}^{d}(\beta_j+\gamma_j S_n) + O\left(\frac{\ln\ln(n+1)}{n}\right),$$

where the implied constant depends on the weights, but is independent of the dimension.

Proof. We have

$$
\begin{aligned}
M_{n,d,\gamma} &= \frac{1}{\varphi(n)^d} \sum_{z \in \mathcal{Z}_n^d} \left(\frac{1}{n} \sum_{k=0}^{n-1} \prod_{j=1}^{d} (\beta_j + \gamma_j C_k(z_j)) - \prod_{j=1}^{d} \beta_j \right) \\
&= \frac{1}{n} \sum_{k=0}^{n-1} \prod_{j=1}^{d} \left(\frac{1}{\varphi(n)} \sum_{z_j \in \mathcal{Z}_n} (\beta_j + \gamma_j C_k(z_j)) \right) - \prod_{j=1}^{d} \beta_j \\
&= \frac{1}{n} \prod_{j=1}^{d} (\beta_j + \gamma_j S_n) + \frac{1}{n} \sum_{k=1}^{n-1} \prod_{j=1}^{d} \left(\beta_j + \frac{\gamma_j}{\varphi(n)} \sum_{z_j \in \mathcal{Z}_n} C_k(z_j) \right) - \prod_{j=1}^{d} \beta_j,
\end{aligned}
$$

where in the last step the $k = 0$ term has been separated out and we have used the fact that $C_0(z) = S_n$. If we denote

$$
T_n(k) = \sum_{z \in \mathcal{Z}_n} C_k(z) = \sum_{z \in \mathcal{Z}_n} \sum_{-n/2 < h \le n/2}{}' \frac{e^{2\pi i h k z/n}}{|h|}, \tag{7}
$$

then we see that the mean can be written as

$$
M_{n,d,\gamma} = \frac{1}{n} \prod_{j=1}^{d} (\beta_j + \gamma_j S_n) + L_{n,d,\gamma} - \prod_{j=1}^{d} \beta_j, \tag{8}
$$

where

$$
L_{n,d,\gamma} = \frac{1}{n} \sum_{k=1}^{n-1} \prod_{j=1}^{d} \left(\beta_j + \frac{\gamma_j}{\varphi(n)} T_n(k) \right). \tag{9}
$$

The rest of this proof follows many of the arguments used in the proof of [Nie92, Theorem 5.10] (see also [Dis90]). Firstly, it may be shown that

$$
T_n(k) = \sum_{a|n} \mu(a) \left(\frac{n}{a}, k \right) S_{a(\frac{n}{a},k)} = \sum_{a|n} \mu \left(\frac{n}{a} \right) (a,k)\, S_{\frac{n(a,k)}{a}}, \tag{10}
$$

where μ denotes the well-known Möbius function from number theory. If n is prime, then we obtain $T_n(k) = -S_n$ for any $1 \le k \le n-1$, which leads to the results obtained in [Joe06]. From [Nie78, Lemmas 1 and 2], we have

$$
S_m = 2 \ln\, m + 2\omega - \ln 4 + \varepsilon(m), \tag{11}
$$

where ω is the Euler-Mascheroni constant given by

$$
\omega = \lim_{\ell \to \infty} \left(\sum_{k=1}^{\ell} \frac{1}{k} - \ln \ell \right),
$$

while

$$|\varepsilon(m)| < 4m^{-2}. \tag{12}$$

Using (10), we now obtain

$$T_n(k) = (2\ln n + 2\omega - \ln 4)B_n(k) - 2H_n(k) + V_n(k), \tag{13}$$

where

$$B_n(k) = \sum_{a|n} \mu\left(\frac{n}{a}\right)(a,k),$$

$$H_n(k) = \sum_{a|n} \mu\left(\frac{n}{a}\right)(a,k)\ln\frac{a}{(a,k)},$$

and

$$V_n(k) = \sum_{a|n} \mu\left(\frac{n}{a}\right)(a,k)\varepsilon\left(\frac{n(a,k)}{a}\right). \tag{14}$$

From the proof of [Nie92, Theorem 5.10], we have $B_n(k) = 0$ for any $1 \le k \le n-1$. Using this result in (13), we get

$$T_n(k) = -2H_n(k) + V_n(k). \tag{15}$$

By combining (9) with (15), we obtain

$$L_{n,d,\boldsymbol{\gamma}} = \frac{1}{n}\sum_{k=1}^{n-1}\prod_{j=1}^{d}\left(\beta_j + \gamma_j\left(-2J_n(k) + \frac{V_n(k)}{\varphi(n)}\right)\right), \tag{16}$$

where

$$J_n(k) = \frac{H_n(k)}{\varphi(n)}.$$

The proof of Theorem 5.10 in [Nie92] yields $V_n(k) = O(1)$ with an absolute implied constant. Hence we have $V_n(k)/\varphi(n) = O(1/\varphi(n))$. This result together with (16) and $\beta_j = 1 + \gamma_j$ yields

$$L_{n,d,\boldsymbol{\gamma}} = \frac{1}{n}\sum_{k=1}^{n-1}\prod_{j=1}^{d}\left(1 + \gamma_j(1 - 2J_n(k)) + \gamma_j O\left(\frac{1}{\varphi(n)}\right)\right). \tag{17}$$

Let us denote by p a prime number and by $e_p(n)$ the largest exponent such that $p^{e_p(n)}$ divides n. Then, from the proof of [Nie92, Theorem 5.10], we obtain

$$H_n(k) = \begin{cases} p^{e_p(k)}\varphi(n/p^{e_p(n)})\ln p, & \text{if } p \text{ is the unique prime with } e_p(n) > e_p(k), \\ 0, & \text{otherwise.} \end{cases}$$

If such a p exists, then by the definition of $e_p(n)$, we have $n/p^{e_p(n)}$ relatively prime with $p^{e_p(n)}$ and hence $\varphi(n/p^{e_p(n)})\varphi(p^{e_p(n)}) = \varphi(n)$. We then obtain

$$J_n(k) = \frac{p^{e_p(k)}\varphi(n/p^{e_p(n)})\ln p}{\varphi(n)} = \frac{p^{e_p(k)}\ln p}{\varphi(p^{e_p(n)})} = \frac{\ln p}{p^{\alpha_k}(p-1)}, \tag{18}$$

where we put $\alpha_k = e_p(n) - e_p(k) - 1$, for $1 \leq k \leq n-1$. For each $1 \leq k \leq n-1$, it is not difficult to check from (18) that $-1 < 1 - 2\ln(p)/(p^{\alpha_k}(p-1)) < 1$ for any prime $p \geq 2$ and for any $\alpha_k \geq 0$. Hence, $1 + \gamma_j(1 - 2J_n(k)) \leq 1 + \gamma_j = \beta_j$ for any $1 \leq j \leq d$. The product in (17) can then be bounded by

$$\begin{aligned}
&\prod_{j=1}^{d}\left(1 + \gamma_j(1 - 2J_n(k)) + \gamma_j O\left(\frac{1}{\varphi(n)}\right)\right)\\
&\leq \prod_{j=1}^{d}\left(\beta_j + \gamma_j O\left(\frac{1}{\varphi(n)}\right)\right)\\
&= \prod_{j=1}^{d}\beta_j + \sum_{\substack{\mathfrak{u}\subseteq\mathcal{D}\\|\mathfrak{u}|\geq 1}}\left(O\left(\frac{1}{\varphi(n)}\right)\right)^{|\mathfrak{u}|}\prod_{j\in\mathfrak{u}}\gamma_j\prod_{j\notin\mathfrak{u}}\beta_j\\
&= \prod_{j=1}^{d}\beta_j + O\left(\frac{1}{\varphi(n)}\right),
\end{aligned}\tag{19}$$

where the implied constant depends on the quantity

$$\sum_{\substack{\mathfrak{u}\subseteq\mathcal{D}\\|\mathfrak{u}|\geq 1}}\prod_{j\in\mathfrak{u}}\gamma_j\prod_{j\notin\mathfrak{u}}\beta_j \leq \prod_{j=1}^{d}(\beta_j + \gamma_j).$$

Next, let us consider

$$\prod_{j=1}^{d}(\beta_j + \gamma_j) = \exp\left(\sum_{j=1}^{d}\ln(\beta_j + \gamma_j)\right) \leq \exp\left(2\sum_{j=1}^{d}\gamma_j\right),$$

where we used that $\beta_j = 1 + \gamma_j$ and $\ln(1+x) \leq x$ for any $x > -1$. Recalling that the weights were assumed to be summable, by denoting $\Gamma := \sum_{j=1}^{\infty}\gamma_j$, it follows that

$$\prod_{j=1}^{d}(\beta_j + \gamma_j) \leq e^{2\Gamma},$$

which shows that the implied constant of (19) is independent of the dimension, but dependent on the weights. From (17), (19) and using that $1/\varphi(n) = O(n^{-1}\ln\ln(n+1))$, we now obtain

$$L_{n,d,\boldsymbol{\gamma}} \leq \frac{n-1}{n}\prod_{j=1}^{d}\beta_j + O\left(\frac{\ln\ln(n+1)}{n}\right).$$

By combining the last inequality with (8), we obtain

$$M_{n,d,\boldsymbol{\gamma}} \leq \frac{1}{n}\prod_{j=1}^{d}(\beta_j + \gamma_j S_n) + O\left(\frac{\ln\ln(n+1)}{n}\right). \qquad \square$$

Corollary 1. *Let $n \geq 2$ be an integer. If the weights $\{\gamma_j\}_{j=1}^{\infty}$ are summable, then there exists a vector $\boldsymbol{z} \in \mathcal{Z}_n^d$ such that*

$$e_{n,d}^2(\boldsymbol{z}) \leq \frac{1}{n} \prod_{j=1}^{d} (\beta_j + \gamma_j S_n) + O\left(\frac{\ln \ln(n+1)}{n}\right),$$

where the implied constant depends on the weights, but is independent of the dimension.

Proof. Clearly, there must be a vector $\boldsymbol{z} \in \mathcal{Z}_n^d$ such that $e_{n,d}^2(\boldsymbol{z}) \leq M_{n,d,\gamma}$ and the result then follows from Theorem 1. □

It is known from [Nie78] or [Nie92] that in an unweighted setting there exist d-dimensional lattice rules having $O(n^{-1}(\ln n)^d)$ star discrepancy with the implied constant depending only on d. Such a bound is widely believed to be the best possible (see [Lar87] or [Nie92] for details). In our situation, from (3), (5) and Corollary 1, together with the observation that $S_n \leq 2 \ln n$ for any $n \geq 2$ (this follows from (11) for $n \geq 3$ and a direct calculation for $n = 2$), it will follow that there exists a vector $\boldsymbol{z} \in \mathcal{Z}_n^d$ such that

$$D_{n,\gamma}^*(\boldsymbol{z}) = O(n^{-1}(\ln\ n)^d),$$

but with the implied constant independent of d. A bound that does not involve $\ln n$ is possible by making use of [HN03, Lemma 3]. This result leads to the conclusion that if the weights are summable, then there exists a generating vector $\boldsymbol{z}$ such that the weighted star discrepancy achieves the error bound

$$D_{n,\gamma}^*(\boldsymbol{z}) = O(n^{-1+\delta}),$$

for any $\delta > 0$, where the implied constant depends on δ and the weights but is independent of n and d.

Let us also remark that corresponding results for a weighted L_p star discrepancy can be deduced, since such a discrepancy is bounded by the discrepancy introduced in (2). Further details can be found in [Joe06].

4 A Component-by-Component Construction

Before presenting the main result regarding the CBC construction, we need the following:

Lemma 1. *There exists a positive constant c independent of n such that*

$$\sum_{k=1}^{n-1} \frac{|T_n(k)|}{\varphi(n)} \leq c \ln n,$$

where $T_n(k)$ has been defined by (7).

Proof. Since $J_n(k) = H_n(k)/\varphi(n) \geq 0$, then from (15), we obtain:

$$\sum_{k=1}^{n-1} \frac{|T_n(k)|}{\varphi(n)} \leq \sum_{k=1}^{n-1} \left(2J_n(k) + \frac{|V_n(k)|}{\varphi(n)} \right). \tag{20}$$

From the proof of [Nie92, Theorem 5.10], we obtain:

$$\sum_{k=1}^{n-1} J_n(k) = \ln n. \tag{21}$$

In order to analyse the second quantity of (20), we see from (14) that

$$|V_n(k)| \leq \sum_{a|n} \left| \mu\left(\frac{n}{a}\right) \right| (a,k) \left| \varepsilon\left(\frac{n(a,k)}{a}\right) \right|.$$

By using (12), we next obtain:

$$|V_n(k)| \leq 4 \sum_{a|n} \left| \mu\left(\frac{n}{a}\right) \right| \left(\frac{a}{n}\right)^2 = 4 \sum_{a|n} \frac{1}{a^2} \leq \frac{2\pi^2}{3}.$$

Recalling that $1/\varphi(n) = O(\ln\ln(n+1)/n)$ with an absolute implied constant, we now deduce that there exists a constant $c_1 > 0$ independent of n such that

$$\sum_{k=1}^{n-1} \frac{|V_n(k)|}{\varphi(n)} \leq (n-1)\frac{2\pi^2 c_1}{3} \frac{\ln\ln(n+1)}{n} \leq \frac{2\pi^2 c_1 \ln n}{3}.$$

From this inequality combined with (20) and (21), we obtain:

$$\sum_{k=1}^{n-1} \frac{|T_n(k)|}{\varphi(n)} \leq \left(2 + \frac{2\pi^2 c_1}{3}\right) \ln n,$$

which leads to the desired result by taking $c = 2 + 2\pi^2 c_1/3$. □

In order to construct the generating vector, we use a component-by-component (CBC) technique, which is essentially a "greedy"-type algorithm, based on finding each component one at a time. This technique has been successfully used in several research papers, for instance [Joe04], [Joe06] or [KJ02]. Here, we are looking to prove that the CBC algorithm produces a generating vector whose corresponding weighted star discrepancy has the same order of magnitude as the bound given in Corollary 1. The CBC algorithm is presented below:

Component-by-component (CBC) algorithm

Assume that $n \geq 2$ is an integer, d is the dimension and all the weights are known. Then the generating vector $\boldsymbol{z} = (z_1, z_2, \ldots, z_d)$ can be constructed as follows:

1. Set the value for the first component of the vector, say $z_1 := 1$.
2. For $m = 2, 3, \ldots, d$, find $z_m \in \mathcal{Z}_n$ such that $e^2_{n,m}(z_1, z_2, \ldots, z_m)$ is minimised.

In the above, we have

$$e^2_{n,m}(z_1, z_2, \ldots, z_m) = \frac{1}{n} \sum_{k=0}^{n-1} \prod_{j=1}^{m} (\beta_j + \gamma_j C_k(z_j)) - \prod_{j=1}^{m} \beta_j .$$

The following theorem and corollary will justify the use of the CBC algorithm.

Theorem 2. *Let $n \geq 2$ be an integer and suppose that the weights $\{\gamma_j\}_{j=1}^{\infty}$ are summable. If there exists a $\boldsymbol{z} \in \mathcal{Z}_n^d$ such that*

$$e^2_{n,d}(\boldsymbol{z}) \leq \frac{1}{n} \prod_{j=1}^{d} (\beta_j + \alpha \gamma_j \ln n) ,$$

where $\alpha = 2 + c$ with c defined by Lemma 1, then there exists $z_{d+1} \in \mathcal{Z}_n$ such that

$$e^2_{n,d+1}(\boldsymbol{z}, z_{d+1}) \leq \frac{1}{n} \prod_{j=1}^{d+1} (\beta_j + \alpha \gamma_j \ln n) .$$

Such a z_{d+1} can be found by minimising $e^2_{n,d+1}(\boldsymbol{z}, z_{d+1})$ over the set $\mathcal{Z}_n$.

Proof. For any $z_{d+1} \in \mathcal{Z}_n$, we see from (6) that

$$\begin{aligned}
e^2_{n,d+1}(\boldsymbol{z}, z_{d+1}) &= \frac{1}{n} \sum_{k=0}^{n-1} \prod_{j=1}^{d} (\beta_j + \gamma_j C_k(z_j)) \left(\beta_{d+1} + \gamma_{d+1} C_k(z_{d+1})\right) - \beta_{d+1} \prod_{j=1}^{d} \beta_j \\
&= \beta_{d+1} e^2_{n,d}(\boldsymbol{z}) + \frac{\gamma_{d+1}}{n} \sum_{k=0}^{n-1} \prod_{j=1}^{d} (\beta_j + \gamma_j C_k(z_j))\, C_k(z_{d+1}) \\
&= \beta_{d+1} e^2_{n,d}(\boldsymbol{z}) + \frac{\gamma_{d+1} S_n}{n} \prod_{j=1}^{d} (\beta_j + \gamma_j S_n) \\
&\quad + \frac{\gamma_{d+1}}{n} \sum_{k=1}^{n-1} \prod_{j=1}^{d} (\beta_j + \gamma_j C_k(z_j))\, C_k(z_{d+1}),
\end{aligned}$$

where in the last step the $k = 0$ term has been separated out. Next we average $e^2_{n,d+1}(\boldsymbol{z}, z_{d+1})$ over all the possible values of z_{d+1} to form

$$\operatorname{Avg}(e^2_{n,d+1}(\boldsymbol{z}, z_{d+1})) = \frac{1}{\varphi(n)} \sum_{z_{d+1} \in \mathcal{Z}_n} e^2_{n,d+1}(\boldsymbol{z}, z_{d+1})$$

$$
\begin{aligned}
&= \beta_{d+1} e_{n,d}^2(\boldsymbol{z}) + \frac{\gamma_{d+1} S_n}{n} \prod_{j=1}^{d} (\beta_j + \gamma_j S_n) \\
&\quad + \frac{\gamma_{d+1}}{n\varphi(n)} \sum_{z_{d+1} \in \mathcal{Z}_n} \sum_{k=1}^{n-1} \prod_{j=1}^{d} (\beta_j + \gamma_j C_k(z_j))\, C_k(z_{d+1}) \\
&= \beta_{d+1} e_{n,d}^2(\boldsymbol{z}) + \frac{\gamma_{d+1} S_n}{n} \prod_{j=1}^{d} (\beta_j + \gamma_j S_n) \\
&\quad + \frac{\gamma_{d+1}}{n} \sum_{k=1}^{n-1} \left(\frac{1}{\varphi(n)} \sum_{z_{d+1} \in \mathcal{Z}_n} C_k(z_{d+1}) \right) \prod_{j=1}^{d} (\beta_j + \gamma_j C_k(z_j)) \\
&\leq \beta_{d+1} e_{n,d}^2(\boldsymbol{z}) + \frac{\gamma_{d+1} S_n}{n} \prod_{j=1}^{d} (\beta_j + \gamma_j S_n) \\
&\quad + \frac{\gamma_{d+1}}{n} \sum_{k=1}^{n-1} \frac{|T_n(k)|}{\varphi(n)} \prod_{j=1}^{d} (\beta_j + \gamma_j S_n).
\end{aligned}
$$

Using Lemma 1 and $S_n \leq 2 \ln n$, we next obtain

$$
\begin{aligned}
\operatorname{Avg}(e_{n,d+1}^2(\boldsymbol{z}, z_{d+1})) &\leq \beta_{d+1} e_{n,d}^2(\boldsymbol{z}) + \frac{\gamma_{d+1} S_n}{n} \prod_{j=1}^{d} (\beta_j + \gamma_j S_n) \\
&\quad + \frac{c\gamma_{d+1} \ln n}{n} \prod_{j=1}^{d} (\beta_j + \gamma_j S_n) \\
&\leq \beta_{d+1} e_{n,d}^2(\boldsymbol{z}) + \frac{(2+c)\gamma_{d+1} \ln n}{n} \prod_{j=1}^{d} (\beta_j + \gamma_j S_n) \\
&\leq \beta_{d+1} e_{n,d}^2(\boldsymbol{z}) + \frac{\alpha\gamma_{d+1} \ln n}{n} \prod_{j=1}^{d} (\beta_j + \alpha\gamma_j \ln n)\,.
\end{aligned}
$$

By making use of the hypothesis, we finally obtain

$$
\begin{aligned}
\operatorname{Avg}(e_{n,d+1}^2(\boldsymbol{z}, z_{d+1})) &\leq \frac{\beta_{d+1}}{n} \prod_{j=1}^{d} (\beta_j + \alpha\gamma_j \ln n) + \frac{\alpha\gamma_{d+1} \ln n}{n} \prod_{j=1}^{d} (\beta_j + \alpha\gamma_j \ln n) \\
&= \frac{1}{n} \prod_{j=1}^{d+1} (\beta_j + \alpha\gamma_j \ln n)\,.
\end{aligned}
$$

It is obvious that the $z_{d+1} \in \mathcal{Z}_n$ chosen to minimise $e_{n,d+1}^2(\boldsymbol{z}, z_{d+1})$ will satisfy

$$
e_{n,d+1}^2(\boldsymbol{z}, z_{d+1}) \leq \operatorname{Avg}(e_{n,d+1}^2(\boldsymbol{z}, z_{d+1})).
$$

This, together with the previous inequality completes the proof. □

Corollary 2. *Let $n \geq 2$ be an integer. If the weights $\{\gamma_j\}_{j=1}^{\infty}$ are summable, then for any $m = 1, 2, \ldots, d$, there exists a $\boldsymbol{z} \in \mathcal{Z}_n^m$ such that*

$$e_{n,m}^2(z_1, z_2, \ldots, z_m) \leq \frac{1}{n} \prod_{j=1}^{m} (\beta_j + \alpha\gamma_j S_n).$$

We can set $z_1 = 1$ and for every $2 \leq m \leq d$, z_m can be chosen by minimising $e_{n,m}^2(z_1, z_2, \ldots, z_m)$ over the set $\mathcal{Z}_n$.

Proof. Recall from Section 2 that $R_n(\boldsymbol{z}, \mathfrak{u}) = 0$ for any subset $\mathfrak{u} \subseteq \mathcal{D}$ with $|\mathfrak{u}| = 1$. Hence for $m = 1$ it follows that $e_{n,1}^2(z_1) = 0$. The result then follows immediately from Theorem 2. □

In order to evaluate the complexity of the CBC construction, we observe first that each $e_{n,m}^2(z_1, z_2, \ldots, z_m)$ can be evaluated in $O(n^2 m)$ operations. This cost can be reduced to $O(nm)$ by using asymptotic techniques as presented in [JS92] (see also [Joe06, Appendix A]) and consequently, the total complexity of the algorithm will be $O(n^2 d^2)$. This can be reduced to $O(n^2 d)$ if we store the products during the construction at an extra expense of $O(n)$. However, this order of magnitude can be further reduced to $O(nd \log n)$ with an approach similar to the one used by Nuyens and Cools in [NC05]. Their approach is essentially based on a fast matrix-vector multiplication and consists of minimising a function of the form

$$\frac{1}{n} \sum_{k=0}^{n-1} \prod_{j=1}^{d} \left(1 + \gamma_j \omega\left(\left\{\frac{kz_j}{n}\right\}\right)\right) - 1,$$

where ω is some function. In our situation we can take

$$\omega(x) = \sum_{-\frac{n}{2} < h \leq \frac{n}{2}}{}' \frac{e^{2\pi i h x}}{|h|}, x \in [0, 1].$$

Thus, with some modifications, the techniques used in [NC05] will also work here.

References

[Dic04] J. Dick. On the convergence rate of the component-by-component construction of good lattice rules. *J. Complexity*, 20: 493–522, 2004.

[Dis90] S. Disney. Error bounds for rank-1 lattice quadrature rules modulo composites. *Monatsh. Math.*, 110: 89–100, 1990.

[DPW07] J. Dick, F. Pillichshammer, and B.J. Waterhouse. The construction of good extensible rank-1 lattices. *Math. Comp.*, 2007. to appear.

[HH97] F.J. Hickernell and H.S. Hong. Computing multivariate normal probabilities using rank-1 lattice sequences. In *Proceedings of the Workshop on Scientific Computing (Hong Kong)*, (G.H. Golub, S.H. Lui, F.T. Luk, and R.J. Plemmons, eds.), Springer-Verlag, Singapore, pp. 209–215, 1997.

[HN03] F.J. Hickernell and H. Niederreiter. The existence of good extensible rank-1 lattices. *J. Complexity*, 19: 286–300, 2003.

[Joe04] S. Joe. Component by component construction of rank-1 lattice rules having $O(n^{-1}(ln\ n)^d)$ star discrepancy. In H. Niederreiter, editor, *Monte Carlo and Quasi-Monte Carlo Methods 2002*, pages 293–298. Springer, 2004.

[Joe06] S. Joe. Construction of good rank-1 lattice rules based on the weighted star discrepancy. In H. Niederreiter and D. Talay, editors, *Monte Carlo and Quasi-Monte Carlo Methods 2004*, pages 181–196. Springer, 2006.

[JS92] S. Joe and I.H. Sloan. On computing the lattice rule criterion *R*. *Math. Comp*, 59: 557–568, 1992.

[KJ02] F.Y. Kuo and S. Joe. Component-by-component construction of good lattice rules with a composite number of points. *J. Complexity*, 18: 943–946, 2002.

[Kuo03] F.Y. Kuo. Component-by-component constructions achieve the optimal rate of convergence for multivariate integration in weighted korobov and sobolev spaces. *J. Complexity*, 19: 301–320, 2003.

[Lar87] G. Larcher. A best lower bound for good lattice points. *Monatsh. Math.*, 104: 45–51, 1987.

[NC05] D. Nuyens and R. Cools. Fast component-by-component construction of rank-1 lattice rules with a non-prime number of points. *J. Complexity*, 22: 4–28, 2005.

[Nie78] H. Niederreiter. Existence of good lattice points in the sense of Hlawka. *Monatsh. Math.*, 86: 203–219, 1978.

[Nie92] H. Niederreiter. *Random number generation and quasi-Monte Carlo methods*. SIAM, Philadelphia, 1992.

[SJ94] I.H. Sloan and S. Joe. *Lattice methods for multiple integration*. Clarendon Press, Oxford, 1994.

[SJ07] V. Sinescu and S. Joe. Good lattice rules based on the general weighted star discrepancy. *Math. Comp.*, 76: 989–1004, 2007.

[SW98] I.H. Sloan and H. Woźniakowski. When are quasi-monte carlo algorithms efficient for high dimensional integrals? *J. Complexity*, 14: 1–33, 1998.

[Zar68] S.K. Zaremba. Some applications of multidimensional integration by parts. *Ann. Polon. Math.*, 21: 85–96, 1968.

Ergodic Simulations for Diffusion in Random Velocity Fields

Nicolae Suciu[1], Călin Vamoş[2], and Karl Sabelfeld[3]

[1] Friedrich-Alexander University of Erlangen-Nuremberg, Institute of Applied Mathematics, Martensstrasse 3, 91058 Erlangen, Germany
suciu@am.uni-erlangen.de

[2] T. Popoviciu Institute of Numerical Analysis, Romanian Academy, 400320 Cluj-Napoca, P. O. Box 68-1, Romania
cvamos@ictp.acad.ro

[3] Weierstrass Institute for Applied Analysis and Stochastics, Mohrenstrasse 39, 10117 Berlin, Germany, and Institute of Computational Mathematics and Mathem. Geophysics, Russian Acad. Sci., Lavrentieva str., 6, 630090 Novosibirsk, Russia
sabelfel@wias-berlin.de

Summary. Ergodic simulations aim at estimating ensemble average characteristics of diffusion in random fields from space averages. The traditional approach based on large supports of the initial concentration in general fails to obtain ergodic simulations. However, such simulations, using single realizations of the velocity, are shown to be feasible if space averages with respect to the location of the initial concentration support are used to estimate ensemble averages.

1 Introduction

Diffusion in random velocity fields is often used to model the transport in random environments as ionized plasmas [MPS93] turbulent flows [MY75] or natural porous media [MM80]. Based on this model, predictions of transport in real life problems, for which no analytical solutions are known, are usually achieved by simple-sampling Monte-Carlo simulations. The latter consist in ensemble averaging over simulations performed for given realizations of the field. This procedure consumes often much computing time, as for instance in the case of large-scale simulations of transport in groundwater [SVV06], where even for reduced size, two-dimensional problems, a sample simulation requires tens of cpu hours on last generation supercomputers. It is therefore desirable to avoid the ensemble averaging. Towards this aim, ergodic hypotheses are usually invoked to justify single realizations approaches.

Ergodicity occurs in stochastic modeling of transport in a broad variety of formulations. The ergodicity of the dynamical system generated by a realization of the velocity field is investigated to assess stochastic average models for actual observables given by time averages. The nonergodicity of this dynamical system is associated with the anomalous diffusion of deterministic transport and with the increase as the square of the Péclet number of the effective coefficient of the diffusion in the corresponding random fields. In a larger sense, ergodicity is also formulated as a measure of the reliability of the stochastic predictions for single realizations of transport. (See [SVV07] for a short review on multiple meanings of ergodicity). But related to numerical simulations of transport, the ergodicity of the random field itself is of main concern [KKS05].

A random field is ergodic if space averages converge to ensemble averages. For instance, this is the case for the mostly used space random functions models in geostatistics, characterized by a finite integral range, for which the ensemble mean can be successfully inferred from space averages [CD99]. Based on this property of the velocity field, a traditional approach was to use extended initial concentration distributions and single realizations of the velocity to estimate ensemble mean properties of transport. It was expected that when the transport simulation starts with a large source and the solute particles experience the variability of the given realization of the velocity field the observables computed from a single simulation behave closely to their average over the statistical ensemble of simulations. This eventuality is the so called "ergodicity in the large sense" [SVV07], investigated numerically in [SVV06]. It was found that the single realization approach using large sources fails in general to reproduce the ensemble averaged results. Even if a numerical evidence was supplied that, for ergodic velocity fields, the diffusion in random fields is asymptotically ergodic in the large sense, irrespective of the source dimensions, considering large sources does not ensure the reliability of single realization simulations for finite times [SVV06, SVV07].

In the present paper we discuss the feasibility of ergodic simulations of diffusion in random fields by a procedure which is different from the traditional approach based on large sources. Using "global random walk" simulations [VSV03], we found that an ergodic property also holds for the sample simulations of the diffusion in fields with finite integral range: the ensemble average of the transport observables can be estimated by the arithmetic mean of the observables resulting from repeated simulations of diffusion, done for the same realization of velocity field and for point-like sources with different locations uniformly distributed over large enough spatial domains. The mainly used observable quantities, dispersion coefficients and concentrations at reference planes across the mean flow, are defined in Sect. 2. In Sect. 3 we show the limitations of the approach based on large source simulations and present a numerical evidence that the transport observables are homogeneous random variables, with respect to the space location of the source. Further, in Sect. 4, we investigate to what extent the observables are also ergodic, in the sense mentioned above. Some concluding remarks are presented in Sect. 5.

2 Observables of Transport Simulations

The mostly used observables which are computed from simulations of diffusion in random velocity fields are the dispersion coefficients, equivalent to the second spatial moments of the solute plume, and the space average of the concentration over cross-sections of the plume. For a two dimensional problem, as considered in the numerical investigation presented in this paper, these quantities are defined as follows.

The effective coefficients for a given realization of the field are given by

$$D_{ll}^{eff}(t) = \frac{s_{ll}(t)}{2t}, \; s_{ll}(t) = \int (x_l - \mu_l(t))^2 c(\mathbf{x}, t) d\mathbf{x}, \tag{1}$$

where $l = 1, 2$, s_{ll} is a diagonal component of the second central moment of the concentration field c and

$$\mu_l(t) = \int x_l c(\mathbf{x}, t) d\mathbf{x}$$

is the first moment or the center of mass of the plume. The effective coefficient (1) is an equivalent representation of the second moment s_{ll} and plays the role of a diffusion coefficient only when the process has a diffusive large time upscaling [SVE06].

Often, to describe the randomness of the center of mass one uses the "center of mass coefficient"

$$D_{ll}^{cm}(t) = \frac{(\mu_l(t) - \langle \mu_l(t) \rangle_\omega)^2}{2t}, \tag{2}$$

where $\langle \mu_l(t) \rangle_\omega$ is the average of the first moment over the realizations ω of the velocity field. The ensemble average of this quantity is equivalent with the variance of the center of mass, $2t\langle D_{ll}^{cm}(t) \rangle_\omega = \langle \mu_l(t)^2 \rangle_\omega - \langle \mu_l(t) \rangle_\omega^2$.

An "ensemble coefficient" can be defined as

$$D_{ll}^{ens}(t) = \frac{\sigma_{ll}(t)}{2t}, \; \sigma_{ll}(t) = \int (x_l - \langle \mu_l(t) \rangle_\omega)^2 c(\mathbf{x}, t) d\mathbf{x}, \tag{3}$$

where σ_{ll} is the second moment with respect to the center of mass of the ensemble averaged plume. σ_{ll} accounts for both diffusive spreading and randomness of the center of mass. The ensemble, effective, and center of mass coefficients are related by the identity

$$D_{ll}^{ens}(t) = D_{ll}^{eff}(t) + D_{ll}^{cm}(t).$$

The ensemble average of this identity is often used in analyses of time behavior of the solute plume [SVV06].

The cross-section concentration at the center of mass of the two-dimensional plume is computed by

$$C(t) = \frac{1}{L_2} \int_0^{L_2} c(\mu_1(t), x_2, t) dx_2, \tag{4}$$

where L_2 is the transverse dimension of the grid [SVV06, SVV07].

The observables (1-4) are random variables depending on the realization ω of the velocity field. For convenience, we define a point source simulation $\mathcal{S}(t; \omega, \mathbf{x}_0)$ by the quadruple

$$\mathcal{S}(t; \omega, \mathbf{x}_0) = \{D_{ll}^{eff}(t; \omega, \mathbf{x}_0), D_{ll}^{ens}(t; \omega, \mathbf{x}_0), D_{ll}^{cm}(t; \omega, \mathbf{x}_0), C(t; \omega, \mathbf{x}_0)\}, \tag{5}$$

where $\mathbf{x}_0$ denotes the space point where the point instantaneous source of the transport simulation is located.

3 Simulations for Extended and Point Sources

We consider an isotropic two-dimensional diffusion in groundwater ($D_1 = D_2 = D = 0.01\ m^2/day$) in a random velocity field $\mathbf{V}$ with ensemble mean $\mathbf{U} = (U, 0)$, $U = 1\ m/day$. The velocity field is generated, with the Kraichnan routine, as a superposition of 6400 random sin modes which approximates a Gaussian field. This Gaussian field is an approximation of the Darcy velocity field in saturated groundwater formations, for a log-hydraulic conductivity field exponentially correlated with a small variance of 0.1 and isotropic correlation length $\lambda = 1\ m$. The transport over 2000 $days$ in given realizations of the velocity field, is simulated by simultaneously tracking $N = 10^{10}$ computational particles with the "global random walk" algorithm (GRW), presented at length in [VSV03, SVV04]. To simulate point instantaneous injection conditions, the particles were released at $t = 0$ from points $\mathbf{x}_0$. For comparisons presented in Figs. 1–4, simulations were also done for an extended initial distribution by releasing the same number N of particles from each of 121 grid points spaced by λ inside a square $10\lambda \times 10\lambda$ centered at the origin of the coordinate system. Details on the implementation of the numerical method used in this paper are presented in [SVV06].

The principle of the GRW algorithm consists in moving all the computational particles lying at a grid site globally, by a numerical procedure based on (exact or approximated) Bernoulli repartitions. The particles undergo displacements proportional to local velocity at the site and diffusion jumps proportional to $\sqrt{2D\delta t}/\delta x$, where δt and δx are respectively the constant time and space steps. Since the latter accounts exactly for the dispersion of the diffusion process of coefficient D, the GRW algorithm has no numerical diffusion. Because the particles are conserved, the algorithm is also stable. The total number of computational particles is not restricted [VSV03] and can be as large as necessary to ensure self-averaging results, i.e. no averages over GRW simulations are necessary to obtain transport observables (for given velocity realization) with the desired accuracy [SVV04]. The overshooting

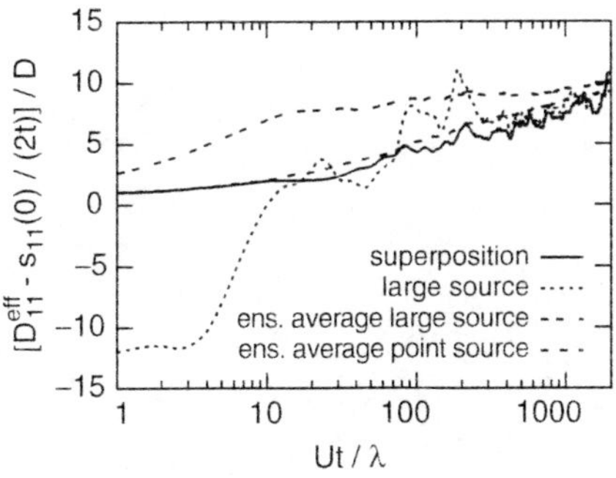

Fig. 1. Longitudinal effective coefficients for a superposition of point sources and an for the corresponding extended source

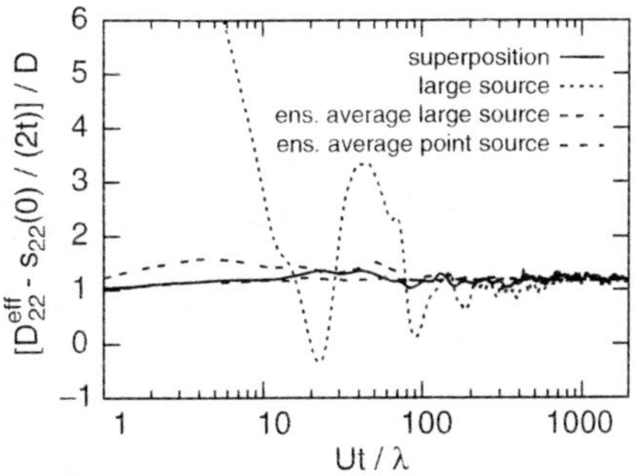

Fig. 2. Transverse effective coefficients for a superposition of point sources and an for the corresponding extended source

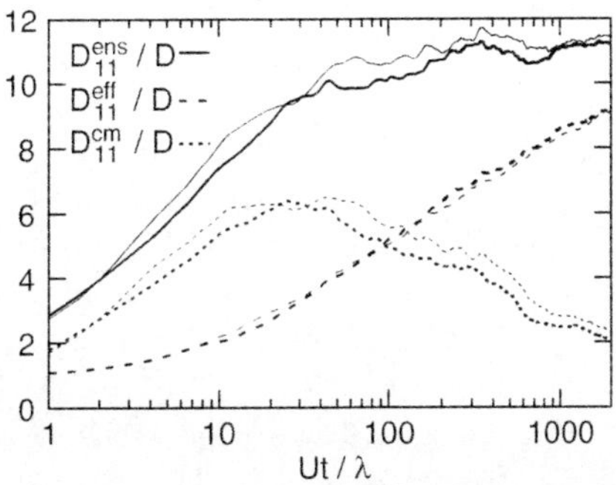

Fig. 3. Longitudinal effective coefficients for two different locations of the point source

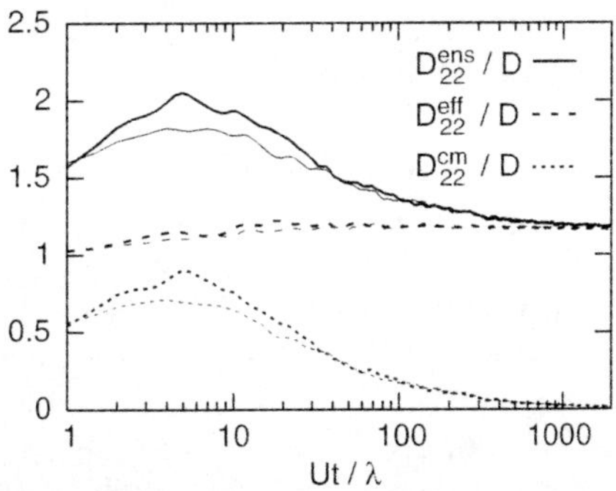

Fig. 4. Transverse effective coefficients for two different locations of the point source

errors, occurring when particles jump over grid points with different velocity values, were limited by a suitable choice of the time and the space steps of $\delta t = 0.5\ days$ and $\delta x = 0.1\ m$, so that the resulting dispersion coefficients were estimated with errors of the order of $D/2$ [SV06, SVE06].

In Fig. 1 and Fig. 2 the longitudinal and transverse effective coefficients are compared for a fixed velocity realization and a large square source $10\lambda \times 10\lambda$ centered at the origin, for the ensemble averages over 121 simulations for different velocity realizations and identical initial conditions consisting of the same square source and of a point source located at the origin, and for a superposition of point source simulations. The latter represents the space average over 121 simulations done for the same fixed velocity realization and for point source locations $\mathbf{x}_0$ uniformly distributed into the same square $10\lambda \times 10\lambda$. To render comparable the effective coefficients for point and extended sources [SVV06], the contribution of the initial second moment $s_{ll}(0)/(2t)$ (which is non-vanishing for extended sources) has been removed from $D_{ll}^{eff}(t)$ and

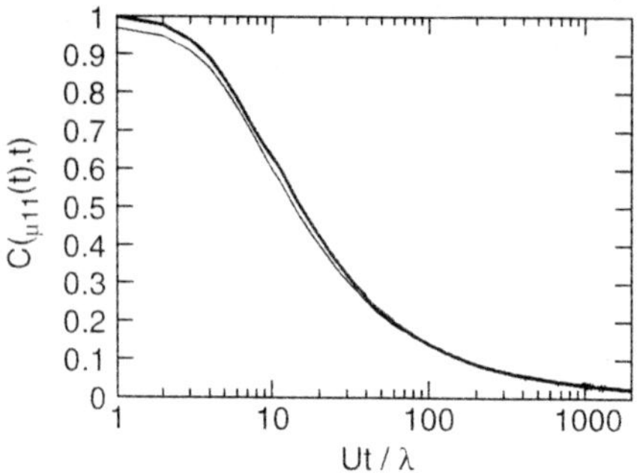

Fig. 5. Time dependence of the cross-section concentration at the plume center of mass for two different locations of the point source

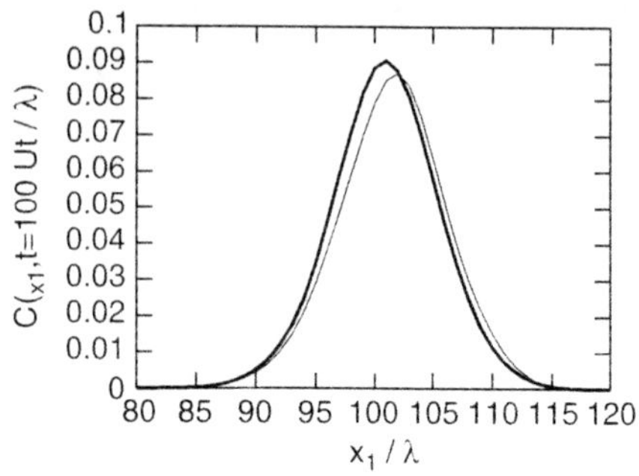

Fig. 6. Space dependence of the cross-section concentration at a fixed time for two different locations of the point source

the result has been normalized by the diffusion coefficient D. One remarks that the superposition results are quite close to those of the ensemble average for point source, which indicate the feasibility of ergodic simulations. On the contrary, the effective coefficients for square source are very different from the results of ensemble averaging (for either square and point sources). This indicates the failure of large source simulations to reproduce ensemble averages. The large early time deviation from ensemble averaged coefficients in the case of large sources is caused by correlations between initial positions and velocity fluctuations on solute particles trajectories [SVV07] (analyzed in detail by Suciu et al., manuscript submitted to Water Resources Research, 2006). However, it was shown that for narrow sources with large extension on the direction perpendicular to that of the simulated coefficient the single realization effective coefficients (1) are close to their ensemble averages and to the ensemble averages of the ensemble coefficients (3) [SVV07]. It is only in this situation that large source conditions yield ergodic simulations.

Since a prerequisite for ergodicity is the statistical homogeneity of the random variable, we check whether this holds true for the point source simulation (5). To do that, we compare the ensemble averages over 256 velocity realizations of simulations for $\mathbf{x}_0 = (0, 0)$ and $\mathbf{x}_0 = (50\lambda, 50\lambda)$. The results for effective coefficients (Fig. 3 and Fig. 4) show differences of the order of D or smaller. The statistical homogeneity is also indicated by the comparisons between time and space behavior of the cross-section concentration (Fig. 5 and Fig. 6). Thus, the ensemble average $\langle \mathcal{S}(t; \omega, \mathbf{x}_0) \rangle_\omega$ can be assumed to be independent of $\mathbf{x}_0$.

4 Ergodic Simulations

A random space function $F(x)$ is ergodic if the space average converges to the ensemble average,

$$\lim_{\mathcal{V}(\Omega) \longrightarrow \infty} \frac{1}{\mathcal{V}(\Omega)} \int_\Omega F(x) \mathrm{d}x = \langle F(x) \rangle ,$$

where $\mathcal{V}(\Omega)$ is the volume of the domain $\Omega \subset \mathbb{R}^3$. The Slutsky's Theorem ensures the mean square convergence if the integral of the correlation function $\langle F(x)F(x+y)\rangle$ is finite. A sufficient condition for that is the existence of a finite correlation range $\int \langle F(x)F(x+y)\rangle dy / \langle F(x)^2\rangle$ of the random function [CD99]. As shown by Chilès and Delfiner [CD99], this is almost always the case for geostatistical models of the velocity field in groundwater. Therefore, space averages can be used to infer the ensemble mean velocity, and for Gaussian fields also to infer the velocity variance.

The ergodicity of the velocity field could induce ergodic properties for simulations starting with extended sources if the observables for large source conditions were expressed as space averages of those for point sources. Since the dispersion coefficients (1-3) depend on squared first moments μ_l, and thus on squared concentrations, they are not linear superpositions of point source quantities [SVV06, SVV07]. The nonlinearity with respect to the initial concentration of the effective coefficient (1) can explain the failure of the traditional approach to yield ergodic simulations (see Fig. 1 and Fig. 2).

As an alternative to the traditional approach, we check the ergodicity of the simulations with respect to the average over the spatial location of the source. In this first paper we shall consider only point-like sources. The simulations $\mathcal{S}(t;\omega,\mathbf{x}_0)$ are ergodic if their ensemble average can be inferred from space averages, i.e.

$$\langle \mathcal{S}(t;\omega,\mathbf{x}_0)\rangle_\omega = \langle \mathcal{S}(t;\omega,\mathbf{x}_0)\rangle_{\mathbf{x}_0}. \tag{6}$$

Even though statistically homogeneous velocity fields with finite correlation lengths are ergodic [CD99] and, as shown by Figs. 3–6, the simulations are almost homogeneous random variables, their ergodicity is extremely difficult to prove theoretically owing to the highly nonlinear dependence on velocity of the particles trajectories, and consequently of $\mathcal{S}(t;\omega,\mathbf{x}_0)$ [SVV07]. Nevertheless, the numerical results presented in the following indicate the ergodicity property (6) for the simulations (5) of diffusion in random velocity fields.

The expectations $E(\cdot)(t)$ and the standard deviations $SD(\cdot)(t)$ of the observables (1-4), defined in Sect. 2, are computed in two ways:

(a) as averages over an ensemble of 121 velocity realizations for the same position of the injection point, and

(b) as averages over simulations of transport in the same realization of the velocity, for 121 distinct positions of the injection point, uniformly distributed in a square of edge equal to 10λ.

The method (a) corresponds to the usual simple-sampling Monte-Carlo simulation, while (b) is the "ergodic simulation method". Figs. 7–12 compare the expectations and the standard deviations of the observables (1-4) computed by ergodic simulations with those obtained by Monte-Carlo simulations (which correspond to smoother curves). The agreement is better for the expectations of the effective coefficients (Fig. 7 and Fig. 8) and for their standard deviations (Fig. 9 and Fig. 10). Less satisfactory is the statistics estimated by ergodic simulations for the center of mass and ensemble coefficients (mainly for longitudinal

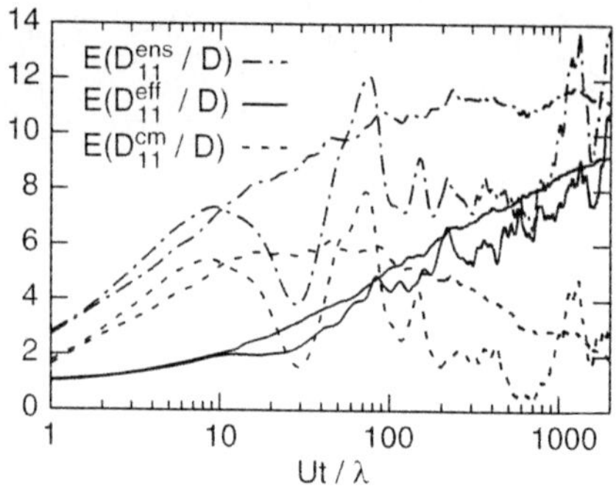

Fig. 7. Expectations for longitudinal dispersion coefficients

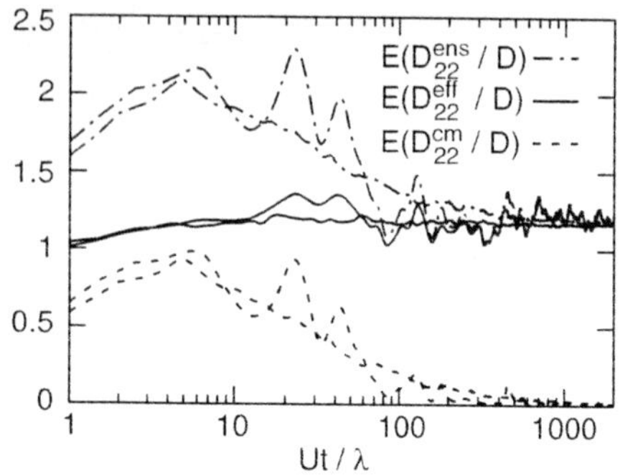

Fig. 8. Expectations for transverse dispersion coefficients

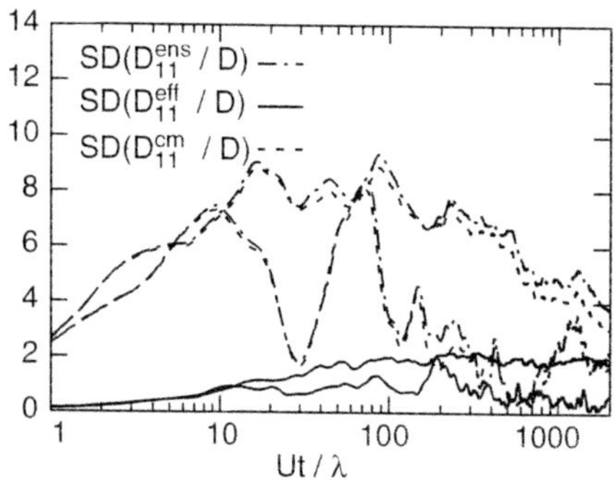

Fig. 9. Standard deviations for longitudinal dispersion coefficients

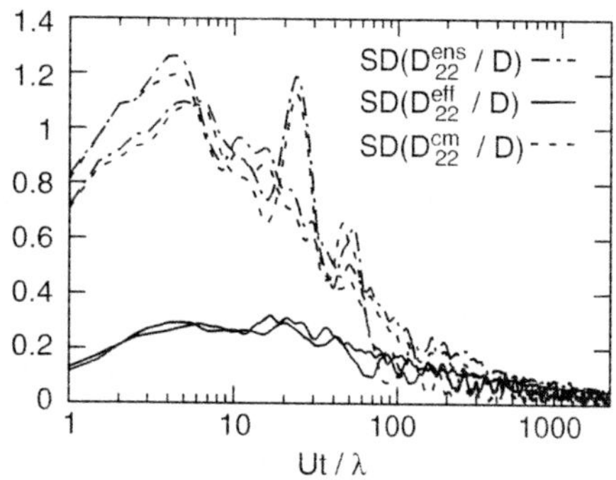

Fig. 10. Standard deviations for transverse dispersion coefficients

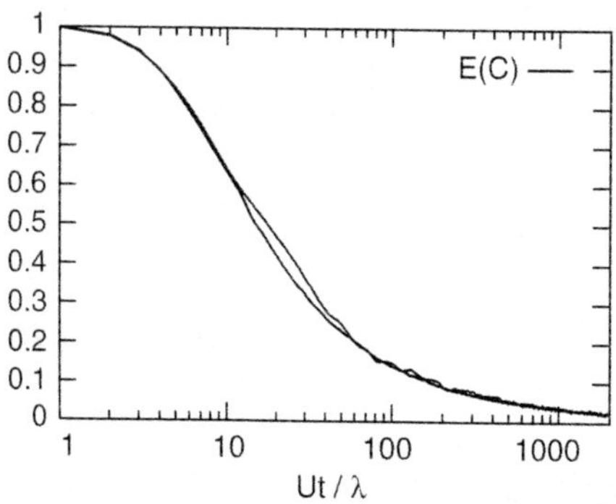

Fig. 11. Expectations for cross-section concentrations

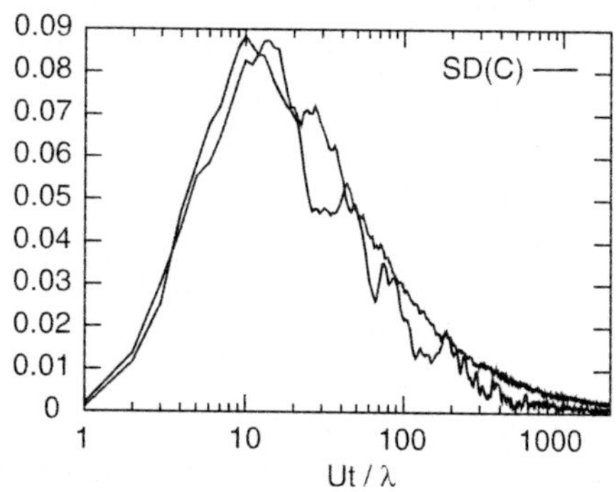

Fig. 12. Standard deviations for cross-section concentrations

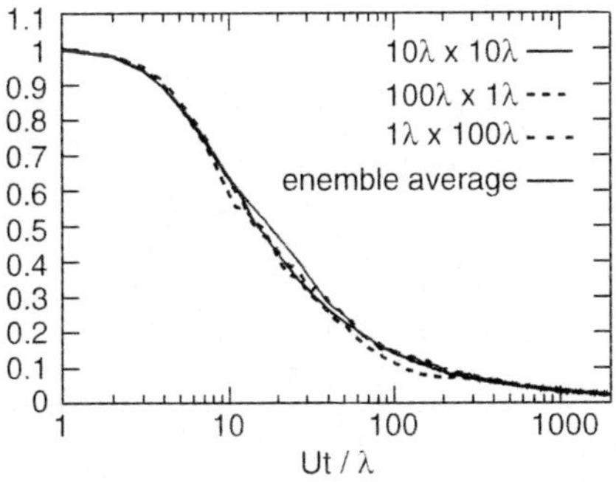

Fig. 13. Expectations of cross-section concentrations for different shapes of the averaging domain

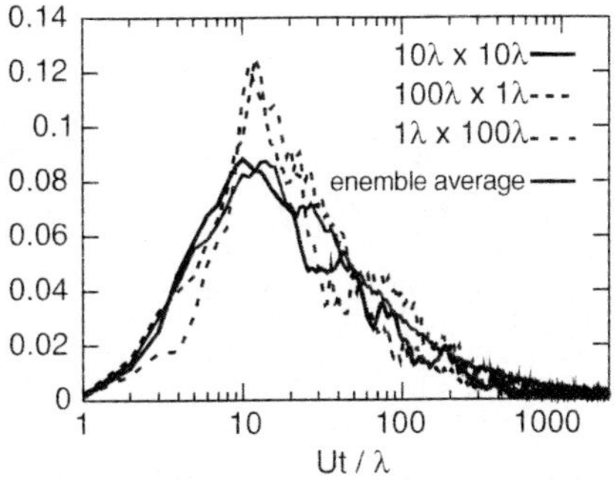

Fig. 14. Standard deviations of cross-section concentrations for different shapes of the averaging domain

coefficients shown in Fig. 7 and Fig. 9). The best results are obtained for the expectation and standard deviation of the cross-section concentration (Fig. 11 and Fig. 12). The comparisons given in Fig. 13 and Fig. 14 further indicate that simulations for symmetric averaging domains have better ergodic properties.

5 Conclusions

The numerical investigation presented in this paper indicate that ergodic simulations of diffusion in random velocity fields with finite correlation range are possible. Ensemble average characteristics of the process can be inferred from space averages with respect to the location of the finite support of the initial concentration distribution. The results are already encouraging for fields generated with the Kraichnan simple-randomization method, in spite of its poor ergodic properties [CD99, KKS05, SVV06].

The less satisfactory results are obtained for the center of mass and ensemble dispersion coefficients. As indicated by their definitions, (2) and (3), these coefficients rather characterize the random velocity field than the spreading of the solute. It is therefore expected that their ergodic simulation also improves when the Kraichnan routine is replaced by a field generator with better ergodic properties [KKS05].

Further refinements of the method of ergodic simulations proposed in this paper require investigations on the role of the averaging domain containing the locations $\mathbf{x}_0$ of the point source and on the relation between the spacing of $\mathbf{x}_0$ and the correlation lengths of the velocity field. By taking different origins $\mathbf{x}_0$ of the coordinate system for the transport simulation, uniformly distributed into the domain Ω of a given realization of the velocity field, we expect that ergodic simulations can be used in investigations of transport for large sources as well.

Acknowledgments. The work presented in this paper was supported by the Deutsche Forschungsgemeinschaft grant SU 415/1-2, the Romanian Ministry of Education and Research grant CEEX-M1-C2-2532, RFBR grant 06-00498, and the NATO Collaborative Linkage Grant ESP.NR.CLG 981426.

References

[CD99] J. P. Chilès and P. Delfiner. Geostatisctics: Modeling Spatial Uncertainty. Wiley, New York (1999)

[KKS05] P. Kramer, O. Kurbanmuradov, and K. Sabelfeld. Extension of multiscale Gaussian random field simulations algorithm. Weierstrass Institute, Berlin, Preprint **1040** (2005)

[MM80] G. Matheron and G. de Marsily. Is transport in porous media always diffusive? Water Resour. Res., **16**, 901–917 (1980)

[MPS93] J. Misguish, G. Pelletier, and P. Schuck (editors). Statistical Description of Transport in Plasma, Astra- and Nuclear Physics. Les Houches Series, Nova Science Publ., Inc., New York (1993)

[MY75] A. S. Monin and A. M. Yaglom. Statistical Fluid Mechanics: Mechanics of Turbulence. MIT Press, Cambridge, M A (1975)

[SVV04] N. Suciu, C. Vamoş, J. Vanderborght, H. Hardelauf, and H. Vereecken. Numerical modeling of large scale transport of contaminant solutes using the global random walk algorithm. Monte Carlo Methods and Appl., **10**(2), 153–177 (2004)

[SV06] N. Suciu and C. Vamoş. Evaluation of overshooting errors in particle methods for diffusion by biased global random walk. Rev. Anal. Num. Th. Approx., **35**, 119-126 (2006)

[SVV06] N. Suciu, C. Vamoş, J. Vanderborght, H. Hardelauf, and H. Vereecken. Numerical investigations on ergodicity of solute transport in heterogeneous aquifers, Water Resour. Res., **42**, W04409, doi:10.1029/2005WR004546 (2006)

[SVE06] N. Suciu, C. Vamoş, and J. Eberhard. Evaluation of the first-order approximations for transport in heterogeneous media. Water Resour. Res., **42**, W11504, doi: 10.1029/2005WR004714 (2006)

[SVV07] N. Suciu, C. Vamoş, and H. Vereecken. Multiple meanings of ergodicity in real life problems. In: Marinoschi, G., Ion, S., Popa, C. (ed) Proceedings of the 5th Workshop on Mathematical Modelling of Environmental and Life Sciences Problems. Rom. Acad. Publishing House, Bucharest (2007, to appear)

[VSV03] C. Vamoş, N. Suciu, and H. Vereecken. Generalized random walk algorithm for the numerical modeling of complex diffusion processes, J. Comp. Phys., **186**(2), 527–544 (2003)

Efficient Simultaneous Simulation of Markov Chains

Carsten Wächter[1] and Alexander Keller[2]

[1] Ulm University, Germany
carsten.waechter@uni-ulm.de
[2] Ulm University, Germany
alexander.keller@uni-ulm.de

Summary. Markov chains can be simulated efficiently by either high-dimensional low discrepancy point sets or by padding low dimensional point sets. Given an order on the state space, both approaches can be improved by sorting the ensemble of Markov chains. We analyze deterministic approaches resulting in algorithmic simplifications and provide intuition when and why the sorting works. Then we discuss the efficiency of different sorting strategies for the example of light transport simulation.

I spent an interesting evening recently with a grain of salt.

Shannon: A Mathematical Theory of Communication, 1948

1 Introduction

A Markov chain describes a memoryless stochastic process, where the transition probabilities do not depend on the history of the process. For example Shannon modeled the English language by a Markov chain, where he computed the relative frequencies of one word following another from an English book. Using these transition probabilities he generated seemingly English sentences. Many more evolution processes, like e.g. the Brownian motion or particle trajectories, can be described as Markov chains.

Properties of processes can be estimated by averaging the contributions of multiple Markov chains. Instead of simulating each trajectory independently, it has been found that simultaneously simulating Markov chains using correlated samples and sorting the ensemble of states after each transition step can notably improve convergence.

2 Simultaneous Simulation of Markov Chains

The idea of improving the simultaneous simulation of Markov chains with quasi-Monte Carlo methods by an intermediate sorting step was originally introduced by Lécot in a series of papers dealing with the Boltzmann equation

[Léc89a, Léc89b, Léc91, LC98] and later on the heat equation [KL99]. This idea was then used and refined for solving the heat equation on a grid by Morokoff and Caflisch [MC93] and recently extended by L'Écuyer, Tuffin, Demers et al. [LL02, LT04a, LT04b, DLT05, LLT05, LDT06, LLT06, HLL07, LDT07] to incorporate randomized versions of the algorithm and splitting for rare event simulation. Independent research, but in a way related to the above approaches, was conducted by [BNN+98] in the field of computer graphics.

In the following we simplify the deterministic version of the scheme. For the derivation we use the algorithm from [LT04b]. The results give practical insight, when and why the scheme is superior to approaches without sorting and how to implement it.

2.1 Analysis of a Deterministic Algorithm in One Dimension

The algorithm presented in [LT04b] simultaneously simulates Markov chains on a discrete state space E with an initial distribution $\mu := (\mu_i)_{i \in E}$ and a transition matrix $P := (p_{i,j})_{i,j \in E}$. Using a $(t,2)$-sequence $(x_i)_{i \in \mathbb{N}_0}$ in base b [Nie92], $N = b^m$ chains are simulated in parallel, where $\mathbf{X}_{n,l}$ is the state of chain l at time step n and $\mathbb{N}_0$ are the natural numbers including zero. Further the algorithm requires that for $m > t$ the $N = b^m$ subsequent components $x_{i,1}$ form a $(0,m,1)$-net in base b. As shown in [LT04b] the algorithm converges if

$$\forall k \in E : \sum_{l=1}^{N-1} \left| \sum_{n=1}^{k-1} p_{l+1,n} - \sum_{n=1}^{k-1} p_{l,n} \right| \leq 1 \tag{1}$$

holds.

Simplification of the Algorithm

We now consider the Sobol' sequence $x_l = (x_{l,1}, x_{l,2}) = (\Phi_2(l), \Phi_S(l)) \in [0,1)^2$, which is a $(0,2)$-sequence in base $b = 2$ and fulfills the assumptions required for the convergence condition to hold. For the definition of the original Sobol' sequence see [Sob67], while a simple code example for $(\Phi_2(l), \Phi_S(l))$ is found in [KK02b].

The simulation itself starts at time step $n = 0$ initializing state $\mathbf{X}_{0,\lfloor N \cdot x_{l,1} \rfloor}$ for $0 \leq l < N$ using $x_{l,2}$ for the realization of μ. The algorithm then continues by sorting the states (this will be discussed in detail in Section 3) and continues the chains by computing $\mathbf{X}_{n,\lfloor N \cdot x_{(l+n \cdot N),1} \rfloor}$ using $x_{(l+n \cdot N),2}$ for $0 \leq l < N$ to realize transitions according to P. The index

$$\sigma(l) := \lfloor N \cdot x_{(l+n \cdot N),1} \rfloor$$

for selecting the next state for transition in fact uses the van der Corput sequence Φ_2 in base 2, which is a $(0,1)$-sequence and thus a sequence of $(0,m,1)$-nets [Nie92]. For example choosing $m = 3 > t = 0$ we have $N = 2^3 = 8$ and

$$(\lfloor 8 \cdot \phi_2(l + n \cdot 8) \rfloor)_{l=0}^{7} \equiv \{0, 4, 2, 6, 1, 5, 3, 7\}.$$

for $n \in \mathbb{N}_0$. Hence all indices used during the different timesteps n are in fact identical for all m.

Assuming uniform probabilities $p_{i,j} = \frac{1}{|E|}$ the convergence theorem still applies, but more important, stable sorting does not change the state order. It thus follows that in fact the index permutation can be chosen as identity without touching the convergence conditions. The same applies for selecting the initial states $\mathbf{X}_{0,l}$ and it results the simplified, but equivalent algorithm

- $n := 0$
- initialize $\mathbf{X}_{0,l}$ using $x_{l,2}$ for $0 \leq l < 2^m$
- loop
  - sort state vector using a suitable order
  - $n := n + 1$
  - continue chain by computing $\mathbf{X}_{n,l}$ using $\Phi_S(l + n \cdot 2^m)$ for $0 \leq l < 2^m$

using only the second component of the $(0, 2)$-sequence x_l.

When and Why it Works

The improved convergence of the scheme, which has been observed in many applications (see the references at the beginning of Section 2), now must be caused by the structure of the samples $\Phi_S(l + n \cdot 2^m)$ used to realize the transitions of $\mathbf{X}_{n,l}$ according to P. This can be understood by decomposing the radical inverse (see also [Kel06])

$$\Phi_S(l + n \cdot 2^m) = \Phi_S(l) + \frac{1}{2^m}\Phi_S(n),$$

which reveals an implicit stratification: $\Phi_S(l)$ is an offset with spacing $\frac{1}{2^m}$ depending on the state number l, while the shift $\Phi_S(n)$ is identical for all the intervals at timestep n.

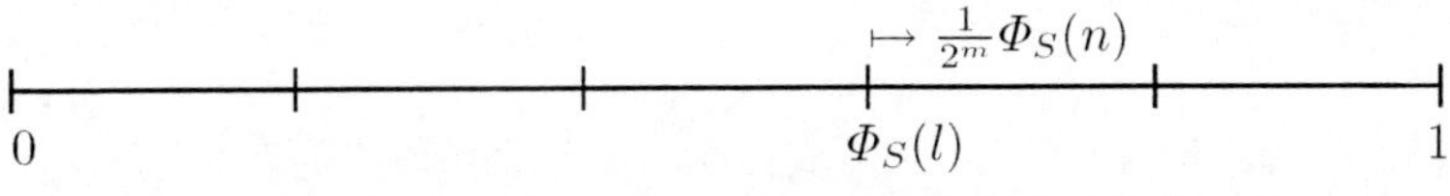

Here the low dimensional setting allows for a misleading interpretation of the samples being a shifted lattice or stratified samples, as the entirety of the $\Phi_S(l)$ for $l = 0, \ldots, 2^m - 1$ in fact must be an $(0, m, 1)$-net and thus an equidistant set of samples.

However, the good performance stems from the property that $\Phi_S(l)$ is a (t, s)-sequence and thus a sequence of (t, m', s)-nets for any m' with $t \leq m' \leq m$. This means that $b^{m'}$ states, that are similar in state space and therefore subsequent by order after sorting, will sample their transition by a (t, m', s)-net, which guarantees for good discrete density approximation. The maximum improvement would be obtained if all 2^m chains were in the same state. The more the states of the chains are separated in state space, the smaller the performance improvements will be.

2.2 Simplified Algorithm in s Dimensions

Using a (t, s)-sequence in base b, which is a sequence of (t, m, s)-nets, the scheme also works in s dimensions: Markov chains, whose states are similar after sorting are guaranteed to sample the transition probability by low discrepancy samples. The simplified algorithm in s dimensions now looks like:

- $n := 0$
- initialize $\mathbf{X}_{0,l}$ using quasi-Monte Carlo points x_l
- loop
  - sort state vector using a suitable order
  - $n := n + 1$
  - continue chain by computing $\mathbf{X}_{n,l}$ using subsequent samples x_l from a (t, s)-sequence

Some simulations require trajectory splitting in order to capture certain local subtle effects. While this already has been addressed in [DLT05, LDT06, LDT07], it in fact can be achieved in a simpler way by just drawing more samples out of the (t, s)-sequence for states to be split.

This is a consequence of the fact that it is not even necessary to simultaneously simulate exactly b^m chains. It is only important to draw subsequent samples from the (t, s)-sequence and to minimize the number b^m of points in the subsequent (t, m, s)-nets in order to enable the maximal performance gain. The choice of the (t, s)-sequence, however, is restricted by the condition that $(0, s)$-sequences only exist for $b \geq s$ and that $m > t$ [Nie92]. Note that other radical inversion based points sets like the Halton sequence or its scrambled variants fulfill properties similar to (t, s)-sequences [Mat98] and will result in similar performance gains.

2.3 Randomization

While there exists no general proof for convergence of the deterministic algorithm in higher dimensions yet, the algorithm becomes unbiased by freshly randomizing the quasi-Monte Carlo points in each time step n [LLT06]. Since this is in fact an instance of padded replications sampling as introduced in [KK02a, KK02b] the argument for unbiasedness becomes simpler than in [LLT06]. Randomization, however, deserves special attention.

The most efficient implementation along the lines of [KK02b] consists of choosing a (t, s)-sequence in base $b = 2$, from which subsequent samples are drawn, which are `XOR`-ed by an s-dimensional random vector. This random vector is freshly drawn after each transition step. However, as random scrambling changes the order in which the points are enumerated, the local properties of the sequences of (t, m, s)-nets are changed, too.

This observation can be taken as an explanation for some of the effects seen in [LLT05]: Sobol and Korobov points used in the array-(R)QMC simulation are worse up to an order of magnitude in variance reduction than their transformed (Gray-code for Sobol, Baker transform for Korobov) counterparts. The explanation for this is found in the structure of the points. The sequence of (t, m, s)-nets extracted from the Sobol sequence is locally worse than its Gray-code variant. The same goes for the Korobov lattice and its transformed variant.

3 Sorting Strategies

In order to have the states as closely together as possible, they have to be enumerated in an order such that the sum of the distances of neighboring states is minimal. This in fact relates to the traveling salesman problem[3], where for a given set of cities and the costs of traveling from one city to another city, the cheapest round trip is sought that visits each city exactly once and then returns to the starting city.

Our problem is very similar except for it is not necessary to return from the last state of the route to the first state. Some techniques are already available to efficiently calculate approximate, but close to optimal, solutions for the traveling salesman problem [DMC91, Rei94]. However, running times of these algorithms are not acceptable in our simulations, as the calculation of the distance matrix alone exhibits an $\mathcal{O}(N^2)$ complexity, while we want to keep the algorithm as close as possible to the $\mathcal{O}(N)$ complexity of classic Monte Carlo methods.

In the following we discuss some possible orders to achieve fast sorting for high-dimensional state spaces.

3.1 Norm of State

The average complexity of quicksort is $\mathcal{O}(N \log N)$, but for certain scenarios even $\mathcal{O}(N)$ algorithms exist, like e.g. radixsort, which, however, requires additional temporary memory. In order to use these one-dimensional sorting algorithms, the multi-dimensional state must be reduced to one dimension. Amongst many choices often some norm $\|\mathbf{X}_{n,l}\|$ is used to define an order on the state space. However, similar norms do not necessarily indicate proximity in state space. A simple example for this is similar energy of particles in a transport simulation that are located far away in space.

3.2 Spatial Hierarchy

A second possibility to enable multidimensional sorting is the usage of a spatial hierarchy to define an order on the states [Wie03]. Efficient data structures for

[3] This problem already was investigated by Euler in the early 18th century and unfortunately can be shown to be NP-hard [Kar72]. For a historical survey on the problem see Lawler et al. [LLKS85].

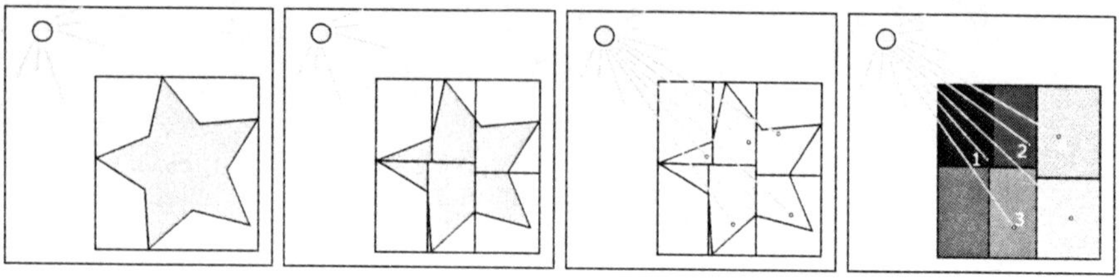

Fig. 1. Sorting the states into the leafs of a spatial hierarchy defines an order of proximity by traversing the hierarchy in in-order.

this purpose are the BSP-tree [SBGS69, Abr95], its specialized axis aligned subset, the kD-tree [Ben75], or bounding volume hierarchies [RW80, KK86, WK06]. The construction of such binary hierarchies is simple: The space is recursively subdivided using planes selected by some heuristic [WK07]. The construction runs in $\mathcal{O}(N \log N)$ on the average. Traversing the hierarchy in in-order enumerates the leaves in an order of proximity. This traversal becomes trivial, if the tree is left-balanced and in consequence can be stored in an array.

If a spatial hierarchy must be used anyway, for example to accelerate ray tracing, there is no additional construction time for the hierarchy. The particles then are stored as linked lists attached to the leaves of the hierarchy (see Figure 1). Unfortunately the quality of this order is strongly determined by the quality of the spatial hierarchy used for simulation, which is especially problematic if the number of leafs in the hierarchy is much smaller than the number of chains N as this results in several states being mapped to the same leaf.

3.3 Bucket Sorting and Space Filling Curves

In order to guarantee linear time complexity, bucket sorting can be used. In the s-dimensional extension [HLL07] of the simple algorithm sketched in Section 2.1, multidimensional states were sorted into buckets by the first dimension of the state, then the states of each bucket were sorted into buckets according to the second dimension, and so forth. This procedure works well, but has the problem that states close in state space can be separated by the sorting procedure. In addition, a stratification of each dimension has to be used, which induces the curse of dimension in the number of Markov chains to be simulated simultaneously.

We therefore divide the state space into equal voxels, which serve as buckets. The bucket of each state is found by truncating the state coordinates according to the resolution of the voxel grid. Note that this applies for continuous as well as discrete state spaces. Enumerating the voxels by proximity yields the desired order on the states and can be done in linear time in the number of voxels.

Orders that enumerate the voxels by proximity are given by space filling curves [Sag94] like e.g. the Peano curve, Hilbert curve, or H-indexing. These curves guarantee every voxel to be visited exactly once and an overall path length being relatively short. For problems with large geometry, which is the

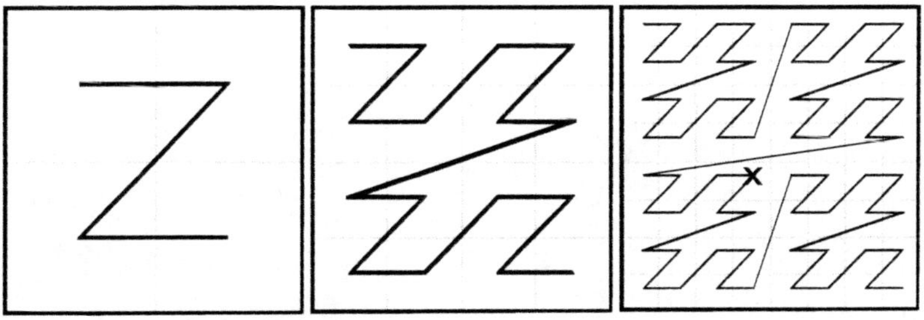

Fig. 2. The Z-curve in two dimensions for 2×2, 4×4, and 16×16 buckets. With the origin $(0,0)$ top left the point marked by $\times$ has the integer coordinates $(3,4)$, which corresponds to $(011, 100)_2$ in the binary system. Its binary Z-curve index 100101_2 is computed by bitwise interleaving the binary coordinates.

case in our own simulations, this can be even one of the few possibilities to generate fast and memory efficient approximate solutions to the traveling salesman problem [Rei94]. However, these curves are rather costly to evaluate, need to be tabulated to be efficient, or are not available for higher dimensions.

Fortunately, the Z-curve, also known as Lebesgue-curve or Morton order, avoids these problems. Given integer coordinates of a bucket in multidimensional space, its one dimensional Z-curve index is simply calculated by bitwise interleaving the coordinate values (see Figure 2). This is very easy and fast to compute for any dimension and problem size, and requires no additional memory. Unfortunately the results are not as good as for example the Hilbert-curve in a global context. However, the average case partitioning quality and average/worst case logarithmic index ranges are comparably good [Wie03]. Problems can arise in highly symmetrical scenes like Shirley's "Scene 6" (see Figure 6) used for our numerical experiments: States on the walls parallel to the (x, y)-plane will be sorted very well, but the states located on the other two walls parallel to the (y, z)-plane will be visited by the curve in an alternating manner, which can lead to correlation artifacts in some scenarios.

4 Application to Light Transport Simulation

For numerical evidence, we apply the algorithm developed in the previous sections to light transport simulation for synthesizing realistic images. The underlying integral equation can be reformulated as a path integral [Vea97]. Sampling path space (see Figure 3) corresponds to simulating Markov chains, where the paths are established by ray tracing and scattering events. The initial distribution is determined by the emission characteristics of the light sources and the transition probabilities are given by bidirectional reflectance distribution functions on the surface.

To solve the path integral, one can think of two basic strategies, which are either using high dimensional low discrepancy points or padding low

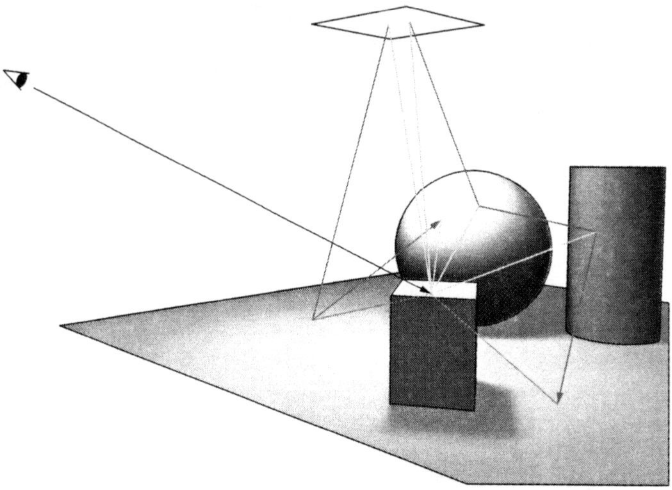

Fig. 3. Sampling transport path space by bidirectional path tracing. Trajectories from the eye and the light sources are generated by Markov chains and connected to determine the transported amount of light.

dimensional low discrepancy points [KK02b]. The latter approach fits our findings in Section 2.2, where high dimensional events are composed as subsequent transitions of a Markov chain. As measured in [KK02a] the difference to using high dimensional sequences for Markov chain simulations is small to none, but using padded sequences is computationally faster and requires less implementation effort. It is also simpler for practitioners in rendering industry.

In addition the low dimensional approach allows for much better results, because the stratification properties of (t, s)-sequences or the Halton sequence and its scrambled variants are much better for small dimensions (see Section 2.2).

4.1 Fredholm or Volterra?

The integral equation underlying light transport can be considered either as a Fredholm or Volterra integral equation, which matters for implementation.

$$L_r(x, \omega) = \int_{S^2_-(x)} f_r(\omega_i, x, \omega) L(x, \omega_i)\, (n(x) \cdot \omega_i) d\omega_i$$

is the radiance reflected off a surface in point x in direction ω, where the domain $S^2_-(x)$ of the integral operator is the hemisphere aligned to the normal $n(x)$ in x (see the illustration in Figure 4). f_r is the bidirectional reflectance distribution function describing the optical surface properties and L is the incident radiance. Using this integral operator results in a Volterra integral equation of the second kind, as the integration domain depends on x.

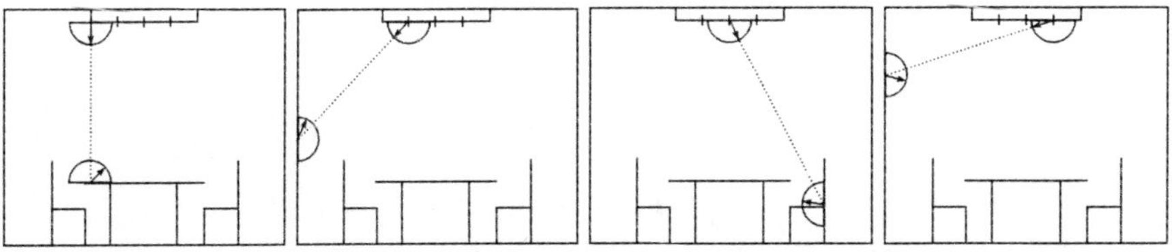

Fig. 4. Photon trajectories are started from the light sources. Upon hitting a surface after tracing a ray, the bidirectional reflectance distribution function is sampled to determine a direction of scattering to continue the path.

$$L_r(x,\omega) = \int_{S^2} f_r(\omega_i, x, \omega) L(x, \omega_i) \max\{n(x) \cdot \omega_i, 0\} d\omega_i$$

on the other hand results in a Fredholm integral equation of the second kind, as we are integrating over all directions of the unit sphere S^2 independent of x.

Using the latter approach of generating global directions [SKFNC97] and rejecting directions with negative scalar product with respect to the surface normal $n(x)$ is computationally attractive, while the first approach requires to generate directions in the hemisphere that have to be transformed into the local surface frame, which is more expensive. Mappings from the unit square to S^2 or $S^2_-(x)$ are found in [SC94, Rus98, Shi00].

An even more important argument for generating global directions is related to our algorithmic approach (see Section 2.2): By sorting it can happen that two close by surface points with different surface normals (e.g. in a corner) use subsequent samples of a (t, s)-sequence to generate scattering directions. Generating global directions now works fine, whereas generating directions in the two different local frames using subsequent samples will destroy the low discrepancy properties. These discontinuities become clearly visible. Using global directions, however, does not allow for importance sampling according to f_r or the cosine term, which often is a disadvantage and deserves further investigation.

4.2 Numerical Evidence

Following the arguments in Sections 2.2 and 2.3, we use the Halton sequence with permutations by Faure [Kel06] randomized by a Cranley-Patterson-rotation [CP76] in order to have unbiased error estimates. For the sorting the Z-curve order (see Section 3.3) worked best in our setting and was used for the following experiments. We further note that we numerically verified that omitting the randomization has no notable effect on the precision of the results. In our numerical experiments we compared four approaches to simulate Markov chains:

MC: Uniform random numbers generated by the Mersenne Twister [SM07] were used for classical Monte Carlo sampling.

RQMC: Used the high-dimensional Halton sequence with permutations by Faure randomized by a Cranley-Patterson rotation, where pairs of components were used to sample the two dimensional emission and scattering events.

lo-dim RQMCS: Used the two-dimensional Halton sequence randomized by a Cranley-Patterson rotation. The Z-curve was used to enumerate the bucket-sorted states.

hi-dim RQMCS: Used the high-dimensional Halton sequence with permutations by Faure randomized by a Cranley-Patterson rotation. The Z-curve was used to enumerate the bucket-sorted states.

In a first experiment the robust global illumination algorithm [KK04, Kol04] was used to compute the path integrals. The resulting graphs are depicted in Figure 6 and display the RMS error to a master solution and the variance averaged over the whole image as well as the pixel-based variance. The numbers were obtained by averaging 10 independent runs for a varying number of Markov chains. The measured numbers only convince for the simple test scene. In the complicated cases even no performance gain over Monte Carlo sampling can be measured, because the number of independent runs is too small and more experiments were not possible due to excessive running times. However, the visual error tells a dramatically different story as can be seen in Figure 5, where a clear superiority of the new algorithm in even very difficult settings becomes obvious. This case is not an exception, but can be observed for many test cases. It only emphasizes that standard error measures are not appropriate error measures for visual quality, which is a known but unsolved problem in computer graphics.

Figure 7 shows measurements for a very difficult light transport problem, where we directly traced photons from the light sources and connected their final path segment to the camera (one technique of bidirectional path tracing [Vea97]). Opposite to the above measurements only one number

Fig. 5. Visual comparison for the test scene "Invisible Date" using 300 chains for simulation. The only lightsource is not visible as it is placed on the ceiling of the neighboring room. Due to the better distribution of the photons randomized quasi-Monte Carlo (RQMC) outperforms Monte Carlo (MC) visually, as can be seen by the reduced shadow artifacts. RQMC with sorting (RQMCS, using 256^3 voxels for the bucket sort) is even superior as more photons made it into the second room and even the back of the door is lit very well.

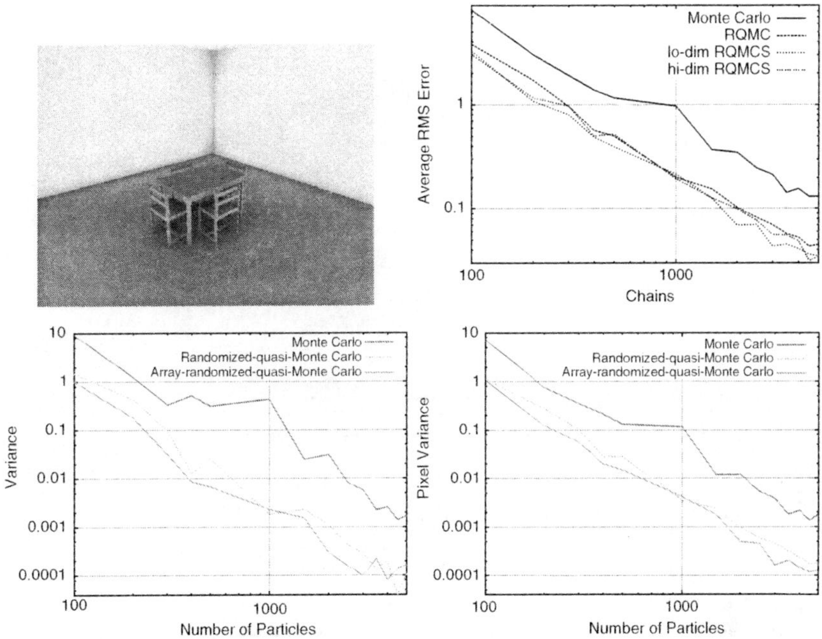

a) Test scene "Shirley 6".

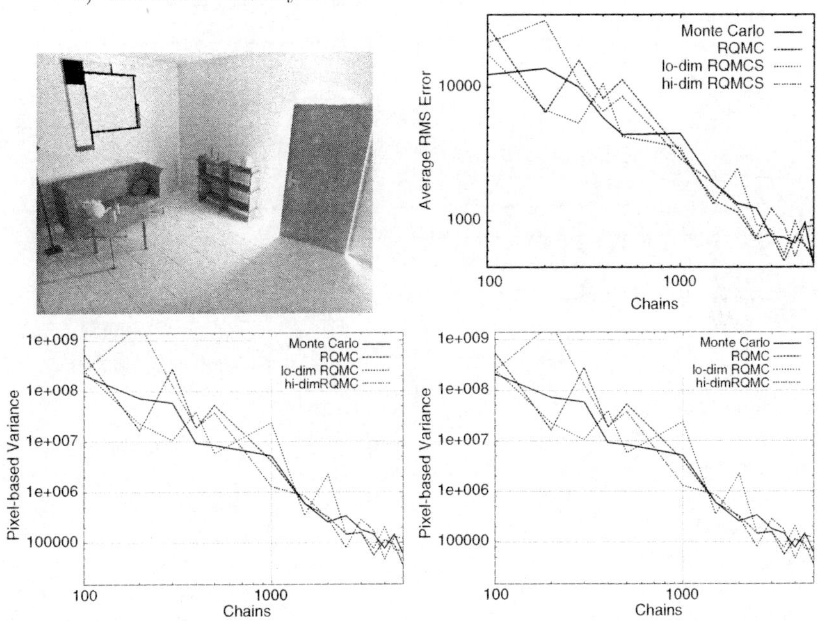

b) Test scene "Invisible date".

Fig. 6. RMS error, global, and per pixel variance for an a) simple and b) more complicated light transport setting.

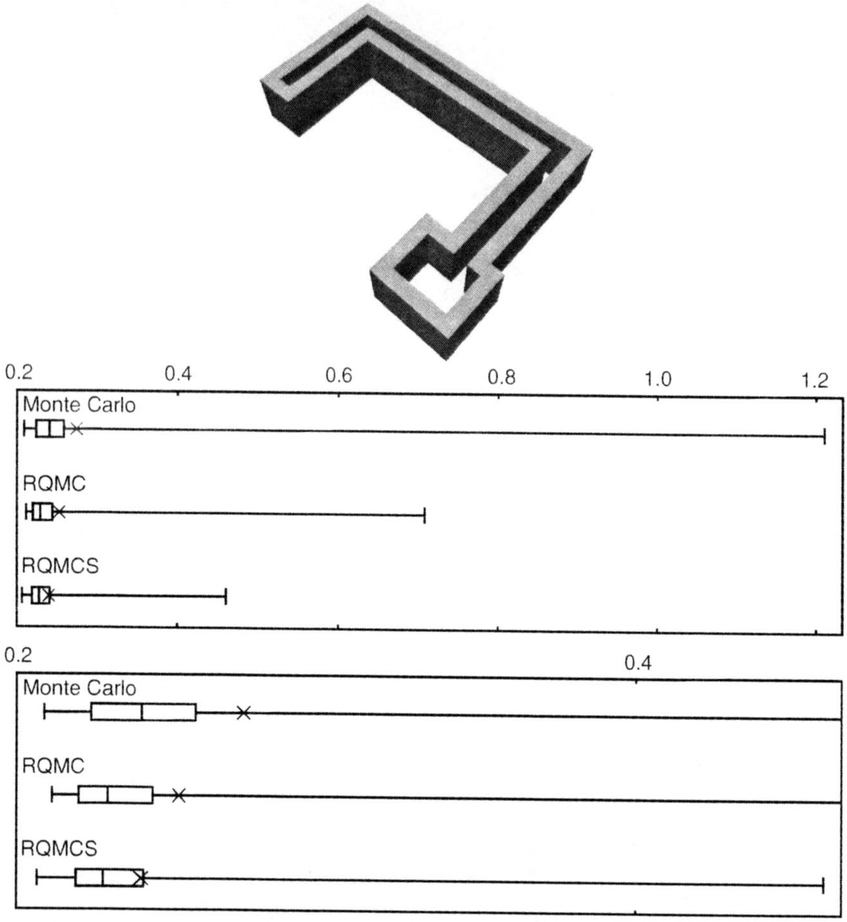

	L_1 per camera	L_2 per camera	L_1 per pixel	L_2 per pixel
Monte Carlo	0.0626524	0.02368870	0.202377	463.397
RQMC	0.0407531	0.00773514	0.162835	62.0994
RQMCS	0.0247592	0.00178907	0.138360	9.79994

Fig. 7. Schematic view of the labyrinth test scene, where floor and roof have been removed for illustration. The camera is situated in the room at the bottom, while a light source is located on the other end of the connecting tunnel. The graphs show the Box-and-Whisker plots and the average amount (marked by $\times$) of the total radiance received by the camera for 1048576 simulated light paths for 50 independent realizations using each technique. The lower graph enlarges the interesting part of the upper graph. The table finally displays the deviation from a master solution using the L_1- and L_2-norm (variance) for the total radiance received by the camera and received by each pixel (256×256 pixels resolution) averaged over the 50 independent realizations. In this difficult setting the new method (RQMCS) is clearly superior.

of simultaneously simulated Markov chains is considered. Now a sufficient amount of experiments was computationally feasible and the superiority of the new algorithm became clearly visible.

5 Conclusion

We successfully simplified the algorithms to simultaneously simulate Markov chains and provided intuition when and why sorting the states can improve convergence. In addition the algorithm no longer is bounded by the curse of dimension and there is no restriction to homogenous Markov chains, because the simulation just can use transition probabilities $P \equiv P_n$ that can change over time.

Our experiments also revealed that not all (t, s)-sequences or radical inversion based points sequences are equally good. This deserves further characterization.

The algorithm would be even simpler, if rank-1 lattice sequences could be applied. The constructions so far, however, lack the properties of (t, s)-sequences that are required for the improved performance. In the future we will investigate whether it is possible to construct suitable rank-1 lattice sequences.

As we are using global directions, i.e. integrate over products of spheres, it is also interesting to establish connections to recent research in that direction [KS05].

Acknowledgments

The authors would like to thank mental images GmbH for support and for funding of this research. Jan Rzehak created the illustration in Figure 3.

References

[Abr95] M. Abrash. BSP Trees. *Dr. Dobbs Sourcebook*, 20(14):49–52, 1995.

[Ben75] J. Bentley. Multidimensional Binary Search Trees used for Associative Searching. *Commun. ACM*, 18(9):509–517, 1975.

[BNN+98] P. Bekaert, L. Neumann, A. Neumann, M. Sbert, and Y. Willems. Hierarchical Monte Carlo Radiosity. *Eurographics Rendering Workshop 1998*, pages 259–268, June 1998.

[CP76] R. Cranley and T. Patterson. Randomization of number theoretic methods for multiple integration. *SIAM Journal on Numerical Analysis*, 13:904–914, 1976.

[DLT05] V. Demers, P. L'Écuyer, and B. Tuffin. A Combination of Randomized quasi-Monte Carlo with Splitting for Rare-Event Simulation. In *Proceedings of the 2005 European Simulation and Modelling Conference*, pages 25–32. SCS Press, 2005.

[DMC91] M. Dorigo, V. Maniezzo, and A. Colorni. Positive Feedback as a Search Strategy. Technical Report 91016, Dipartimento di Elettronica e Informatica, Politecnico di Milano, Italy, 1991.

[HLL07] R. El Haddad, C. Lécot, and P. L'Écuyer. Quasi-Monte Carlo Simulation of Discrete-Time Markov Chains in Multidimensional State Spaces. In A. Keller, S. Heinrich, and H. Niederreiter, editors, *Monte Carlo and Quasi-Monte Carlo Methods 2006*, page in this volume. Springer, 2007.

[Kar72] R. Karp. Reducibility among Combinatorial Problems. In R. Miller and J. Thatcher, editors, *Complexity of Computer Computations, Advances in Computing Research*, pages 85–103. Plenum Press, 1972.

[Kel06] A. Keller. Myths of Computer Graphics. In H. Niederreiter, editor, *Monte Carlo and Quasi-Monte Carlo Methods 2004*, pages 217–243. Springer, 2006.

[KK86] T. Kay and J. Kajiya. Ray Tracing Complex Scenes. *Computer Graphics (Proceedings of SIGGRAPH 86)*, 20(4):269–278, August 1986.

[KK02a] T. Kollig and A. Keller. Efficient Bidirectional Path Tracing by Randomized Quasi-Monte Carlo Integration. In H. Niederreiter, K. Fang, and F. Hickernell, editors, *Monte Carlo and Quasi-Monte Carlo Methods 2000*, pages 290–305. Springer, 2002.

[KK02b] T. Kollig and A. Keller. Efficient Multidimensional Sampling. *Computer Graphics Forum*, 21(3):557–563, September 2002.

[KK04] T. Kollig and A. Keller. Illumination in the Presence of Weak Singularities. In D. Talay and H. Niederreiter, editors, *Monte Carlo and Quasi-Monte Carlo Methods 2004*, pages 245–257. Springer, 2004.

[KL99] F. El Khettabi and C. Lécot. Quasi-Monte Carlo Simulation of Diffusion. *J. Complexity*, 15(3):342–359, 1999.

[Kol04] T. Kollig. *Efficient Sampling and Robust Algorithms for Photorealistic Image Synthesis*. PhD thesis, University of Kaiserslautern, Germany, 2004.

[KS05] F. Kuo and I. Sloan. Quasi-Monte Carlo Methods can be efficient for Integration over Products of Spheres. *J. Complexity*, 21(2):196–210, 2005.

[LC98] C. Lécot and I. Coulibaly. A quasi-Monte Carlo Scheme using Nets for a linear Boltzmann Equation. *SIAM J. Numer. Anal.*, 35(1):51–70, 1998.

[LDT06] P. L'Écuyer, V. Demers, and B. Tuffin. Splitting for Rare-Event Simulation. In *Proceedings of the 2006 Winter Simulation Conference*, pages 137–148, 2006.

[LDT07] P. L'Écuyer, V. Demers, and B. Tuffin. Rare Events, Splitting, and quasi-Monte Carlo. *ACM Transactions on Modeling and Computer Simulation*, 17(2): Art. No. 9, 2007.

[Léc89a] C. Lécot. A Direct Simulation Monte Carlo Scheme and Uniformly Distributed Sequences for Solving the Boltzmann Equation. *Computing*, 41(1–2):41–57, 1989.

[Léc89b] C. Lécot. Low Discrepancy Sequences for solving the Boltzmann Equation. *Journal of Computational and Applied Mathematics*, 25(2):237–249, 1989.

[Léc91] C. Lécot. A quasi-Monte Carlo Method for the Boltzmann Equation. *Mathematics of Computation*, 56(194):621–644, 1991.

[LL02] P. L'Écuyer and C. Lemieux. Recent Advances in Randomized quasi-Monte Carlo methods. In M. Dror, P. L'Ecuyer, and F. Szidarovszky, editors, *Modeling Uncertainty: An Examination of Stochastic Theory, Methods, and Applications*, pages 419–474. Kluwer Academic Publishers, 2002.

[LLKS85] E. Lawler, J. Lenstra, A. Rinnooy Kan, and D. Shmoys. *The Traveling Salesman Problem: A Guided Tour of Combinatorial Optimization*. John Wiley, 1985.

[LLT05] P. L'Écuyer, C. Lécot, and B. Tuffin. A Randomized quasi-Monte Carlo Simulation Method for Markov Chains. Technical report g-2006-64, to appear in Operations Research, GERAD, Université de Montréal, 2005.

[LLT06] P. L'Écuyer, C. Lécot, and B. Tuffin. Randomized quasi-Monte Carlo Simulation of Markov Chains with an Ordered State Space. In H. Niederreiter and D. Talay, editors, *Monte Carlo and quasi-Monte Carlo Methods 2004*, pages 331–342. Springer, 2006.

[LT04a] C. Lécot and B. Tuffin. Comparison of quasi-Monte Carlo-Based Methods for the Simulation of Markov Chains. *Monte Carlo Methods and Applications*, 10(3–4):377–384, 2004.

[LT04b] C. Lécot and B. Tuffin. Quasi-Monte Carlo Methods for Estimating Transient Measures of Discrete Time Markov Chains. In H. Niederreiter, editor, *Monte Carlo and Quasi-Monte Carlo Methods in Scientific Computing 2002*, pages 329–344. Springer, 2004.

[Mat98] J. Matoušek. On the L_2-discrepancy for anchored boxes. *J. Complexity*, 14(4):527–556, 1998.

[MC93] W. Morokoff and R. Caflisch. A Quasi-Monte Carlo Approach to Particle Simulation of the Heat Equation. *SIAM Journal on Numerical Analysis*, 30(6):1558–1573, 1993.

[Nie92] H. Niederreiter. *Random Number Generation and Quasi-Monte Carlo Methods*. SIAM, Philadelphia, 1992.

[Rei94] G. Reinelt. *The Traveling Salesman - Computational Solutions for TSP Applications*. Springer, 1994.

[Rus98] D. Rusin. Topics on Sphere Distributions, http://www.math.niu.edu/~rusin/known-math/95/sphere.faq, 1998.

[RW80] S. Rubin and J. Whitted. A 3-dimensional representation for fast rendering of complex scenes. *Computer Graphics (Proceedings of SIGGRAPH 80)*, 14(3):110–116, 1980.

[Sag94] H. Sagan. *Space-Filling Curves*. Springer, 1994.

[SBGS69] R. Schumacker, R. Brand, M. Gilliland, and W. Sharp. Study for Applying Computer-Generated Images to Visual Simulation. Technical report AFHRL-TR-69-14, U.S. Air Force Human Resources Laboratory, 1969.

[SC94] P. Shirley and K. Chiu. Notes on Adaptive Quadrature on the Hemisphere. Technical Report TR-411, Dept. of Computer Science Indiana University, 1994.

[Shi00] P. Shirley. *Realistic Ray Tracing*. AK Peters, Ltd., 2000.

[SKFNC97] L. Szirmay-Kalos, T. Fóris, L. Neumann, and B. Csébfalvi. An Analysis of Quasi-Monte Carlo Integration Applied to the Transillumination Radiosity Method. *Computer Graphics Forum*, 16(3):271–282, 1997.

[SM07] M. Saito and M. Matsumoto. SIMD-oriented Fast Mersenne Twister: A 128-bit Pseudorandom Number Generator. In A. Keller, S. Heinrich, and H. Niederreiter, editors, *Monte Carlo and Quasi-Monte Carlo Methods 2006*. Springer, in this volume.

[Sob67] I. Sobol'. On the Distribution of Points in a Cube and the approximate Evaluation of Integrals. *Zh. vychisl. Mat. mat. Fiz.*, 7(4):784–802, 1967.

[Vea97] E. Veach. *Robust Monte Carlo Methods for Light Transport Simulation*. PhD thesis, Stanford University, 1997.

[Wie03] J.-M. Wierum. *Anwendung diskreter raumfüllender Kurven - Graphpartitionierung und Kontaktsuche in der Finite-Elemente-Simulation*. PhD thesis, Universität Paderborn, 2003.

[WK06] C. Wächter and A. Keller. Instant Ray Tracing: The Bounding Interval Hierarchy. In T. Akenine-Möller and W. Heidrich, editors, *Rendering Techniques 2006 (Proc. of 17th Eurographics Symposium on Rendering)*, pages 139–149, 2006.

[WK07] C. Wächter and A. Keller. Terminating Spatial Partition Hierarchies by A Priori Bounding Memory. Technical report, Ulm University, 2007.

Part III

Appendix

Conference Participants

Eloy Anguiano
Centro de Referencia Linux
(CRL, UAM-IBM) Despacho B-206
Escuela Politécnica Superior
C/Francisco Tomás y Valiente n°11
Universidad Autónoma de Madrid
Campus Cantoblanco
28049 Madrid
Spain
eloy.anguiano@uam.es

Tatiana Averina
Institute of Computational Mathematics
and Mathematical Geophysics SB RAS
Prospekt Lavrentjeva, 6
Novosibirsk 630090
Russian Federation
ata@osmf.sscc.ru

Michael Beil
Department of Internal Medicine I,
University Ulm
Robert-Koch-Str. 8, 89079 Ulm
Germany
michael.beil@uni-ulm.de

Hans Braxmeier
Abteilung Stochastik, Universität Ulm
Helmholtzstrasse 18, 89069 Ulm
Germany

Evelyn Buckwar
Fakultät für Mathematik
Otto-von-Guericke-Universität
Magdeburg
Universitätsplatz 2, 39106 Magdeburg
Germany

Alexander Burmistrov
Institute of Computational Mathematics
and Mathematical Geophysics SB RAS
Prospekt Lavrentjeva, 6
Novosibirsk 630090
Russian Federation
burm@osmf.sscc.ru

Ronald Cools
Department of Computer Science,
K.U.Leuven
Celestijnenlaan 200A, 3001 Heverlee
Belgium
Ronald.Cools@cs.kuleuven.be

Jakob Creutzig
Stochastics and Operations Research
Fachbereich Mathematik, TU Darmstadt
Schlossgartenstr. 7, 64295 Darmstadt
Germany
creutzig@mathematik.
tu-darmstadt.de

Ligia Loreta Cristea
Institut für Finanzmathematik
Johannes Kepler Universität Linz
Altenbergerstrasse 69, 4040 Linz
Austria
ligia-loretta.cristea@jku.at

Holger Dammertz
Abteilung Medieninformatik
Universität Ulm
Albert-Einstein-Allee 11, 89069 Ulm
Germany
holger.dammertz@uni-ulm.de

Sabrina Dammertz
Abteilung Medieninformatik
Universität Ulm
Albert-Einstein-Allee 11, 89069 Ulm
Germany
sabrina.dammertz@uni-ulm.de

Fred Daum
Raytheon
318 Acton Street Carlisle MA 01741
USA
frederick_e_daum@raytheon.com

Valérie Demers
Université de Montréal
1168 Pierre-Cognac, #102 Chambly
Quebec
Canada H3L 5W8
demersv@iro.umontreal.ca

Lih-Yuan Deng
Department of Mathematical Sciences
University of Memphis
Memphis, TN 38152
USA
lihdeng@memphis.edu

Steffen Dereich
TU Berlin
Institut für Mathematik, MA 7-5
Fakultät II
Straße des 17. Juni 136, 10623 Berlin
Germany
dereich@math.tu-berlin.de

Josef Dick
School of Mathematics and Statistics
University of New South Wales
Sydney NSW 2052
Australia
josef.dick@unsw.edu.au

Stefanie Eckel
University of Ulm
Department of Stochastics
Helmholtzstrasse 22, 89069 Ulm
Germany
stefanie.eckel@uni-ulm.de

Yves Edel
Mathematisches Institut der Universität
Im Neuenheimer Feld 288, 69120
Heidelberg
Germany
y.edel@mathi.uni-heidelberg.de

Sergej Ermakov
Saint-Petersburg State
University Faculty of Mathematics
and Mechanics
Bibliotechnaya Sq. 2 198904, St.
Petersburg
Russia
sergej.ermakov@pobox.spbu.ru

Greg Fasshauer
Department of Applied Mathematics
Illinois Institute of Technology Chicago
IL 60616
USA
fasshauer@iit.edu

Henri Faure
CNRS Marseille
Institut de Mathématiques de Luminy
UMR 6206, 163 Avenue de Luminy, case
907 13288 Marseille cedex 9
France
faure@iml.univ-mrs.fr

Bernhard Finkbeiner
Abteilung Medieninformatik
Universität Ulm
Albert-Einstein-Allee 11, 89069 Ulm
Germany
bernhard.finkbeiner@uni-ulm.de

Frank Fleischer
Department of Stochastics
University of Ulm
Germany
frank.fleischer@uni-ulm.de

Piero Foscari Widmann Rezzonico
Sanpaolo AM Luxembourg
35, Rue Michel Rodange Luxembourg Ville
Luxembourg
piero.foscari@gmail.com

Stefan Geiss
Department of Mathematics and Statistics
University of Jyväskylä
P.O. Box 35 (MaD) 40014
Finland
geiss@maths.jyu.fi

Alan Genz
Department of Mathematics
Washington State University
P.O. Box 643113 Pullman
WA 99164-3113
USA
alangenz@wsu.edu

Michael Giles
Oxford University Computing Laboratory
Wolfson Building, Parks Road
Oxford OX1 3QD
U.K.
giles@comlab.ox.ac.uk

Michael Gnewuch
Institut für Informatik
Christian-Albrechts-Universitaet zu Kiel
Christian-Albrechts-Platz 4 24098 Kiel
Germany
mig@informatik.uni-kiel.de

Leonhard Grünschloß
Abteilung Medieninformatik
Universität Ulm
Albert-Einstein-Allee 11, 89069 Ulm
Germany
leonhard.gruenschloss@googlemail.com

Flavius Guiaş
Universität Dortmund
Fachbereich Mathematik
Vogelpothsweg 87, 44221 Dortmund
Germany
flavius.guias@mathematik.uni-dortmund.de

Johannes Hanika
Abteilung Medieninformatik
Universität Ulm
Albert-Einstein-Allee 11, 89069 Ulm
Germany
johannes.hanika@uni-ulm.de

Hiroshi Haramoto
Hiroshima University
Department of Mathematics
Kagamiyama 1-3-1, Higashi-Hiroshima 739-8526
Hiroshima
Japan
haramoto@hiroshima-u.ac.jp

Erika Hausenblas
University Salzburg
Hellbrunnerstr. 34, 5020 Salzburg
Austria
erika.hausenblas@sbg.ac.at

Daniela Hauser
Abteilung Medieninformatik
Universität Ulm
Albert-Einstein-Allee 11, 89069 Ulm
Germany
daniela.hauser@uni-ulm.de

Carole Hayakawa
Department of Chemical Engineering and Materials Science
University of California, Irvine
916 Engineering Tower Irvine, CA 92697
USA
hayakawa@uci.edu

Stefan Heinrich
Fachbereich Informatik
Universität Kaiserslautern
67653 Kaiserslautern
Germany
heinrich@informatik.uni-kl.de

Peter Hellekalek
Department of Mathematics
University of Salzburg
Hellbrunner Strasse 34, 5020 Salzburg
Austria
peter.hellekalek@sbg.ac.at

Fred J. Hickernell
Department of Applied Mathematics
Illinois Institute of Technology
Room 208, Bldg. E1 10 W. 32nd St.
Chicago, IL 60616
USA
hickernell@iit.edu

Heinz Hofbauer
Department of Computer Sciences
University of Salzburg
Jakob Haringer Str. 2, 5020 Salzburg
Austria
hhofbaue@cosy.sbg.ac.at

Tito Homem-de-Mello
Northwestern University
USA
tito@northwestern.edu

Raquel Iniesta
Cancer Epidemiology Service
Catalan Insititute of Oncology
Avda. Gran Via Km 2,7 s/n 08907
L'Hospitalet de Llobregat, Barcelona
Spain
riniesta@iconcologia.net

Boleslaw Kacewicz
Faculty of Applied Mathematics
AGH University of Science
and Technology
A3/A4, p. 301 Al. Mickiewicza 30 30-059
Cracow
Poland
kacewicz@uci.agh.edu.pl

Reinhold Kainhofer
Financial and Actuarial Mathematics
Vienna University of Technology
Wiedner Hauptstrasse 8/105-1 1040
Wien
Austria
reinhold@kainhofer.com

Alexander Kalouguine
Image Processing Systems Institute
of RAS
Samara State Aerospace University
151, Molodoguardeyskaya St.
Samara 443001
Russia
alexklg@mercdev.com

Jonathan Keith
Department of Mathematics
University of Queensland
St. Lucia Qld. 4072
Australia
j.keith@qut.edu.au

Alexander Keller
Abteilung Medieninformatik
Universität Ulm
Albert-Einstein-Allee 11, 89069 Ulm
Germany
alexander.keller@uni-ulm.de

Christian Kempter
Abteilung Medieninformatik
Universität Ulm
Albert-Einstein-Allee 11, 89069 Ulm
Germany
christian.kempter@uni-ulm.de

Adam Kolkiewicz
Department of Statistics and Actuarial
Science
University of Waterloo, 200 University
Ave. West Waterloo, Ontario
Canada N2L 3G1
wakolkie@uwaterloo.ca

Hyungwoon Koo
Department of Mathmatics
Korea University
Anam-dong, Sungbuk-gu Seoul, 136-701
Korea
koohw@korea.ac.kr

Peter Kritzer
Department of Mathematics
University of Salzburg
Hellbrunnerstr. 34, 5020 Salzburg
Austria
peter.kritzer@sbg.ac.at

Manuel Kugelmann
Abteilung Medieninformatik
Universität Ulm
Albert-Einstein-Allee 11, 89069 Ulm
Germany
dragoon@bitcraft.org

Frances Kuo
School of Mathematics and Statistics
University of New South Wales
Sydney NSW 2052
Australia
f.kuo@unsw.edu.au

Elisabeth Larsson
Department of Information Technology
Uppsala University
Box 337 SE-751 05 Uppsala
Sweden
Elisabeth.Larsson@it.uu.se

Claudia Lautensack
Fraunhofer Institut für Techno-und Wirtschaftsmathematik
Fraunhofer-Platz 1, 67663 Kaiserslautern
Germany
claudia.lautensack@itwm.fraunhofer.de

Christian Lécot
Laboratoire de Mathematiques
UMR 5127 CNRS and Université de Savoie
73376 Le Bourget du-Lac cedex
France
Christian.Lecot@univ-savoie.fr

Pierre L'Ecuyer
DIRO, Université de Montréal
C.P. 6128, Succ. Centre-Ville, Montréal,
Canada H3C 1J7
lecuyer@iro.umontreal.ca

Josef Leydold
Department of Statistics and Mathematics
Vienna University of Economics and Business Administration
Augasse 2-6, 1090 Vienna
Austria
leydold@statistik.wu-wien.ac.at

Vitaliy Lukinov
Institute of Computational Mathematics and Mathematical Geophysics SB RAS
Prospekt Lavrentjeva, 6
Novosibirsk 630090
Russian Federation
Vitaliy.Lukinov@ngs.ru

Karin Lunde
Hochschule Ulm
Prittwitzstr. 10, 89073 Ulm
Germany
k.lunde@hs-ulm.de

Roman Makarov
Department of Mathematics
Wilfrid Laurier University
75 University Avenue West Waterloo
Ontario
Canada N2L 3C5
rmakarov@wlu.ca

Peter Mathé
Weierstrass Institute
Mohrenstrasse 39, 10117 Berlin
Germany
mathe@wias-berlin.de

Makoto Matsumoto
Department of Mathematics
Hiroshima University
1-3-1 Kagamiyama, Higashi-Hiroshima-shi, Hiroshima Prefecture
Japan
m-mat@math.sci.hiroshima-u.ac.jp

Ilya Medvedev
Institute of Computational Mathematics and Mathematical Geophysics SB RAS
Prospekt Lavrentjeva, 6
Novosibirsk 630090
Russian Federation
medvedev79@ngs.ru

Bernhard Milla
Technische Universität Kaiserslautern
Fachbereich Informatik
Postfach 3049, 67653 Kaiserslautern
Germany
milla@informatik.uni-kl.de

Kenichi Miura
National Institute of Informatics
Center for Grid Research and Development, NII 1-105 Kanda-Jimbocho, Jimbocho Mitsui Building #1402
Chiyoda-ku, Tokyo
Japan
kenmiura@grid.nii.ac.jp

Thomas Müller-Gronbach
University of Magdeburg
Germany
gronbach@mail.math.uni-magdeburg.de

Carmen Blanca Navarrete
Centro de Referencia Linux (CRL, UAM-IBM)
Despacho B-206. Escuela Politécnica Superior. C/Francisco Tomás y Valiente n°11. Universidad Autónoma de Madrid. Campus Cantoblanco.
28049 Madrid
Spain
carmen.navarrete@uam.es

Sehera Nawaz
Abteilung Medieninformatik
Universität Ulm
Albert-Einstein-Allee 11, 89069 Ulm
Germany
sehera.nawaz@uni-ulm.de

Andreas Neuenkirch
TU Darmstadt
FB Mathematik/AG Stochastik
Schlossgartenstr. 7, 64289 Darmstadt
Germany
neuenkirch@mathematik.tu-darmstadt.de

Harald Niederreiter
Department of Mathematics
National University of Singapore
2 Science Drive 2 Singapore
Singapore, 117543
nied@math.nus.edu.sg

Erich Novak
Uni Jena
Mathematisches Institut, Ernst-Abbe-Platz 2, 07743 Jena
Germany
novak@math.uni-jena.de

Dirk Nuyens
Department of Computer Science
K.U.Leuven
Celestijnenlaan 200A, 3001 Heverlee
Belgium
dirk.nuyens@cs.kuleuven.be

Thomas Önskog
Department of Mathematics and Mathematical Statistics, Umera University S-901 87 Umera Sweden
Sweden
thomas.onskog@math.umu.se

Gilles Pagès
LPMA-Université Paris 6
case 188, 4, pl. Jussieu
F. 75252 Paris Cedex 5
France
gpa@ccr.jussieu.fr

Robert Patterson
Department of Chemical Engineering
University of Cambridge
Pembroke St CB2 3RA Cambridge
U.K.
riap2@cam.ac.uk

Leif Persson
Swedish Defence Research Agency FOI
Cementvägen 20 SE-901 82 Umera
Sweden
leiper@foi.se

Friedrich Pillichshammer
Institut für Finanzmathematik
Universität Linz
Altenbergerstrasse 69, 4040 Linz
Austria
friedrich.pillichshammer@jku.at

Leszek Plaskota
Faculty of Mathematics Informatics and Mechanics Warsaw University ul. Krakowskie Przedmiescie 26/28 00-927 Warszawa
Poland
leszekp@mimuw.edu.pl

Marco Pollanen
Department of Mathematics
Trent University Peterborough
Ontario
Canada, K9J 7B8
marcopollanen@trentu.ca

Matthias Raab
Ulmer Zentrum für Wissenschaftliches Rechnen
Abteilung Medieninformatik, Universität Ulm
Albert-Einstein-Allee 11, 89069 Ulm
Germany
matthias.raab@uni-ulm.de

Klaus Ritter
TU Darmstadt
Fachbereich Mathematik
Schlossgartenstrasse 7 64289 Darmstadt
Germany
ritter@mathematik.tu-darmstadt.de

Andreas Rößler
TU Darmstadt
Fachbereich Mathematik
Schlossgartenstr. 7, 64289 Darmstadt
Germany
roessler@mathematik.tu-darmstadt.de

Anna Rukavishnikova
Faculty of Mathematics & Mechanics
S-Petersburg State University
Srednegavansky 3,10; S-Petersburg
Russia
anyaruk@mail.ru

Jonas Rumpf
University of Ulm
Department of Stochastics
Helmholtzstr. 18, 89069 Ulm
Germany
jonas.rumpf@uni-ulm.de

Karl Sabelfeld
WIAS, Berlin
Mohrenstrasse 39, 10117 Berlin
Germany
sabelfel@wias-berlin.de

Mutsuo Saito
Department of Mathematics
Hiroshima University
1-3-1 Kagamiyama, Higashi-Hiroshima-shi, Hiroshima Prefecture
Japan
saito@math.sci.hiroshima-u.ac.jp

Charles Sanvido
DIRO, Université de Montréal
C.P. 6128, Succ. Centre-Ville, Montréal
Canada, H3C 1J7
sanvidoc@iro.umontreal.ca

Thorsten Sauder
Bankgesellschaft Berlin
Alt-Stralau 31, 10245 Berlin
Germany
Thorsten.Sauder@bankgesellschaft.de

Wolfgang Ch. Schmid
Department of Mathematics
University of Salzburg
Hellbrunnerstraße 34, 5020 Salzburg
Austria
wolfgang.schmid@sbg.ac.at

Hendrik Schmidt
Abteilung Stochastik
Universität Ulm
Helmholtzstrasse 18, 89069 Ulm
Germany
hendrik.schmidt@uni-ulm.de

Rudolf Schürer
Department of Mathematics
University of Salzburg
Hellbrunnerstraße 34, 5020 Salzburg
Austria
rudolf.schuerer@sbg.ac.at

Ronnie L. Schwede
Eawag, Swiss Federal Institute
of Aquatic Science and Technology
Ueberlandstrasse 133, 8600 Dübendorf
Switzerland
Ronnie.Schwede@eawag.ch

Daniel Seibert
mental images GmbH
Fasanenstr. 81, 10623 Berlin
Germany
daniel@mental.com

Raffaello Seri
Dipartimento di Economia via Monte
Università degli Studi dell'Insubria
Generoso 71
Italy
raffaello.seri@uninsubria.it

Peter Shirley
University of Utah
50 S Central Campus Dr, Rm 3190
Salt Lake City, UT 84112
USA
shirley@cs.utah.edu

Nikolai Simonov
Institute of Computational Mathematics
and Mathematical Geophysics SB RAS
Prospekt Lavrentjeva, 6
Novosibirsk 630090
Russian Federation
simonov@scs.fsu.edu

Vasile Sinescu
Department of Mathematics
University of Waikato
Private Bag 3105 Hamilton
New Zealand
vs27@waikato.ac.nz

Ian Sloan
School of Mathematics and Statistics
University of New South Wales
Sydney NSW 2052
Australia
i.sloan@unsw.edu.au

Daria Spivakovskaya
Delft Institute of Applied Mathematics
Delft University of Technology
Office 05.270 Mekelweg 4 2628 CD Delft
Netherlands
D.Spivakovskaya@ewi.tudelft.nl

Nicolae Suciu
Institute for Applied Mathematics
Friedrich Alexander University
Erlangen-Nuremberg
Martensstr. 3, 91058 Erlangen
Germany
suciu@am.uni-erlangen.de

Ali Tarhini
Laboratoire de Mathematiques
Université de Savoie Campus Scientifique
73376 Le Bourget du Lac
France
atarh@univ-savoie.fr

Tomas Tichy
Department of Finance
Faculty of Economics VSB-Technical
University Ostrava Sokolska 33 701 21
Ostrava
Czech Republic
tomas.tichy@vsb.cz

Tatiana Tovstik
St. Petersburg University
Botanicheskaya 18-4-40 Stary Peterhof
198504 St. Petersburg Russia
Russia
peter.tovstik@mail.ru

Seth Tribble
Stanford University
444 Webster St. Palo Alto, CA 94301
USA
stribble@stanford.edu

Horst Trinker
Department of Mathematics
University of Salzburg Hellbrunnerstr.
34, 5020 Salzburg
Austria
horst.trinker@sbg.ac.at

Olga Ukhinova
Institute of Computational Mathematics
and Mathematical Geophysics SB RAS
Prospekt Lavrentjeva, 6
Novosibirsk 630090
Russian Federation
olsu@osmf.sscc.ru

Karsten Urban
Department of Numerical Mathematics
University of Ulm
Helmholtzstr. 18, 89069 Ulm
Germany
karsten.urban@uni-ulm.de

Bart Vandewoestyne
Department of Computer Science
K.U.Leuven
Celestijnenlaan 200A, 3001 Heverlee
Belgium
Bart.Vandewoestyne@cs.kuleuven.be

Jochen Voß
Mathematics Institute
University of Warwick
Coventry CV4 7AL
U.K.
voss@maths.warwick.ac.uk

Anton Voytishek
Institute of Computational Mathematics
and Mathematical Geophysics SB RAS
Prospekt Lavrentjeva, 6
Novosibirsk 630090
Russian Federation
vav@osmf.sscc.ru

Carsten Wächter
Abteilung Medieninformatik
Universität Ulm
Albert-Einstein-Allee 11, 89069 Ulm
Germany
carsten.waechter@uni-ulm.de

Tim Wagner
TU Darmstadt, Fachbereich Mathematik
Stochastics and Operations Research
Schloßgartenstr. 7, 64289 Darmstadt
Germany
twagner@mathematik.tu-darmstadt.de

Wolfgang Wagner
Weierstrass Institute for Applied
Analysis and Stochastics
Mohrenstrasse 39, 10117 Berlin
Germany
wagner@wias-berlin.de

Vasant Waikar
Department of Mathematics
and Statistics
Miami University Oxford, OH 45056
USA
waikarvb@muohio.edu

Patrick G. Walch
Graduiertenkolleg 1100
Fakultät für Mathematik und
Wirtschaftswissenschaften, Uni Ulm
Helmholtzstr. 22, Room E16, 89069 Ulm
Germany

Xiaoqun Wang
Department of Mathematical Sciences
Tsinghua University Beijing 100084
P. R. China
xwang@math.tsinghua.edu.cn

Grzegorz Wasilkowski
Department of Computer Science 773
Anderson Hall University of Kentucky
Lexington, KY, 40506-0046
USA
greg@cs.uky.edu

Ben Waterhouse
School of Mathematics and Statistics
University of New South Wales
Sydney NSW 2052
Australia
benjw@maths.unsw.edu.au

Hirotake Yaguchi
Department of Mathematics
Faculty of Education
Mie University 514-8507 Tsu, Mie
Japan
yaguchi@edu.mie-u.ac.jp

Author Index

Boulos, Solomon: 167
Burmistrov, Alexander: 181

Cools, Ronald: 3
Creutzig, Jakob: 195

Dammertz, Holger: 205
Dammertz, Sabrina: 205, 217
Debrabant, Kristian: 237
Deng, Lih-Yuan: 251, 263
Dick, Josef: 275
Doerr, Benjamin: 299

Edel, Yves: 313
Edwards, Dave: 167
El Haddad, Rami: 413
Ermakov, Sergej: 23

Faure, Henri: 327

Genz, Alan: 35
Giles, Mike: 343
Gnewuch, Michael: 299, 359
Grünschloß, Leonhard: 397
Guiaş, Flavius: 383

Hanika, Johannes: 397
Hayakawa, Carole: 431
Heinrich, Stefan: 445
Hickernell, Fred: 461
Hofbauer, Heinz: 495, 511

Iniesta, Raquel: 529

Joe, Stephen: 645

Keith, Jonathan: 537
Keller, Alexander: 205, 217, 397, 591, 669
Kritzer, Peter: 275, 549
Kroese, Dirk: 537
Kuo, Frances Y: 275

L'Ecuyer, Pierre: 313, 413
Lécot, Christian: 413
Li, Huajiang: 263
Lifshits, Mikhail: 195

Müller-Gronbach, Thomas: 53, 577
Matsumoto, Makoto: 607
Medvedev, Ilya: 561
Mikhailov, Gennadii: 561
Moreno, Victor: 529

Niederreiter, Harald: 83
Nuyens, Dirk: 3

Pagès, Gilles: 101
Pillichshammer, Friedrich: 549

Rößler, Andreas: 237
Raab, Matthias: 591
Ritter, Klaus: 53, 577

Sabelfeld, Karl: 143, 659
Saito, Mutsuo: 607
Schürer, Rudolf: 623
Schwede, Ronnie: 397
Seibert, Daniel: 591

Shiau, Jyh-Jen Horng: 263
Shirley, Peter: 167
Simonov, Nikolai: 633
Sinescu, Vasile: 645
Sofronov, George: 537
Spanier, Jerome: 431
Suciu, Nicolae: 659

Tsai, Gwei-Hung: 263

Uhl, Andreas: 495, 511

Vamoş, Călin: 659
Venugopalan, Vasan: 431

Wächter, Carsten: 669
Wagner, Tim: 577
Wasilkowski, Greg: 461
Woźniakowski, Henryk: 359, 461

Zinterhof, Peter: 495, 511